SYNTHETIC METHODS IN STEP-GROWTH POLYMERS

SYNTHETIC METHODS IN STEP-GROWTH POLYMERS

Edited by

Martin E. Rogers
Luna Innovations
Blacksburg, VA

Timothy E. Long
Department of Chemistry
Virginia Tech
Blacksburg, VA

A JOHN WILEY & SONS, INC., PUBLICATION

Library of Congress Cataloging-in-Publication Data:

Synthetic methods in step-growth polymers / edited by Martin E. Rogers and Timothy Long.
    p. cm.
    Includes index.
    ISBN 0-471-38769-X (cloth)
    1. Polycondensation. 2. Plastics. I. Rogers, Martin E. II. Long, Timothy E., 1969–
    QD281.P6S96 2003
    668.4—dc21   2002011134

Printed in the United States of America

10 9 8 7 6 5 4 3 2 1

CONTRIBUTORS

A. CAMERON CHURCH.   Department of Chemistry, University of Florida, Gainesville, FL 32611-7200

JEFF DODGE.   Bayer Corporation, Pittsburgh, PA 15205

ALAIN FRADET.   Chimie des Polymères, Université Pierre et Marie Curie, Paris, France

REINOUD J. GAYMANS.   Twente University, Chemistry and Technology of Engineering Plastics, 7500 AE Enschede, The Netherlands

S. LIN-GIBSON.   Polymers Division, NIST, Gaithersburg, MD 20899-8543

QIAO-SHENG HU.   Department of Chemistry, City University of New York, College of Staten Island, Staten Island, NY 10314

TIMOTHY E. LONG.   Department of Chemistry, Virginia Polytechnic Institute and State University, Blacksburg, VA 24061

R. MERCIER.   LMOPS, 69390 Vernaison, France

J. E. McGRATH.   Department of Chemistry, Virginia Polytechnic Institute and State University, Blacksburg, VA 24061

JAMES H. PAWLOW.   Department of Chemistry, University of Florida, Gainesville, FL 32611-7200

D. PICQ.   LMOPS, 69390 Vernaison, France

MALCOLM B. POLK.   Georgia Institute of Technology, School of Textile and Fiber Engineering, Atlanta, GA 30332-0295

J. S. RIFFLE.   Department of Chemistry, Virginia Polytechnic Institute and State University, Blacksburg, VA 24061

MARTIN E. ROGERS.   Luna Innovations, Blacksburg, VA

B. SILLION.   SCA 69390 Vernaison, France

JASON A. SMITH.   University of Florida, Department of Chemistry, Gainesville, FL 32611-7200

MARTINE TESSIER.   Chimie des Polymères, Université Pierre et Marie Curie, Paris, France

S. RICHARD TURNER.   Eastman Chemical Company, Kingsport, TN

KENNETH B. WAGENER.   Department of Chemistry, University of Florida, Gainesville, FL 32611-7200

SHENG WANG.   Department of Chemistry, Virginia Polytechnic Institute and State University, Blacksburg, VA 24061

CONTENTS

PREFACE

Step-growth polymerization continues to receive intense academic and industrial attention for the preparation of polymeric materials used in a vast array of applications. Polyesters used in fibers, containers and films are produced globally at a rate of millions of metric tons per year. Polyamides (1.7M metric tons) and polycarbonates (1.6M metric tons) led the global engineering polymers marketplace in 2000. High temperature engineering liquid crystalline polyesters were projected to grow an amazing 13 to 15% per year from 2001–2006. A step-wise polymerization mechanism serves as the fundamental basis for these polymer products, and future discoveries will require fundamental mechanistic understanding and keen awareness of diverse experimental techniques.

This text was not intended to be comprehensive, but serve as a long-standing resource for fundamental concepts in step-growth polymerization processes and experimental methodologies. Ten invited chapters provide a review of major classes of macromolecules prepared via step-growth polymerization, including polyesters, polyamides, polyurethanes, polyimides, poly(arylene ethers), and phenolic resins. Moreover, recent advances in acyclic diene metathesis polymerization and transition metal coupling represent exciting new directions in step-growth processes. The final chapter describes processes for subsequent recycling and depolymerization of step-growth polymers, which are important considerations as we attempt to minimize the negative impact of step-growth polymers on our environment. In addition to providing a literature review of this rapidly evolving research area, special attention was devoted to the incorporation of detailed experimental methodologies enabling researchers with limited polymerization experience to quickly impact this field. We would like to express our gratitude to the chapter authors for their valuable contributions, and we hope that this text will cultivate new ideas and catalyze discoveries in your laboratory.

MARTIN E. ROGERS
TIMOTHY E. LONG

1 Introduction to Synthetic Methods in Step-Growth Polymers

Martin E. Rogers

Luna Innovations, Blacksburg, Virginia 24060

Timothy E. Long

Department of Chemistry, Virginia Polytechnic Institute and State University, Blacksburg, Virginia 24061

S. Richard Turner

Eastman Chemical Company, Kingsport, Tennessee 37662

1.1 INTRODUCTION

1.1.1 Historical Perspective

Some of the earliest useful polymeric materials, the Bakelite resins formed from the condensation of phenol and formaldehyde, are examples of step-growth processes.[1] However, it was not until the pioneering work of Carothers and his group at DuPont that the fundamental principles of condensation (step-growth) processes were elucidated and specific step-growth structures were intentionally synthesized.[2,3] Although it is generally thought that Carothers' work was limited to aliphatic polyesters, which did not possess high melting points and other properties for commercial application, this original paper does describe amorphous polyesters using the aromatic diacid, phthalic acid, and ethylene glycol as the diol. As fundamental as this pioneering research by Carothers was, the major thrust of the work was to obtain practical commercial materials for DuPont. Thus, Carothers and DuPont turned to polyamides with high melting points and robust mechanical properties. The first polymer commercialized by DuPont, initiating the "polymer age," was based on the step-growth polymer of adipic acid and hexamethylene diamine—nylon 6,6.[4] It was not until the later work of Whinfield and Dickson in which terephthalic acid was used as the diacid moiety and the benefits of using a para-substituted aromatic diacid were discovered that polyesters became commercially viable.[5]

Synthetic Methods in Step-Growth Polymers.   Edited by Martin E. Rogers and Timothy E. Long
© 2003 John Wiley & Sons, Inc.   ISBN: 0-471-38769-X

In these early days of polymer science, the correlation of structure and property in the newly synthesized structures was a daunting challenge. As Carothers said, "problem of the more precise expression of the relationships between the structures and properties of high polymers is complicated by the fact that some of the properties of this class of substances which are of the greatest practical importance and which distinguish them most sharply from simple compounds can not be accurately measured and indeed are not precisely defined. Examples of such properties are toughness and elasticity" (ref. 6, p. 317).

Today, step-growth polymers are a multi-billion-dollar industry. The basic fundamentals of our current understanding of step-growth polymers from monomer functionality to molecular weight distribution to the origins of structure–property relationship all had their beginnings in the pioneering work of Carothers and others at DuPont. A collection of these original papers offers an interesting and informative insight into the development of polymer science and the industry that it spawned.[7]

1.1.2 Applications

In general, step-growth polymers such as polyesters and polyamides possess more robust mechanical properties, including toughness, stiffness, and higher temperature resistance, than polymers from addition polymerization processes such as polyolefins and other vinyl-derived polymers. Even though many commercial step-growth polymerization processes are done on enormous scale using melt-phase processes, most step-growth-based polymers are more expensive than various vinyl-based structures. This is, at least in part, due to the cost of the monomers used in step-growth polymerizations, which require several steps from the bulk commodity petrochemical intermediates to the polymerizable monomer, for example, terephthalic acid from the xylene stream, which requires oxidation and difficult purification technology. These cost and performance factors are key to the commercial applications of the polymers.

Most of the original application successes for step-growth polymers were as substitutes for natural fibers. Nylon-6,6 became an initial enormous success for DuPont as a new fiber. Poly(ethylene terephthalate) (PET) also found its initial success as a textile fiber. An examination of the polymer literature in the 1950s and 1960s shows a tremendous amount of work done on the properties and structures for new fibers. Eventually, as this market began to mature, the research and development community recognized other commercially important properties for step-growth polymers. For example, new life for PET resulted from the recognition of the stretch-blow molding and barrier properties of this resin. This led to the huge container plastics business for PET, which, although maturing, is still fast growing today.

The remainder of this introductory chapter covers a few general but important parameters of step-growth polymerization. References are provided throughout the chapter if further information is desired. Further details of specific polymers made by step-growth polymerization are provided in subsequent chapters within this book.

1.2 STRUCTURE–PROPERTY RELATIONSHIPS IN STEP-GROWTH POLYMERS

1.2.1 Molecular Weight

Polymers produced by step-growth polymerization are composed of macro-molecules with varying molecular weights. Molecular weights are most often reported as number averages, $\overline{Mn}$, and weight averages, $\overline{Mw}$. Rudin, in *The Elements of Polymer Science and Engineering*, provides numerical descriptions of molecular weight averages and the derivation of the molecular weight averages.[8] Other references also define molecular weight in polymers as well as methods for measuring molecular weights.[8-11] Measurement techniques important to step-growth polymers include endgroup analysis, size exclusion chromatography, light scattering, and solution viscometry.

The physical properties of polymers are primarily determined by the molecular weight and chemical composition. Achieving high molecular weight during polymerization is critical if the polymer is to have sufficient thermal and mechanical properties to be useful. However, molecular weight also influences the polymer melt viscosity and solubility. Ease of polymer processing is dependent on the viscosity of the polymer and polymer solubility. High polymer melt viscosity and poor solubility tend to increase the difficulty and expense of polymer processing.

The relationship between viscosity and molecular weight is well documented.[12-14] Below a critical molecular weight, the melt viscosity increases in proportion to an increase in molecular weight. At this point, the viscosity is relatively low allowing the material to be easily processed. When the molecular weight goes above a critical value, the melt viscosity increases exponentially with increasing molecular weight. At higher molecular weights, the material becomes so viscous that melt processing becomes more difficult and expensive.

Several references discuss the relation between molecular weight and physical properties such as the glass transition temperature and tensile strength.[15-17] The nature of thermal transitions, such as the glass transition temperature and crystallization temperature, and mechanical properties are discussed in many polymer texts.[8,17,18] Below a critical molecular weight, properties such as tensile strength and the glass transition temperature are low but increase rapidly with increasing molecular weight. As the molecular weight rises beyond the critical molecular weight, changes in mechanical properties are not as significant. When developing polymerization methods, knowledge of the application is necessary to determine the target molecular weight. For example, polymers used as rigid packaging or fibers require high strength and, consequently, high molecular weights.

Thermoplastic commercial step-growth polymers such as polyesters, polycarbonates, and polyamides are generally made with number-average molecular weights in the range of 10,000–50,000 g/mol. Polymers within this molecular weight range are generally strong enough for use as structural materials yet low enough in melt viscosity to be processable at a reasonable cost.

Thermosetting resins are combined with fibers and other fillers to form composites.[19] Thermosetting resins with low viscosities are necessary to wet fibers or other fillers and to allow efficient processing and application prior to curing. When preparing thermosetting resins, such as unsaturated polyesters, phenolics, and epoxides, it is necessary to minimize viscosity by severely limiting molecular weight.

For example, the molecular weight of unsaturated polyesters is controlled to less than 5000 g/mol. The low molecular weight of the unsaturated polyester allows solvation in vinyl monomers such as styrene to produce a low-viscosity resin. Unsaturated polyesters are made with monomers containing carbon–carbon double bonds able to undergo free-radical crosslinking reactions with styrene and other vinyl monomers. Crosslinking the resin by free-radical polymerization produces the mechanical properties needed in various applications.

Step-growth polymerizations can produce polymers with a wide range of physical properties. Polysiloxanes made from the step-growth polymerization of silanols have among the lowest glass transition temperatures. Polydimethyl siloxanes have a glass transition temperature near $-125°C$. On the other hand, step-growth polymerization produces polyimides and polybenzoxazoles with glass transition temperatures of 300°C to over 400°C.[20,21]

Even within a particular class of polymers made by step-growth polymerization, monomer composition can be varied to produce a wide range of polymer properties. For example, polyesters and polyamides can be low-T_g, amorphous materials or high-T_g, liquid crystalline materials depending on the monomer composition.

The dependence of polymer properties on chemical compositions is reviewed in basic polymer texts.[9,10] The backbone structure of a polymer defines to a large extent the flexibility and stability of a polymer molecule. Consequently, a great range of polymer properties can be achieved within each class of step-growth polymers by varying the backbone structure using different monomers.

The most common backbone structure found in commercial polymers is the saturated carbon–carbon structure. Polymers with saturated carbon–carbon backbones, such as polyethylene, polypropylene, polystyrene, polyvinyl chloride, and polyacrylates, are produced using chain-growth polymerizations. The saturated carbon–carbon backbone of polyethylene with no side groups is a relatively flexible polymer chain. The glass transition temperature is low at $-20°C$ for high-density polyethylene. Side groups on the carbon–carbon backbone influence thermal transitions, solubility, and other polymer properties.

Nearly all of the polymers produced by step-growth polymerization contain heteroatoms and/or aromatic rings in the backbone. One exception is polymers produced from acyclic diene metathesis (ADMET) polymerization.[22] Hydrocarbon polymers with carbon–carbon double bonds are readily produced using ADMET polymerization techniques. Polyesters, polycarbonates, polyamides, and polyurethanes can be produced from aliphatic monomers with appropriate functional groups (Fig. 1.1). In these aliphatic polymers, the concentration of the linking groups (ester, carbonate, amide, or urethane) in the backbone greatly influences the physical properties.

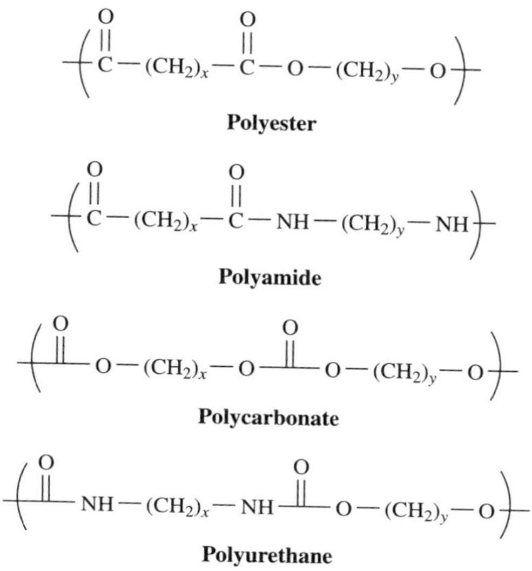

$$\left[\!\!\begin{array}{c}O\\ \|\\ C\end{array}\!-(CH_2)_x-\!\!\begin{array}{c}O\\ \|\\ C\end{array}\!-O-(CH_2)_y-O\right]$$

Polyester

$$\left[\!\!\begin{array}{c}O\\ \|\\ C\end{array}\!-(CH_2)_x-\!\!\begin{array}{c}O\\ \|\\ C\end{array}\!-NH-(CH_2)_y-NH\right]$$

Polyamide

$$\left[\!\!\begin{array}{c}O\\ \|\\ \end{array}\!O-(CH_2)_x-O-\!\!\begin{array}{c}O\\ \|\\ \end{array}\!O-(CH_2)_y-O\right]$$

Polycarbonate

$$\left[\!\!\begin{array}{c}O\\ \|\\ \end{array}\!NH-(CH_2)_x-NH-\!\!\begin{array}{c}O\\ \|\\ \end{array}\!O-(CH_2)_y-O\right]$$

Polyurethane

Figure 1.1 Aliphatic step-growth polymers.

Increasing the methylene content increases the melting point, eventually tending toward the T_m of polyethylene at low linking group concentrations. The linear aliphatic polyesters and polycarbonates have relatively low T_g's (-70 to $-30°C$) and melting points below $100°C$. The linear aliphatic polyesters and polycarbonates are not used as structural materials due to the low melting temperatures and limiting hydrolytic stability. Aliphatic polyesters are used as soft-segment polyols in polyurethane production.

In contrast to the polyesters and polycarbonates, the linear aliphatic polyamides and polyurethanes have high melting points and higher glass transition temperatures as the amide and urethane linking groups participate in intermolecular hydrogen bonding. In Chapter 3 of *Polymer Chemistry*, Stevens discusses the influence of hydrogen bonding in polyamides compared with polyesters.[9] Stevens notes that poly(hexamethylene adipamide) melts at $265°C$ compared to $60°C$ for poly(hexamethylene adipate).[9]

Aromatic groups in the polymer backbone bring rigidity and thermal stability to the polymer molecule (Fig. 1.2). Consequently, the demands of high-strength and high-temperature applications are met by polymers with a high aromatic content in the backbone. Polymers with a particularly high aromatic content can show main-chain liquid crystallinity.

Aromatic polymers are often more difficult to process than aliphatic polymers. Aromatic polyamides have to be processed from very aggressive solvents such as sulfuric acid. The higher melting temperatures and viscosity also make melt processing more difficult. Thermal stability and processing of aromatic polymers can be balanced by the use of flexible spacing groups in between aromatic rings

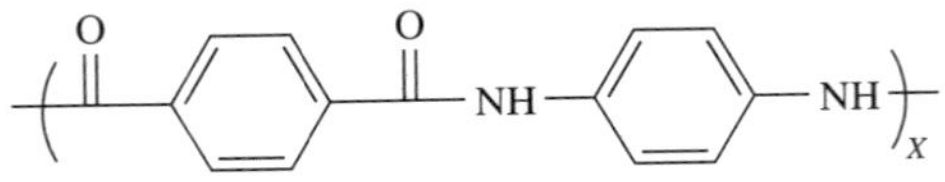

Polyamide

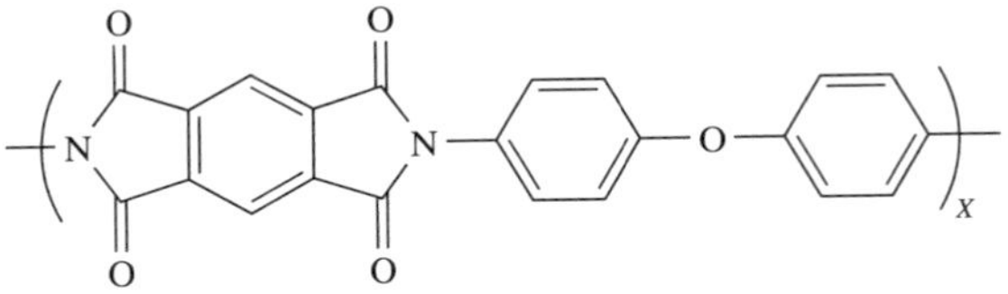

Polyimide

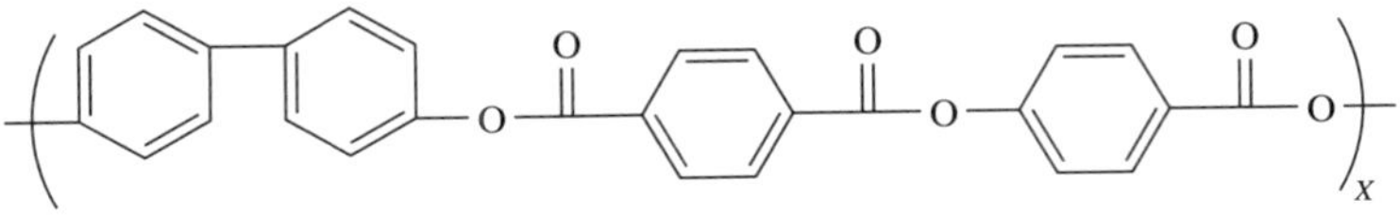

Polyester

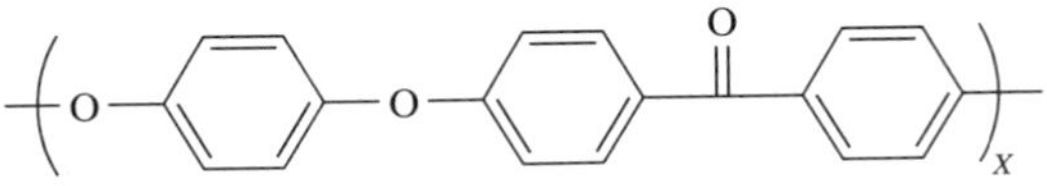

Polyetheretherketone

Figure 1.2   Aromatic step-growth polymers.

on a polymer backbone. Hexafluoroisopropylidene, isopropylidene, oxygen, car-bonyl, and sulfonyl bridging groups between rings increase opportunities for bond rotation, which decreases T_g's and increases solubility. Also, incorporating non-symmetrical monomers with meta and ortho linkages causes structural disorder in the polymer chain, improving processability. Flexible groups pendant to an aromatic backbone will also increase solubility and processability.

The following chapters will provide detailed discussions of the structure–property relations with various classes of step-growth polymers.

1.2.2 Polymer Architecture

Block copolymers are composed of two different polymer segments that are chemically bonded.[23,24] The sequential arrangement of block copolymers can vary from diblock or triblock copolymers, with two or three segments respectively, to multiblock copolymers containing many segments. Figure 1.3 is a schematic representation of various block copolymer architectures. The figure also includes graft and radial block copolymers. Step-growth polymerization can be used effectively to produce segmented or multiblock copolymers and graft copolymers. Well-defined diblock and triblock copolymers are generally only accessible by chain-growth polymerization routes.

A variety of morphologies and properties can be achieved with microphase-separated block copolymers. Copolymers of hard and soft polymer segments have

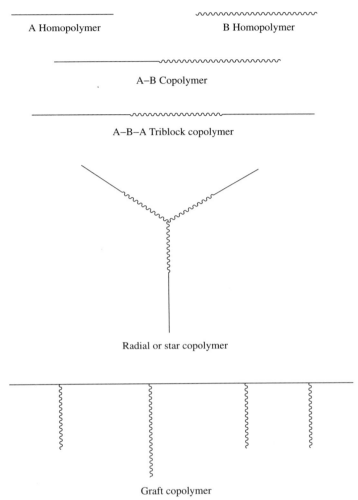

Figure 1.3   Various block copolymer architectures.

a variety of properties depending on their composition. Copolymers with small amounts of a soft segment will behave as a toughened glassy polymer while copolymers made predominately of the soft segment will act as a thermoplastic elastomer.

The thermal properties of block copolymers are similar to physical blends of the same polymer segments. Each distinct phase of the copolymer displays unique thermal transitions, such as a glass transition and/or a crystalline melting point. The thermal transitions of the different phases are affected by the degree of intermixing between the phases.

Segmented or multiblock copolymers can be made by combining a functionally terminated oligomer or prepolymer with at least two monomers. To form a segmented copolymer, the backbone oligomer must not be able to participate in interchange reactions with the monomers. For example, combining a polyester oligomer with a diacid and diamine in a melt polymerization might result in interchange reactions between the monomers and the ester linking groups in the oligomer backbone. In this case a random polyesteramide copolymer would be produced instead of a segmented copolymer. Commercial examples of segmented copolymers produced by step-growth polymerization include polyester–polyether, polyurethane–polyether, and polyurethane–polyester copolymers.

Multifunctional monomers with functionality greater than 2 can be used to form three-dimensional polymer structures during step-growth polymerization. Incorporating multifunctional monomers, A_x, with AA and BB monomers results in crosslinking between polymer chains and eventual gelation. The point at which gelation occurs depends on the average functionality of the monomer mixture and the conversion of functional groups.[25]

Adding small amounts of multifunctional monomers results in branching of the main polymer chain. The branched polymer will have a higher polydispersity and melt viscosity than analogous linear polymers. Branching agents are often used to modify the melt viscosity and melt strength of a polymer. Branching in step-growth polymers also changes the relationship between melt viscosity and the shear applied to a melt. Branched polymers tend to undergo a greater degree of shear thinning than unbranched linear polymers.

Monomers of the type $A_x B_y$ are used in step-growth polymerization to produce a variety of polymer architectures, including stars, dendrimers, and hyperbranched polymers.[26-28] The unique architecture imparts properties distinctly different from linear polymers of similar compositions. These materials are finding applications in areas such as resin modification, micelles and encapsulation, liquid crystals, pharmaceuticals, catalysis, electroluminescent devices, and analytical chemistry.

Dendrimers are characterized by highly regular branching following a strict geometric pattern (Fig. 1.4). Dendrimers are prepared in a multistep synthesis often requiring purification between steps. One method of producing dendrimers is known as the divergent method.[29] Using the divergent approach, dendrimer growth starts at the core and proceeds radially out from the center. Each layer is built in a stepwise addition process.

In the convergent method, dendrimer growth begins with chain ends of "surface functional groups" coupling with an AB_y building block.[30] This leads to the next-generation dendron. The process can be repeated to build larger dendrons. Finally, the dendrons can be attached to a polyfunctional core producing a dendritic macromolecule.

Dendrimers produced by divergent or convergent methods are nearly perfectly branched with great structural precision. However, the multistep synthesis of dendrimers can be expensive and time consuming. The treelike structure of dendrimers can be approached through a one-step synthetic methodology.[31] The step-growth polymerization of AB_x-type monomers, particularly AB_2, results in a randomly branched macromolecule referred to as hyperbranch polymers.

Figure 1.4 Dendrimer structure.

The hyperbranch polymers differ from dendrimers in that perfect branching is not achieved and additional linear units are present in the molecule (Fig. 1.5). The extensive branching in hyperbranched polymers prevents crystallization and results in amorphous materials. Hyperbranched materials are generally brittle with low melt viscosity due to the lack of long chains to form entanglements. These properties can be exploited as functional modifiers in crosslinking resins,[32] thermoplastic processing aids,[33] as well as components in adhesives and coatings.

1.3 SYNTHESIS OF STEP-GROWTH POLYMERS

Many synthetic methodologies have been investigated for the synthesis of high-molecular-weight step-growth polymers. However, only organic reactions that proceed in a quantitative fashion (>99%) are suitable for the preparation of high-molecular-weight linear polymers. The susceptibility of the electrophilic carbonyl to nucleophilic attack has received significant attention in step-growth polymerization processes and is widely utilized in commercially important

Figure 1.5 Hyperbranch structure.

families of polymeric materials, including polyesters,[34] polyamides, polyimides, polyurethanes, polycarbonates,[35] epoxy resins,[36] and phenol-formaldehyde polymers.[37] Nucleophilic and electrophilic substitution reactions are also employed in the synthesis of many other classes of step-growth polymers. For example, poly(arylene ethers) are synthesized via the nucleophilic substitution of an aryl halide with a diphenol in the presence of a basic catalyst. Diverse polymer families are prepared using nucleophilic and electrophilic substitution reactions in a step-growth polymerization, including aromatic poly(ketones), poly(arylates), poly(phenylene sulfides), poly(sulfones), and poly(siloxanes).[38,39] Transition metal coupling has also received recent attention for the synthesis of high-performance poly(arylenes) or poly(aryl alkenes).[40,41] In addition, nonpolar polymers are readily prepared via recent advances in step-growth polymerization using ADMET polymerization.[42]

A diverse array of polymeric compositions are attainable using step-growth polymerization processes; however, many experimental criteria must be addressed in order to achieve well-defined compositions and predictable molecular weights. In order to achieve high molecular weight in a step-growth polymerization process, the synthetic methodologies described above must meet certain well-established criteria. The following *essential criteria* are often cited for the successful preparation of high-molecular-weight linear polymers:

1. high reaction conversions (>99.9%) as predicted using the Carothers' equation,
2. monomer functionality (f) equal to 2.0,
3. functional group stoichiometry equal to 1.0,
4. absence of deleterious side reactions that result in loss of monomer functionality,
5. efficient removal of polymerization condensates, and
6. accessibility of mutually reactive groups.[43]

Most introductory polymer textbooks discuss the growth of molecular weight for a step-growth polymerization process. High molecular weight is not achieved until high monomer conversions are reached.[44] This is in sharp contrast to free-radical addition polymerizations where high-molecular-weight polymers are produced at relatively low conversions.

The Carothers equation relates the number-average degree of polymerization to the extent of reaction and average functionality of a step-growth polymer. In the Carothers equation, the number-average degree of polymerization, X_n, relates to the extent of reaction, p, and average functionality, f_{avg}, of the polymer system:

$$X_n = \frac{2}{2 - pf_{avg}}$$

The molecular weight of a polymer will be reduced if either the extent of conversion or the average functionality is decreased. At 95% conversion of difunctional monomers, for example, X_n is only 20.[25] The molecular weight is also related to a stoichiometric imbalance, r, which is normally defined to be less than 1.0:

$$X_n = \frac{1 + r}{1 - r} \quad \text{or} \quad r = \frac{X_n - 1}{X_n + 1}$$

The number-average molecular weight of a polymer may be controlled by offsetting the stoichiometry of two dissimilar mutually reactive difunctional monomers. The polymer will have the same endgroup functionality as that of the monomer used in excess. For a generic polymer made from a difunctional monomer AA with A functional groups and an excess of difunctional monomer BB with B functional groups, r is defined as

$$r = \frac{N_A}{N_B}$$

where N_A is the moles of A functional groups and N_B is the moles of B functional groups. The amount of AA and BB monomer used is then $\frac{1}{2}N_A$ and $\frac{1}{2}N_B$, respectively.

The molecular weight can also be controlled by adding a monofunctional monomer. The monofunctional endgroup, B, has the same functionality as monomer BB. In this case, the moles of A functional groups in the difunctional monomer, AA, is given as N_A and the moles of AA is $\frac{1}{2}N_A$. The moles of B functional group in the difunctional monomer, BB, is given as N_B and the moles of BB is $\frac{1}{2}N_B$. The moles of B functionality in the monofunctional endgroup, B, is given as N'_B which is also equal to the moles of B. The moles of monomers, both mono- and difunctional, containing B functional groups is $\frac{1}{2}N_B + N'_B$. Thus, the stoichiometric imbalance is defined as

$$r = \frac{\frac{1}{2}N_A}{\frac{1}{2}N_B + N'_B}$$

and simplifies to

$$r = \frac{N_A}{N_B + 2N'_B}$$

The derivation of these important equations is described in detail in earlier introductory texts.[25,41–45]

Generally, N_A is assigned an arbitrary value and the values of N_B and N'_B must be calculated. To determine N_B and N'_B, two equations must be solved. The first comes from the above equation, which rearranges to

$$N_B + 2N'_B = \frac{N_A}{r}$$

In order to obtain polymers that are only end capped with the monofunctional end group, the moles of B functional groups must equal the moles of A functional groups. This is expressed in a second equation as

$$N_B + N'_B = N_A$$

By solving these two equations simultaneously, N_B and N'_B can be determined. Figure 1.6 summarizes the impact of the functional group conversion on the molecular weight.[46] High reaction conversion (p) is required to achieve high molecular weight for linear step-growth polymerization processes.[47]

Although most step-growth polymerizations involve the formation of a volatile condensate, this is not a prerequisite for step-growth polymerization, and polyurethane formation is a classic example of a step-growth polymerization that does not form a low-molar-mass condensate.[48] Thus, step growth defines the polymerization process in terms of the basic mechanism, and step-growth polymerization is preferred terminology compared to earlier terms such as condensation polymerization. However, in most instances when a condensate is formed, efficient removal of the condensate using either low pressures (typically 0.1–0.5 mm Hg) or a dry nitrogen purge at high temperatures is required. In addition, efficient agitation and reactor engineering have received significant attention in order to facilitate removal of condensates and ensure accessibility of mutually reactive functional groups. This is especially important in melt polymerization processes where the zero shear melt viscosity (η_o) is proportional to the 3.4 power of the

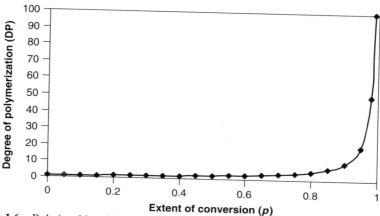

Figure 1.6   Relationship of degree of polymerization to conversion of functional groups in step-growth polymerizations.

weight-average molecular weight.[49] Thus, as molecular weight increases with conversion, the melt viscosity increases dramatically and the requirement for efficient agitation and condensate removal becomes more important.

Linear step-growth polymerizations require exceptionally pure monomers in order to ensure 1 : 1 stoichiometry for mutually reactive functional groups. For example, the synthesis of high-molecular-weight polyamides requires a 1 : 1 molar ratio of a dicarboxylic acid and a diamine. In many commercial processes, the polymerization process is designed to ensure perfect functional group stoichiometry. For example, commercial polyesterification processes often utilize dimethyl terephthalate (DMT) in the presence of excess ethylene glycol (EG) to form the stoichiometric precursor bis(hydroxyethyl)terephthalate (BHET) in situ.

Step-growth polymerization processes must be carefully designed in order to avoid reaction conditions that promote deleterious side reactions that may result in the loss of monomer functionality or the volatilization of monomers. For example, initial transesterification between DMT and EG is conducted in the presence of Lewis acid catalysts at temperatures (200°C) that do not result in the premature volatilization of EG (neat EG boiling point 197°C). In addition, polyurethane formation requires the absence of protic impurities such as water to avoid the premature formation of carbamic acids followed by decarboxylation and formation of the reactive amine.[50] Thus, reaction conditions must be carefully chosen to avoid undesirable consumption of the functional groups, and 1 : 1 stoichiometry must be maintained throughout the polymerization process.

As mentioned previously, the use of multifunctional monomers results in branching. The introduction of branching and the formation of networks are typically accomplished using trifunctional monomers, and the average functionality of the polymerization process will exceed 2.0. As the average functionality increases, the extent of conversion for network formation decreases. In many instances, the trifunctional or higher functional monomers contain reactive groups that are identical to the difunctional monomers. For example, pentaerythritol ($f = 4$) and 1,3,5-benzene tricarboxylic acid ($f = 3$) and trimellitic anhydride ($f = 3$) are commonly used in polyesterification. Many novel families of step-growth polymers are attained through the judicious combination of controlled endgroup functionality, extent of branching, and molecular weight. Hyperbranched step-growth polymers have received significant review in the literature and are an exquisite example of controlled functionality and topology using well-defined monomer functionality.[51]

REFERENCES

1. L. H. Beakland, U.S. Patent. 942, 699, 1907; L. H. Beakland, *Ind. Eng. Chem.*, **1**, 149 (1909).

2. W. H. Carothers, *J. Am. Chem. Soc.*, **51**, 2548 (1929).

3. W. H. Carothers, *J. Am. Chem. Soc.*, **51**, 2560 (1929).

4. W. H. Carothers, U.S. Patents 2,130,947 and 2,130,948 (to DuPont), 1938.

5. J. R. Whinfield and J. T. Dickson, British Patent 578,079, June 14, 1946.

6. W. H. Carothers and F. J. van Natta, *J. Am. Chem. Soc.*, **52**, 314 (1930).

7. H. Mark and G. S. Whitby (Eds.), *Collected Papers of Wallace Hume Carothers on High Polymeric Substances*, Interscience Publishers, New York, 1940.

8. A. Rudin, *The Elements of Polymer Science and Engineering*, 2nd ed., Academic, San Diego, 1998.

9. M. P. Stevens, *Polymer Chemistry: An Introduction*, Addison-Wesley Publishing, Reading, MA, 1975.

10. R. J. Young and P. A. Lovell, *Introduction to Polymers*, 2nd ed., Chapman and Hall, London, 1991.

11. C. S. Wu, *Handbook of Size Exclusion Chromatography*, Dekker, New York, 1995.

12. M. Alger (Ed.), *Polymer Science Dictionary*, 2nd ed., Chapman and Hall, London, 1997.

13. C. K. Schoff, in *Encyclopedia of Polymer Science and Engineering*, Vol. 14, Wiley, New York, 1988, p. 466.

14. L. J. Fetters, D. J. Lohse, and S. T. Milner, *Macromolecules*, **32**, 6847–6851 (1999).

15. J. R. Martin, J. F. Johnson, and A. R. Cooper, *J. Macromol. Sci. Revs. Macromol. Chem.*, **C8**(1), 57–199 (1972).

16. R. W. Nunes, J. R. Martin, and J. Johnson, *Polym. Eng. Sci.*, **22**(4), 205 (1982).

17. R. J. Young, in *Comprehensive Polymer Science: The Synthesis, Characterization, Reactions and Application of Polymers*, Vol. 2: *Polymer Properties*, Pergamon, New York, 1989.

18. J. E. Mark, A. Eisenberg, W. W. Graessley, L. Mandelkern, E. T. Samulski, J. L. Koenig, and G. D. Wignall, *Physical Properties of Polymers*, 2nd ed., American Chemical Society, Washington, DC, 1993.

19. S. H. Goodman, *Handbook of Thermoset Plastics*, 2nd ed., Noyes Publication, Westwood, NJ, 1998.

20. M. E. Rogers, M. H. Brink, J. E. McGrath, and A. Brennan, *Polymer*, **34**(4), 849 (1993).

21. W. D. Joseph, J. C. Abed, R. Mercier, and J. E. McGrath, *Polymer*, **35**(23), 5046–5050 (1994).

22. K. B. Wagener, D. Valenti, and S. F. Hahn, *Macromolecules*, **30**(21), 6688–6690 (1997).

23. A. Noshay and J. E. McGrath (Eds.), *Block Copolymers*, Academic, New York, 1977.

24. R. P. Quirk, D. J. Kinning, and L. J. Fetters, *Comprehensive Polym. Sci.*, **7**, 1 (1989).

25. G. Odian, *Principles of Polymerization*, 3rd ed., Wiley, New York, 1991.

26. H. R. Kricheldorf, *Pure Appl. Chem*, **70**(6), 1235 (1998).

27. H. R. Kricheldorf, *Macromol. Symp.* **122**, 12–23 (1997).

28. K. Inoue, *Prog. Polym. Sci.*, **25**, 453 (2000).

29. C. N. Moorefield and G. R. Newkome, in *Advances in Dendritic Macromolecules*, Vol. 1, G. R. Newkome (Ed.), JAI Press, Greenwich, CT, 1994.

30. C. J. Hawker and K. L. Wooley, in *Advances in Dendritic Macromolecules*, Vol. 2, G. R. Newkome (Ed.), JAI Press, Greenwich, CT, 1995.

31. A. Sunder, J. Heinemann, and H. Frey, *Chem. Eur. J.*, **6**(14), 2499 (2000).

32. R. Mezzenga, L. Boogh, and J. E. Manson, *Composites Sci. Technol.*, **61**, 787–795 (2001).

33. P. Froehling and J. Brackman, *Macromol. Symp.*, **151**, 581–589 (2000).

34. I. Goodman and J. A. Rhys, *Polyesters*, Elsevier, New York, 1975.

35. D. G. LeGrand and J. T. Bendler, *Handbook of Polycarbonate Science and Technology*, Marcel Dekker, New York, 2000.

36. B. Ellis, *Chemistry and Technology of Epoxy Resins*, Blackie, Glasgow, United Kingdom, 1993.

37. J. Ferguson, *Macromol. Chem.*, **3**, 76–92 (1984).

38. I. Yilgor and J. E. McGrath, *Adv. Polym. Sci.*, **86**, 1 (1988).

39. R. W. Dyson (Ed.), *Siloxane Polymers*, Chapman and Hall, New York, 1987.

40. E. E. Braunsteiner and J. K. Stille, *Macromolecules*, **12**, 1033 (1979).

41. G. K. Noren and J. K. Stille, *J. Polym. Sci. (D), Macromol. Rev.*, **5**, 209 (1984).

42. K. B. Wagener and T. A. Davidson, "Non-Conjugated and Conjugated Dienes in Acrylic Diene Metathesis (ADMET) Chemistry," in *New Macromolecular Architecture and Functions. Proceedings OUMS '95*, M. Kamachi and A. Nakamura (Eds.), Springer Verlag, New York, 1996.

43. M. Chanda, *Advanced Polymer Chemistry*. Marcel Dekker, New York, 2000.

44. P. C. Hiemenz, *Polymer Chemistry: The Basic Concepts*, Marcel Dekker, New York, 1984.

45. P. J. Flory, *Principles of Polymer Chemistry*, Cornell University Press, Ithaca, NY, 1953.

46. P. E. Slade, Jr., (Ed.), *Polymer Molecular Weights*, Part 2, Marcel Dekker, New York, 1975.

47. H. R. Allock and F. W. Lampe, *Contemporary Polymer Chemistry*, 2nd ed., Prentice-Hall, Englewood Cliffs, NJ, 1990.

48. D. L. Lyman, *Rev. Macromol. Chem.*, **1**, 191 (1966).

49. J. R. Fried, *Polymer Science and Technology*, Prentice-Hall, Upper Saddle River, NJ, 1995.
50. D. J. Sparrow and I. G. Walton, *Macromol. Chem.*, **3**, 93–97 (1984).
51. P. E. M. Allen and C. R. Patrick, *Kinetics and Mechanisms of Polymerization Reactions*, Wiley, New York, 1974.

2 Polyesters

Alain Fradet and Martine Tessier
Chimie des Polymères, Université Pierre et Marie Curie, Paris, France

2.1 INTRODUCTION

Polyesters are defined as polymers containing at least one ester linking group per repeating unit. They can be obtained by a wide range of reactions, the most important being polyesterifications between dibasic acids and diols or their derivatives (Scheme 2.1).

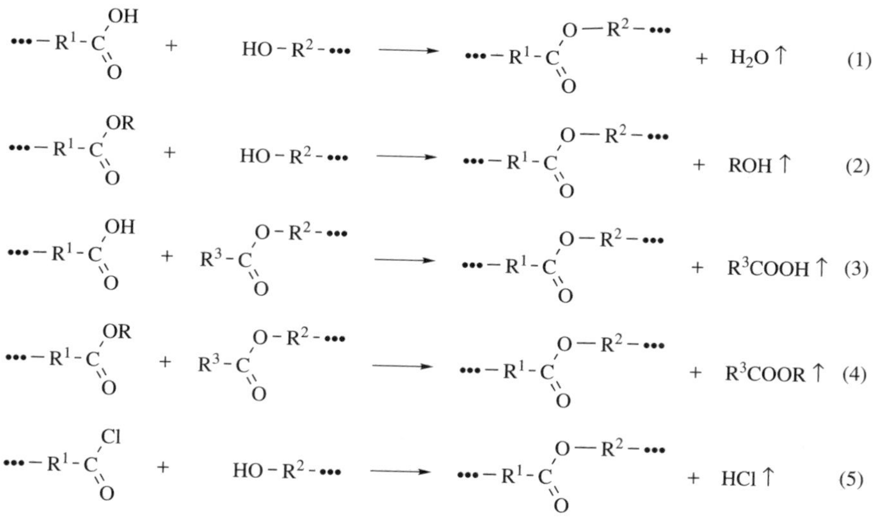

Scheme 2.1

Many other reactions have been reported for the synthesis of polyesters, such as reactions between dicarboxylic acid salts and dialkyl halides, reactions between

Synthetic Methods in Step-Growth Polymers. Edited by Martin E. Rogers and Timothy E. Long
© 2003 John Wiley & Sons, Inc. ISBN: 0-471-38769-X

chlorocarbonyloxy-terminated monomers and diacids, or reactions between bisketenes and diols. These reactions, however, cannot be applied to the synthesis of high-molar-mass polyesters under economically viable conditions and are limited to very specific laboratory-scale syntheses. Two notable exceptions are the ring-opening polymerization of lactones and lactides for the production of degradable polyesters and the biosynthesis of aliphatic polyesters by bacteria or genetically modified plants.

The polyester family is extremely large and, depending on the nature of R^1 and R^2 (Scheme 2.1), exhibits an enormous variety of structures, architectures, properties, and, therefore, applications. Aliphatic polyesters comprise (i) linear low-molar-mass hydroxy-terminated macromonomers used in the synthesis of polyurethanes, (ii) biodegradable and bioabsorbable thermoplastic linear polyesters such as polylactides and polylactones, and (iii) the recently described hyperbranched polyesters applied as crosslinkers in coatings or as rheology modifiers in thermoplastics. Aliphatic–aromatic polyesters such as poly(ethylene terephthalate) (PET) and poly(butylene terephthalate) (PBT) are processed into economically important fibers, films, and engineering thermoplastics. Wholly aromatic copolyesters exhibit superior mechanical properties and heat resistance and have found a number of applications as high-performance thermoplastics. Unsaturated polyesters and alkyd resins (glyptal resins) are thermosetting resins widely used in the coating and composite industries. The polyester family comprises not only thermoplastic polymers and thermosetting resins but also rubberlike polymers: Polyester thermoplastic elastomers (ester TPEs)—copolyesters of terephthalic acid, 1,4-butanediol, and dihydroxy aliphatic polyethers—associate the mechanical properties of rubber with the processing characteristics of thermoplastics. Many other ester-containing polymers could be added to the polyester family, such as polyesteramides, obtained by reaction between diacids, diols and diamines; polyestercarbonates, prepared by reaction of ester-containing monomers; or block copolyesters, prepared by reactions of polyester macromonomers. Polycarbonates, sometimes considered as a special class of polyesters, are described in Chapter 3.

Polyesters are now one of the economically most important classes of polymers, with an overall world production between 25 and 30 million tons in 2000, consisting mostly of PET. This production is rapidly increasing and is expected to continue to do so during the next decade, driven by packaging applications, due to a very favorable image of environmentally friendly and recyclable polymers in western countries, and by textile applications, due to a strong demand in the far-east area to satisfy the needs of an increasing population.

The synthesis and properties of polyesters have been treated in numerous encyclopedia and general review articles.[1–9] This chapter, therefore, does not discuss these aspects extensively but focuses more on recent academic and industrial research developments in polyesters and on the practical aspects of polyester synthesis.

2.1.1 History

It has been known since the early nineteenth century that heating carboxylic polyacids and glycerol resulted in resinous compounds, the exact composition of which remained unknown. In the 1910–1920s, the General Electric Company led extensive research on the chemistry of phthalic anhydride–glycerol reaction and developed the technology of alkyd resins (glyptal resins), which are essentially polyesters of phthalic anhydride, glycerol, and monocarboxylic unsaturated fatty acids. These resins, readily soluble in aromatic or aliphatic hydrocarbons, are still widely used in coatings, varnishes, and paints. However, the modern history of polyesters began in the 1930s when Carothers[10-12] proved the macromolecular theory of Staudinger from experimental studies on reactions between aliphatic dibasic acids and diols and established the relationships between degree of polymerization, conversion, functionality, and gel point, that is, the base relationships of step-growth polymerization. These polyesters had low melting points, were sensitive to hydrolysis, and, therefore, were not suitable for commercial applications. They could not compete with aliphatic polyamides (nylons), also discovered in the 1930s by Carothers at DuPont Company.

To increase polyester melting point and to approach the thermomechanical properties obtained with nylons, it was necessary to stiffen the polyester chain by using rigid aromatic monomers instead of flexible aliphatic ones. This was realized in the early 1940s in the laboratories of the Calico Printers Association in the United Kingdom when Whinfield synthesized high-melting-point fiber-forming polyesters from terephthalic acid and aliphatic diols.[13,14] After World War II the patent rights on these aliphatic–aromatic polyesters were shared between I.C.I. and DuPont, and several members of this family became — and are still today — major commercial polymers. Poly(ethylene terephthalate) (PET) is now one of the most produced polymers, primarily for textile and packaging applications. Poly(butylene terephthalate) (PBT) finds uses as solid-state molding resin. Poly(trimethylene terephthalate) (PTT), though described in Whinfield's original patent, is a newcomer in the commercial polyester family and has found its first applications in the textile industry.

At the end of the 1930s, when Carothers and Flory were studying aliphatic polyesters, a new type of thermosetting resin appeared,[15] based on unsaturated polyesters. Unsaturated polyesters are synthesized by reacting mixtures of saturated and unsaturated dibasic acids or anhydrides with aliphatic 1,2-diols. The thermosetting resin is obtained by dissolving this linear polyester in an unsaturated monomer, such as styrene, capable of undergoing free-radical copolymerization with the unsaturations in polyester chains. The liquid resin is transformed into a rigid, insoluble, and infusible crosslinked polymer network after radical polymerization in the presence of heat or catalysts. Unsaturated polyester (UP) resins found their first applications in combination with glass fibers for protective radar domes during World War II,[8] but the technology was widely commercially available only in the mid-1950s. UPs are now one of the most important matrix resin for glass-fiber-reinforced composite materials.

Although they were the first step-growth polymers fully characterized, it is interesting to note that the first commercial applications of aliphatic polyesters appeared only in the late 1950s and the 1960s, with the development of polyurethane foams and elastomers (see Chapter 5).

The 1970s were a period of intensive research on block copolymers, especially in step-growth polymerization. A number of polyester block copolymers were reported in the literature, but only one found commercial success: the polyesterether block copolymers marketed by DuPont under the tradename Hytrel and exhibiting the characteristics of thermoplastic elastomers.[16] During this period and later in the 1980s, much attention was also focused on high-performance wholly aromatic polyesters. Although the first publications on these polyesters appeared at the end of the 1950s, the commercial introduction of the first one, the amorphous poly(bisphenol-A isophthalate-terephthalate) (Union Carbide's Ardel), took place only in the mid-1970s.[6] In spite of relatively high cost, liquid crystalline thermotropic polyesters, such as poly(6-hydroxy-2-naphthoic acid-co-4-hydroxybenzoic acid) (Ticona's Vectra), described at the end of the 1970s[17], now find a number of applications in high-technology markets.

In the 1990s, environmental concerns began to be gaining ground. The versatility of the ester linkage, able to undergo hydrolysis, alcoholysis, and acidolysis in some conditions, makes polyesters the polymers of choice to fulfill the increasing demand for recyclable and/or biodegradable polymers. This has resulted in a renewed interest in aliphatic polyesters, such as poly(lactones), poly(lactides), or copolyesters containing aliphatic moieties. PET production is also strongly driven by the demand of recyclable polymers.

Polyester chemistry is the same as studied by Carothers long ago, but polyester synthesis is still a very active field. New polymers have been very recently or will be soon commercially introduced: PTT for fiber applications; poly(ethylene naphthalate) (PEN) for packaging and fiber applications; and poly(lactic acid) (PLA), a biopolymer synthesized from renewable resources (corn syrup) introduced by Dow-Cargill for large-scale applications in textile industry and solid-state molding resins. Polyesters with unusual hyperbranched architecture also recently appeared and are claimed to find applications as crosslinkers, surfactants, or processing additives.

2.1.2 Applications

Polyesters are one of the most versatile classes of polymers ever produced, covering a wide range of properties and applications. Polyesters are present in fibers, engineering thermoplastics, and high-performance polymers as well as in thermosetting resins and elastomers. Table 2.1 lists the chemical structure, abbreviations, and uses of some commercially important thermoplastic polyesters.

2.1.2.1   Thermoplastic Polyesters

In terms of volume and economic importance, thermoplastic polyesters are dominated by PET, which has experienced a tremendous development in fibers and molding resins due to a strong demand for textile applications and in food packaging and bottle markets for glass replacement. PET has a very favorable environmentally friendly image, and recycling is a supplementary force driving PET consumption. The 2000 world PET production can be estimated to be ca. 26 million tons and, according to the forecasts of PET industry, is expected to reach a total of 55 million tons by 2010.[18] By comparison with the enormous volume of PET produced, the production of other polyesters appears quite modest (Table 2.2).

2.1.2.1.1   Poly(ethylene terephthalate)

PET FIBERS   Fibers are the largest end use for PET. The world production of PET fibers surpassed that of nylon fibers in the beginning of the 1970s and has now reached a level close to that of cotton, with about 18 million tons in 2000.[19,20] PET fiber production is expected to grow during the next few years at a yearly rate close to 7%, reaching 25 million tons by 2005.[19] The growth of PET fiber production may be assigned to the combination of many factors: low production costs, excellent properties, ability to be blended with natural fibers such as cotton, growing population and increasing wealth in the far-east area, and, depending on their relative price, substitution for cotton or other man-made fibers such as acrylic and polyamides.

Polyester fibers are available in a wide variety of fiber products: staple for blending, filament yarns, high-strength industrial yarns, and nonwoven fabrics. They are present in all segments of the textile market: garments, carpets, nonwoven, and upholstery. Industrial applications include ropes, filters, conveyor belts, and tire cords. The emergence of new high-technology polyester fibers such as silklike very fine PET microfibers is also worth mention.

PET SOLID-STATE RESIN   PET resin, the second largest application of PET, experienced a double-digit yearly growth rate during the last decade[21] and is expected to continue to do so during the next one. The 2000 world production of PET resin was close to 6.5 million tons, driven by food packaging applications, mainly reheat stretch blow-molded bottles (87% of total resin) for carbonated and non-carbonated soft drinks and water.[22,23] PET bottles are also used for a number of other beverages and foods, including syrups, pickled food, and salad dressing. PET containers are also used in toiletries and cosmetics. The excellent price-to-performance ratio of PET recently pushed the development of new, sophisticated applications: Multilayer and coated high-barrier PET containers are reaching the beer bottle market in Europe and the United States. This market is expected to be a great opportunity of growth in PET demand during the next decade.[23] Apart blow-molding applications, PET solid-state resin applications are extremely diverse, ranging from sheet thermoforming applications to glass-fiber-reinforced resin for injection-molding applications.

PET FILM   Due to a slow crystallization rate in the unoriented state but rapid crystallization upon orientation, PET is extremely well suited for the biaxial film orientation process. In this process, the amorphous film obtained after extrusion

TABLE 2.1 Names, Structures, and Applications of Some Commercial Thermoplastic Polyesters

Polyester	Trade Names	Chemical Structure	Applications
Poly(ethylene terephthalate) (PET)	Rynite, Mylar, and Dacron (DuPont) Eastapak (Eastman) Ertalyte (DSM) Impet (Ticona) Tergal (Tergal Fibres)	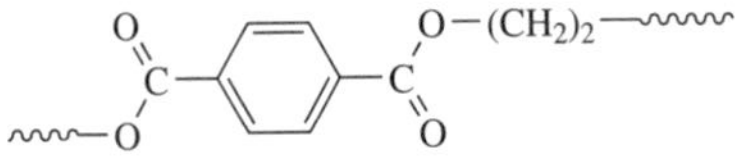 $O-(CH_2)_2$ benzene ring diester	Fibers, engineering thermoplastic resins, bottle and container resins, films
Poly(trimethylene terephthalate) (PTT)	Corterra (Shell) Sorona (DuPont)	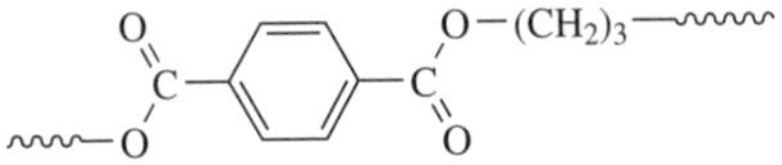 $O-(CH_2)_3$ benzene ring diester	Fibers, engineering thermoplastic resins
Poly(butylene terephthalate) (PBT)	Ektar (Eastman) Pocan (Bayer) Valox (G.E.) Crastin (DuPont)	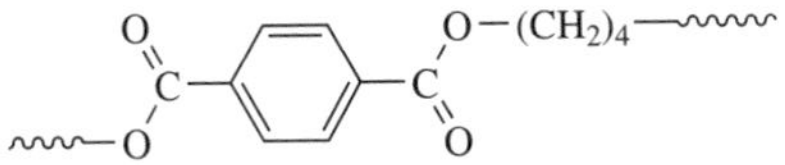 $O-(CH_2)_4$ benzene ring diester	Engineering thermoplastic resins
Poly(ethylene 2,6-naphthalate) (PEN)	Kaladex (DuPont) Pentex (Honeywell-AlliedSignal)	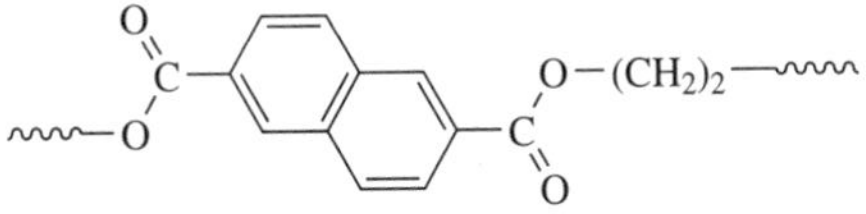 $O-(CH_2)_2$ naphthalene diester	High-performance films and fibers, hot-fill containers

Name	Trade name (Company)	Structure	Applications
Poly(2,2'-bis(1,4-phenylene)propane terephthalate-*co*-isophthalate) (polyarylate, PAR)	Ardel (Amoco)	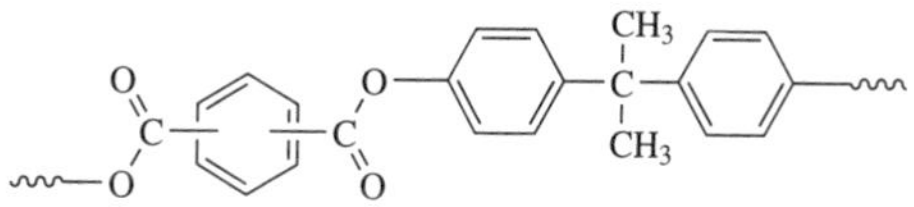 *(structure: bisphenol-A terephthalate/isophthalate)*	Injection-molding resins for UV-resistant, transparent parts
Poly(2,6-naphthalate-*co*-1,4-benzoate) (polyester LCP)	Vectra (Ticona)	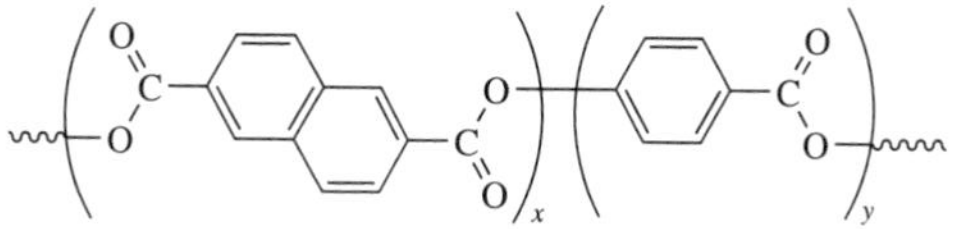 *(structure: naphthalate/benzoate copolymer)*	Heat-resistant parts, precision parts, miniature connectors
Poly(poly(oxytetramethylene)-*block*-poly(1,4-butylene terephthalate)) (polyester TPE)	Hytrel (DuPont)	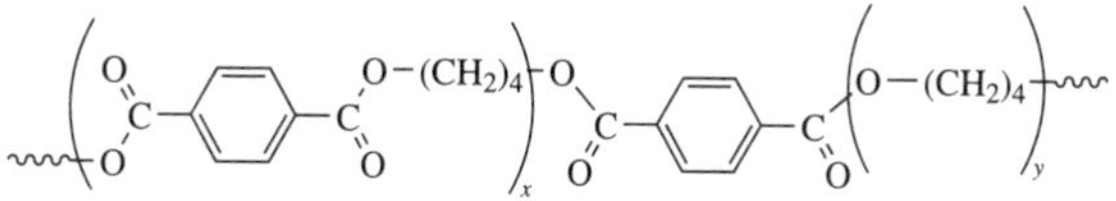 *(structure: poly(oxytetramethylene)-block-poly(1,4-butylene terephthalate))*	Elastomeric thermoplastic resins, tubing, pipes, rubber like parts
Poly(L-lactate) (PLLA)	NatureWorks (Cargill-Dow)	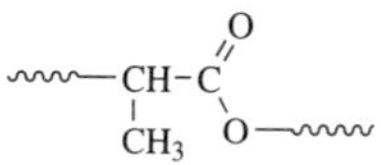 *(structure: poly(L-lactate))*	Fibers, engineering thermoplastic resins, bioresorbable/biodegradable thermoplastic resins

TABLE 2.2   World Production and Prices of Some Commercially Important Polyesters

Polyesters	World Production[a] (1000 tons)	Prices[b] (U.S.$/kg)
Thermoplastics		
PET (total)	25,900[c]	1.30–2.60
PET (fibers)	18,000[c]	
PET (solid-state resin)	6,300[c]	
PET (films)	1,000[c]	
PBT	400[d]	2.20–2.60
PEN	40[e]	8.90[e]–38.00[f]
Polyarylate (PAR)	—[g]	3.90–4.70
Polyester LCPs	12[h]	21.00–26.20
Thermoplastic elastomers	65[i]	4.70–6.00
Macromonomers		
Polyester polyols	730[j]	—[g]
Thermosetting resins		
Alkyd resin	—[g]	1.90–2.10
Unsaturated polyester	1700[k]	1.65–3.75

[a] Estimated 2000 world production unless otherwise specified.
[b] From www.plasticsnews.com for pure resin unless otherwise specified.
[c] Refs 18–20.
[d] Ref. 443.
[e] Ref. 24.
[f] PEN film price, ref. 444.
[g] Data not available.
[h] Ref. 35.
[i] Ref. 37.
[j] Ref. 47.
[k] Ref. 48.

and quenching is successively drawn in two orthogonal directions, leading to stress-induced crystallization with biaxial orientation of polymer chains in the film plane. The resulting transparent film is ca. 50% crystalline and exhibits excellent mechanical, thermal, and electrical insulating properties. PET film is the base film for many photographic films and magnetic tapes. It is used as dielectric film in capacitors and finds a number of applications in food and pharmaceutical packaging. The world production of PET film is estimated to be 1 million tons, increasing at a rate of ca. 5% per year.[18]

2.1.2.1.2 Other Aromatic–Aliphatic Thermoplastic Polyesters

PBT is mainly used as glass-fiber-reinforced engineering thermoplastic, although PBT fibers can also be made. PBT crystallizes much more rapidly than PET and does not require nucleating agents for extrusion or injection-molding applications. PBT crystallizes rapidly in mold at essentially room temperature. Moreover, due to a lower melting point than PET, it can be molded at comparatively lower molding temperature. This permits fast processing and rapid production cycles. However, larger volumes and lower prices (Table 2.2) make PET a serious PBT competitor for solid-state resin applications.

PEN presents many advantages over PET in terms of thermomechanical performances. However, the cost of dimethyl 2,6-naphthalene dicarboxylate monomer limits its present use to specific applications, for example, high performance long-playing magnetic tapes, photographic film for advanced photo system (APS), high-strength fibers, and aerospace-related products. Taking advantage of outstanding barrier properties, PEN and PEN/PET copolymers are also expected to find new applications in the pasteurizable, hot-fill, and returnable/refillable container and bottle markets.[24]

Poly(1,4-cyclohexylenedimethylene terephthalate) (PCT) finds some niche applications in the packaging market as amorphous, clear, extrusion, or injection-molding material for films and bottles.

PTT was already mentioned in Whinfield and Dickson's 1941 original patent[13] that described PET and PBT. However, no economically viable route to 1,3-propanediol monomer was available until the 1990s when Shell developed a route based on ethylene oxide hydroformylation and Degussa a route based on acrolein hydration/hydrogenation.[25] Shell and DuPont launched respectively their Corterra and Sorona PTT polymers in 2000. DuPont uses the acrolein route but plans to produce 1,3-propanediol from cornstarch by a biotechnological process in 2003.[26] PTT fibers present a combination of properties, including stretch recovery, resilience, softness of touch, dyeability, and easy care which, make them very attractive for the fabric, carpet, and apparel markets. PTT is also intended to compete for film and engineering thermoplastic applications with PET, PBT, and polyamides.[27] According to Shell Chemicals, the PTT demand is expected to exceed 1 million tons by 2010.[28]

2.1.2.1.3 Wholly Aromatic Polyesters

This polyester class comprises amorphous high-T_g copolyesters, known as amorphous polyarylates, and semicrystalline polyesters that often exhibit anisotropic liquid crystalline (LC) melts. Liquid crystalline polyesters are often termed as polyester LCPs.

Copolyesters of bisphenol-A and iso- and terephthalic acid are amorphous engineering thermoplastics with excellent heat and ultraviolet (UV) light resistance and general properties that compare well with other engineering resins. Their amorphous morphology imparts on them properties of transparency and dimensional stability with somewhat low solvent resistance. Amorphous polyarylates are light-yellow to amber transparent resins that find a number of niche applications for injection-molded parts in automotive, electronics, safety and building equipment — for example, headlight housings, fire helmets, face shields, and transparent exterior parts such as solar energy components and glazing.

Polyarylates can be blended with a wide range of commercially available thermoplastics, including polyamides, polycarbonates, polyetherketones, polyesters, and poly(phenylene sulfide), thus broadening their application domain.

Liquid crystalline polyesters are copolyesters of rigid para-oriented monomers such as 4-hydroxybenzoic acid, hydroquinone, terephthalic acid, or 2,6-hydroxynaphthoic acid. They exhibit an LC mesophase in the molten state, usually a nematic one: Due to high stiffness, polymer chains tend to line up parallel to each other, instead of forming random coils and isotropic melts. This unique structure is retained in the semicrystalline solid forming materials termed as "self-reinforced" thermoplastic resins, by analogy with fiber-reinforced polymers. Polyester LCPs offer a combination of properties that makes them extremely well suited for precision molding of small parts: extremely high melt fluidity, high heat resistance and dimensional stability, exceptional tensile strength and modulus in the flow direction, and low-temperature expansion coefficient. Their major market is electronics in applications such as miniature connectors with thin walls and surface-mount interconnection devices where LCP flow properties allow high yield molding.[29,30] Other applications comprise precision parts for audiovisual equipment and under-hood parts for automotive industry. Owing to their outstanding barrier properties to oxygen, carbon dioxide, and water vapor, applications as barrier layer in multilayer packaging materials are also under consideration.[30-33] Following analysts, LCP demand is growing at a yearly rate higher than 20% from a 2000 production estimated to be ca. 12,000 tons.[34,35]

2.1.2.1.4 Polyester Thermoplastic Elastomers

Thermoplastic elastomers (TPEs) are polymers that present the typical mechanical properties of rubber and the processing ease of thermoplastics. Polyester TPEs are multiblock copolymers containing blocks of amorphous low-T_g polyether covalently bonded to blocks of semicrystalline PBT polyester. At service temperature, the material exhibits biphasic morphology with microphase separation between a low-T_g polyether-rich phase and a semicrystalline PBT-rich phase. The semicrystalline microphase acts as physical crosslinks for the amorphous low-T_g regions imparting the material properties of an elastomer. At processing temperature, the PBT phase melts, allowing the polyester TPE to be processed by conventional methods such as extrusion or injection-molding. Polyesterethers are structurally related to linear polyurethanes (Chapter 5).[16,36] A typical member of this class of polyesters is DuPont's Hytrel, a poly(poly(oxytetramethylene)-block-poly(butylene terephthalate)) obtained by replacing some of 1,4-butanediol by dihydroxy-poly(oxytetramethylene) polyether in PBT synthesis. Depending on the polyether–PBT ratio, polyester TPE properties range from that of rubbery polymers to that of rigid engineering thermoplastics. They replace thermoplastics when improved impact resistance, flexural and compressive properties, and spring characteristics are required: industrial and automotive hydraulic tubing, hoses, gaskets and bellows; autoclavable medical tubing; and jacketing for electrical and fiber-optics cable. Because of their high water vapor permeability, some polyoxyethylene polyether-based grades are used as waterproof breathable films in the manufacture of outer garments, tents, and footwear.[37] Polyester TPE production and prices are reported in Table 2.2.

2.1.2.1.5 Degradable Polyesters

The ester linkage of aliphatic and aliphatic–aromatic copolyesters can easily be cleaved by hydrolysis under alkaline, acid, or enzymatic catalysis. This feature makes polyesters very attractive for two related, but quite different, applications: (i) bioresorbable, bioabsorbable, or bioerodible polymers and (ii) environmentally degradable and recyclable polymers.

BIORESORBABLE POLYESTERS Bioresorbable implants, for instance orthopedic fixations or sutures, are devices designed to slowly degrade in the body after implantation so that a second surgical intervention is not required for implant removal after healing. Polyesters and copolyesters of lactic acid and glycolic acid (Table 2.3) have been used as bioresorbable sutures since the 1960s. These polyesters exhibit a set of properties that makes them the ideal bioresorbable polymers: They have good mechanical and processing properties, can be sterilized, give no toxic or inflammatory response, and are slowly and completely hydrolyzed into natural metabolic by-products that are eliminated by the body. Since then, other polymers have been accepted for use as bioresorbable polymers: poly(dioxanone) (PDO), poly(trimethylene carbonate) (PTC), and poly(ε-caprolactone) (PCL) and their copolymers.[38] Copolymer composition can be adjusted to fulfill a large range of properties and degradation times. Polyesters produced by microorganisms, such as poly(hydroxyalkanoic acid) (PHA), have also been considered for use in biomedical devices.[39,40] However, synthetic biodegradable polymers are generally preferred for such applications since they offer a more reliable source of raw material with respect to antigenicity concerns.[38] Bioresorbable polymers can be melt processed by conventional methods such as injection-molding, compression-molding, and extrusion. Sterilization is achieved by irradiation or by exposure to ethylene oxide. They are used in wound closure (sutures, staples), osteosynthetic materials (orthopedic fixation devices: pins, screws, rods, bone plates), cardiovascular surgery (stents, grafts), and intestinal surgery (anastomosis rings).[41] They also find interesting applications as matrix materials for implanted drug release devices or drug-containing microspheres or microcapsules.[41,42]

ENVIRONMENTALLY DEGRADABLE POLYESTERS The use of environmentally biodegradable/compostable harmless plastics — ecofriendly plastics — that disappear after a few weeks or months in soil is an elegant way of dealing with solid waste disposal — a crucial problem for large modern cities. Environmentally degradable polymers are defined as polymers that are hydro-biodegradable and completely converted by microorganisms to carbon dioxide (or methane in anaerobic conditions), water, and biomass. To be attractive, such polymers should present the thermomechanical properties of common plastics, typically between those of low-density polyethylene and polypropylene, and should be melt processable by

TABLE 2.3 Names, Composition, and Applications of Some Commercial Degradable Polyesters

Composition	Trademarks	Company	Uses	Origin of Raw Materials
Polyesteramide	BAK	Bayer	Environmental	Petrochemistry
Copolyester of terephthalic acid, aliphatic acids, and diols	Biomax	DuPont	Environmental	Petrochemistry
Poly(hydroxyalkanoic acid)s	Biopol[a]	Monsanto	Environmental	Biosynthesis by bacteria or transgenic plants
Copolyesters of succinic acid, adipic acid, 1,2-ethanediol, and 1,4-butanediol	Bionolle	Showa HighPolymer	Environmental	Petrochemistry
Poly(ε-caprolactone)	CAPA	Solvay	Environmental	Petrochemistry
Poly(glycolic acid)	Dexon	Davis & Geck	Biomedical	—[b]
Copolyester of terephthalic acid, adipic acid, and 1,4-butanediol	Eastar Bio	Eastman	Environmental	Petrochemistry
Copolyester of terephthalic acid, aliphatic acids, and diols	Ecoflex	BASF	Environmental	Petrochemistry
Poly(L-lactic acid)	EcoPLA, NatureWorks	Dow-Cargill	Environmental, biomedical	Lactic acid from corn starch fermentation
Poly(L-lactic acid)	Lacty	Shimadzu	Environmental, biomedical	—[b]
Poly(ethylene succinate)	Lunare SE	Nippon Shokubai	Environmental	Petrochemistry

[a]Originally developed by ICI-Zeneca Bioproducts.
[b]Information not available.

conventional methods. Aliphatic polyesters and aliphatic–aromatic copolyesters are ideally suited to fulfill these specifications. During the last few years, a number of companies have put biodegradable polymers on the market. Almost all these polymers are polyesters or copolyesters: aliphatic polyesters such as poly(ε-caprolactone), poly(butylene-*co*-ethylene succinate), poly(lactic acid) and microbial poly(hydroxyalkanoic acid), copolyesters of terephthalic acid with various aliphatic diols and diacids, and polyesteramides (Table 2.3).

Price is of course the determining factor for applications including fast food service cups, containers and cutlery, paper coatings, disposable films for food packaging, ground covers, garden bags, and seed mats. The biodegradable plastics market is expected to reach 90,000 tons/year by 2003 with an estimated price of U.S. \$3–\$4/kg,[43] making microbial polyesters by far too expensive for these applications.[44]

Special mention must be made of poly(lactic acid), a biodegradable/bio-resorbable polyester, obtained from renewable resources through fermentation of corn starch sugar. This polymer can compete with conventional thermoplastics such as PET for conventional textile fibers or engineering plastics applications. The first Dow-Cargill PLA manufacturing facility is scheduled to produce up to 140,000 tons of NatureWorks PLA per year beginning in 2002[45] at an estimated price close to that of other thermoplastic resins: U.S. \$1/kg.[46] Other plants are planned to be built in the near future.[45]

2.1.2.2   *Polyester Polyols*

The most representative member of this class of polyesters is the low-molar-mass ($\overline{M_n} \approx 1000–3000$) hydroxy-terminated aliphatic poly(2,2'-oxydiethylene adipate) obtained by esterification between adipic acid and diethylene glycol. This oligomer is used as a macromonomer in the synthesis of polyurethane elastomers and flexible foams by reaction with diisocyanates (see Chapter 5). Hydroxy-terminated poly(ε-caprolactone) and copolyesters of various diols or polyols and diacids, such as *o*-phthalic acid or hydroxy acids, broaden the range of properties and applications of polyester polyols.

Polyester polyols account for only ca. 10% of the total polyol market, which is dominated by polyether polyols such as hydroxy-terminated polyoxyethylene or polyoxypropylene. Polyester polyols are preferred for applications where better mechanical properties, wear resistance, and UV stability are required. The largest application of polyester polyols is flexible specialty polyurethane foam in the furniture, packaging, and automotive industries. Polyester polyols are also used for nonfoam applications such as coatings, paints, sealants, and adhesives.[47]

2.1.2.3   *Thermosetting Polyester Resins*

2.1.2.3.1   *Unsaturated Polyester Resins*

Unsaturated polyesters are low-molar-mass polyesters obtained by melt poly-condensation of 1,2-diols with saturated and unsaturated anhydrides or dibasic acids.[7] A typical unsaturated polyester composition consists of maleic anhydride, phthalic anhydride and/or isophthalic acid, and 1,2-propanediol. For almost all applications, unsaturated polyesters are diluted with a copolymerizable monomer such as styrene. The resulting syruplike unsaturated polyester resin (UPR) can be compounded with fillers and/or glass fibers and cured in the presence of free-radical initiators. It yields thermoset articles with a wide range of properties, depending on the composition of the initial unsaturated polyester, the nature and concentration of crosslinking monomer, the presence of reinforcing fibers, and the choice of initiator and catalytic systems.

Cast UPRs are not reinforced with glass fibers but are filled with large amounts (up to 90%) of low-cost fillers. Their applications include polyester cements and mastics, simulated marble and wood, and various encapsulation compounds. UPR-laminating resins are sprayed on glass fiber or laid on glass cloth with an initiator and used for molding, continuous lamination, filament winding, or pultrusion applications. Gel coats are protective layers primarily used in the marine industry and consisting of pigmented, filled UPRs that afford high gloss surfaces with good environmental resistance. They are usually sprayed with a peroxide initiator onto the mold before reinforcement with glass fiber and lami-nating resins. The largest applications of UPRs are fiberglass-reinforced molding resins for high-temperature and/or high-pressure cure. Semifinished compounds containing highly reactive UPR, high-temperature initiators, reinforcing glass fiber, fillers, mold release agents, pigments, and thermoplastics to prevent shrink-age, are required for fast molding cycles in automated industries: the so-called sheet molding compounds ("SMCs") and bulk molding compounds ("BMCs") are ready-to-use compounds for compression-molding (SMC and BMC) or injection-molding (BMC) applications.

The primary UPR end markets are construction, automotive, and marine industries with applications such as house paneling, tub and shower applications, chemical-resistant storage tanks, pultruded profiles, and fiberglass composite boat hulls. The UPR industry is mature, with a world production close to 1.7 million tons (Table 2.2), but must face two important issues: increasingly strict regulations for styrene emissions and poor recycling potential for polyester thermosets.[48,49]

2.1.2.3.2 Alkyd Resins

Alkyd resins are branched polyesters made by reacting dicarboxylic acids or anhydrides, polyols such as glycerol or pentaerythritol, and long-chain unsatu-rated monocarboxylic fatty acids derived from natural oils (e.g., linseed oil, soya oil, or dehydrated castor oil).[50] The unsaturations present in fatty acid chains are susceptible to undergo oxidative crosslinking by air oxygen in the presence of catalytic amounts of siccatives, such as metal salt of organic acids. Alkyd resins are therefore well suited to applications such as air-drying varnishes or archi-tectural paints, typically in solvent-based formulations. Their low cost and great versatility make them the largest type of coatings produced worldwide. Because of regulations for reduction of volatile organic compounds (VOCs) emissions, their use is, however, slowly decreasing.[51] To address this issue, waterborne alkyd resin formulations have been developed, but their application is limited by the tendency of ester groups to hydrolyze in the basic aqueous systems used for solubilization or dispersion.

2.1.3 Issues and Research Trends

The most important trend in industrial and academic research on polyesters is perhaps due to the increasing environmental awareness of end consumers and the resulting strong demand for biodegradable or recyclable materials made by ecofriendly production processes (see Chapter 11). Most thermoplastic polyesters are versatile, fully recyclable polymers, harmless by nature, and this is one of the reasons for the good health and bright future of the polyester industry. A number of biodegradable/compostable copolyester compositions have been developed in the recent years by the polyester industry to satisfy the demand for biodegradable packaging materials (Table 2.3). Even the newly introduced PTT, which is otherwise a quite conventional aromatic–aliphatic polyester, is presented as a "green" polymer because one of its monomers (1,3-propanediol) can be produced from corn starch, a renewable resource. This seems more justified for PLA, which, as mentioned above, is a biodegradable/bioresorbable recyclable polyester exhibiting quite interesting thermomechanical properties that can be produced from renewable resources. Other attempts are the biosynthesis of aliphatic polyesters by transgenic plants to produce poly(hydroxyalkanoic acid)s (PHAs).[44] Biosynthetic processes based on transgenic potatoes or soybeans are expected to be economically viable in the near future, with production costs as low as U.S. $0.5/kg for PHA.[39] On the other hand, many research groups are involved in studies on the conventional ring-opening polymerization of cyclic esters such as glycolide, lactide, or ε-caprolactone for the synthesis of biodegradable/bioresorbable polyesters.[52]

Ester groups can undergo a variety of interchange reactions that are used for the chemical modification of polyesters. The chemical recycling of thermoplastic polyesters into low-molar-mass reactive oligomers by alcoholysis or glycolysis is described in Chapter 11 (e.g., the synthesis of hydroxy-terminated oligomers for polyurethane applications). A number of efforts have also been put in the reactive extrusion of polyesters with various reactants, including polymeric ones, for the synthesis of copolyesters, block copolyesters, or modified polyesters.[53–56]

Polyester thermosets, both UPR and alkyd resins, are facing severe regulatory issues for limiting VOC and toxic compound emissions. Although these issues are not new, only partly satisfying solutions have been proposed until now, and further research is clearly needed in this domain.

The polymerization of cyclic low-molar-mass polycarbonates, polyarylates, and PBT to high-molar-mass thermoplastics has been extensively studied by the General Electric Company during the last decade.[57,58] Due to very low viscosity, cyclic oligoesters can be processed like thermosetting resins but retain thermoplastic properties in the final state, after polymerization in the presence of suitable catalysts. This permits the fabrication of thermoplastic composites that present a number of advantages over their thermosetting counterparts: very fast processing speed, ability to thermoform and recycle parts, and higher damage tolerance. This innovative technology is now commercialized by Cyclics Corporation.[59]

A new type of polyester appeared recently exhibiting unusual hyperbranched architecture instead of the linear and tridimensional architectures of thermoplastics and thermosets, respectively. Hyperbranched polymers belong to the class of dendritic polymers, which present a tridimensional, highly branched structure with a great number of endgroups.[60-63] Hyperbranched polyesters are prepared by the step-growth polymerization of AB_x-type monomers, where A and B are ester-forming, mutually reactive endgroups.[64] A polyfunctional B-type chain limiter is often added to limit molar mass and to reduce the polydispersity index, which would otherwise be extremely high. By analogy with dendrimers, this chain limiter is often referred to as "core molecule." The bulk polyesterification of 2,2-dimethylolpropionic acid (bis-MPA) and a polyhydroxy core molecule in the presence of acid catalyst yields hydroxy-terminated hyperbranched polyesters[65,66] that have recently been commercially introduced under the brand name Boltorn by Perstorp Polyols (see Section 2.4.5.1). Hyperbranched polyesters, like other hyperbranched polymers, have low viscosity, a large number of endgroups, high solubility, and high molar mass and are nonentangled molecules. This combination of properties makes them brittle and unusable as structure thermoplastic materials. They are well suited for use as rheology modifiers or processing aids in thermoplastics, as blend components, and as crosslinking macromonomers in thermosets, after endgroup modification if necessary.[60,64] The use of hyperbranched polyesters has also been suggested for a rapidly growing number of applications such as nonlinear optics, drug delivery systems, catalysis, or metal surface modification.[60,64]

2.2 STRUCTURE–PROPERTY RELATIONSHIPS

The nature — aliphatic or aromatic — of the bivalent $-R^1-$ and $-R^2-$ radicals in polyester chains (Scheme 2.1) exerts a profound influence on the properties of polyesters and define four main classes of linear polyesters:

1. Aliphatic polyesters are low-melting (40–80°C) semicrystalline polymers or viscous fluids and present inferior mechanical properties. Notable exceptions are poly(α-hydroxy acid)s and poly(β-hydroxy acid)s.

2. Aromatic–aliphatic polyesters, in which either R^1 or R^2 is aromatic, are generally high-melting (150–270°C) semicrystalline materials that find applications as engineering thermoplastics, films, or fibers.

3. Wholly aromatic polyesters, in which both R^1 and R^2 are aromatic, are either high-T_g amorphous polymers or very high melting semicrystalline polymers that often exhibit liquid crystalline properties.

4. Polyester thermoplastic elastomers, which are obtained by replacing a part of the R^2 diol by dihydroxy polyether macromonomer, present biphasic morphology and rubberlike properties.

The other structural parameters, such as the ester group content, the presence, nature, and position of substituents on monomer units, and the number and nature of comonomers, are of less importance and do not modify this classification of linear polyesters.

Additional parameters should be taken into account for polyester networks and hyperbranched polyesters, for example, crosslink density and degree of branching.

2.2.1 General Considerations

2.2.1.1 Thermal Properties

There are close correlations between melting temperature and a number of important polymer properties: High-melting polymers generally present high tensile strength and modulus, high deflection temperature under load, high thermal resistance, low elongation at break, and low solubility in common solvents.

Since melting is a thermodynamic equilibrium, melting temperature (T_m) is given by the classical relationship $T_m = \Delta H_m / \Delta S_m$, where ΔH_m and ΔS_m are the enthalpy and entropy of melting, respectively. Therefore, T_m can be regarded at first approximation as a measure of both intermolecular interactions, which govern ΔH_m, and chain flexibility, which govern ΔS_m, parameters that obviously control many of the polymer properties. Examining melting temperature, therefore, indirectly provides useful information on polymer general properties. However, it is to be noted that the melting temperature of semicrystalline polymers might also be influenced by other factors, for example, molar mass, degree of crystallinity, and crystallite size.

Linear polyesters do not have strongly interacting groups, for example, H-bonded ones. Consequently, ΔH_m is not dramatically different from a given polyester to another one, and the predominant factor affecting melting temperature is ΔS_m, that is, the number of conformations available in the molten state, in other words chain flexibility. Higher flexibility leads to higher entropy of melting and lower melting temperature. Therefore, T_m is very sensitive to the nature of R^1 and R^2 moieties in polyester chains and as a result varies over a very broad range, from room temperature to temperatures above 500°C.

At equivalent ester group content, the T_m of aliphatic polyesters is significantly lower than that of aromatic–aliphatic ones (Table 2.4). When the length of the alkyl chain decreases, the rigidifying influence of $-CO-O-$ groups predominates, resulting in the remarkably high melting point of poly(glycolic acid) $[-(CH_2-COO-)_n-]$ with respect to that of other poly(ω-hydroxyacids) (Table 2.5).

In the poly(alkylene arylate) series, T_m decreases with increasing length of flexible $-(CH_2)_n-$ moieties and, as in the aliphatic series, approaches the limiting value of polyethylene melting point for large n values (Table 2.6). Aromatic–aliphatic polyesters with even numbers of methylene groups melt at higher

TABLE 2.4   Melting Point, T_m (°C), and Glass Transition Temperature, T_g (°C), of Poly(alkylene adipate)s and Poly(alkylene terephthalate)s

Repeating-Unit Formula	$-\overset{\text{O}}{\underset{\text{}}{\text{C}}}-(CH_2)_4-\overset{\text{O}}{\underset{\text{}}{\text{C}}}-O-(CH_2)_n-O-$		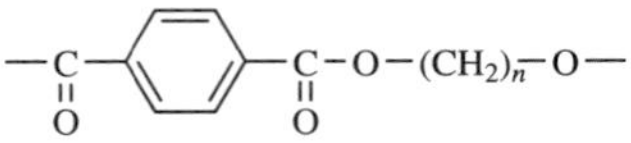 $-\overset{\text{O}}{\underset{\text{}}{\text{C}}}\!\!-\!\!\bigcirc\!\!-\!\!\overset{\text{O}}{\underset{\text{}}{\text{C}}}-O-(CH_2)_n-O-$	
n	T_g^a	T_m^b	T_g^c	T_m^c
1			101	269
2	−63	50	69[a]	265[b]
3	−59	45	35	233
4	−74[d]	55.8[d]	17[a]	232[b]
5	—	—	10	134
6	−73[e]	61[e]	−9	154
7	—	—	3	85
8	—	—	—	132
9	—	—	−3	95
10	−56	80	−5	125

[a]Ref. 445.
[b]Ref. 448.
[c]Ref. 4.
[d]Ref. 446.
[e]Ref. 447.

TABLE 2.5   Melting Point, T_m (°C), and Glass Transition Temperature, T_g (°C), of Poly(ω-hydroxyacids)[a]

Repeating-Unit Formula	$-O-(CH_2)_n-\overset{\text{O}}{\underset{\text{}}{\text{C}}}-$	
n	T_g	T_m
1	45	228
2	−24	93
3	−59	64
4	−66	58
5	−64	69.2
9	—	80[b]
10	−46	92
12	−44	95
14	−22	97.5
15	—	94[c]
—	−36	114.6

[a]Ref. 449.
[b]Ref. 448.
[c]Ref. 450.

TABLE 2.6   Melting Point, T_m (°C), and Glass Transition Temperature, T_g (°C), of Poly(alkylene arylate)s

Repeating-Unit Formula	$-\overset{\text{O}}{\underset{\text{O}}{\text{C}}}-\text{Ar}-\overset{\text{O}}{\underset{\text{O}}{\text{C}}}-\text{O}-(CH_2)_n-\text{O}-$					
Ar	⬡		⬡⬡		⬡⬡	
n	T_g^a	T_m^a	T_g^b	T_m^a	T_g^c	T_m^c
1	101	269	—	340	—	—
2	69[b]	265[d]	117[e]	337[e]	—	>350[a]
3	35	233	73	199	—	263
4	17[b]	232[d]	76	241	—	296
5	10	134	38	135	42	176
6	−9	154	44	211	—	240
7	3	85	—	130	—	168[f]
8	—	132	—	185	—	193[d]
9	−3	95	—	124	—	128[d]
10	−5	125	14	144	—	164[d]

[a] Ref. 4.
[b] Ref. 445.
[c] Ref. 452.
[d] Ref. 448.
[e] Ref. 451.
[f] Ref. 453.

temperature than odd-numbered ones. This odd–even effect has been reported for a number of polyesters[4] and copolyesters[67,68] as well for many other step-growth polymers.[69,70] It is generally assigned to variations in the planar zigzag packing of polymer chains. Chain stiffness is enhanced by resonance interactions between phenylene groups and ester carbonyl as exemplified by the much higher melting point of poly(alkylene terephthalate) compared to their 1,4-phenylene-bis acetate-based isomers (Table 2.7). The biphenylene dicarboxylate series forms anisotropic liquid crystalline melts due to the presence of rigid biphenylene mesogenic groups.[71] As a rule, T_m increases with increasing proportion of aromatic or trans-cycloaliphatic rigid units (see Table 2.8 and compare Tables 2.4 and 2.6).

Aromatic polyesters that do not contain any flexible structural units are often nonmeltable or extremely high melting polymers that cannot be processed. Copolymerization is a way to obtain processable wholly aromatic polyesters: The T_m versus copolyester composition curve is a U-shaped curve exhibiting a minimum that is generally well below the T_m of corresponding homopolymers. Liquid crystalline aromatic polyesters, for instance, are usually copolymers.[72] An example is Ticona's Vectra, a random copolyester containing 4-oxybenzoyl and 6-oxy-2-naphthoyl units in ca. 70 : 30 mol ratio. This copolymer melts at ca.

TABLE 2.7   Melting Temperatures of Isomeric Poly(alkylene terephthalate)s and Poly(alkylene 1,4-phenylene-bisacetate)s

Repeating-Unit Formula	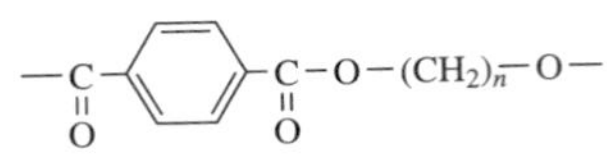 $-\overset{\underset{\|}{O}}{C}$—⟨benzene⟩—$\overset{\underset{\|}{O}}{C}$—O—$(CH_2)_n$—O—		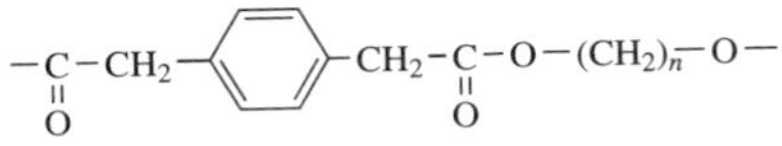 $-\overset{\underset{\|}{O}}{C}-CH_2$—⟨benzene⟩—$CH_2-\overset{\underset{\|}{O}}{C}$—O—$(CH_2)_n$—O—	
	n	T_m (°C)[a]	n	T_m (°C)[b]
	4	232[b]	2	137
	5	134	3	58
	6	154	4	73[c]

[a]Ref. 4.
[b]Ref. 445.
[c]Ref. 420.

TABLE 2.8   Melting Point, T_m (°C), and Glass Transition Temperature, T_g (°C), of Aliphatic, Aromatic–Aliphatic, and Cycloaliphatic Polyesters

Repeating-Unit Formula	T_g	T_m
$-\overset{\underset{\|}{O}}{C}-(CH_2)_4-\overset{\underset{\|}{O}}{C}-O-(CH_2)_4-O-$	-74[a]	55.8[a]
$-\overset{\underset{\|}{O}}{C}$—⟨benzene⟩—$\overset{\underset{\|}{O}}{C}-O-(CH_2)_4-O-$	17[b]	232[c]
$-\overset{\underset{\|}{O}}{C}-(CH_2)_4-\overset{\underset{\|}{O}}{C}-O$—⟨benzene⟩—$O-$	—	230[d]
$-\overset{\underset{\|}{O}}{C}-(CH_2)_4-\overset{\underset{\|}{O}}{C}-O$—⟨cyclohexane, *trans*⟩—$O-$	—	225[d]

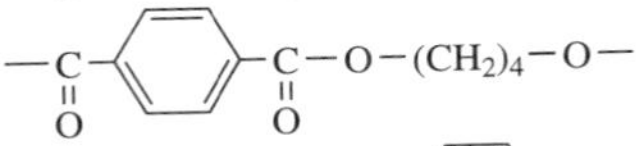

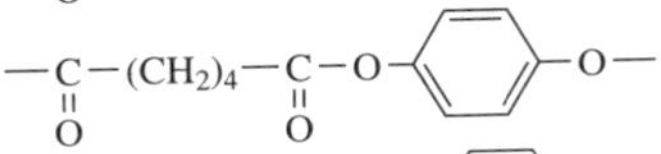

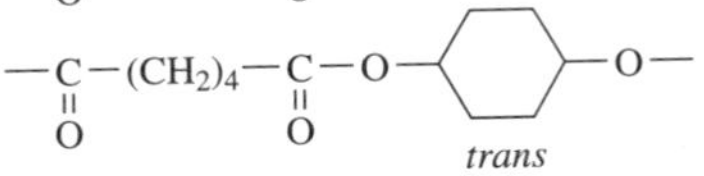

[a]Ref. 446.
[b]Ref. 445.
[c]Ref. 448.
[d]Ref. 4.

275°C and is easily processable, whereas the homopolymers do not melt before the onset of thermal degradation, at temperatures as high as 500°C.[73,74] Varying copolymer composition permits the adjustment of melting temperature and of other properties (e.g., solubility) to desired values. This method is frequently used for aliphatic and aromatic–aliphatic polyesters as well.

The disruption of chain regularity by the introduction of lateral substituents or kinks on repeating units is a supplementary means to decrease the melting temperature of aromatic polyesters.[72] This is illustrated in Table 2.9, where the melting temperatures of unsubstituted and methyl-substituted aromatic–aliphatic and aliphatic acids are reported. Regularity disruptions often cause significant

TABLE 2.9 Melting Point, T_m (°C), of Unsubstituted and Methyl-substituted Aliphatic and Aromatic–Aliphatic Polyesters

Repeating units:

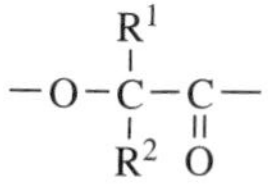

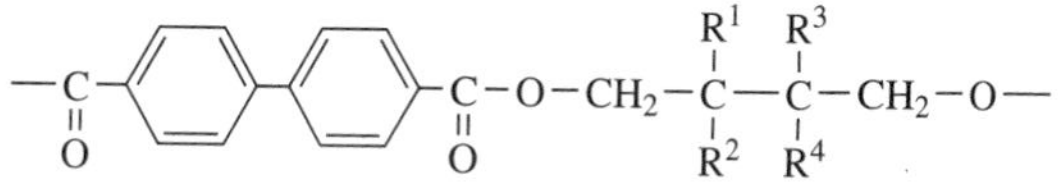

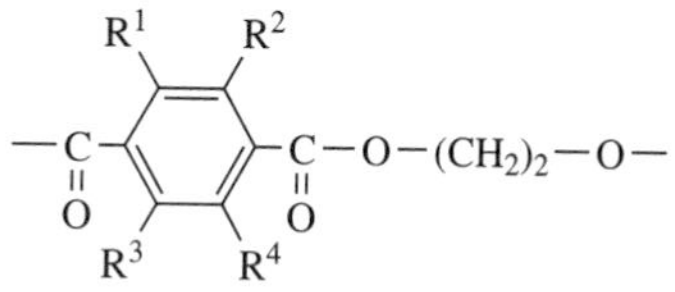

Structure 1 (aliphatic):

$$-O-\underset{R^2}{\overset{R^1}{C}}-\underset{O}{\overset{\|}{C}}-$$

Structure 2 (aromatic–aliphatic):

$$-\underset{O}{\overset{\|}{C}}-\!\!\left\langle\!\!\bigcirc\!\!\bigcirc\!\!\right\rangle\!\!-\underset{O}{\overset{\|}{C}}-O-CH_2-\underset{R^2\;R^4}{\overset{R^1\;R^3}{C\!-\!C}}-CH_2-O-$$

Structure 3 (aromatic):

$$-\underset{O}{\overset{\|}{C}}-\!\!\left\langle\!\!\overset{R^1\quad R^2}{\underset{R^3\quad R^4}{\bigcirc}}\!\!\right\rangle\!\!-\underset{O}{\overset{\|}{C}}-O-(CH_2)_2-O-$$

Repeating Unit	Structure 1			Structure 2				Structure 3		
R^1	H	H	CH$_3$	H	H	H	H	H	CH$_3$	CH$_3$
R^2	H	CH$_3$	CH$_3$	H	CH$_3$	H	H	H	H	CH$_3$
R^3	—	—	—	H	CH$_3$	H	H	H	H	H
R^4	—	—	—	H	H	H	H	H	H	CH$_3$
T_m	228[a]	175[b]	240[c]	277[d]	277[d]	217[d]	183[d]	265[e]	70[e]	180[e]

[a]Ref. 449.
[b]Ref. 375.
[c]Ref. 4.
[d]Ref. 454.
[e]Ref. 448.

decreases in crystallinity and may even lead to the complete loss of crystalline character. Aliphatic polyesters obtained from unsymmetrical diols [e.g., poly(1,2-propylene adipate)] and aromatic copolyesters containing iso- or orthophthalate units, 2,2′-bis(1,4-phenylene)propane units, or unsymmetrical 1,4-phenylene units with bulky substituents are usually amorphous polymers. On the other hand, symmetrical substitution does not hinder crystallization. Due to rotational constraints induced by the presence of substituents, chain stiffness increases, explaining why symmetrically substituted polyesters present melting points that may be higher than that of their unsubstituted homologues (Table 2.9).

Glass transition temperature (T_g) reflects the segmental mobility in the amorphous phase and follows similar trend as that observed for T_m, increasing chain rigidity leading to increasing T_m and T_g. For semicrystalline polyesters, T_g can be found close to $0.6T_m$ to $0.65T_m$ (T values in Kelvin).[4] Amorphous aliphatic polyesters are usually low-T_g viscous oils while amorphous aromatic polyesters are high-T_g polymers, some of which find applications as engineering thermoplastics. The T_g of random binary copolyesters vary monotically with composition between the T_g of homopolymers and have been reported to satisfy Fox[75] or Gordon–Taylor equations.[76–78]

2.2.1.2   Thermal and Thermo-Oxidative Degradation

The thermal resistance of polyesters increases with the aromatic group content: Aliphatic polyesters are stable up to 220°C, while aromatic–aliphatics begin to undergo thermal degradation above 280°C only. The primary degradation process involves random ester scission through a six-center concerted mechanism with a β-CH hydrogen transfer. It results in the formation of carboxyl and vinyl endgroups[79–83] (Scheme 2.2).

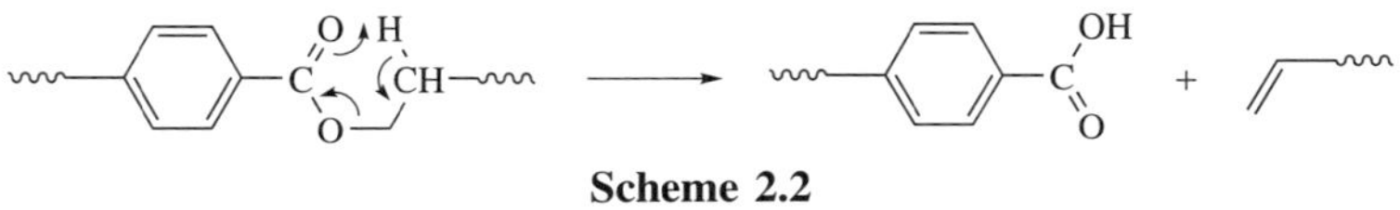

Scheme 2.2

Depending on the nature of the alkyl chain attached to the vinyl endgroups, the degradation further proceeds following various reaction pathways leading to the release of dienes, linear or cyclic ethers, and aldehydes. The formation of acetaldehyde, dioxane, and diethylene glycol has been reported for PET degradation above 290°C,[83] dienes, and tetrahydrofuran (THF) in the case of PBT degradation, above 240°C.[84] At the end of the degradation process, CO, CO_2, and aromatics are released while polydienes and crosslinked structures are formed.[83,85] Polyesters without β-CH on the diol moiety, for example polyesters of 2,2-dimethyl-1,3-propanediol or polyesters from aromatic diols, cannot degrade following this path and consequently present improved thermal stability.[86,87] Wholly aromatic liquid crystalline polyesters have been reported to degrade at 450–550°C through the homolytic scission of ester linkages releasing CO, CO_2, phenol, and other aromatics. This process is accompanied by char formation.[85]

The decarboxylation of carboxy endgroups taking place during high-temperature syntheses is another important polyester degradation reaction (Scheme 2.3). Decarboxylation takes place at significantly lower temperature for polyesters deriving from aliphatic acids than from aromatic ones. This temperature is particularly low for polyesters of oxalic and malonic acids, which, therefore, cannot be synthesized by the conventional diacid–diol bulk reaction.[88]

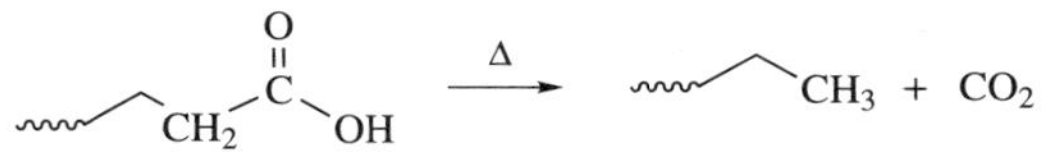

$$\text{(Scheme 2.3)} \quad \text{CH}_2\text{—C(=O)—OH} \xrightarrow{\Delta} \text{CH}_3 + CO_2$$

Scheme 2.3

In addition to the pathways depicted above, a 4-center concerted mechanism yielding ketenes has been reported during the vacuum pyrolysis of aliphatic polyesters (Scheme 2.4).[89,90]

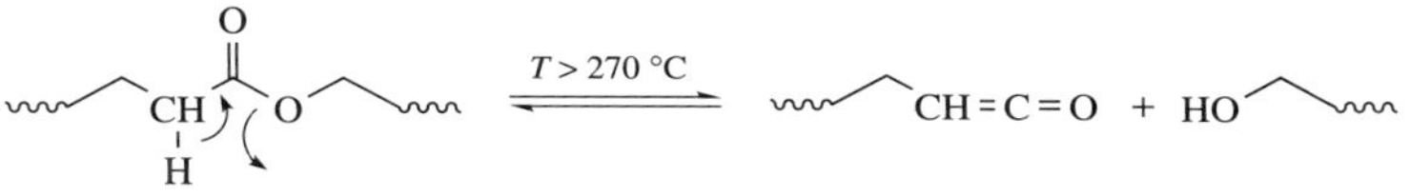

$$\text{CH(H)—C(=O)—O} \xrightleftharpoons{T > 270\,°C} \text{CH=C=O} + HO\text{—}$$

Scheme 2.4

In polyesters, as in many other step-growth polymers, the formation of linear species is always accompanied by the formation of small amounts of cyclic oligomers. At high temperature, low-boiling-point cyclic oligomers are distilled off, thus displacing the ring-chain equilibrium toward the formation of cyclics. The thermal decomposition of polyesters, therefore, generates nonnegligible amounts of cyclic oligomers, which may even be the major degradation products, as in the case of polylactones or poly(hydroxy acids) such as PLA. The formation of cyclics has also been reported for various aromatic–aliphatic and wholly aromatic polyesters.[91,92]

In the presence of oxygen, more complex thermo-oxidative processes occur in polyesters containing aliphatic moieties. They result in crosslinked products and in the formation of compounds such as aldehydes, carboxylic acids and vinyl esters, as reported in the case of PET.[93,94] On the other hand, the presence of oxygen has little effect on the thermal resistance of wholly aromatic polyesters below 550°C. Above this temperature a char combustion process takes place.[85]

2.2.1.3 Chemical Reactions Involving Ester Linkage

Most ester-forming reactions are reversible. Depending on circumstances, these reactions may be either undesirable side reactions, for example hydrolytic chain scissions occurring during processing, or useful reactions when chemical modification or polymer recycling is considered.

2.2.1.3.1 Hydrolysis

The ester linkage is susceptible to hydrolytic cleavage giving back carboxy and hydroxy endgroups. Most polyesters undergo degradation when heated in the presence of water. The degradation reaction is accelerated in the presence of acid or base catalysts (Scheme 2.5). In order to avoid chain scission during processing, solid-state polyester resins (e.g., PET and PBT) must be carefully dried before utilization, usually at 120–150°C for several hours, so that water content is below 200 ppm. The structural parameters that govern the hydrolytic degradation of aliphatic polyesters have been examined by several authors for the design of biodegradable materials.[95,96] As a rule, polyesters from sterically hindered diols, diacids, or hydroxy acids and polyesters with aromatic units offer superior hydrolytic resistance. 2,2-Dimethyl-1,3-propanediol, hydrogenated bisphenol-A, α,α'-disubstituted hydroxyacids, and aromatic monomers are often used for this purpose. On the other hand, hydrolyzable/biodegradable polyesters are unhindered polymers: aliphatic–aromatic copolyesters or polyesters from α- or β-hydroxyacids, from nonsubstituted aliphatic lactones, or from diols and diacids (Table 2.3). The design, synthesis, and properties of degradable polyesters have recently been reviewed and discussed by many authors.[41,52,97,98]

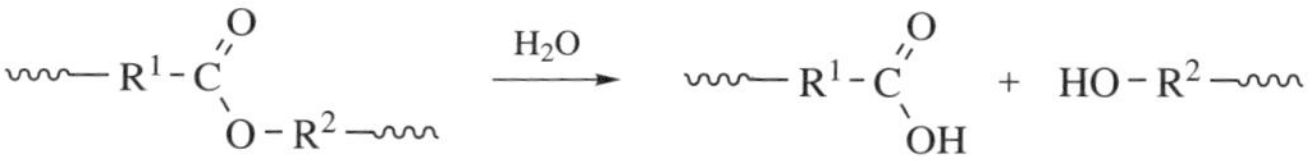

$$\text{www}-R^1-\underset{\underset{O-R^2-\text{www}}{|}}{\overset{\overset{O}{\|}}{C}} \quad \xrightarrow{H_2O} \quad \text{www}-R^1-\underset{\underset{OH}{|}}{\overset{\overset{O}{\|}}{C}} \quad + \quad HO-R^2-\text{wwv}$$

Scheme 2.5

2.2.1.3.2 Interchange Reactions

Interchange reactions are important ester-forming reactions, widely employed in polyester synthesis (Scheme 2.1, reactions 2–4). These reactions are also involved in the redistribution and randomization of monomer units during melt synthesis, processing, or blending.[99] Melt blending provides an attractive way of obtaining new copolymers. Depending on reaction conditions, block to random copolyesters can be obtained. Structural randomization has been reported, for instance, in blends of PBT and bisphenol-A polycarbonate,[100] of PBT and aromatic polyesters,[101] of poly(ethylene adipate) and poly(trimethylene adipate),[102] and of poly(6-hydroxy-2-naphthoic acid) and poly(4-hydroxybenzoic acid).[103] Similarly, aliphatic polyester–aromatic polyester block copolymers undergo randomization during melt processing and, for this reason, have not found any commercial applications as thermoplastic elastomers, unlike polyesterurethanes or polyesterethers.

2.2.2 Structure and Properties of Important Polyesters

2.2.2.1 Aliphatic Polyesters

Due to low hydrolytic and chemical resistance and to low melting point, aliphatic polyesters have long been considered to be limited to applications such as plasticizers or macromonomers for the preparation of polyurethane foams, coatings, or elastomers (see Chapter 5). Hydrolytic sensitivity, however, is a requested feature for the design of environmentally degradable plastics and biomedical polymers. Aliphatic degradable polyesters or copolyesters, either of natural or synthetic origin, have been the subject of intense academic and industrial attention during the past two decades.[41,52,97,98,104,105]

Depending on their structure, properties, and synthetic methods, degradable polyesters can be divided into four groups: poly(α-esters), poly(β-esters), poly(lactones), and polyesters of aliphatic diols and diacids.

2.2.2.1.1   Poly(α-esters)

Most poly(α-esters) are high-melting, semicrystalline polymers. Poly(glycolic acid) (PGA) has the highest melting point of all aliphatic polyesters (220–225°C). Tacticity has a strong influence on the morphology, melting point, and properties of poly(lactic acid) (PLA) (Table 2.10). Pure enantiomeric L-lactic acid and L,L-lactide produce isotactic poly(L-lactic acid) (PLLA), which is semicrystalline and melts at 170–190°C. L-Lactic acid is the most frequent form of lactic acid and is industrially produced by the fermentation of glucose or hydrolyzed starch. D-Lactic acid exists in some bacterial systems[104] and leads to isotactic poly(D-lactic acid) (PDLA), a polymer with similar properties as isotactic PLLA. Syndiotactic PDLLA has been obtained by the stereoselective polymerization of D,L-lactide (*meso*-lactide).[106] Due to ester interchange in the melt,[107] however, redistribution of stereosequences takes place to some extent during processing, leading on prolonged heating to atactic PDLLA, a polymer also obtained from the polycondensation of racemic lactic acid. Depending on microstructure, PDLLA may be amorphous or semicrystalline. Amorphous PDLLA has significantly lower tensile strength and modulus than PLLA (Table 2.10). When crystallized, it presents a much lower melting point than PLLA (130–135°C). Consequently, the properties of PLA are strongly affected by the optical purity of starting monomer, by polymerization and processing conditions, and by thermal history. A wide range of properties can also be achieved by copolymerization with glycolide or ε-caprolactone. These general properties compare well with those of commodity thermoplastics such as polyethylene, polystyrene, or poly(vinyl chloride) (PVC).[104]

2.2.2.1.2   Poly(β-esters)

Poly(3-hydroxy alkanoic acid)s (PHAs), also known as poly(3-hydroxyalkanoate)s, are aliphatic polyesters naturally produced by a wide range of bacteria as energy storage material. They have been investigated as biodegradable polymers from renewable resources. Various PHA homo- and copolymers, such as poly(3-hydroxybutanoic acid) (PHB) and poly(3-hydroxybutanoic-*co*-3-hydroxyvaleric acid) (PHBV), Monsanto's Biopol polyesters, can be obtained by the fermentation of glucose or of various feedstock materials. The high cost of the fermentation process made the bacterial production of PHAs uneconomical (U.S. $12/kg)[52] compared to other biodegradable plastics, and the current trend in PHA research is directed toward biosynthesis by genetically modified plants.

TABLE 2.10 Properties of Biodegradable Aliphatic Polyesters

Polyester	PGA[a]	PLLA[a]	PDLLA[a]	PHB[b]	PHBV[b]	CAPA[c]	Bionolle[d]		Eastar Bio[e]
							PBSu	PESu	
T_m(°C)	225–230	170–190	—	177	135	58	114	104	108
T_g(°C)	40	50–60	50–60	4[f]	—	−60[a]	−32	−10	−30
Density (g/cm^3)	1.50–1.69	1.25–1.29	—	1.25[f]	—	1.15	1.26	1.32	1.22
Melt flow index, g/10 min (190°C, 2.16 kg)	—	10–30[g]	—	—	—	2.9	—	—	28
Tensile strength (MPa)	80–980[h]	120–2300[h]	40–50[i]	40	20	40/43 (MD/TD)[j]	—	—	22/20 (MD/TD)[j]
Tensile modulus (MPa)	3900–13,800[h]	6900–980[h]	1500–1900[i]	—	—	—	550[k]	580[k]	107/106 (MD/TD)[j]
Elongation at break (%)	30–40[h]	12–26[h]	5–10[i]	3.0	100	—	560	200	700/730 (MD/TD)[j]
Stress at yield (MPa)	—	—	—	—	—	18/16 (MD/TD)[j]	33	20.5	—
Notched Izod impact strength (J/m)	—	16[g]	—	35	300	—	294	98	—

[a]PGA: poly(glycolic acid); PLLA: poly(L-lactic acid), PDLLA: poly(DL-lactic acid), ref. 41.
[b]PHB: poly(3-hydroxybutanoic acid); PHBV: poly(3-hydroxybutanoic acid-co-3-hydroxyvaleric acid), ref. 105.
[c]CAPA: poly(ε-caprolactone) FB100, ref. 456.
[d]Bionolle PBSU: poly(butylene succinate); Bionolle PESU: poly(ethylene succinate), ref. 457.
[e]Eastar Bio: poly(tetramethylene adipate-co-terephthalate), ref. 110.
[f]Ref. 52.
[g]Ref. 455.
[h]Oriented fiber.
[i]Nonoriented fiber.
[j]MD: machine direction; TD: transverse direction.
[k]Stiffness following JIS-K7203.

PHAs obtained from microbial synthesis present always the D(−) configuration and are isotactic. The tacticity of chemically synthesized PHB depends on the enantiomeric configuration of starting monomer and the nature of the catalyst. Atactic, syndiotactic, and isotactic PHBs have been reported.[108,109] The properties of naturally occurring PHAs depend on the composition: PHB is a highly crystalline, brittle polymer with $T_m = 177°C$ while PHBV copolymers exhibit lower melting point (135–150°C) and higher impact resistance and flexibility. Selected properties of PLA and PHB are compared to those of biodegradable polymers in Table 2.10.

2.2.2.1.3 Poly(Lactones)

Poly(ε-caprolactone) (PCL), the most representative member of this polyester family, is obtained by the ring-opening polymerization of ε-caprolactone. It is a low-T_m (60°C), low-T_g (−60°C) semicrystalline polyester that presents mechanical properties resembling those of low-density polyethylene (Table 2.10).

2.2.2.1.4 Copolyesters from Diacids and Diols

Various environmentally degradable copolyesters obtained from reactions between diols and diacids have recently been commercially introduced.

Bionolle (Showa Highpolymer) is a semicrystalline polyester produced by the polycondensation of 1,4-butanediol or 1,2-ethanediol with succinic acid or mixtures of succinic and adipic acids.[110] Molar mass can be increased to $\overline{M}_n = 200,000$ by using diisocyanates as chain coupling agents.[111] Depending on composition, the properties of Bionolle vary between those of low-density polyethylene and those of high-density polyethylene[110] (Table 2.10).

A different approach consists of reacting dimethyl terephthalate and dimethyl adipate with 1,4-butanediol in the experimental conditions of poly(butylene terephthalate) (PBT) synthesis. The resulting copolyester (Eastman's Eastar Bio) presents properties close to those of other degradable materials (Table 2.10).

2.2.2.1.5 Polyester Biodegradation

The predominant biodegradation reactions of polyesters are chemical and microbial hydrolysis. Degradation rate is influenced by the molar mass, crystallinity, hydrophobicity, and T_g of the polyester, by the degradation conditions, and by the presence of degradation products. PGA, PLA, and their copolymers are hydrophilic polymers and degrade either in vivo or in vitro through an abiotic bulk hydrolysis mechanism, which first affects the amorphous regions. The degree of crystallinity has been reported as the major controlling factor in the enzyme-catalyzed hydrolysis of PLA.[112] PHAs and PCL are more hydrophobic than poly(α-esters), and their abiotic hydrolysis occurs more slowly, through surface erosion rather than bulk hydrolysis. Low-molar-mass polyesters are known to undergo hydrolytic degradation in the presence of lipases and esterase.[113] The degradation products of polyesters are bioassimilated by microorganisms, irrespective of enantiomeric composition — as shown on lactide oligomers[114] — and the nature of monomers. Polymer biodegradation mechanisms and testing methods have recently been reviewed.[52]

2.2.2.2   Aromatic–Aliphatic Polyesters

This class of polyesters consists of four major commercial polymers and their copolymers, namely PET, PTT, PBT, and PEN (see Table 2.1). They compete for engineering thermoplastics, films, and fibers markets with other semicrystalline polymers, such as aliphatic polyamides, and for some other applications with amorphous engineering plastics such as polycarbonate. The syntheses of PET and PBT, detailed in numerous reviews and books,[2–5] are described in Sections 2.3.2.2 and 2.3.2.1.

PET, PTT, and PBT have similar molecular structure and general properties and find similar applications as engineering thermoplastic polymers in fibers, films, and solid-state molding resins. PEN is significantly superior in terms of thermal and mechanical resistance and barrier properties. The thermal properties of aromatic–aliphatic polyesters are summarized in Table 2.6 and are discussed above (Section 2.2.1.1).

The crystalline structure of these polymers, all triclinic, reveal some interesting differences.[115–117] In crystalline regions, PET adopts a planar zigzag conformation with almost completely extended chains (98–99% extended). Two crystalline forms, α and β, are known for PBT. The α-form is stable in the relaxed unstretched state at ambient temperature and reversibly transforms into the more extended β-form when stretched (96% extended instead of 86%). The crystalline structure of PTT is helical with successive monomer units lying at approximately 60° to one another about the helix axis, polymer chains being only 76% extended. This springlike structure explains the low tensile modulus and the excellent elastic recovery of PTT compared to PET, since the initial part of the strain–stress curve is thought to correspond to the elastic uncoiling of polymer chains. The semicrystalline PEN yarns obtained by melt spinning contain β-crystals that present a sinusoidal 94% extended chain conformation. Under stress or hot drawing, they irreversibly transform into the fully extended α-form by rotation of the naphthalene ring.[118] The elastic modulus of crystalline regions, measured from X-ray measurements, vary in the order PEN α (145 GPa) > PET (108 GPa) > PBT α (13.5 GPa) > PTT (2.6 GPa). This is explained by the higher rigidity of naphthalene with respect to phenylene for PEN and PET, which are both in fully extended conformation, and by the less extended conformations of chains in PBT and PTT crystals.[119]

Crystallization rate is an important parameter for the processing of semicrystalline polymers. PET[120] and PEN[121–123] crystallize relatively slowly in the absence of nucleating agents. They find applications where crystallization can be induced by mechanical orientation (e.g., fibers, films and stretch blow-molded bottles). The as-spun PET[124] and PEN[125] fibers are amorphous. As a result, polymer chains are easily reoriented parallel to the fiber axis during the drawing process, causing stress-induced crystallization, while, on the other hand, crystallites present in the original material would hinder molecular reorientation.

High degrees of orientation are thus achieved both in crystalline and amorphous regions, resulting in enhanced strength in the drawing direction.[124,126] Analogous chain orientation and mechanical strengthening process takes place during the manufacture of high-strength biaxially oriented PET[127] and PEN[128,129] films and PET stretch blow-molded bottles.[130]

PBT and PTT are fast crystallizing polymers and, therefore, well suited for extrusion and injection-molding applications with high cycle times. The cold drawing temperature window for PTT and PBT (temperature range between glass transition temperature and cold-crystallization temperature) is rather narrow. Careful temperature control must be exerted when drawing PTT and PBT fibers or films and high-speed lines are generally preferred for PTT fibers.[27]

The number-average molar mass of thermoplastic polyesters varies from about 20,000 for film and fiber applications to 40,000 for injection-molding or blow-molding resins. Relationships between intrinsic or molten viscosity and molar mass have been published for PET,[131–136] PBT,[135,137] and PEN.[138]

Semicrystalline aromatic–aliphatic polyesters present mechanical properties (e.g., tensile strength and modulus) comparable to that of other engineering plastics, slightly lower than aliphatic polyamides (nylon-6 and nylon-6,6) but higher than polycarbonate. Heat deflection temperature (HDT), within the 55–85°C range, is relatively low, making reinforcement by glass fiber required to reach sufficient levels for engineering molding resins applications. Grades containing 30% glass fiber exhibit HDT in the 210–230°C range. Altogether, the properties of PTT are intermediate between those of PET and PBT. The mechanical and electrical properties of unfilled PET, PBT, and PTT engineering plastics are reported in Table 2.11.

Compared to PET, PEN has higher melting point and glass transition temperature and possesses higher modulus — twice as high when processed as a fiber — and lower elongation to break, properties that are typical of a more rigid polymer (Table 2.12). Among other properties shared by the thermoplastic polyesters are good dimensional stability and gas barrier properties. The naphthalate polymer family possesses excellent barrier properties,[27] which, combined with high T_g, make PEN suitable for applications such as pasteurizable hot-filled containers.

The semicrystalline polyesters of the terephthalate and naphthalate family are resistant to a wide range of chemicals at room temperature, including water, alcohols, ketones, ethers, glycols, chlorinated solvents, aliphatic hydrocarbons, and oils. They are slowly hydrolyzed in boiling water and rapidly degraded in strongly basic or acidic medium.

It must finally be kept in mind that it is extremely easy to adjust the properties of polyesters to desired values by adding small quantities (usually less than 10%) of comonomers in starting monomer feed. Isophthalic, adipic, dodecanedioic, p-hydroxybenzoic acids or esters and diethylene glycol, cyclohexanedimethanol, or bisphenol-A are often used for this purpose. Examples of property adjustment are the modification of solvent diffusivity of PET membranes by the addition of low levels of isophthalate or naphthalene dicarboxylate units in polymer chains[139]

TABLE 2.11   Typical Properties of Unfilled Poly(ethylene terephthalate) (PET), Poly(trimethylene terephthalate) (PTT), and Poly(butylene terephthalate (PBT) Solid-State Polyester Resins

Properties	PET[a]	PTT[b]	PBT[a]
T_m (°C)	265[c]	233[c]	232[c]
T_g (°C)	69[c]	35[c]	22[c]
Specific gravity	1.37	1.35	1.31
Tensile strength (MPa)	53	59.3	52
Flexural modulus (GPa)	2.83	2.76	2.34
HDT at 1.8 MPa (°C)	85	59	55
Notched Izod impact (J/m)	43	48	53
Dielectric strength (V/μm)	23.6	21	23.2
Dielectric constant at 1 MHz	3.4	3.0	3.1
Dissipation factor at 1 MHz	0.02	0.015	0.02
Volume resistivity ($\Omega \cdot cm \times 10^{16}$)	3.5	1.0	4.0

[a]Ref. 120.
[b]Ref. 27.
[c]Ref. 4.

TABLE 2.12   Physical Properties of Poly(ethylene terephthalate) (PET) and Poly(ethylene 2,6-naphthalate) (PEN) Polyester Fibers

Properties	PET[a]	PEN[b]
T_m (°C)	258–263	274
T_g (°C)	69[c]	117[d]
Specific gravity	1.38–1.39	1.36
Initial modulus (N/Tex)	6.6–10.6	16.8–19.0; 22[e]
Breaking tenacity (N/Tex)	0.35–0.85	0.57–0.67; 0.9[e]
Breaking elongation (%)	10–50	6

[a]Data range for regular–high tenacity yarns, ref. 124.
[b]Ref. 260.
[c]Ref. 4.
[d]Ref. 451.
[e]Data for Honeywell's PentexTM, ref. 458.

or the modification of PET crystallization kinetics by copolymerization with hydroquinone[140] or isophthalic acid.[141] Copolymerizing comonomers in larger proportions is a means to obtain copolymers with properties completely different from that of the corresponding homopolymers. Thus, 50 : 50 (mol) poly(ethylene 2,6-naphthalate-*co*-terephthalate) is soluble in dimethyl sulfoxide (DMSO) at room temperature while homopolymers are insoluble[142]; most terephthalate-*co*-

isophthalate copolyesters are amorphous in the 20 : 80–80 : 20 mol composition range[143-145]; copolymers of PET and poly(4-hydroxybenzoic acid) (PHBA) containing 60 mol % HBA units (Unitika's Rodrun) exhibit liquid crystalline properties while PHBA does not melt and PET is not liquid crystalline[146,147]; poly(1,4-cyclohexylenedimethylene terephthalate-*co*-isophthalate) (PCTI, 50 : 50 mol) is an amorphous ($T_g = 80°C$) thermoplastic while the terephthalate homopolymer is a high-melting (280–320°C) semicrystalline, brittle polymer.[120] Melt blending is another widely used approach for polyester property modification. In the case of polyester–polyester or polyester–polycarbonate blending, ester interchange reactions result in the rapid formation of small amounts of block copolymer which facilitate blend compatibilization. Random copolymers are produced on prolonged heating in the presence of ester interchange catalysts.[100,148-153]

2.2.2.3 Aromatic Polyesters

2.2.2.3.1 Amorphous Polyarylates

Although many other monomers have been utilized, the typical commercial amorphous polyarylate is a polyester containing bisphenol-A units and nearly equal proportions of isophthalate and terephthalate units (Formula 2.1).[6]

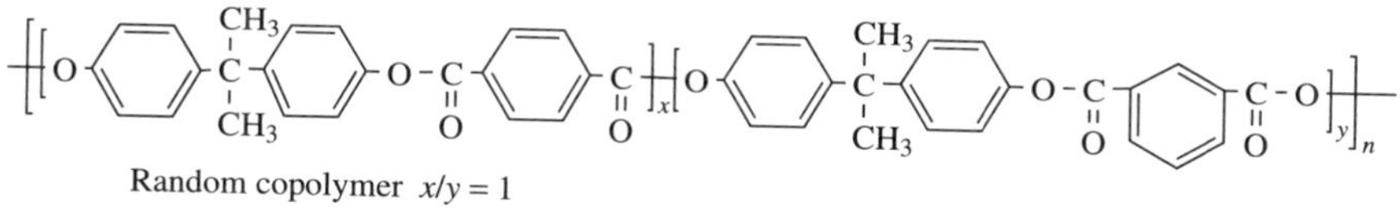

Random copolymer *x/y* = 1

Formula 2.1

Amorphous polyarylates are light-amber transparent materials which exhibit mechanical properties comparable to that of unfilled PET in terms of tensile or flexural strength and modulus (Table 2.13) but are notably superior in terms of heat resistance (HDT = 174°C vs. 85°C for PET) and impact strength.

Upon UV light exposure, they undergo a photo-Fries rearrangement which results in the formation of *o*-hydroxy-benzophenone moieties in polymer chains[6,154,155] (Scheme 2.6). These moieties are very efficient UV absorbers that stabilize the polymer against further degradation. The resulting light-yellow protective layer formed on the polymer surface imparts polyarylates with excellent retention of mechanical properties upon UV light exposure. Combined to their transparency, weather resistance, and heat stability, this makes them suitable for applications such as tinted parts in automotive lighting, outer glazing, and solar energy devices.

Numerous blend formulations with amorphous polyarylate as one of the components have been described[6] and are generally claimed to exhibit improved UV and weather resistance, improved molding performance characteristics, and

TABLE 2.13   Typical Properties of Ardel D-100 Bisphenol-A Polyarylate and of Liquid Crystalline Polyesters Vectra A-950 and Xydar SRT-500

Properties	Ardel D-100[a]	Vectra A950[b]	Xydar SRT-500[c]
T_m (°C)	—	280	358
T_g (°C)	190	—	—
Tensile strength (MPa)	65	126	126
Tensile modulus (GPa)	2.0	8	8.3
Tensile elongation at break (%)	50	3.9	5
Tensile elongation at yield (%)	8.0	—	—
Flexural strength (MPa)	76[d]	137[e]	131
Flexural Modulus (GPa)	2.13	7.4	13.1
Notched Izod impact strength (J/m)	234	684	208
Density	1.21	1.4	1.35
Heat deflection temperature (HDT) at 1.8 MPa (°C)	174	178	335
Dielectric strength (V/μm)	15.7	47	31
Volume resistivity (Ω·cm)	1.2×10^{16}	10^{16}	—

[a] Ref. 6.
[b] Ref. 459.
[c] Ref. 161.
[d] Value at 5% strain.
[e] Value at 3.5% strain.

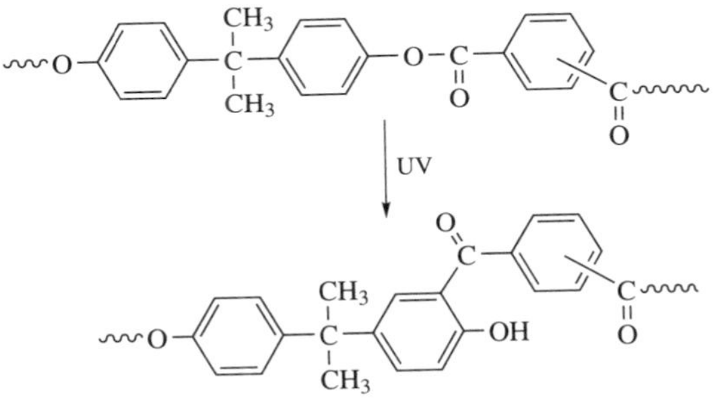

Scheme 2.6

better impact properties. Studies dealing with polyamides,[156] polycarbonates,[77] aromatic–aliphatic polyesters,[78,101,157] and polyetherimides[158] have more recently been published. The compatibilizing effect of polyarylate homo- or block copolymers in various blends has also been investigated.[159,160]

2.2.2.3.2   Liquid Crystalline Polyesters

The liquid crystalline (LC) state is an intermediary state between isotropic liquid and crystalline solid. It arises from the tendency of rod like or disc like rigid molecules to align parallel to each other in the molten state (thermotropic compounds) or in solution (lyotropic compounds), forming anisotropic ordered phases, termed mesophases. LC polymers are prepared by combining rigid units, termed mesogenic units, into polymeric structures. Depending on their architecture, liquid crystalline polymers (LCPs) can be classified into three principal categories (Table 2.14): (i) main-chain LCPs, with the mesogenic units incorporated into the polymer backbone; (ii) side-chain LCPs, with the mesogenic units linked to polymer chains as side groups; and (iii) combined LCPs with both types of structures.[161] Side-chain mesogens are decoupled from polymer backbone by flexible spacers and, therefore, can easily be aligned in electric or magnetic fields. This imparts side-chain LCP properties that are of particular interest for electrooptics or nonlinear optics applications. On the other hand, main-chain LCPs are high-strength/high-modulus rigid polymers applied as materials for structural applications.

LC polyesters belong to the class of thermotropic main-chain LCPs, which also comprises polymers such as polycarbonates, polyethers, polyphenylenes, polyester-imides, polymers containing azo- or azo N-oxide linking groups, some cellulose derivatives, and polypeptides such as poly(γ-benzyl-L-glutamate). Both from the academic and industrial points of view, polyesters are by far the most important representatives of this class of polymers.

Only a few building blocks are available for the synthesis of ester-containing rigid mesogenic units, namely dihydroxy-, dicarboxy-, or (hydroxy, carboxy)-1,4-phenylenes, 4,4'-biphenylenes, and 2,6-, 1,4-, or 1,5-naphthylenes (Table 2.15).

The first rigid-rod polyester poly(4-hydroxybenzoic acid) (PHBA, Ekonol) was commercially introduced by 1970.[162] However, this polyester, like the

TABLE 2.14 Schematic Representation of Three Main Categories of Liquid Crystalline Polymers (LCPs)

LCP Types	Schematic Structures[a]
Main-chain LCPs	
Side-chain LCPs	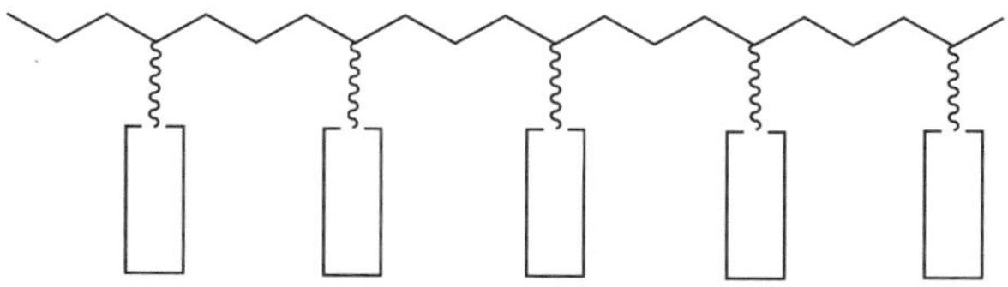
Combined LCPs	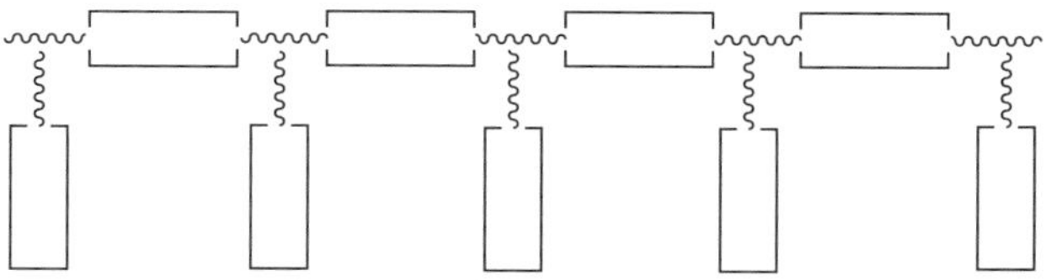

[a] ☐: mesogenic unit; ⌃⌃⌃: chain backbone; ∿∿∿: flexible spacer.

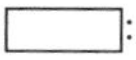

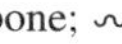

TABLE 2.15 Building Blocks Used in Main-Chain
Polyester LCP Mesogenic Units

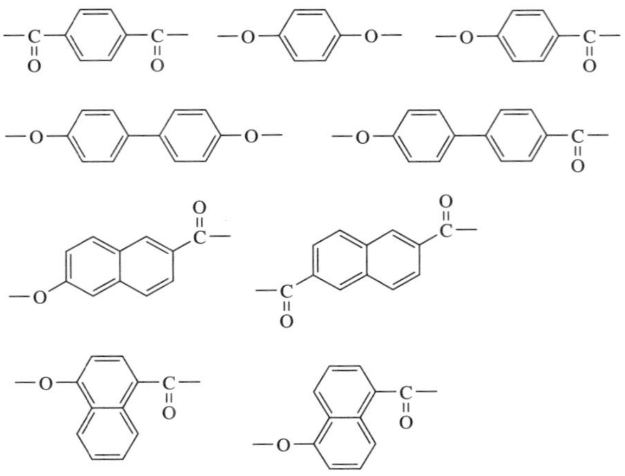

structurally resembling poly(1,4-phenylene terephthalate), melts after degradation starts (above 600°C)[163] and can be processed only by sintering-compression, metal-forming processes, or plasma coating.[162,164] During the last two decades, a great deal of effort has been devoted to the study of structure–property relationships in thermotropic polyesters in order to direct synthesis toward copolyesters exhibiting lower melting temperature, better solubility, and better processability. For instance, copolyesters of PHBA and comonomers such as terephthalic acid and 4,4'-diphenol are LC melt-processable copolymers. These efforts have recently been reviewed[163–166] and are briefly summarized below.

Several approaches have been used, all aiming at disruption of the structural regularity of aromatic polyester chains:

1. *Insertion of flexible aliphatic spacers between mesogenic units in polymer backbone.* This leads to appreciable decrease in melting and isotropization temperatures but also reduces the mechanical strength and the thermal stability of resulting polymers unless spacer length is kept short.[167] This approach is applied in PET/PHBA random copolyesters[168–170] (now Unitika's Rodrun), which can be regarded as thermotropic copolyesters containing short 1,2-ethylene spacers between rigid terephthalate and 4-oxybenzoyl moieties.[146,171–173]

2. *Introduction of bulky lateral substituents on monomer units to increase interchain distance and prevent close packing in polymer crystal.* The use of unsymmetrically substituted monomers, resulting in a random distribution of head-to-head and head-to-tail structures in polymer chains, further helps in disrupting regularity. Some examples of substituent effects are given in Table 2.16.

TABLE 2.16 Melting Point, T_m (°C), and Isotropization Temperature, T_i (°C), of Poly(1,4-phenylene terephthalate) LCPs Containing Various Substituents

Structure	T_m	T_i	References
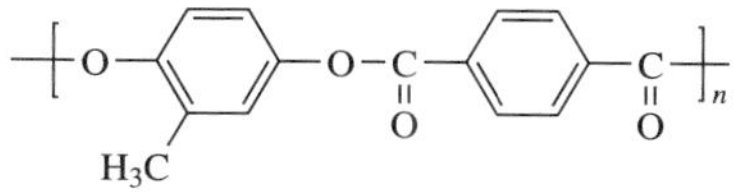 (1,4-phenylene terephthalate with CH_3 substituent)	>500	—	460
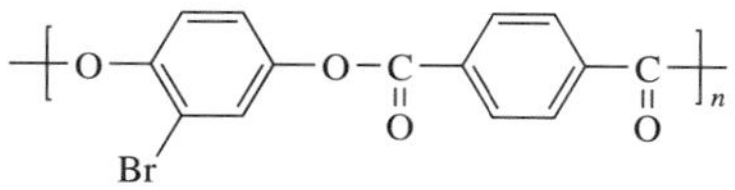 (1,4-phenylene terephthalate with Br substituent)	353	475	174
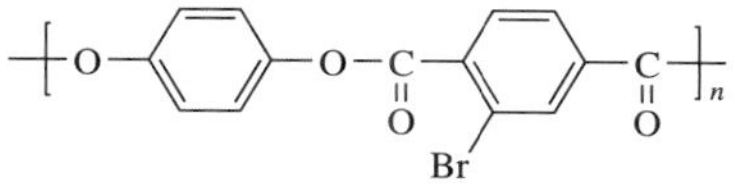 (1,4-phenylene terephthalate with Br on acid ring)	405	—	174
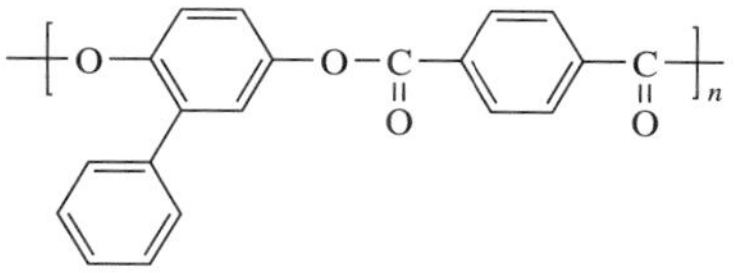 (1,4-phenylene terephthalate with phenyl substituent)	346	475	174
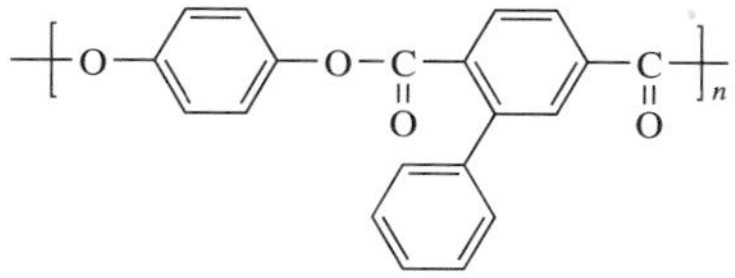 (1,4-phenylene terephthalate with phenyl on acid ring)	265	343	174
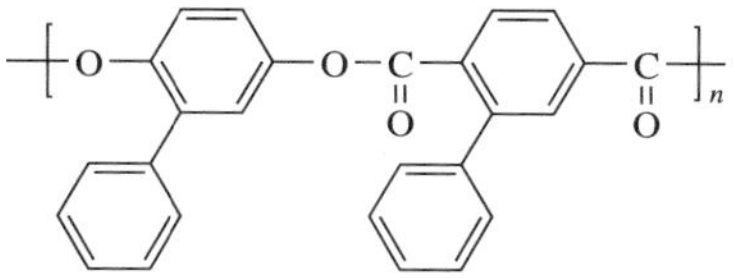 (1,4-phenylene terephthalate with two phenyl substituents)	157	231	174
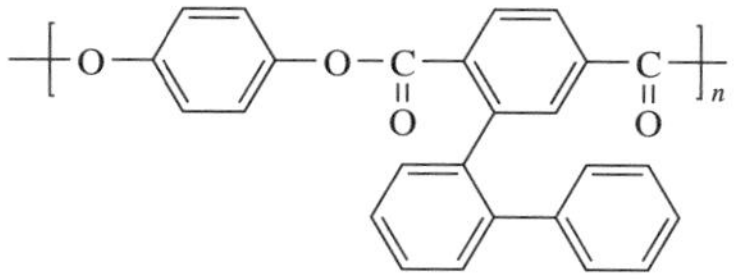 (1,4-phenylene terephthalate with biphenyl substituent)	285	303	461
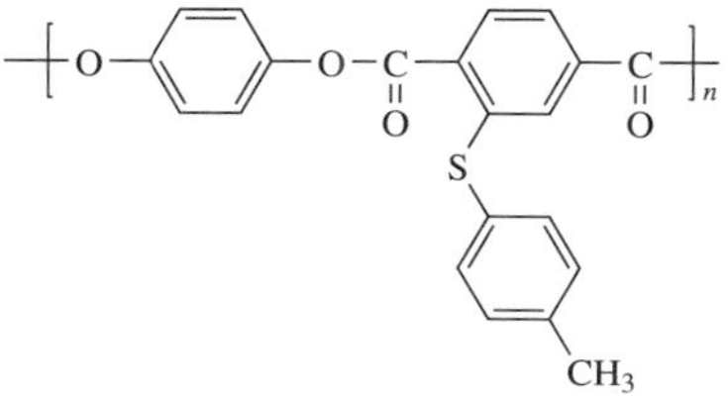 (1,4-phenylene terephthalate with S–C$_6$H$_4$–CH$_3$ substituent)	303	—	462
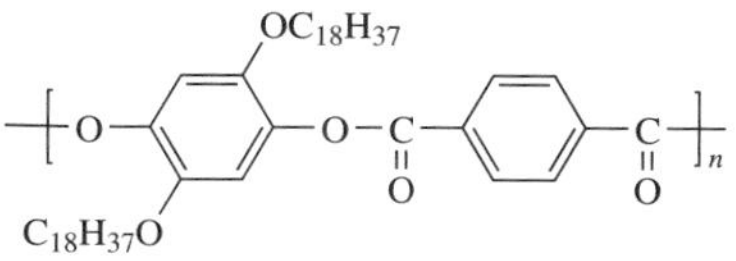 (1,4-phenylene terephthalate with two $OC_{18}H_{37}$ substituents)	95–107	120	463

This method is attractive, since polymers with good thermal stability are obtained, especially with aryl or halogeno substituents.[165,174] Moreover, a number of substituted polyester LCPs exhibit solubility in common organic solvents, thus facilitating their structural characterization. However, the cost of starting monomers has hampered the commercial development of thermotropic polyesters based on substituted monomers.

3. *Introduction of nonmesogenic units in polymer chains.*   (E.g., using meta-substituted aromatic monomers such as isophthalic acid or resorcinol). This results in the formation of kinks in polymer chain, which disrupt lateral interactions.

4. *Copolymerization of two or more mesogenic monomers yielding copolymers with a random distribution of repeating units such as those given in Table 2.15.*   This method gives convenient polymers in terms of mechanical properties, thermal resistance, and LC properties. The use of monomers of unequal length generally favors the formation of LC nematic mesophases, in which mesogens are orientally ordered but lack positional order, over smectic mesophases in which positional ordering gives rise to layered structures.[165] This approach is used in several commercial polyester LCPs such as 4-hydroxybenzoic acid–2,6-hydroxynaphthoic acid copolyesters[17] and hydroxybenzoic acid, terephthalic acid, and 4,4′-diphenol copolyesters,[175,176] now Ticona's Vectra and Solvay's Xydar, respectively.

Polyester LCPs exhibit mechanical properties (e.g., tensile strength and modulus) which are notably greater than conventional thermoplastics and comparable to glass-filled PET (Table 2.13). Melt-spun fibers with tensile properties equivalent to those of high-performance aramides can be obtained.[177] These superior properties originate from the high degree of molecular orientation in the flow direction achieved by these stiff molecules during melt processing. The ease of orientation in the machine direction (flow direction), however, adversely affects the properties in the transverse direction. This makes difficult the production of films with balanced properties, which usually require special techniques such as multiaxial stretching, cross-lamination, or rotation compression injection-molding.[178–182]

The melt rheology of LCPs in the LC state is quite complex.[173,183,184] LCPs generally present a non-Newtonian behavior with a rapid decrease in viscosity with increasing shear rates. As a consequence, the bulk viscosity of polyester LCPs can be two orders of magnitude less than that of PET under normal processing conditions.[185] The morphology and properties of resulting products, fibers or moldings, are strongly dependent on the processing conditions.[186,187] Skin-core and microfibrillar morphologies are generally observed in melt-processed polyester LCPs, with highly oriented skin and an underlying hierarchy of micro- to macrofibrillar structures ranging from 50 nm to 5 μm in diameter.[164,188]

Besides high mechanical strength, polyester LCPs exhibit excellent thermal and chemical resistance, high heat deflection temperature, low mold shrinkage, low thermal expansion coefficient, and outstanding gas barrier properties. Glass-reinforced grades are generally used for molding resin applications in order to improve transverse properties and compressive strength.

2.2.3 Role of Microstructure and Architecture

2.2.3.1 Block Copolymers: Polyester Thermoplastic Elastomers

Polyester block copolymers can be defined as $(AB)_n$-type alternating multi-block copolymers composed of flexible aliphatic polyester or polyether blocks (A-type blocks) and rigid high-melting aromatic–aliphatic polyester blocks (B-type blocks) (Formula 2.2).

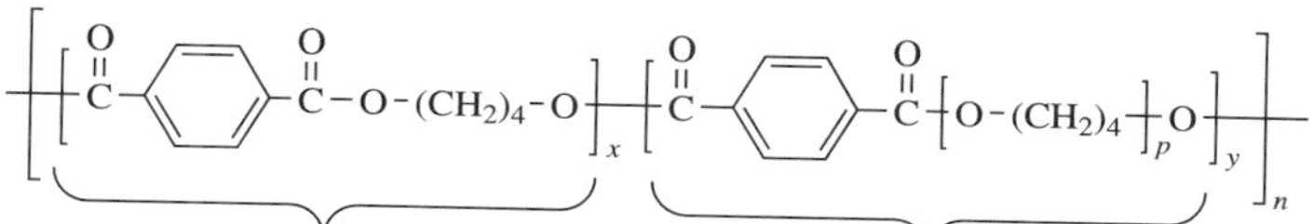

Rigid block   Soft block

Formula 2.2

The low-T_g flexible blocks and the semicrystalline rigid blocks are incompatible at the service temperature and segregate, giving rise to materials that exhibit biphasic morphology. This morphology is generally described as a dispersion of microdomains formed by partially crystallized rigid blocks in a continuous amorphous matrix containing flexible blocks and noncrystallized rigid blocks[36,189] (Fig. 2.1). The crystallized microdomains act as physical crosslinks for the soft amorphous matrix and impart both elastomeric and thermoplastic properties to the material. The resulting copolymer behaves as rubber at the service temperature but can be processed as a conventional thermoplastic above the melting point of the rigid domains.

The rigid and soft blocks used in polyesterether thermoplastic elastomers (polyesterether TPEs) are typically PBT and poly(oxytetramethylene) (PTMO), respectively, with block number-average molar mass varying between 1000 and 3000. They are obtained by the melt reaction between dimethyl terephthalate, butanediol, and dihydroxy-terminated PTMO in the conditions typical of a PBT synthesis.

As shown in Table 2.17, the properties of polyester TPEs are strongly dependent on the soft–rigid block ratio and vary from those of soft rubbery polymers to those of rigid thermoplastics. The softer grades (60% mass PTMO, 40 Shore D hardness) have very low Young's modulus (53 MPa), do not break in the notched Izod impact test even at −40°C, and present high elongation at break (700%). The rigid grades (10% PTMO, 75 Shore D hardness) exhibit properties close to that of the parent PBT homopolymer. The stress–strain curves (Fig. 2.2) present a yield point at ca. 5–20% elongation and samples strained to higher levels exhibit

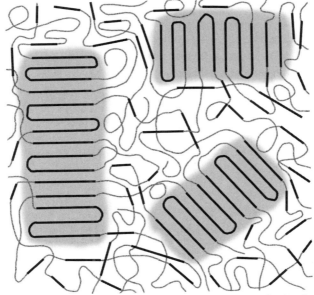

Figure 2.1   Model of microstructure in polyester thermoplastic elastomers.

TABLE 2.17   Physical Properties of Poly(poly(butylene terephthalate)-*block*-poly(oxytetramethylene)) Polyester–Ether TPEs[a]

PTMO[b] in Copolymer (mass %)	0	10	25	35	50	60
Shore D hardness	80	74	63	55	45	38
Density	1.3	1.27	1.23	1.2	1.16	1.12
T_g (°C)[c]	55	44	2	−23	−50	−60
T_m (°C)	223	221	210	202	183	192
Ultimate elongation (%, ASTM 412)	200	360	540	570	700	650
Tensile strength (MPa, Din 53504)	52	45	38	32	21	17
Young's modulus (MPa)	2600	830	300	170	90	53

[a]Reproduced from ref. 464 with permission. Copyright © 1999 John Wiley & Sons, Inc.
[b]$\overline{M_n}$ = 1000 except for the 60% mass copolymer in which $\overline{M_n}$ = 2000.
[c]Taken at peak tan δ.

considerable permanent deformation. This behavior, quite different from that of conventional vulcanized rubber, is assigned to the progressive and irreversible disruption of PBT crystalline matrix and extensive crystallite orientation during the drawing process (Region II, Fig. 2.2).

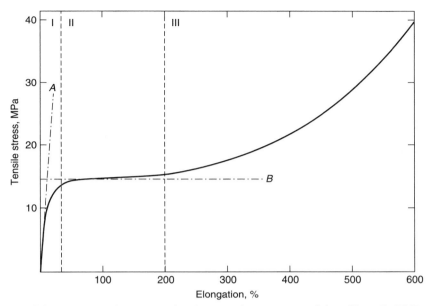

Figure 2.2 Stress–strain curve of polyester elastomer containing 58 wt % PBT. *A*, slope = Young's modulus; *B*, yield stress. (Reprinted with permission from ref. 190, p. 96. Copyright © 1988 John Wiley & Sons, Inc.)

At constant PBT/PTMO composition, when the molar mass of PTMO block is ≥2000, partial crystallization of the polyether phase leads to copolymer stiffening. The properties of polyesterether TPEs are not dramatically different when PTMO is replaced by polyethers such as poly(oxyethylene) (PEO) or poly(oxypropylene). PEO-based TPEs present higher hydrophilicity, which may be of interest for some applications such as waterproof breathable membranes but which also results in much lower hydrolysis resistance. Changing PBT into a more rigid polymer by using 2,6-naphthalene dicarboxylic acid instead of terephthalic acid results in compounds that exhibit excellent general properties but poorer low-temperature stiffening characteristics.

TPEs associating both rigid and soft polyester blocks have also been described. They cannot be obtained by the melt polyesterification used for polyesterether TPEs, since interchange reactions would yield random — rather than block — copolyesters. The preferred method involves the reaction of OH-terminated aliphatic and aromatic–aliphatic polyesters with chain extenders such as diisocyanates and results in copoly(ester–ester–urethane)s.

2.2.3.2 Hyperbranched Polyesters

Hyperbranched polyesters are prepared by the step-growth polymerization of AB_x-type monomers where A and B are $-OH$ and $-COOH$ groups or derivatives such as CH_3COO-, $HO-CH_2CH_2-O-$, $(CH_3)_3SiO-$, $-COOCH_3$, or $-COCl$

TABLE 2.18 Monomers Used for Synthesis of Hyperbranched Polyesters

No.	Monomer	References	No.	Monomer	References
1	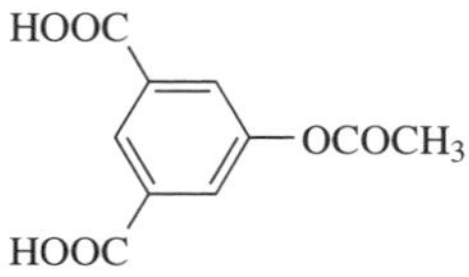 HOOC, HOOC—benzene—OCOCH$_3$	269	7	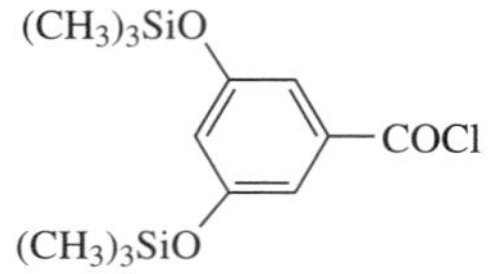 (CH$_3$)$_3$SiO, (CH$_3$)$_3$SiO—benzene—COCl	202, 442
2	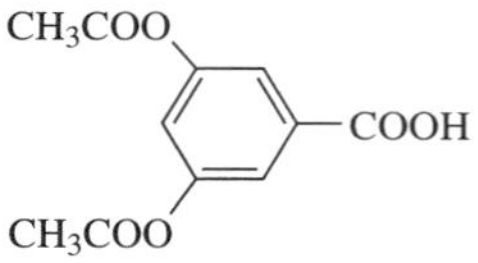 CH$_3$COO, CH$_3$COO—benzene—COOH	198, 442	8	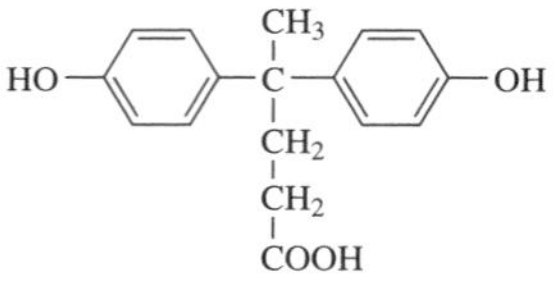 HO—benzene—C(CH$_3$)(CH$_2$CH$_2$COOH)—benzene—OH	465
3	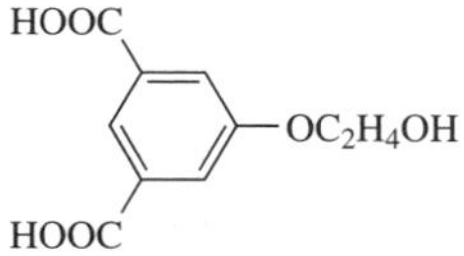 HOOC, HOOC—benzene—OC$_2$H$_4$OH	198, 466	9	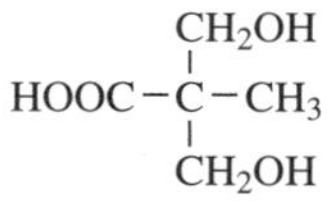 HOOC—C(CH$_2$OH)(CH$_2$OH)—CH$_3$	65, 197
4	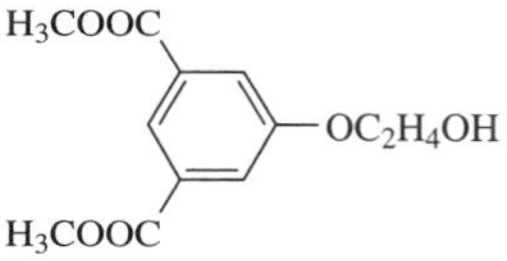 H$_3$COOC, H$_3$COOC—benzene—OC$_2$H$_4$OH	466	10	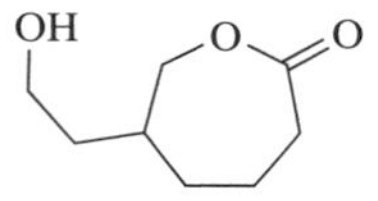 OH—CH$_2$CH$_2$—(caprolactone ring)	467
5	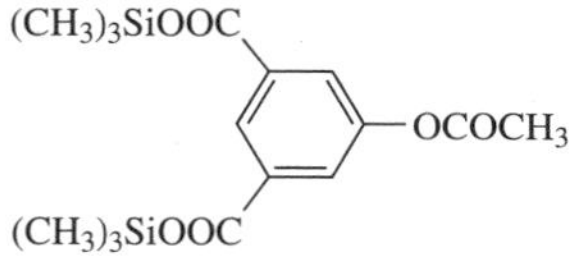 (CH$_3$)$_3$SiOOC, (CH$_3$)$_3$SiOOC—benzene—OCOCH$_3$	269	11	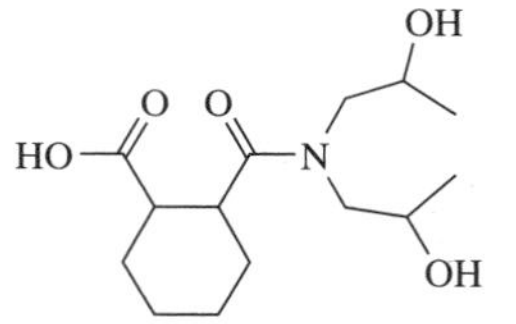 HO—CO—cyclohexane—CO—N(CH$_2$CH(OH)CH$_3$)$_2$	468
6	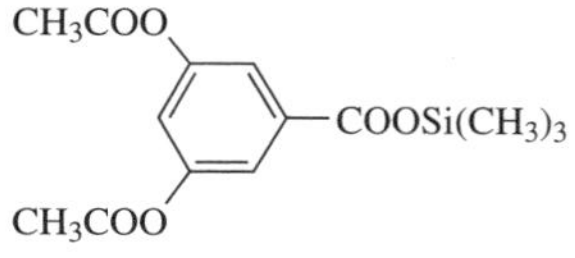 CH$_3$COO, CH$_3$COO—benzene—COOSi(CH$_3$)$_3$	268			

groups. Most reported hyperbranched polyesters are wholly aromatic or aromatic–aliphatic polyesters obtained from readily available 1,3,5-trisubstituted-benzene AB$_2$ monomers (Table 2.18). It is, however, noteworthy that only aliphatic hyperbranched polyesters (Perstorp's Boltorn obtained from dimethylol propionic acid, monomer 9, Table 2.18) and aliphatic polyesteramides (DSM's Hybrane obtained from monomer 11, Table 2.18) have been commercially introduced so far. The branched architecture and the presence of a large number of endgroups impart to hyperbranched polymers properties which are not found in linear polymers. The synthesis, structure, and properties of hyperbranched polymers have been recently reviewed by many authors.[60-63]

Hyperbranched polymers are characterized by their degree of branching (DB). The DB of polymers obtained by the step-growth polymerization of AB_2-type monomers is defined by Eq. (2.1) in which dendritic units have two reacted B-groups, linear units have one reacted B-group, and terminal units have two unreacted B-groups[191]:

$$DB = \frac{\sum(\text{dendritic units}) + \sum(\text{terminal units})}{\sum(\text{linear units}) + \sum(\text{dendritic units}) + \sum(\text{terminal units})} \quad (2.1)$$

Another definition, taking into account polymerization conversion, has been more recently proposed.[192] Perfect dendrimers present only terminal- and dendritic-type units and therefore have DB = 1, while linear polymers have DB = 0. Linear units do not contribute to branching and can be considered as structural defects present in hyperbranched polymers but not in dendrimers. For most hyperbranched polymers, nuclear magnetic resonance (NMR) spectroscopy determinations lead to DB values close to 0.5, that is, close to the theoretical value for randomly branched polymers. Slow monomer addition[193,194] or polycondensations with nonequal reactivity of functional groups[195] have been reported to yield polymers with higher DBs (0.6–0.66 range).

Assuming that no intramolecular or side reactions take place and that all groups are equireactive, the polydispersity index, I_p, of hyperbranched polymers obtained by step-growth polymerization of AB_x monomers is given by Eq. (2.2), where p_A is the conversion in A groups.[196] Note that the classical Flory relationship $\overline{DP_n} = 1/(1 - p_A)$ holds for AB_x monomer polymerizations:

$$I_p = \frac{\overline{DP_w}}{\overline{DP_n}} = \frac{1 - p_A^2/x}{1 - p_A} \quad (2.2)$$

The molar mass distribution of hyperbranched polymers is, therefore, always larger than that of their linear homologues and tends toward infinity when conversion becomes close to 1. The use of a B_y comonomer, acting as a chain limiter and "core molecule," helps in reducing polydispersity and controlling the molar mass of the final polymer.[197]

Due to their compact, branched structure and to the resulting lack of chain entanglement, dendritic polymers exhibit much lower melt and solution viscosity than their linear counterparts. Low α-values in the Mark–Houwink–Sakurada intrinsic viscosity–molar mass equation have been reported for hyperbranched polyesters.[198,199] Dendrimers do not obey this equation, a maximum being observed in the corresponding log–log viscosity–molar mass curves.[200] The lack of chain entanglements, which are responsible for most of the polymer mechanical properties, also explains why hyperbranched polymers cannot be used as thermoplastics for structural applications. Although some crystalline or liquid crystalline hyperbranched polymers have been reported, most hyperbranched polyesters described so far are amorphous. Glass transition temperature is strongly influenced by the nature of endgroups, increasing endgroup polarity generally leading to T_g increase. For instance, T_g values of 150 and 250°C have been reported respectively for acetate- and carboxy-terminated aromatic hyperbranched polyesters,[201] while the analogue terminated with long alkyl chains exhibited a T_g close to 10°C.[202,203] A modified Flory equation taking into account the nature and number of endgroups has been proposed to predict the T_g of dendritic molecules.[204] The melt viscosity of hyperbranched polymers is also strongly influenced by the nature of endgroups. Increase in melt viscosity by several orders of magnitude has been reported for aliphatic polyesters when ester endgroups are replaced by more polar hydroxy endgroups.[62] High solubility in common organic solvents is another remarkable feature of hyperbranched polymers, which strongly depends on the nature of endgroups. Polar endgroups such as metal carboxylate may confer solubility in water to nonpolar hyperbranched polymers. The term unimolecular micelle has been coined by Kim and Webster[205] for lithium carboxylate-terminated hyperbranched polyphenylenes, which can be described as globular macromolecules with a hydrophobic inner part and a polar outer shell.

2.2.3.3 Polyester-Based Networks

The polyesterification of carboxy and hydroxy polyfunctional monomers such as glycerol and phthalic acid yields crosslinked, hard materials. Since gelation starts at relatively low conversion, it is difficult to prepare prepolymers having viscosity and other properties appropriate for applications as thermosetting resin in coatings, laminates, or moldings. Moreover, the high reaction temperature and long reaction time required to achieve fully crosslinked materials and the formation of water during crosslinking make this chemistry of very limited practical use.

Polyester-based networks are typically prepared from polyester prepolymers bearing unsaturations which can be crosslinked. The crosslinking process is either an autoxidation in the presence of air oxygen (alkyd resins) or a copolymerization with unsaturated comonomers in the presence of radical initiators (unsaturated polyester resins). It should also be mentioned that hydroxy-terminated saturated polyesters are one of the basis prepolymers used in polyurethane network preparation (see Chapter 5).

2.2.3.3.1 Unsaturated Polyester Resins

Unsaturated polyesters are low-molar-mass polymers (1500–2500) obtained by the polyesterification of stoichiometric mixtures of diols and mixtures of saturated and unsaturated diacids or anhydrides (see Section 2.4.2.1).

For almost all applications unsaturated polyesters are dissolved in an unsaturated monomer capable of free-radical polymerization with the unsaturations in polyester chains. This polymerizable comonomer is generally styrene, but other compounds, such as methyl methacrylate, vinyl toluene, α-methylstyrene, and diallylphthalate, are also used in some applications. Upon heating and in the presence of initiators such as peroxides, free-radical copolymerization takes place and transforms the low-viscosity thermosetting polyester resin into a hard, insoluble, and infusible crosslinked network. Curing temperature, usually in the 80–160°C range, depends on initiator decomposition temperature. In the case of cast polyester resins, accelerators, mainly metal salts, catalyze peroxide decomposition and promote room temperature cure.

Maleic anhydride is the most widely used unsaturated condensation monomer for polyester resin preparation. During high-temperature polyesterification, maleate units are partly or wholly converted into their fumarate trans isomer. Since fumarate double bonds are much more reactive with styrene than maleate, the extent of maleate–fumarate isomerization exerts a great influence on crosslink density and, therefore, should be carefully controlled. Other unsaturated dibasic acid monomers, such as itaconic acid or cyclopentadiene–maleic anhydride adducts, can also be employed. General-purpose resins are obtained by reacting maleic anhydride and phthalic anhydride with 1,2-propanediol and/or 1,2-ethanediol. The use of isophthalic or terephthalic acid and of diols derived from bisphenol-A yields resins with higher rigidity and thermal resistance. On the other hand, saturated aliphatic acids such as adipic or sebacic acid and long-chain diols such as diethylene glycol, triethylene glycol, or poly(ethylene glycol) improve flexibility but reduce the water resistance of final material.

The final properties depend not only on unstaturated polyester structure but also on a number of other parameters, such as the nature and proportion of unsaturated comonomer, the nature of the initiator, and the experimental conditions of the crosslinking reaction. Moreover, since polyester resins are mainly used as matrices for composite materials, the nature and amount of inorganic fillers and of reinforcing fibers are also of considerable importance. These aspects have been discussed in many reviews and book chapters and are beyond the scope of this chapter.[7–9]

2.2.3.3.2 Alkyd Resins

Alkyd resins are polyesters of polyacids, polyols, and monobasic unsaturated fatty acids which can undergo crosslinking with oxygen from the air in the presence of promoters, generally metal salts such as cobalt naphthenate. They yield high-gloss, flexible, crosslinked coatings with good durability. Typical monomers for the preparation of alkyd resins are phthalic anhydride, glycerol, and/or pentaerythritol and drying oils such as soya or castor oil, that is, mixtures of saturated and unsaturated fatty acid triglycerides. Due to the presence of mono-, di-, tri-, and tetrafunctional monomers, alkyd resins are highly branched, polydisperse polyesters. It is important that the monomer composition be adjusted so that (i) gelation does not take place during resin synthesis and (ii) the final carboxyl content is low. Carothers equation is generally used to determine the theoretical conversion at the gel point, although this method suffers many limitations. Since hydroxyl groups are used in slight excess to ensure low carboxyl group content at the end of reaction, the nongelation conditions are given by the equation.

$$(p_{COOH})_{gel} = \frac{1}{\overline{f_{n,COOH}}} \geq 1 \Rightarrow \overline{f_{n,COOH}} \leq 1 \qquad (2.3)$$

where $\overline{f_{n,COOH}}$ is the number-average functionality in carboxyl groups, defined as the total number of carboxyl groups over the total number of molecules in starting monomer mixture. Since both ester interchange and direct esterification take place during the synthesis, triglyceride molecules should be taken as mixtures of glycerol (one molecule, $f_{OH} = 3$) and fatty acids (three molecules, $f_{COOH} = 1$) in these equations.

The properties of alkyd resins are controlled by the unsaturated fatty acid content, expressed as the weight percent of corresponding triglyceride ("oil length"). High triglyceride contents (55–60%) give resins that are soluble in aliphatic hydrocarbons and rapidly crosslink by autoxidation, yielding films with high gloss and good mechanical properties. Resins with low triglyceride content (below 40%) are soluble in aromatic hydrocarbons, present longer drying times, and result in films with inferior properties. Properties also markedly depend on the conjugated polyunsaturated fatty acid content of the formulation. High contents yields resins with short drying times which, however, exhibit a notable yellowing tendency. As usual in polyester chemistry, properties can be finely adjusted by varying monomer composition. The addition of isophthalic acid gives faster drying resins and tougher films with higher thermal stability, while aliphatic acids or long-chain diols impart flexibility to the final material.

The chemistry and structure–properties relationships of alkyd resins are well established and discussed in many textbooks.[50]

2.3 SYNTHETIC METHODS

2.3.1 General Considerations

The main polyesterification reactions are assumed to proceed through a substitution reaction at the carbonyl carbon with the formation of a tetrahedral intermediate, according to the $A_{AC}2$ addition–elimination mechanism of Ingold's classification[206] (Scheme 2.7). This scheme applies to (i) hydroxy–carboxy direct esterification ($X = -OH$, $Y = -H$), (ii) hydroxy–ester interchange (alcoholysis, $X = -OR^3$, $Y = -H$), (iii) carboxy–ester interchange (acidolysis, $X = -OH$, $Y = R^4CO-$), (iv) ester–ester interchange (transesterification, $X = -OR^3$, $Y = R^4CO-$), (v) acid chloride–hydroxy reaction ($X = -Cl$, $Y = -H$), and (vi) anhydride–hydroxy reaction ($X = R^3COO-$, $Y = -H$), where R^1 to R^4 represent either monomer or growing-chain ends. Negatively charged nucleophiles such as alcoholates or phenolates are also suitable for solution or interfacial polyesterifications with acid chlorides.

According to Scheme 2.7, when X is a good leaving group and $-OY$ a strong nucleophile, the reaction proceeds at high rate in mild conditions. On the other hand, reactions involving poor leaving groups and weak nucleophiles must be carried out at high temperature during a long reaction time. Increasing the

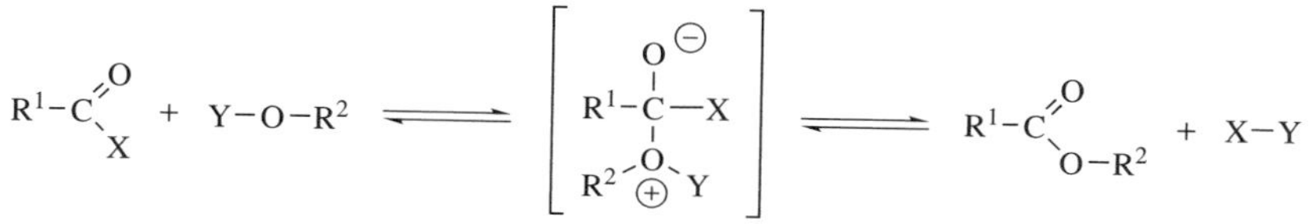

$$R^1-C{\overset{O}{\underset{X}{\diagup}}} + Y-O-R^2 \rightleftharpoons \left[\begin{array}{c} O^{\ominus} \\ | \\ R^1-C-X \\ | \\ \underset{R^2 \overset{\oplus}{\diagdown} Y}{O} \end{array} \right] \rightleftharpoons R^1-C{\overset{O}{\underset{O-R^2}{\diagup}}} + X-Y$$

Scheme 2.7

electron-withdrawing character of R and/or X decreases the electron density on the carbonyl carbon atom and favors the formation of reaction intermediate, thus increasing the reaction rate. On the other hand, resonance interactions between the carbonyl group and either R or X have a stabilizing effect on the starting carboxylic acid derivative, which decreases the rate of formation of the reaction intermediate. The presence of bulky groups close to the reaction site also does not favor the formation of ester, explaining why tertiary alcohols cannot be used in polyesterification reactions.

Most esterifications are equilibrium processes. The equilibrium constant depends on the nature of monomers and is generally low for common polyesterification reactions ($K \approx 4$ for carboxylic acid–alcohol reactions). This means that high-molar-mass polyesters can be obtained only if the by-product, water, alcohol, or acid (X–Y in Scheme 2.7) is continuously and efficiently removed from reaction medium in order to shift the equilibrium to the right. On the other hand, reactions involving acid chlorides or activated carboxylic acids are generally regarded as "nonequilibrium" polyesterifications, due to the high value of their equilibrium constant ($K = 4700$ for acid chloride–phenol polyarylate synthesis).[207] Polyesterifications can, therefore, be divided into two main categories: (i) high-temperature bulk polyesterifications, mainly hydroxy–carboxy reactions and hydroxy–ester and carboxy–ester interchange reactions, and (ii) nonequilibrium polyesterifications, generally low-temperature solution reactions involving highly reactive monomers such as acid chlorides or activated carboxylic acids. A third class of polyester syntheses consists of reactions that are generally not regarded as polyesterifications: epoxy–carboxy reaction, ring-opening polymerization of cyclic esters, and some miscellaneous, sometimes exotic, reactions.

2.3.2 High Temperature Bulk Polyesterifications

From the preceding discussion, it is easily understood that direct polyesterifications between dicarboxylic acids and aliphatic diols (Scheme 2.8, $R^3 = H$) and polymerizations involving aliphatic or aromatic esters, acids, and alcohols (Scheme 2.8, $R^3 = $ alkyl group, and Scheme 2.9, $R^3 = H$) are rather slow at room temperature. These reactions must be carried out in the melt at high temperature in the presence of catalysts, usually metal salts, metal oxides, or metal alkoxides. Vacuum is generally applied during the last steps of the reaction in order to eliminate the last traces of reaction by-product (water or low-molar-mass alcohol, diol, or carboxylic acid such as acetic acid) and to shift the reaction toward the formation of high-molar-mass polyester. Ester–ester interchange (Scheme 2.9, R^3 = alkyl group) is much slower than hydroxy–ester and carboxy–ester interchange and is seldom applied to polyester synthesis.

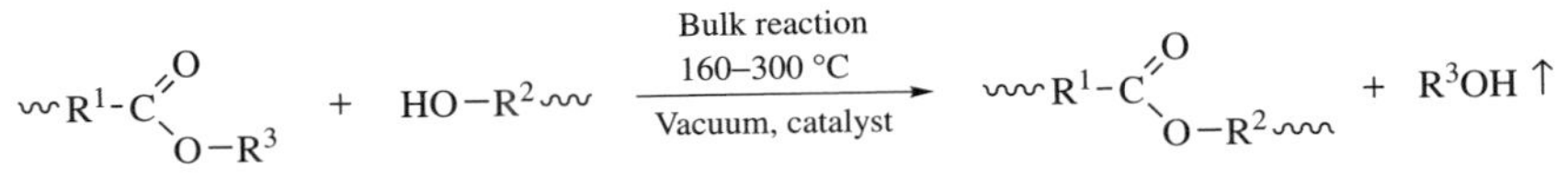

R^1 = Alkyl or aryl group; R^2 = Alkyl group; R^3 = H or alkyl group

Scheme 2.8

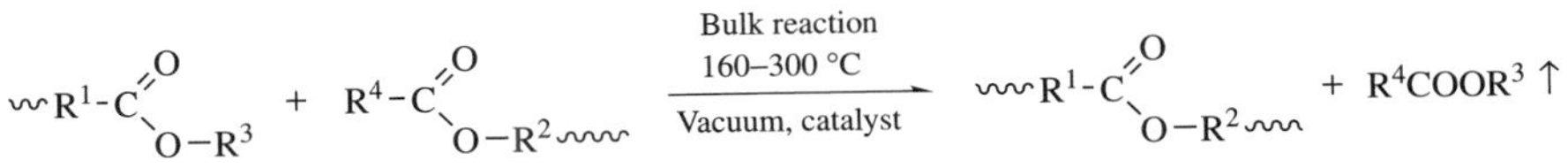

R^1 = Alkyl or aryl groups; R^2, R^4 = Alkyl groups; R^3 = H or alkyl group

Scheme 2.9

Phenolic compounds are weaker nucleophiles and better leaving groups than aliphatic alcohols. They do not yield polyesters when reacted with carboxylic acids or alkyl carboxylates. The synthesis of polyesters from diphenols is, therefore, generally carried out through the high-temperature carboxylic acid–aryl acetate or phenyl ester–phenol interchange reactions with efficient removal of reaction by-product (Schemes 2.10 and 2.11, respectively).

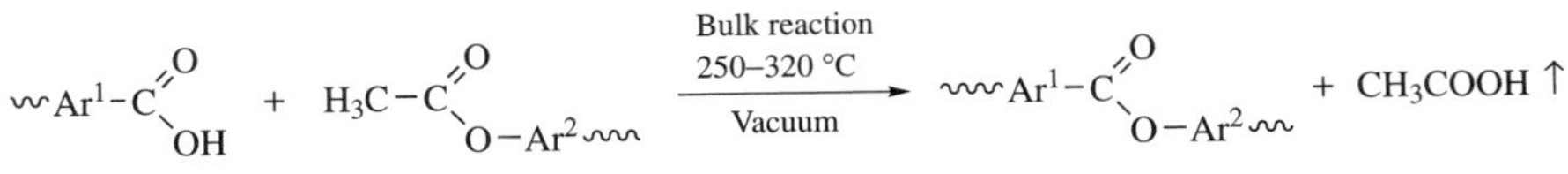

Ar^1, Ar^2 = Aryl groups

Scheme 2.10

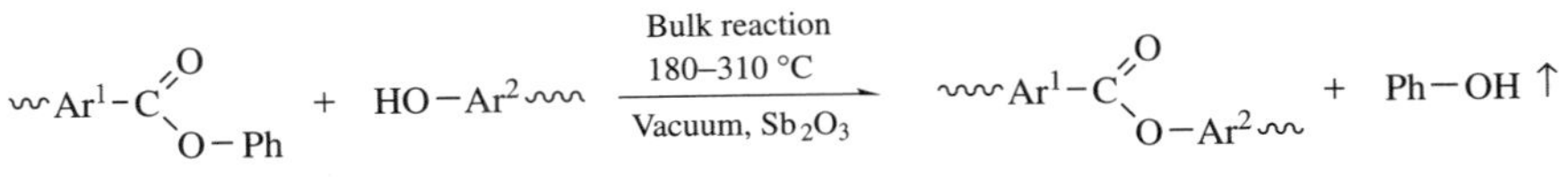

Ar^1, Ar^2 = Aryl groups; Ph = Phenyl group

Scheme 2.11

Since polyesterifications are step-growth processes, the synthesis of high-molar-mass polyesters requires high-purity monomers in precisely adjusted stoichiometric ratio and with conversion close to unity (say above 99%). Side reactions (see Section 2.2.1.2) which may cause imbalance of reactive groups and limit the molar mass or may lead to the formation of undesired changes in polymer structure, must also be carefully controlled. This may be difficult in the case of high-temperature bulk reactions, especially those involving aliphatic monomers, which begin to degrade above 220°C.

It should be stressed that the various high-temperature ester interchange reactions play a very important role in achieving high-molar-mass polyesters. Monomer excess adventitiously or deliberately introduced in reaction mixture can be eliminated at high temperature and under vacuum through hydroxy–ester, carboxy–ester, or ester–ester interchanges, either during the last steps of the reaction or during a subsequent high-temperature solid-state postpolymerization stage. Such a procedure is applied in the synthesis of PET from terephthalic acid and 1,2-ethanediol (see Sections 2.3.2.1, 2.4.3.1.2, and 2.4.3.1.3) and can be extended to almost any kind of polyester.

As already discussed (Section 2.2.1.3), interchange reactions are also implicated in the formation of random copolyesters exhibiting the "most probable" molar mass distribution when polyester blends are melt mixed. They are also involved in the randomization of block copolyesters taking place in the melt upon heating.[208–211]

Cyclic oligomers are formed during step-growth polymerization,[212–214] either by reactions between endgroups of the same chain or by hydroxy–ester and carboxy–ester "back-biting" reactions between endgroups and ester groups of the same chain. The presence of cyclics is, therefore, unavoidable in most step-growth polymers and polyesters are not an exception.[92,215] Since the ring-chain equilibrium is displaced toward the formation of cyclics in dilute solutions, solution processes are likely to yield higher amounts of cyclics than reactions carried out in the bulk.

2.3.2.1   Direct Polyesterification

2.3.2.1.1   Applications

The carboxy–hydroxy reaction (direct esterification) is the most straightforward method of polyester synthesis. It was first reported in the 1930s by Carothers[10–12] and is still a very widely used method for the synthesis of polyesters from diacids and diols (Scheme 2.12) or from hydroxy acids (Scheme 2.13). Direct

$$n \, \text{HOOC}-\text{R}^1-\text{COOH} + n \, \text{HO}-\text{R}^2-\text{OH} \xrightarrow[\text{Bulk}]{160-290\,°\text{C}} \text{H}-\left(\text{OOC}-\text{R}^1-\text{COO}-\text{R}^2\right)_n\text{OH}$$

$$+$$

$$(2n-1)\,\text{H}_2\text{O}\uparrow$$

Scheme 2.12

$$n \ HO-R-COOH \xrightarrow[\text{Bulk}]{160-290\ ^{\circ}C} HO-\!\!\left(\!R-COO\!\right)_{\!\!n}\!\!-H \ + \ (n-1)\,H_2O \uparrow$$

Scheme 2.13

polyesterification is rather slow at room temperature and must be carried out at high temperature (150–290°C depending on monomers) and high endgroup concentration, preferably in the bulk. Vacuum is generally applied during the last steps of reaction to distill off reaction water and to continuously shift the reaction toward the formation of high-molar-mass polyester. In the case of alkyd resins or unsaturated polyesters, toluene or xylene (5–15% mass) is sometimes added in reaction medium to facilitate water removal through the formation of toluene– or xylene-water azeotropic mixture (see Section 2.4.2.3). Direct polyesterification is particularly well suited to the syntheses of aliphatic polyesters, unsaturated polyesters, and aromatic–aliphatic polyesters such as PET and PTT, which involve reactions between aliphatic or aromatic carboxylic acids and primary or secondary aliphatic alcohols. It is not applicable to the synthesis of polyesters from monomers containing phenol or tertiary alcohol groups, which exhibit a very low reactivity with carboxylic acids, even at high temperature and in the presence of catalysts. In this case (e.g., for the preparation of wholly aromatic polyesters), methods involving hydroxy–ester or carboxy–ester interchange reactions are generally preferred.

Although low-molar-mass aliphatic polyesters and unsaturated polyesters can be synthesized without added catalyst (see Sections 2.4.1.1.1 and 2.4.2.1), the presence of a catalyst is generally required for the preparation of high-molar-mass polyesters. Strong acids are very efficient polyesterification catalysts but also catalyze a number of side reactions at elevated temperature (>160°C), leading to polymers of inferior quality. Acid catalysts are, therefore, not much used. An exception is the bulk synthesis of hyperbranched polyesters reported in Section 2.4.5.1, which is carried out at moderate temperature (140°C) under vacuum in the presence of p-toluene sulfonic acid catalyst. The use of strongly acidic oil-soluble catalysts has also been reported for the low-temperature synthesis of polyester oligomers in water-in-oil emulsions.[216]

Metal salts and oxides, such as zinc acetate and diantimony trioxide, and organometallic compounds, mainly titanium and zirconium alkoxides, are the preferred catalysts for bulk polyesterifications carried out in the 160–290°C range.[217] Thus, the most important method of PET synthesis involves the bulk polyesterification of purified terephthalic acid (PTA) with 1,2-ethanediol (ethylene glycol, EG). The reaction is performed in two stages. In the first stage PTA is reacted with excess EG (ca. 1 : 1.2 ratio) under pressure at 230–270°C with elimination of water (direct esterification), yielding a low-molar-mass hydroxyethyl terephthalate–terminated oligomer. The medium is then stepwise heated to 270–290°C under a progressively reduced pressure (10–50 Pa) until excess EG is eliminated and high-molar-mass PET ($\overline{DP_n} \approx 200$) is obtained. No catalyst is usually required for the first reaction stage. The second stage involves hydroxy–ester interchange reactions for which Sb_2O_3 is an efficient catalyst (see Sections 2.3.2.2.1 and 2.4.3.1.2.) (Scheme 2.14). The molar mass of PET can be further increased by a solid-state postpolymerization under inert-gas flow or under vacuum at 220–250°C (see Section 2.4.3.1.3). The synthesis of poly(trimethylene terephthalate), PTT, is essentially the same, starting from 1,3-propanediol instead of 1,2-ethanediol.[27] Due to 1,4-butanediol dehydration to THF in acidic medium, PBT cannot be prepared by direct polyesterification (see Section 2.3.2.2.1).

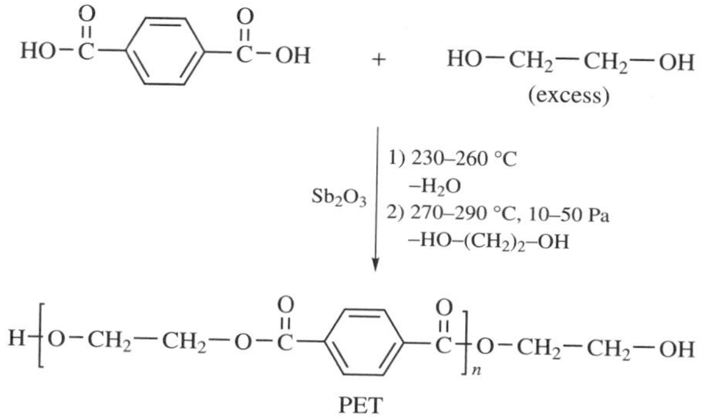

Scheme 2.14

The reaction between acid anhydrides and diols is another convenient method of polyester synthesis. The reaction proceeds in two steps with the formation of an intermediate hydroxy acid (Scheme 2.15).

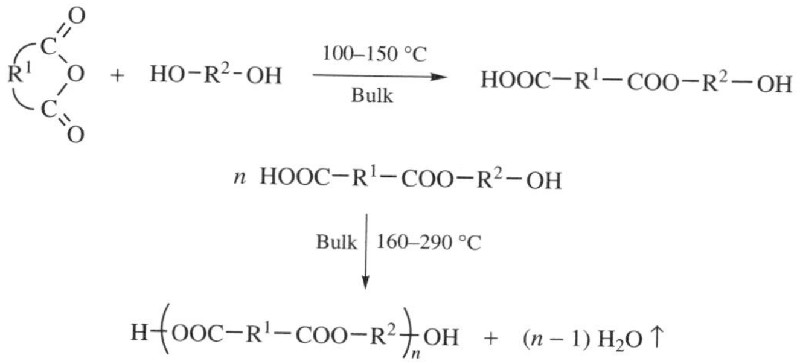

Scheme 2.15

Since anhydrides are much more reactive than carboxylic acids, reaction kinetics is controlled by the second step. The scope and applications of this reaction are the same as direct polyesterification but are practically limited to the synthesis of unsaturated polyesters and alkyd resins from phthalic and maleic anhydrides (see Sections 2.4.2.1 and 2.4.2.3).

2.3.2.1.2   Reaction Mechanisms and Catalysis

The kinetics of uncatalyzed and catalyzed bulk direct polyesterifications has been extensively studied, both for theoretical considerations — the elucidation of reaction mechanisms — and for practical purposes — the optimization of industrial processes. The polymeric nature of reaction medium complicates the interpretations of kinetic studies carried out on bulk polyesterification reactions: The polarity of reaction medium dramatically changes during polyesterification, from that of a relatively polar diol–diacid mixture to that of a relatively nonpolar high-molar-mass polyester. Moreover, reaction kinetics becomes diffusion controlled at high extents of reaction, due to increasing viscosity. Since the devolatilization of reaction water becomes increasingly difficult as the reaction proceeds, ester hydrolysis may also contribute to the overall kinetics in the last steps of the reaction. Some studies have been carried out taking into account a contribution of the reverse reaction (hydrolysis) at the end of the reaction,[218–224] but it is, however, generally considered that water is efficiently removed from reaction medium in the conditions used and that this side reaction does not significantly contribute to the overall kinetics. In the case of some reaction systems (e.g., monomer mixtures containing terephthalic acid), the kinetics of the early stages may also be perturbed by initial heterogeneity. It must finally be remarked that ester–ester, hydroxy–ester, and carboxy–ester interchange reactions, which also take place during diacid–diol polyesterifications, do not contribute to chain growth and polyesterification kinetics provided that low-molar-mass diols or diacids are not distilled off from reaction medium.

In his early studies on polyesterification kinetics, Flory[225–227] demonstrated that reaction stages above ca. 80% conversion only should be taken into account when discussing the mechanisms of high-temperature bulk polyesterifications. In these conditions polarity no longer changes with conversion, and the reaction medium can be considered as an "ideal" diluted solution of reactive groups in polyester — provided that viscosity is not too high and the reaction is not diffusion controlled. Although these limits were clearly stated by Flory, many authors have put forward reaction mechanisms based on studies where the whole course of the reaction was taken into account, resulting in many controversial interpretations of polyesterification kinetics. The kinetics and mechanisms of diacid–diol polyesterifications and the conditions in which kinetic studies can be discussed in terms of reaction mechanisms have been reviewed.[2,217]

REACTIONS WITHOUT ADDED CATALYST   Flory[225–227] showed that both mono- and polyesterifications are third-order processes similar to the processes found for solution esterifications and that the reactivity of hydroxyl and carboxyl end-groups does not depend on chain length. These early findings have been confirmed by many other studies,[228–231] and order 2 in acid and order 1 in alcohol were confirmed in polymeric medium.[232,233] The proposed mechanism involves an autocatalysis of the reaction by the carboxylic acid, the rate-determining step being the nucleophilic addition of the alcohol on protonated carboxylic acid

(Scheme 2.16). Assuming that protonation is an equilibrium and the concentration of protonated carboxylic acid is very low, the rate equation is as

$$v = k[\text{B}][\text{R}'\text{OH}] = kK[\text{RCOOH}]^2[\text{R}'\text{OH}] \tag{2.4}$$

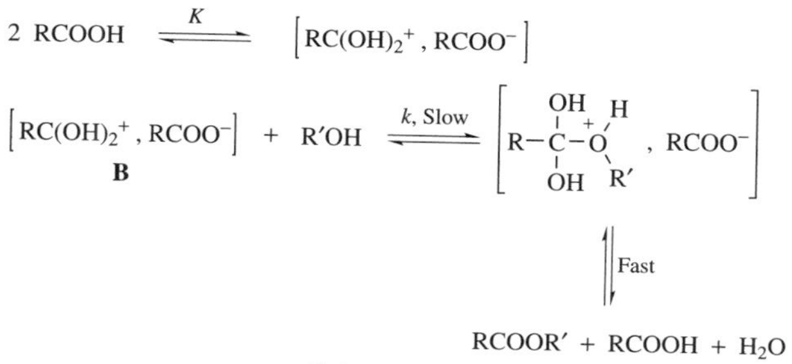

Scheme 2.16

Although the third-order mechanism is the most commonly accepted, many other have been proposed, including 1st order (in the presence of excess glycol),[234] 2nd order,[235,236] 2.5 order,[235,237–239] 3rd order with order 1 in acid and order 2 in alcohol,[240–242] succession of 2nd and 3rd orders[219,243] or varying reaction order,[244] and more complex mechanisms such as a 6th overall order.[245] Most of these studies, however, were carried out in reaction media which do not fulfill Flory's ideality conditions or in reaction media that are quite different from polyesters (e.g., nonpolar long-chain fatty acid esters).[239] It must also be mentioned that many studies are devoted to the simulation of polyesterifications taking place in technological conditions rather than to the determination of reaction mechanisms. In these studies, reaction orders are taken from plausible reaction mechanisms or are only semiempirical parameters not directly connected to reaction mechanisms.[237,244,246–249]

REACTIONS CATALYZED BY STRONG PROTONIC ACIDS The kinetics of polyesterifications catalyzed by strong protonic acids, mainly H_2SO_4 and benzene, toluene, and naphthalene sulfonic acids, has also been extensively examined. In contrast to the reactions carried out without added catalyst, nearly all results agree with a third-order mechanism, first order in acid, alcohol, and catalyst. At constant catalyst concentration, the reaction can be considered as a second-order one.[220,226,227,233,238,239,250,251] In spite of this concordance, the various authors suggest mechanisms that involve either ion pairs or free ions. The generally accepted reaction scheme has a close resemblance to the preceding one, with protonation of carboxylic acid by the added protonic acid and formation of ion pairs (Scheme 2.17). From this scheme, reaction rate can be expressed by Eq. (2.5), assuming that the concentration of the dissociated form of protonic acid catalyst is negligible with respect to initial catalyst concentration $[\text{AH}]_0$:

$$v = k[\text{C}][\text{R}'\text{OH}] = kK[\text{AH}]_0[\text{RCOOH}][\text{R}'\text{OH}] \tag{2.5}$$

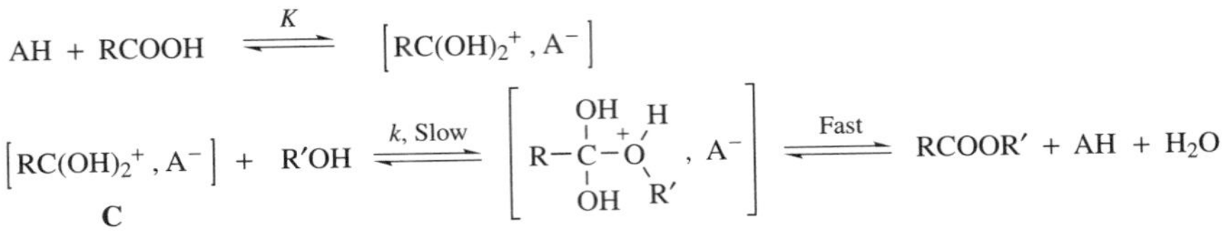

$$AH + RCOOH \xrightarrow{K} \left[RC(OH)_2^+, A^- \right]$$

$$\left[RC(OH)_2^+, A^- \right] + R'OH \xrightarrow{k, \text{Slow}} \left[\begin{array}{c} OH \quad H \\ | \quad + \, / \\ R-C-O \\ | \quad \backslash \\ OH \quad R' \end{array}, A^- \right] \xrightarrow{\text{Fast}} RCOOR' + AH + H_2O$$

$$\quad C$$

Scheme 2.17

CATALYSIS BY LEWIS ACIDS AND METAL ALKOXIDES Most industrial processes relative to direct polyesterification require the use of metal salts or organometallic derivatives as catalysts, especially for the synthesis of high-melting aromatic–aliphatic polyesters such as PET. An enormous number of references relative to polyesterification catalysts, mainly patents, have appeared in the literature, and derivatives of practically all elements have been claimed to be esterification or polyesterification catalysts.[217] The number of references relative to reaction mechanisms and to reactivity comparison is, however, much lower and the information available on metal-catalyzed reaction mechanisms is often controversial. The catalytic efficiency of various metal derivatives has been estimated by Malek et al.[252,253] from reactions of aromatic acids in aliphatic diols. These studies, as well as others, show that the most efficient catalysts are titanium and zirconium alkoxides and tin derivatives such as Bu_2SnO. Other catalysts such as Zn, Pb, Al, Mn, Mg, Cd, Co acetates, Cu and La acetylacetonates, and Sb_2O_3 exhibit much lower reactivity.[252–254] It is generally accepted that the metal ligands are exchanged with the carboxyl and hydroxyl groups present in reaction medium. Tetrabutyltin,[253] which does not exchange its ligands, and titanium alkoxides with strongly attached ligands such as 1,2-phenylenedioxy groups,[254] display no or little catalytic activity. On the other hand, when ligands can be exchanged, the catalytic activity does not depend on initial ligands. A reasonable reaction scheme has been proposed for tetraalkoxyzirconium-catalyzed reactions[255] involving (i) the exchange of initial ligands (*i*-propoxy in the present case), (ii) the reversible formation of carboxylatozirconium species, (iii) the nucleophilic attack of hydroxyl groups leading to the formation of the ester and Zr–OH groups, and (iv) the re-formation of catalytic species, alkoxy- and/or carboxylatozirconium compounds (Scheme 2.18).

According to this scheme, the catalytic effect arises from the complexation of carbonyl group on the metal atom, which induces a positive charge on the carbonyl carbon atom and favors the nucleophilic attack of hydroxyl groups. The nature of catalytic species, therefore, changes during the course of reaction, and the results suggest that the catalytic activity depends on the composition of reaction medium. Moreover, the presence of reaction water, which is unavoidable in direct esterification, may lead to the formation of metal–O–metal bonds and of condensed forms of metal alkoxides. This decreases the number of active exchangeable sites and, therefore, lowers catalytic activity.[256,257] In view of the complexity of such reactions, it is rather difficult to discuss the kinetic results in terms of reaction orders.

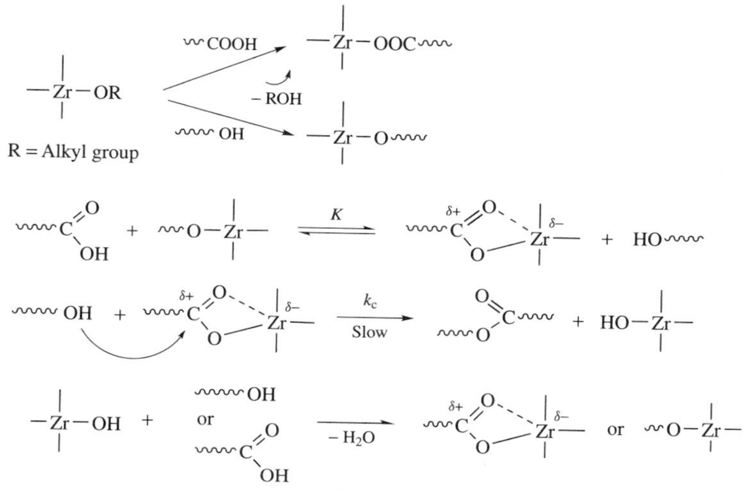

$$-\overset{|}{\underset{|}{Zr}}-OR \quad \overset{\text{\footnotesize\raise2pt\hbox{~COOH}}}{\underset{\text{~~OH}}{\xrightarrow{\hspace{2cm}}}} \quad \begin{array}{l} -\overset{|}{\underset{|}{Zr}}-OOC\text{~~} \\[4pt] -\overset{|}{\underset{|}{Zr}}-O\text{~~} \end{array}$$

$-ROH$

R = Alkyl group

Scheme 2.18

2.3.2.2   *Transesterification Polymerizations*

2.3.2.2.1   *Applications*

Transesterifications, also termed ester exchange or ester interchange reactions, include hydroxy–ester, carboxy–ester, and ester–ester reactions. Hydroxy–ester reaction is the most important one and is used for many aromatic–aliphatic and wholly aromatic polyester syntheses. Carboxy–ester interchange is restricted to the synthesis of wholly aromatic polyesters while the ester–ester route is rarely used for polyester preparation due to slow kinetics. All these reactions may take place in comparable experimental conditions and can be catalyzed by similar classes of compounds.

SYNTHESIS OF AROMATIC–ALIPHATIC POLYESTERS Hydroxy–ester interchange (alcoholysis) reactions play a predominant role in most industrial preparations of aliphatic and aromatic–aliphatic polyesters. The original industrial synthesis of PET[13] is based on the reaction between dimethyl terephthalate (DMT) and excess EG (Scheme 2.19). The reaction is carried out in two stages. During the first stage, methanol is distilled off and a low-molar-mass hydroxyethyl terephthalate–terminated oligomer is produced. The second stage involves intermolecular reactions between hydroxy and ester endgroups with elimination of EG. Vacuum is applied at the end of the reaction until high-molar-mass PET is obtained. This stage is similar to the second stage of the direct esterification method (see Section 2.3.2.1.1). Catalysts such as acetates of calcium, zinc, magnesium, or manganese are generally used in the first stage and Sb_2O_3 in the second one. Phosphorus-containing compounds may be added at the end of the first stage to deactivate the metal acetate catalyst, which may promote yellow discoloration during the second stage. Titanium compounds are very efficient catalysts for

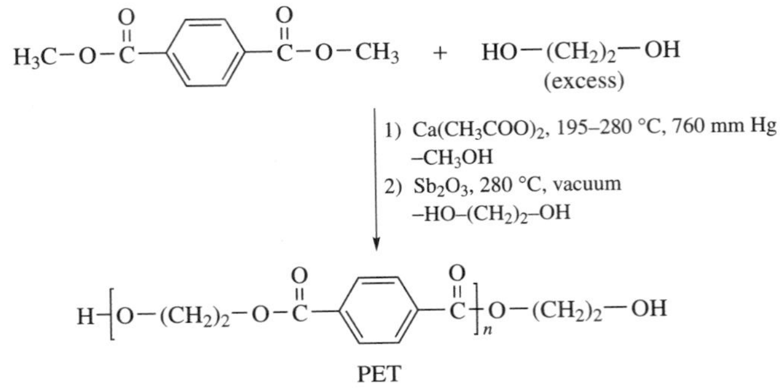

PET

Scheme 2.19

both stages but are not generally used because they generate undesirable yellow discoloration in the presence of EG. However, complex mixtures comprising titanium alkoxides, phosphorus compounds, and complexing agents[258] or cobalt salts[259] have been claimed to yield high-molar-mass low-color PET. Although the modern PET industrial synthesis is based on the direct polyesterification of purified terephthalic acid with EG (Section 2.3.2.1.1), the DMT/EG method discussed here is still used for ca. 10% of PET production. Experimental details are given in Section 2.4.3.1.1.

Since 1,4-butanediol (BD) undergoes dehydration side reaction in the presence of acid resulting in THF formation, the hydroxy–ester interchange reaction is the preferred method for the preparation of PBT. The first stage of reaction is carried out at 150–200°C and consists of a hydroxy–ester interchange between DMT and excess butanediol with elimination of methanol. In the second stage, temperature is raised to 250°C and BD excess is eliminated under vacuum. Tetraisopropoxy- and tetrabutoxytitanium are efficient catalysts for both stages (Scheme 2.20).

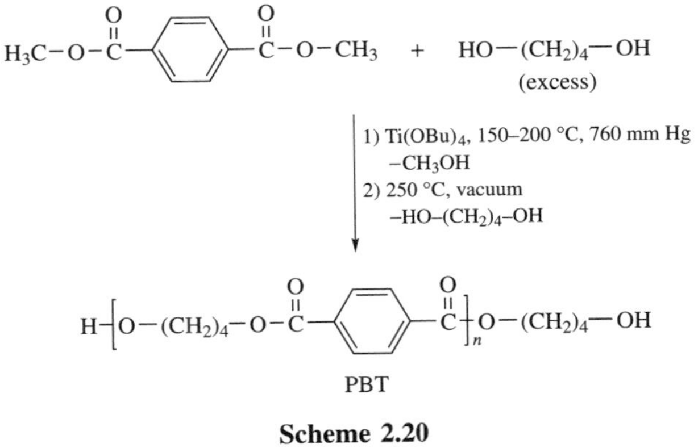

PBT

Scheme 2.20

Because naphthalene-2,6-dicarboxylic acid has a very low solubility in EG and is not easily available, PEN is also prepared using the two-stage hydroxy–ester interchange procedure, starting from dimethyl naphthalene-2,6-dicarboxylate (DMN) and excess EG. Zinc or manganese diacetate catalysts are used for the first stage and Sb_2O_3 for the second one, with final temperature close to 290°C.[118,260–263] Titanium compounds are very efficient ester interchange catalysts, but as observed for PET synthesis, discoloration in the presence of EG prevents their use in PEN preparation. It should finally be mentioned than hydroxy-terminated aliphatic polyesters can also be prepared following this reaction scheme.

Due to the high reaction temperatures required during the last stages of these syntheses, side reactions cannot be avoided. Acetaldehyde, carboxyl endgroups, and vinyl endgroups are formed during PET and PEN synthesis. The formation of 2,2′-oxydiethylene moieties in polymer chains by etherification of hydroxyl endgroups is also a well-known side reaction of EG polyester syntheses.[264] These reactions should be carefully controlled since they can exert an important influence on polymer properties such as T_g, mechanical properties, hydrolytic stability, and discoloration.

SYNTHESIS OF WHOLLY AROMATIC POLYESTERS Besides the acid chloride/phenol reaction (see below Section 2.3.3.1), high-temperature melt reactions between bisphenol diacetates and aromatic dicarboxylic acids (Scheme 2.21) or between bisphenols and aromatic dicarboxylic acid diphenyl esters (Scheme 2.22) are the preferred routes for the synthesis of wholly aromatic polyesters. The presence of ester interchange catalysts such as antimony oxide or titanium alkoxide is generally required for the phenol–ester route but is optional for the carboxy–ester one, for which acetates of various metals such as zinc or manganese are the most effective catalysts. As usual, vacuum must be applied during the last steps of reaction to efficiently remove the by-products, either acetic acid or phenol, in order to achieve high molar mass and good thermal stability for the final polymer. The bisphenol diacetate route, which produces acetic acid by-product, is the preferred method for the synthesis of liquid crystalline aromatic polyesters.

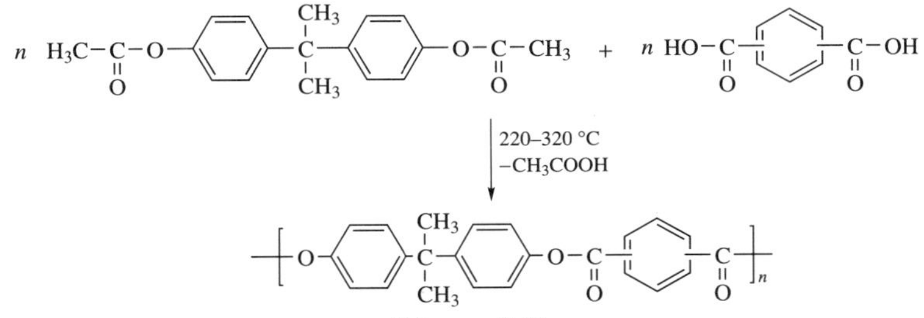

Scheme 2.21

Carrying out reactions in such a temperature range, however, leads to a number of difficulties and limitations that are not easy to overcome: (i) the volatilization

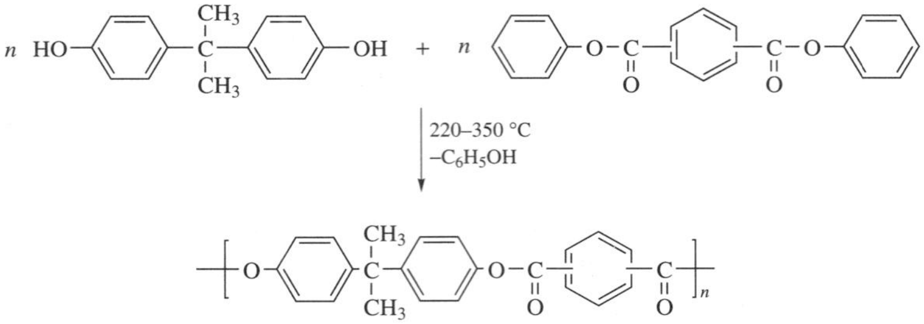

Scheme 2.22

of dicarboxylic acid monomers during the first steps of reaction may cause stoichiometric imbalance of reactants; (ii) polymers melts are often very viscous, resulting in diffusion-controlled kinetics and slowed reaction rates; and (iii) side reactions occurring at high temperature lead to structural defects such as *o*-hydroxybenzophenone and quinonoid structures which result in yellow- or amber-colored products. The addition of high-boiling-point inert solvents, diluents, or swelling agents such as diphenyl ether, aromatic sulfones, chlorinated biphenyls, cyclohexylbiphenyls, and dibenzylbenzenes often helps in reducing melt viscosity and eliminating reaction by-products, thus enhancing reaction rate. In a final stage, the molar mass of high-melting semicrystalline polyesters can be increased by solid-state postpolymerization carried out under vacuum or nitrogen flow at a temperature below T_m.

The use of silylated monomers is an interesting alternative method of aromatic polyester synthesis since the silylated gaseous by-products cannot participate in the reverse reaction, shifting polyesterification toward polymer formation. Reactions between silyl esters and acetates (Scheme 2.23) and reactions between silyl ethers and acid chlorides (Scheme 2.24) have been applied to the synthesis of linear[265–267] and hyperbranched wholly aromatic polyesters[202,268,269] (see Section 2.4.5.2.2).

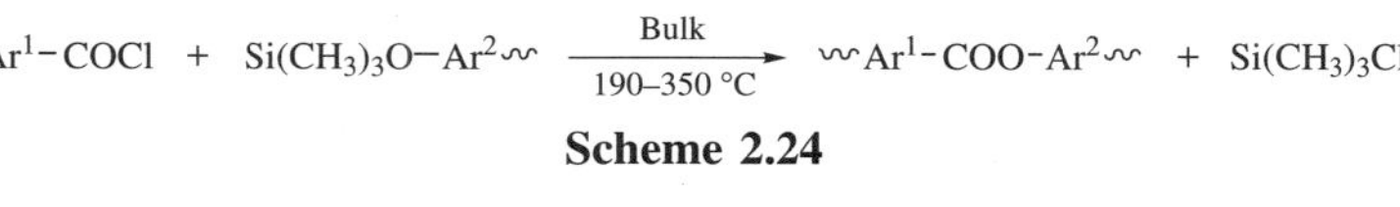

Scheme 2.23

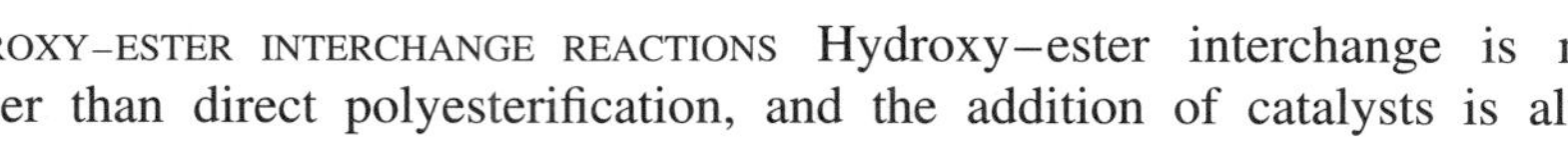

Scheme 2.24

2.3.2.2.2 Reaction Mechanisms and Catalysis

HYDROXY–ESTER INTERCHANGE REACTIONS   Hydroxy–ester interchange is much slower than direct polyesterification, and the addition of catalysts is always required for the synthesis of high-molar-mass polyesters. Within the very large number of derivatives of almost all existing elements that have been claimed to catalyze hydroxy–ester interchange reactions, metal salts, oxides, and alkoxides are the most often cited compounds. The activity of a given catalyst during the first step of polymerization (reaction of methyl ester groups) may be quite different from that observed during the polycondensation step (reaction of hydroxyalkyl ester groups) (Scheme 2.19).

Lead, Zn, and Mn derivatives are generally considered as the most efficient catalysts for the first stage of reaction, followed by a group of catalysts comprising Co, Mg, Ti, Ni, Sn, and Ca derivatives. The catalytic activity of Sb compounds is much lower. Thus, the following decreasing orders in catalytic activity were found for the dimethyl 2,6-naphthalate–1,2-ethanediol reaction taking place during the first stage of PEN synthesis: Pb(II) > Zn(II) > Mn(II) > Co(II) > Ti(IV) > Sn(II) > Mg(II) > Ca(II) > Na(I) > Sb(III)[270] and Pb(II) $\geq$ Zn(II) > Co(II) > Mg(II) > Ni(II) $\geq$ Sb(III).[271] Many resembling results have been published on the first stage of PET synthesis. It is finally worth mentioning that synergistic effects can be obtained using mixtures of catalysts such as Mg, Mn, and Zn acetates.[272]

On the other hand, Sb_2O_3 and metal oxides such as GeO_2 exhibit a good catalytic activity for the polycondensation step. This explains why the associations of metal acetates with Sb_2O_3 are often reported catalytic systems. The following order in catalytic activity was found for the PEN polycondensation step: Ti(IV) > mixtures of Ti(IV) and Sb(III) > Sn(II) > Sb(III) > Co(II) > Zn(II) > Pb(II) > Mn(II) > Mg(II).[270] It should be underlined that titanium and zirconium alkoxides are efficient catalysts for both stages of reaction. Lanthanide compounds such as 2,2′-bipyridyl, acetylacetonate, and o-formyl phenolate complexes of Eu(III), La(III), Sm(III), Er(III), and Tb(III) appear to be even more efficient than titanium alkoxides, Ca or Mn acetates, Sb_2O_3, and their mixtures.[273] Moreover, PET produced with lanthanides has been reported to exhibit better thermal and hydrolytic stability as compared to PET synthesized with the conventional Ca acetate–Sb_2O_3 catalytic system.[273]

Kinetic studies in such reaction media are experimentally difficult, and the reaction mechanisms are not yet fully understood. It has been shown that the reactivity of hydroxyl endgroups is not affected by chain length but that the reactivity of 1,2-ethanediol hydroxyls can be very different from that of hydroxyl endgroups.[274–276] The experimental data generally fit second-order theoretical kinetic curves, and the reaction is assumed to be first order in both hydroxyl and ester groups. This allows the determination and the comparison of catalytic activities.[2,270,276–279] Decreasing orders in catalyst are often reported at high catalyst concentrations or high conversions. This is usually assigned to the formation of catalyst dimers or oligomers.[277,280,281] The catalytic activity of metal acetates has been examined with respect to their acidity factor β, which corresponds to the stability constant of dibenzoylmethane complex for the considered metal species.[282] The results indicate that very strong or very weak carbonyl–metal interactions are unfavorable factors with regard to catalytic activity.[282,283] The absence of interactions obviously results in poor catalytic activity. On the other hand, very strong carbonyl–metal interactions would result in the formation of stable complexes, slowing the ligand exchanges responsible for the catalytic effect. The formation of strong complexes between EG and antimony has frequently been proposed to explain the low catalytic activity of this metal in alcohol-rich medium.[284,285] The presence of carboxyl groups has also been reported to inhibit the catalytic activity of various acetates and alkoxides. A reasonable mechanism should involve the coordination of the ester carbonyl to the metal atom, favoring the nucleophilic attack of alkoxy ligands or hydroxyl endgroups on the corresponding carbon. It could be written as a succession of (1) ester interchange and (2) ligand exchange, yielding the low-molar-mass alcohol or diol by-product (Scheme 2.25).

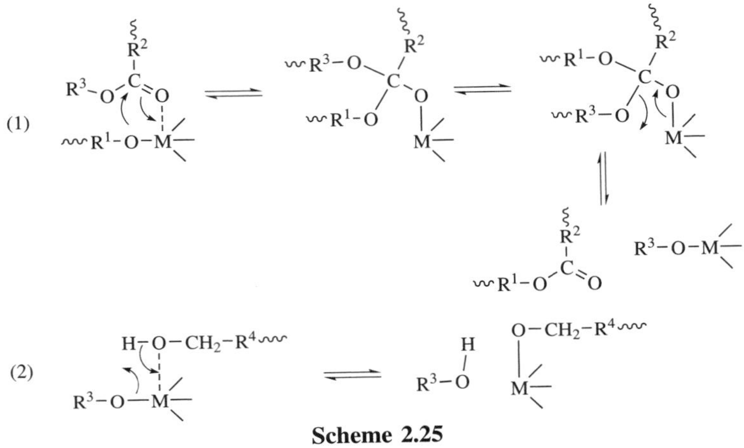

Scheme 2.25

This scheme is remarkably close to the coordination insertion mechanism believed to operate in the metal alkoxide–catalyzed ring-opening polymerization of cyclic esters (see Section 2.3.6). It shares many features with the mechanism proposed above for the metal alkoxide–catalyzed direct polyesterification (Scheme 2.18), including the difficulty of defining reaction orders.

CARBOXY–ESTER INTERCHANGE REACTION   Carboxy–ester interchange (also termed acidolysis) has been much less studied than hydroxy–ester interchange and little is known about the reaction mechanisms.[286] The reaction of terephthalic acid with aromatic diacetates without added catalyst has been examined assuming first order in each reactant.[287] The kinetics is rather complex since reaction takes place in a multiphase system, terephthalic acid being insoluble in reaction medium. Similar results were found for the polyesterification of 4-acetoxybenzoic acid,[288] yielding nonmeltable poly(4-hydroxybenzoic acid), or for its copolyesterification in PET, forming poly(ethylene terephthalate-co-4-oxybenzoyl), through ester interchange reactions.[289]

2.3.3 Nonequilibrium Polyesterifications

Nonequilibrium polyesterifications are characterized by their irreversible nature, which drives the reaction toward the formation of polyester.[207] These reactions generally involve highly reactive monomers such as acid chlorides or activated carboxylic acids and are conducted either in solution or in heterogeneous medium. They present some marked advantages over equilibrium polyesterifications since they can yield high-molar-mass polymers in mild conditions in a short time without ester interchange or degradative side reactions. Therefore, they are well suited for the synthesis of very high melting aromatic polyesters[290] and for the synthesis of polyesters bearing functional groups (e.g., unsaturations) which would react during melt synthesis.[291] They are also convenient for the synthesis of polyesters with controlled architecture or microstructure, as block or alternating copolyesters.[292,293] The disadvantages comprise uneasy control of polymer molar mass and the formation of higher amounts of cyclic oligomers with comparison to melt polycondensations. In addition, environmental concerns connected to the use of acid chlorides and organic solvents, difficulties in getting rid of traces of catalysts or of reaction by-products, and the use of relatively expensive starting materials have somewhat limited the interest of researchers in these reactions. Very little new information has appeared in the literature on the mechanisms and catalysis of these reactions since the end of the 1980s.

2.3.3.1 Acid Chloride–Alcohol Reaction

The reaction between acid chlorides and aliphatic alcohols or phenolic compounds commonly takes place at low to moderate temperature (−10 to 100°C) in solution or by interfacial or dispersion polymerization techniques. It has been widely applied to polyester synthesis. Reviews and books are available on the polymerization techniques as well as on the chemistry of this reaction.[2,207,294]

2.3.3.1.1 Solution Reactions

Solution reactions between diacid chlorides and diols or diphenols are carried out in THF or CH_2Cl_2 at −10 to 30°C in the presence of tertiary amines such as triethylamine or pyridine, which play a role of both reaction catalyst and HCl acceptor (Scheme 2.26). This synthetic method is also termed acceptor–catalytic polyesterification.[295-297] High-temperature solution reactions have also been reported for a number of less soluble, generally semicrystalline, aromatic polyesters.[6] They yield high-molar-mass polyesters exhibiting good mechanical properties and thermal stability.

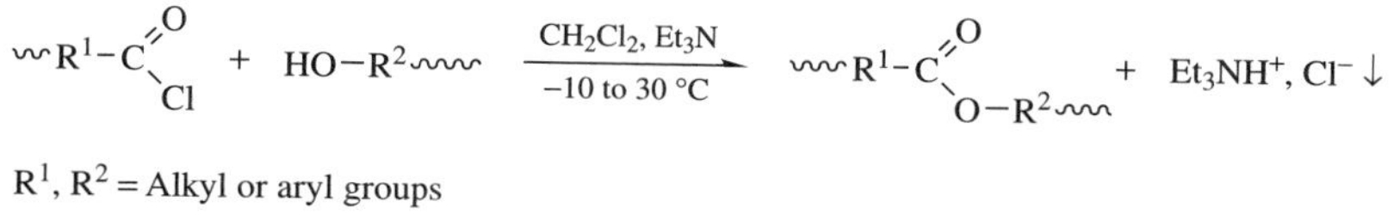

R^1, R^2 = Alkyl or aryl groups

Scheme 2.26

The by-product (HCl) precipitates in the form of its quaternary ammonium salt and cannot participate in the reverse reaction. This also prevents possible H^+-catalyzed side reactions during polymerization or polymer processing. The existence of such side reactions is the reason high-temperature bulk or solid-state processes involving acid chlorides, although possible, are not commonly applied to polyester synthesis. It is also worth mentioning that highly exothermic reactions between tertiary amines and the dichlorides of maleic, fumaric, or succinic acids, leading to black tarry compounds, have been reported.[291,298]

The reaction is generally believed to proceed via the formation of ionic acylammonium intermediate compounds (Reaction 1, Scheme 2.27). The equilibrium constant of the acylammonium formation depends mostly on steric and resonance factors, while the basicity of the tertiary amine seems to play a secondary role.[297] In the case of the less basic compounds, such as acidic phenols, and of strong tertiary amines, such as trialkylamines, the reaction has been reported to proceed through a general base mechanism via the formation of hydroxy–amine H-bonded complexes (Reaction 2, Scheme 2.27).[297]

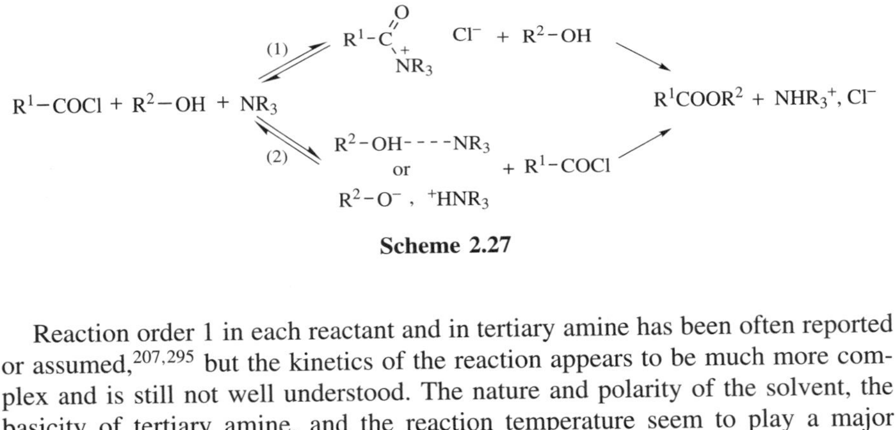

$$R^1-COCl + R^2-OH + NR_3 \; \rightleftharpoons \;
\begin{cases}
(1) & R^1-\overset{O}{\underset{NR_3^+}{C}} \;\; Cl^- + R^2-OH \\[2ex]
(2) & R^2-OH\text{-}\text{-}\text{-}\text{-}NR_3 \;\; \text{or} \;\; R^2-O^-,\, {}^+HNR_3 \;\; + R^1-COCl
\end{cases}
\; \rightarrow \; R^1COOR^2 + NHR_3^+,\, Cl^-$$

Scheme 2.27

Reaction order 1 in each reactant and in tertiary amine has been often reported or assumed,[207,295] but the kinetics of the reaction appears to be much more complex and is still not well understood. The nature and polarity of the solvent, the basicity of tertiary amine, and the reaction temperature seem to play a major role in reaction kinetics: Increasing solvent polarity results in a dramatic increase in reaction rate,[297] but it has also been reported that reaction rate decreases in solvents which are able to form H bonds with phenols.[296] The rate constants for the first and second COCl of the acid chloride monomer may be quite different, depending on the acid chloride structure and the nature of bisphenol monomer.[207] This feature has been applied to the synthesis of polyesters with predominant head-to-head or tail-to-head microstructures.[297] Polymer precipitation during the course of reaction is often observed, preventing the formation of high-molar-mass compounds unless the solvent exhibits a high swelling power with respect to the polymer.[290] Dropwise addition of the acid chloride into a solution of the other monomer is recommended to achieve high-molar-mass polyester: The acid chloride is completely consumed before the addition of a new portion, and high-molar-mass polymer is formed when the stoichiometric amount has been added. Since excess acid chloride cannot react, stoichiometry is much more easily controlled than in the case where monomers are added all at once (see Section 2.4.4.1.1).

2.3.3.1.2 Interfacial Reactions

Interfacial reactions between acid chlorides and alkali metal salts of diphenols can be carried out at $0-35°C$ in systems such as water–CH_2Cl_2 or water–chlorobenzene. The acid chloride, which can be either aliphatic or aromatic, is dissolved in the water-immiscible organic phase and is reacted with the aqueous alkaline bisphenolate under high-speed agitation. Aliphatic diols do not form alcoholate ions in aqueous solutions and are, therefore, not suitable monomers for the preparation of polyesters by the interfacial technique. On the other hand, this route is very attractive for organic phase-soluble or swellable aromatic polyesters since good yields and high molar mass can easily be achieved in this case. It is commercially applied to the synthesis of bisphenol-A polyarylates such as Unitika's U-Polymer and is useful for laboratory-scale preparation of block copolyesters,[292,299] of double-bond-containing aromatic polyesters,[291,300,301] and of polythioesters.[302]

Since reaction takes place at the interface or near the interface, factors such as stirring speed, the relative volume of organic and aqueous phases, monomer concentrations, and the nature and concentration of surfactant exert a marked influence on reaction kinetics and on the resulting polymer yield and molar mass.[294,303] The surfactant increases the total interfacial area and, consequently, the overall reaction rate and also acts as a phase transfer agent, facilitating the transportation of phenolate ions in the organic phase.[294,304] Ionic surfactants, mainly quaternary ammonium salts and fatty acid sulfates, at ca. $1-2\%$ mass concentration and stirring speed in the $700-1000$-rpm range are generally preferred. In contrast to polyesterifications carried out in homogeneous media, the maximum molar mass is not necessarily obtained at the equimolar ratio of reactants. The actual optimum stoichiometric ratio depends on the individual diffusion rate of reactants to the reaction zone.[294] The synthesis of high-molar-mass polyesters has been reported for acid chloride–bisphenol stoichiometric ratios ranging between 0.58 and 2, depending on reactants and experimental conditions.[305,306] As with solution polyesterification, the gradual addition of one monomer helps in achieving high-molar-mass polymers (see Section 2.4.1.1.2).

2.3.3.2 Activation Polyesterification

Activating agents, such as trifluoroacetic anhydride; 1,1'-carbonyldiimidazole; carbodiimides; sulfonyl, tosyl, and picryl chlorides; and a range of phosphorus derivatives can promote direct solution reactions between dicarboxylic acids and diols or diphenols in mild conditions. The activating agents are consumed during the reaction and, therefore, do not act as catalysts. These so-called direct polycondensation or activation polycondensation reactions proceed via the in situ transformation of one of the reactants, generally the carboxylic acid, into a more reactive intermediate. This intermediate is not isolated from the reaction medium and immediately reacts with the other monomer. They have been the subject of intensive academic research during the 1980s, mainly for the synthesis of high-melting aromatic polyesters and polyamides in mild conditions. Due to the lower nucleophilicity of phenols with respect to amines, activation polyesterifications, however, are much less efficient than the corresponding polyamidations. It is also worth mentioning that activation polyesterifications have not gained commercial acceptance, cheaper processes and reactants being available for synthesizing similar polyesters.

Trifluoroacetic anhydride, which activates the polyesterification of 4-hydroxybenzoic acid via the formation of a mixed anhydride,[307] and 1,1'-carbonyldiimidazole[308] were the first reported activating agents. The reaction between 1,1'-carbonyldiimidazole and carboxylic acids proceeds through the formation of N-acylimidazoles, which react with aliphatic diols in the presence of sodium ethoxide catalyst (Scheme 2.28).

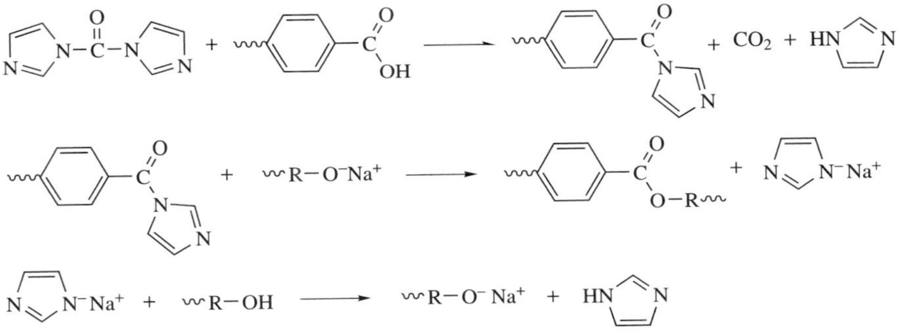

Scheme 2.28

2,4,6-Trinitrochlorobenzene (picryl chloride) in pyridine–N-methylpyrrolidinone (NMP) solutions were later used for the preparation of polyesters from dicarboxylic acids and diphenols or aliphatic diols,[309] but better results have been obtained with sulfonyl chlorides and phosphorus compounds.

2.3.3.2.1 Phosphorus Compounds

Within the wide range of phosphorus compounds described as activating agents for polyesterification reactions,[2,310] triphenylphosphine dichloride and diphenylchlorophosphate (DPCP) were found to be the most effective and convenient ones. In pyridine solution, DPCP forms a N-phosphonium salt which reacts with the carboxylic acid giving the activated acyloxy N-phosphonium salt. A favorable effect of LiBr on reaction rate and molar masses has been reported and assumed to originate from the formation of a complex with the N-phosphonium salt. This decreases the electron density of the phosphorus atom and favors the reaction with the aromatic acid. Good results were obtained using pyridine–LiBr solutions of DPCP for the polyesterification of aromatic dicarboxylic acids with diphenols and for the polyesterification of hydroxybenzoic acids[311] (Scheme 2.29). N,N-dialkylamides, such as dimethyl formamide (DMF), have also been reported to enhance the activating behavior of DPCP during the polyesterification of hydroxybenzoic acids, probably through the formation of a Vilsmeier adduct.[312]

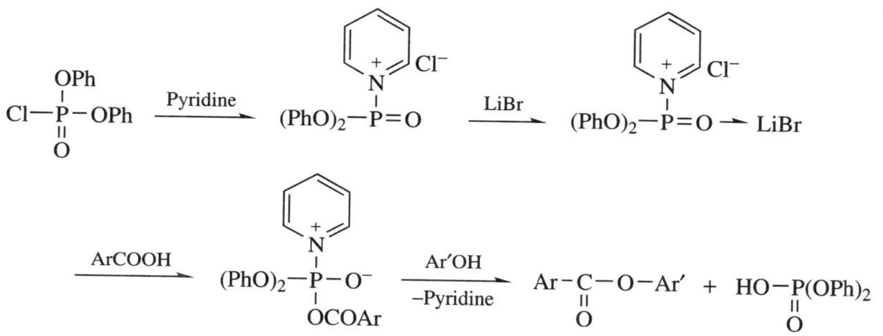

Scheme 2.29

Similarly, triphenylphosphine dichloride (TPPCl$_2$) can activate aromatic carboxylic acids in pyridine through the formation of acyloxyphosphonium salts (Scheme 2.30).[313] A side reaction between the intermediate acyloxyphosphonium species and a second carboxyl endgroup leading to the formation of anhydrides has been reported.[313] This chain-limiting reaction decreases the molar mass, while the presence of anhydride linkages in the chains adversely affects the thermal and hydrolytic stability of the final polyester.

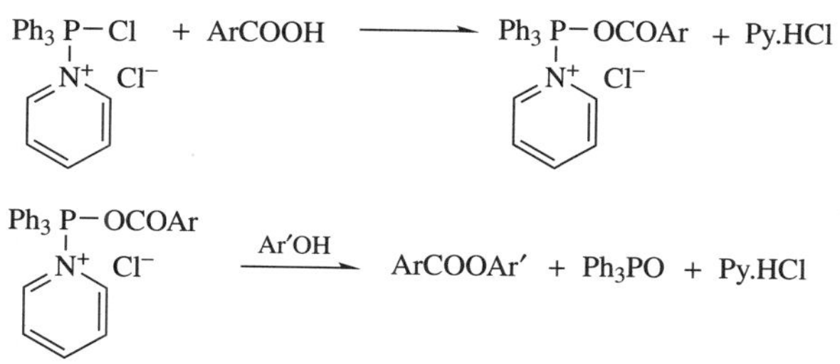

Scheme 2.30

Interestingly, the final inactive by-products of these phosphorus-activated reactions, triphenylphosphine oxide or diphenylphosphate, can be converted back to starting TPPCl$_2$ or DPCP by reaction with COCl$_2$ and SOCl$_2$, respectively.

2.3.3.2.2 Sulfur Compounds

Sulfur compounds have also been widely studied as activating agents for polyesterification reactions. *p*-Toluenesulfonyl chloride (tosyl chloride) reacts with DMF in pyridine to form a Vilsmeir adduct which easily reacts with carboxylic acids at 100–120°C, giving highly reactive mixed carboxylic–sulfonic anhydrides.[312] The reaction is efficient both for aromatic dicarboxylic acid–bisphenol[312] and hydroxybenzoic acid[314] polyesterifications (Scheme 2.31). The formation of phenyl tosylates as significant side products of this reaction has been reported.[315]

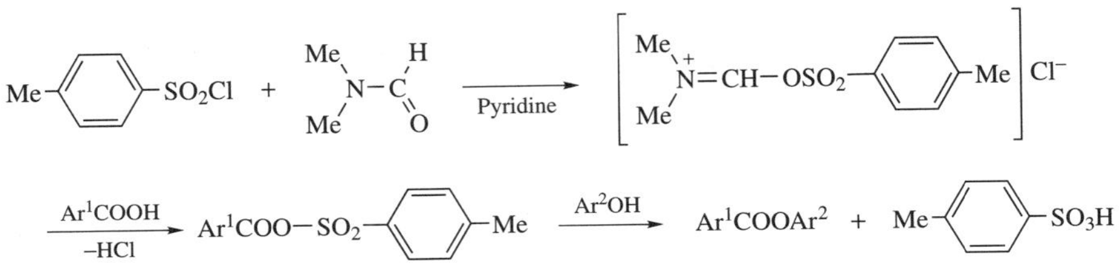

Scheme 2.31

Thionyl chloride is another activating agent employed for reactions between aromatic carboxylic acids and phenols in pyridine solution. The mechanism suggested does not involve the formation of an acid chloride but assumes the existence of an intermediary mixed sulfinic anhydride which undergoes reaction with phenolic endgroups (Scheme 2.32).[311]

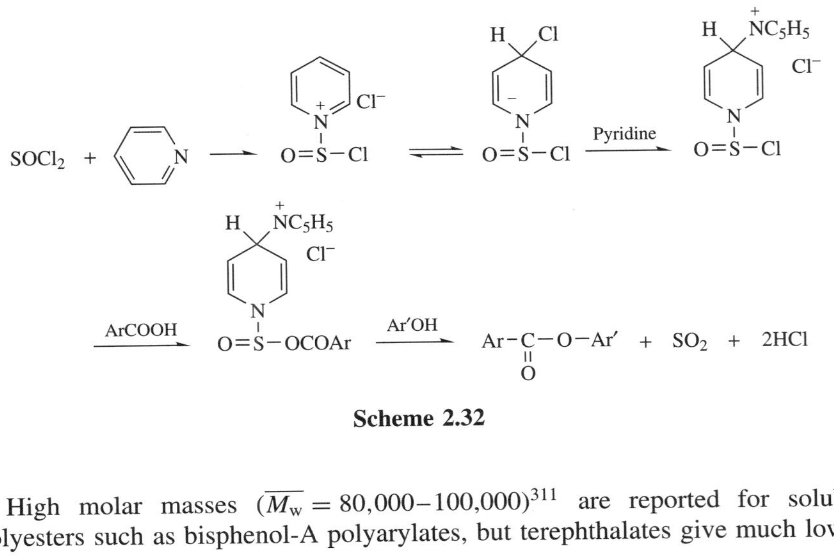

Scheme 2.32

High molar masses ($\overline{M_w} = 80,000-100,000$)[311] are reported for soluble polyesters such as bisphenol-A polyarylates, but terephthalates give much lower molar masses, probably due to premature oligomer precipitation during the course of reaction. The same remark can be made for any type of activation polyesterifications.

An example of thionyl chloride–activated polyesterification is detailed in Section 2.4.4.1.3.

2.3.3.2.3 N,N'-Dialkylcarbodiimides

N,N'-dialkylcarbodiimides, particularly N,N'-dicyclohexylcarbodiimide (DCC), are well-known condensing agents used in the synthesis of peptides. Carbodiimides have also been utilized for the synthesis of some polyesters and, in contrast to the sulfur or phosphorus activating agent discussed above, are suitable for the synthesis of aliphatic polyesters. Low-molar-mass $(\overline{M}_n = 1000–5000)$ polyesters of malic acid[316] and polyesters of various alkanedicarboxylic acids with aliphatic diols[317] were thus prepared by reacting the monomer mixture with DCC in THF solution in the presence of pyridine and p-toluene sulfonic acid. Poly(lactic acid) has been synthesized by direct polyesterification in CH_2Cl_2 in the presence of DCC and 4-(N,N-dimethylamino)pyridine.[318] Improved yields and high-molar-mass products have been reported for the polyesterification of various aromatic–aliphatic hydroxy acids in methylene chloride in the presence of the 4-(N,N-dimethylamino)pyridinium salt of p-toluenesulfonic acid (Scheme 2.33).[315] This salt is assumed to catalyze both the formation of an active N-acylpyridinium species (reaction 1) and the reaction of phenol endgroups with anhydride groups (reaction 2) arising from a possible side reaction of activated carboxylic acid (reaction 3). The formation of inactive N-acylurea (reaction 4) is, therefore, suppressed. This route is, of course, only suitable to soluble polyesters.

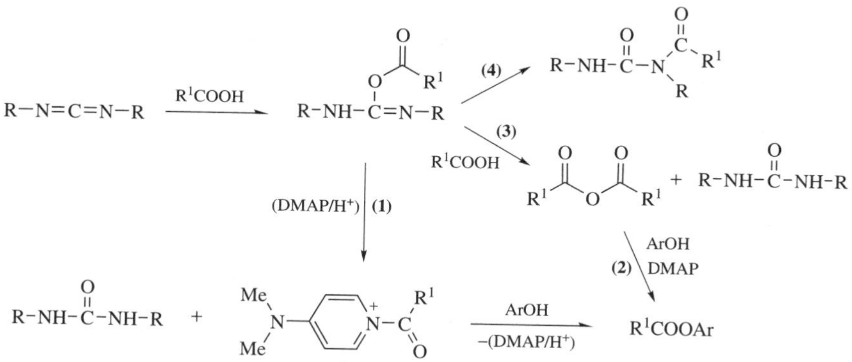

Scheme 2.33

2.3.3.3 Reaction between Alkali Metal Carboxylates and Dihaloalkanes

The reaction of dicarboxylic acid cesium or potassium salts with alkylene dihalides is another method of polyester preparation in relatively mild conditions

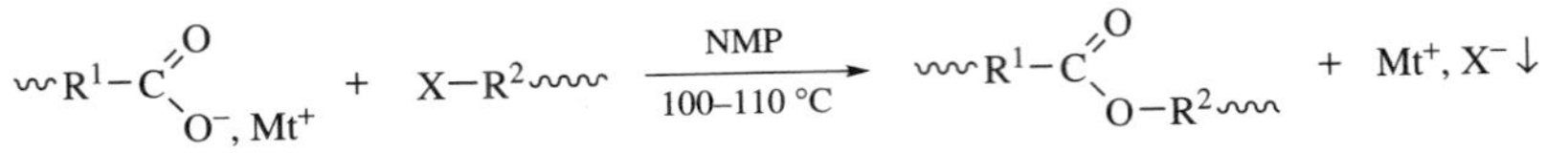

$$\text{\textasciitilde\textasciitilde} R^1 - C\overset{O}{\underset{O^-,\,Mt^+}{\diagdown}} \quad + \quad X - R^2 \text{\textasciitilde\textasciitilde} \quad \xrightarrow[\text{100–110 °C}]{\text{NMP}} \quad \text{\textasciitilde\textasciitilde} R^1 - C\overset{O}{\underset{O - R^2 \text{\textasciitilde\textasciitilde}}{\diagdown}} \quad + \quad Mt^+, X^- \downarrow$$

R^1 = Alkyl or aryl group, R^2 = Alkyl group, X = Cl, Br, Mt = K, Cs

Scheme 2.34

(Scheme 2.34).[319–322] Since the alkali metal salt formed does not react with ester groups, the molar mass is not limited by the reverse reaction and this polyesterification can be regarded as a nonequilibrium one.

Dibromo derivatives are much more reactive than their dichloro counterparts, which generally need prolonged heating at higher temperature to yield only moderately high molar mass polymers. A carboxylate salt excess, (ca. 10%) is favorable for the polyesterification reaction[319] while larger excess (50%) does not cause any important changes in yield or molar mass.[323] This reflects the heterogeneous nature of reaction medium. As with other polyesterifications carried out in solution, the amount of cyclic oligomers is quite high, especially in the case of reactions involving cesium carboxylates[320] or in the case of polyesters containing maleate[321], o-phthalate,[321] or pimelate units.[320] Thus, the amount of cyclic products reported in maleate, phthalate, and fumarate polyesters obtained by reacting the corresponding potassium dicarboxylates with 1,4-dibromobutane was respectively 33, 49, and 10% mass.[321] Since the polymer is isolated by precipitation in a nonsolvent in which cyclics are soluble, usually methanol, the reported yields decrease accordingly.

Using this method, it was possible to prepare high-molar-mass (80,000), configurationally pure poly(alkylene maleate) without any isomerized fumarate units in the chains.[324] This could not be achieved by the conventional high-temperature reaction between maleic anhydride and diols due to the simultaneous cis/trans maleate/fumarate isomerization taking place in these conditions. This method has also proven to be useful for the laboratory-scale preparation of alternating- or multiblock unsaturated polyesters as well as hydroxyl group — containing polyesters from tartaric or malic acids.[322]

2.3.4 Enzyme-Catalyzed Polyesterifications

Enzymes are proteins that exhibit a high catalytic efficiency and specificity for a given chemical reaction. Lipases and esterases belong to a class of enzymes referred to as hydrolases that specifically catalyze the in vivo hydrolysis of esters. These enzymes — mainly lipases — were also found to catalyze a number of in vitro esterification and polyesterification reactions in organic medium and have consequently been employed as biocatalysts in various nonbiosynthetic processes. These reactions proceed in very mild conditions with a high substrate-, stereo-, regio-, and chemioselectivity without forming undesirable side products.[325] Enzyme-catalyzed polyesterifications are carried out at 20–80°C in solution or in the bulk using free or immobilized enzymes, which are usually removed from reaction medium by filtration at the end of the reaction. Solvent hydrophobicity plays an important role in enzymatic activity, particularly for lipases which normally act at oil–water interfaces in living cells.[326] The best reactivity is found for hydrophobic solvents such as hexane, toluene, diisopropylether, or diphenylether. Hydrophilic polar solvents such as DMSO or methanol lead to deep significant modifications in enzyme conformation and, therefore, to a dramatic decrease in catalytic activity.[327]

Polyesters have been obtained in organic medium by polyesterification of hydroxy acids,[328,329] hydroxy esters,[330] stoichiometric mixtures of diols and diacids,[331–333] diols and diesters,[334–339] and diols and cyclic anhydrides.[340] Lipases have also been reported to catalyze ester–ester interchanges in solution or in the bulk at moderate temperature.[341] Since lipases obviously catalyze the reverse reaction (i.e., hydrolysis or alcoholysis of polyester), lipase-catalyzed polyesterifications can be regarded as equilibrium polycondensations taking place in mild conditions (Scheme 2.35).

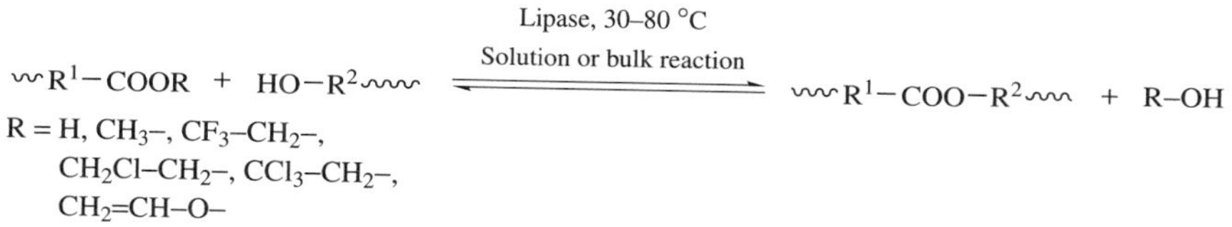

$$\text{\textasciitilde R}^1\text{–COOR} + \text{HO–R}^2\text{\textasciitilde} \xrightleftharpoons[\text{Solution or bulk reaction}]{\text{Lipase, 30–80 °C}} \text{\textasciitilde R}^1\text{–COO–R}^2\text{\textasciitilde} + \text{R–OH}$$

R = H, CH$_3$–, CF$_3$–CH$_2$–,
   CH$_2$Cl–CH$_2$–, CCl$_3$–CH$_2$–,
   CH$_2$=CH–O–

Scheme 2.35

Various methods have been employed to remove the reaction by-product, water or alcohol, in order to displace the equilibrium toward the formation of polyester: addition of molecular sieve,[328,332] Dean–Stark distillation of toluene–water azeotrope,[329] or bubbling inert gas in reaction medium.[336] In the case of reactions carried out without solvent, vacuum can be applied to distill off reaction water or alcohol.[331] The use of vinyl esters is a convenient way to displace esterification equilibrium: The theoretical by-product, vinyl alcohol, tautomerizes into acetaldehyde, which cannot participate in the reverse reaction.[342–344] Activated esters such as 2,2,2-trifluoroethyl-,[345] 2-chloroethyl, and 2,2,2-trichloroethyl esters[346] have also been reported to yield high-molar-mass polyesters. From the preceding discussion, it can be seen that the method is limited to soluble or low-melting-point polyesters, mainly aliphatic ones. Although the synthesis of some aromatic–aliphatic polyesters of isophthalic acid has also been reported,[339,347–349] lipases exhibit a very low catalytic activity for rigid aromatic monomers such as terephthalates[348] or diphenols.[349]

The reported molar masses of polyesters obtained by enzymatic catalysis are relatively low, generally below 8000, except for polymers recovered by precipitation.[336] This procedure results in the elimination of a soluble fraction consisting of low-molar-mass linear and cyclic oligomers.[336] An $\overline{M_w}$ as high as 46,400 has thus been reported for a poly(tetramethylene decanedioate) obtained in diphenylether using *Mucor miehei* lipase as catalyst and recovered by precipitation in methanol.[345]

The catalytic site of lipases is known to contain a serine residue which plays a key role in esterification catalysis.[326] A stable acyl-enzyme species is formed by the reaction of the serine residue with ester or acid groups. This intermediate species further reacts with hydroxyl groups, yielding the final ester and regenerating the serine residue. The generally proposed kinetic scheme for lipase-catalyzed acid–alcohol or hydroxy–ester esterifications is a conventional two-substrate enzymatic reaction scheme with release of the first product (H_2O or alcohol respectively) before combination of the second substrate. Inhibition by alcohol groups has been reported in the case of hydroxy–ester interchange esterifications[350] and polyesterifications.[351] A reasonable reaction pathway for the dimethyl ester–diol polyesterification is illustrated in Scheme 2.36, where Lip–OH and Lip–OOCR represent free and acyl-lipase species, respectively. A similar scheme has been proposed for the lipase-catalyzed ester interchange reactions taking place between poly(ε-caprolactone) and poly(ω-pentadecalactone) in solution or in the bulk.[341]

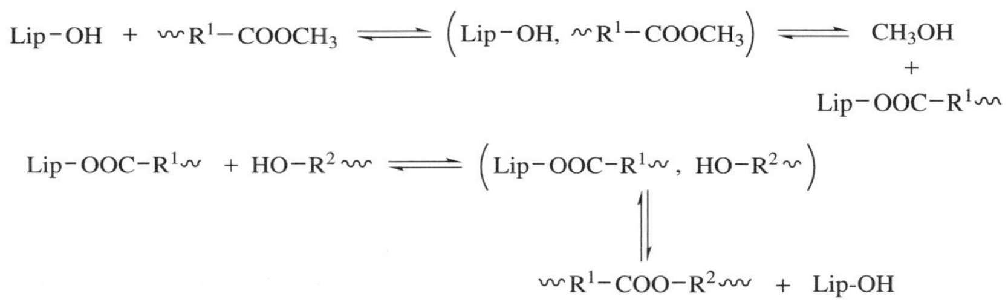

$$\text{Lip-OH} + \sim\!\!\text{R}^1\!\!-\!\text{COOCH}_3 \rightleftharpoons \left(\text{Lip-OH}, \sim\!\!\text{R}^1\!\!-\!\text{COOCH}_3\right) \rightleftharpoons \begin{array}{c}\text{CH}_3\text{OH}\\ +\\ \text{Lip-OOC-R}^1\!\!\sim\end{array}$$

$$\text{Lip-OOC-R}^1\!\!\sim + \text{HO-R}^2\!\!\sim \rightleftharpoons \left(\text{Lip-OOC-R}^1\!\!\sim, \text{HO-R}^2\!\!\sim\right)$$

$$\sim\!\!\text{R}^1\!\!-\!\text{COO-R}^2\!\!\sim + \text{Lip-OH}$$

Scheme 2.36

Lipases have also been used as initiators for the polymerization of lactones such as β-butyrolactone, δ-valerolactone, ε-caprolactone, and macrolides.[341,352–357] In this case, the key step is the reaction of lactone with the serine residue at the catalytically active site to form an acyl-enzyme hydroxy-terminated activated intermediate. This intermediate then reacts with the terminal hydroxyl group of a *n*-mer chain to produce an ($n + 1$)-mer.[325,355,358,359] Enzymatic lactone polymerization follows a conventional Michaelis–Menten enzymatic kinetics[353] and presents a "controlled" character, without termination and chain transfer,[355] although more or less controlled factors, such as water content of the enzyme, may affect polymerization rate and the nature of endgroups.[360]

2.3.5 Polyesters from Polyaddition Reactions

Provided that they are carried out at temperatures below 100–120°C, polyaddition reactions between epoxides and carboxylic acids or anhydrides lead to linear polyesters bearing pendant secondary hydroxyl groups,[361–363] (Scheme 2.37). At higher temperatures (150–200°C) the reaction is accompanied by a number of side reactions, such as the esterification of secondary hydroxyl groups or the ring-opening polymerization of epoxy groups, yielding branched or crosslinked materials.[362,363] The high-temperature acid anhydride–epoxy reaction is a well-known cure reaction of epoxy resins.

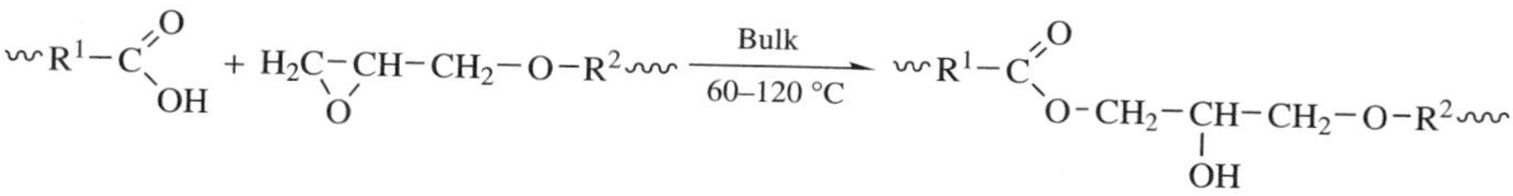

Scheme 2.37

2.3.6 Polyesters from the Ring-Opening Polymerization of Cyclic Esters

The ring-opening polymerization of cyclic esters, mainly lactones and lactides (Scheme 2.38) is a very convenient method of preparation of degradable aliphatic polyesters (See also Section 2.2.2.1). Poly(lactic acid) (PLA) and poly(glycolic acid) (PGA)—also named poly(lactide) and poly(glycolide) respectively—and their copolymers are usually synthesized by the ring-opening polymerization of their cyclic dimers, lactide and glycolide, respectively (Scheme 2.39).

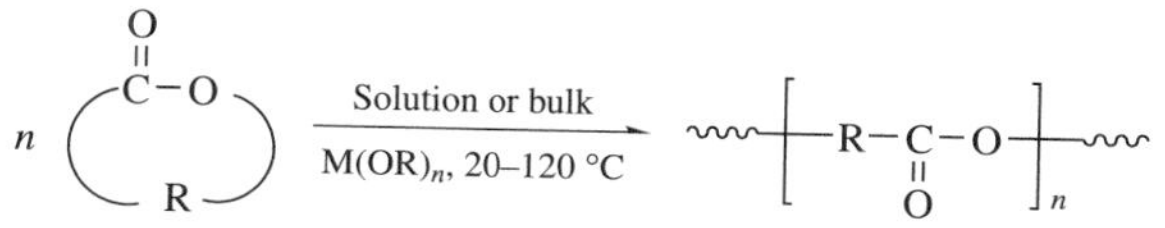

Scheme 2.38

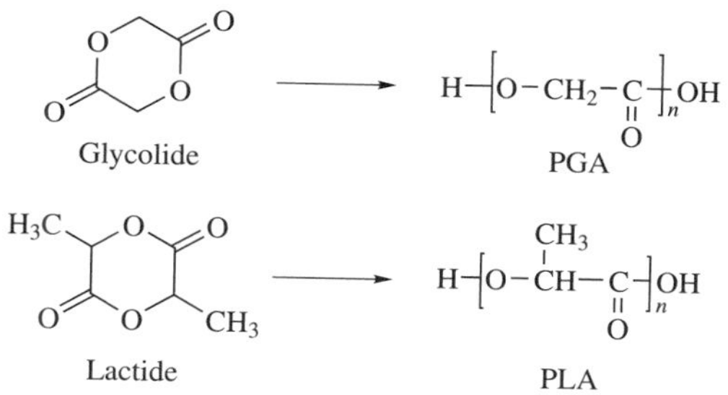

Glycolide   PGA

Lactide   PLA

Scheme 2.39

A wide range of initiators have been proposed for this reaction, mainly salts or organocompounds of tin,[364–367] zinc,[368,369] aluminum,[369] and, more recently, yttrium and lanthanides.[370-372] In spite of potential toxicity, especially when biomedical applications are considered, the most versatile and commonly used initiators are tin compounds, such as stannous(II) bis(2-ethylhexanoate) (tin octoate), although less toxic zinc catalysts have also proven to be efficient for this reaction.[368] High-molar-mass polymers, from several thousands to several millions, have been reported for PLA synthesized in these conditions.[373] The synthetic procedure involves a first step where the α-hydroxy acid is condensed to a low-molar-mass oligomer, which is then depolymerized at high temperature to give the cyclic dimer. The dimer must carefully be purified by solvent extraction/crystallization procedure to achieve high-molar-mass polymers by the following ring-opening polymerization (Scheme 2.40).

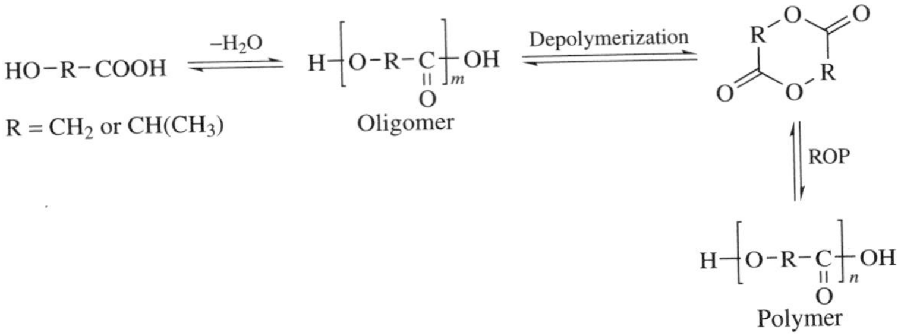

$$\text{HO}-\text{R}-\text{COOH} \underset{-\text{H}_2\text{O}}{\rightleftharpoons} \text{H}\left[\text{O}-\text{R}-\overset{\text{O}}{\underset{\|}{\text{C}}}\right]_m\text{OH} \overset{\text{Depolymerization}}{\rightleftharpoons}$$

$$\text{R} = \text{CH}_2 \text{ or } \text{CH(CH}_3)$$

Oligomer

ROP

$$\text{H}\left[\text{O}-\text{R}-\overset{\text{O}}{\underset{\|}{\text{C}}}\right]_n\text{OH}$$

Polymer

Scheme 2.40

The synthesis of high-molar-mass PLA and PGA by two-step polycondensations of lactic and glycolic acids, respectively, has recently been reported.[374,375] It involves the formation of a low-molar-mass oligomer followed by a polycondensation step either in the solid state[374] or in the melt under vacuum.[375] The procedures are detailed in Section 2.4.1.5.2.

Poly(ε-caprolactone) (PCL), other poly(lactones), and their copolymers are synthesized by the ring-opening polymerization of the corresponding lactones. Various organometallic initiators have proven to be efficient, such as oxides, alkoxides, or carboxylates of titanium, zirconium, tin, aluminum and various metals,[371,376-378] as well as enzymatic catalysts (See Section 2.3.4). It must be noted that PHB and PHB copolymers, produced by bacteria or genetically modified plants, can also be chemically synthesized by the ring-opening polymerization of 3-butyrolactone in the presence of tin derivatives following procedures similar to that described for lactides and lactones.[52]

There is still a great deal of effort being made on the study of the mechanisms and catalysis of lactide, glycolide, and lactone polymerizations,[367,379-383] and many reviews have been published on this topics.[365,384-386] Since it is basically a chain-growth, rather than a step-growth, process, it is beyond the scope of this book and will not be extensively discussed here. Nevertheless, it is worth mentioning that very close resemblances exist between the mechanisms reported in the literature for the coordination ring-opening polymerization of cyclic esters initiated by metal derivatives and those reported for transesterification polymerizations. Most of the ring-opening polymerization initiators used, especially titanium, zirconium, and tin alkoxides, are also very efficient catalysts for the hydroxy–ester or ester–ester interchange reactions. The synthesis of poly(butylene terephthalate) from cyclic oligomers in the presence of titanium alkoxides[387] is clearly a ring-opening polymerization of a cyclic ester but also closely resembles the ring-chain equilibria taking place during the synthesis of most step-growth polymers.

2.3.7 Miscellaneous Synthetic Methods

Many ester-forming reactions reported in the literature cannot be applied to the synthesis of polyesters due to side reactions, incomplete conversions, or non-quantitative yields. Some examples of nonconventional polyester syntheses are listed below. Most of them lead to oligomers rather than polymers and require expensive reactants or special reaction conditions, which make them of little practical interest.

The reaction between bischloroformates and dicarboxylic acids has been applied to the synthesis of polyesters and polyestercarbonates. Thus, mixtures of bisphenol and dicarboxylic acid react with phosgene in pyridine, yielding high-molar-mass copolymers with bisphenol carbonate and bisphenol dicarboxylate structural units in the chains. The carbonate–carboxylate ratio is controlled by varying the initial dicarboxylic acid–bisphenol monomer ratio. The reaction proceeds via the formation of chloroformate endgroups, which further react either with remaining phenol endgroups or with carboxylic acid endgroups with evolution of CO_2. The reaction is assumed to proceed through the formation of an intermediate carboxylic–carbonic anhydride[388] (Scheme 2.41). Aliphatic polyesters were also prepared following this method by reacting adipic acid with the bischloroformates of aliphatic diols.[389]

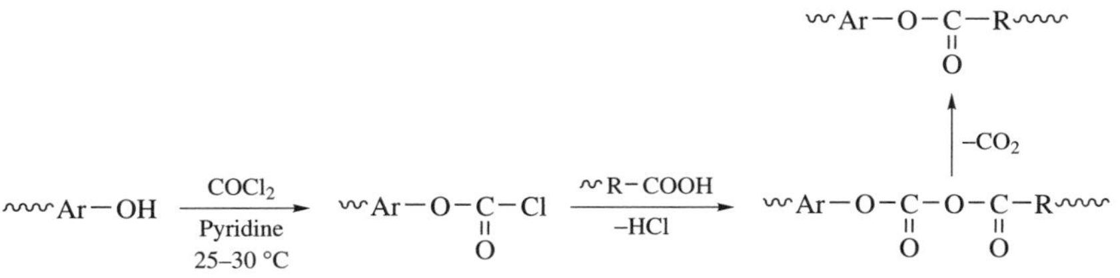

HO–Ar–OH = bisphenol-A   HOOC–R–COOH = isophthalic, terephthalic, adipic, azelaic or nadic acid

Scheme 2.41

Aliphatic dinitriles, such as adipodinitrile, react with aliphatic diols at low temperature ($0-5°C$) in the presence of HCl and form poly(iminoesters) which

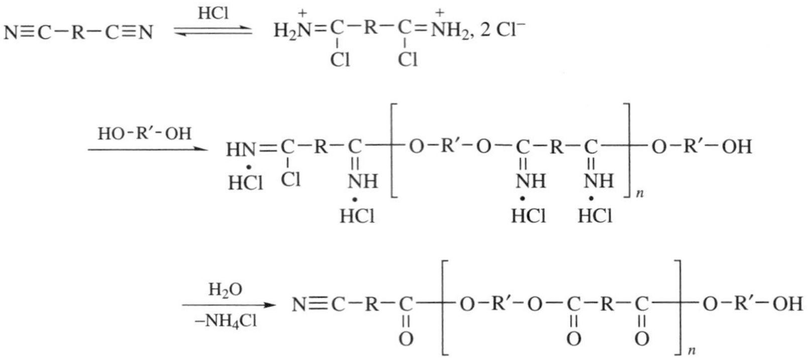

$$N\equiv C-R-C\equiv N \; \underset{}{\overset{HCl}{\rightleftharpoons}} \; H_2\overset{+}{N}=\underset{Cl}{C}-R-\underset{Cl}{C}=\overset{+}{N}H_2, \; 2\,Cl^-$$

Scheme 2.42

can be hydrolyzed to low-molar-mass polyesters ($\overline{M}_n = 700–1700$).[390] Reaction temperature should be kept low in order to avoid the substitution of OH groups by chlorine atoms during the poly(iminoester) formation step (Scheme 2.42).

The polymerization of terephthaldehyde in the presence of triethoxyaluminum[391] or trialkylaluminum[392] (Tischchenko–Claisen reaction) yields random copolyesters containing ca. 1 : 1 mol ratio of *p*-methylenebenzoate and *p*-xyleneterephthalate units (Scheme 2.43). The reaction does not take place with sterically hindered aldehydes.[392]

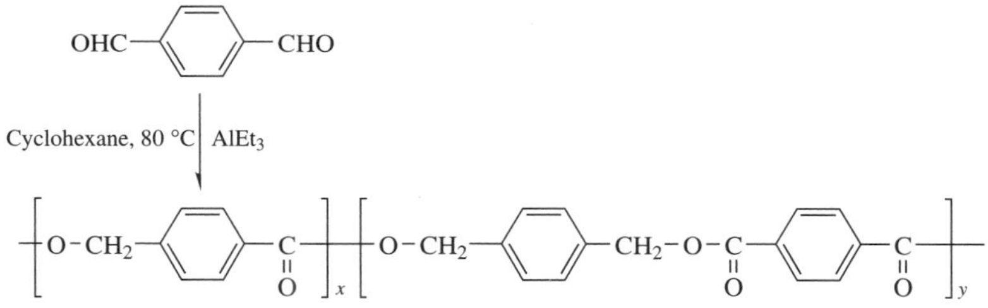

Random copolymer ; $x/y = 1$

Scheme 2.43

The anionic copolymerization of ketenes with aldehydes[393,394] or ketones[395] is a synthetic method allowing the preparation of sterically hindered polyesters which cannot easily be prepared by conventional ways. Thus, dimethylketene copolymerizes with aromatic aldehydes, such as benzaldehyde, in the presence of anionic initiators (e.g., BuLi or NaOEt), yielding polyesters with stereoregular structure[393] (Scheme 2.44). Ketenes are also intermediate compounds formed during the reaction of bisdiazoketoalkanes with diols, claimed to give polyesters in good yields.[396]

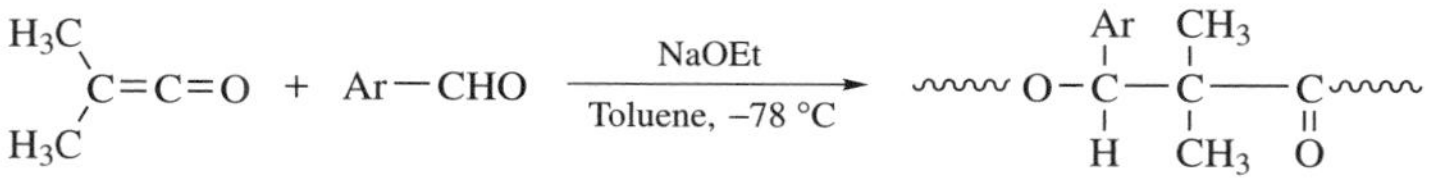

$$H_3C\diagdown \underset{H_3C\diagup}{C}=C=O \;+\; Ar-CHO \xrightarrow[\text{Toluene, } -78\,°C]{\text{NaOEt}} \;\text{\textasciitilde\textasciitilde} O - \underset{\underset{H}{|}}{\overset{\overset{Ar}{|}}{C}} - \underset{\underset{CH_3}{|}}{\overset{\overset{CH_3}{|}}{C}} - \underset{\underset{O}{\|}}{C}\text{\textasciitilde\textasciitilde}$$

Scheme 2.44

Besides these reactions, ester-containing copolymers can be obtained through non-ester-forming reactions by reacting ester-containing monomers or oligomers having suitable endgroups. The interfacial synthesis of copolyesteramides by polyamidification of ester-containing bis(acid chlorides) with diamines,[397] the formation of alternating copolyesters by reaction of 1,2-ethylene-bis(4-hydroxy-benzoate) with terephthalic acid,[398] and the synthesis of polyesterthioethers by polyaddition between bisacrylates or methacrylates and hydrogen sulfide[399] (Scheme 2.45) or dithiols[400] are examples of this approach.

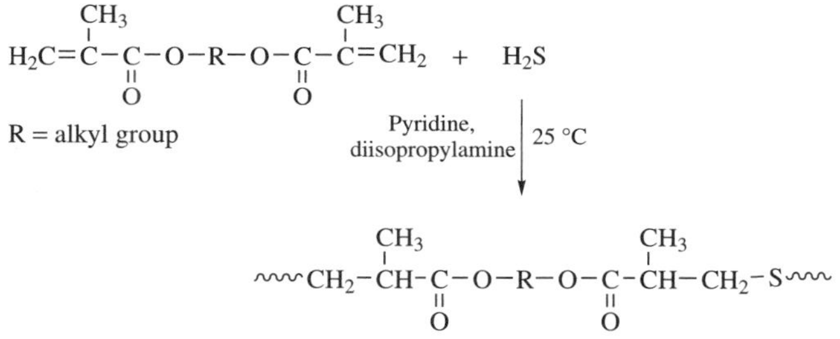

$$H_2C=\overset{\overset{CH_3}{|}}{C}-\underset{\underset{O}{\|}}{C}-O-R-O-\underset{\underset{O}{\|}}{C}-\overset{\overset{CH_3}{|}}{C}=CH_2 \;+\; H_2S$$

R = alkyl group

Pyridine, diisopropylamine $\Big|$ 25 °C $\downarrow$

$$\text{\textasciitilde\textasciitilde}CH_2-\overset{\overset{CH_3}{|}}{CH}-\underset{\underset{O}{\|}}{C}-O-R-O-\underset{\underset{O}{\|}}{C}-\overset{\overset{CH_3}{|}}{CH}-CH_2-S\text{\textasciitilde\textasciitilde}$$

Scheme 2.45

In a similar way, polyester block copolymers were obtained by reacting acid chloride–terminated polyesters with hydroxy-terminated polyethers,[401] or by reacting polyester-containing polymer mixtures with coupling agents such as diisocyanates.[402]

It should also be noted that the treatment of poly(alkylene terephthalates) with bisacyllactams, bisoxazinones, or bisoxazolinones results in polyesters with higher molar mass and improved properties. These coupling agents react in the bulk at high temperature (>240°C) with the hydroxy endgroups of polyesters, resulting in a rapid chain extension (molar mass boosting).[402] Similar results have been reported for the reactions between polyester carboxy endgroups and bisoxazolines.[402] An example of PET chain extension by reaction with 2,2′-bis(4H-3,1-benzoxazine-4-one)[403] is given in Scheme 2.46.

Many random copolyesters and polyester–polycarbonates have also been prepared by ester interchange reactions in the molten state. Thus, poly(ethylene terephthalate-co-isophthalates) can be obtained by simple melt blending of PET and poly(ethylene isophthalate) (PEI) homopolyesters at 270°C. The copolymer changes gradually from a block type at the beginning of reaction to a random-type

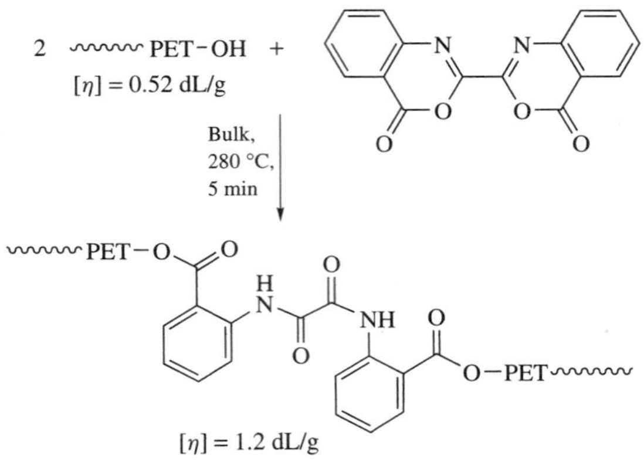

Scheme 2.46

copolymer after 20 min heating.[404] This method is quite general and has been applied to a large number of polyesters.[100,148–153] Ester interchange catalysts such as those described in Section 2.3.1.2.2 may be added to reduce reaction time.

Finally, the synthesis of biodegradable poly(hydroxyalkanoic acids) (PHA) by bacteria or genetically modified plants should be mentioned. The microbiology, biochemistry, and genetics of PHA biosynthesis have been reviewed by several authors[52,105,405–407] and are beyond the scope of this chapter.

2.3.8 Characterization of Polyesters

Polyesters are not different from other polymers, and any of the characterization methods commonly used in polymer science can obviously be applied to polyesters and provide information on their structure and properties. In this section, some data specific to polyesters — solubility information, COOH and OH endgroup titration, and infrared (IR) and NMR spectra assignments — are briefly summarized. Most of these data originate from the authors' laboratory. References are provided on some particular points only.

2.3.8.1 Solvents

Aliphatic polyesters are soluble in most organic solvents and insoluble only in apolar solvents (alkanes, diethyl ether) or in polar protic solvents such as methanol and water. THF is generally used for size exclusion chromatography (SEC) analyses and CH_2Cl_2 for viscosimetric studies. NMR spectra can be recorded in common solvents such as $CDCl_3$ or DMSO-d_6. Poly(glycolic acid) (PGA) and poly(lactic acid) (PLA) are exceptions. Since they are not soluble in THF, SEC analyses must be carried out in chlorinated solvents (e.g., $CHCl_3$) for PLA[375] and hexafluoroisopropanol (HFIP)/CF_3COONa (1 mM) for PGA.[374]

Poly(alkylene terephthalate)s are soluble only in solvents such as 1,1,2,2-tetrachloroethane–phenol or o-cresol–$CHCl_3$ mixtures, o-chlorophenol, and 2,4-dichlorophenol. Viscosimetric studies are generally carried out in phenol–tetrachloroethane (60/40 mass) at 35°C. The NMR spectra of PET can be recorded in DMSO-d$_6$ at 130°C[408] or, more conveniently, in CD$_2$Cl$_2$/CF$_3$CO$_2$D (4/1 vol.)[409] or CDCl$_3$/CF$_3$CO$_2$D (2/1 vol.) solutions.

Amorphous bisphenol-A polyarylates are soluble in dioxane and in chlorinated solvents such as CH$_2$Cl$_2$, 1,2-dichlororethane, 1,1,2-trichloroethane, and 1,1,2,2-tetrachloroethane while semicrystalline and liquid crystalline wholly aromatic polyesters are only sparingly soluble in solvents such as tetrachloroethane–phenol mixtures or pentafluorophenol, which is often used for inherent viscosity determinations.

2.3.8.2 IR Spectroscopy

The carbonyl group of polyesters of aliphatic diols gives a characteristic and very strong absorption (stretching vibration) in the 1715–1740-cm^{-1} range. Polyesters deriving from phenols present a C=O absorption at somewhat higher frequency, at 1740–1750 cm^{-1}. The carboxylic endgroups absorption at 1700–1710 cm^{-1} (C=O stretching vibration of COOH) appears as a shoulder on the intense ester carbonyl peak of carboxy-terminated polyesters. The corresponding OH stretching vibration of COOH appears as a series of overlapping broad peaks between 2500 and 3000 cm^{-1}, while hydroxyl endgroups give rise to a broad absorption centered around 3540 cm^{-1}. Unsaturated polyesters containing fumarate or maleate units present a characteristic absorption in the 1650-cm^{-1} region assigned to the C=C stretching vibration. The C—O stretching vibration of esters gives two intense absorptions in the region 1000–1300 cm^{-1} which are very dependent on polyester structure, often overlap with other resonances, and are, therefore, difficult to assign unambiguously.

2.3.8.3 NMR Spectroscopy

2.3.8.3.1 ^{13}C NMR Spectroscopy

The ester C—O resonance of aliphatic polyesters appears in the 170–175-ppm range, while that of polyesters deriving from aromatic acids appears in the 165–170-ppm range, depending on polyester structure and on the solvent. Thus, the ester carbonyl resonance of methylhydroquinone/isophthalic polyester appears at 168.3 ppm (163 ppm in solid-state cross-polarization magic angle spinning (CP-MAS) ^{13}C NMR spectra) while the corresponding poly(ε-caprolactone) and poly(ethylene adipate) ester carbonyl resonances are close to 173 ppm. When present, carboxylic acid endgroups give slightly deshielded C=O resonances with respect to the corresponding ester. The COO—CH$_2$ methylene and COO—CH(R) methine carbons give easily assigned resonances in the 60–65- and 65–70-ppm regions, respectively, while the —CH$_2$—COO methylenes give rise to a peak close to 35 ppm. The assignments of the ^{13}C NMR spectra of some typical polyesters are reported in Table 2.19.

TABLE 2.19 **Assignments of Polyester ^{13}C NMR Spectra (ref. TMS)**

Repeating-Unit Formulas and Assignments (ppm)	Experimental Conditions	References

64.10 28.90 25.95 24.97 34.30 172.67
—O—CH$_2$—CH$_2$—CH$_2$—CH$_2$—CH$_2$—C—
                                                                  ‖
                                                                  O

C$_6$D$_6$, 60°C, 88 MHz — 469

61.7   33.2   23.8   172.8
—O—CH$_2$—CH$_2$—O—C—CH$_2$—CH$_2$—CH$_2$—CH$_2$—C—
                        ‖   ‖
                        O   O

HO—CH$_2$—CH$_2$—O—C—   —CH$_2$—C—OH
60.3   65.5   ‖   178.0 O
                 O

CDCl$_3$, r.t., 62.9 MHz — 421

                              28.1 ; 28.3
68.8   71.2   ⌒   171.12 ; 171.41
—O—CH$_2$—CH—O—C—CH$_2$—CH$_2$—C—
           | ‖   ‖
           CH$_3$ O   O
           15.4

64.1   67.7   171.63   64.5 65.3   170.80
HO—CH$_2$—CH—O—C—   HO—CH—CH$_2$—O—C—
           | ‖ | ‖
           CH$_3$ O   CH$_3$ O
           15.3   18.4

CDCl$_3$, r.t., 62.9 MHz — 470

                     129.5 (cis)
                     134.0 (trans)
62.40   ⌒   165.0
—O—CH$_2$—CH$_2$—O—C—CH=CH—C—
                    ‖   ‖
                    O   O

CDCl$_3$, r.t., 62.9 MHz — 470

                   127.6
          132.3   ⌒
61.4   ↘   163.4
—O—CH$_2$—CH$_2$—O—C—⟨benzene⟩—C—
                    ‖   ‖
                    O   O

DMSO, 130°C, 88 MHz — 408

            130.25   128.56
62.87   ↘
—O—CH$_2$—CH$_2$—O—C—⟨benzene⟩—133.74
                    ‖
                    O
                    ↗   C—
               130.69   O 165.29

CDCl$_3$, r.t., 62.9 MHz — 470

16.83 125.65
H$_3$C   ↓   148.60 131.05   131.05
134.01 →⟨benzene⟩—O—C—⟨benzene⟩—137.45
                         ‖
                         O
149.95 124.4 121.59
                     ↗   C—
                133.65   O 168.32

CDCl$_3$/CF$_3$CO$_2$D — 471
(1/1 vol.), r.t. 62.9 MHz

Note: r.t. = room temperature.

TABLE 2.20   Assignments of Polyester ^{1}H NMR Spectra (ref. TMS)

Repeating-Unit Formulas and Assignments (ppm)	Experimental Conditions	References
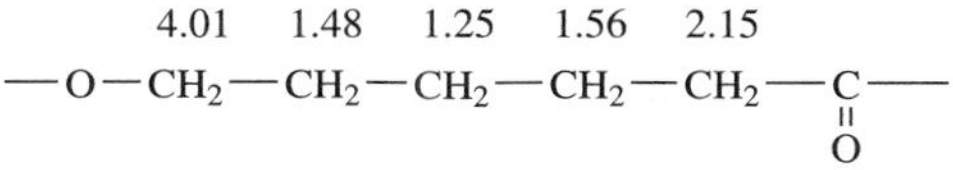 4.01 1.48 1.25 1.56 2.15 —O—CH$_2$—CH$_2$—CH$_2$—CH$_2$—CH$_2$—C— ‖ O	C$_6$D$_6$, 50°C, 350 MHz	469
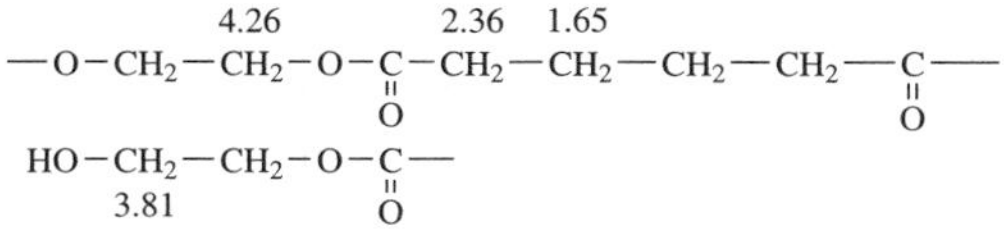 4.26 2.36 1.65 —O–CH$_2$–CH$_2$–O–C–CH$_2$–CH$_2$–CH$_2$–CH$_2$–C— ‖ ‖ O O HO–CH$_2$–CH$_2$–O–C— 3.81 ‖ O	CDCl$_3$, r.t., 250 MHz	421
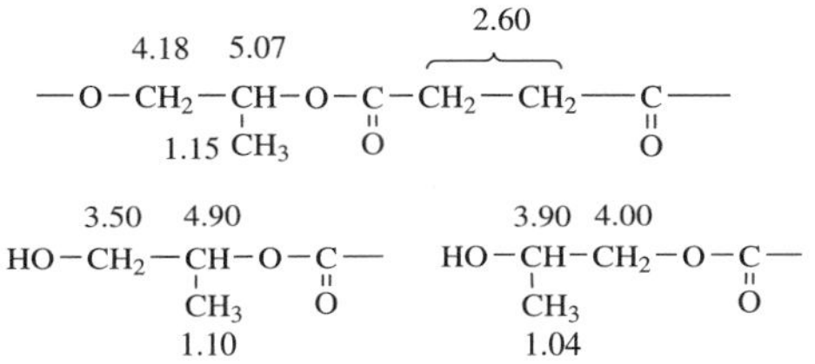 2.60 4.18 5.07 ⏜ —O–CH$_2$–CH–O–C–CH$_2$–CH$_2$—C— | ‖ ‖ 1.15 CH$_3$ O O 3.50 4.90 3.90 4.00 HO–CH$_2$–CH–O–C— HO–CH–CH$_2$–O–C— | ‖ | ‖ CH$_3$ O CH$_3$ O 1.10 1.04	CDCl$_3$, r.t., 250 MHz	470
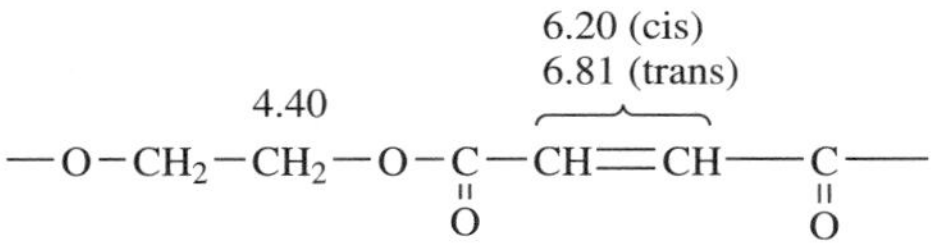 6.20 (cis) 6.81 (trans) 4.40 ⏜ —O–CH$_2$–CH$_2$–O–C—CH═CH—C— ‖ ‖ O O	CDCl$_3$, r.t., 250 MHz	470
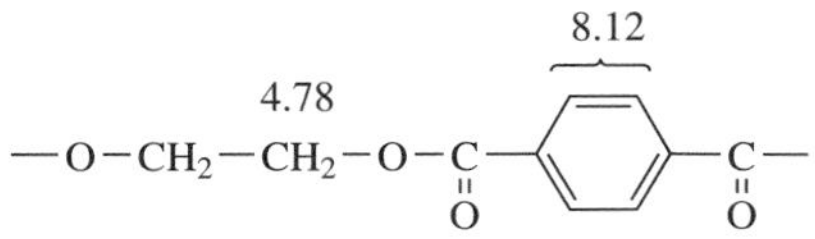 8.12 4.78 ⏜ —O–CH$_2$–CH$_2$–O–C⟨benzene ring⟩C— ‖ ‖ O O	CD$_2$Cl$_2$/CF$_3$CO$_2$D (4/1 vol), r.t., 400 MHz	409
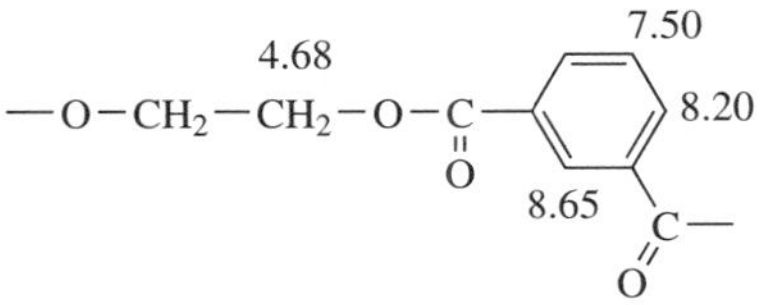 7.50 4.68 ⟨benzene ring⟩ —O–CH$_2$–CH$_2$–O–C⟨ring⟩8.20 ‖ O 8.65 C— 8.65 ‖ O	CDCl$_3$, r.t., 250 MHz	470

Note: r.t. = room temperature.

2.3.8.3.2 ^{1}H NMR Spectroscopy

The —CH$_2$–COO and —COO–CH$_2$ methylene protons present resonances in the 2.1–2.4- and 4.0–4.3-ppm ranges, respectively. The peaks of —CH$_2$OH endgroup methylenes appear close to 3.8 ppm in most aliphatic polyesters and can be used for the determination of the OH endgroup content. In the spectra of PET recorded in CD$_2$Cl$_2$/CF$_3$CO$_2$D, these peaks (a triplet) are shifted to 4.2 ppm due to esterification by trifluoroacetic acid.[409] In unsaturated polyesters, the —OOC–CH=CH–COO methines give resonances close to 6.7 (fumarate units) and 6.3 ppm (maleate units). The assignments of the ^{1}H NMR spectra of some typical polyesters are given in Table 2.20.

2.3.8.4 Endgroup Determination

The endgroup content allows the calculation of the number-average molar mass of linear polyesters and its accurate measurement is, therefore, of great importance. Many methods can be applied to the determination of endgroups in polyesters,[410,411] such as ^{1}H NMR[412] or IR[413-415] direct estimations or methods involving the derivatization of endgroups by proper reactants. The OH group content determination in polyesters can thus be carried out by ^{19}F NMR spectroscopy after reaction with trifluoroacetic anhydride or by UV or IR spectrophotometry after reaction with aromatic isocyanates or 3,5-dinitrobenzoyl chloride.[416]

The widely used −COOH and −OH endgroup chemical titrations are detailed below.

2.3.8.4.1 Carboxyl Endgroup Chemical Titration

A sample of polyester (ca. 1 g, exactly weighed) is dissolved in 20 mL toluene−ethanol mixture (1/1 vol.) and titrated by a solution of KOH in ethanol (0.05 mol/L) using a potentiometric titrator. A blank titration must be performed under the same conditions. Hardly soluble polyesters (e.g., PET) must be dissolved in an o-cresol−chloroform mixture or in hot benzyl alcohol.[417] The result (acid content) is normally expressed in mmol COOH/g polyester but may also be given as the "acid number," defined as the number of milligrams of KOH required to neutralize 1 g of polyester. [Acid number = (number of mmol COOH/g polyester) × 56.106.]

2.3.8.4.2 Hydroxyl Endgroup Chemical Titration

Hydroxyl endgroups are usually titrated by acetylation or phthalylation in pyridine using an excess acetyl chloride, acetic anhydride, or phthalic anhydride. The excess of reactant is hydrolyzed and back titrated by aqueous sodium or potassium hydroxide.

Thus, a polyester sample (1−3 g, exactly weighed) is dissolved in 25 mL of a titrated solution of acetic anhydride in dry pyridine (10% mass). The solution is heated to reflux for 1 h. After cooling, 50 mL pyridine and 10 mL water are added. The excess acetic acid present in the resulting solution is titrated by aqueous potassium hydroxide (0.5 mol/L) using a potentiometric titrator. The determination must be carried out in duplicate and a blank titration must be performed under the same conditions. The mass of polyester and the concentration of reactants should be adjusted to ensure that at least a fourfold excess of acetic anhydride is used. The final result (OH content) is expressed in mmol OH/g polyester or as the "hydroxyl number," defined as the number of milligrams of KOH required to neutralize the acetic acid consumed per gram of polyester. [Hydroxyl number = (number of mmol OH/g polyester) × 56.106.]

The OH endgroups of hardly soluble polyesters of the PET family can be esterified by excess 3,5-dinitrobenzoic acid in nitrobenzene and back titrated by a solution of NaOH in ethylene glycol[418] or derivatized by dichloroacetyl chloride in dichloroacetic acid and the Cl content determined by X-ray fluorescence after precipitation and purification of modified PET.[419]

2.4 POLYESTER SYNTHESES

Unless otherwise specified, all monomers, reactants, and solvents were reagent grade 99+% (from Aldrich or Fluka) and used without further purification. In polymer formulas, n, x, y, and z represent number-average numbers of repeating units.

(*Caution*: The reactions described below should be carried out in a hood behind a protective shield.)

2.4.1 Aliphatic Polyesters

2.4.1.1 Poly(ethylene adipate) by Diacid–Diol Reaction

2.4.1.1.1 COOH-Terminated Poly(ethylene adipate)

This dicarboxy-terminated oligomer is prepared by reacting excess adipic acid with 1,2-ethanediol in the bulk until hydroxyl group conversion is complete (Scheme 2.47). The molar mass of final polymer depends on the initial molar ratio of monomers (procedure similar to that described in ref. 401).

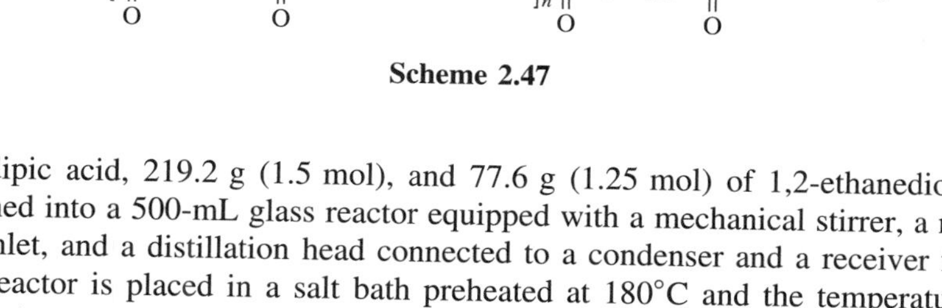

$$(1 + 1/n) \ HO-\underset{\underset{O}{\|}}{C}-(CH_2)_4-\underset{\underset{O}{\|}}{C}-OH \quad + \quad 1 \ HO-CH_2-CH_2-OH$$

1) 180–220 °C, 1 bar, N_2
2) 160–220 °C, 0.07 mbar

$$1/n \ HO\left[\underset{\underset{O}{\|}}{C}-(CH_2)_4-\underset{\underset{O}{\|}}{C}-O-CH_2-CH_2-O\right]_n\underset{\underset{O}{\|}}{C}-(CH_2)_4-\underset{\underset{O}{\|}}{C}-OH \quad + \quad 2 \ H_2O \uparrow$$

Scheme 2.47

Adipic acid, 219.2 g (1.5 mol), and 77.6 g (1.25 mol) of 1,2-ethanediol are weighed into a 500-mL glass reactor equipped with a mechanical stirrer, a nitrogen inlet, and a distillation head connected to a condenser and a receiver flask. The reactor is placed in a salt bath preheated at 180°C and the temperature is then raised gradually to 220°C (see note at end of procedure) until the greater part of water has been removed (3 h). The reactor is cooled down to 160°C and vacuum is applied slowly to ca. 0.07 mbar (30 min). Temperature is ramped to 220°C (see note below) at a rate of 1°C/min and reaction is continued for an additional 90 min. At the end of reaction, the carboxylic acid endgroup content is close to 1.90 mol/kg. No purification of final polyester is carried out.

Differential scanning calorimetry (DSC): $T_g = -49°C$, $T_m = 36°C$. SEC (in THF, calibrated with commercial polystyrene standards): $\overline{M_n} = 860$, $\overline{M_w} = 1660$, $\overline{M_w}/\overline{M_n} = 1.9$. Proton NMR (CDCl$_3$, ref. tetramethylsilane (TMS): δ 1.65 (m, CH$_2$−CH_2−CH$_2$), 2.36 (m, CH_2−COO), 4.26 ppm (s, CH_2−OCO). Carbon-13 NMR (CDCl$_3$, ref. TMS): δ 23.9 (CH$_2$−CH$_2$−CH$_2$), 33.3 (CH$_2$−COO), 61.8 (CH$_2$−OCO), 172.8 (COO−CH$_2$), 178.0 ppm (COOH), $\overline{M_n} = 950$ (from quantitative ^{13}C NMR).

(*Note*: Heating above 220°C should be avoided because of polymer degradation Zncl$_2$ catalyst (0.1% mass) may be added to increase reaction rate.)

2.4.1.1.2 OH-Terminated Poly(ethylene adipate)

A large excess of 1,2-ethanediol (threefold) is reacted with adipic acid until carboxyl groups have reacted. The excess of 1,2-ethanediol is distilled off under vacuum in the presence of Ti(OBu)$_4$ as hydroxy−ester interchange catalyst. The reaction can be followed by SEC analysis of samples withdrawn from reaction medium and stopped when the desired molar mass has been reached (Scheme 2.48).

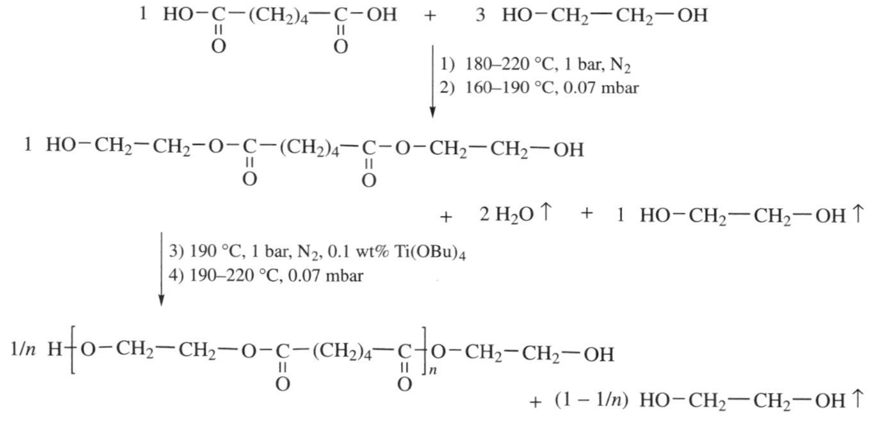

Scheme 2.48

Adipic acid, 146.1 g (1.0 mol), and 186.2 g (3 mol) of 1,2-ethanediol are weighed into a 500-mL glass reactor equipped with a mechanical stirrer, a nitrogen inlet, and a distillation head connected to a condenser and a receiver flask. The reactor is placed in a salt bath preheated at 180°C and the temperature is then raised gradually to 220°C (see note at end of procedure) until the greater part of water has been removed (3 h). The reactor is cooled down to 160°C and vacuum is applied slowly to about 0.07 mbar (30 min). Excess free 1,2-ethanediol begins to distill and temperature is raised gradually to 190°C (90 min). The reaction is continued for an additional 1 h. Pressure is then ramped to atmospheric with a nitrogen flow and 0.115 g (1.15 mL of a 10% solution in CH_2Cl_2) of $Ti(OBu)_4$, as catalyst, is added to the reaction medium. Vacuum is applied slowly to about 0.07 mbar (30 min) and temperature is raised to 220°C (30 min) (see note below). Additional 1,2-ethanediol arising from ester interchange is removed. At the end of the reaction, the carboxylic acid endgroup content is lower than 10^{-3} mol/kg. The hydroxyl endgroup content varies within 0.6–2.0 mol/kg depending on the duration of the last heating step.

DSC, SEC, and NMR analyses of a sample with [OH] = 0.65 mol/kg: DSC: $T_g = -47°C$, $T_m = 45.5°C$. SEC (THF, calibrated with polystyrene standards): $\overline{M_n} = 2050$, $\overline{M_w} = 4250$, $\overline{M_w}/\overline{M_n} = 2.1$. Proton NMR ($CDCl_3$, ref. TMS): δ 1.64 (m, $CH_2-CH_2-CH_2$), 2.34 (m, CH_2-COO), 3.81 (t, CH_2-OH), 4.21 (t, CH_2- CH_2-OH), 4.26 ppm (s, CH_2-OCO); $\overline{M_n} = 3050$. Carbon-13 NMR ($CDCl_3$, ref. TMS): δ 23.8 ($CH_2-CH_2-CH_2$), 33.2 (CH_2-COO), 65.5 (CH_2-CH_2-OH) 61.7 (CH_2-OCO), 60.3 (CH_2-OH) 172.8 ppm (COO).

(*Note*: Heating above 220°C should be avoided because of beginning of polymer degradation.)

2.4.1.2 OH-Terminated Poly(ethylene adipate-co-maleate) by Reaction between Diacid, Anhydride, and Diol

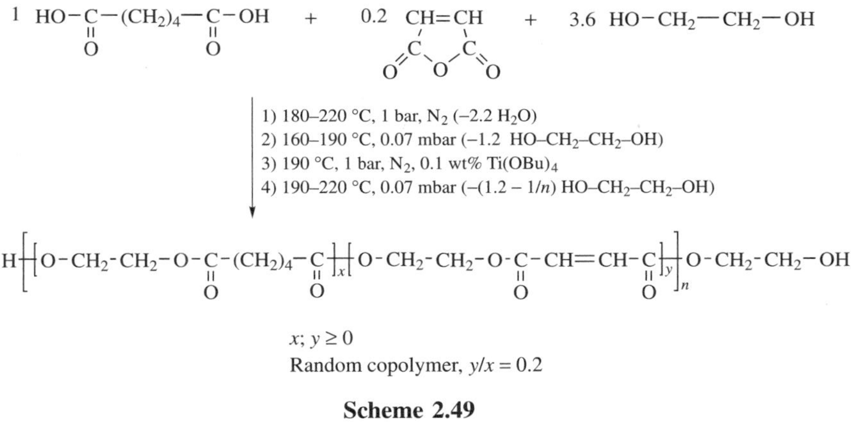

Scheme 2.49

Adipic acid, 146.14 g (1.0 mol), 19.61 g (0.2 mol) of maleic acid, and 223.45 g (3.6 mol) of 1,2-ethanediol are reacted according to the procedure given in Section 2.4.1.1.2. At the end of the reaction, the carboxylic acid endgroup content is lower than 10^{-2} mol/kg. The hydroxyl endgroup content varies within 0.6–2.0 mol/kg depending on the duration of the last heating step (Scheme 2.49).

Vapor pressure osmometry (VPO) analysis (toluene, 60°C) of a sample with [OH] = 1.41 mol/kg: $\overline{M_n} = 1500$.

2.4.1.3   High-Molar-Mass Poly(tetramethylene octanedioate) by Diester–Diol Reaction

This procedure has been adapted from ref. 420.

Dimethyl octanedioate (dimethyl suberate), 71.2 g (0.352 mol), 1,4-butanediol (5% excess: 0.370 mol, 33.3 g), and 0.02 g of tetraisopropoxytitanium (0.025% of final polyester mass) are placed in a three-necked round-bottomed flask fitted with a mechanical stirrer. The medium is slowly heated to 150°C within 4 h under nitrogen atmosphere while methanol is distilled off. Vacuum is then slowly applied and the reaction continued at 0.01 mbar and 150°C for 48 h. The resulting polyester is cooled down, dissolved in chloroform (50 g polyester/200 mL chloroform), and slowly added to a 10-fold volume of methanol under high-speed agitation (1000 rpm). The precipitated polyester is filtered off and dried at 30°C under vacuum (0.1 mbar).

Melting temperature is 60°C; SEC (THF, polystyrene standards): $\overline{M_n} = 21{,}500$, $I_p = 1.75$.

Note: Purification by precipitation results in the elimination of low-molar-mass oligomers and cyclics and yields polyesters with narrower distributions than the theoretical one ($I_p = 2$).

2.4.1.4   OH-Terminated Poly(ε-caprolactone) by ε-Caprolactone–1,6-Hexanediol Reaction

The preparation of poly(ε-caprolactone) given below is a bulk ring-opening polymerization of ε-caprolactone initiated by Ti(OBu)$_4$ in the presence of 1,6-hexanediol as chain limiter (Scheme 2.50). The resulting polymer consists of a random distribution of macromolecules in which hexamethylene moieties can be either terminal or internal units.

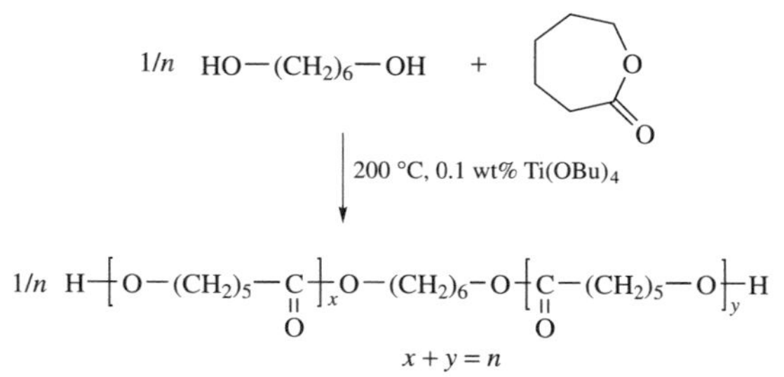

$1/n$ HO—(CH$_2$)$_6$—OH   +

200 °C, 0.1 wt% Ti(OBu)$_4$

$1/n$ H$\left[\!\!\left[\text{O}-(\text{CH}_2)_5-\overset{\text{O}}{\underset{\|}{\text{C}}}\right]_x\!\!\text{O}-(\text{CH}_2)_6-\text{O}\left[\overset{\text{O}}{\underset{\|}{\text{C}}}-(\text{CH}_2)_5-\text{O}\right]_y\right]$H

$x + y = n$

Scheme 2.50

ε-Caprolactone, 88.2 g (0.77 mol), and 11.8 g (0.1 mol) of 1,6-hexanediol are charged into a 250-mL three-necked round-bottom flask equipped with a magnetic stirrer, a reflux condenser, and a nitrogen inlet. The reaction mixture is heated to 200°C under nitrogen stream. As catalyst, 0.1 g (1 mL of a 10% solution in CH$_2$Cl$_2$) Ti(OBu)$_4$ is added. The mixture is reacted for 4 h at 200°C, cooled down, and analyzed without further purification.

DSC: $T_g = -68°C$; $T_m = 45°C$. Proton NMR (CDCl$_3$, ref. TMS): δ 1.29 (m, γ-CH_2 to ester group), 1.56 (m, β-CH_2 to ester group), 2.22 (t, CH_2–COO), 3,53 (t, CH_2–OH), 3.97 ppm (t, CH_2–OCO), $\overline{M}_n = 1160$.

2.4.1.5 Bioresorbable Polyesters

2.4.1.5.1 Poly(glycolic Acid) (PGA)

The preparation of PGA given below (Scheme 2.51) is a polycondensation of glycolic acid followed by a solid-state postpolycondensation.[374] (Reproduced from ref. 374. Copyright © 2000 Elsevier Science Ltd, by permission of the copyright owner.) This method is different from the commonly used ring-opening polymerization of lactide (see Section 2.3.6).

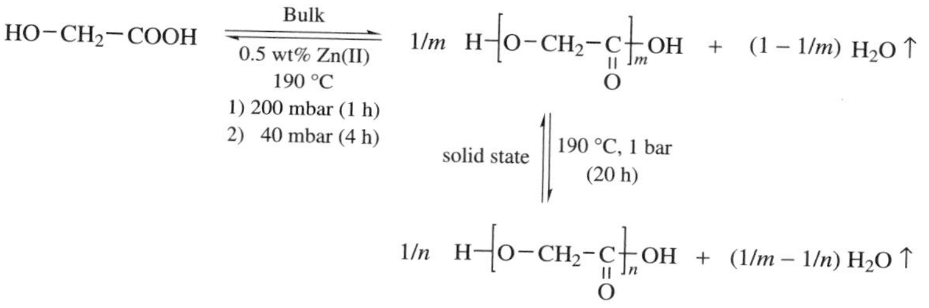

Scheme 2.51

Glycolic acid (GA), 50 g (0.66 mol), is weighed into a 200-mL reaction flask equipped with a magnetic stirrer. Zinc acetate dihydrate [Zn (CH$_3$CO$_2$)$_2$,2H$_2$O] (0.5% mass relative to GA) is added to reaction mixture. The mixture is heated to 190°C with stirring under vacuum, 200 mbar for 1 h, then 40 mbar for 4 h. During this step the medium turns from liquid to solid. The condensation product formed (oligomer state) is crushed into powder in a nitrogen atmosphere, and the powder is heated at the solid state under stirring at 190°C for 20 h.

DSC: $T_g = 40°C$, $T_m = 220°C$, degree of crystallinity 44%. SEC (in HFIP containing 1 mM sodium trifluoroacetate and relative to poly(methylmethacrylate) standards): $\overline{M}_w = 44{,}000$, $\overline{M}_w/\overline{M}_n = 1.8$

2.4.1.5.2 Poly(L-lactic Acid) (PLA)

Poly(lactic acid)s are generally produced by the ring-opening polymerization of D-, L-and/or D, L-lactide using various initiators in solution or in the bulk.[364,366,422] Recently, Moon et al.[375] described the synthesis of high-molar-mass PLA using a melt polycondensation of L-lactic acid in the presence of Sn(II) catalysts and protonic acids (Scheme 2.52). Their procedure is given below. (Reproduced from ref. 375. Copyright © 2000 John Wiley & Sons, Inc., by permission of the copyright owner.)

A 300-mL three-necked flask is equipped with a mechanical stirrer and a reflux condenser connected to a vacuum system through a cold trap.

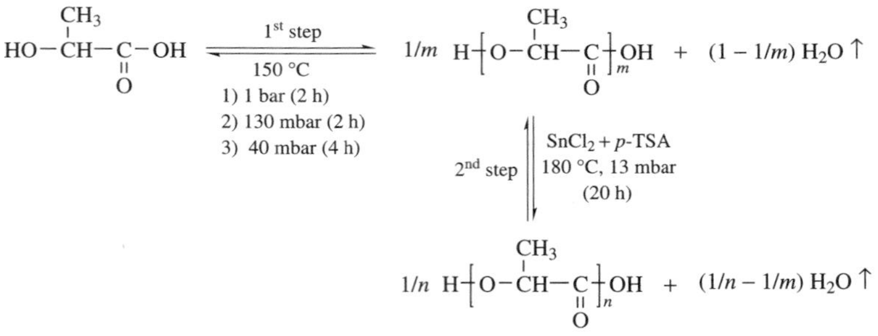

$$\text{HO}-\overset{\overset{\displaystyle CH_3}{|}}{\underset{\underset{\displaystyle O}{||}}{CH-C}}-\text{OH} \quad \underset{\substack{150\,^{\circ}\text{C} \\ 1)\ 1\ \text{bar (2 h)} \\ 2)\ 130\ \text{mbar (2 h)} \\ 3)\ 40\ \text{mbar (4 h)}}}{\overset{1^{st}\ \text{step}}{\rightleftharpoons}} \quad 1/m\ \text{H}\!\!\left[\text{O}-\overset{\overset{\displaystyle CH_3}{|}}{\underset{\underset{\displaystyle O}{||}}{CH-C}}\right]_{m}\!\!\text{OH} \ + \ (1-1/m)\ H_2O \uparrow$$

$$2^{nd}\ \text{step} \quad \begin{array}{c} \text{SnCl}_2 + p\text{-TSA} \\ 180\ ^{\circ}\text{C},\ 13\ \text{mbar} \\ (20\ \text{h}) \end{array}$$

$$1/n\ \text{H}\!\!\left[\text{O}-\overset{\overset{\displaystyle CH_3}{|}}{\underset{\underset{\displaystyle O}{||}}{CH-C}}\right]_{n}\!\!\text{OH} \ + \ (1/n-1/m)\ H_2O \uparrow$$

Scheme 2.52

Oligomerization: Two hundred grams of a 90% aqueous solution of L-lactic acid is charged into the flask and is dehydrated at 150°C, first at atmospheric pressure for 2 h, then at a reduced pressure (130 mbar) for 2 h, and finally under a pressure of 40 mbar for another 4 h. A viscous oligo(L-lactic acid) is formed quantitatively.

Polycondensation: At room temperature, 0.4% mass of Sn(II) chloride dihydrate ($SnCl_2 \cdot 2H_2O$) and 0.4% mass of *p*-toluenesulfonic acid monohydrate (*p*-TSA) are introduced into the mixture. The mixture is heated to 180°C under mechanical stirring. The pressure is reduced stepwise to reach 13 mbar, and the reaction is continued for 20 h. The reaction system becomes gradually viscous, and a small amount of L-lactide is formed and refluxed through the reflux condenser. At the end of the reaction, the flask is cooled down, the product is dissolved in chloroform and subsequently precipitated into diethyl ether. The resulting white fibrous solids are filtered and dried under vacuum (average yield 67%).

DSC: $T_m = 145°C$; degree of racemization 5%. SEC (in $CHCl_3$ at 35°C and calibrated according to polystyrene standards): $\overline{M_w} = 100,000$.

When the polymerization is carried out in the same experimental conditions but without addition of *p*-TSA ($SnCl_2 \cdot 2H_2O = 0.4\%$ mass, 180°C and 20 h), the resulting polymer is amorphous with $\overline{M_w} = 40,000$ and a degree of racemization equal to 75% (determined from ^{13}C NMR measurements).

2.4.1.6 Poly(hexamethylene fumarate) by Diester–Diol Enzyme-Catalyzed Solution Reaction

The procedure given below is taken from ref. 337 with permission. (Copyright © 1995 Hüthig & Wepf Verlag. Reprinted by permission of John Wiley & Sons, Inc.) (Scheme 2.53).

1,6-Hexanediol (0.5 mol/L), dimethyl fumarate (0.5 mol/L), toluene and Novozyme (33.3 g/L) are introduced in a thermostatted double-jacketted reactor fitted with a short thermostatted distillation column and a nitrogen inlet. The temperature is set at 60°C and a nitrogen flow (0.2 L/min) is bubbled into reaction medium. Methanol and toluene are collected in a flask and the volume of the solution is held constant by addition of toluene. After reaction (15 days) the catalyst is removed by filtration and the solvent is evaporated under reduced

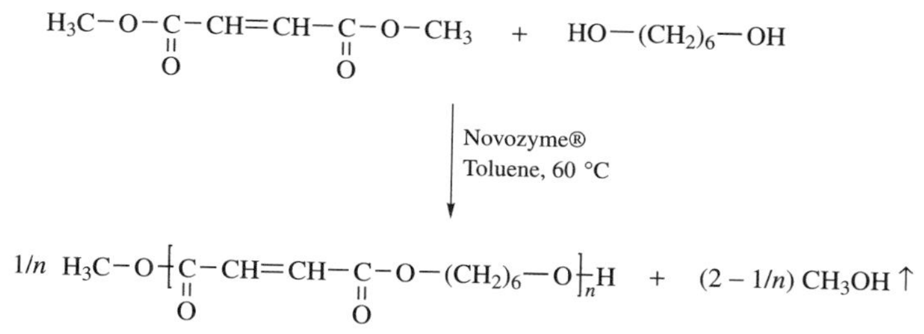

$$H_3C-O-\underset{\underset{O}{\|}}{C}-CH{=}CH-\underset{\underset{O}{\|}}{C}-O-CH_3 \quad + \quad HO-(CH_2)_6-OH$$

Novozyme®
Toluene, 60 °C

$$1/n \ H_3C-O\!\!\left[\underset{\underset{O}{\|}}{C}-CH{=}CH-\underset{\underset{O}{\|}}{C}-O-(CH_2)_6-O\right]_n\!\!H \quad + \quad (2-1/n)\,CH_3OH\uparrow$$

Scheme 2.53

pressure to give a quantitative yield of polyester. DSC: $T_m = 127°C$. NMR ^{1}H (CDCl$_3$, ref. TMS): δ 1.30 (m, γ-CH_2 to ester group), 1.58 (m, β-CH_2 to ester group), 3.52 (t, CH_2−OH), 3.68 (s, CH_3), 4.08 (t, CH_2−OCO), 6.70 ppm (s, $CH{=}$).

Note: Poly(hexamethylene maleate) can be synthesized according to the same procedure using dimethyl maleate instead of dimethyl fumarate. In this case, a mixture of linear chain (76% mass) and cyclic species (24% mass) are obtained.

2.4.2 Unsaturated Polyester Thermosetting Resins

2.4.2.1 Unsaturated Maleate/O-Phthalate/1,2-Propanediol Polyester Prepolymer

Maleic anhydride, 98 g (1.0 mol), 148 g (1.0 mol) of phthalic anhydride, and 160 g (2.1 mol) of 1,2-propanediol are polycondensed in a three-necked flask equipped with a mechanical stirrer, a nitrogen inlet, and a distillation head connected to a condenser and a receiver flask. The flask is placed in a salt bath preheated at 160°C. Water begins to distill and the temperature is then raised gradually to 190°C. The polycondensation is stopped (after about 15 h) when the reaction mixture has an acid number of 50 (see Section 2.3.8.4.1) (Scheme 2.54). A slightly different procedure is described in ref. 423.

2.4.2.2 Preparation and Cure of Unsaturated Polyester/Styrene Resin

Forty grams of unsaturated polyester (see above, Section 2.4.2.1), 26.7 g of styrene, and 10 mg of hydroquinone are weighed into a plastic beaker. The mixture is stirred and slightly warmed, if necessary, until complete dissolution. Then, at room temperature, 0.25 mL Co(II) 2-ethylhexanoate (note a below) and 1.0 mL of 2-butanone peroxide (MEKP, note b below) are added to the stirred mixture. The mixture gels after a 10–20-min period (stir during this period) and solidifies after ca. 1 h (Scheme 2.55)[424].

Notes: (a) Cobalt(II) 2-ethylhexanoate, 65% mass solution in mineral spirits; (b) 2-butanone peroxide, ca. 32% mass solution in dimethyl *o*-phthalate.

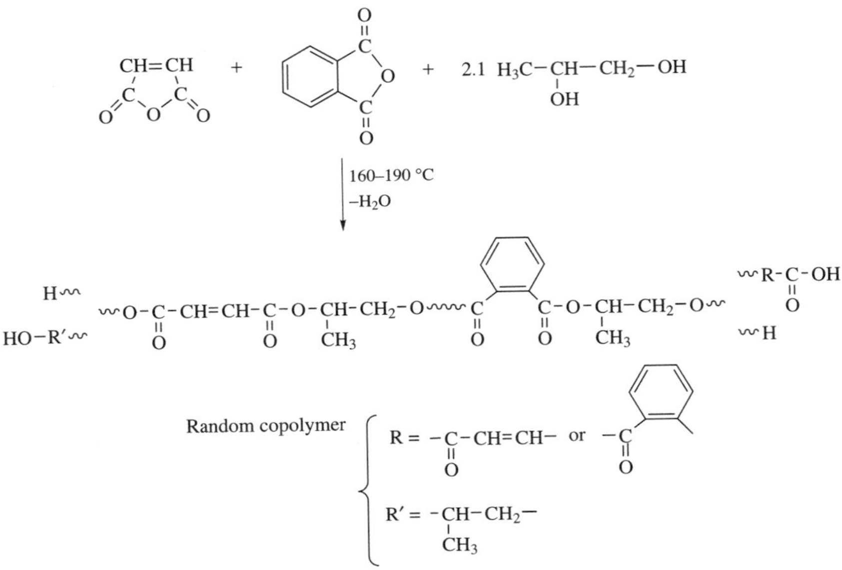

Scheme 2.54

2.4.2.3 Alkyd Resin Synthesis and Cure

Alkyd resin synthesis: This synthesis consists of two steps. In the first step, a triglyceride oil is reacted at ca. 250°C with polyols, such as glycerol or pentaery-thritol, in the presence of a basic catalyst to form a monoglyceride. In the second step, phthalic anhydride, with or without another dibasic acid such as maleic anhydride, is added to the reaction medium and reacted at high temperature. The resulting product is a branched polyester (Scheme 2.56).

Soya oil, 88.6 g, 20.0 g of pentaerythritol, and 0.06 g of lithium hydroxide monohydrate are weighed into a 250-mL three-necked round-bottom flask fitted with a magnetic stirrer, a Dean–Stark separator, and nitrogen inlet and outlet. The reaction mixture is heated to 250°C for 30 min under nitrogen (note a below), then cooled to 200°C. Phthalic anhydride, 34.6 g, 0.70 g of maleic anhydride, and 8.0 g of xylene are added. The reaction mixture is heated to 230°C and the toluene–water azeotrope is removed by distillation. The reaction is stopped when the acid number of reaction medium is lower than 10 (note b below). After cooling to room temperature, 52 g of white spirit is added.

Notes: (a) The reaction must be carried out until a sample withdrawn from the reaction medium becomes completely soluble in methanol (ca. 30 min). (b) See Section 2.3.8.4.1.

Cure of alkyd resin: A varnish is prepared by mixing 100 g of the alkyd resin prepared above, 0.05 g of Co(II) 2-ethylhexanoate, 0.05 g of calcium 2-ethyl-hexanoate, and 0.02 g of lead 2-ethylhexanoate. This mixture is held for 24 h in a closed flask before use. A thin layer of this varnish, applied on a metal plate, forms a hard, glossy coating after 24 h drying at room temperature.

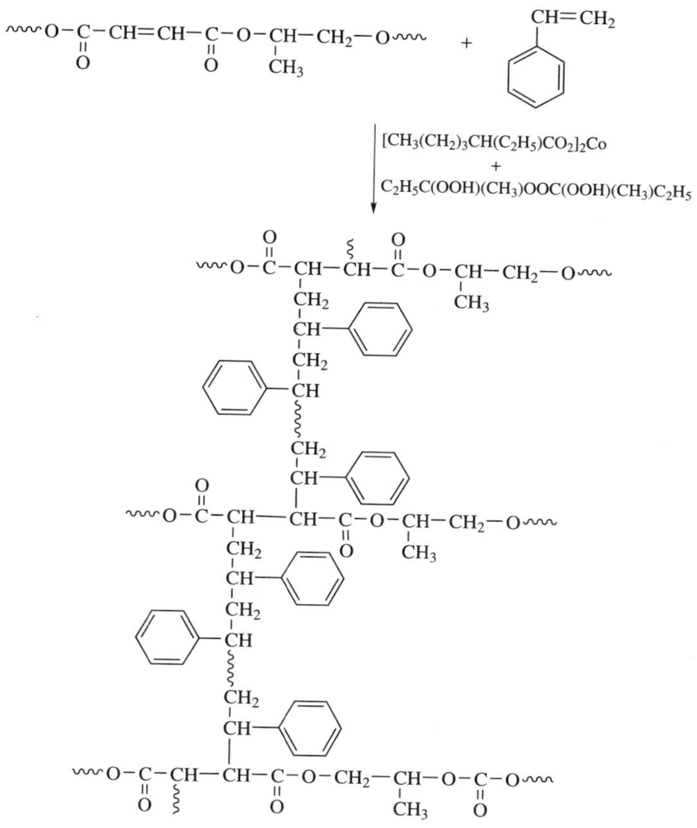

Scheme 2.55

2.4.3 Aromatic–Aliphatic Polyesters

2.4.3.1 *Poly(ethylene terephthalate) (PET)*

2.4.3.1.1 *Dimethyl Terephthalate–1,2-Ethanediol Reaction*

The preparation of PET is given below (Scheme 2.57). (Reproduced from ref. 425 with permission. Copyright © 1992 Academic Press, Inc. Reprinted by permission of Elsevier Science Ltd.)

To a weighed thick-walled glass cylindrical reactor fitted with a mechanical stirrer, nitrogen inlet, distillation side arm, and a vacuum line are added 15.5 g (0.08 mol) of dimethyl terephthalate, 11.8 g (0.19 mol) of 1,2-ethanediol, 0.025 g of calcium acetate dihydrate [$(CH_3CO_2)_2 \cdot H_2O$], and 0.006 g of antimony trioxide (Sb_2O_3). The tube is warmed gently in a salt bath to melt the mixture. While heating to 197°C, a slow stream of nitrogen is passed through the melt to help eliminate the methanol. The tube is heated for 2–3 h at 197°C or until all the methanol has been removed. The side arm is also heated to prevent clogging by

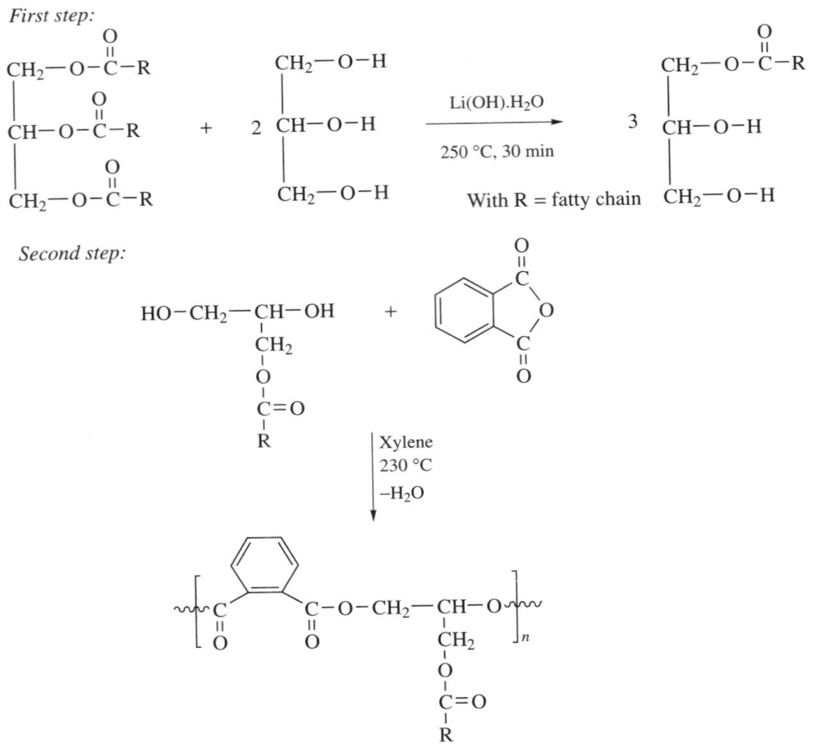

First step:

$$CH_2-O-\overset{\overset{\displaystyle O}{\|}}{C}-R$$
$$CH-O-\overset{\overset{\displaystyle O}{\|}}{C}-R$$
$$CH_2-O-\overset{\overset{\displaystyle O}{\|}}{C}-R$$

$$+ \quad 2 \quad \begin{matrix} CH_2-O-H \\ CH-O-H \\ CH_2-O-H \end{matrix}$$

$$\xrightarrow[\text{250 °C, 30 min}]{\text{Li(OH).H}_2\text{O}}$$

$$3 \quad \begin{matrix} CH_2-O-\overset{\overset{\displaystyle O}{\|}}{C}-R \\ CH-O-H \\ CH_2-O-H \end{matrix}$$

With R = fatty chain

Second step:

$$HO-CH_2-\underset{\underset{\underset{\underset{R}{C=O}}{O}}{CH_2}}{CH}-OH \quad + \quad$$

$$\Big| \begin{matrix} \text{Xylene} \\ \text{230 °C} \\ -\text{H}_2\text{O} \end{matrix}$$

Scheme 2.56

the condensation of some dimethyl terephthalate. The polymer tube is next heated to 222°C for 20 min and then at 283°C for 10 min. The side arm is connected to a vacuum pump and the pressure is reduced to 0.4 mbar or less while heating at 283°C for 3.5 h. The resulting polymer melts at ca. 270°C.

The inherent viscosity of a 0.5% solution in 1,1,2,2-tetrachloroethane–phenol (40/60) is ca. 0.6–0.7 dL/g at 30°C. NMR ^{1}H [CDCl$_3$–trifluoroacetic acid (70/30 v/v)], ref. TMS): δ 4.17 (t, CH_2–OH), 4.61 (m, CH_2–CH$_2$–OH), 4.78 (s, CH_2–OOC), 8.12 ppm (s, ArH). The most important side reaction during PET synthesis is the formation of 2–4 mol% of oxydiethylene units in polymer chains: δ 4.10 (t, COO–CH$_2$–CH$_2$–O–CH_2–CH$_2$–OOC), 4.64 ppm (t, COO–CH_2–CH$_2$–O–CH$_2$–CH$_2$–OOC).

2.4.3.1.2 Terephthalic Acid–1,2-Ethanediol Reaction

(Adapted from the procedure described in ref. 426 for a batch pilot plant equipment.)

A 250 mL stainless steel polycondensation reactor is charged with 90 g (0.54 mol) of terephthalic acid, 40.5 g (0.65 mol) of 1,2-ethanediol, 0.025 g of

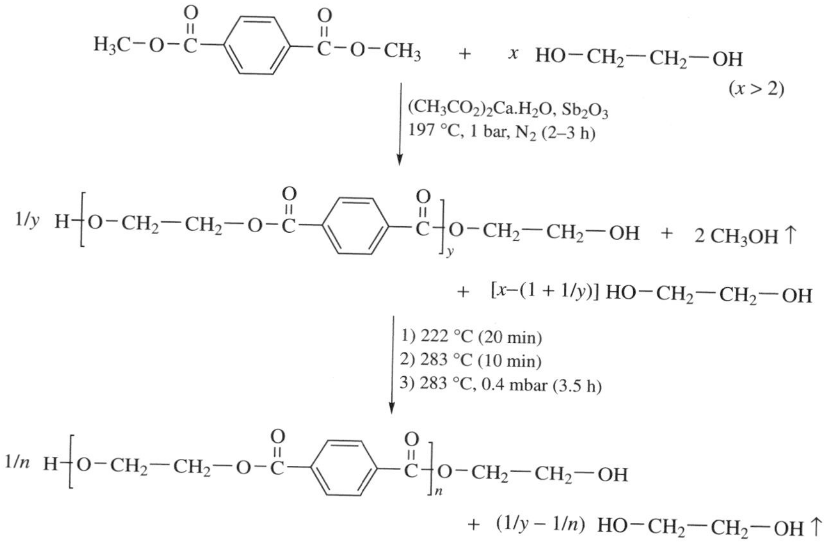

Scheme 2.57

antimony trioxide (catalyst) and 0.02 mmol of choline hydroxide (diethylene glycol suppressant) (45% mass in methanol). The pressure is set at 3.5 bar with nitrogen and the mixture is heated at 93°C for 40 min under stirring (60 rpm). Then, the reactor temperature is increased very slowly to 266°C (150 min) while eliminated water is collected in a condensate receiver. At this point, the pressure is slowly decreased to atmospheric at a rate of 0.7 bar/min. After 30 min reaction at atmospheric pressure, vacuum is slowly applied (30 mbar/min) until pressure reaches 1.3 mbar. At the same time, the reaction temperature is slowly increased to 285°C (55 min). The agitator speed is decreased stepwise in 10-rpm increments to 30 rpm and the reaction is stopped after further 95 min reaction. The intrinsic viscosity of the polymer is close to 0.67 dL/g (o-chlorophenol solution at 25°C).

2.4.3.1.3 Solid-State Post Polymerization of PET

This procedure has been adapted from ref. 427.

The PET prepolymer utilized in the solid-state polymerization can be in the form of pellets or chips. However, it will preferably be crushed into powder (particle size 125–250 μm)[428] and then vacuum dried overnight at 110°C.

Ninety grams of PET prepolymer having an inherent viscosity of 0.59 dL/g (see note below) is weighed into a laboratory-size solid-state polymerization reactor equipped with a cindered glass dispersing plate. A nitrogen flow is applied through the reactor, which is placed in an oil bath at 230°C. The prepolymer is left at this temperature for 5 h. The inherent viscosity of the resulting polymer is of about 0.9 dL/g (see note).

Note: Viscosity measurement: 30°C, 0.4 g/dL in 60/40 phenol–tetrachloro-ethane.

2.4.3.2 *Poly(butylene terephthalate) (PBT) by Dimethyl Terephthalate–1,4-Butanediol Reaction*

This procedure has been adapted from ref. 429 (Scheme 2.58).

1,4-Butanediol, 50.47 g (0.56 mol), 77.68 g (0.4 mol) dimethyl terephthalate, and tetrabutoxytitanium ester interchange catalyst (150 ppm Ti) are placed in a stainless steel reactor equipped with a thermocouple, stirring motor, distillation column, and nitrogen inlet. The reaction mixture is stirred and heated to 150°C under nitrogen at atmospheric pressure. The ester interchange reaction starts and the reaction medium is progressively (160 min) heated to 200°C while methanol distills off. The distillation column is then replaced by a condenser and an additional 270 ppm of Ti is added. The temperature is raised to 205°C under vacuum (0.3 mbar) for 45 min, then to 251°C to continue the polymerization for another 65 min or longer, until polymer intrinsic viscosity reaches 0.8 dL/g (*o*-chlorophenol at 25°C). The carboxylic acid endgroup content is equal to 0.013 mol/kg.

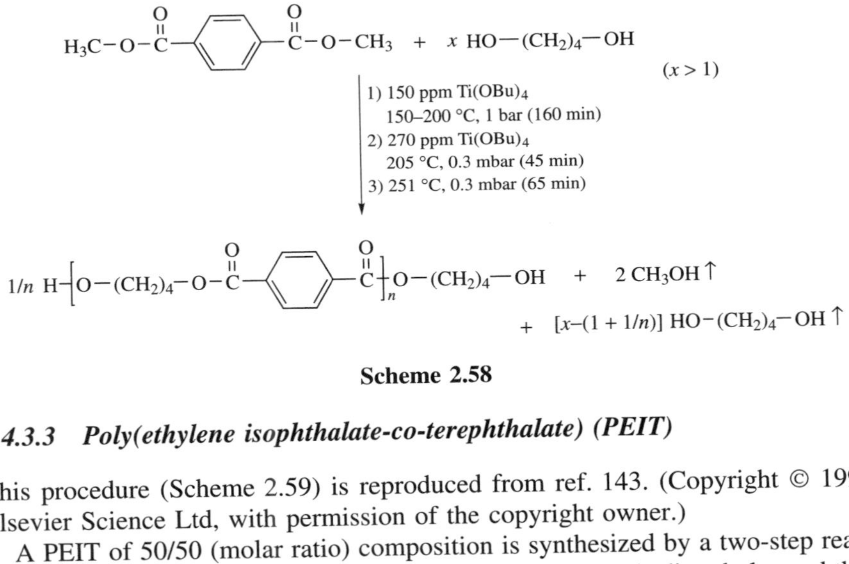

Scheme 2.58

2.4.3.3 *Poly(ethylene isophthalate-co-terephthalate) (PEIT)*

This procedure (Scheme 2.59) is reproduced from ref. 143. (Copyright © 1999 Elsevier Science Ltd, with permission of the copyright owner.)

A PEIT of 50/50 (molar ratio) composition is synthesized by a two-step reaction sequence as follows. In the first step, 97.10 g (0.5 mol) dimethyl terephthalate (DMT), 97.10 g (0.5 mol) dimethyl isophthalate (DMI), 136.55 g (2.2 mol) 1,2-ethanediol, and zinc acetate dihydrate ester interchange catalyst ($2.7 \times 10^{-4}\%$ mass of the total amount of DMI and DMT mixture) are weighed into a three-necked flask fitted with a mechanical stirrer, a nitrogen inlet, and a condenser. The medium is stirred for 2.0–2.5 h at 180–210°C under nitrogen. Ninety-two percent of the theoretical amount of methanol is removed by distillation. In the second step, antimony acetate polycondensation catalyst and trimethyl phosphate thermal stabilizer (9.9×10^{-4} and $1.5 \times 10^{-3}\%$ mass of the total amount of DMI

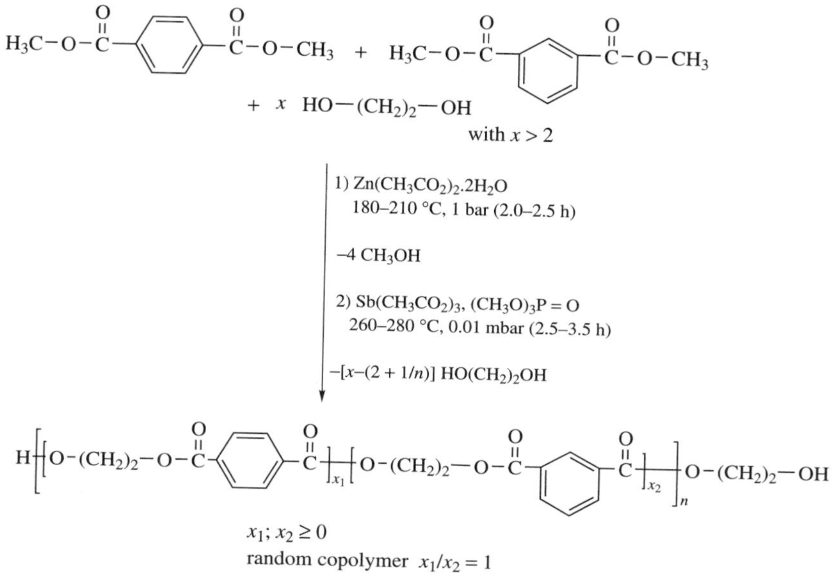

$$H_3C-O-\overset{\overset{\displaystyle O}{\|}}{C}-\!\!\!\!\!\raisebox{-0.5ex}{\text{(ring)}}\!\!\!\!\!-\overset{\overset{\displaystyle O}{\|}}{C}-O-CH_3 \ + \ H_3C-O-\overset{\overset{\displaystyle O}{\|}}{C}-\!\!\!\!\!\raisebox{-0.5ex}{\text{(ring)}}\!\!\!\!\!-\overset{\overset{\displaystyle O}{\|}}{C}-O-CH_3$$

$$+ \ x \ HO-(CH_2)_2-OH$$

with $x > 2$

1) $Zn(CH_3CO_2)_2 \cdot 2H_2O$
   180–210 °C, 1 bar (2.0–2.5 h)

$-4\ CH_3OH$

2) $Sb(CH_3CO_2)_3$, $(CH_3O)_3P = O$
   260–280 °C, 0.01 mbar (2.5–3.5 h)

$-[x-(2 + 1/n)]\ HO(CH_2)_2OH$

$$1/n\ \ H\!\left[\!\!\left[O-(CH_2)_2-O-\overset{\overset{\displaystyle O}{\|}}{C}-\!\!\!\!\!\raisebox{-0.5ex}{\text{(ring)}}\!\!\!\!\!-\overset{\overset{\displaystyle O}{\|}}{C}\right]_{x_1}\!\!\!\!\left[O-(CH_2)_2-O-\overset{\overset{\displaystyle O}{\|}}{C}-\!\!\!\!\!\raisebox{-0.5ex}{\text{(ring)}}\!\!\!\!\!-\overset{\overset{\displaystyle O}{\|}}{C}\right]_{x_2}\right]_n\!\!\!O-(CH_2)_2-OH$$

$x_1; x_2 \geq 0$
random copolymer $x_1/x_2 = 1$

Scheme 2.59

and DMT mixture, respectively) are added. The polymerization is carried out with stirring for 2.5–3.5 h at 260–280°C under vacuum (ca. 10^{-2} mbar). The intrinsic viscosity of the resulting copolymer is 0.504 dL/g (measured in CF_3COOH at 30°C). This copolyester is amorphous ($T_g = 64.1$°C). A series of other PEIT copolyesters can be prepared at various DMI–DMT monomer ratios.

2.4.3.4 Aromatic–Aliphatic Alternating Copolyesteramide

The dimethyl ester of a carboxy-terminated diamide is reacted with 1,2-ethanediol in the conditions of PBT synthesis. The amide functions are unaffected while the hydroxy–ester interchange reaction proceeds to the formation of an alternating copolyesteramide (Scheme 2.60). The procedures given below are reproduced from ref. 430. (Copyright © 1997 Elsevier Science Ltd, with permission of the copyright owner.)

Synthesis of N,N'-bis(p-carbomethoxybenzoyl) butanediamine (T4T-dimethyl). Dimethyl terephthalate (DMT) (275 g, 1.42 mol) is dissolved at 65°C in a mixture of 1100 mL anhydrous toluene and 85 mL anhydrous methanol in a 2-L flask equipped with a stirrer, condenser, calcium chloride tube, and nitrogen inlet. When all the DMT is dissolved, 15 mL lithium methoxide (1.25 M) is added. 1,4-Butanediamine (BDA) (34 mL, 0.34 mol), dissolved in 150 mL anhydrous toluene and 10 mL anhydrous methanol, is added dropwise in 4–6 h while the nitrogen flow is stopped. Three hours after the start of the BDA addition, the temperature is gradually increased up to 90°C (5°C/h) while the methanol is stripped off. After a total reaction time of 24 h the reaction is stopped. The white

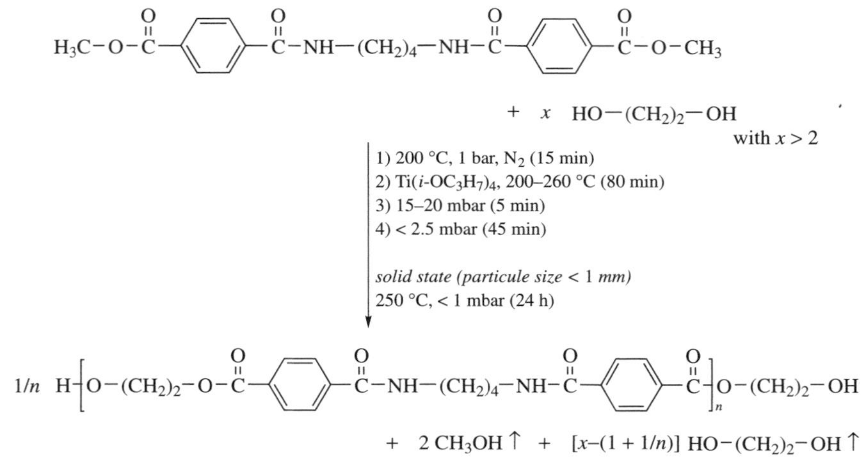

$$H_3C-O-\overset{\overset{O}{\|}}{C}-\langle\!\!\!\bigcirc\!\!\!\rangle-\overset{\overset{O}{\|}}{C}-NH-(CH_2)_4-NH-\overset{\overset{O}{\|}}{C}-\langle\!\!\!\bigcirc\!\!\!\rangle-\overset{\overset{O}{\|}}{C}-O-CH_3$$

$$+\quad x\quad HO-(CH_2)_2-OH$$
$$\text{with } x > 2$$

1) 200 °C, 1 bar, N_2 (15 min)
2) Ti(i-OC$_3$H$_7$)$_4$, 200–260 °C (80 min)
3) 15–20 mbar (5 min)
4) < 2.5 mbar (45 min)

solid state (particule size < 1 mm)
250 °C, < 1 mbar (24 h)

$$1/n\;\;H\!\left[O-(CH_2)_2-O-\overset{\overset{O}{\|}}{C}-\langle\!\!\!\bigcirc\!\!\!\rangle-\overset{\overset{O}{\|}}{C}-NH-(CH_2)_4-NH-\overset{\overset{O}{\|}}{C}-\langle\!\!\!\bigcirc\!\!\!\rangle-\overset{\overset{O}{\|}}{C}\right]_n\!\!O-(CH_2)_2-OH$$

$$+\;\; 2\, CH_3OH\uparrow\;\;+\;\;[x-(1+1/n)]\, HO-(CH_2)_2-OH\uparrow$$

Scheme 2.60

precipitate is filtered through Soxhlet extraction thimbles and washed twice with hot *m*-xylene to remove DMT. The product is subsequently dried overnight in a vacuum oven at 70°C. T4T-dimethyl is recrystallized from hot 1-methyl-2-pyrrolidinone (50 g/L at 160°C) and washed with hot acetone twice. The purified T4T-dimethyl is dried overnight in a vacuum oven at 70°C.

Melt polycondensation: The reaction is carried out in a 250-mL stainless steel vessel with nitrogen inlet and mechanical stirrer. The vessel containing T4T-dimethyl (30 g, 72.8 mmol) and ethanediol (30 g, 0.48 mol) is heated up in an oil bath at 200°C. After 15 min reaction Ti(i-OC$_3$H$_7$)$_4$ (1.5 mL of 0.1 *M* solution in CH$_2$Cl$_2$) is added and subsequently the temperature is gradually raised to 260°C (1°C/min). After 10 min at 260°C the pressure is reduced (15–20 mbar) for 5 min. Then the pressure is reduced further (<2.5 mbar) for 45 min. The vessel is cooled down slowly to room temperature, maintaining the low pressure. After solidification, the polymer is ground (particle size <1 mm) and subsequently dried in a vacuum oven at 80°C.

Solid-state postpolycondensation: The postpolycondensation is carried out at reduced pressure (<1 mbar) in a glass tube which is placed in an oven at a temperature of 250°C for 24 h. The inherent viscosity of the resulting polymer is 0.61 dL/g (0.1 g/dL solution in *p*-chlorophenol at 45°C). DSC: $T_g = 124°C$, $T_m = 316°C$.

2.4.3.5   Polyester–Polyether Thermoplastic Elastomers: Dimethyl Terephthalate/1,4-Butanediol/Dihydroxy–Poly(oxytetramethylene) Reaction

Poly(poly(alkylene terephthalate)-*block*-poly(oxyalkylene)) thermoplastic elastomers are prepared by the bulk polycondensation of dimethyl terephthalate with a mixture of 1,4-alkanediol and hydroxy-terminated poly(oxyalkylene) in the conditions of PBT synthesis (Section 2.4.3.2). The procedure reported below (Scheme 2.61) describes the synthesis of a PBT–poly(oxytetramethylene) multi-block copolymer (57.8% mass hard block).[431]

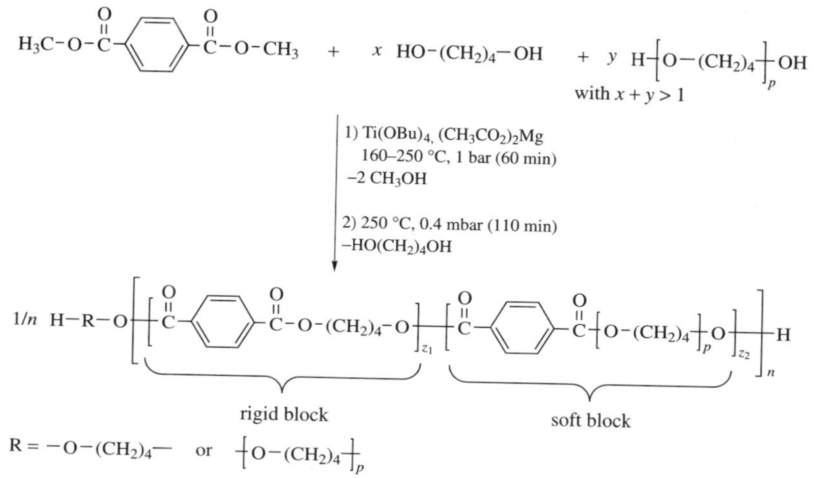

$$H_3C-O-\overset{O}{\overset{\|}{C}}-\!-\overset{O}{\overset{\|}{C}}-O-CH_3 \quad + \quad x\ HO-(CH_2)_4-OH \quad + \quad y\ H\!-\!\!\left[O-(CH_2)_4\right]_p\!\!-OH$$

with $x+y>1$

1) Ti(OBu)$_4$, (CH$_3$CO$_2$)$_2$Mg
160–250 °C, 1 bar (60 min)
–2 CH$_3$OH

2) 250 °C, 0.4 mbar (110 min)
–HO(CH$_2$)$_4$OH

$$1/n\ H-R-O-\left[\!\left[\overset{O}{\overset{\|}{C}}-\!\!\!\!\!\!\!\!-\overset{O}{\overset{\|}{C}}-O-(CH_2)_4-O\right]_{z_1}\!\!\left[\overset{O}{\overset{\|}{C}}-\!\!\!\!\!\!\!\!-\overset{O}{\overset{\|}{C}}\!\!\left[O-(CH_2)_4\right]_p\!\!O\right]_{z_2}\right]_n\!\!-H$$

rigid block   soft block

$R = -O-(CH_2)_4-$ or $\left[O-(CH_2)_4\right]_p$

Scheme 2.61

Hydroxy-terminated poly(oxytetramethylene) (number average molar mass $\approx$ 975) 38.5 g (0.0395 mol), 36.5 g (0.405 mol) of 1,4-butanediol, 60.0 g (0.309 mol) of dimethyl terephthalate, and 0.298 g (0.828 mmol) of N,N'-bis(β-naphthyl)p-phenylenediamine as stabilizer are weighed into a 250-mL agitated flask fitted with nitrogen inlet and a short distillation column. The flask is placed in an oil bath at 160°C, agitated for 5 min and then 0.7 mL of a catalyst solution (see note at end of procedure) is added. Methanol distills from the reaction mixture as the temperature is slowly raised to 250°C, over a period of 1 h. When the temperature reaches 250°C, the pressure is gradually reduced to 0.4 mbar within 20 min. The polymerization mass is agitated at 250°C/0.4 mbar for 90 min. The inherent viscosity of the resulting product (0.1 g/dL in m-cresol) is 1.65 dL/g. A series of thermoplastic elastomer polyester with other hard/soft block ratios can be prepared following a similar procedure

Note: The catalyst solution is prepared by mixing 20 mL of a solution of 11.10 mL of tetrabutoxytitane in 90 mL dry butanol with 10 mL of a solution of 0.3 g anhydrous magnesium acetate in 10 mL dry methanol.

2.4.4 Aromatic Polyesters

2.4.4.1 Bisphenol-A Polyarylates

2.4.4.1.1 Terephthaloyl Chloride/Isophthaloyl Chloride/Bisphenol-A Solution Polyesterification

This procedure has been adapted from that described in ref. 432.

Terephthaloyl chloride, 20.32 g (0.1 mol), and 20.32 g (0.1 mol) of isophthaloyl chloride are charged into a 1-L three-necked Morton flask equipped with a mechanical stirrer, thermometer, reflux condenser with attached drying tube, an inert gas inlet, and a pressure-equalized addition funnel. Methylene chloride (600 mL, distilled and dried over molecular sieve) is added to the flask. The flask is immersed in an ice bath, but the temperature is kept at 15°C or higher to prevent the precipitation of terephthaloyl chloride. The solution is kept under constant dry nitrogen flow with stirring. A solution of *p*-tertiobutylphenol (0.12 g, 0.8 mmol), used as a chain stopper, bisphenol-A (45.57 g, 0.1996 mol), distilled dry triethylamine (42.5 g, 0.42 mol), and distilled dry methylene chloride (60 mL) is added to the diacid chloride solution over a 10-min period with stirring at 15–21.5°C. The addition funnel is rinsed with methylene chloride (40 mL). The reaction mixture is stirred for an additional 1 h. The polymer is recovered by precipitation in distilled water in a Waring blender. The precipitated polymer is washed once with hot distilled water and once with cold distilled water in the Waring blender. The final wash is free of chloride ions. The polyarylate is dried in a vacuum oven. The intrinsic viscosity of the polyester is close to 0.9 dL/g (determined in 1,1,2,2-tetrachoroethane at 30°C).

2.4.4.1.2   Terephthaloyl Chloride–Isophthaloyl Chloride–Bisphenol-A Interfacial Polyesterification

This procedure has been adapted from ref. 433 (Scheme 2.62).

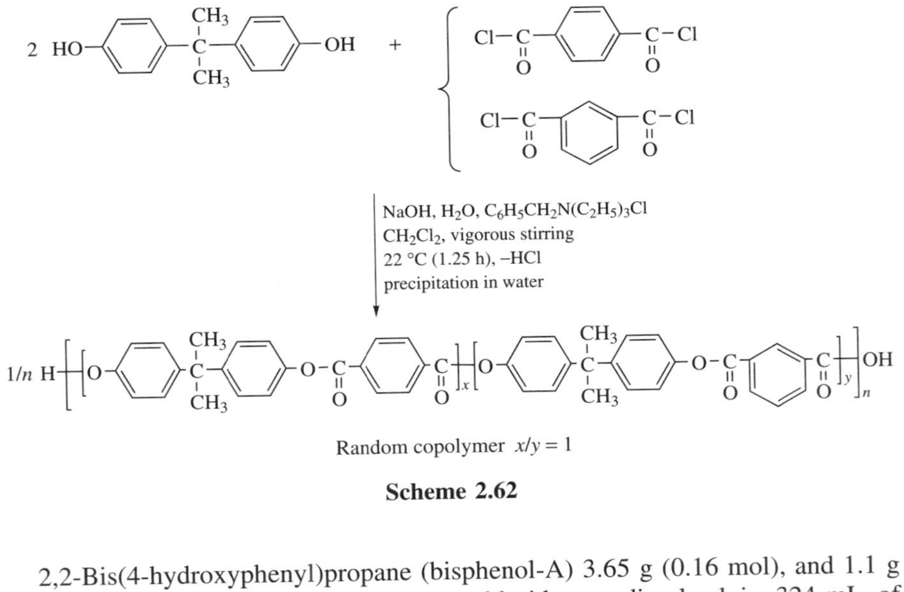

Random copolymer $x/y = 1$

Scheme 2.62

2,2-Bis(4-hydroxyphenyl)propane (bisphenol-A) 3.65 g (0.16 mol), and 1.1 g (4.8 mmol) of benzyltriethylammonium chloride are dissolved in 324 mL of NaOH (1 mol/L). Methylene chloride, 270 mL, is added to this solution. At a temperature not exceeding 22°C, a solution of 16.25 g (0.08 mol) of terephthaloyl chloride and 16.25 g (0.08 mol) of isophthaloyl chloride in 80 mL of methylene chloride is added over a period of 15 min to the very vigorously stirred mixture. Stirring is continued for an additional 1 h. The upper aqueous layer is decanted and replaced by 300 mL of water, and the mixture is again stirred for 15 min. This procedure is repeated once more. Upon pouring the polymer solution into water, a fibrous white polymer precipitates which can easily be isolated. The inherent viscosity of the resulting polyarylate is 1.4 dL/g [0.5% mass in 1,1,2,2-tetrachloro-ethane/phenol (40/60 mass) at 30°C].

DSC: $T_g = 194$°C, $T_m = 260$°C. A series of polyarylates can be prepared following the same procedure at various terephthaloyl chloride–isophthaloyl chloride ratios.[434]

2.4.4.1.3 Terephthalic Acid–Isophthalic Acid–Bisphenol-A Reaction Using Thionyl Chloride Activation

This procedure is reproduced from ref. 435. (Copyright © 1986 John Wiley & Sons, Inc., by permission of the copyright owner.)

To cold thionyl chloride (1.31 g, 11 mmol) in an ice-water bath, pyridine (10 mL) is added slowly for 10 min to keep the reaction temperature low. The reaction medium is stirred for 30 min. Then, a mixture of isophthalic acid (0.41 g, 2.5 mmol) and terephthalic acid (0.41 g, 2.5 mmol) in pyridine (10 mL) is added slowly for 10–20 min to control the reaction temperature. The cooling bath is then removed and the reaction mixture is stirred at room temperature for 20 min. 2,2-Bis(4-hydroxyphenyl)propane (bisphenol-A, 1.14 g, 5 mmol) in pyridine (10 mL) is added all at once to the mixture, and the whole solution is heated to 80°C (bath temperature) for 4 h. The resulting viscous solution is diluted with pyridine and poured into methanol to precipitate the polymer, which is washed in boiling methanol and dried. The inherent viscosity of polymer is 2.2 dL/g (determined in 60/40 phenol–1.1.2.2-tetrachloroethane at 30°C)

Note: Several condensing agents for the direct polycondensation reaction such as diphenyl chlorophosphate–LiBr and tosyl chloride–DMF have also been developed by Higashi et al.[311,312] (see Section 2.3.3.2).

2.4.4.1.4 Diphenyl Terephthalate–Diphenyl Isophthalate–Bisphenol-A Melt Polyesterification

This synthesis involves the hydroxy–ester interchange reaction between phenyl ester and phenol endgroups[436] (Scheme 2.63) (see Section 2.3.2.2).

Diphenyl terephthalate, 31.8 g (0.1 mol), and 31.8 g (0.1 mol) of diphenyl isophthalate are reacted under nitrogen at a temperature of 180°C with 45.6 g (0.2 mol) of 2,2-bis(4-hydroxyphenyl)propane (bisphenol-A) in the presence of 64 mg of Sb_2O_3. After 3 h, the temperature is increased to 230°C for 1 h, then to 250°C for 1 h, and the reaction is continued with distillation of phenol. The temperature is then raised to 310°C, the condensation vessel is removed, and vacuum (0.3 mbar) is applied. The polycondensation continues very rapidly with

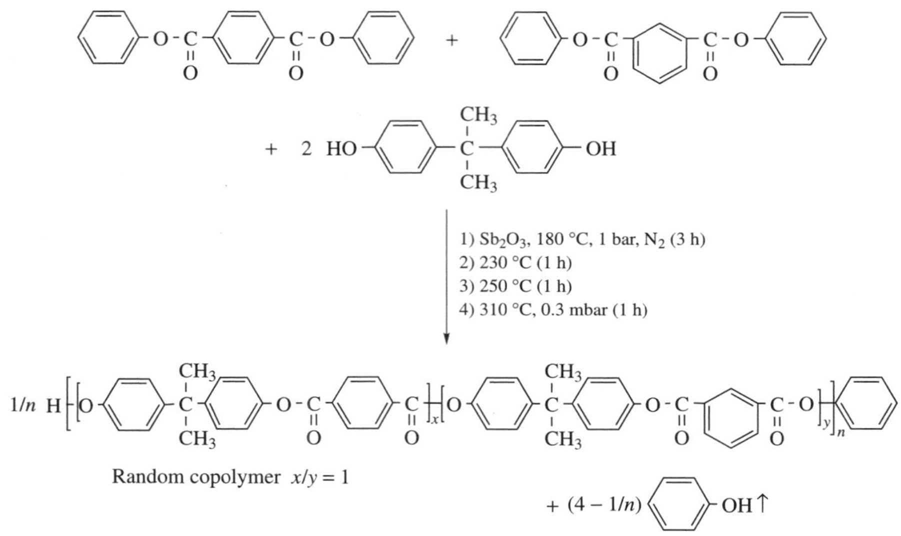

Scheme 2.63

a strong increase in the viscosity of the melt. The reaction is terminated after 1 h. A brownish, noncrystalline, somewhat brittle polycondensate is obtained with an inherent viscosity equal to 0.44 dL/g (1 g/dL in phenol−1,1,2,2-tetrachloroethane (60/40) at 25°C).

Note: Bisphenol-A and the diaryl esters of terephthalic acid and isophthalic acid are nonvolatile compounds, so that any excess of these components cannot completely be removed, resulting in a low-molar-mass, unusable polyester. More-over, excess bisphenol-A causes a strong discoloration of the polyester melt due to thermal degradation at the high reaction temperature used. This can be avoided if the diaryl esters are mixed with 5 mol % of diphenyl carbonate. Any excess of this compound can easily be removed in vacuum at the polycondensation temperature.

Experiment with addition of diphenyl carbonate: Diphenyl terephthalate, 31.8 g (0.1 mol), 28.62 g (0.09 mol) of diphenyl isophthalate, and 2.37 g (0.011 mol) of diphenyl carbonate are polycondensed with 45.6 g (0.2 mol) of 2,2-bis(4-hydroxyphenyl)propane (bisphenol-A) under the preceding conditions. A slightly brownish, extremely tough, noncrystalline polyester is obtained with an inherent viscosity equal to 0.56 dL/g. The softening point of the polyester is equal to 200°C and the melting range is 215−285°C.

2.4.4.1.5 *Terephthalic Acid–Isophthalic Acid–Bisphenol-A Diacetate Melt Polyesterification*

A procedure similar to that given below for wholly aromatic liquid crystalline polyesters can be used (see Section 2.4.4.2). The inherent viscosity of the resulting polymers were ca. 0.6 dL/g (0.5 g/dL in phenol−*o*-dichlorobenzene at 25°C).[437,438]

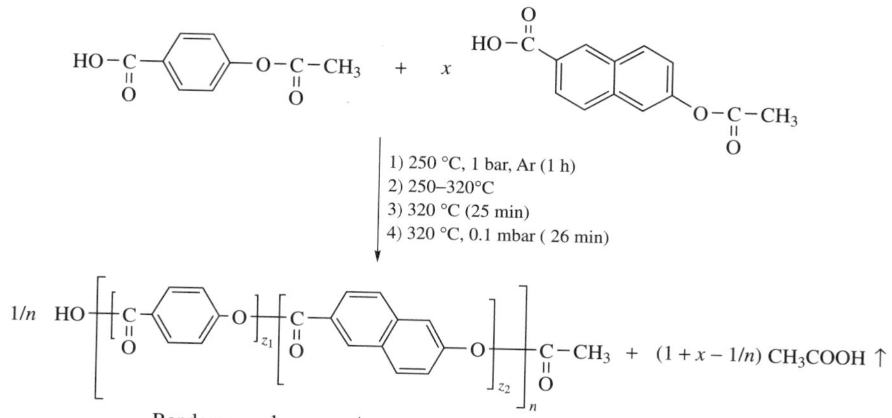

Scheme 2.64

2.4.4.2 Liquid Crystalline Polyesters

2.4.4.2.1 Poly(4-oxybenzoyl-co-6-oxy-2-naphthoyl)

This procedure has been adapted from ref. 17 (Scheme 2.64).

4-Acetoxybenzoic acid, 101 g (0.56 mol), and 55 g (0.24 mol) of 6-acetoxy-2-naphthoic acid are weighed into a three-necked round-bottom flask equipped with a stirrer, argon inlet tube, and a heating tape wrapped around a distillation head connected to a condenser. The charged flask is purged with argon and placed in a salt bath preheated at 250°C. The reaction mixture melt is stirred rapidly under a stream of dry argon while acetic acid is distilled from the polymerization vessel. The mixture is stirred for 3 h at 250°C and then for 1 h 15 min at 280°C. The temperature is then increased gradually to 320°C. The viscous polymer melt is held for 25 min at 320°C. Then, vacuum is applied slowly to 0.1 mbar and held for 26 min. During these stages polymer melt viscosity increases and stirring should be slowed down. After cooling, the polymer is finely ground and dried at 150°C for 1 h.

Characterization[17]: Inherent viscosity 5.7 dL/g (1% mass in pentafluorophenol at 60°C). DSC: $T_m = 275°C$ (nematic melt).

2.4.4.2.2 Poly(p-oxybenzoyl-co-(p-phenylene isophthalate))

The carboxy–ester interchange melt polyesterification used for the preparation of this wholly aromatic copolyester is followed by a solid-state postpolyesterification (Scheme 2.65). (Reprinted from ref. 439. Copyright © 1990 John Wiley & Sons, Inc., by permission of the copyright owner.)

Into a 1-L three-necked glass reaction kettle equipped with a mechanical stirrer, nitrogen inlet tube, and a distillation head connected to a condenser are

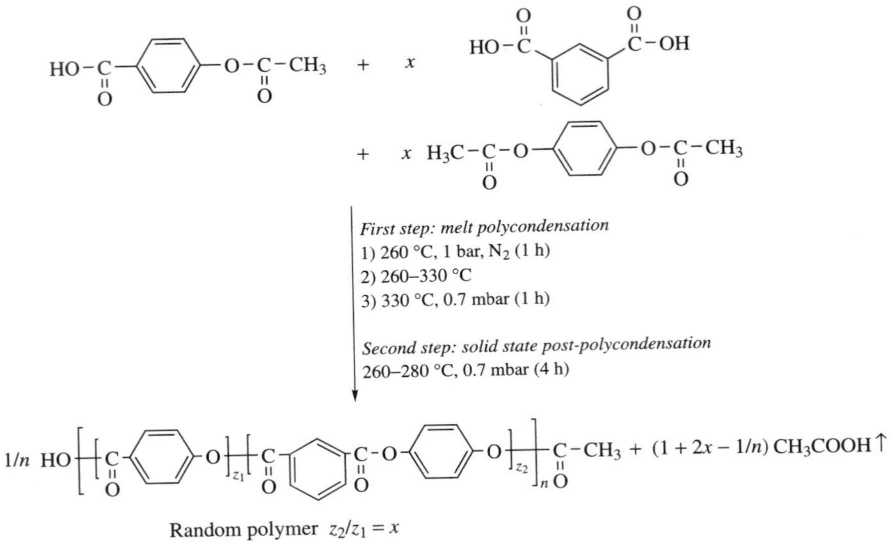

First step: melt polycondensation
1) 260 °C, 1 bar, N_2 (1 h)
2) 260–330 °C
3) 330 °C, 0.7 mbar (1 h)

Second step: solid state post-polycondensation
260–280 °C, 0.7 mbar (4 h)

$$1/n \; HO \left[\overset{O}{\underset{O}{C}} \!\!-\!\!\!\! \begin{array}{c} \end{array} \!\!\!\!-\!\!O \right]_{z_1} \!\!\! \left[C \begin{array}{c} \end{array} C\!-\!O \begin{array}{c} \end{array} O \right]_{z_2} \!\!\! C\!-\!CH_3 \right]_n + (1 + 2x - 1/n)\, CH_3COOH \!\uparrow$$

Random polymer $z_2/z_1 = x$

Scheme 2.65

added 45.0 g (0.25 mol) of *p*-acetoxybenzoic acid, 48.5 g (0.25 mol) of 4.4′-phenylenediacetate, and 41.5 g (0.25 mol) of isophthalic acid. The monomers are melted under nitrogen and heated to 260°C for 1 h; acetic acid distills out during this stage. The reaction mixture is then heated to 320–330°C, and vacuum is applied slowly (ca. 0.7 mbar) and held for one hour. The polymer product is dried and ground. The ground sample is then charged into a solid-sate polymerization tube equipped with a vacuum line. Vacuum (0.7 mbar) is applied and the tube is heated to 260–280°C and held for 4 h at this temperature.

Characterization[439]: Inherent viscosity before and after solid-sate polymerization is 0.46 and 3.20 dL/g, respectively (0.5 g/dL in pentafluorophenol at 25°C). DSC: $T_g = 135°C$, $T_m = 317°C$. A copolyester of similar composition[440] exhibited a liquid crystalline behavior with crystal–nematic and nematic–isotropic transition temperatures at 307 and 410°C, respectively (measured by DSC and hot-stage polarizing microscopy). The high-resolution solid-state ^{13}C NMR study of a copolyester with a composition corresponding to $z_2/z_1 = 1.35$ has been reported.[441]

2.4.5 Hyperbranched Polyesters

2.4.5.1 *Hyperbranched Aliphatic Polyester Based on 2,2-Bis(hydroxymethyl) Propionic Acid*

This aliphatic hyperbranched polyester is prepared by the bulk polycondensation of 2,2-bis(hydroxymethyl)propionic acid (bis-MPA) as AB_2 monomer and 1,1,1-tris(hydroxymethyl)propane (TMP) as B_3 core molecule, according to a procedure published by Malmström et al.[65] A typical example corresponding to the synthesis of the third-generation polyester (see Scheme 2.66) is detailed below. (From ref. 65. Copyright © 1995 American Chemical Society, by permission of the copyright owner.)

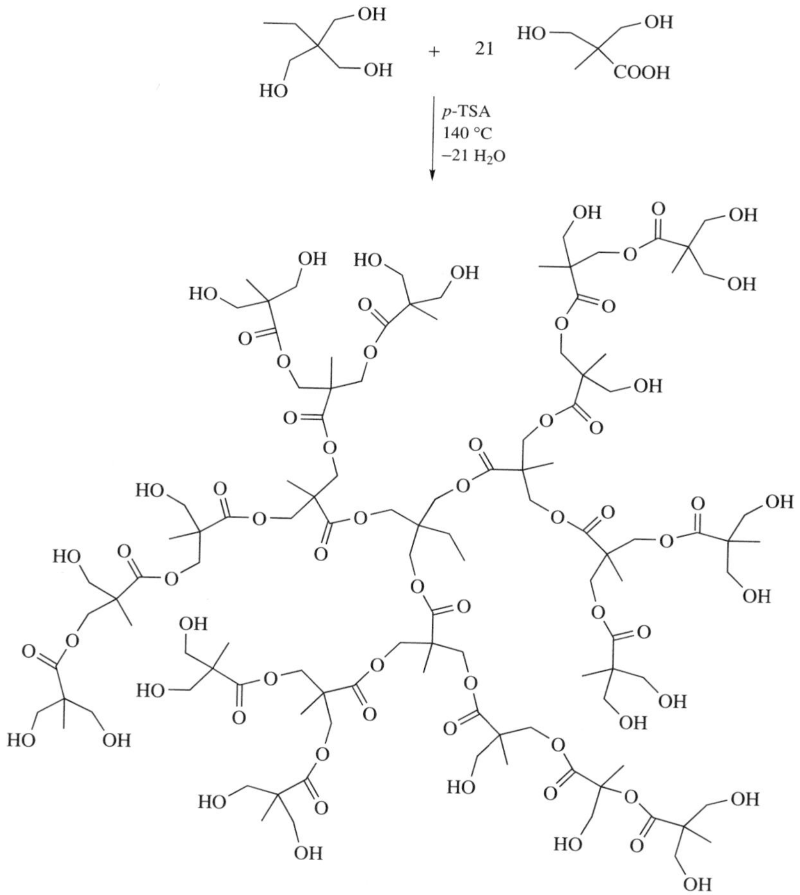

Scheme 2.66

Bis-MPA (6.71 g, 50.0 mmol), TMP (0.745 g, 5.55 mmol) (in stoichiometric correspondence to a perfect dendrimer of second generation), and p-toluene sulfonic acid monohydrate (p-TSA, 33.6 mg) are carefully mixed in a three-necked flask equipped with an argon inlet, a drying tube, and a stirrer. The flask is placed in an oil bath previously heated to 140°C. The mixture is left to react under a stream of argon, removing the water formed during the reaction. After 2 h, the argon stream is turned off and the flask sealed and connected to a vacuum line (12 mbar, cooling trap) for 1 h. The pressure is increased to atmospheric with argon, bis-MPA corresponding to the third generation (8.94 g, 66.7 mmol) and *p*-TSA (44.8 mg) are added, and the argon flow is started. After 2 h of reaction at normal pressure, vacuum is applied for 1 h before the reaction mixture is removed from the flask.

Fourier transform infrared (FTIR) spectroscopy (NaCl) shows no remaining carboxylic acid (1696 cm^{-1}, carbonyl) but only ester groups (1736 cm^{-1}, carbonyl): $\overline{M_n}$(SEC) = 6530; $\overline{M_n}$(^{1}H NMR) = 1640; theory for third generation: $\overline{M_n}$ = 2570.[65] Polyesters of higher generation were synthesized according to this pseudo-one-step procedure and were analyzed by SEC, VPO, and ^{1}H NMR.[65]

2.4.5.2 Hyperbranched Aromatic Polyesters

Wholly aromatic hyperbranched polyesters based on 3,5-dihydroxyisophthalic acid can be prepared either from bis(acetoxy)isophthalic acid or from its silylated derivative (Scheme 2.67).

2.4.5.2.1 Hyperbranched Aromatic Polyester Prepared from 3,5-Diacetoxybenzoic Acid

This procedure is reproduced from ref. 442. (Copyright © 1993 American Chemical Society, by permission of the copyright owner.)

Polymerization of 3,5-diacetoxybenzoic acid: 3,5-Diacetoxybenzoic acid (20 g, 0.084 mol) is placed in a dry flask under argon. The reaction vessel consists of a 100-mL round-bottom flask with a long neck, a mechanical stirrer, a nitrogen inlet with a long glass tube going to the bottom of the flask, and a ground joint for connection to a vacuum pump. The reaction vessel is also connected with a condenser and a receiver to distill acetic acid. *p*-Toluene sulfonic acid monohydrate (0.6–2 mol %), as catalyst, is added into the reaction mixture. The reaction flask is placed in a salt bath at 230°C. A slow inert gas (nitrogen) stream is applied during the first stage of the reaction until viscosity is built up. Then vacuum is applied (about 10^{-2} mbar) to drive the reaction to high conversion (0.5–3 h). At the end of the reaction, stirring with the mechanical glass stirrer is no longer possible due to the high viscosity of the polymer. At room temperature the polymer is very brittle and can be easily broken into small pieces. The polymer is dissolved in THF overnight, filtered, and precipitated into methanol to yield a white powder. The yield is 80–90%.

Characterization[442]: Intrinsic viscosity 0.095 dL/g (determined at 25°C in THF). Proton NMR (DMSO-d6): δ = 2.25 (s, 3H), four broad signals between 7.25 and 8.18 (3H), no OH protons. IR (KBr): ν = 3045, 2940, 1765 (COOR), 1600 cm^{-1}. DSC: The polymer is amorphous with T_g = 149°C. SEC: $\overline{M_w}$ = 62,000, $\overline{M_w}/\overline{M_n}$ = 12.0 (determined in THF from universal calibration or directly from low angle laser light scattering measurements).

Note: Kricheldorf et al.[268] synthesized the same polyester by a similar procedure using Ti(OPr)$_4$ as polymerization catalyst. The characteristics of the resulting polymer were inherent viscosity 0.22 dL/g (determined at 20°C in 4/1 CH$_2$Cl$_2$– trifluoroacetic acid at a concentration of 0.2 g/dL). DSC: T_g = 164°C. Degree of branching (DB) 0.48 (determined by ^{1}H NMR).

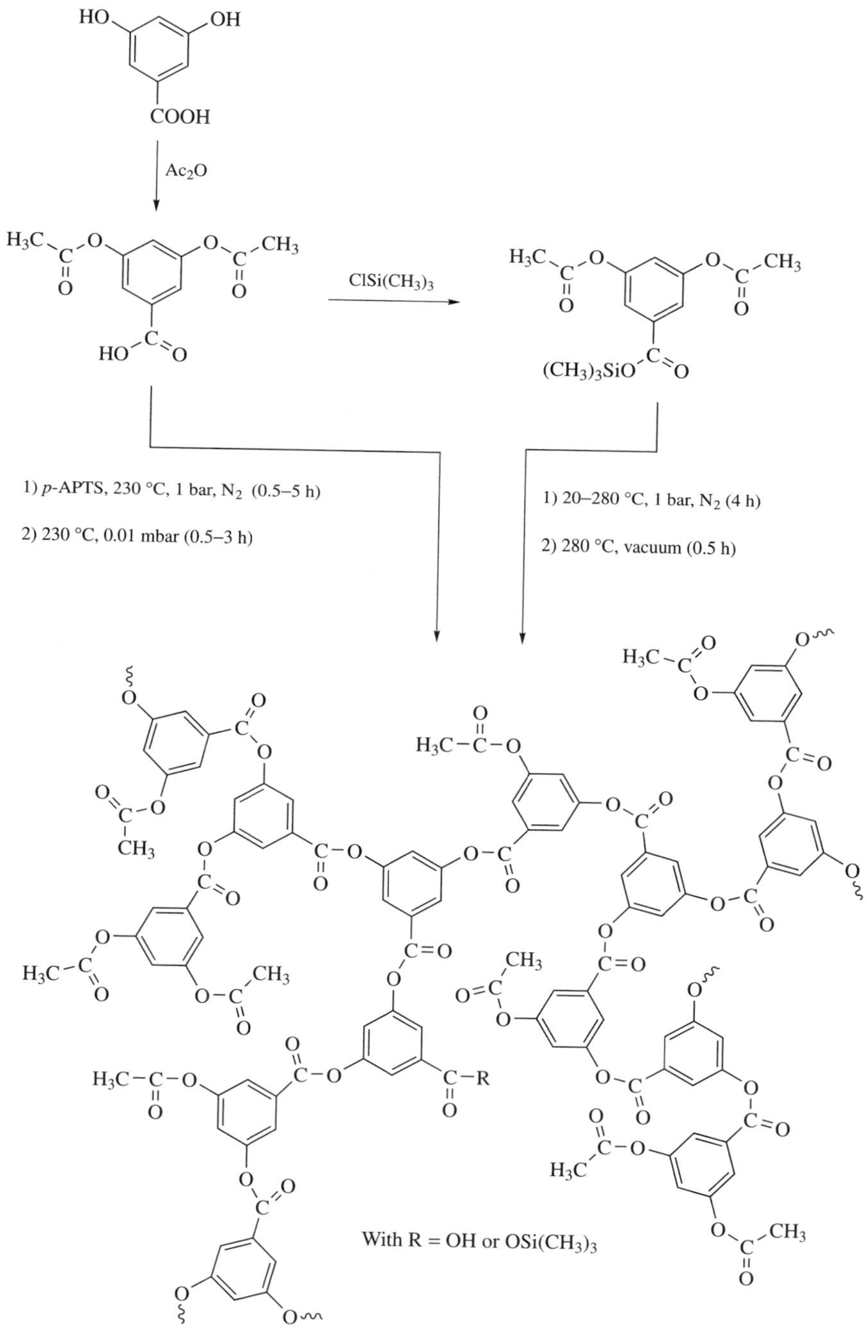

Scheme 2.67

2.4.5.2.2 Hyperbranched Aromatic Polyester Prepared from Trimethylsilyl 3,5-Diacetoxybenzoate

This procedure is reprinted from ref. 268. (Copyright © 1995 American Chemical Society, by permission of the copyright owner.)

Synthesis of trimethylsilyl 3,5-diacetoxybenzoate: 3,5-Diacetoxybenzoic acid (119.11 g, 0.5 mol) and chlorotrimethylsilane (65.18 g, 0.6 mol) are dissolved in dry toluene (1 L), and triethylamine (60.71 g, 0.6 mol) is added dropwise with stirring and heating. The reaction mixture is refluxed for 2 h, cooled with ice, and filtered under exclusion of moisture. The product is isolated from the filtrate by distillation in vacuum over a short-path apparatus. Yield 79%; $T_m = 40-42°C$. Proton NMR ($CDCl_3$/TMS): δ 0.42 (s, 9H), 2.31 (s, 6H), 7.10 (t, 1H), 7.70 (d, 2H).

Polycondensations of trimethylsilyl 3,5-diacetoxybenzoate: Trimethylsilyl 3,5-diacetoxybenzoate (15.52 g, 50 mmol) is weighed into a cylindrical reactor equipped with a glass stirrer and gas inlet and outlet tubes. The reaction vessel is placed into a metal bath preheated to 200°C. The temperature is raised in 20°C steps over a period of 1 h and finally maintained at 280°C for 3 h. Vacuum is then applied for an additional 0.5 h. Finally, the cold reaction product is powdered, dissolved in CH_2Cl_2–trifluoroacetic acid (volume ratio 4 : 1), and precipitated into cold methanol.

Characterization[268]: Inherent viscosity 0.36 dL/g (measured at 20°C in 4 : 1 CH_2Cl_2–trifluoroacetic acid at a concentration of 0.2 g/dL). DSC: $T_g = 162°C$. SEC: $\overline{M_w} = 27,000$, $\overline{M_w}/\overline{M_n} = 10$ (determined in THF, calibrated with commercial polystyrene standards). Degree of branching (DB) 0.49 (calculated from the ^1H NMR spectrum).

Note: This hyperbranched polyester was also prepared by the bulk polymerization of 3,5-bis(trimethylsilyloxy)benzoyl chloride.[202] DSC: $T_g = 190°C$. DB = 0.55–0.60 (^1H NMR measurements).

REFERENCES

1. M. Arroyo, *Plast. Eng.* (*New York*), **41**, 417 (1997).

2. F. Pilati, in *Comprehensive Polymer Science*, Vol. 5, G. Allen and J. C. Bevington (Eds.), Pergamon, Oxford, 1989, p. 275.

3. H. Köpnick, M. Schmidt, W. Brügging, J. Rüter, and W. Kaminsky, in *Ullmann's Encyclopedia of Industrial Chemistry*, Vol. A21, VCH, Weinheim, 1992, p. 227.

4. I. Goodman, in *Encyclopedia of Polymer Science and Engineering*, Vol. 12, 2nd ed., H. F. Mark, N. M. Bikales, C. G. Overberger, and G. Menges (Eds.), Wiley Interscience, New York, 1988, p. 1.

5. D. B. G. Jaquiss, W. F. H. Borman, and R. W. Campbell, in *Kirk-Othmer Encyclopedia of Chemical Technology*, Vol. 18, H. F. Mark, D. F. Othmer, C. G. Overberger, and G. T. Seaborg (Eds.), Wiley, New York, 1982, p. 549.

6. B. D. Dean, M. Matzner, and J. M. Tibbitt, in *Comprehensive Polymer Science*, Vol. 5, G. Allen and J. C. Bevington (Eds.), Pergamon, Oxford, 1989, p. 317.

7. A. Fradet and P. Arlaud, in *Comprehensive Polymer Science*, Vol. 5, G. Allen and J. C. Bevington (Eds.), Pergamon, Oxford, 1989, p. 331.

8. J. Selley, in *Encyclopedia of Polymer Science and Engineering*, Vol. 12, 2nd ed., H. F. Mark, N. M. Bikales, C. G. Overberger, and G. Menges (Eds.), Wiley Interscience, New York, 1988, p. 256.

9. J. Makhlouf, in *Kirk-Othmer Encyclopedia of Chemical Technology*, Vol. 18, H. F. Mark, D. F. Othmer, C. G. Overberger, and G. T. Seaborg (Eds.), Wiley, New York, 1982, p. 575.

10. W. H. Carothers, *Chem. Rev.*, **8**, 353 (1931).

11. W. H. Carothers and J. W. Hill, *J. Am. Chem. Soc.*, **54**, 1577 (1932).

12. W. H. Carothers, *Trans. Faraday Soc.*, **32**, 43 (1936).

13. J. R. Whinfield and J. T. Dickson, British Patent 578,079 (to ICI), June 16, 1946.

14. J. R. Whinfield, *Nature (London)*, **158**, 930 (1946).

15. C. Ellis, U.S. Patent 2,225,313 (to Ellis-Foster Co), August 6, 1937.

16. G. Holden, N. R. Legge, R. Quirk, and H. E. Schroeder, *Thermoplastic Elastomers*, 2nd ed., Carl Hanser Verlag, Munich, 1996.

17. G. W. Calundann, U.S. Patent 4,161,470 (to Celanese Corp.), October 20, 1977.

18. N. Coleman, "PTA, PET Supply/Demand," paper presented at Polyester '99, 4th Annual World Congress, Zurich, Switzerland, 1999.

19. F. Charaf, *Int. Fiber J.*, **15**(5), 81. (2000).

20. D. Osman, *Int. Fiber J.*, **15**(4), 38 (2000).

21. J. Schumacher, M. Jäckel, A. Leder, M. Petesch, and Y. Sakuma, *Polyethylene Terephthalate Solid-State Resins*, CEH Report, SRI Consulting, Menlo Park, CA, July 1997.

22. R. Brown, *Chem. Mark. Rep.*, **257**(24), 27 (2000).

23. R. Brown, *Chem. Mark. Rep.*, **257**(16), 3 (2000).

24. A. Tullo, *Chem. Mark. Rep.*, **256**(5), 13 (1999).

25. H. H. Chuah, "Corterra Poly(trimethylene) terephthalate — New Polymeric Fiber for Carpets," Paper presented at Tifcon '96, Blackpool, U.K. (1966). http://www.shellchemicals.com.

26. E. I. Du Pont de Nemours, "DuPont Sorona™ Announces New Licensee," July 21, 2000. http://www.dupont.com.

27. C. Hwo, T. Forschner, R. Lowtan, D. Gwyn, and B. Cristea, *J. Plast. Film Sheeting*, **15**, 219 (1999).

28. K. Greenberg, *Chem. Mark. Rep.*, **256**(23), 3 (1999).

29. A. Tullo, *Chem. Mark. Rep.*, **256**(1), 1 (1999).

30. L. C. Sawyer, H. C. Linstid, and M. Römer, *Plast. Eng.*, **54**(12), 37 (1998).

31. S. Kenig, A. Ophir, F. Wiener, and M. Omer, *Proc. 58th Annu. Tech. Conf. Soc. Plast. Eng.*, **1**, 438 (2000).

32. B. Sparenberg, *Kunststoffe*, **89**, 136 (1999).

33. B. Sparenberg, *Kunststoffe*, **88**, 1734 (1998).

34. R. Westervelt, *Chem. Week*, **162**(24), 37 (2000).

35. R. Westervelt, *Chem. Week*, **162**(25), 14 (2000).

36. R. W. M. van Berkel, R. J. M. Borggreve, C. L. Van der Sluijs, and G. H. Werumeus Buning, *Plast. Eng.* (*New York*), **41**, 397 (1997).

37. M. S. Reisch, *Chem. Eng. News*, **74**, 10 (1996).

38. J. C. Middleton and A. J. Tipton, *Med. Plast. Biomater. Mag.*, **1998**, 30 (1998).

39. A. Steinbüchel and B. Füchtenbusch, *TIBTECH*, **16**, 419 (1998).

40. R. H. Marchessault, *Trends Polym. Sci.*, **4**, 163 (1996).

41. Y. Ikada and H. Tsuji, *Macromol. Rapid Commun.*, **21**, 117 (2000).

42. R. Arshady, *J. Control. Release*, **17**, 1 (1991).

43. G. Bohlmann and Y. Yoshida, *Biodegradable Polymers*, CEH Report, SRI Consulting, Menlo Park, CA, February 2000.

44. O. Baker, *Science News*, **156**, 246 (1999).

45. R. Rudie, *Int. Fiber J.*, **15**(1), 8 (2000).

46. A. Wood, *Chem. Week*, **160**(4), 51 (1998).

47. H. Chinn, S. Cometta, and K. Sakota, *Polyester Polyols*, CEH Report, SRI Consulting, Menlo Park, CA, June 1999.

48. K.-L. Ring, A. DeBoo, and M. Yamazaki, *Unsaturated Polyester Resins*, CEH Report, SRI Consulting, Menlo Park, CA, April 1999.

49. A. Tullo, *Chem. Mark. Rep.*, **255**(18), 3 (1999).

50. H. J. Lanson, in *Encyclopedia of Polymer Science and Engineering*, Vol. 1, 2nd ed., H. F. Mark, N. M. Bikales, C. G. Overberger, and G. Menges (Eds.), Wiley Interscience, New York, 1985, p. 644.

51. C. Boswell, *Chem. Mark. Rep.*, **254**(15), 22 (1998).

52. W. Amass, A. Amass, and B. Tighe, *Polym. Int.*, **47**, 89 (1998).

53. R. Mani, M. Bhattacharya, and J. Tang, *J. Polym. Sci., Part A: Polym. Chem.*, **37**, 1693 (1999).

54. T. G. Gopakumar, S. Ponrathnam, A. Lele, C. R. Rajan, and A. Fradet, *Polymer*, **40**, 357 (1998).

55. V. N. Ignatov, C. Carraro, V. Tartari, R. Pippa, M. Scapin, F. Pilati, C. Berti, M. Toselli, and M. Fiorini, *Polymer*, **38**, 195 (1997).

56. W. M. Stevels, A. Bernard, P. Van De Witte, P. J. Dijkstra, and J. Feijen, *J. Appl. Polym. Sci.*, **62**, 1295 (1996).

57. M. L. Todt and J. W. Carbone, U.S. Patent 5,663,282 (to General Electric Company), September 2, 1997.

58. D. J. Brunelle, U.S. Patent 5,498,651 (to General Electric Company), March 12, 1996.

59. Cyclics Corporation, *CBT XBO-1 Datasheet*, 2001. www.cyclics.com.

60. B. Voit, *J. Polym. Sci. Part A: Polym. Chem. Ed.*, **38**, 2505 (2000).

61. C. J. Hawker, *Adv. Polym. Sci.*, **147**, 113 (1999).

62. A. Hult, M. Johansson, and E. Malmström, *Adv. Polym. Sci.*, **143**, 1 (1999).

63. Y. H. Kim, *J. Polym. Sci. Part A: Polym. Chem.*, **36**, 1685 (1998).

64. E. Malmström, M. Johansson, and A. Hult, *Polym. News*, **22**, 128 (1997).

65. E. Malmström, M. Johansson, and A. Hult, *Macromolecules*, **28**, 1698 (1995).

66. E. Malmström, M. Johansson, and A. Hult, *Macromolecules*, **29**, 1222 (1996).

67. E. J. Tijsma, L. van der Does, and A. Bantjes, *Macromolecules*, **27**, 179 (1994).

68. P. J. M. Serrano, E. Thüss, and R. J. Gaymans, *Polymer*, **38**, 3893 (1997).

69. R. M. Versteegen, R. P. Sijbesma, and E. W. Meijer, *Angew. Chem.*, **111**, 3095 (1999).

70. R. Hill and E. E. Walker, *J. Polym. Sci.*, **3**, 609 (1948).

71. J. Watanabe, M. Hayashi, A. Morita, and T. Niiori, *Mol. Cryst. Liq. Cryst. Sci. Technol., Sect. A*, **254**, 221 (1994).

72. V. Percec and D. Tomazos, in *Comprehensive Polymer Science*, First Supplement Vol., G. Allen and J. C. Bevington (Eds.), Pergamon, Oxford, 1992, p. 299.

73. A. Mühlebach, J. Lyerla, and J. Economy, *Macromolecules*, **22**, 3746 (1989).

74. G. Schwarz and H. R. Kricheldorf, *Macromolecules*, **24**, 2829 (1991).

75. T. G. Fox, *Bull. Am. Phys. Soc.*, **2**, 123 (1956).

76. D. Ma, G. Zhang, Z. Huang, and X. Luo, *J. Polym. Sci. Part A: Polym. Chem.*, **36**, 2961 (1998).

77. T. Nagasawa, Y. Murata, K. Tadano, R. Kawai, K. Ikehara, and S. Yano, *J. Mater. Sci.*, **35**, 3077 (2000).

78. S.-H. Hwang, *J. Appl. Polym. Sci.*, **76**, 1947 (2000).

79. G. Montaudo and C. Puglisi, in *Comprehensive Polymer Science*, First Supplement Vol., G. Allen and J. C. Bevington (Eds.), Pergamon, Oxford, 1992, p. 227.

80. M. E. Bednas, M. Day, K. Ho, R. Sander, and D. M. Wiles, *J. Appl. Polym. Sci.*, **26**, 277 (1981).

81. T. Ohtani, T. Kimura, and S. Tsuge, *Anal. Sci.*, **2**, 179 (1986).

82. N. Grassie and G. Scott, *Polymer Degradation and Stabilization*, Cambridge University Press, Cambridge, 1985.

83. K. C. Khemani, *Polym. Degrad. Stab.*, **67**, 91 (2000).

84. R. M. Lum, *J. Polym. Sci. Polym. Chem. Ed.*, **17**, 203 (1979).

85. H. Sato, T. Kikuchi, N. Koide, and K. Furuya, *J. Anal. Appl. Pyrol.*, **37**, 173 (1996).

86. F. D. Trischler and J. Hollander, *J. Polym. Sci., Part A-1*, **7**, 971 (1969).

87. H. A. Pohl, *J. Am. Chem. Soc.*, **73**, 5660 (1951).

88. V. V. Korshak and S. V. Vinogradova, *Polyesters*, Pergamon, Oxford, 1965, p. 25.

89. I. Lüderwald, *J. Macromol. Sci.-Chem.*, **A13**, 869 (1979).

90. I. Luderwald and M. Urrutia, *Makromol. Chem.*, **177**, 2093 (1976).

91. M. Bounekhel and I. C. McNeill, *Polym. Degrad. Stab.*, **51**, 35 (1996).

92. P. Maravigna and P. Montaudo, in *Comprehensive Polymer Science*, Vol. 5, G. Allen and J. C. Bevington (Eds.), Pergamon, Oxford, 1989, p. 63.

93. L. H. Buxbaum, *Angew. Chem. Int. Ed.*, **7**, 182 (1968).

94. K. Yoda, A. Tsuboi, M. Wada, and R. Yamadera, *J. Appl. Polym. Sci.*, **14**, 2357 (1970).

95. M. Vert, S. Li, H. Garreau, J. Mauduit, M. Boustta, G. Schwach, R. Engel, and J. Coudane, *Angew. Makromol. Chem.*, **247**, 239 (1997).

96. M. Mochizuki and M. Hirami, *Polym. Adv. Technol.*, **8**, 203 (1997).

97. K. Sudesh and Y. Doi, *Polym. Adv. Technol.*, **11**, 865 (2000).

98. S. J. Huang, *Polym. Mater. Sci. Eng.*, **80**, 286 (1999).

99. S. R. Porter and L. H. Wang, *Polym. Rev.*, **33**, 2019 (1992).

100. J. Devaux, P. Godard, and J. P. Mercier, *J. Polym. Sci. Polym. Phys. Ed.*, **20**, 1901 (1982).

101. M. J. Fernandez-Berridi, J. J. Iruin, and I. Maiza, *Polymer*, **36**, 1357 (1995).

102. H. G. Ramjit and R. D. Sedgwick, *J. Macromol. Sci. Chem.*, **A10**, 815 (1976).

103. A. Mühlebach, J. Economy, R. D. Johnson, T. Karis, and J. Lyerla, *Macromolecules*, **23**, 1803 (1990).

104. G. B. Kharas, F. Sanchez-Riera, and D. K. Severson, in *Plastics from Microbes* D. P. Mobley (Ed.), Carl Hanser Verlag, Munich, 1994, p. 94.

105. D. Byrom, in *Plastics from Microbes*, D. P. Mobley (Ed.), Carl Hanser Verlag, Munich, 1994, p. 5.

106. T. M. Ovitt and G. W. Coates, *J. Am. Chem. Soc.*, **121**, 4072 (1999).

107. F. Chabot, M. Vert, S. Chapelle, and P. Granger, *Polymer*, **24**, 53 (1983).

108. P. J. Hocking and R. H. Marchessault, *Polym. Bull.*, **30**, 163 (1993).

109. S. Bloembergen, D. A. Holden, T. L. Bluhm, G. K. Hamer, and R. H. Marchessault, *Macromolecules*, **22**, 1656 (1989).

110. T. Fujimaki, *Polym. Degrad. Stab.*, **59**, 209 (1998).

111. E. Takiyama, T. Hokari, S. Seki, T. Fujimaki, and Y. Hatano, U.S. Patent 5,310,782 (to Showa Highpolymer), December 1, 1993.

112. M. S. Reeve, S. P. McCarthy, M. J. Downey, and R. A. Gross, *Macromolecules*, **27**, 825 (1994).

113. T. Tokiwa, T. Ando, T. Suzuki, and T. Takeda, *Polym. Mat. Sci. Eng.*, **62**, 988 (1990).

114. A. Torres, S. M. Li, S. Roussos, and M. Vert, *J. Environ. Polym. Degrad.*, **4**, 213 (1996).

115. M. G. Brereton, C. R. Davies, R. Jakeways, T. Smith, and I. M. Ward, *Polymer*, **19**, 17 (1978).

116. I. M. Ward and M. A. Wilding, *J. Polym. Phys.: Polym. Phys. Ed.*, **14**, 263 (1976).

117. R. Jakeways, I. M. Ward, M. A. Wilding, I. H. Hall, I. J. Desborough, and M. G. Pass, *J. Polym. Phys.: Polym. Phys. Ed.*, **13**, 799 (1975).

118. C. J. M. van der Heuvel and E. A. Klop, *Polymer*, **41**, 4249 (2000).

119. K. Nakamae, T. Nishino, K. Tada, T. Kanamoto, and M. Ito, *Polymer*, **34**, 3322 (1993).

120. J. Y. Jadhav and S. W. Kantor, in *Encyclopedia of Polymer Science and Engineering*, Vol. 12, 2nd ed., H. F. Mark, N. M. Bikales, C. G. Overberger, and G. Menges (Eds.), Wiley Interscience, New York, 1988, p. 217.

121. A. M. Ghanem and R. S. Porter, *J. Polym. Sci.: Part B: Polym. Phys.*, **27**, 2587 (1989).

122. S. Buchner, D. Wiswe, and H. G. Zachmann, *Polymer*, **30**, 480 (1989).

123. Y. Uelcer and M. Cakmak, *Polymer*, **35**, 5651 (1994).

124. G. W. Davis and J. R. Talbot, in *Encyclopedia of Polymer Science and Engineering*, Vol. 12, 2nd ed., H. F. Mark, N. M. Bikales, C. G. Overberger, and G. Menges (Eds.), Wiley Interscience, New York, 1988, p. 118.

125. A. Suzuki, T. Kuwabara, and T. Kunugi, *Polymer*, **39**, 4235 (1998).

126. G. Wu, Q. Li, and J. A. Cuculo, *Polymer*, **41**, 8139 (2000).

127. E. Werner, S. Janocha, M. J. Hopper, and K. J. Mackenzie, in *Encyclopedia of Polymer Science and Engineering*, Vol. 12, 2nd ed., H. F. Mark, N. M. Bikales, C. G. Overberger, and G. Menges (Eds.), Wiley Interscience, New York, 1988, p. 133.

128. M. Cakmak, Y. D. Wang, and M. Simhambhatla, *Polym. Eng. Sci.*, **30**, 721 (1990).

129. J. C. Kim, M. Cakmak, and X. Zhou, *Polymer*, **39**, 4225 (1998).

130. R. A. Bubeck and M. A. Barger, *Int. Polym. Process.*, **15**, 337 (2000).

131. M. L. Wallach, *Polym. Prepr. (Am. Chem. Soc., Div. Polym. Chem.)*, **6**(2), 860 (1965).

132. W. R. Moore and D. Sanderson, *Polymer*, **9**, 153 (1968).

133. W. L. Hergenrother and C. J. Nelson, *J. Polym. Sci., Polym. Chem. Ed.*, **12**, 2905 (1974).

134. K. Kamide, Y. Miyazaki, and H. Kobayashi, *Polym. J.*, **9**, 317 (1977).

135. A. Horbach, R. Binsack, and H. Mueller, *Angew. Makromol. Chem.*, **98**, 35 (1981).

136. M. M. Sang, N. N. Jin, and E. F. Jiang, *J. Liq. Chromatogr.*, **5**, 1665 (1982).

137. H. Iida, H. Ikeuchi, S. Nakazawa, H. Suganuma, H. Komatsu, and M. Tanaka, *Sen'i Gakkaishi*, **43**, 251 (1987).

138. R. Po, E. Tampellini, E. Occhiello, and F. Garbassi, *Polym. Bull.*, **30**, 551 (1993).

139. C. C. McDowell, J. M. Partin, B. D. Freeman, and G. W. G. W. McNeely, *J. Membr. Sci.*, **163**, 39 (1999).

140. Y. Sakaguchi, *Polymer*, **38**, 2201 (1997).

141. Y. Zhang and L. Gu, *Eur. Polym. J.*, **36**, 759 (2000).

142. Y.-M. Sun, J.-Y. Shieh, and C.-S. Wang, *Eur. Polym. J.*, **33**, 317 (1997).

143. S. W. Lee, M. M. Ree, C. E. Park, Y. K. Jung, C.-S. Park, Y. S. Jin, and D. C. Bae, *Polymer*, **40**, 7137 (1999).

144. B. Li, J. Yu, S. Lee, and M. Ree, *Eur. Polym. Sci.*, **35**, 1607 (1999).

145. B. Li, J. Yu, S. Lee, and M. Ree, *Polymer*, **40**, 5371 (1999).

146. J. I. Suenaga and T. Okada, *Mol. Cryst. Liq. Cryst.*, **169**, 97 (1989).

147. T.-H. Shinn, C.-C. Lin, and D. C. Lin, *Polymer*, **36**, 283 (1995).

148. Y. J. Kim and O. O. Park, *J. Environ. Polym. Degrad.*, **7**, 53 (1999).

149. J.-C. Ho, Y.-S. Lin, and K.-H. Wei, *Polymer*, **40**, 3843 (1999).

150. M. Okamoto and T. Kotakaa, *Polymer*, **38**, 1357 (1997).

151. V. N. Ignatov, C. Carraro, V. Tartari, R. Pippa, M. Scapin, F. Pilati, C. Bertid, M. Tosellid, and M. Fiorinie, *Polymer*, **38**, 195 (1997).

152. S. H. Jang and B. S. Kim, *Polym. Eng. Sci.*, **35**, 538 (1995).

153. E. Andresen and H. G. Zachmann, *Colloid Polym. Sci.*, **272**, 1352 (1994).

154. S. R. Maerov, *J. Polym. Sci.*, **3**, 487 (1965).

155. D. Bellus, Z. Manasek, P. Hrdlovic, and P. Slama, *J. Polym. Sci., Part C*, **16**, 267 (1967).

156. B. D. Dean, U.S. Patent 5,115,046 (to Amoco Corp.), May 19, 1992.

157. J. Runt, D. M. Miley, K. P. Gallagher, X. Zhang, C. A. Barron, and S. K. Kumar, *Polym. Adv. Technol.*, **5**, 333 (1994).

158. S. Bastida, J. I. Eguiazábal, and J. Nazábal, *Polymer*, **37**, 2317 (1996).

159. T. O. Ahn, S. Lee, H. M. Jeong, and S. W. Lee, *Polymer*, **37**, 3559 (1996).

160. S. Bastida, J. I. Eguiazábal, and J. Nazábal, *Polymer*, **42**, 1157 (2001).

161. S. I. Kwolek, P. W. Morgan, and J. R. Schaefgen, in *Encyclopedia of Polymer Science and Engineering*, Vol. 9, 2nd ed., H. F. Mark, N. M. Bikales, C. G. Overberger, and G. Menges (Eds.), Wiley Interscience, New York, 1987, p. 1.

162. J. Economy, R. S. Storm, V. I. Matkovitch, S. G. Cottis, and B. E. Novak, *J. Polym. Sci.: Polym. Chem. Ed.*, **14**, 2207 (1976).

163. J.-I. Jin and C.-S. Kang, *Prog. Polym. Sci.*, **22**, 937 (1997).

164. J. Economy and K. Goranov, *Adv. Polym. Sci.*, **117**, 221 (1994).

165. H. Han and P. K. Bhowmik, *Prog. Polym. Sci.*, **22**, 1431 (1997).

166. J.-I. Jin, *Plast. Eng.*, **40**, 669 (1997).

167. W. J. Jackson, *Br. Polym. J.*, **12**, 154 (1980).

168. W. J. Jackson and H. F. Kuhfuss, *J. Polym. Sci., Polym. Chem. Ed.*, **14**, 2043 (1976).

169. H. F. Kuhfuss and W. J. Jackson, U.S. Patent 3,778,410 (to Eastman Kodak Co), December 11, 1973.

170. H. F. Kuhfuss and W. J. Jackson, U.S. Patent 3,804,805 (to Eastman Kodak Co), April 16, 1974.

171. P. Driscoll, S. Hayase, and T. Masuda, *Polym. Eng. Sci.*, **34**, 519 (1994).

172. K. Nakamae, T. Nishio, and T. Kuroki, *Polymer*, **36**, 2681 (1995).

173. M. Tormes, M. E. Munoz, J. J. Pena, and A. Santamaria, *J. Polym. Sci., Part B: Polym. Phys.*, **36**, 253 (1998).

174. W. R. Krigbaum, H. Hakemi, and R. Kotek, *Macromolecules*, **18**, 965 (1985).

175. S. G. Cottis, J. Economy, and B. E. Nowak, U.S. Patent 3,637,595, January 25, 1972.

176. S. G. Cottis, J. Economy, and L. C. Wohrer, U.S. Patent 3,975,487 (to Carborundum Co.), August 17, 1976.

177. D. Acierno, F. P. La Mantia, G. Polizzotti, A. Ciferri, and B. Valenti, *Macromolecules*, **15**, 1455 (1982).

178. A. J. Cesaroni and E. L. Fletcher, U.S. Patent 6,159,412 (to Du Pont Canada), December 12, 2000.

179. R. D. Jester and J. A. Penoyer, U.S. Patent 5,789,042 (to Hoechst Celanese), August 4, 1998.

180. A. C. Harvey, R. W. Lusignea, and J. L. Racich, U.S. Patent 4,966,807 (to Foster Miller), October 20, 1990.

181. A. Zachariades and J. Economy, *Polym. Eng. Sci.*, **23**, 266 (1983).

182. T. Urasaki, Y. Hirabayashi, M. Ogasawara and H. Inata, U.S. Patent 4,333,907 (to Teijin Ltd), June 24, 1980.

183. K. F. Wissbrun and A. C. Griffin, *J. Polym. Sci.: Polym. Phys. Ed.*, **20**, 1835 (1982).

184. K. F. Wissbrun, *Br. Polym. J.*, **12**, 163 (1980).

185. S. I. Kwolek, P. W. Morgan, and J. R. Schaefgen, in *Encyclopedia of Polymer Science and Engineering*, Vol. 9, 2nd ed., H. F. Mark, N. M. Bikales, C. G. Overberger, and G. Menges (Eds.), Wiley Interscience, New York, 1987, p. 31.

186. T. N. Nguyen, K. Geiger, and T. Walther, *Polym. Eng. Sci.*, **40**, 1643 (2000).

187. D. Beery, S. Kenigs, and A. Siegmann, *Polym. Eng. Sci.*, **31**, 451 (1991).

188. L. Sawyer and M. Jaffe, *J. Mater. Sci.*, **21**, 1897 (1986).

189. R. K. Adams, G. K. Hoeschele, and W. K. Witsiepe, in *Thermoplastic Elastomers*, G. Holden, N. R. Legge, R. P. Quirk, and H. E. Schroeder (Eds.), Hanser Publishers, Munich, 1996, p. 191.

190. H. Schroeder and R. J. Cella, in *Encyclopedia of Polymer Science and Engineering*, Vol. 12, 2nd ed., H. F. Mark, N. M. Bikales, C. G. Overberger, and G. Menges (Eds.), Wiley Interscience, New York, 1988, p. 75.

191. C. J. Hawker, R. Lee, and J. M. J. Fréchet, *J. Am. Chem. Soc.*, **113**, 4583 (1991).

192. D. Hölter, A. Burgath, and H. Frey, *Acta Polym.*, **48**, 30 (1997).

193. C. Lach and H. Frey, *Macromolecules*, **31**, 2381 (1998).

194. H. Frey and D. Hölter, *Acta Polym.*, **50**, 67 (1999).

195. D. Schmalljohan, J. G. Baratt, H. Komber, and B. I. Voit, *Macromolecules*, **33**, 6284 (2000).

196. P. J. Flory, *J. Am. Chem. Soc.*, **75**, 2718 (1952).

197. M. Johansson, E. Malmström, and A. Hult, *J. Polym. Sci. Part A: Polym. Chem.*, **31**, 619 (1993).

198. S. R. Turner, F. Walter, B. I. Voit, and T. H. Mourey, *Macromolecules*, **27**, 1611 (1994).

199. S. R. Turner, B. I. Voit, and T. H. Mourey, *Macromolecules*, **26**, 4617 (1993).

200. T. H. Mourey, S. R. Turner, M. Rubinstein, J. M. J. Fréchet, C. J. Hawker, and K. L. Wooley, *Macromolecules*, **25**, 2401 (1992).

201. B. I. Voit, *Acta Polymer.*, **46**, 87 (1995).

202. C. J. Hawker, R. Lee, and J. M. J. Fréchet, *J. Am. Chem. Soc.*, **113**, 4583 (1991).

203. K. L. Wooley, C. J. Hawker, R. Lee, and J. M. J. Fréchet, *Polym. J.*, **26**, 187 (1994).

204. K. L. Wooley, C. J. Hawker, J. M. Pochan, and J. M. J. Fréchet, *Macromolecules*, **26**, 1514 (1993).

205. Y. H. Kim and O. W. Webster, *J. Am. Chem. Soc.*, **112**, 4592 (1990).

206. C. K. Ingold, *Structure and Mechanisms in Organic Chemistry*, 2nd ed., Cornell University Press, Ithaca, NY, 1969, p. 1129.

207. S. V. Vinogradova, *Polym. Sci. USSR*, **19**, 769 (1977).

208. H. R. Kricheldorf and Z. Denchev, in *Transreactions in Condensation Polymers*, S. Fakirov (Ed.), Wiley-VCH Verlag GmbH, Weinheim, Germany, 1999, p. 1.

209. J. Devaux, in *Transreactions in Condensation Polymers*, S. Fakirov (Ed.), Wiley-VCH Verlag GmbH, Weinheim, Germany, 1999, p. 125.

210. J. Economy, L. A. Schneggenburger, and D. Frich, in *Transreactions in Condensation Polymers*, S. Fakirov (Ed.), Wiley-VCH Verlag GmbH, Weinheim, Germany, 1999, p. 195.

211. G. Montaudo, C. Puglisi, and F. Samperi, in *Transreactions in Condensation Polymers*, S. Fakirov (Ed.), Wiley-VCH Verlag GmbH, Weinheim, Germany, 1999, p. 159.

212. F. R. Jones, L. E. Scales, and J. A. Semlyen, *Polymer*, **15**, 738 (1974).

213. R. E. Heath, B. R. Wood, and J. A. Semlyen, *Polymer*, **41**, 1487 (1999).

214. J. A. Semlyen, J. J. L. Bryant, S. C. Hamilton, and B. R. Wood, *Polym. Prepr. (Am. Chem. Soc., Div. Polym. Chem.)*, **38**(2), 189 (1997).

215. U. W. Suter, in *Comprehensive Polymer Science*, Vol. 5, G. Allen and J. C. Bevington (Eds.), Pergamon, Oxford, 1989, p. 91.

216. L. Jong and J. C. Saam, *ACS Symp. Ser.*, **624**, 332 (1996).

217. A. Fradet and E. Maréchal, *Adv. Polym. Sci.*, **43**, 51 (1982).

218. D. Beigzadeh, S. Sajjadi, and F. A. Taromi, *J. Polym. Sci., Part A: Polym. Chem.*, **33**, 1505 (1995).

219. Y.-R. Fang, C.-G. Lai, J.-L. Lu, and M.-K. Chen, *Sci. Sin.*, **18**, 72 (1975).

220. C. T. Kuo and S. Chen, *J. Polym. Sci., Part A: Polym. Chem.*, **27**, 2793 (1989).

221. E. Makay-Bodi and I. Vancso-Szmercsanyi, *Eur. Polym. J.*, **5**, 145 (1969).

222. S. Chen and K. C. Wu, *J. Polym. Sci., Polym. Chem. Ed.*, **20**, 1819 (1982).

223. S. A. Chen and J. C. Hsiao, *J. Polym. Sci., Polym. Chem. Ed.*, **19**, 3123 (1981).

224. I. Vancso-Szmercsanyi, E. Makay-Bodi, E. Szabo-Rethy, and P. Hirschberg, *J. Polym. Sci., Part A-1*, **8**, 2861 (1970).

225. P. J. Flory, *J. Am. Chem. Soc.*, **62**, 2261 (1940).

226. P. J. Flory, *J. Am. Chem. Soc.*, **59**, 466 (1937).

227. P. J. Flory, *J. Am. Chem. Soc.*, **61**, 3334 (1939).

228. W. L. Chang and T. Karalis, *J. Polym. Sci., Part A: Polym. Chem.*, **31**, 493 (1993).

229. P. Laporte, A. Fradet, and E. Maréchal, *J. Macromol. Sci. Chem.*, **A24**, 1269 (1987).

230. M. Gordon and C. G. Leonis, *J. Chem. Soc., Faraday Trans. 1*, **71**(2), 161 (1975).

231. M. Gordon and G. R. Scantlebury, *J. Chem. Soc. B*, **1**, 1 (1967).

232. A. Fradet and E. Maréchal, *J. Polym. Sci., Polym. Chem. Ed.*, **19**, 2905 (1981).

233. S. D. Hamann, D. H. Solomon, and J. D. Swift, *J. Macromol. Sci., Chem.*, **2**, 153 (1968).

234. D. E. V. Sickle and R. I. Garrity, *Polym. Mater. Sci. Eng.*, **70**, 163 (1993).

235. I. Vancso-Szmercsanyi and E. Makay-Bodi, *J. Polym. Sci., Part C*, **16**, 3709 (1968).

236. M. Davies, *Research (London)*, **2**, 544 (1949).

237. R. Bacaloglu, M. Maties, C. Csunderlik, L. Cotarca, A. Moraru, J. Gros, and N. Marcu, *Angew. Makromol. Chem.*, **164**, 1 (1988).

238. A. C. Tang and K. S. Yao, *J. Polym. Sci.*, **35**, 219 (1959).

239. A. Fradet and E. Maréchal, *J. Macromol. Sci., Chem.*, **A17**, 859 (1982).

240. C. C. Lin and P. C. Yu, *J. Polym. Sci., Polym. Chem. Ed.*, **16**, 1005 (1978).

241. C. C. Lin and K. H. Hsieh, *J. Appl. Polym. Sci.*, **21**, 2711 (1977).

242. C. C. Lin and P. C. Yu, *J. Appl. Polym. Sci.*, **22**, 1797 (1978).

243. A. J. Amass, *Polymer*, **20**, 515 (1979).

244. E. Paatero, K. Naerhi, T. Salmi, M. Still, P. Nyholm, and K. Immonen, *Chem. Eng. Sci.*, **49**, 3601 (1994).

245. K. Uberreiter and W. Hager, *Makromol. Chem.*, **189**, 1697 (1979).

246. R. Bacaloglu, M. Maties, L. Cotarca, I. Gros, A. Moraru, and A. Tarnaveanu, *Angew. Makromol. Chem.*, **165**, 9 (1989).

247. R. Bacaloglu, M. Fisch, and K. Biesiada, *Polym. Eng. Sci.*, **38**, 1014 (1998).

248. I. S. Dairanieh, F. Rasoul, and N. Khraishi, *Chem. Eng. Commun.*, **96**, 57 (1990).

249. J. Lehtonen, K. Immonen, T. Salmi, E. Paatero, and P. Nyholm, *Chem. Eng. Sci.*, **51**, 2799 (1996).

250. F. G. Baddar, M. H. Nosseir, G. G. Gabra, and N. E. Ikladious, *J. Polym. Sci., Part A-1*, **9**, 1947 (1971).

251. A. F. Shaaban, M. A. Salem, and N. N. Messiha, *Indian J. Chem., Sect. A*, **22A**, 767 (1983).

252. L. Nondek and J. Malek, *Makromol. Chem.*, **178**, 2211 (1977).

253. O. M. O. Habib and J. Malek, *Collect. Czech. Chem. Commun.*, **41**, 2724 (1976).

254. F. Leverd, A. Fradet, and E. Marechal, *Eur. Polym. J.* **23**, 695 (1987).

255. B. Paul, *Modélisation de la catalyse de polyestérification en masse par les composés à liaisons métal-oxygène*, Thèse de l'Université P. et M. Curie, Paris, 1998.

256. F. Leverd, A. Fradet, and E. Marechal, *Eur. Polym. J.*, **23**, 699 (1987).

257. F. Leverd, A. Fradet, and E. Marechal, *Eur. Polym. J.*, **23**, 705 (1987).

258. D. E. Putzig, E. F. McBride, H. Q. Do, J. A. Trainham, H. L. Jaeger and H. Schulte, PCI Int. Appl. WO 0062927 (to E. I. Du Pont de Nemours & Co), October 26, 2000.

259. R. Hagen, E. Schaaf, and H. Zimmermann, U.S. Patent 6,013,756 (to Karl Fischer Industrieanlagen GmbH), January 11, 2000.

260. J. Jager, J. A. Juijin, C. J. M. van den Heuvel, and R. A. Huijts, *J. Appl. Polym. Sci.*, **57**, 1429 (1995).

261. C. S. Wang and Y. M. Sun, in *CRC Polymeric Materials Encyclopedia*, Vol. 7, J. C. Salamone, (Ed.), CRC Press, New York, 1996, p. 5355.

262. C. S. Wang and Y. M. Sun, *Polym. Prepr. (Am. Chem. Soc., Div. Polym. Chem.)*, **35**(2), 723 (1994).

263. C. S. Wang and Y. M. Sun, *J. Polym. Sci.: Part A: Polym. Chem.*, **32**, 1295 (1994).

264. J. M. Besnoin and K. Y. Choi, *Macromol. Phys.*, **C29**, 55 (1989).

265. H. R. Kricheldorf and G. Schwarz, *Polm. Bull.*, **1**, 383 (1979).

266. H. R. Kricheldorf, G. Schwarz, and F. Ruhser, *Macromolecules*, **24**, 3485 (1991).

267. H. R. Kricheldorf and D. Lübbers, *Makromol. Chem. Rapid Commun.*, **12**, 691 (1991).

268. H. R. Kricheldorf, O. Stöber, and D. Lübbers, *Macromolecules*, **28**, 2118 (1995).

269. H. R. Kricheldorf and O. Stöber, *Macromol. Rapid Commun.*, **15**, 87 (1994).

270. S. S. Park, S. S. Im, and D. K. Kim, *J. Polym. Sci., Part A: Polym. Chem.*, **32**, 2873 (1994).

271. C.-S. Wang, Y.-M. Sun, and L.-C. Hu, *J. Polym. Res.*, **1**, 131 (1994).

272. M. Di Serio, R. Tesser, F. Trulli, and E. Sàntacesaria, *J. Appl. Polym. Sci.*, **62**, 409 (1996).

273. V. N. Ignatov, F. Pilati, C. Bert, V. Tartari, C. Carraro, G. Nadali, M. Fiorini, and M. Toselli, *J. Appl. Polym. Sci.*, **58**, 771 (1995).

274. J. Yamanis and M. Adelman, *J. Polym. Sci. Polym. Chem. Ed.*, **14**, 1945, 1961 (1976).

275. K. V. Datye and H. M. Raje, *J. Appl. Polym. Sci.* **30**, 205 (1985).

276. E. Santacesaria, F. Trulli, L. Minervini, M. Di Serio, R. Tesser, and S. Contessa, *J. Appl. Polym. Sci.*, **54**, 1371 (1994).

277. A. Khrouf, S. Boufi, R. El Gharbi, and A. Gandini, *Polym. Int.*, **48**, 649 (1999).

278. C. Gutierrez, A. Alvarez, M. Chumacero, J. Flores-Estrada, G. Hernandez-Padron, and V. M. Castano, *Int. J. Polym. Mater.*, **45**, 29 (2000).

279. C. S. Wang and Y. M. Sun, *J. Polym. Sci., Part A: Polym. Chem.*, **32**, 1295 (1994).

280. B. Fortunato, P. Manaresi, A. Munari, and F. Pilati, *Polymer*, **27**, 29 (1986).

281. F. Pilati, P. Manaresi, B. Fortunato, A. Munari, and V. Passalacqua, *Polymer*, **22**, 799 (1981).

282. K. Tomita and H. Ida, *Polymer*, **16**, 185 (1975).

283. M. Di Serio, B. Apicella, G. Grieco, P. Iengo, L. Fiocca, R. Po, and E. Santacesaria, *J. Mol. Catal. A: Chem.*, **130**, 233 (1998).

284. J. Otton, S. Ratton, G. D. Markova, K. M. Nametov, V. I. Bakhmutov, S. V. Vinogradova, and V. V. Korshak, *J. Polym. Sci., Part A: Polym. Chem.*, **27**, 3535 (1989).

285. J. Otton and S. Ratton, *J. Polym. Sci., Part A: Polym. Chem.*, **29**, 377 (1991).

286. I. Vulic and T. Schulpen, *J. Polym. Sci., Polym. Chem. Ed.*, **30**, 2725 (1992).

287. X. Han, A. B. Padias, and H. K. J. Hall, *Macromolecules*, **30**, 8205 (1997).

288. J. Mathew, R. V. Bahukelar, R. S. Ghadage, C. R. Rajan, S. Ponrathnam, and S. D. Prasad, *Macromolecules*, **25**, 7338 (1992).

289. J. Mathew, R. S. Ghadage, S. Ponrathnam, and S. D. Prasad, *Macromolecules*, **27**, 4021 (1994).

290. V. V. Korshak, S. V. Vinogradova, V. A. Vasnev, and T. I. Mithaishvili, *Polym. Sci. USSR*, **11**, 89 (1969).

291. T. Galcéra, A. Fradet, and E. Maréchal, *Eur. Polym. J.*, **31**, 733 (1995).

292. T. Y. Tseng, N. J. Chu, and Y. D. Lee, *J. Appl. Polym. Sci.*, **41**, 1651 (1990).

293. D. Jehnichen, J. Tobisch, P. Friedel, and D. Pospiech, *Polymer*, **37**, 1463 (1996).

294. S. C. Temin, in *Interfacial Synthesis*, Vol. II: *Polymer Applications and Technology*, F. Millich and C. E. J. Carraher (Eds.), Marcel Dekker, New York, 1977, p. 27.

295. V. A. Vasnes, V. N. Ignatov, S. V. Vinogradova, and H. M. Tseitlin, *Makromol. Chem.*, **191**, 1759 (1990).

296. V. A. Vasnev, A. I. Tarasov, S. V. Vinogradova, and V. V. Korshak, *Polym. Sci. USSR*, **17**, 828 (1975).

297. V. A. Vasnev and S. V. Vinogradova, *Russ. Chem. Rev.*, **48**, 16 (1979).

298. J. Urbanski and H. Urbanska, *Polish J. Chem.*, **52**, 307, 523 (1978).

299. H. B. Tsai and Y. D. Lee, *J. Polym. Sci., Part A: Polym. Chem.*, **25**, 3405 (1987).

300. D. Acierno, E. Amendola, S. Concilio, R. Fresa, P. Iannelli, and P. Vacca, *Macromolecules*, **33**, 9376 (2000).

301. D. Acierno, R. Fresa, P. Iannelli, and P. Vacca, *Polymer*, **41**, 4179 (2000).

302. W. Podkoscielny and B. Tarasiuk, *Angew. Makromol. Chem.*, **207**, 173 (1993).

303. H. B. Tsai and Y. D. Lee, *J. Polym. Sci., Part A: Polym. Chem.*, **25**, 2195 (1987).

304. C. Y. Wang, D. C. Wang, W. Y. Chiu, and L. W. Chen, *Angew. Makromol. Chem.*, **248**, 123 (1997).

305. L. B. Sokolov, *Polymer Sci. USSR*, **12**, 1097 (1970).

306. G. Borissov and H. Sivriev, *Makromol. Chem.*, **158**, 215 (1972).

307. P. W. Morgan, *Polym. Rev.*, **10**, 366 (1965).

308. H. A. Staab, *Angew. Chem. Int. Ed. Engl.*, **1**, 351 (1962).

309. H. Tanaka, Y. Iwanaga, G.-C. Wu, K. Sanui, and N. Ogata, *Polym. J.*, **14**, 643 (1982).

310. E. Maréchal, *Bull. Soc. Chim. Fr.*, **4**, 713 (1987).

311. F. Higashi, A. Hoshio, Y. Yamada, and M. Ozawa, *J. Polym. Sci. Polym. Chem. Ed.*, **23**, 69 (1985).

312. F. Higashi, N. Akiyama, I. Takahashi, and T. Koyama, *J. Polym. Sci. Polym. Chem. Ed.*, **22**, 1653 (1984).

313. G.-C. Wu, H. Tanaka, K. Sanui, and N. Ogata, *Polym. J.*, **14**, 797 (1982).

314. F. Higashi, I. Takahashi, N. Akiyama, and T.-C. Chang, *J. Polym. Sci. Polym. Chem. Ed.*, **22**, 3607 (1984).

315. J. S. Moore and S. I. Stupp, *Macromolecules*, **23**, 65 (1990).

316. N. Belcheva, C. Tsvetanov, I. M. Panayotov, and S. Lazarova, *Makromol. Chem.*, **191**, 213 (1990).

317. I. M. Panayotov, N. Belcheva, and C. Tsvetanov, *Makromol. Chem.*, **188**, 2821 (1987).

318. F. Akutsu, M. Inoki, H. Uei, M. Sueyoshi, Y. Kasashima, K. Naruchi, Y. Yamaguchi, and M. Sunahara, *Polym. J.*, **30**, 421 (1998).

319. G. C. East and M. Morshed, *Polymer*, **23**, 168 (1982).

320. A. L. Cimecioglu and G. C. East, *Makromol. Chem., Rapid. Commun.*, **8**, 141 (1987).

321. J. Djonlagic, M.-O. Sepulchre, M. Sepulchre, and N. Spassky, *Makromol. Chem.*, **191**, 1529 (1990).

322. N. Spassky, M.-O. Sepulchre, M. Sepulchre, B. Dunjic, and J. Djonglagic, *J. Serb. Chem. Soc.*, **58**, 285 (1993).

323. A. L. Cimecioglu and G. C. East, *J. Polym. Sci., Part A: Polym. Chem.*, **30**, 313 (1992).

324. M. S. Jacovic, J. Djonlagic, M. Sepulchre, M.-O. Sepulchre, A. Le Borgne, and N. Spassky, *Makromol. Chem.*, **189**, 1353 (1988).

325. S. Kobayashi, *J. Polym. Sci. Part A: Polym. Chem. Ed.*, **37**, 3041 (1999).

326. L. Brady, A. M. Brzozowski, Z. S. Derewenda, E. Dodson, G. Dodson, S. Tolley, J. P. Turkenburg, L. Christiansen, B. Huge-Jensen, L. Norskov, L. Thim, and U. Menge, *Nature*, **343**, 767 (1990).

327. T. Lalot, M. Brigodiot, and E. Maréchal, in *Comprehensive Polymer Science*, Second Supplement Vol., G. Allen, S. L. Aggarwal, and S. Russo (Eds.), Pergamon, Oxford, 1996, p. 29.

328. X. Shuai, Z. Jedlinski, M. Kowalczuk, J. Rydz, and H. Tan, *Eur. Polym. J.*, **35**, 721 (1999).

329. K. F. Brandstadt, J. C. Saam, and A. Sharma, *Polym. Prepr. (Am. Chem. Soc., Div. Polym. Chem.)*, **41**(2), 1857 (2000).

330. A. L. Gutman, D. Knaani, and T. Bravdo, *Macromol. Symp.*, **122**, 39 (1997).

331. F. Binns, P. Harffey, S. M. Roberts, and A. Taylor, *J. Chem. Soc., Perkin Trans. 1*, (19), 2671 (1999).

332. F. Binns, S. M. Roberts, A. Taylor, and C. F. Williams, *J. Chem. Soc., Perkin Trans. 1*, (8), 899 (1993).

333. F. Binns, P. Harffey, S. M. Roberts, and A. Taylor, *J. Polym. Sci., Part A: Polym. Chem.*, **36**, 2069 (1998).

334. S. Geresh and Y. Gilboa, *Biotechnol. Bioeng.*, **37**, 883 (1991).

335. J. S. Wallace and C. J. Morrow, *J. Polym. Sci., Part A: Polym Chem.*, **27**, 3271 (1989).

336. C. Berkane, G. Mezoul, T. Lalot, M. Brigodiot, and E. Maréchal, *Macromolecules*, **30**, 7729 (1997).

337. G. Mezoul, T. Lalot, M. Brigodiot, and E. Maréchal, *Macromol. Rapid Commun.*, **16**, 613 (1995).

338. G. Mezoul, T. Lalot, M. Brigodiot, and E. Maréchal, *J. Polym. Sci., Part A: Polym. Chem.*, **33**, 2691 (1995).

339. G. Mezoul, T. Lalot, M. Brigodiot, and E. Maréchal, *Polym. Bull.*, **36**, 541 (1996).

340. S. Kobayashi and H. Uyama, *Makromol. Chem., Rapid Commun.*, **14**, 841 (1993).

341. A. Kumar and R. A. Gross, *J. Am. Chem. Soc.*, **122**, 11767 (2000).

342. A. K. Chaudhary, B. J. Kline, E. J. Beckman, and A. J. Russell, *Polym. Prepr. (Am. Chem. Soc., Div. Polym. Chem.)*, **38**(2), 396 (1997).

343. H. Uyama, S. Yaguchi, and S. Kobayashi, *J. Polym. Sci., Part A: Polym. Chem.*, **37**, 2737 (1999).

344. H. Uyama, K. Inada, and S. Kobayashi, *Macromol. Biosci.*, **1**, 40 (2001).

345. Y.-Y. Linko, Z.-L. Wang, and J. Seppala, *Enz. Microb. Technol.*, **17**, 506 (1995).

346. Z.-L. Wang, K. Hiltunen, P. Orava, J. Seppala, and Y.-Y. Linko, *J. Macromol. Sci. Pure Appl. Chem.*, **A33**, 599 (1996).

347. H. G. Park, H. N. Chang, and J. S. Dordick, *Biocatalysis*, **11**, 263 (1994).

348. X. Y. Wu, Y. Y. Linko, J. Seppala, M. Leisola, and P. Linko, *J. Ind. Microbiol. Biotechnol.*, **20**, 328 (1998).

349. R. L. Rodney, B. T. Allinson, E. J. Beckman, and A. J. Russell, *Biotechnol. Bioeng.*, **65**, 485 (1999).

350. M. Rizzi, P. Stylos, A. Rieck, and M. Reuss, *Enz. Microb. Technol.*, **14**, 709 (1992).

351. A. Lavalette, *Etude cinétique de la polytransestérification catalysée par une lipase en milieu organique*, Thèse, de l'Université P. et M. Curie, Paris, 2000.

352. A. Kumar, B. Kalra, A. Dekhterman, and R. A. Gross, *Macromolecules*, **33**, 6303 (2000).

353. S. Kobayashi and H. Uyama, *Macromol. Symp.*, **144**, 237 (1999).

354. W. Xie, J. Li, D. Chen, and P. G. Wang, *Macromolecules*, **30**, 6997 (1997).

355. L. A. Henderson, Y. Y. Svirkin, R. A. Gross, D. L. Kaplan, and G. Swift, *Macromolecules*, **29**, 7759 (1996).

356. D. Knani, A. L. Gutman, and D. H. Kohn, *J. Polym. Sci. Part A: Polym. Chem.*, **31**, 1221 (1993).

357. K. S. Bisht, L. A. Henderson, R. A. Gross, D. L. Kaplan, and G. Swift, *Macromolecules*, **30**, 2705 (1997).

358. R. T. MacDonald, S. K. Pulapura, Y. Y. Svirkin, R. A. Gross, D. L. Kaplan, J. Akkara, G. Swift, and S. Wolk, *Macromolecules*, **28**, 73 (1995).

359. H. Uyama, S. Namekawa, and S. Kobayash, *Polym. J.*, **29**, 299 (1997).

360. F. Deng, L. Henderson, and R. A. Gross, *Polym. Prepr. (Am. Chem. Soc., Div. Polym. Chem.)*, **39**(2), 144 (1998).

361. N. M. Mang, J. E. White, A. P. Haag, S. L. Kram, and C. N. Brown, *Polym. Prepr. (Am. Chem. Soc., Div. Polym. Chem.)*, **36**(2), 180 (1995).

362. J. E. Klee, F. Claussen, and H. H. Hoerhold, *Polym. Bull.*, **35**, 79 (1995).

363. P.-J. Madec and E. Marechal, *Adv. Polym. Sci.*, **71**, 153 (1985).

364. G. Schwach, J. Coudane, R. Engel, and M. Vert, *Polym. Bull.*, **32**, 617 (1994).

365. A. Duda and S. Penczek, *ACS Symp. Ser.*, **764**, 160 (2000).

366. H. R. Kricheldorf, C. Boettcher, and K.-U. Tönnes, *Polymer*, **33**, 2817 (1992).

367. H. R. Kricheldorf, I. Kreiser-Saunders, and A. Stricker, *Macromolecules*, **33**, 702 (2000).

368. G. Schwach, J. Coudane, R. Engel, and M. Vert, *Polym. Int.*, **46**, 177 (1998).

369. M. Bero, J. Kasperczyk, and Z. Jedlinski, *Makromol. Chem.*, **191**, 2287 (1990).

370. B. M. Chamberlain, Y. Sun, J. R. Hagadorn, E. W. Hemmesch, M. A. Hillmyer, and W. B. Tolman, *Macromolecules*, **32**, 2400 (1999).

371. D. Mercerreyes, R. Jérôme, and P. Dubois, *Adv. Polym. Sci.*, **147**, 1 (1999).

372. V. Simic, N. Spassky, and L. G. Hubert-Pfalzgraf, *Macromolecules*, **30**, 7338 (1997).

373. S.-H. Hyon, K. Jamshidi, and Y. Ikada, *Biomaterials*, **18**, 1503 (1997).

374. K. Takahashi, I. Taniguchi, M. Miyamoto, and Y. Kimura, *Polymer*, **41**, 8725 (2000).

375. S. I. Moon, C. W. Lee, M. Miyamoto, and Y. Kimura, *J. Polym. Sci. Part A: Polym. Chem.*, **38**, 1673 (2000).

376. F. Schue, C. Jaimes, R. Dobreva-Schue, O. Giani-Beaune, W. Amass, and A. Amass, *Polym. Int.*, **49**, 965 (2000).

377. M. Möller, R. Kange, and J. L. Hedrick, *J. Polym. Sci., Part A: Polym. Chem.*, **38**, 2067 (2000).

378. J. Gimenez, P. Cassagnau, and A. Michel, *J. Rheol.*, **44**, 527 (2000).

379. H. R. Kricheldorf and I. Kreiser-Saunders, *Polymer*, **41**, 3957 (2000).

380. A. Kowalski, A. Duda, and S. Penczek, *Macromol. Rapid Commun.*, **19**, 567 (1998).

381. A. Kowalski, A. Duda, and S. Penczek, *Macromolecules*, **33**, 689 (2000).

382. S. Penczek, A. Duda, R. Szymanski, J. Baran, J. Libiszowski, and A. Kowalski, *NATO Sci. Ser., Ser. E*, **359**, 283 (1999).

383. H. Dong, S.-G. Cao, Z.-Q. Li, S.-P. Han, D.-L. You, and J.-C. Shen, *J. Polym. Sci., Part A: Polym. Chem.*, **37**, 1265 (1999).

384. R. Jérôme and P. Teyssié, in *Comprehensive Polymer Science*, Vol. 3, G. Allen and J. C. Bevington (Eds.), Pergamon, Oxford, 1989, p. 501.

385. S. Penczek and S. Slomkowski, in *Comprehensive Polymer Science*, Vol. 3, G. Allen and J. C. Bevington (Eds.), Pergamon, Oxford, 1989, p. 813.

386. S. Penczek, A. Duda, and J. Libiszowski, *Macromol. Symp.*, **128**, 241 (1998).

387. D. J. Brunelle and J. Serth-Guzzo, *Polym. Prepr. (Am. Chem. Soc., Div. Polym. Chem.)*, **40**(1), 566 (1999).

388. E. P. Goldberg, S. F. Strause, and H. E. Munro, *Polym. Prepr. (Am. Chem. Soc., Div. Polym. Chem.)*, **5**(1), 233 (1964).

389. K. Adam and H.-G. Trieschmann, Ger. Offen. DE 890,792 (to B.A.S.F.-AG), September 21, 1953.

390. E. N. Zil'berman, A. E. Kulikova, and N. M. Teplyakov, *J. Polym. Sci.*, **56**, 417 (1962).

391. Y. V. Mitin, N. Sazanov, and G. P. Vlasov, *Vysokomol. Soedin.*, **2**, 716 (1960).

392. W. Sweeny, *J. Appl. Polym. Sci.*, **7**, 1983 (1963).

393. G. Natta, G. Mazzanti, G. Pregaglia, and G. Pozzi, *J. Polym. Sci.*, **58**, 1201 (1962).

394. D. Nagai, A. Sudo, F. Sanda, and T. Endo, *J. Polym. Sci. Part A: Polym. Chem.*, **39**, 2078 (2001).

395. G. Natta, G. Mazzanti, G. Pregaglia, and M. Binaghi, *J. Am. Chem. Soc.*, **82**, 5511 (1960).

396. H. J. Hagemeyer, U.S. Patent 2,533,455 (to Eastman Kodak Co.), December 12, 1950.

397. J. L. R. Williams, J. M. Carson, and G. A. Reynolds, *Makromol. Chem.*, **65**, 54 (1963).

398. I. Goodman and E. Haddock, British Patent 968,390 (to I.C.I.), September 2, 1964.

399. J. G. Erickson, *J. Polym. Sci. Part A1*, **4**, 519 (1966).

400. K. Sugiyama and K. Shiraishi, *J. Polym. Sci., Part A: Polym. Chem.*, **28**, 1983 (1990).

401. A. Fradet and E. Maréchal, *Eur. Polym. J.*, **14**, 749 (1978).

402. A. Fradet, in *Comprehensive Polymer Science*, Second Supplement Vol., G. Allen, S. L. Aggarwal, and S. Russo (Eds.), Pergamon, Oxford, 1996, p. 133.

403. H. Inata and S. Matsumura, *J. Appl. Polym. Sci.*, **34**, 2609 (1987).

404. W. S. Ha, Y. K. Chun, S. S. Jang, D. M. Rhee, and C. R. Park, *J. Polym. Sci., Part B: Polym. Phys.*, **35**, 309 (1997).

405. A. J. Anderson and E. A. Dawes, *Microbiol. Rev.*, **54**, 450 (1990).

406. S. Y. Lee, J.-i. Choi, and H. H. Wong, *Int. J. Biol. Macromol.*, **25**, 31 (1999).

407. N. Yoshie and Y. Inoue, *Int. J. Biol. Macromol.*, **25**, 193 (1999).

408. R. Pétiaud, H. Waton, and Q.-T. Pham, *Polymer*, **33**, 3155 (1992).

409. A. M. Kenwright, S. K. Peace, R. W. Richards, A. Bunn, and W. A. MacDonald, *Polymer*, **40**, 2035 (1999).

410. V. P. Boiko and V. K. Grishchenko, *Acta Polym.*, **36**, 459 (1985).

411. R. G. Garmon, *Tech. Methods Polym. Eval.*, **4**(1), 31 (1975).

412. J. C. Bevington, J. R. Ebdon, and T. N. Huckerby, in *NMR Spectroscopy of Polymers*, R. N. Ibbett (Ed.), Blackie, Glasgow, United Kingdom, 1993, p. 80.

413. H. Zhang, A. Rankin, and I. M. Ward, *Polymer*, **37**, 1079 (1996).

414. K. A. B. Lee, S. M. Hurley, R. A. Siepler, R. D. Mills, K. A. Handrich, and J. J. Conway, *Appl. Spectrosc.*, **44**, 1719 (1990).

415. R. L. Addleman and V. J. I. Zichy, *Polymer*, **13**, 391 (1972).

416. S. Mori, *Anal. Chim. Acta*, **189**, 17 (1986).

417. M. J. Maurice and F. Huizinga, *Anal. Chim. Acta*, **22**, 363 (1960).

418. H. Zimmermann and A. Tryonadt, *Faserforsch. Textiltech.*, **18**, 487 (1967).

419. G. D. B. Van Houwelingen, M. W. M. G. Peters, and W. G. B. Huysmans, *Fresenius' Z. Anal. Chem.*, **293**, 396 (1978).

420. E. Marten, *Korrelationen zwischen der Struktur und der Enzymatischen Hydrolyse von Polyestern*, Dr. Rer. Nat. Dissertation, Teschnische Universität Carolo-Wilhelmina, Braunschweig, 2000.

421. A. Fradet and M. Tessier, unpublished work.

422. N. Spassky, M. Wisnieswski, C. Pluta, and A. Le Borgne, *Macromol. Chem. Phys.*, **197**, 2627 (1996).

423. A. Bodnar, M. Solymoss, I. Vancson-Szmercsanyi, I. Szigeti, and L. Szommer, U.S. Patent 4,902,773 (to Nitrokemia Ipartelepek), February 20, 1990.

424. G. M. Lampman, W. R. Hale, A. Pinkers, and C. G. Sewel, *J. Chem. Educ.*, **56**, 626 (1979).

425. S. R. Sandler and W. Karo, in *Polymer Synthesis*, Vol. 1, H. H. Wasserman (Eds.), Academic, London, 1992, p. 69.

426. D. E. James and L. G. Packer, *Ind. Eng. Chem. Res.*, **34**, 4049 (1995).

427. P. R. Wendling, U.S. Patent 4,532,319 (to Goodyear Tire and Rubber Co.), July 30, 1985.

428. B. Gantillon, *Préparation du poly(éthylene téréphtalate) à l'état solide en milieu dispersé*, Thèse de l'Université Claude Bernard, Lyon, 1999.

429. M. L. Doerr, U.S. Patent 4,824,930 (to Hoechst-Celanese Corp.), July 25, 1990.

430. M. C. E. J. Niesten, K. Bouma, and R. J., Gaymans, *Polymer*, **39**, 93 (1998).

431. W. K. Witsiepe, U.S. Patent 3,763,109 (to Du Pont), October 2, 1973.

432. J. A. Pawlak, A. Lemper and V. A. Pattison, U.S. Patent 4,049,629 (to Hooker Chemicals & Plastics), September 20,1977.

433. A. J. Conix, U.S. Patent 3,216,970 (to Gevaert Photo-Produkten), September 11, 1965.

434. W. M. Eareckson, *J. Polym. Sci.*, **40**, 399 (1959).

435. F. Higashi, T. Mashimo, and I. Takahashi, *J. Polym. Sci. Polym. Chem. Ed.*, **24**, 97 (1986).

436. F. Blaschke and W. Ludwig, U.S. Patent 3,395,119 (to Witten GmbH Chem. Werke), July 30, 1968.

437. M. Levine and S. C. Temin, *J. Polym. Sci.*, **28**, 179 (1958).

438. P. Tacke, U. Grogp, and K.-J. Idel, U.S. Patent 4,841,000 (to Bayer Aktiengesellschaft), June 20, 1989.

439. H.-B. Tsai, C. Lee, N.-S. Chang, M.-S. Chen, and S.-J. Chang, *J. Appl. Polym. Sci.*, **440**, 1499 (439).

440. A. B. Erdemir, D. J. Johnson, and J. G. Tomka, *Polymer*, **27**, 441 (1986).

441. U. Wiesner, F. Lauprêtre, and L. Monnerie, *Macromolecules*, **27**, 3632 (1994).

442. S. R. Turner, B. I. Voit, and T. H. Mourey, *Macromolecules*, **26**, 4617 (1993).

443. S. Milmo, *Chem. Mark. Rep.*, **257**(20), 1 (2000).

444. Eldib Research Report, *Year 2000 Market Survey on PEN Film in USA*, http://www.eldib.com, 2000.

445. R. J. Andrews and E. A. Grulke, in *Polymer Handbook*, 4th ed., J. Brandrup, E. H. Immergut, and E. A. Grulke (Eds.), Wiley-Interscience, New York, 1999, p. VI/221.

446. M. Varma-Nair and B. Wunderlich, *J. Phys. Chem. Ref. Data*, **20**, 349 (1991).

447. M. Gilbert and F. J. Hybart, *Polymer*, **15**, 407 (1974).

448. R. L. Miller, in *Polymer Handbook*, 4th ed., J. Brandrup, E. H. Immergut, and E. A. Grulke (Eds.), Wiley-Interscience, New York, 1999, p. VI/40.

449. M. Varma-Nair, R. Pan, and B. Wunderlich, *J. Polym. Sci. Part B, Polym. Phys.*, **29**, 1107 (1991).

450. Y. Tokiwa, and S. Komatsu, Jpn. Kokai Tokkyo, Koho JP 8,295,726 (to Nippon Oil & Fats Co. Ltd.), November 12, 1996.

451. S. Z. D. Cheng and B. Wunderlich, *Macromolecules*, **21**, 789 (1988).

452. L. Rozes, *Synthèse de poly(polyester-bloc-polyéther)s thermotropes. Approche des propriétés thermiques, viscoélastiques et rhéologiques*, Thèse, de l'Université P. et M. Curie, Paris, 1996.

453. E. Perez, A. Bello, M. H. Marugan, and J. M. Perena, *Polym. Comm.*, **31**, 386 (1990).

454. J. B. Loman, L. Van der Does, A. Bantjes, and I. Vuluc, *J. Polym. Sci. Part A, Polym. Chem.*, **33**, 493 (1995).

455. Cargill Dow LLC, *P L A Polymers — Injection Molding Process Guide*, 1999. http://www.cdpoly.com.

456. Solvay, *Biodegradable Blown Film Grades of CAPA Polycaprolactones*, January 2001. http://www.solvaycaprolactones.com.

457. Eastman, *Eastar BIO GP Copolyester Product Data Sheet*, 2001. http://www.eastman.com.

458. Honeywell Performance Fibers, *Pentex*® *Product Data Sheet*, 2001. http://www.performan-cefibers.com.

459. Ticona, *Vectra*® *A-950 Data Sheet*, 2001. http://www.ticona-us.com.

460. R. Cai, J. Preston, and E. T. Samulski, *Macromolecules*, **25**, 563 (1992).

461. W. Hatke, H. T. Land, H. W. Schmidt, and W. Heitz, *Makromol. Chem. Rapid Commun.*, **12**, 235 (1991).

462. H. R. Kricheldorf, V. Döring, and V. Eckhardt, *Makromol. Chem.*, **189**, 1425 (1988).

463. K. Berger and M. Ballauff, *Mol. Cryst. Liq. Cryst.*, **157**, 109 (1988).

464. K. B. Wagener, in *Polymer Handbook*, 4th ed., J. Brandrup, E. H. Immergut, and E. A. Grulke (Eds.), Wiley-Interscience, New York, 1999, p. V/119.

465. P. Kambouris and C. J. Hawker, *J. Chem. Soc. Perkin Trans 1.*, (22), 2717 (1993).

466. W. J. Feast, *Mater. Chem.*, **5**, 404 (1995).

467. M. Liu, N. Vladimirov, and J. M. J. Fréchet, *Macromolecules*, **32**, 6881 (1999).

468. R. A. T. M. van Benthem, D. Muscat, and D. A. W. Stanssens, *A.C.S. Polym. Mat. Sci. Eng. Prepr.*, **80**, 72 (1999).

469. Q.-T. Pham, R. Pétiaud, H. Waton, and M.-F. Llauro-Darricades, *Proton and Carbon NMR Spectra of Polymers*, Penton, London, 1991.

470. A. Piras, *Edude par résonance magnétique nucléaire haute résolution en solution de la structure des polyesters insaturés*, Thèse de, l'Université P. et M. Curie, Paris, 1988.

471. J.-P. Leblanc, *Synthèse et caractérisation de polycondensats renfermant des motifs stilbéniques*, Thèse, de l'Université P. et M. Curie, Paris, 1988.

3 Polyamides

Reinoud J. Gaymans

Twente University, 7500 AE Enschede, The Netherlands

3.1 INTRODUCTION

Polyamides (PAs) are high-molecular-weight materials containing amide units; the hydrocarbon segments between the amide groups can be aliphatic, partially aromatic, or wholly aromatic. The type of hydrocarbon segment used has an effect on the chain flexibility and structural regularity; this is important for the formation of the crystalline phase. Polyamides, in common with polypeptides, contain an amide group and are often called Nylons, the trade name given to them by DuPont.

PAs can be divided into polymers synthesized from diamines and diacids, called AA–BB-type polymers as in (3.1) and polymers based on amino acids, called AB-type polymers as in (3.2):

$$-\left(N-(CH_2)_{\overline{x}}N-C-(CH_2)_{\overline{y}}C\right)_{\overline{n}} \qquad (3.1)$$

$$-\left(N-(CH_2)_{\overline{x}}C\right)_{\overline{n}} \qquad (3.2)$$

Polymers are further defined based on the number of carbon atoms they contain. Thus, the PA based on 1,6-hexamethylene diamine and sebacic acid is named PA-6,10 and the polymer based on caprolactam, PA-6. Terephthalic and isophthalic acid units are coded T and I. Thus, PA-6,T represents the PA from hexamethylene diamine and terephthalic acid.

3.1.1 Historical Perspective

Many types of PAs have been studied, and in one of the first patents on PAs, filed by of Carothers of DuPont in 1937, 31 were described.[1] Of these 31 polyamides,

Synthetic Methods in Step-Growth Polymers. Edited by Martin E. Rogers and Timothy E. Long
© 2003 John Wiley & Sons, Inc. ISBN: 0-471-38769-X

nylon-6,6 is still very important, nylon-6,10 is being produced on a smaller scale and nylon-4,6 has only been on the market for a few years since it is now possible to avoid degradation during polymerization.[2] In the 1930s, Schlack at I. G. Farben in Germany[3] developed the other important aliphatic polyamide, nylon-6. Sisters of PA-6 which are also on the market are PA-11 and PA-12. Since this time, partially aromatic and wholly aromatic PAs have been successfully developed.

PA-6,6 was the first synthetic condensation polymer on the market. The announcement of the start of industrial production of PA-6,6, the birth of a new kind of material, was a milestone event. PA-6,6 was chosen for industrial development as the starting materials were easily prepared and the polymer exhibited excellent thermal stability and ductile deformation behavior. In the 1930s, not only was the synthesis of the polymer studied and large-scale production routes developed, but monomer synthesis, polymerization equipment, melt processing apparatuses, and the analytical techniques for studying the polymers were also developed. This major engineering achievement was combined with a strong marketing effort. It is remarkable that after many decades since their discovery, PA-6,6 in the United States and PA-6 in Europe are still dominant polymers in the marketplace. A number of excellent reviews have been written on polyamides.[4-13]

3.1.2 Applications

The first use of PAs was in fiber applications, in which fibers were produced by melt spinning. These materials have a high strength and good wear resistance and can be easily dyed. The tension stiffening effect of the PA melt made the production of fibers with homogeneous thickness possible.

PAs as engineering plastics came onto the market following the development of the injection-molding machine, and most PAs as engineering plastics are processed by this method. The major part of PAs as engineering plastics are either a reinforced or a rubber-modified grade. The glass-fiber-filled PAs have a high modulus, high strength, and good dimensional stability up to their melting temperature. These PAs are used in highly demanding applications such as in the transport sector "under the hood," in housing for electrical equipment, and in sporting goods. With caprolactam and laurolactam, casting is also possible, which is an interesting technique for thick-walled articles. High-molecular-weight materials have good mechanical properties, but for ease of processing, the melt viscosity should be low. For melt spinning, the melt viscosity can be very low, whereas for extrusion and blow-molding grades the higher viscosities are desirable (Table 3.1).

Partial aromatic PAs have also been well studied and are of industrial importance. The later developed, wholly aromatic PAs poly(*m*-phenyleneisophthalamide) in (3.3) and poly (*p*-phenyleneterephthalamide) in (3.4) are interesting fiber-forming materials with high maximum-use temperature[7,9,14].

TABLE 3.1   Typical PA-6 Granulate Specifications[a]

Relative Viscosity[b]	Application
2.4–2.6	Textile filament
2.5–2.7	Staple fibers
2.7–2.9	Carpet yarns
2.8–3.0	Industrial yarn
3.2–3.6	Tire yarn, film
2.7–4.0	Engineering plastic, film, foil

[a] From ref. 13.
[b] Determined on 1% solutions in 96% sulfuric acid at 25°C.

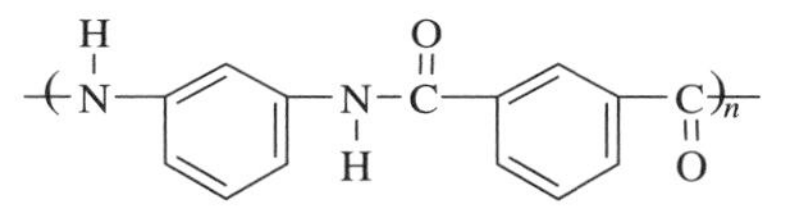

$$\underset{H}{\overset{}{+}N}\!\!-\!\!\!\!\!\overset{}{\bigcirc}\!\!\!\!\!-\!\!\underset{H}{N}\!\!-\!\!\overset{O}{\underset{}{C}}\!\!-\!\!\!\!\!\overset{}{\bigcirc}\!\!\!\!\!-\!\!\overset{}{\underset{O}{C}}\!\!\!+_n \tag{3.3}$$

$$\underset{H}{\overset{}{+}N}\!\!-\!\!\!\!\!\overset{}{\bigcirc}\!\!\!\!\!-\!\!\underset{H}{N}\!\!-\!\!\overset{O}{\underset{}{C}}\!\!-\!\!\!\!\!\overset{}{\bigcirc}\!\!\!\!\!-\!\!\overset{}{\underset{O}{C}}\!\!\!+_n \tag{3.4}$$

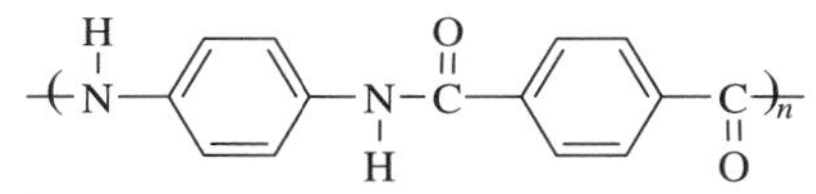

Poly(*p*-phenyleneterephthalamide) forms a liquid crystalline solution and can be spun into a fiber with a very high orientation; these fibers have excellent tensile and thermal properties. These high-modulus fibers are suitable as reinforcing materials in technical applications.

3.1.3 Research Trends and Critical Issues

PA-6,6 is still often synthesized in a one-pot batch process with a prepolymerization stage under pressure. Continuous processes have long been described, and for prepolymerization, in particular, they seem to work well. However, with this method, there does seem to be a problem with quality control.

PA-6,6 is made from the relatively expensive materials hexamethylene diamine and adipic acid. An alternative synthesis of PA-6,6 from adiponitrile and hexamethylene diamine utilizing water is under investigation.[16] PA-6 can be synthesized in a continuous process at atmospheric pressure, but reaction times are very long as the ring-opening initiation step is particularly slow. The reaction time can be shortened considerably by carrying out prepolymerization in the presence of excess water at pressure; however, this makes the continuous polymerization process more complex. Copolymers with amide units of uniform length (diamides) are relatively new; the diamide units are able to crystallize easily and have a thermally stable crystalline structure.

3.2 STRUCTURE–PROPERTY RELATIONSHIPS

PAs have good mechanical properties due to their hydrogen bonding. The hydrogen bonding increases the chain interaction resulting in a higher yield stress, fracture stress, impact strength, tear strength, and abrasion resistance. Most PAs on the market are semicrystalline polymers, which have a high maximum-use temperature and a good solvent resistance. The aliphatic PA types $-(AA-BB)_n-$ in (3.1) and $-(AB)_n-$ in (3.2) are named after the number of carbon atoms in each unit. The polyamides $-(AA-BB)_n-$ are called polyamide-x ($y+2$), for example, polyamide-4,6, -6,6, -6,10, and -6,12. The x in the name stands for the number of methylene units in the diamine unit, and y stands for the number of methylene units in the diacid unit. The total number of carbon atoms in the diacid unit including the two acid carbons is thus $y+2$. The $(AB)_n$-type PAs such as PA-6, -11, and -12 are called by one number, $x+1$, where x stands for the number of methylene units and $x+1$ for the number of carbon atoms including the acid carbon in this amino acid unit.

PAs have several thermal transitions. The melting temperature, the crystallization temperature, and the rate of crystallization can all be studied with differential scanning calorimetry (DSC). However, at the T_g the change in the C_p value for aliphatic PAs is not sharp, and thus the transition is often not resolved by this method. The T_g and other thermal transitions can be studied by dynamic mechanical analysis (DMA) (Fig. 3.1).[5–7,11,12,17] The observed transitions are the flow temperature (melting temperature) (T_{fl} or T_m), the glass transition temperature (T_g), a β-transition temperature (T_β) of a non-hydrogen-bonded chain at about $-50°C$, and a γ-transition temperature (T_γ) of the mobile of methylene units at about $-120°C$. PAs are often semicrystalline and in the region between the T_g and T_m have a reasonably high modulus. Thus, semicrystalline PAs are dimensionally stable up to their melting temperature.

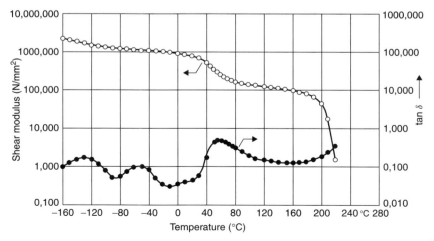

Figure 3.1   Influence of temperature on shear modulus and tan δ at 1 Hz of PA-6 (dry).[17]

3.2.1 General Considerations

PAs are used neat in fibers, films, and engineering plastics and may contain fillers in engineering plastics applications.[5–7,11,15] Reinforced PA accounts for about 80% of the engineering plastic market. Most of the reinforced PA is filled with glass fibers and to a lesser extent with particles like talc, kaolin, and mica. For engineering plastics applications, dimensional stability at high temperatures is often sought. Solvent resistance to hydrocarbons is also important for use in automotive applications such as PA-11 in fuel lines. The processing of engineering plastics is achieved by injection molding and to a lesser extent by extrusion; to ensure good processability, a low melt viscosity is required and rapid crystallization is important.

Semicrystalline partial aromatic PAs have a very high melting temperature, often too high for melt synthesis and melt processing. Amorphous partial aromatic PAs do not have a melting temperature and can thus be synthesized and processed at lower temperatures. They are transparent and have good solvent resistance and relatively low water absorption.[18] Wholly aromatic PA can be processed from solution into film and fibers. These polymers have very good dimensional stability and excellent heat stability. All para wholly aromatic PAs form liquid crystalline solutions, and fibers can be obtained which are highly oriented and which have a para-crystalline structure.[7,9,14] These all-aromatic PA fibers have a very high modulus and fracture strength.

3.2.2 Role of Chemical Structure

3.2.2.1 Amide Concentration

Linear polymers have crystalline order, which is a function of the amide content and the regularity of the spacing of the amide units. The melting temperature (T_m) increases with amide concentration and is highest for the even–even PAs (Fig. 3.2). The glass transition temperature (T_g) also increases with amide content (Fig. 3.3), and this steady increase in T_g with amide content is due to the strong interchain hydrogen bonding of amide groups.

The melting temperature is a function of the melting enthalpy and entropy:

$$T_m = \Delta H_m / \Delta S_m \qquad (3.5)$$

Although the amide groups allow the formation of strong interchain hydrogen bonding, the increase in melting temperature is generally not due to an increase in ΔH_m but due to a decrease in ΔS_m.[19] The minimal effect of amide content on ΔH_m is due to the fact that, in melting, the hydrogen bonding is not broken.

3.2.2.2 Structural Regularity

As PA can be regarded as polyethylene (PE) containing amide units, it is obvious that the positioning of the amide units in the chain is important for structural

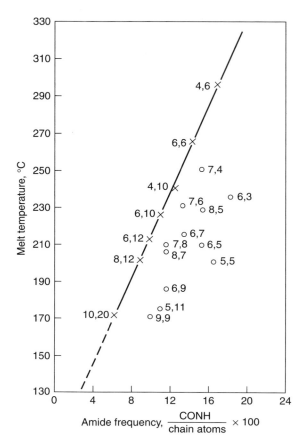

Figure 3.2 Influence of amide content (amide units/chain atoms × 100) on melting temperature of PA AA-BB.[6]

order.[6,13] The main consideration is that hydrogen bonding between amide units occurs in the melt. For optimal packing in the crystalline phase, the methylene sequence must be fully extended in a planar zigzag. In even–even PA (as in PA-6,6), the chains order in a parallel fashion (Fig. 3.4a) with an extended planar zigzag conformation of the ethylene units.[6,20] The unit cell of this stacking is triclinic and, with wide-angle X-ray spectroscopy (WAXS), two strong signals are observed at spacings of 0.44 and 0.37 nm. These spacings represent the projected interchain distance within a hydrogen-bonded sheet and the intersheet distance, respectively. In uneven chain segments (as in PA-7,7), the olefinic chain segments have to twist in order to accommodate full hydrogen bonding and the order is much lower. PA-6 stacks best in an antiparallel fashion (Fig. 3.4c). However, in the antiparallel structure there is a small mismatch in stacking of the chains due to the angle of the hydrogen bond with the chain axis. This results in a lower order

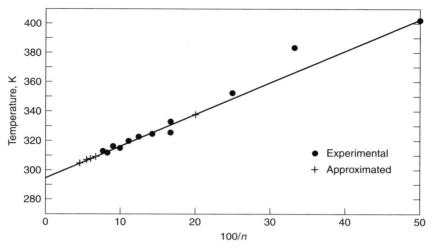

Figure 3.3 Influence of amide content [100 chain atoms/amide unit (n)] on glass transition temperature of PA AB.[12]

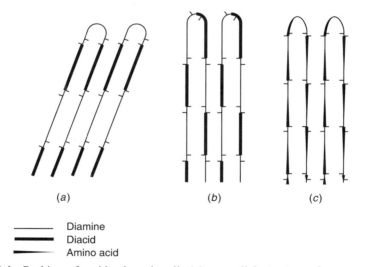

Diamine
Diacid
Amino acid

Figure 3.4 Packing of amides in unit cell: (*a*) α-parallel structure of even–even PA; (*b*) β-antiparallel structure of even–even PAs with equal methylene sequence length in amine and acid unit, as in PA-4,6[20]; (*c*) antiparallel PA-6-type polymer.

and a lower melting temperature than in PA-6,6. The structure of crystallized PA-6-type material is monoclinic with two spacings. Poorly crystallized PA-6 does not have an extended methylene sequence at room temperature and the unit cell structure is pseudohexagonal with single spacing. In this structure, the direction of the amide bonds can be in both planes.[12]

A special case is if the methylene units in the even–even PA are in the diamine as long as in the diacid (as in PA-4,6, -6,8, and -8,10) so that an ordering in both a parallel and an antiparallel fashion is possible (Figs. 3.4a, b).[20] Because of this feature, these polymers have the highest order.

If the well-crystallized PAs are heated, the two characteristic diffraction signals move together and meet at what is called the Brill temperature (Fig. 3.5).[6,20] Above this Brill temperature, there is only one spacing for both directions; this implies a pseudohexagonal packing, such as that observed in the poorly crystallized PA-6 at room temperature.

The melting temperatures of the even–even PAs are higher than for other AA–BB-type PAs and A–B-type PAs.[6,11,21] However, the amide order has little effect on the glass transition temperature (Fig. 3.3).

A higher structural regularity of the chains makes possible a more rapid crystallization and, as a result, a higher crystallinity of the polymer. A higher rate of crystallization increases the rate of solidification on cooling and eases the processability of the polymer. In injection molding, with a rapidly crystallizing polymer system, the cycle times can be lowered; the greater the crystallinity of the polymer during processing, the lower the shrinkage afterward. The crystalline structure formed is comprised of unit cells which are stacked into lamellae 6–10 nm thick which can grow into spherulites. Secondary crystallization can take place in between the lamella formed. The lamellae are the physical crosslinks for the amorphous chain segments. The effect of the crystalline phase

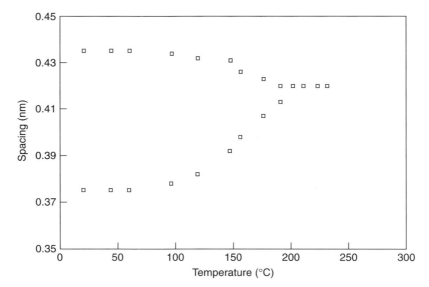

Figure 3.5   Changing spacing of unit cell with temperature as measure with WAXS.[20] Transition from triclinic (α-structure, with two spacings) to pseudohexagonal (γ-structure, with one spacing) can be seen.

is most important in the region between the T_g and T_m. It is generally observed that in this T_g-T_m region the modulus gradually decreases with temperature (Fig. 3.1). The higher the crystallinity of the system, the higher the modulus is in this region.

3.2.2.3 *Water Absorption*

Water absorption in PA increases exponentially with amide content (Fig. 3.6). The water absorption equilibration can be carried out at room temperature and also in boiling water. The latter method gives comparable results as the crystalline phase under these conditions is not dissolved. Water drastically lowers the T_g of the PA and with that the modulus and the yield strength at room temperature (Fig. 3.7).[17] With the lowering of the T_g, a β-transition at $-30°C$ is observed. A fully saturated PA has a T_g of $-30°C$. The T_g of a fully water saturated sample approaches that of a non-hydrogen-bonded system such as an N-methylated PA or an aliphatic polyester. Surprisingly, the fracture strength is little affected by the water content. It is typical to observe that a water-containing PA has a higher modulus below its T_g than a dry sample (Fig. 3.7). Thus, in this region, the water has an antiplasticization effect. Curiously, the first percent of water is absorbed without increasing the volume of the system; this water in PA is therefore present in the free space between the polymer chains. At temperatures higher than the T_g, the effect of water on the modulus is small. There is also little effect of the water on the T_m if the sample is warmed gradually. This is probably due to the evaporation of the water during the warming process. However, in an autoclave, the aliphatic PAs dissolve in water well below their melting temperature.

Partial aromatic PAs have, when dry, a T_g of well over $100°C$ and maintain a T_g above room temperature when wet.[18] Thus their properties at room temperature are only marginally affected by their water content.

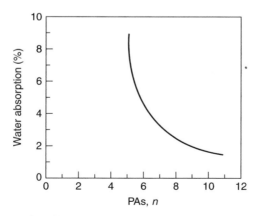

Figure 3.6 Influence of amide content in n PAs on water absorption at 100% relative humidity.[21]

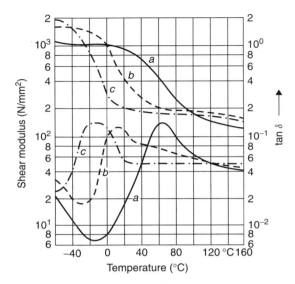

Figure 3.7 Shear modulus (at 1 Hz) as function of temperature of PA-6 at different water contents: *a*, dry; *b*, 3.5%; *c*, 10% water.[17]

3.2.3 Copolymers

3.2.3.1 Copolyamides

If an aliphatic, semicrystalline PA is blended with another aliphatic PA, then initially the polymers mix and crystallize separately. With longer residence times in the melt, transamidation takes place and the melting temperatures change (Fig. 3.8).[6] With even longer mixing times, complete randomization takes place where a random copolyamide is formed having a poor structural order. These random copolyamides do not crystallize quickly; they also have a low crystallinity and their melting temperatures are low. The glass transition temperature of the copolymer is unaffected by this lower order. Similar copolymers can also be prepared in a reactor starting from monomers. A special case is if the copolymers are made from structural units of similar length, such as adipamide and terephthalamide as in PA-6,6/PA-6,T. The adipic and terephthalic units are almost the same in length and, as a result, the melting temperature of the random polymers PA-6,6/PA-6,T is only somewhat lower than can be expected from the rule of mixtures (Fig. 3.9). However, sebacic and terephthalic acid have a dissimilar length, and thus the PA-6,10/PA-6,T copolymers have a poor order. A closer look at the PA-6,6/PA-6,T copolymers by WAXS reveals that the adipamide and terephthalamide units form their own unit cells. Thus, although very similar in length, they are not miscible in the crystalline phase and thus do not form isomorphous structures. The glass transition temperature in these random copolymers increases according to the Fox rule.[6]

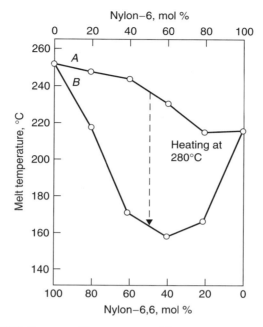

Figure 3.8   PA-66/PA-6 composition versus melting temperature: *A*, polymer blend; *B*, equilibrated random copolymer.[6]

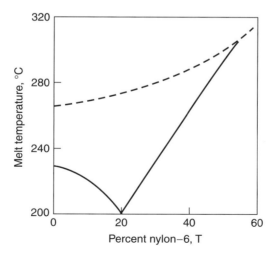

Figure 3.9   Melting temperatures of PA-6,6 and PA-6,10 as function of PA6,T content: (− − −) PA-6,6/PA-6T; (—) PA-6,10/PA-6,T.[6]

3.2.3.2   *Copolyesteramides*

Another important type of condensation polymer are the linear polyesters, such as poly(ethylene terephthalate) (PET) and poly(butylene terephthalate) (PBT). Copolymers of polyesters and PA have been studied in detail, and it has been shown that random copolyesteramides have a low structural order and a low melting temperature. This is even the case for structurally similar systems such as when the group between the ester unit is the same as that between the amide unit, as in caprolactam–caprolactone copolymers (Fig. 3.10).[22] Esters and amide units have different cell structures and the structures are not therefore isomorphous. If block copolymers are formed of ester and amide segments, then two melting temperatures are present.

Alternating ester–amide copolymers behave differently. These copolymers have one T_g and one T_m and can crystallize quickly. Typical examples of these are the 4NT6 polymers in (3.6), the properties of which compare well with PA-6,6

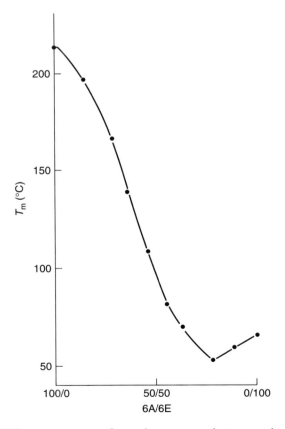

Figure 3.10   Melting temperatures of caprolactam–caprolactone random copolymers as function of molar composition.[22]

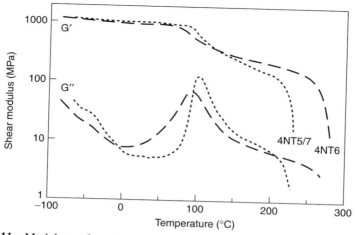

Figure 3.11   Modulus as function of temperature of alternating ester–amide copolymers based on tetramethylenediamine–erephthalic acid: 4NT6 with hexanediol; 4NT5/7 with pentanediol–heptanediol mixture.[23]

and PET (Fig. 3.11):

$$\text{+}(O\text{+}CH_2\text{)}_6\,O—\overset{O}{\overset{\|}{C}}—\!\!\!\!\!\!\!\!\!\!\!\!\!\!—\overset{O}{\overset{\|}{C}}—N\text{+}CH_2\text{)}_4\,N—\overset{O}{\overset{\|}{C}}—\!\!\!\!\!\!\!\!\!\!\!\!\!\!—\overset{O}{\overset{\|}{C}}\text{)}_n \qquad (3.6)$$

An unusual alternating copolyesteramide is 4NT5,7, made with a penta-diol–heptadiol mixture. Although the mixture of uneven diols in this copolymer is expected to drastically lower the crystallinity, the polymer crystallizes quickly, has a high modulus in the T_g–T_m region and, rather surprisingly, is transparent. In this polymer the spherulites must be so small or imperfect that they do not scatter light.

Another ordered polyesteramide is polyester containing a small amount of diamides (amide units of uniform length):

$$\text{+}(O\text{+}CH_2\text{)}_4\,O—\overset{O}{\overset{\|}{C}}—\!\!\!\!\!\!\!\!\!\!\!\!\!\!—\overset{O}{\overset{\|}{C}}\text{)}_9\,N\text{+}CH_2\text{)}_4\,N—\overset{O}{\overset{\|}{C}}—\!\!\!\!\!\!\!\!\!\!\!\!\!\!—\overset{O}{\overset{\|}{C}}\text{)}_n \qquad (3.7)$$

The diamide unit acts as a nucleation initiator for the polyester segments.[24] The presence of these diamide units at low concentrations results in an increased crys-tallization temperature. With increasing diamide concentrations in the polyester, the T_g and T_m also increase.

3.2.3.3   *Copolyetheramide, Thermoplastic Elastomers (TPEs)*

Copolymers can also be made with other polymers. Typically these are the block copolymers with polyether segments in (3.8) which have thermoplastic elastomeric properties:

$$+\{O\!+\!CH_2\!+\!_4O\}_n\!-\!\overset{O}{\underset{\|}{C}}\!+\!CH_2\!+\!_{10}\overset{O}{\underset{\|}{C}}\!+\!\underset{H}{N}\!-\!\!+\!CH_2\!+\!_{10}\overset{O}{\underset{\|}{C}}\}_x\}_m \qquad (3.8)$$

The PA blocks used are often made of PA-11 and PA-6. One difficulty with using these copolymers is that the ether and amide blocks are often incompatible, resulting in melt phasing taking place during the polymerization. If melt phasing takes place, then attaining a high-molecular-weight polymer is more difficult and the copolymer has many transitions. Melt phasing can be avoided by reducing the segment length. However, at a small polyether segment length the materials do not have a very low T_g and low modulus above the T_g. In addition, if the amide segment length is reduced, then the melting temperature, crystallinity, and crystallization rates are lowered. A good combination of properties can be obtained by using high-melting diamide units, which are small segments of uniform length[25,26]:

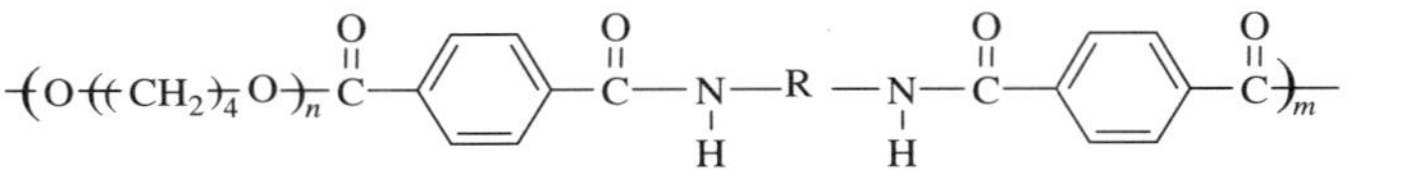

$$+\{O\!+\!CH_2\!+\!_4O\}_n\!-\!\overset{O}{\underset{\|}{C}}\!-\!\!\bigcirc\!\!-\!\overset{O}{\underset{\|}{C}}\!-\!\underset{H}{N}\!-\!R\!-\!\underset{H}{N}\!-\!\overset{O}{\underset{\|}{C}}\!-\!\bigcirc\!\!-\!\overset{O}{\underset{\|}{C}}\}_m \qquad (3.9)$$

By using diamide units, melt phasing can be avoided and at the same time the crystallization rate is rapid. Surprisingly, these segmented copolymers with diamide units not only crystallize quickly but also have a high fracture strain (>1500%) and good elasticity.

3.2.4 Applications

3.2.4.1 Fibers

The PAs were first produced for fiber applications. The PA melt is easily spinnable due to the tension stiffening effect of the melt.[27] PA fibers have a high strength, good wear resistance, can easily be dyed, and are often used as carpet yarn and technical fibers. The all-para wholly aromatic fibers are spun from a liquid crystalline solution.[7,14] The polymer chains can, in this way, easily be oriented and the resulting fibers have a very high modulus and fracture strength. The melting temperature of these materials is also quite high, allowing them to be used in high-temperature applications. PAs with aromatic meta links also have good fire resistance.

3.2.4.2 Engineering Plastics

The high strength, good wear resistance, high ductility, low melt viscosity, and high crystallization rate makes PA the most important engineering plastic.[11–13] The major drawback in this application is the effect of wetting. On wetting, the modulus is lowered and the material swells a few percent; the dimensional stability can be improved by filling the material with glass fibers or particles

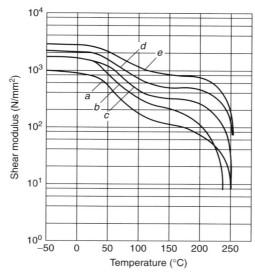

Figure 3.12 Influence of fillers on modulus–temperature relationship: *a*, PA-6 neat; *b*, 30 wt % mineral; *c*, 15 wt % glass fibers; *d*, 30 wt % glass fibers; *e*, 50 wt % glass fibers.[17]

(Fig. 3.12).[11,12,17] In this way, a material with a higher modulus, higher strength, and less swelling is obtained and the maximum-use temperature is considerably increased.

If high impact resistance is required, the PA can be modified with rubber particles.[11,15] The blends are usually made by reactive compounding from maleic-anhydride-modified rubbers, such as, EPDM, EPR, polybutadiene, or SEBS. Partial amorphous PA with a high T_g combines to give a high dimensional stability and good solvent resistance with transparency.

3.2.4.3   *Reaction-in-Mold Nylon*

In water-free conditions, lactam can polymerize anionically using alkali metals such as sodium.[10,12] This anionic polymerization is rapid and can be carried out well below the melting temperature of the polymer, and the molecular weights obtained can be very high. This rapid polymerization allows the reaction to be carried out during processing with the "reaction in mold" (RIM) technique. In the RIM system, polyethers are often added resulting in an ether–amide block copolymer with a very high ductility. The anionic polymerization can also be used for casting thick-walled, large-size products.

3.3 OVERVIEW OF CHEMISTRY AND ANALYTICAL TECHNIQUES

PA belongs to the group of condensation polymers about which many excellent reviews have been written about.[4–13] The PA $(AB)_n$ can be synthesized from cyclic amides (lactams) by the initiation step in (3.10) followed by the addition process in (3.11) and the condensation step in (3.12):

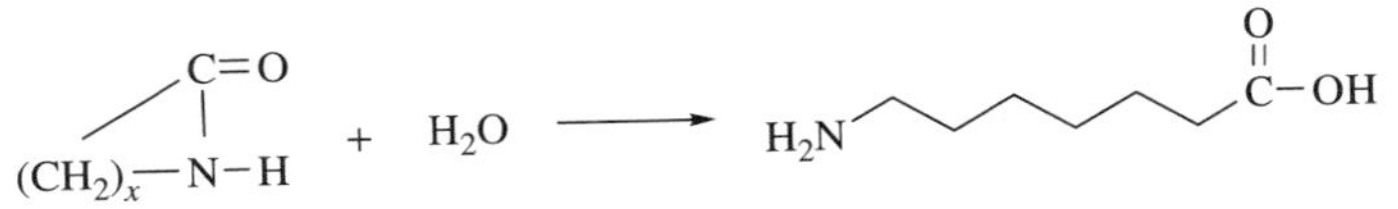

$$
\underset{(CH_2)_x-N-H}{\overset{C=O}{\diagup}} + H_2O \longrightarrow H_2N\diagdown\diagup\diagdown\diagup\overset{O}{\underset{\|}{C}}-OH \qquad (3.10)
$$

$$
\diagup\diagdown\diagup NH_2 + \underset{(CH_2)_x-N-H}{\overset{C=O}{\diagup}} \rightleftharpoons \diagup\diagdown\diagup\overset{H}{\underset{}{N}}-\overset{}{\underset{\|}{\underset{O}{C}}}-CH_2\text{+}NH_2)_x \quad (3.11)
$$

$$
\diagup\diagdown\diagup NH_2 + HO-\overset{O}{\underset{\|}{C}}\diagdown\diagup\diagdown \rightleftharpoons \diagup\diagdown\diagup\overset{H}{\underset{}{N}}-\overset{}{\underset{\|}{\underset{O}{C}}}\diagdown\diagup\diagdown + n\,H_2O
$$

$$
(3.12)
$$

It is also possible to prepare them from amino acids by the self-condensation reaction (3.12). The PAs $(AABB)_n$ can be prepared from diamines and diacids by hydrolytic polymerization [see (3.12)]. The polyamides can also be prepared from other starting materials, such as esters, acid chlorides, isocyanates, silylated amines, and nitrils. The reactive acid chlorides are employed in the synthesis of wholly aromatic polyamides, such as poly(p-phenyleneterephthalamide) in (3.4). The molecular weight distribution (M_w/M_n) of these polymers follows the classical theory of molecular weight distribution and is nearly always in the region of 2. In some cases, such as PA-6,6, chain branching can take place and then the M_w/M_n ratio is higher.

3.3.1 Chemistry and Catalysis

3.3.1.1 *Hydrolytic Process*

The main polymerization method is by hydrolytic polymerization or a combination of ring opening as in (3.11) and hydrolytic polymerization as in (3.12).[5,7,9,11,28] The reaction of a carboxylic group with an amino group can be noncatalyzed and acid catalyzed. This is illustrated in the reaction scheme shown in Fig. 3.13. The kinetics of the hydrolytic polyamidation-type reaction has the form shown in (3.13). In aqueous solutions, the polycondensation can be described by second-order kinetics.[29] Equation (3.13) can also be expressed as (3.14) in which B is the temperature-independent equilibrium constant and ΔH_a the enthalpy change of the reaction[5,6,8,12,28,29]:

$$
-\frac{d[-CO_2H]}{dt} = k[-CO_2H][-NH_2] - \frac{K}{[-CONH-][H_2O]} \qquad (3.13)
$$

$$
\frac{[-CONH-][H_2O]}{[-CO_2H][-NH_2]} = K = B\,\exp\frac{-\Delta H_a}{RT} \qquad (3.14)
$$

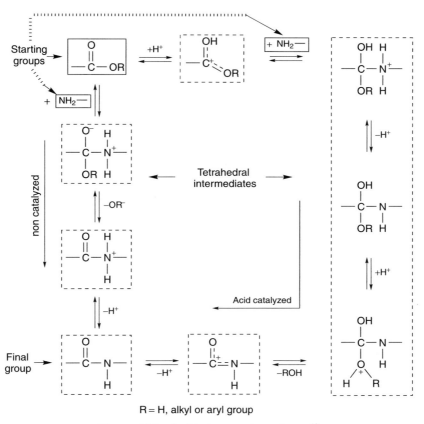

R = H, alkyl or aryl group

Figure 3.13 Amidation reaction scheme.[10]

The equilibrium constant K (also called K_{III} in the synthesis of PA-6) has also been found to be independent of the water concentration. Giori and Hayes[30] found that with up to 0.5 mol water per kilogram of reaction mass (0.9 wt %), K increases linearly with water content (Fig. 3.14). At water concentrations higher than 1 mol $\cdot$ kg^{-1} (1.8 wt %), K decreases again. At very high water concentrations, K seems to be independent of water content.

With increasing water content, the dielectric constant of the medium and the degree of endgroup ionization will increase.[30] This is likely to influence the endgroup activity coefficients, depending on whether the polycondenzation reaction involves the condensation of predominantly neutral or ionized species.

If the equilibrium constant is calculated with activity constants derived from Raoul's law instead of concentrations, then K is virtually independent of the water concentration.[30] The reported values of ΔH_a for hydrolytic polyamidation are in the order of 25–29 kJ $\cdot$ mol^{-1}.[29] This means that on decreasing the temperature at a constant water concentration, the equilibrium molecular weight shifts to a higher value.

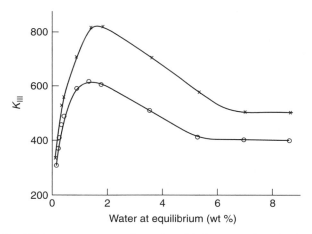

Figure 3.14 Equilibrium constant K for polyamidations reaction (3.13) as function of water content at different temperatures: ($\times$) 240°C; ($\circ$) 260°C.[29]

Others have presented the kinetics of polyamidation differently. At high water concentrations ($5-10$ mol $\cdot$ kg^{-1}), a second-order reaction is given with an activation energy of approximately 86 kJ $\cdot$ mol^{-1}.[5,6,12,28] At low water concentrations in the final stages of the polymerization, a mixed uncatalyzed second-order reaction and an acid-catalyzed third-order reaction are observed. The rate constant k in (3.13) can then be written as

$$k = k' + [-COOH]K'' \tag{3.15}$$

The calculated activation energy is now 96 kJ $\cdot$ mol^{-1} for the uncatalyzed second-order reaction and 88 kJ $\cdot$ mol^{-1} for the third-order reaction. From hydrolysis data using very low water concentrations ($0.005-0.1$ mol $\cdot$ kg^{-1}), the reaction was found to be second order but exhibited a dependence on water concentration in the rate constants (Fig. 3.14). With 1.1 mol $\cdot$ kg^{-1} water, a combination of second- and third-order reactions was observed with activation energies of 109 and 63 kJ $\cdot$ mol^{-1}, respectively.[8]

By removing the water, the reaction equilibrium in (3.12) is shifted to higher molecular weights. However, synthesis under high vacuum (anhydrous) conditions and high temperatures was found to result in colored polymers, and in the case of PA-6,6 also led to branched (gelled) materials.[8,28] Under these conditions, significant degradation reactions take place. The color formation under anhydrous conditions is a result of the formation of Schiff base structures[6]:

$$\tag{3.16}$$

The branching of the polymer is the result of the dimerization of hexamethylene diamine to a tri functional amine[6,28]:

$$\text{(3.17)}$$

In the polymerization of PA-4,6, a side reaction is the cyclization of the diaminobutane endgroup[2,32]:

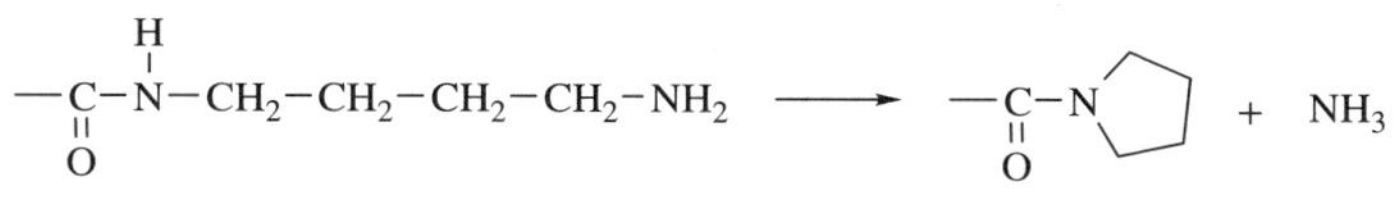

$$\text{(3.18)}$$

The diaminobutane as endgroup is more susceptible to cyclization than as free diamine or as diamide.

3.3.1.2 Ring-Opening Polymerization

Starting with the cyclic lactams, by the hydrolytic polymerization method, three steps can be observed:

- the ring opening to amino acids as in (3.10),
- the lactam addition to prepolymers as in (3.11), and
- the hydrolytic polymerization of the prepolymers as in (3.12).

The concentration of the lactam in the final product is determined by (3.11). Cyclic dimers can also form, and these also take part in the polymerization[12]; the reactions are acid catalyzed. The kinetics of this ring-opening polymerization with the three reactions in (3.10)–(3.12) is complex. The reaction rate constants and equilibrium constants have been described by several authors,[5,6,8,12,28] and more pragmatic approaches for describing the reaction kinetics have also been given.[28,31,33]

The polymerization of PA is carried out between 220 and 280°C, and water is added to initiate the ring-opening reaction. The effect of the water content on the degree of polymerization at equilibrium is given in Fig. 3.15. At a high water content, the degree of polymerization is low but the rate of ring opening (initiation) is high. At low temperatures, the degree of polymerization is higher but the rate of polymerization is lower. Thus, initially, one wants to have a high water content at a high temperature and, later, a low water content at a somewhat lower temperature. For a high water content at high temperatures, the ring-opening step has to be carried out under pressure. The ring opening with water to aminocaprolic acid is, at atmospheric pressure, slow. One can avoid this ring-opening step by starting the lactam reaction with aminocaprolic acid whereby the overall reaction rate increases (Fig. 3.16).[28] Various catalysts have been recommended; however, in practice, PA-6 is normally produced without a catalyst. In the case where a very high molecular weight is aimed for, phosphoric

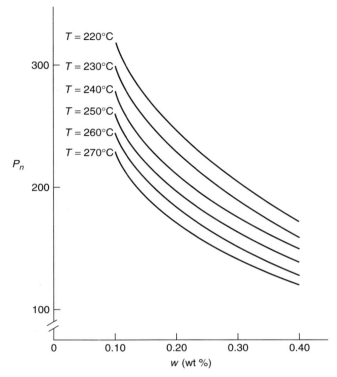

Figure 3.15 Influence of water content (w) on degree of polymerization (P_n) at chemical equilibrium.[31]

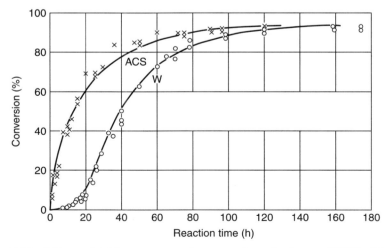

Figure 3.16 Effect of initiator on caprolactam polymerization at 260°C: (○) water 0.01 mol %; (×) aminocaprolic acid 0.01 mol %.[28]

TABLE 3.2   Amino Acid Polymers[a]

Polymer	T_m[b]	Used[c]	Amino Acid T_m	Pka1	Pka2	Lactam T_m	Equilibrium Lactam Concentration at 280°C
PA-2	>350	−C	—	2.43	9.76	—	—
PA-3	∼340	−C	320–340	3.60	10.36	74–76	—
PA-4	278	−C	198	4.23	10.43	127–130	100
PA-5	298	−C	162–163	4.27	10.77	—	15
PA-6	232	+	208	4.37	10.8	68–71	10
PA-7	237	+	196	—	—	∼30	3
PA-8	211	−	192	—	—	—	2.5
PA-9	209	−	191–192	—	—	—	1.5
PA-10	192	−	188	—	—	—	0.8
PA-11	190	+	190–192	—	—	—	1.2
PA-12	187	+	184	—	—	150–153	0.7
3-Aminobenzoic acid		−	178	3.07	9.27		
4-Aminobenzoic acid		−	187	2.30	9.36		

[a] From ref. 10.
[b] Perfect crystalline structure.
[c] Used on a large scale: (−), no; (+), yes; C, easy cyclization.

acid is sometimes used; however, this is at the cost of polymer quality.[28] For the lower $-(AB)_n-PA$ the ring stability of the lactam is unfavorable for melt polymerization (Table 3.2).[5,6,12,28] The equilibrium lactam concentrations of the seven-membered lactam ring and the larger rings are relatively low, so that melt polymerization and melt processing are possible.

3.3.1.3   Acid Chloride Method

PAs can also be prepared with acid chlorides:

$$H_2N-R_1-NH_2 \;+\; Cl-\overset{\overset{O}{\|}}{C}-R_2-\overset{\overset{O}{\|}}{C}-Cl \;\longrightarrow\; \left(\!N-R_1-\overset{H}{\underset{H}{N}}-\overset{\overset{O}{\|}}{C}-R_2-\overset{}{\underset{\|}{C}}\!\right)_{\!n} \;+\; 2\,HCl$$

$$(3.19)$$

Acid chlorides are very reactive and have as a condensation product hydrochloric acid.[4,7,9] This hydrochloric acid can form an amine salt with unreacted amine groups, which should be avoided. To prevent this happening, acid binders, which are more reactive than the amines, are added. Polyamidation can be carried out using a solution and with an interfacial method. With the interfacial method one has the choice between a stirred and an unstirred process. In an unstirred process, the polymerization is at the interface and a rope can be drawn from the interface,

Collapsed film

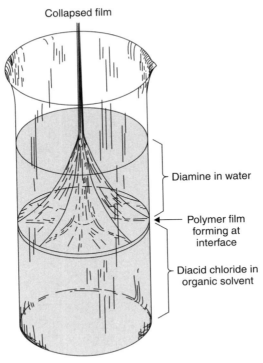

Diamine in water

Polymer film
forming at
interface

Diacid chloride in
organic solvent

Figure 3.17   Setup for interfacial polymerization with pulling a rope.[34]

which is spectacular to view (Fig. 3.17).[4,6,34] In the stirred process, the interface
is much larger and the production faster. Acid binders used in this interfacial
process are potassium hydroxide, sodium hydroxide, and calcium oxide. With
the interfacial method, high-molecular-weight PA can be obtained with PA-6,10.
The shorter diamines and diacids have lower molecular weights (Table 3.3).[4]

The polymerization of aromatic diamines with acid chlorides in solution works
well.[7,9,14,35] The basicity of the aromatic diamine is low and acid binding can be
achieved with several compounds and even solvents such as N-methylpyrrolidone
(NMP) and dimethylacetamide (DMAc). The all-para aromatic amide poly(p-
phenyleneterephthalamide) can be synthesized in DMAc.[7,9,14] To prevent precipitation of the polymer, a salt such as calcium chloride or lithium chloride can be
added. It is also possible to react the acid chloride with a silylated diamine:

$$
\underset{\substack{|\\H}}{Me_3Si-N}-R_1-\underset{\substack{|\\H}}{N}-SiMe_3 \;+\; Cl-\overset{\substack{O\\||}}{C}-R_2-\overset{\substack{O\\||}}{C}-Cl \;\longrightarrow\; \left(\underset{\substack{|\\H}}{N}-R_1-\underset{\substack{|\\H}}{N}-\overset{\substack{O\\||}}{C}-R_2-\overset{\substack{O\\||}}{C}\right)_{\!n}
$$

$$
+
$$

$$
2\,Me_3SiCl
$$

$$(3.20)$$

TABLE 3.3   Inherent Viscosities of Interfacial Polymers[a]

Polyamide	Inherent Viscosity[b]
PA-2,10	1.2
PA-4,10	1.0
PA-6,10	2.06
PA-10,10	0.51
PA-*m*-phenylene,10	0.73
PA-*p*-phenylene,10	0.56
PA-6,2	0.56
PA-6,4	0.28
PA-6,6	1.07
PA-6,10	2.06
PA-6,I	0.81
PA-6,T	0.90

[a] From ref. 4.
[b] 1% solutions.

This reaction is simple and qualitative[36,37]; the diamine can be both an aromatic and an aliphatic diamine. With this method, even star-shaped PAs have been synthesized.[37] Solution polymerization from acid chlorides and aliphatic diamines is more difficult due to the strong basicity of the aliphatic amine groups. Acid binders which have been used with aliphatic diamines are the tertiary amines with high k_b values; these include dimethylbenzylamine and diisopropylethylamine.[4,38]

3.3.1.4   Other Methods

PAs can also be prepared from diisocyanate and diacids:

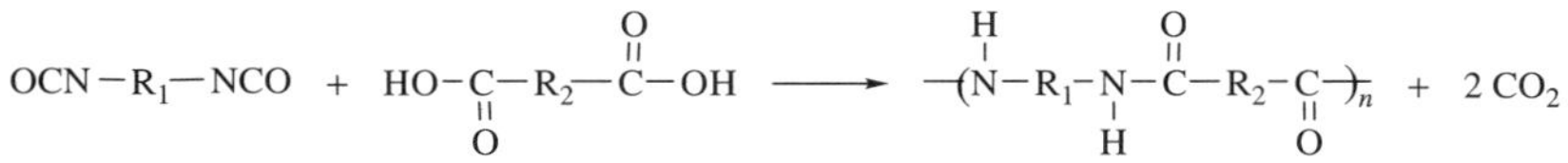

$$OCN-R_1-NCO \ + \ HO-\overset{O}{\underset{O}{\overset{\|}{C}}}-R_2-\overset{O}{\overset{\|}{C}}-OH \ \longrightarrow \ -\!\!\left(\!N-R_1-\overset{H}{\underset{H}{N}}-\overset{O}{\overset{\|}{C}}-R_2-\overset{O}{\underset{O}{C}}\!\right)_{\!\!n} \ + \ 2\,CO_2$$

$$(3.21)$$

This reaction is relatively fast but can have side reactions, as the isocyanate group can also react with an amide group. Esteramide copolymers are prepared from polyesters with acid endgroups and a diisocyanate.[39]

Polyamides can also be prepared from diesters and diamines:

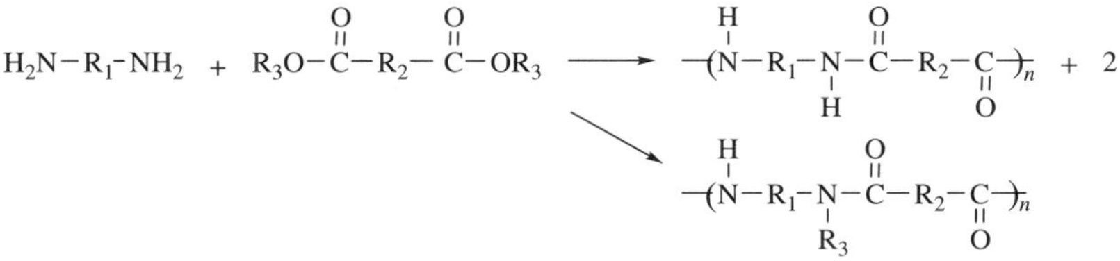

$$H_2N-R_1-NH_2 \ + \ R_3O-\overset{O}{\overset{\|}{C}}-R_2-\overset{O}{\overset{\|}{C}}-OR_3 \ \longrightarrow \ -\!\!\left(\!N-R_1-\overset{H}{\underset{H}{N}}-\overset{O}{\overset{\|}{C}}-R_2-\overset{O}{\overset{\|}{C}}\!\right)_{\!\!n} \ + \ 2\,R_3OH$$

$$(3.22)$$

$$-\!\!\left(\!N-R_1-\overset{H}{\underset{R_3}{N}}-\overset{O}{\overset{\|}{C}}-R_2-\overset{O}{\overset{\|}{C}}\!\right)_{\!\!n}$$

$$(3.23)$$

This reaction is rapid and can, under anhydrous conditions, be carried out at mild temperatures (60–120°C). The type of leaving group has a strong effect on attainable molecular weights. The polyamide melt synthesis with dimethyl terephthalate has however not been so successful, because N-methylation takes place at high temperatures. This N-methylation is due to the methyl ester alkylation of the amines and not due to the presence of methanol.[28,40] This N-methylation reaction is significant at temperatures over 200°C. With other esters, N-alkylation takes place to a much lower extent.

PAs can also be formed from dinitriles and diamines in the presence of water[18]:

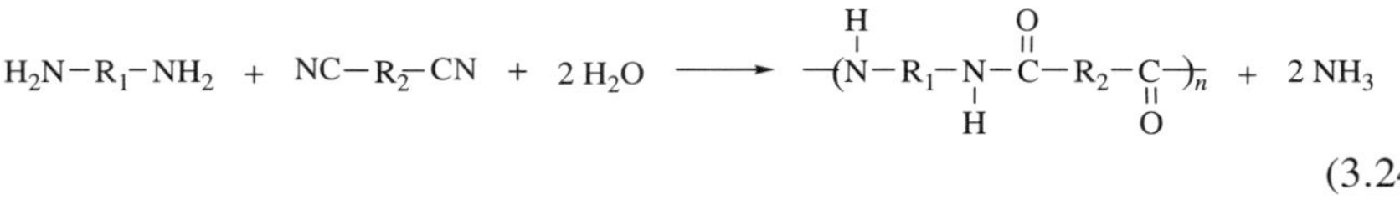

$$H_2N-R_1-NH_2 \ + \ NC-R_2-CN \ + \ 2\,H_2O \ \longrightarrow \ -(N-R_1-N-C-R_2-C)_n \ + \ 2\,NH_3$$

$$(3.24)$$

The reaction proceeds via the hydrolysis of nitrile groups to an amide. The amides may also be *N*-formyl amines, which react with acid groups whereby volatile formic acid is stripped[10]:

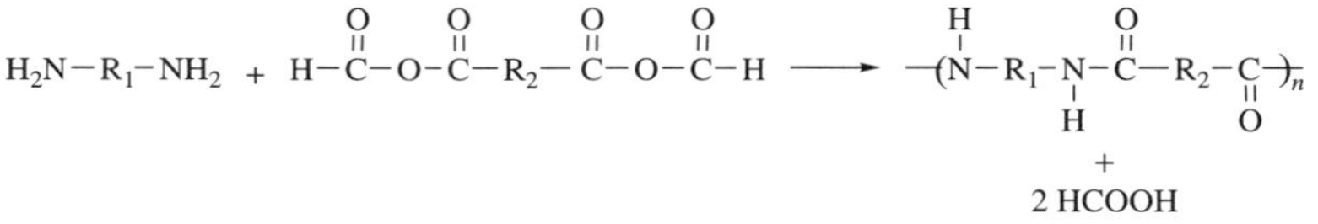

$$H_2N-R_1-NH_2 \ + \ H-C-O-C-R_2-C-O-C-H \ \longrightarrow \ -(N-R_1-N-C-R_2-C)_n$$

$$+ \ 2\,HCOOH$$

$$(3.25)$$

3.3.1.5 *Transamidation*

Next to direct polymerization reactions, interchange reactions between two amide groups can also take place.[37] The effect of the transamidation depends on whether two PAs are mixed or if a PA of two different molecular weights is mixed in the melt. If two types of amides are mixed for 3 min at 260°C, 5% of the amide groups undergo amide–amide interchange [as measured by nuclear magnetic resonance (NMR)] and a block copolymer is formed. By allowing the reaction to proceed for 120 min, a completely random copolyamide is formed. When polyamides of differing molecular weights are mixed, the initial bimodal distribution changes in 3 min at 260°C to a single distribution of almost normal width. The amide–amide interchange reaction is acid catalyzed. Interchange also takes place by reacting an amine with an amide forming a new amide and amine. Therefore, two PAs with amine endgroups can react together to form a larger PA and a diamine, and the latter can be stripped if sufficiently volatile:

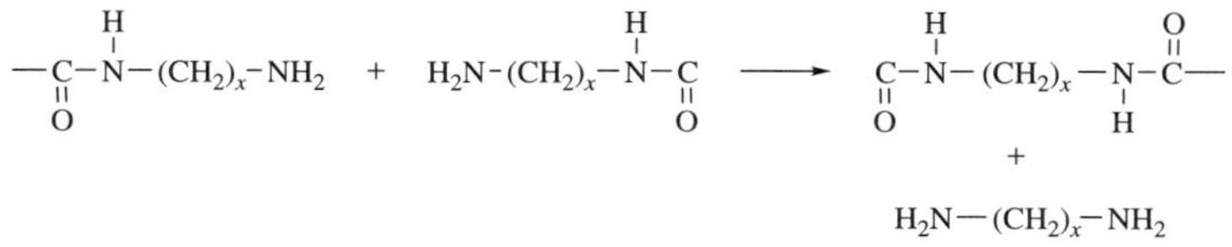

$$-C-N-(CH_2)_x-NH_2 \ + \ H_2N-(CH_2)_x-N-C \ \longrightarrow \ C-N-(CH_2)_x-N-C-$$

$$+ \ H_2N-(CH_2)_x-NH_2$$

$$(3.26)$$

3.3.1.6   Solid-State Polymerization

High-molecular-weight polymers can easily be obtained by solid-state polymerization carried out just below the melting temperature of the polymer.[2,5,6,8,28,41,42] In solid-state polymerization

- stripping of water from small granules is easy,
- thermal degradation is limited as the reaction temperatures are relatively low, and
- equilibrium molecular weights are higher than in the melt.

The higher equilibrium molecular weights in solid-state polymerization are due to the lower reaction temperature and the higher chain-end concentration in the amorphous phase. The endgroups are situated in the amorphous phase and part of the chain segments are in the crystalline phase. This means that due to the crystalline phase, the effective endgroup concentration in the amorphous phase is higher in the solid state than in the melt.[5,43] Another observed effect is that the mobility of the chain ends is limited by the crystalline phase.[42] It was found that it was more difficult to obtain a high-molecular-weight polymer from a low-molecular-weight starting material unless the polymer was remelted in the course of the process. With remelting, randomization of the chain ends can take place and a higher molecular weight material can be obtained.

The starting materials for solid-state polymerization can be monomers (PA salts), prepolymers, and polymers.[6] Solid-state polymerization is carried out on granules at $10-40°C$ below their melting temperature for 24 h; the size of the granules seems to have little effect on the reaction rate. Solid-state polymerization can be carried out in a high vacuum with a nitrogen sweep; however, at high temperatures ($T > 220°C$), in the absence of water, the polymer becomes colored.[2] The use of superheated steam prevents granules from turning yellow, and the steam does not seem to reduce the reaction rate. In the solid state, postcondensation phosphoric acid compounds were found to have a catalytic effect.[12,28,42]

3.3.2   Experimental Methods

The main method of PA synthesis is by melt polymerization. The polymerization of PA-6,6 occurs in two stages, a prepolymerization of the PA salt at elevated pressures followed by a melt polymerization at atmospheric pressure. The prepolymerization stage requires an autoclave, preferably with a glass insert. The glass insert allows easy extraction of the polymer. PA-6 polymerization is simple; it can be carried out at atmospheric pressure, and the evaporating water stirs the reaction medium.

At atmospheric or reduced pressure, a melt polymerization on a small scale (25–100 g) can easily be carried out using straight-wall flange flasks (250 mL) (Fig. 3.18a). The straight wall often allows the extraction of the polymer from

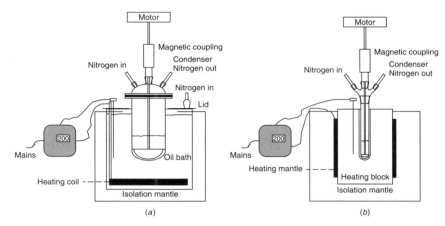

Figure 3.18 Laboratory melt polymerization setup: (*a*) 250 mL; (*b*) 50-mL reactor in either oil bath or heating block.

the flask without breaking it. Flange flasks on the market are made of glass and stainless steel and are sealed by a fluoro rubber ring. The top of the flange flask is comprised of glass and can be three or four necked; for the stirrer, there is the nitrogen inlet, the condenser/nitrogen outlet, and an opening for adding reactants. The stirrer is preferably mechanical with a magnetic coupling. For small quantities (5–15 g) a wide-bore test tube (50 mL) which has been fitted with a three- or four-neck setup can be used (Fig. 3.18*b*). A good heating medium is a silicon–oil bath containing special grade high-temperature oil suitable for temperatures up to 250°C (Fig. 3.18*a*), in which the temperature is controlled by a thermocouple and an electronic controller. To prevent degradation of the oil, the bath is partially enclosed with a glass lid and a stream of nitrogen is fed over the oil. As the silicon oil expands considerably on warming, the bath should not be filled too high. At higher temperatures, sand or salt baths should be used. It is also possible to use a heating block (Fig. 3.18*b*).

For very high melting polymers ($T_m > 300°C$), a solution polymerization is normally employed. If this is started from the reactive acid chloride, the reaction temperature can be low. Polymers from acid chlorides can also be prepared by the interfacial method. Semicrystalline PA can be postcondensed in the solid state to higher molecular weights. To do this, the polymer powder/particles are heated for many hours below their melting temperature in an inert atmosphere.

3.3.3 Characterization

Using IR spectroscopy and NMR, one can analyze the chemical structure of PA. The molecular weight and molecular weight distribution can be analyzed by endgroup analysis, viscometry, and high-pressure liquid chromatography (HPLC). The crystalline order can be analyzed by WAXS, small-angle X-ray spectroscopy

(SAXS), IR spectroscopy, NMR, transmission electron microscopy (TEM), or atomic force microscopy (AFM) and the thermal transitions by DSC and DMA.

3.3.3.1   Solubility of PA[44,45]

Due to their strong hydrogen bonding and high crystallinity, PAs have a high degree of solvent resistance and are thus difficult to dissolve. Organic solvents used to dissolve PA include formic acid, *m*-cresol, phenol–water mixtures, benzyl alcohol (at elevated temperatures), hexafluroisopropanol, trifluoroethanol, trifluoroacetic acid, trifluoroacetic acid anhydride, and aprotic solvents such as DMAc, dimethylformamide (DMF), and NMP containing lithium chloride or calcium chloride. Inorganic solvents include concentrated sulfuric acid and phosphoric acid.

3.3.3.2   Molecular Weight Determination

The molecular weights of PAs are often not very high ($M_n > 20,000$); in this range M_n can be determined by endgroup analysis or, less frequently, by osmometry, M_w can be determined by light scattering. Both M_n and M_w can indirectly be determined by HPLC.

3.3.3.3   Endgroups[45]

Polyamides are mostly linear polymers having reactive amine and/or acid endgroups, unless the polymer chains are capped with monofunctional groups. Monofunctional groups (such as benzoic acid) can be added to control the molecular weight or can be produced as reaction by-products, such as Pyrrolidine, in the synthesis of PA-4,6 in (3.18). Amine endgroups can potentiometrically be titrated in phenol–water solutions[46] while carbonyl endgroups can be titrated in a 1% solution in benzyl alcohol.[47]

Other endgroups can indirectly be quantified by first hydrolyzing the polymer in diluted chloric acid solution and then determining the composition of the compound by HPLC, reverse-phase chromatography, or gas chromatography (GC).[45,48]

3.3.3.4   Light Scattering[45]

For the determination of the molecular weight (M_w) by light scattering, the number of solvent systems is limited. The refractive index difference should be at least 0.1 and the solvent should not have an electrolytic effect. Useful solvents include formic acid containing KCl salt and fluorinated alcohols.

3.3.3.5   Viscometry[11,12,45,48]

Aliphatic PAs dissolve well in *m*-cresol, formic acid (85–90%), and concentrated sulfuric acid (96–98%). Industry usually determines the relative viscosity (η_{rel}) of a 1% solution in concentrated sulfuric acid. The inherent viscosity (η_{inh}) is often used, as this value is relatively independent of concentration. Intrinsic viscosities $[\eta]$ are sometimes determined and correlations with the molecular weight are made.

3.3.3.6 High-Performance Liquid Chromatography[48,49]

The molecular weight and molecular weight distribution can be studied by HPLC. For obtaining accurate molecular weights, calibration is required. The HPLC analysis of PA can be achieved in a solution of *m*-cresol; however, as the viscosity of *m*-cresol is high, the analysis is carried out at elevated temperatures. Other powerful solvents for HPLC analysis are hexafluroisopropanol and trifluoroethanol. A technique typical for PA is acetylating the amide hydrogen with trifluoroacetic acid anhydride.[49] In this way the hydrogen bonds are broken and the modified PA dissolves readily in many suitable solvents for HPLC, such as tetrahydrofuran (THF).

3.3.3.7 Structure Analysis

3.3.3.7.1 Nuclear Magnetic Resonance

Using proton NMR of solutions, the composition of polymers can be analyzed.[47] Carbon-13 NMR spectroscopy is a useful tool for studying the sequence length of segments in copolymers and thereby determining the blockiness of the copolymer. With solid-state NMR, the mobility of chain segments can be studied and the crystallinity determined.

3.3.3.7.2 Infrared Spectroscopy

Infrared spectroscopy can be used to study the structure of PAs and, in particular, their hydrogen bonding. It can best be studied on a 30-μm film preferably obtained from the melt or from solution. An easy method is to take a KBr disc, place on top of that a drop of a PA solution, and evaporate the solvent. The amide group has several specific bonds; the 3300-cm^{-1} band is the N—H stretch and is sensitive to hydrogen bond interactions. In non-H-bonded conditions (dilute solutions), it is at 3450 cm^{-1}; the stronger the H bond, the lower the value. The amide I band (carbonyl stretch) shifts from 1720 cm^{-1} for non-H-bonded conditions to lower values with increasing H bonding and is usually found at 1640 cm^{-1}. The methylene sequence, which changes position on crystallization in the amorphous phase, is at 1118 cm^{-1} and in the crystalline phase is at 836 cm^{-1}.[11]

3.3.3.7.3 Differential Scanning Calorimetry

Thermal transitions can be studied by DSC. The crystallization transition is usually sharp with a good baseline. The melting transition is more complex and often not a single transition (Fig. 3.19)[48] as it depends on the thermal history of the sample and the structural changes that can take place upon heating. In warming, solid-state transitions can take place in the unit cell, the lamellae can thicken, and secondary crystallization can also take place. The heats of crystallization and

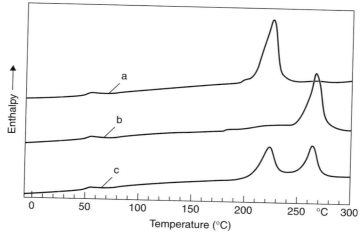

Figure 3.19   DSC trace of PAs: *a*, PA-6; *b*, PA-6,6; *c*, PA-6/PA-6,6 mixture.[48]

fusion are a measure of crystallinity. The glass transition temperature for PA does often not give a good measurable change in C_p value.

3.3.3.7.4   Dynamic Mechanical Analysis

With DMA the effect of temperature on the modulus can be studied. By increasing the temperature from -150 to $300°C$, one encounters several transitions in PA (Fig. 3.1). There is a transition at about $-120°C$, the γ-transition, which is due to the mobilization of methylene units. There is also a transition at $-30°C$, which is present in wetted aliphatic PA; this is due to non-H-bonded amide units and is termed the β-transition. At about $50°C$ the glass transition (T_g) (α-transition) of the aliphatic polyamides PA-6 and PA-6,6 occurs. At this transition, the modulus is lowered considerably. For partially aromatic PA, the T_g occurs above $100°C$. The last transition is the flow temperature, at which temperature the material melts; the flow temperature and the melt temperature, as measured by DSC, correspond well. The modulus is a measure of dimensional stability and increases with crystallinity and filler content (Fig. 3.12).

3.3.3.7.5   X-Ray Diffraction

The order in PA can be studied by WAXS and SAXS. WAXS gives insights into the unit cell structure, the crystallinity, the crystallite size (if not too large), and the crystallite orientation. SAXS gives a more accurate measure of the lamella thickness. In a temperature sweep, the changes in structure with temperature can be followed.

3.3.3.7.6   Electron Microscopy

The lamella and spherulitic structures can be studied using electron microscopy. The most informative technique is TEM, although a more recently developed technique is AFM. For TEM, three methods of staining are frequently reported: osmium tetraoxide, ruthenium tetraoxide and phosphor tungsten acid.

3.4 SYNTHETIC METHODS

3.4.1 Introduction

Amides are formed by the reaction of an amine with an acid group [see (3.1)]. PAs can consist of either alternating diamines (AA) and diacids (BB) or amino acids (AB). In order to obtain high-molecular-weight polymers, the amine and acid functional groups must be present stoichiometrically. This stoichiometry is inherently present in the amino acid AB-type monomers. In $-(AA-BB)_n-$polymers, the diamines and diacids have to be present in equal amounts. PAs can be synthesized in the melt phase, in solution, or using an interfacial method. If the aliphatic PA is thermally stable to above its melting temperature, a melt synthesis method is almost exclusively used. Many PAs have been studied, but only a handful are sufficiently thermally stable to be melt synthesizable and melt processable. On an industrial scale, most PAs are prepared in the melt phase. In order to obtain high molecular weights, the polymers are postcondensed in the solid state. This occurs above the T_g of the polymer but 10–40°C below its melting temperature. The reaction times vary from a few hours to up to 24 h.

3.4.2 Aliphatic AA–BB-Type Polyamides

The polyamides $-(AA-BB)_n-$can be prepared from diamines (AA) and diacids (BB), for example, PA-4,6, PA-6,6, PA-6,10, and PA-6,12; possible starting diamines and diacids are given in Table 3.4. The lower diamines readily cyclize and the lower diacids decarboxylize. The polymers are formed by the equilibrium condensation reaction (3.12) where the condensation product, usually water, has to be removed. The smallest cyclic product (AA–BB) that can form is quite large $(x + y + 4)$, and thus the concentration of cyclic products in these polymers will be small. These aliphatic AA–BB polymers are made in the melt phase, starting with a watery solution of the diamine and the diacid. It is also possible to synthesize these polymers from diacid chlorides in an organic solvent or by using an interfacial method. Synthesis from a diamine and a diester has little been reported. However, synthesis from dinitriles and diamines together with water has recently been described [see (3.24)].[15] For obtaining very high molecular weight polymers, a postcondensation step is carried out in the melt phase in an extruder or in the solid state.[6,12,28]

The diamine and the diacid form a PA salt that is soluble in water at elevated temperatures. The polymerization from the PA salt solution occurs in two or three stages. In the first stage a prepolymer is made. This step is carried out under pressure to prevent the evaporation of the more volatile diamine. In the second stage, a polymer is made in the melt phase at atmospheric or slightly reduced pressure.

TABLE 3.4   Diamines and Diacids[a]

Monomer	n^b	Used[c]	T_m	T_b	PKa1	PKa2
Diamines						
Hydrazine	0	−C	2.0	113.5	8.11	>12
1,2-Ethylene	2	−C	9.95	116.7	6.98	9.98
1,3-Propylene	3	−C	−22.5	136.5	8.58	10.62
1,4-Tetramethylene	4	+	27	158	9.32	10.36
1,5-Pentamethylene	5	+	9	178	9.74	11.00
1,6-Hexamethylene	6	+	39	196	10.76	11.86
1,8-Octamethylene	8	−	53	122		
1,10-Decamethylene	10	−	63			
1,12-Dodecamethylene	12	+	70			
1,4-Cyclohexylene	—	−	68–72			
1,3-Phenylene	—	+	64		9.02	11.5
1,4-Phenylene	—	+	147		7.84	11.5
1,5-Naphthylene	—	−	197–190			
1,8-Naphthylene	—	−	63–65			
Carboxylic diacids						
Ethanedioic (oxalic)	2	−D	189		1.46	4.4
Propanedioic (malonic)	3	−D	135		2.8	5.85
Butanedioic (succinic)	4	−D	185		4.17	5.64
Pentanedioic (glutaric)	5	−D	97.5		4.33	5.57
Hexanedioic (adipic)	6	+	151		4.43	5.52
Heptanedioic (pimelic)	7	+	105		4.47	5.52
Octanedioic (suberic)	8	−	142		4.52	5.52
Nonanedioic (azalic)	9	−	106		4.54	5.52
Decanedioic (sebacic)	10	−	134		4.55	5.52
Undecanedioic	11	−	111			
Dodecanedioic	12	+	112			
Phthalic	—	−C	191		3.0	5.40
Isophthalic	—	+	348		3.54	4.62
Terephthalic	—	+	>300		3.54	4.46
1,4-Naphthalic	—	−	>300			
2,6-Naphthalic	—	−	>300			

[a] From ref. 28.
[b] Number methylene units.
[c] Used on a large scale: (−), no; (+), yes; C, easy cyclization.

At moderately high molecular weights, the reaction mass becomes highly viscous, which limits heat transfer and evaporation of the condensation water. This high viscosity limits further melt polymerization in the bulk. In the literature, the melt polymerization stage is sometimes omitted and the prepolymers are condensed to high molecular weights in the solid state.[6,28,41] The polymerizations can easily be carried out without a catalyst.

The AA–BB polyamides are nearly always prepared using bulk polymerization with a prepolymerization step at higher pressure, although not every laboratory has the facility to carry out a polymerization in an autoclave. Nielinger[50] has reported bulk polymerization of PA-6,I from its salt solution at atmospheric pressure (Example 11). This method may also be usable for other laboratory polymerizations like PA-6,6.

3.4.2.1   PA-6,6

The most important AA–BB-type polymer is PA-6,6. It is a semicrystalline material and has a high melting temperature (265°C). PA-6,6 is prepared from 1,6 hexamethylenediamine and adipic acid (Eq. 3.27):

The diamine part has six methylene units and the diacid part four methylene units and two acid carbon atoms. Thus, it has an even number of methylene units in both the diamine and the diacid part, which gives the polymer chain regularity so it is able to crystallize easily.

The polymer is industrially prepared using a melt — polymerization method starting with a 50% watery solution of the diamide and the diacid[5,6,8,28]; this concentrated solution can be pumped and transported. Water can be omitted, but then the PA salt must be isolated first. In methanol and ethanol the PA salt precipitates in equal molar amine–acid amounts. In preparing the PA salt, the reactants are also purified and a less volatile starting material is made. The dry salt is a good starting material for synthesis on a small scale. In the polymer, the diamine and the diacid should be present in equal amounts. However, hexamethylenediamine is more volatile than adipic acid, and if care is not taken, an imbalance occurs during the polymerization. The loss of the diamine can be limited by carrying out the prepolymerization under pressure or in a closed system. A closed system or an autoclave reaction limits the loss of the diamine and, at the same time, the evaporation of water. Water dissolves not only the PA salt but also the prepolymer and the polymer. The presence of water makes it possible to keep the reaction mass molten well below the melting temperature of the (pre)polymer. The complex situation is that water is not only a solvent for the PA prepolymer but also a condensation product of the polymerization. Too much water will therefore limit the attainable molecular weight. In the prepolymer, the diamine is bound and the polymerization can then be continued at atmospheric pressure above the melting temperature of the polymer. As the molecular weight increases, the viscosity of the polymer melt also increases. In a tank reactor, stirring then becomes difficult, resulting in a slower evaporation of the condensation product and a buildup of heat at the reactor wall.

A logical step to overcome the difficulty of the evaporation of water seems to be to apply a vacuum, as done in polyester synthesis. A vacuum has the effect of increasing the molecular weight. However, under a vacuum at high temperatures ($T > 220°C$), degradation of the PA takes place.[5,6,8,28] The degradation at reduced pressure takes the form of branching of the polymer[17] and, at the same time, color formation.[16]

A typical polymerization process as carried out on an industrial scale is given in Fig. 3.20.[5] An autoclave (about 4 m^3) is charged with a concentrated PA salt solution in water (50% at 50°C). The autoclave is flushed with nitrogen and heated to 210°C while the pressure is allowed to increase to 18 bar. In this heating step, the salt concentration is increased to 75%. At 210°C, the reaction is carried out for 60 min. In the second step, the temperature is slowly increased to 280°C while maintaining the pressure at 18 bar. Subsequently, the reaction mass is warmed to 290°C and, at the same time, the pressure is lowered to atmospheric. At 290°C and at a slight pressure, the polymerization is continued for 1 h before the polymer is discharged.

Continuous polymerization processes for PA-6,6 have been reported for over 30 years.[5,6,28] Prepolymerization in tubular (Fig. 3.21) or baffled reactors is particularly well suited to continuous polymerization. The polymerization of prepolymers to high-molecular-weight materials in a continuous process is more difficult to control as small differences is molecular weights result in large differences in viscosities. Viscosity differences result in different hold-up times in the reactor and thus nonhomogeneous products.

To reduce the chance of side reactions, such as the dimerization of the diamine to ω,ω'-diaminodihexylamine in (3.17) and the degradation of adipic acid to Schiff bases in (3.16), the precondensation can be carried out either for 30–60 min below 250°C or very rapidly (seconds) at 250–290°C. In the (pre)polymerization step, a concentrated PA salt solution is pumped into a set of heating tubes. These tubes have several heating zones as their diameter gradually increases in size.[5,6,28]

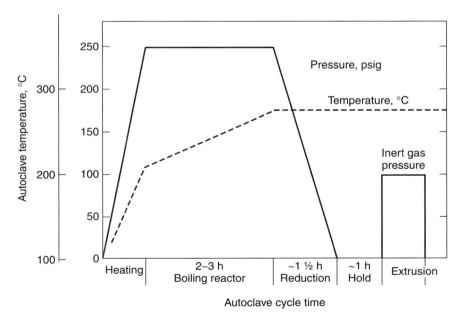

Figure 3.20   Typical autoclave cycle for batch polymerization of PA-6,6.[5]

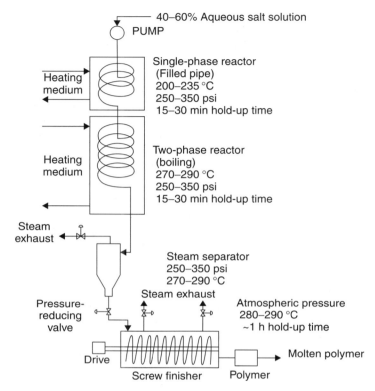

Figure 3.21   Early continuous PA-6,6 polymerization setup: (*a*) salt concentrator; (*b*) vent; (*c*) pump; (*d*) flash tubes; (*e*) finisher.[5]

The process starts with a one-phase flow of a low-viscosity PA salt solution and ends with a two-phase flow with the viscous PA melt along the tube wall and steam through the middle. A stream separator and a finishing reactor are attached to the polymerization tubes.

In a polymerization from a watery solution, the added water has to be removed at some stage during the reaction. This evaporation uses a considerable amount of energy and reaction time. An alternative method is to directly react the molten diamine with the molten diacid.[51] In this process, the mixing of the reactants is critical; this mixing can be carried out using a static mixer. It has also been proposed that polymerization can be started with molten diamine-rich and diacid-rich streams.[52] The advantage of using diamine-rich and diacid-rich streams is that the reactants are less volatile and therefore the reaction pressure can be lower. The metering also appears to be easier to control. On an industrial scale, PA-6,6 is synthesized from a concentrated watery salt solution and on a small scale from the dry PA salt.[53,54]

Example 1. PA-66 from PA salt.

1a. Hexamethylenediamine-adipic acid salt [53] :

$$H_2N(CH_2)_6NH_2 + HO-\overset{O}{\underset{\|}{C}}(CH_2)_4\overset{O}{\underset{\|}{C}}-OH \longrightarrow {}^{\oplus}H_3N(CH_2)_6NH_3^{\oplus} {}^{\ominus}O-\overset{O}{C}(CH_2)_4\overset{O}{C}{}^{\ominus}O$$

$$2\ H_2O + \left(\underset{H}{\overset{H}{N}}(CH_2)_6\underset{H}{N}-\overset{O}{\underset{\|}{C}}(CH_2)_4\overset{O}{\underset{\|}{C}}\right)_n$$

$$(3.27)$$

In a 500-mL Erlenmeyer flask, 29.2 g (0.2 mol) of adipic acid is dissolved in 250 mL of warm ethanol, and the solution is cooled to room temperature. A solution of 23.43 g (0.202 mol) of hexamethylenediamine in 50 mL of ethanol is added to the adipic acid solution with gentle stirring (rinse the flask containing the diamine with ethanol). The solution warms up by the exothermic reaction and immediate precipitation of the salt takes place. The suspension is allowed to cool, then filtered, washed with methanol, and air dried. The pH of a 1% solution of the salt in water is measured and should be 7.6.

The relationship between the composition and pH is given in Fig. 3.22.

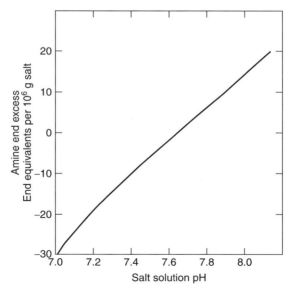

Figure 3.22   The pH of PA-6,6 salt as function of excess amine group concentration.[5]

1b. PA-6,6 polymerization from dry PA salt in a small autoclave. A 100-mL autoclave with a glass insert is filled with 20 g of hexamethylenediamine–adipic acid salt. The autoclave is closed and flushed with nitrogen by alternately evacuating with a vacuum pump and filling with nitrogen (four to five cycles). It is then filled with nitrogen to a starting pressure of 5 bar. The autoclave is warmed to 210°C and kept at this temperature for 2 h, whereby the pressure increases to about 14 bar. Subsequently, the autoclave is warmed to 280°C over a period of 30 min while at the same time gradually lowering the pressure to atmospheric. The polymerization is continued for 1 h at 290°C at atmospheric pressure without letting air in. After cooling the autoclave, the PA-6,6 is removed. The polymer can be crushed into small particles. For this, the polymer is cooled in liquid nitrogen, placed in a PE or polypropylene (PP) bag and crushed in a vice while still cold.

The polymer obtained is white, opaque, and tough and has a crystalline melting temperature of 265°C. The inherent viscosity (η_{inh}) is 1.0–1.2 in 96% sulfuric acid (1.0 % solution, 25°C). In the past, prepolymers were prepared from dry PA salt on a small scale in sealed heavy-walled glass tubes. As these heavy-walled tubes are not safe to handle, they should no longer be used.

1c. Solid-state postcondensation of (pre)polymers. To increase the molecular weight further, the polymer can be postcondensed in the solid state. The granular (pre)polymer (10 g) is placed in a 100-mL flask. The flask is flushed with nitrogen and placed in an oil bath. The oil bath is heated to 200°C over a 1-h period and maintained at this temperature for 24 h. The polymer obtained has an η_{inh} of 1.4–1.6 in 98% sulfuric acid.

An oven can be used instead of an oil bath. If the amount of polymer to be post-condensed is large, a tumble dryer (rotavapor apparatus) or moving bed reactor can be used.

3.4.2.2   PA-6,10 and PA-6,12

PA-6,10 is synthesized from 1,6-hexamethylenediamine and sebacic acid, and PA-6,12 from 1,6-hexamethylenediamine and dodecanedioic acid. The melt synthesis from their salts is very similar to PA-6,6 (see Example 1). These diacids are less susceptible to thermal degradation.[55] PA-6,10 can also be synthesized by interfacial methods at room temperature starting with the very reactive sebacyl dichloride.[4,35] A demonstration experiment for interfacial polycondensation without stirring can be carried out on PA-6,10. In this nice classroom experiment, a polymer rope can be pulled from the polymerization interface.[34]

Example 2. PA-6,10 interfacially.[34]

$$H_2N \text{---}(CH_2)_6\text{---}NH_2 + Cl\text{---}\overset{O}{\underset{\|}{C}}\text{---}(CH_2)_8\text{---}\overset{O}{\underset{\|}{C}}\text{---}Cl \ + \ 2\,Na_2CO_3 \ \longrightarrow \ \text{---}(\overset{H}{\underset{|}{N}}\text{---}(CH_2)_6\text{---}\overset{H}{\underset{|}{N}}\text{---}\overset{O}{\underset{\|}{C}}\text{---}(CH_2)_8\text{---}\overset{O}{\underset{\|}{C}})_n\text{---}$$

$$+$$

$$2\,NaHCO_3 + 2\,NaCl$$

(3.28)

In 50 mL of perchloroethylene is dissolved 1.5 mL of sebacyl chloride; 2.2 g of hexamethylenediamine and 4.0 g of sodium carbonate are dissolved in 50 mL water. The aqueous solution is placed carefully on the perchloroethylene solution in a 200-mL tall-form beaker or similar vessel (Fig. 3.17). A film is immediately formed at the interface. The film is grasped with tweezers and raised as a rope of continuously forming polymer film. The rope is guided over two glass rods with a 1-mm drop. With this setup, speeds up of 5 m/min can be attained for 8 min.

The collapsed film should be thoroughly washed with water.

3.4.2.3 PA-4,6

PA-4,6 is synthesized from 1,4-tetramethylenediamine and adipic acid. Due to its chain regularity and high amide content, PA-4,6 has a melting temperature of 290°C. A melt synthesis without degradation is therefore difficult. The cyclization of the 1,4-tetramethylenediamine endgroup to a nonfunctional endgroup is typical for PA-4,6 [see (3.18)]. A high pyrrolidine content limits the attainable molecular weight.

A good method for making PA-4,6 is a two-step process involving prepolymerization of the nylon salt at 180–220°C and 8–18 bar for 1–2 h followed by postcondensation of the powder in the solid state at 220–260°C.[2] It has been found that at these high reaction temperatures colorless products can be best obtained by conducting the solid-state polymerization in an atmosphere of superheated steam. As 1,4-tetramethylenediamine has a low boiling temperature, some diamine loss during the high-temperature polymerization is unavoidable, and this has to be corrected for. A small excess of diamine does not seem to be a problem due to transamidation where the excess diamine can be freed and removed, as in (3.26).

The precondensation can be carried out continuously with the use of a tubular reactor at a temperature of 290–310°C.[56] The tubular reactor is a 4-m-long coiled pipe with a diameter of 4 mm which is heated at 300°C. At the end of the pipe is a valve which is regulated so that the pressure is 1.5 bar. The residence time in the pipe is only seconds. The prepolymer obtained can be postcondensed in the solid state to a high molecular weight.

Example 3. PA-4,6 from dry PA salt.[2]

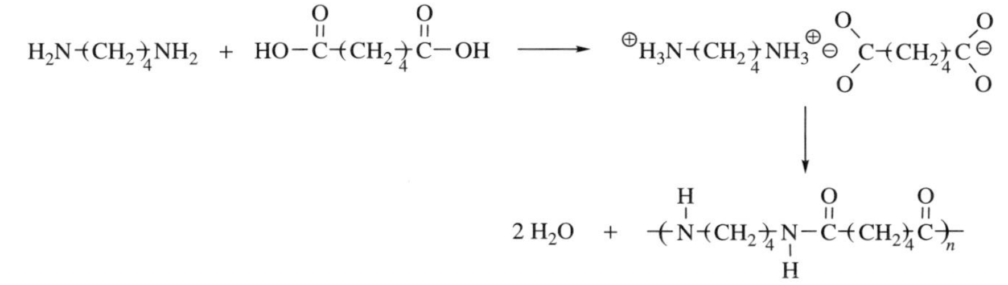

$$(3.29)$$

*PA-4,6 salt is prepared from adipic acid and 1,4-tetramethylenediamine as describ-
ed for the PA-6,6 salt (Example 1a). PA-4,6 salt (20 g), 2 mL water, and 0.2 mL
1,4-tetramethylenediamine (2.1 mol % excess) are added to a 100-mL glass container in an autoclave. The autoclave is flushed with nitrogen, closed, and given
a starting nitrogen pressure of 5 bar. The autoclave is heated over a period of
60 min to 180°C and maintained at that temperature for 100 min, when the pressure is increased to about 8 bar. The pressure is then gradually released, the
reaction mass cooled, and the material removed from the autoclave. The prepolymer is crushed into small particles (0.1–0.2 mm) (see Example 1b). This
prepolymer has a relative viscosity (η_{rel}) of 1.3 as measured in 96% sulfuric acid
(1% solution at 25°C).*

*The prepolymer is subjected to further condensation by heating for 4 h at 260°C
while passing superheated steam at a pressure of 1 bar through the solid bed. The
white PA-4,6 obtained has a η_{rel} of 4.6 as measured in 96% sulfuric acid.*

3.4.2.4   PA-10,2

Polymers containing oxalic acid can be prepared from their esters.[57] With the
reactive esters the prepolymers can be synthesized at low temperatures. Polymerization of these prepolymers is possible at 270°C.

Example 4. PA-10,2 from oxalic ester.[57]

$$H_2N\text{-}(CH_2)_{10}\text{-}NH_2 \;+\; C_4H_9O-\overset{\overset{O}{\|}}{C}-\overset{\overset{O}{\|}}{C}-OC_4H_9 \longrightarrow \text{-}(N\text{-}(CH_2)_{10}\text{-}\overset{\overset{H}{|}}{N}-\overset{\overset{O}{\|}}{C}-\overset{\|}{C})_n$$

$$+$$

$$2\, C_4H_9OH$$

$$(3.30)$$

*Dibutyloxalate (20.22 g, 0.10 mol) is added to a solution of 17.23 g (0.10 mol)
decamethylenediamine in 25 mL toluene (dried over sodium) in a 250-mL three-
necked flask equipped with stirrer, nitrogen inlet, and drying tube in a heating
block (Fig. 3.18a). The residual oxalate ester is quickly washed into the flask
with another portion of dry toluene. Heat is liberated and a white solid begins to
form within a very short period of time. Stirring is continued until the suspension
becomes too viscous (at this stage the η_{inh} is 0.15–0.25). Two hours after the initial
addition, the flask is heated to 270°C. A current of nitrogen is continually passed
over the reaction mass and toluene and butanol are distilled off. The reaction
is continued for 1 h at 270°C after which the polymer is permitted to cool. A
tough white polymer is obtained having an η_{inh} of 0.7–0.8 (0.5% solution in 96%
sulfuric acid) and a melting temperature of 240°C.*

3.4.2.5   PA-4,2 from Oxalic Ester

PA-4,2 is an even–even PA with a very high melting temperature (390°C). It is prepared from an oxalic ester and 1,4-tetramethylenediamine in a two-step procedure: a prepolymerization in solution and a polymerization of the prepolymers in the solid state.

Example 5. PA-4,2 from oxalic ester.[58]

$$H_2N\text{-}(CH_2)_4\text{-}NH_2 \; + \; C_2H_5O\text{-}\overset{O}{\underset{||}{C}}\text{-}\overset{O}{\underset{||}{C}}\text{-}OC_2H_5 \; \longrightarrow \; \left(\text{N-}(CH_2)_4\text{-}\underset{H}{\overset{H}{N}}\text{-}\overset{O}{\underset{||}{C}}\text{-}\overset{}{\underset{O}{C}}\right)_n \; + \; 2\,C_2H_5OH$$

$$(3.31)$$

To a dry-nitrogen-flushed 1-L flange flask fitted with a mechanical stirrer, a nitrogen inlet, and a distilling head, 250 g phenol and 250 g trichlorobenzene are added and warmed to 65°C. To this solvent mixture, 22.44 g (0.255 mol) tetramethylenediamine and 36.5 g (0.250 mol) diethylene oxalate are added. The solution is warmed to 140°C over 30 min and maintained at this temperature for 2 h. During the polymerization, ethanol is liberated and the initial homogeneous solution gradually turns cloudy and viscous. The reaction mixture is then cooled to 80°C and poured into a glass beaker with 2 L of petroleum ether. The precipitated prepolymer is filtered, washed twice with diethyl ether, and dried in a vacuum oven. The material has an η_{inh} of 0.16 in 96% sulfuric acid (0.5% solution, 25°C).

Fifteen grams of this prepolymer powder in a wide-bore reaction tube (Fig. 3.18b) which is flushed with nitrogen is placed in a heating block. The heating block is warmed over a period of 1 h to 270°C and maintained at this temperature for 4 h, after which the reaction vessel is removed. The yellow polymer obtained has an η_{inh} of 1.9. The polymer has a melting temperature of 391°C, a heat of fusion of 148 J/g, a T_g dry at 120°C, and a T_g wet at −15°C.

3.4.3 Aliphatic AB-Type PAs

The AB polyamides are made from either ω-amino acids or cyclic lactams, derivatives of the ω-amino acids (Table 3.1). In these polymers, the amino and acid groups are inherently balanced and the polymer also contains one amino and one acid endgroup. There are a number of different routes available for polymerizing these AB-type polyamides:

- hydrolytic process, starting with amino acids;
- ring-opening polymerization of lactams, and
- anionic and cationic polymerizations.

If prepared from amino acids, the polymers are made in a condensation reaction whereby water is split off:

$$H_2N\text{-}(CH_2)_5\text{-}\overset{O}{\underset{||}{C}}\text{-}OH \; \longrightarrow \; \left(\text{N-}(CH_2)_5\text{-}\overset{O}{\underset{||}{C}}\right)_n \; + \; H_2O \qquad (3.32)$$

This is a typical equilibrium reaction, which proceeds relatively quickly. Next to the water–polymer equilibrium, a ring-chain lactam–polymer equilibrium also exists:

$$\underset{(CH_2)_5-N-H}{\overset{C=O}{\diagup}} \quad + \quad H_2N-(CH_2)_5-\overset{O}{\overset{\|}{C}}-OH \quad \longrightarrow \quad -(N-(CH_2)_5-\overset{O}{\overset{\|}{C}})_n- \quad (3.33)$$

The ω-amino acids are available at high purity but are generally more expensive than their lactams and only suitable for the small-scale preparation of polymers. As they are bulk polymerizations, the polymerization temperature is preferably above the melting temperature of the polymer.

Another method is to start with lactams. The cyclic lactams have a lower melting temperature compared to ω-amino acids and are therefore more easily purified and easier to handle. ε-Caprolactam has a melting temperature of 69°C and can be transported in the molten state in heated tanks. The energy consumption of the lactam polymerization is also low as little water is added by the polymerization process and therefore there is little to evaporate.

The melt polymerizations of PA-AB from PA-6 onward (PA-6 to PA-12) are straightforward. The polymerization is an exothermic reaction, meaning that equilibrium conditions for the lactam–polymer in (3.11) and condensation reaction prepolymer–polymer in (3.12) are most favorable at low temperatures. The reaction rates, however, are increased at higher temperatures. Very high molecular weights can be obtained by including a postcondensation step in the solid state. Of the lower PA (PA-2 to PA-5), the "ring-chain equilibrium" is unfavorable at high temperatures and melt polymerization is not effective (Table 3.2). A rapid, low-temperature polymerization method for lactams is anionic polymerization. Anionic polymerization is also employed for PA-6 and PA-12 in cast PA and RIM nylon.

The most important AB-type PA is PA-6, having six carbon atoms in the repeat unit and therefore five methylene groups. In addition to PA 6, PA-11 and PA-12 are also commercially available. The synthesis of many AB-type PAs, PA-1 to PA-22, have been studied,[5,6,12,28] but the quality of the resultant polymers is highly dependent on the purity of the starting materials.

3.4.3.1   Polyamide-6

On an industrial scale, PA-6 is synthesized from ε-caprolactam with water as the initiator. The process is very simple if the reaction is carried out at atmospheric pressure. The polymerization is carried out in a VK-reactor (Fig. 3.23), which is a continuous reactor without a stirrer, with a residence time of 12–24 h at temperatures of 260–280°C.[5,28] Molten lactam, initiator (water), and chain terminator (acetic acid) are added at the top and the polymer is discharged at the bottom to an extruder. In this extruder, other ingredients such as stabilizers, whiteners, pigments, and reinforcing fillers are added. The extruded thread is cooled in a water bath and granulated. The resultant PA-6 still contains 9–12%

lactam, which can be largely extracted by hot water or methanol. For this type of polymerization, there is a simple relationship between the solution viscosity as a function of temperature (T) and acetic acid chain stopper concentration (c)[28]:

$$\eta_{rel} = 0.02 \cdot T - (2.37 + 1.6 \cdot c) \tag{3.34}$$

where η_{rel} is the viscosity in 96% sulfuric acid (1% solution, 25°C) after extraction.

The initiation of the lactam polymerization with water is directly dependent on the water concentration:

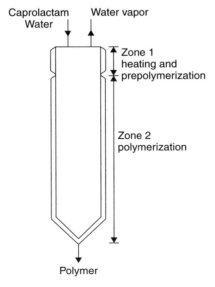

$$\tag{3.35}$$

At higher water concentrations (at higher pressure) (Fig. 3.24), the reaction is considerably more rapid. The prepolymer obtained is further polymerized in another reactor, which has a working pressure of 1 bar or less. The total reaction time of the prepolymerization can be considerably shortened by processing it in an autoclave.[8,12,28] In a laboratory, PA-6 can be synthesized in several ways: from ω-aminocaprolic acid, from lactam and ω-aminocaprolic acid, from lactam and water, and anionically.

Example 6. PA-6 from ω-aminocaprolic acid in (3.32).[59] *To a wide-bore test tube (Fig. 3.18b) with a volume of 50 mL fitted with a nitrogen inlet and an outlet with a water-cooled condensing unit and distilled water, ω-aminocaprolic acid*

Caprolactam
Water

Water vapor

Zone 1
heating and
prepolymerization

Zone 2
polymerization

Polymer

Figure 3.23   Conventional VK column reactor for PA-6 synthesis.[31]

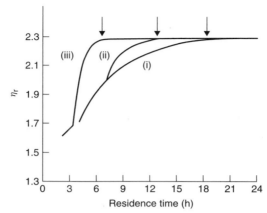

Figure 3.24 Influence of PA-6 polymerization process on relative viscosity as function of reaction time: (i) atmospheric in VK column; (ii) prepolymerization atmospheric followed by water removal; (iii) prepolymerization at elevated pressure followed by water removal.[31]

(20 g) is added. The condensation product, water, can be collected in a small flask. The system is flushed with nitrogen for 10 min before the reaction is started and the nitrogen flow rate during the reaction can be low. The flask is placed in a silicon oil-heating bath at 220°C. Once all of the ω-aminocaprolic acid is molten, the temperature of the oil bath is raised to 250°C. The polymerization is allowed to take place for 4 h at 250°C and atmospheric pressure under a slow nitrogen bleed. Stirring of the reaction mass is not necessary as the condensation product, water, forms gas bubbles which float to the surface.

The reaction mass is then taken from the oil bath and allowed to cool. The polymer is extracted by breaking the glass flask in a towel and can then be crushed into small particles. To do this, the polymer is cooled in liquid nitrogen, placed in a PE or PP bag, and crushed in a vice while still cold.

The lactam and higher oligomers, which are formed during the polymerization, can be extracted by boiling (refluxing) methanol (6 h), and the polymer is then dried in a vacuum oven at 80°C for 6 h. The resultant PA-6 has an η_{inh} of 0.8 as measured in 85% formic acid (1% solution, 25°C).

The progress of the reaction can be followed by measuring the amount of water condensed. The water can be collected in a tube with calcium chloride or alternatively collected in a small flask.[59] The reaction can also be followed by taking samples (~0.5 g, with a spatula) at different intervals and analyzing them. In this way, the change in solution viscosity (molecular weight) can be determined as a measure of reaction progression.

3.4.3.2 PA-6 Hydrolytic Melt Polymerization of Caprolactam

Starting with caprolactam, an initiator is required for the hydrolytic melt polymerization; this can be any of the following: water, ω-amino acids, acids, diacids, amines, or diamines. The concentration of initiator required is at least one per polymer chain — thus at least 0.5%. If acids, diacids, amines, or diamines are used, then a higher initiator concentration limits the maximum molecular weight obtainable. If water or ω-amino acids are used, then the amine–amide balance is not disturbed and higher concentrations are possible. A higher initiator concentration allows the preparation of prepolymers, which can be further polymerized by the hydrolytic reaction. However, the water concentration at high temperatures and atmospheric pressure is low and therefore the initiation is slow. The polymerization with water is also slow, as the ω-amino acid must be formed first (Fig. 3.16).[28] For laboratory-scale experiments, a good initiator is ω-aminocaprolic acid; the concentration of ω-amino acids can range between 1 and 10 wt %.

Example 7. PA-6 from ε-caprolactam and ω-aminocaprolic acid in (3.33). *To a wide-bore test tube (Fig. 3.18b) with a volume of 50 mL fitted with a nitrogen inlet and an outlet with a water-cooled condensing unit and a water lock, 15 g of ε-caprolactam and 1.5 g of ω-aminocaprolic acid are added. The system is flushed with nitrogen for 10 min before the reaction is started and the nitrogen flow rate during the reaction can be low. The flask is immersed in a silicon oil-heating bath at 200°C. When all of the ingredients are molten, the temperature of the oil bath is raised to 250°C and allowed to react for 5 h at this temperature. After cooling the flask, the polymer is extracted as described above (see Example 1). The η_{inh} is 0.8 as measured in 85% formic acid (1% solution, 25°C).*

If polymers with a molecular weight over 15,000 are required, then a postcondensation step is needed. The postcondensation step is carried out at low water concentrations in the melt or in the solid state. The typical solid-state postcondensation conditions for PA-6 are given below.

3.4.3.3 PA-6 Solid-State Postcondensation

If a very high viscosity is required, the granulated polymer can be postcondensed in the solid state at 160–190°C for 4–24 h. The postcondensation step can be done batchwise in large revolving reactors or can be carried out in a continuous manner using tall moving-bed reactors. Surprisingly, the water concentration is not critical to the rate of postcondensation. The method employed for PA-6 is similar to that for PA-6,6 (Example 1c). As a result of a postcondensation step for 24 h at 190°C, the η_{inh} in 85% formic acid is increased from 0.8 to 1.35 ($M_n = 18,000-35,000$).

3.4.3.4 PA-6 by Anionic Polymerization of Lactam

The anionic polymerization of caprolactam is rapid and can be carried out below the melting temperature of the PA. Care should be taken to work under anhydrous conditions.

Example 8. PA-6 by anionic polymerization.[60]

$$
\underset{(CH_2)_x-N-H}{\diagdown C=O} \quad + \quad \underset{\underset{\oplus}{(CH_2)_x-N-H_2}}{\diagdown C=O} \quad \longrightarrow \quad \underset{n}{+N+CH_2+_5\,\overset{O}{\overset{\|}{C}}+} \tag{3.36}
$$
$$
CH_3COO^{\ominus}
$$

To a thoroughly dried (24 h in a oven at 80°C) 50-mL flask 12.5 g (0.11 mol) ε-caprolactam and 0.3 g (0.0125 mol) sodium hydride are added (Fig. 3.18b). The flask is made oxygen free by alternating evacuation with a vacuum pump and filling with nitrogen (three cycles). The flask is warmed to 80°C in an oil bath to melt the ε-caprolactam. The small amounts of water still present are reacted away by sodium hydride resulting in the formation of hydrogen gas. After the evolution of hydrogen gas has ceased, 0.33 g (0.002 mol) N-acetylcaprolactam is added. The mixture is homogenized with a magnetic stirring device or by shaking the flask and is then placed in an oil bath, which is heated to 140°C. This temperature is well below the melting temperature ($T_m \sim 218°C$). The reaction mass solidifies quickly and can be cooled after 30 min. The reaction product can be worked up as described above (see Example 1). The polymer has an η_{inh} of >1.5 as measured in 85% formic acid (1% solution, 25°C).

3.4.3.5 Lower PA

PA-3 can be prepared from β-propiolactam (2-azetidinone) either by thermally initiated ring opening or by anionic polymerization.[12] However, the melting temperature (340°C) is too high for melt synthesis and melt processing. The 3,3-dimethyl-substituted PA-3 has a lower melting temperature (250°C) and is melt synthesizable and melt processable. PA-4 can be prepared from 2-pyrrolidone (γ-butyrolactam). The polymer has a melting temperature of 265°C but is thermally unstable in the melt.

Example 9. PA-4 synthesis.[61]

$$
\underset{(CH_2)_3-N-H}{\diagdown C=O} \;+\; SiCl_4 \;\xrightarrow[-NaCl]{NaOH}\; \underset{(CH_2)_3-N-SiCl_3}{\diagdown C=O} \;+\; \underset{(CH_2)_3-N-H}{\diagdown C=O}
$$
$$
\tag{3.37}
$$
$$
\underset{n}{+N+CH_2+_3\,\overset{O}{\overset{\|}{C}}+}
$$

A 250-mL, three-necked, round-bottomed flask equipped with stirrer and distillation head suitable for vacuum distillation is filled with 120 g of freshly distilled 2-pyrrolidone. The reaction mass is heated under nitrogen to 80°C. Flake potassium hydroxide (3.4 g) is then added. The water formed is rapidly distilled from the flask under 1 mbar pressure. The hot solution is then rapidly *transferred to a 100-mL polyethylene bottle previously purged with nitrogen. While the solution is still hot, 0.5 g of silicon tetrachloride is added. After 10 min and at a temperature of about 50°C, precipitation of the polymer starts. After 24 h at room temperature, the mixture becomes very hard, the bottle may be cut open to allow removal of the polymer, and it can be broken up using a hammer. After cooling the lumps with liquid nitrogen, they can be transferred to a polyethylene bag and crushed in a vice. The fine particles are then blended with 150 mL of water containing 0.1% of formic acid in a high-speed blender. The powdered product is filtered and washed on the filter with 150 mL of 0.1% formic acid solution, followed by three washes with distilled water. It is finally washed with methanol and dried. A white PA-4 (73 g) product is obtained which has an η_{inh} of 2.38 in m-cresol (0.5% solution, 25°C).*

3.4.3.6   PA-5

PA-5 can be synthesized by the anionic ring-opening polymerization of α-piperidone.[12] First the catalyst, N-acetyl piperidone, is prepared. Then the polymerization is performed anionically, well below its melting temperature.

3.4.3.7   *Higher PA*

The higher PA can all be prepared from either amino acids or lactams,[12] depending on the availability of the starting material. PA-7 may be prepared from either 7-aminoheptanoic acid or from the corresponding 7-heptanelactam (ζ-enantholactam). The polymerization from 7-aminoheptanoic acid is a hydrolytic polymerization similar to the polymerization from 6-aminohexanoic acid (see Example 6). The polymerization temperature is 280°C, which is well above its melting temperature (230°C).[12] PA-11 is made from 11-aminoundecanoic acid suspended in water.[12,62,63] This hydrolytic polymerization has three stages: the melting of the amino acid, the prepolymerization, and the polymerization. In the first stage, most of the water is expelled from the system and later the remainder is evaporated.

Example 10. PA-11 from 11-aminoundecanoic acid.[63]

$$H_2N{+}CH_2{\underset{10}{\Large)}}\overset{O}{\overset{\|}{C}}{-}OH \longrightarrow {+}\overset{H}{\overset{|}{N}}{+}CH_2{\underset{10}{\Large)}}\overset{O}{\overset{\|}{C}}{\underset{n}{\Large)}} + H_2O \qquad (3.38)$$

To a 50-mL three-necked flask (Fig. 3.18b) equipped with a stirrer (comprised of a stainless steel shaft and paddle), a head for the distillation of water, and a nitrogen inlet is added 20 g of purified 11-aminoundecanoic acid. The flask is then purged with nitrogen for 5 min. The flask is warmed in a silicon oil bath to 220°C and maintained at this temperature for 10 h. After raising the stirrer from the molten mass, the reaction is cooled under nitrogen and the resultant polymer removed by breaking the glass. The T_m of the polymer is 185–190°C and the η_{inh} in m-cresol (0.5% at 35°C) is 0.6–0.7.

3.4.3.8 PA-12

In common with all the higher AB polyamides, PA-12 can be made from either the amino acid or the lactam.[12] In practice, PA-12 is made from the cheaper 12-laurolactam (12-dodecane lactam or ω-laurolactam). Water is less soluble in laurolactam compared to caprolactam, and the initiation with water is slower. For laboratory-scale synthesis it is advisable to start with the amino acid or a combination of amino acid and lactam.

3.4.4 Partially Aromatic PAs

Partially aromatic PAs have a higher glass transition temperature and a higher melting temperature than their aliphatic counterparts; thus these polymers have a better dimensional stability.[18] Due to the high T_g and the low water absorption as a result of the presence of the aromatic unit, the effect of water on the properties at room temperature is small. The barrier properties of partially aromatic PAs are also better. However, despite all these favorable properties, their commercial market is relatively small.

The partially aromatic PAs are exclusively made of the diamine–diacid type and not the amine–acid type. The aromatic diamines, similar to phenylene diamines, color easily and their polymers are conjugated, having a golden brown color. The aromatic diacids used in the formation of partially aromatic PAs are mainly terephthalic and isophthalic acids. Starting with the diacids, the PA salt is made first and with this the salt prepolymers are prepared. The prepolymerization is usually carried out in an autoclave to prevent the sublimation of the reactants. In a laboratory synthesis it would be preferable to avoid this autoclave step as one is not always available. It is possible to start with the more reactive esters, such as diphenyl isophthalate, or with the acid chlorides; starting with the reactive isocyanates is, in principle, also possible. The terephthalic and isophthalic acids are also used to modify PA-6,6 and PA-4,6 to more dimensionally stable copolymers.[6,18]

3.4.4.1 Diacid Method

In the diacid method, the PA salt is made first. A solution of this PA salt in water can be used for the polymerization. In the temperature range where the reaction rates are high, the diamines are volatile, and thus, it is preferable to carry out the prepolymerization under pressure. The prepolymerization can be carried out either at 220–250°C for 1 h or at 280–320°C in a matter of seconds. In the latter case, the reaction is carried out in a small-diameter tubular reactor.[64] Although a prepolymerization under pressure is preferred, Nielinger[28] has described a polymerization at atmospheric pressure at 210°C, whereby the loss in diamine is compensated for.

Example 11. PA-6,I from PA salt[51]:

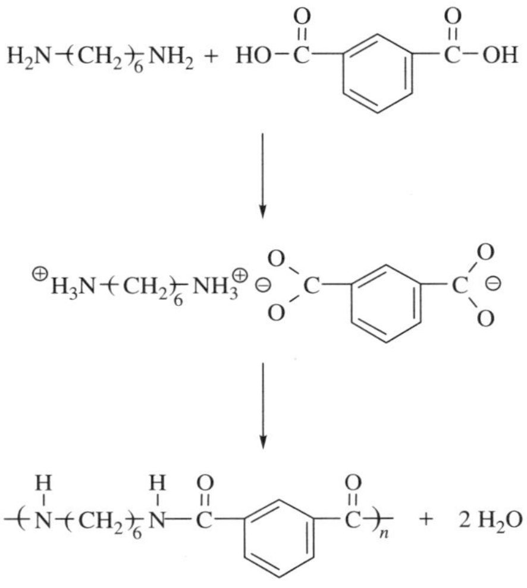

$$H_2N-(CH_2)_6-NH_2 \ + \ HO-\overset{\displaystyle O}{\overset{\|}{C}}\!\!\!\!\!\!\!\!\!\!\!\!\!\!\!\!-\!\!\!\!\!\!\!\!\!\!\!\!\!\!\!\!\overset{\displaystyle O}{\overset{\|}{C}}-OH$$

$$^{\oplus}H_3N-(CH_2)_6-NH_3^{\oplus} \ \overset{\displaystyle O}{\underset{\displaystyle O}{C}}\!\overset{\displaystyle O}{\underset{\displaystyle O}{C}}\ominus \qquad (3.39)$$

$$-(\overset{H}{\overset{|}{N}}-(CH_2)_6-\overset{H}{\overset{|}{N}}-\overset{\displaystyle O}{\overset{\|}{C}}\!\!\!\!\!\!\!\!\!\!\!\!\!\!\!\!-\!\!\!\!\!\!\!\!\!\!\!\!\!\!\overset{\displaystyle O}{\overset{\|}{C}})_n \ + \ 2\,H_2O$$

PA-6,I salt is made by adding 33.2 g of isophthalic acid to a reaction vessel with a reflux cooler filled with a 100-mL mixture of methanol and water (9 : 1). To this suspension, which is under nitrogen, 23.2 g of hexamethylene diamine in 20 mL of methanol is added slowly. The reaction mass is then allowed to cool whereby the nylon salt crystallizes. The precipitated salt is filtered, washed with methanol, and air dried.

To a 50-mL straight-wall three-necked flask (Fig. 3.18b) equipped with a magnetic stirrer, nitrogen inlet/outlet, and condenser unit in a heating block are added 11.29 g of nylon salt and 0.093 g of hexamethylene diamine (2 mol % excess). This mixture is reacted for 2 h at 210°C and 3 h at 270°C. The resultant polymer is colorless and transparent and has a η_{rel} of 2.54 (1% solution in m-cresol, 25°C).

3.4.4.2 Diester Method

The reaction of a diamine with a diester under anhydrous conditions is reasonably rapid. However, a side reaction at high temperatures ($>200°C$) is N-substitution. Unfortunately, this N-substitution is particularly strong with methyl esters,[28,37,65] and therefore, methyl esters such as dimethyl terephthalate or dimethyl isophthalate cannot be used for thermal polyamidations. Other esters, such as ethyl, butyl, and phenyl ester, do not seem to have this problem.

Example 12. PA-6,I from diphenyl isophthalate.[65] *To a well-dried 50-mL straight-wall three-necked flask with nitrogen inlet/outlet, condenser unit, and magnetic stirrer in a heating block (Fig. 3.18b), 15.1 g of diphenyl isophthalate and 6.15 g of 1,6-hexamethylene diamine are added. The mixture is heated to 190°C over a 1-h period and to 253°C over a further 40 min, a vacuum is then* *applied (1 mbar), and the mass warmed to 285°C and reacted for 1 h at this temperature. The resultant polymer has an η_{inh} of 0.83 as determined on a 1% solution in a 40 : 60 tetrachloroethane–phenol mixture at 30°C.*

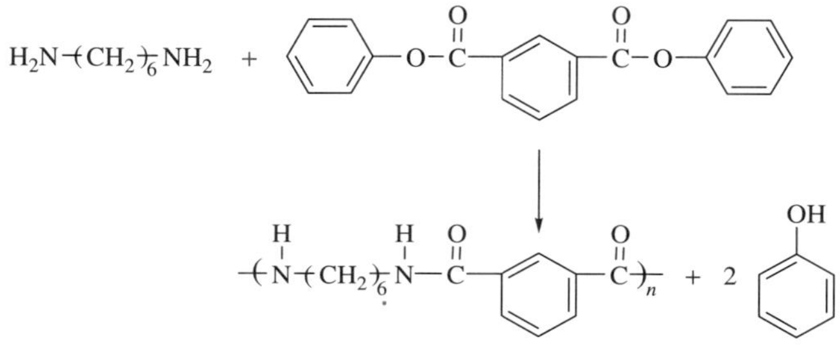

$$H_2N\text{-}(CH_2)_6\text{-}NH_2 \; + \; \text{(structure)}$$

$$\text{(3.40)}$$

$$\text{(polymer structure)} \; + \; 2 \; \text{(phenol)}$$

3.4.4.3   Diacid Chloride Methods

Acid chlorides are very reactive and at room temperature react readily with amines. Synthesis by interfacial and solution methods is possible. However, care should be taken that the hydrochloric acid produced does not react with unreacted amine groups. With the strong basic aliphatic diamines, the acid binder must preferably be even more basic. The attainable molecular weights are strongly dependent on the concentrations; this is particularly the case for easily precipitated terephthalamide polymers. Possible problems with the acid binder can be overcome by starting with silylated diamines.[33,34] A typical example for interfacial polymerization of terephthalamides is PA-2,T.[66]

Example 13. PA-2,T by interfacial polymerization.[66]

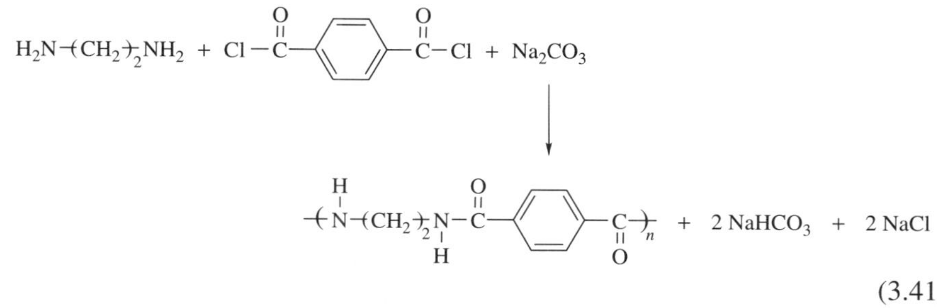

$$H_2N\text{-}(CH_2)_2\text{-}NH_2 \; + \; Cl\text{-}C\text{(...)}C\text{-}Cl \; + \; Na_2CO_3$$

$$\text{(polymer structure)} \; + \; 2\,NaHCO_3 \; + \; 2\,NaCl$$

$$\text{(3.41)}$$

To an 8-L beaker a solution of 3.78 g (0.630 mol) of ethylenediamine and 0.126 mol of potassium hydroxide in 4.5 L of distilled water are added and stirred with a high-speed mixer. To this, 12.79 g (0.634 mol) of terephthaloyl chloride dissolved in 1 L of methylene chloride (a safer solvent is THF) is added and stirred for 10 min. The suspension is filtered and washed twice with methanol. After drying, the polymer has an η_{inh} of 1.0 as measured in 96% sulfuric acid (0.5% solution at 30°C). The melting temperature of the polymer is 455°C.

The molecular weights obtained by a solution polymerization starting with acid chlorides are highly dependent not only on the type of solvent but also on the acid binder used. An example of a solution polymerization is given for the high melting PA-4,T.

Example 14. PA-4,T by solution method.[67]

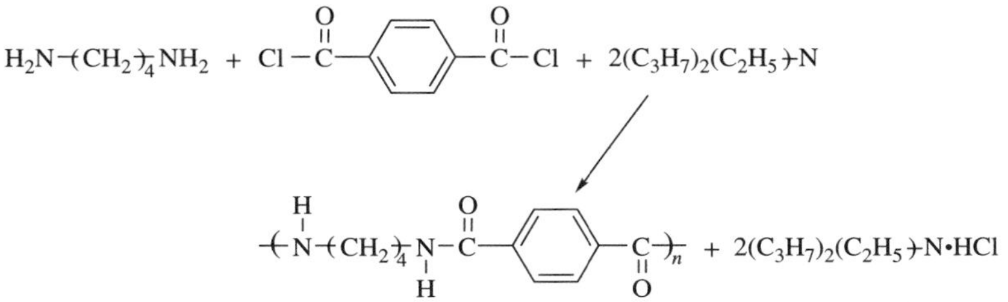

$$H_2N\text{-}(CH_2)_4\text{-}NH_2 \ + \ Cl\text{-}\overset{\overset{O}{\|}}{C}\text{-}\underset{}{\bigcirc}\text{-}\overset{\overset{O}{\|}}{C}\text{-}Cl \ + \ 2(C_3H_7)_2(C_2H_5)\text{-}N$$

$$\text{-}(N\text{-}(CH_2)_4\text{-}\underset{H}{N}\text{-}\overset{\overset{O}{\|}}{C}\text{-}\underset{}{\bigcirc}\text{-}\underset{\overset{\|}{O}}{C})_n \ + \ 2(C_3H_7)_2(C_2H_5)\text{-}N\cdot HCl$$

$$(3.42)$$

To terephthaloyl chloride (0.015 mol) in 100 mL of THF cooled to $-15°C$, tetramethylenediamine (0.015 mol) and diisopropylethylamine (0.030 mol) in 20 mL THF are added while stirring vigorously. A suspension is formed directly. Stirring is continued for 15 min, after which the precipitated polymer is filtered and washed with boiling water. The resultant polymer has an η_{inh} of 0.39 in 96% sulfuric acid (1% solution at 25°C).

This polymer is postcondensed in the solid state. For this, 5 g of material is placed in a glass flask which has been flushed with a stream of nitrogen. The flask is placed in a tube oven at 290°C and kept at that temperature for 1 h. The resultant polymer now has an η_{inh} of 1.52. The polymer is highly crystalline and has a melting temperature of 475°C.

3.4.4.4 PA from Aromatic Diamines and Aliphatic Diacids

Morgan and Kwolek[33] have described a large number of PA derived from phenylenediamines and aliphatic diacids by low-temperature solution polymerization starting with aliphatic diacid chlorides.

Example 15. PA from *m*-phenylenediamine and adipoyl chloride by solution polymerization.[2,32] *To a well-dried 250-mL three-necked straight-wall flange flask with nitrogen inlet/outlet, dropping funnel, and magnetic stirrer (Fig. 3.18b), 5.4 g (0.05 mol) of m-phenylenediamine and 37 mL of dimethylacetamide are added. The solution is cooled by ice to about 10°C and 7.35 mL (0.05 mol) of adipoyl chloride is then added. The resulting mixture is stirred for 1 h at approximately 10–15°C. The polymer precipitates on the addition of water and is washed several times with water and alcohol. After drying in a vacuum oven at 80°C for 8 h, the resultant polymer has an η_{inh} of 1.66 in m-cresol (1%*

solution at 25°C).

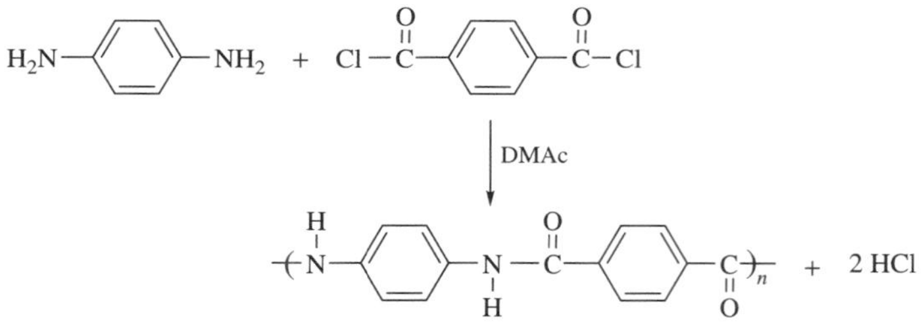

$$(3.43)$$

3.4.4.5   PAs from Diisocyanates

Diisocyanates are highly reactive and readily available compounds. The diisocyanates and acids form amides with the liberation of carbon dioxide; side reactions are possible as diisocyanates can also react with an amide group.

Example 16. Adipic acid and toluene diisocyanate.[68]

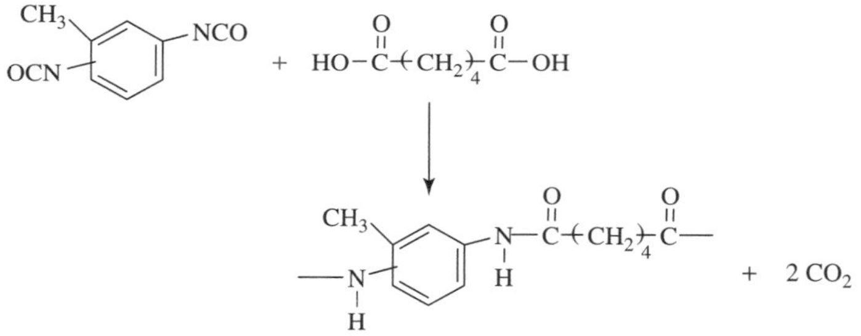

$$(3.44)$$

To a well-dried 250-mL four-necked straight-wall flange flask with nitrogen inlet/ outlet, dropping funnel, and magnetic stirrer (Fig. 3.18a), 3.65 g (25 mmol) of adipic acid and 75 ml of sulfolane are added at room temperature. The mixture is then heated in an oil bath to 40°C and the adipic acid dissolved. Aluminum chloride (915 mg, 6.87 mmol) and toluene diisocyanate (80% 2,4-isomer and 20% 2,6-isomer)(4.35 g, 25 mmol) are added. Carbon dioxide gas generation begins immediately, while the temperature rises to about 45–50°C. After 30 min the reaction mixture is very viscous and appears muddy, while the generation of gas should have virtually stopped. Subsequently, the reaction mass is heated for 1.5 h at 75°C and 1 h at 100°C. Upon cooling, the reaction mass is poured into 500 mL of methanol and a light yellow suspension is formed, which is filtered and washed four times with methanol. After drying, 3.8 g of polymer is obtained (65% yield) having an η_{inh} of 1.12 in m-cresol (1% solution at 25°C).

3.4.5 Wholly Aromatic PAs

Wholly aromatic PAs have a high glass transition temperature (>200°C) and, if crystalline, a very high melting temperature (>500°C). High-molecular-weight polymers cannot be prepared in the melt and melt processing is also not possible. Their synthesis is usually carried out in solution, and due to the very low solubility, special solvents are required to obtain high-molecular-weight polymers. The first wholly aromatic PA on the market (1961) was a low-crystallinity PA [poly(*m*-phenyleneisophthalamide)] [see (3.11)] based on isophthalic acid and *m*-phenylene diamine. This polymer has excellent flame-resistant properties. Later the para-aromatic polyamide [poly(*p*-phenylene terephthalamide)] [see (3.4)] based on terephthalic acid and *p*-phenylenediamine was introduced. This polymer is highly crystalline and as a fiber has a high modulus and a high strength.[7,9,14] Many modifications of these aramides have been studied.[7,9]

With the solution polymerization method, high-molecular-weight aromatic PAs are more easily obtained; however, the interfacial method can also be employed. Since the reactivity of the aromatic acids toward the aromatic amines is low, the very reactive-acid chloride system is generally employed. With acid chlorides the condensation product is hydrochloric acid, which has to be bound so it does not react with the aromatic diamine. In the solution method, the reaction takes place principally in solution, but gelation by crystallization of the main chain is allowed.[7,9,14] For the solution polymerization of aromatic polyamides it is advantageous to use aprotic polar amide solvents, such as DMAc, and NMP. DMF and dimethyl sulfoxide (DMSO) are not suitable because they react with acid chlorides. Hexamethylphosphoramide (HMPT) is a powerful solvent but should not be used, as it has been shown to be carcinogenic in rats.[9] The dissolving power of the solvents can be increased by the addition of salts, such as lithium chloride and calcium chloride. The amide solvents further serve as acid acceptors. Although the amide solvents are weaker bases than the aromatic amines, they are effective as their concentration is much higher. Calcium hydroxide may be added to the amine solvent solution as this forms, with the hydrochloric acid, the calcium chloride salt, which increases the solubility of the polymer. For high-modulus fiber applications, very high molecular weight compounds are required. There are several alternatives to the dichloride–diamine solution route.[9] The methods that give high molecular weights are polymerization from terephthalic acid with phosphorylation and polymerization using N-silylated diamines and diacid chlorides.

3.4.5.1 Poly(m-phenylene isophthalamide)

This polymer can be synthesized from *m*-phenylenediamine and isophthaloyl chloride. It can be prepared by interfacial polymerization or solution polymerization.[4,7,9,14]

Example 17. Poly(*m*-phenylene isophthalamide) in solution.[14] *To a well-dried 250-mL four-necked straight-wall flange flask with nitrogen inlet/outlet, dropping funnel, and magnetic stirrer (Fig. 3.18a). 2.163 g of 1,3-phenylenediamine, 5.62 g of triethylamine, 5.506 g of triethylamine hydrochloride, and 36 mL of dry chloroform are added. Isophthaloyl chloride (4.06 g) in 14 mL of chloroform is then added through the dropping funnel over a 15-min period at 30°C with slow* *stirring. An additional 3 mL of chloroform is used to rinse any residue from the funnel. After 20 min, the reaction mixture, which is a clear, extremely viscous solution, is poured into a large volume of petroleum ether yielding a fibrous precipitate. This is filtered and washed thoroughly with hot water. A 99% yield of the polymer, having an η_{inh} of 1.9 in 96–98% sulfuric acid (1% solution at 30°C), is obtained.*

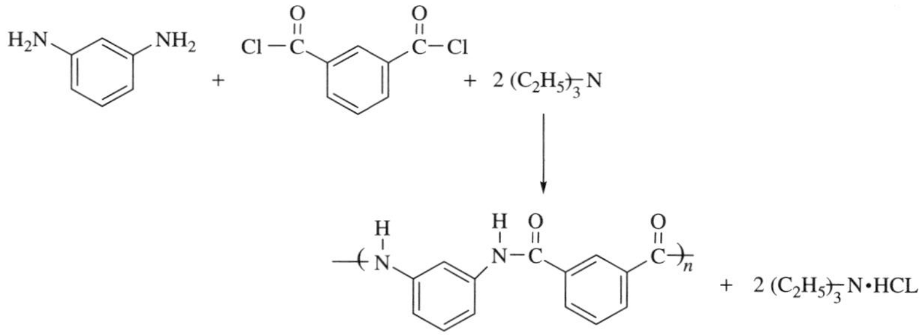

$$(3.45)$$

3.4.5.2   Poly(p-phenylene terephthalamide)

This polymer can be prepared from *p*-phenylenediamine and terephthaloyl chloride. The polymer is highly crystalline and, thus, difficult to keep in solution. Sufficiently high molecular weight polymers can be obtained by solution polymerization using a special solvent system. This ridged rod polymer can form a liquid crystalline solutions.[7,9,14]

Example 18. Poly(p-phenylene terephthalamide) from terephthaloyl chloride.[69]

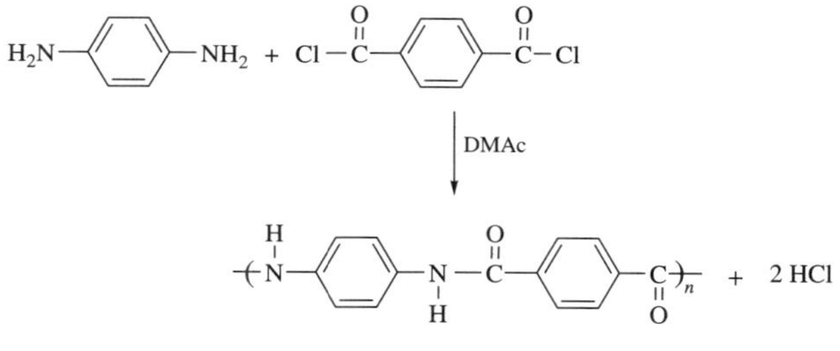

$$(3.46)$$

Into a Waring blender with a 1-L mixing beaker with a cooled bottom connected to a cryostat, 24 g of finely ground and dried calcium chloride suspended in 200 g of N-methyl pyrrolidone (12 wt% calcium chloride) is added. The calcium chloride is partially present in the solid state. To this suspension, 7.3 g of powdered p-phenylene diamine is added with stirring. The resulting mixture is cooled to 0°C. Subsequently, with continued cooling and vigorous stirring, 13.9 g of terephthaloyl chloride is added rapidly. Stirring is continued for 30 min while the temperature rises to about 30°C. A crumbled mass is formed which contains 16 g *of poly(p-phenylene terephthalamide) (8 wt %). A suspension of the polymer is obtained by precipitation with water under vigorous stirring. Following filtration, washing, and drying, a poly(p-phenylene terephthalamide) with an η_{inh} of 4.02 is obtained, as measured in 98% sulfuric acid (1% solution, 25°C).*

3.4.5.3 Phosphorylation

This method involves the direct polycondensation of aromatic diamines with aromatic diacids in the presence of an aryl phosphite (triphenyl phosphite) and an organic base like pyridine.[7,9,70,71] The addition of salts improves the solubility of the polymer and, with this, the maximum attainable molecular weight.[71] The concentrations are, however, lower than by the dichloride method.

Example 19. Poly (p-phenylene terephthalamide) by phosphorylation.[71]

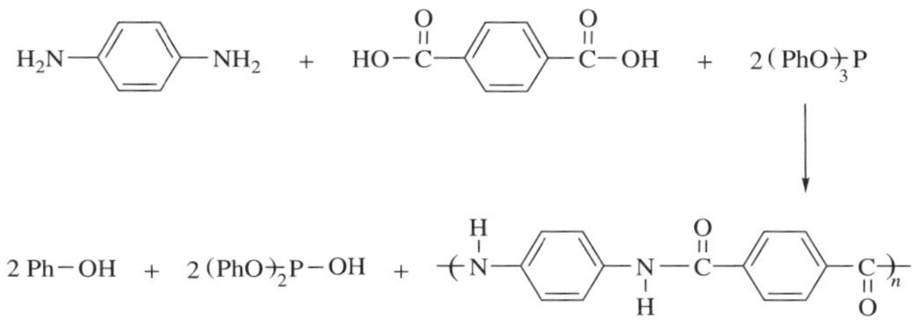

$$(3.47)$$

To a well-dried, straight-wall three-necked flask fitted with nitrogen inlet/outlet, magnetic stirrer, and condenser unit (Fig. 3.18b), 25 mL of NMP, 5 mL of pyridine, 0.4154 g (0.0025 mol) of terephthalic acid, 0.2704 g (0.0025 mol) of p-phenylene diamine, 1.70 g (0.0055 mol) of triphenyl phosphite, 1.5 g of $CaCl_2$, and 0.5 g of LiCl are added under inert gas (nitrogen or argon). Oxygen traces in the above solvents are removed by several cycles of evacuation with a vacuum pump and filling with argon (or nitrogen). All operations are carried out in a dry box in an inert atmosphere in order to avoid humidity and air. The vessel is placed in an oil bath heated to 115°C. The reaction mixture becomes a clear gel in 15–30 min. After 4 h, the reaction is stopped and the gel ground in a blender in the presence of methanol. The polymer is then washed with boiling methanol and dried in a vacuum oven for 18 h at 80°C. The resultant polymer has an η_{inh} of 9.0 as measured in 98% sulfuric acid (1% solution at 25°C).

3.4.5.4 Silylated Diamines

Aromatic polyamides can be prepared by starting with acid chlorides and a hydrogen chloride scavenger. Another way is to first silylate the diamine.[33,34] On reaction of the silylated diamine with the acid chloride, the polyamide is formed along with a silylchloride. In this way high-molecular-weight polymers can be easily prepared. Star-branched polymers have also been prepared using the silyl method[34]; an example is given for the synthesis of poly(p-phenylene terephthalamide).

Example 20. Poly(p-phenylene terephthalamide) by silylated diamines.[33]

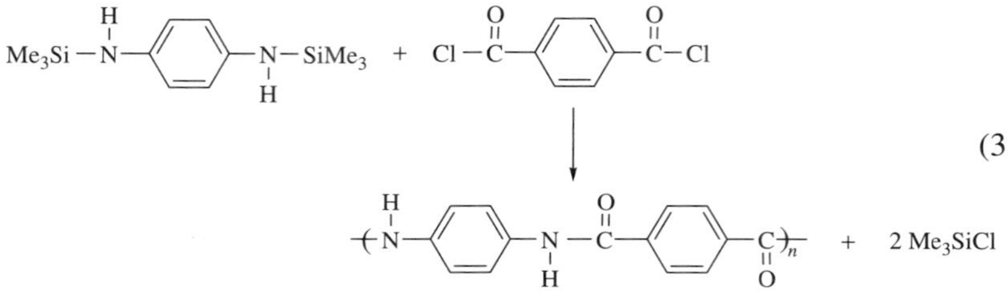

$$Me_3Si-N(H)-\langle aryl \rangle-N(H)-SiMe_3 \quad + \quad Cl-\overset{O}{\underset{}{C}}-\langle aryl \rangle-\overset{O}{\underset{}{C}}-Cl$$

$$(3.48)$$

$$\leftarrow N(H)-\langle aryl \rangle-N(H)-\overset{O}{\underset{}{C}}-\langle aryl \rangle-\overset{}{\underset{O}{C}}\rightarrow_n \quad + \quad 2\,Me_3SiCl$$

To a well-dried, straight-wall three-necked flask fitted with with nitrogen inlet/outlet, magnetic stirrer, and condenser unit (Fig. 3.18b), 1.263 g (5 mmol) of N,N'-bis(trimethylsilyl)-p-phenylenediamine and 1.33 g of lithium chloride are dissolved in 25 mL of N-methyl-2-pyrrolidone (NMP) under inert gas. The solution is brought to −15°C using an ice–salt mixture, and 1.015 g (5 mmol) of powdered terephthaloyl chloride is then added. The mixture is stirred at −10 to −5°C under nitrogen. The polymerization proceeds in this homogeneous solution and becomes a gel after 6 h. The reaction mixture is worked up by agitating with methanol. The polymer is collected by filtration, washed thoroughly with hot methanol, and dried at 80°C in vacuum. The resultant polymer has an η_{inh} of >4 as measured in 98% sulfuric acid (0.5% solution at 30°C).

3.4.5.5 Poly(p-benzamide)

This fully aromatic amide, based on the amino acid p-aminobenzoic acid, can be spontaneously synthesized from p-aminobenzoic chloride.[7,9,72] To prevent this occurring at an unwanted moment, the amine group is masked by forming the hydrochloric acid salt with hydrochloric acid.

Example 21. Poly(p-benzamide) from acid chloride.[72]

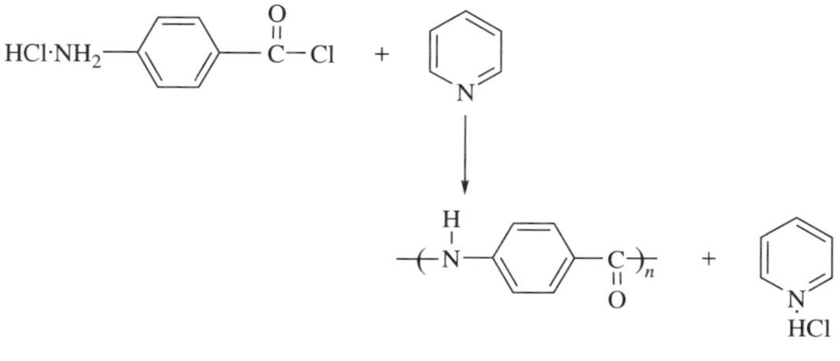

$$HCl\cdot NH_2-\langle aryl \rangle-\overset{O}{\underset{}{C}}-Cl \quad + \quad \langle pyridine \rangle$$

$$(3.49)$$

$$\leftarrow N(H)-\langle aryl \rangle-\overset{O}{\underset{}{C}}\rightarrow_n \quad + \quad \langle pyridine \rangle \cdot HCl$$

In a 250-mL straight-wall three-necked flask equipped with a magnetic stirrer and nitrogen inlet/outlet (Fig. 3.18a), 50 mL of dry hexane and 6.4 g of 4-aminobenzoyl chloride hydrochloride are added. Under a flow of dry nitrogen, the stirred mixture is cooled in a pack of solid carbon dioxide. To the cooled *solution, 10 mL of pyridine is added quickly. After 15 min the flask is allowed to warm to room temperature and is reacted for 2 h. The polymer is collected by filtration, washed thoroughly with water, and dried. The resultant polymer has an η_{inh} of 3.7 in 98% sulfuric acid (1% solution at 25°C).*

3.4.6 Copolymers Containing Amide Units

PAs are made with relatively few starting materials; copolymers can also be made from the same starting materials and have been extensively studied. Random copolymers have an inherently lower structural regularity and thus a lower order and are interesting for hot-melt adhesive applications. The simplest random copolymers are of PA-6 and PA-6,6 (Fig. 3.8). Partially aromatic random copolyamides, such as PA-6,T/PA-6,I, are amorphous and have a high T_g. These polymer systems combine a high dimensional stability up to their T_g and a good solvent resistance to hydrocarbons with transparency. Copolymers that have a high order include PA-6,6/PA-6,T copolymers where an adipic unit is interchanged for a terephthalic unit of a similar length (Fig. 3.9).[6]

PAs have also been copolymerized with other polymer systems and, in particular, with polyesters and polyethers. In the copolyesteramides the crystallinity is decreased by copolymerization, as the crystalline structure of the amide unit is very different from the ester unit. However, alternating polyesteramides behave as homopolymers with a glass transition temperature and a melting temperature intermediate to the polyester and the PA polymer (Figs. 3.10 and 3.11).[23,24] Polyesters, such as PBT and PET, modified with a small amount of diamide are also copolymers that have a high order.[24,73]

3.4.6.1 Alternating Polyesteramide

Strictly alternating polyesteramides behave as homopolymers, and if their structure is regular, they can crystallize quickly.[21,22,74] The best way of synthesizing these materials is by forming first the diamide unit and later the diester unit. This procedure is followed because an ester is readily transamidated by an amine but an amide is not easily transesterified by an alcohol. A typical alternating polymer is 6NT6 [see (3.51) below], based on terephthalic acid, hexamethylenediamine, and hexamethylenediol, which are all readily available monomers. With this synthesis, bisesterdiamide (e.g., T6T-dimethyl) is first formed from hexamethylenediamine and dimethyl terephthalate. Starting from the bisesterdiamide and a diol, alternating polyesteramides can be obtained. The polymerization procedure is the same as that for polyester synthesis.

Example 22. Alternating polyesteramide 6NT6.

22a. T6T-dimethyl[75]. Into a reactor having a volume of 500 mL fitted with a stirrer and a N_2 supply are fed 160 mL of dry toluene, 16 mL of dry methanol, 46 g of dimethyl terephthalate, and 5 mmol of $LiOCH_3$. The mixture is heated to 70°C and 8.4 g of hexamethylenediamine is added to the solution. After 1 h a precipitate *is formed. The reaction is continued for 7 h. During this time a major part of the methanol is gradually distilled off as an azeotrope with toluene. At the end of the reaction, the hot solution is filtered and the precipitate washed on the filter with 120 mL of a mixture of methanol and toluene having a volume composition of 75/25. The bisester diamide obtained is dried in a vacuum oven at 80°C for 6 h. The bisesterdiamide [N,N'-bis(p-carbomethoxybenzoyl)hexamethylenediamine] has a melting temperature of 232°C.*

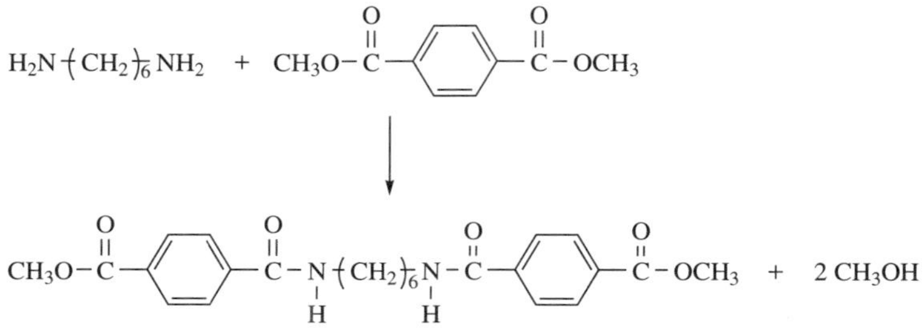

$$(3.50)$$

22b. Polymerization of 6NT6 with hexanediol[74] *:*

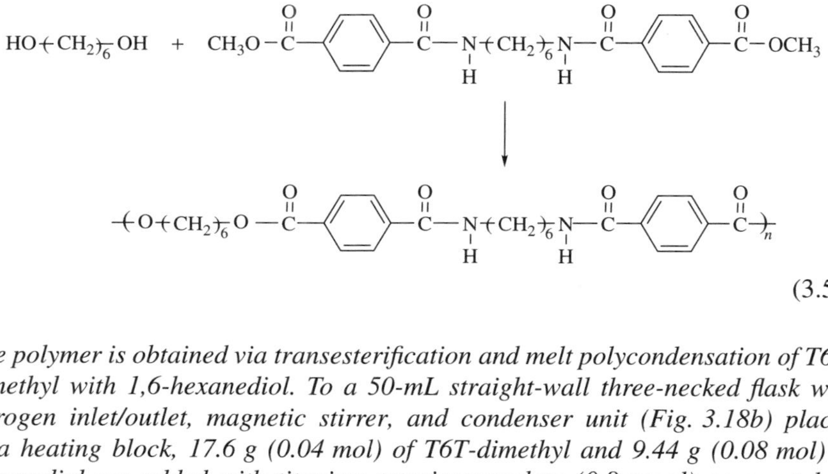

$$(3.51)$$

The polymer is obtained via transesterification and melt polycondensation of T6T-dimethyl with 1,6-hexanediol. To a 50-mL straight-wall three-necked flask with nitrogen inlet/outlet, magnetic stirrer, and condenser unit (Fig. 3.18b) placed in a heating block, 17.6 g (0.04 mol) of T6T-dimethyl and 9.44 g (0.08 mol) of hexanediol are added with titanium tetraisopropylate (0.8 mmol) as a catalyst. After flushing with nitrogen, the reaction mass is heated to 240°C. After 1 h at this temperature, the temperature is further increased to 270°C and a vacuum applied (1 mbar) for 1 h. The resultant polymer had an η_{inh} of 1.5 as measured at 25°C in phenol–tetrachloroethane (50/50 by weight).

3.4.6.2   Polyesters Modified with Diamides

If one wants to modify an existing polyester without losing its high order, copolymerizing it with diamide units is an option. PBT can be modified with a diamide based on 1,4-butane diamine or *p*-phenylene diamine and PET with 1,2-ethylene diamine.[21,76] Small quantities $(0.1-2\%)$ of the diamide increase the rate of crystallization of the ester and higher concentrations also increase the T_g and T_m. An example is given for the amide modification of PBT.

Example 23. Poly(butyleneterephthalate) with 10 mol % diamide units.[24]

$$-\{O-(CH_2)_4\,O-\overset{\overset{\text{O}}{\|}}{C}-\langle\bigcirc\rangle-\overset{\overset{\text{O}}{\|}}{C}\}_9\,\underset{H}{N}-(CH_2)_4\underset{H}{N}-\overset{\overset{\text{O}}{\|}}{C}-\langle\bigcirc\rangle-\overset{\overset{\text{O}}{\|}}{C}\}_n \qquad (3.52)$$

From DMT and 1,4-tetramethylene diamine, a dimethylesterdiamide unit (T4T-dimethyl) is prepared as described above for T6T-dimethyl (Example 22a). The T4T-dimethyl has a T_m of 265°C and a ΔH_m of 152 J/g.

To a 250-mL three-necked straight-wall flange flask with nitrogen inlet/outlet, magnetic stirrer, and condenser unit (Fig. 3.18a), 46.8 g (221.4 mmol) DMT, 4.40 g (10.06 g mmol) T4T-dimethyl, 37 mL (369 mmol) butanediol, and 0.62 mL Ti(i-propylate)$_3$ are added. The reaction flask is placed in an oil-heated bath, warmed to 180°C, and reacted for 30 min. During this time methanol boils off from the reaction mass. The reaction mass is then warmed to 250°C over a 1-h period and a vacuum is also applied: 15 min at 15–20 mbar and 60 min at 0.1–0.4 mbar. Subsequently, the flask is cooled and the polymer removed. The resultant polymer has an η_{inh} of 1.16 as measured in tetrachloroethane–phenol (1 : 1) at 25°C, 0.5% solution. Compared to PBT, the effect of adding 10% diamide includes a T_g increase from 47 to 60°C, a T_m increase from 222 to 230°C, and a crystallization temperature increase from 186 to 198°C (on cooling at 20°C/min by DSC).

3.4.6.3 Polyether–PA Segmented Copolymers

Segmented copolymers, such as PA–polyether segmented copolymers, have also been extensively studied. Polyether–amide segmented copolymers have interesting thermoplastic elastomer properties.[76] In these copolymers the polyether segments give the material a low T_g and the polyamide segments give it a high-melting crystalline phase. As the polarity difference between the ether and amide segment is large, the miscibility of the segments is poor and melt phasing easily takes place. The longer the segment's length, the poorer the miscibility. Melt phasing should be avoided as it limits the maximal attainable molecular weight.

Example 24. PA-11–polyether segmented block copolymer.[76]

$$HO-(CH_2)_4O)_n\,H \;+\; HO-\overset{\overset{\text{O}}{\|}}{C}-(CH_2)_{10}\overset{\overset{\text{O}}{\|}}{C}-OH \;+\; X\;\underset{(CH_2)_{10}-N-H}{\overset{C=O}{\diagup}}$$

$$\downarrow \qquad\qquad (3.53)$$

$$-\{O-(CH_2)_4O)_n-\overset{\overset{\text{O}}{\|}}{C}-(CH_2)_{10}\overset{\overset{\text{O}}{\|}}{C}-(\underset{H}{N}-(CH_2)_{10}\overset{\overset{\text{O}}{\|}}{C})_x\}_m$$

In a 100-mL autoclave with magnetic stirrer, 15 g of laurolactam, 3.94 g poly(tetramethylene oxide) with a molecular weight of 860, 1.06 g of decanedicarboxylic acid, and 0.7 mL of water are added. The autoclave is closed, flushed with nitrogen, and given a starting nitrogen pressure of 5 bar. The system is warmed to 270°C and then the pressure is lowered to atmospheric pressure over a period of 120 min. The reaction is continued for 11 h at 270°C with a stream of nitrogen over the reaction mass. After cooling, the polymer is taken out of the autoclave and granulated. The polymer has an η_{inh} of 2.0 as measured in m-cresol (0.5% solution at 25°C).

It is expected that, if instead of laurolactam, 11-aminoundecanoic acid is used, then the reaction can be carried out at atmospheric pressure, as in Example 10.

3.4.6.4 Polyether–Diamide Segmented Copolymers

A special case of segmented copolymers is if the crystallizable unit is of uniform length. If the amide unit is uniform in length, then a faster and more complete crystallization is possible.[22,23] If the amide unit is short, such as with a diamide unit, then melt phasing is less likely to occur. With these polyether–diamide segmented copolymers, one can combine a low modulus, a high elasticity, and a rapid crystallization with a good processability. The type of diamine in the diamide strongly influences the melting temperature of these segmented copolymers. Changing tetramethylene diamine for p-phenylene diamine results in a polymer with 60°C higher melting temperature.[22,23]

Example 25. Diamide–poly(tetramethylene oxide) segmented copolymer (T4T–PTMO$_{1000}$).[25]

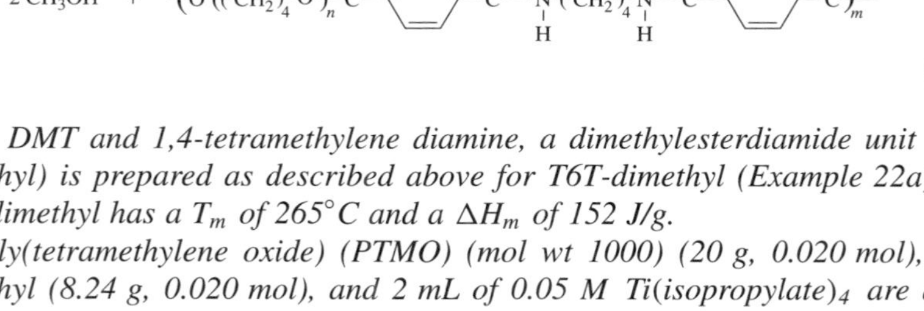

$$(3.54)$$

From DMT and 1,4-tetramethylene diamine, a dimethylesterdiamide unit (T4T-dimethyl) is prepared as described above for T6T-dimethyl (Example 22a). The T4T-dimethyl has a T_m of 265°C and a ΔH_m of 152 J/g.

Poly(tetramethylene oxide) (PTMO) (mol wt 1000) (20 g, 0.020 mol), T4T-dimethyl (8.24 g, 0.020 mol), and 2 mL of 0.05 M Ti(isopropylate)$_4$ are added to a 50-mL straight-wall three-necked flask with nitrogen inlet/outlet, magnetic stirrer, and condenser unit (Fig. 3.18b). The reaction flask is heated in an oil bath to 170°C. After 30 min at this temperature, the temperature is raised to 250°C. At 250°C a vacuum of 10–15 mbar is applied for 30 min and a vacuum *of 0.1 mbar is applied for a further 60 min. Following this, the reaction mass is allowed to cool. The resultant polymer is highly elastic, has an η_{inh} of 1.18 as determined in tetrachloroethane–phenol (1 : 1) (0.25% solution at 25°C), and a melting temperature of 153°C.*

REFERENCES

1. W. H. Carothers, U.S. Patent 2,130,948 (to DuPont), 1937.
2. R. J. Gaymans and E. H. J. P. Bour, U.S. Patent 4,408,036, 1981.
3. P. Schlack, German Patent 748,253 (to Farbenindustrie A.G.), 1938.
4. P. W. Morgan, *Condensation Polymers: By Interfacial and Solution Methods*, Wiley-Interscience, New York, 1965.
5. D. B. Jacobs and J. Zimmermann, in *High Polymers*, Vol. 29, C. E. Schieldknecht and I. Skeist (Eds.), Wiley-Interscience, New York, 1977, p. 424.
6. J. Zimmerman, in *Encyclopedia of Polymer Science and Engineering*, Vol. 11, H. F. Mark, N. M. Bikales, C. G. Overberger, and G. Menges (Eds.), Wiley-Interscience, New York, 1989, p. 315.
7. J. Preston, in *Encyclopedia of Polymer Science and Engineering*, Vol. 11, H. F. Mark, N. M. Bikales, C. G. Overberger, and G. Menges (Eds.), Wiley-Interscience, New York, 1989, p. 381.
8. R. J. Gaymans and D. J. Sikkema, in *Comprehensive Polymer Science*, Vol. 5: *Step Polymerization*, G. Allan, J. C. Bevington, G. C. Eastmond, A. Ledwith, S. Russo, and P. Sigwald (Eds.), Pergamn, Oxford, 1989, p. 357.
9. L. Vollbracht, in *Comprehensive Polymer Science*, Vol. 5: *Step Polymerization*, G. Allan, J. C. Bevington, G. C. Eastmond, A. Ledwith, S. Russo, and P. Sigwald (Eds.), Pergamon, Oxford, 1989, p. 373.
10. H. Sekiguchi and B. Coutin, in *Handbook of Polymer Synthesis*, Part A, H. R. Kricheldorf (Ed.), Marcel Dekker, New York, 1992, p. 807.
11. M. I. Kohan, *Nylon Plastics Handbook*, Hanser, New York, 1995.
12. S. M. Aharoni, *n-Nylons: Their Synthesis, Structure and Properties*, Wiley, New York, 1997.
13. L. Bottenbruch and R. Binsack, *Kunststoff Handbuch*, Vol. 3/4: *Polyamide*, Hanser Verlag, Munich, 1998.
14. P. W. Morgan, *Macromolecules*, **10**, 1381 (1977).
15. H. Hofmann, in *Kunststoff Handbuch*, Vol. 3/4: *Polyamide*, L. Bottenbruch and R. Binsack (Eds.), Hanser Verlag, Munich, 1998, p. 57.
16. W. Gotz, U.S. Patent 5,596,070 (to BASF), 1995.
17. R. Greiner, in *Kunststoff Handbuch*, Vol. 3/4: *Polyamide*, L. Bottenbruch and R. Binsack (Eds.), Hanser Verlag, Munich, 1998, p. 206.
18. H. Daebel, W. Götz, G. Oenbrink, E. Roerdink, U. Roh-Liebenau, and F. A. M. Schenkels, in *Kunststoff Handbuch*, Vol. 3/4: *Polyamide*, L. Bottenbruch and R. Binsack (Eds.), Hanser Verlag, Munich, 1998, p. 803.
19. D. W. van Krevelen, *Properties of Polymers*, Elsevier, Amsterdam, 1990, p. 109.
20. N. A. Jones, E. D. T. Atkins, M. J. Hill, S. J. Cooper, and L. Franco, *Macromolecules*, **30**, 3569 (1977).

21. W. Nielinger, in *Kunststoff Handbuch*, Vol. 3/4: *Polyamide*, L. Bottenbruch and R. Binsack (Eds.), Hanser Verlag, Munich, 1998, p. 1.

22. I. Goodman and R. N. Vachon, *Eur. Polym. J.*, **20**, 529 (1984).

23. P. J. M. Serrano, E. Thüss, and R. J. Gaymans, *Polymer*, **38**, 3893 (1997).

24. A. C. M. van Bennekom and R. J. Gaymans, *Polymer*, **37**, 5439 (1996).

25. R. J. Gaymans and J. L. de Haan, *Polymer*, **34**, 4360 (1993).

26. M. C. E. J. Niesten, J. Feijen, and R. J. Gaymans, *Polymer*, **41**, 8487 (2000).

27. J. H. Saunders, in *Encyclopedia of Polymer Science and Engineering*, Vol. 11, H. F. Mark, N. M. Bikales, C. G. Overberger, and G. Menges (Eds.), Wiley-Interscience, New York, 1989, p. 410.

28. W. Nielinger, in *Kunststoff Handbuch*, Vol. 3/4: *Polyamide*, L. Bottenbruch and R. Binsack (Eds.), Hanser Verlag, Munich, 1998, p. 22.

29. C. Giori and B. T. Hayes, *J. Polym. Sci., Part A-1*, **9**, 335 (1970).

30. C. Giori and B. T. Hayes, *J. Polym. Sci., Part A-1*, **9**, 351 (1970).

31. H. Jacobs and C. Scheigman, in *Proceedings of the Fifth European Symposium on Chemical Reaction Engineering*, Vol. B7, Elsevier, London, 1972, p. 1.

32. E. Roerdink and J. M. M. Warnier, *Polymer*, **26**, 1582 (1985).

33. S. K. Gupta and M. Tjahjadi, *J. Appl. Polym. Sci.*, **33**, 933 (1987).

34. P. W. Morgan and S. L. Kwolek, *J. Chem. Educ.*, **36**, 182 (1959).

35. P. W. Morgan and S. L. Kwolek, *Macromolecules*, **8**, 104 (1975).

36. Y. Oishi, M.-A. Kakimoto, and Y. Imai, *Macromolecules*, **20**, 703 (1987).

37. H. R. Kricheldorf and Z. Denchev, in *Transreactions in Condensation Polymers*, S. Fakirov (Ed.), Wiley-VCH, New York, 1998, p. 1.

38. R. J. Gaymans, *J. Polym. Sci., Polym. Chem. Ed.*, **23**, 1599 (1985).

39. R. C. Nelb and A. T. Chen, in *Thermoplastic Elastomers*, G. Holden, N. R. Legge, R. Quirk, and H. E. Schroeder (Eds.), Hanser, New York, 1996, p. 227.

40. J. E. Flannigan and G. A. Mortimer, *J. Polym. Sci., Polym. Chem ed.*, **16**, 1221 (1978).

41. F. Pilati, in *Comprehensive Polymer Science*, Vol. 5: *Step Polymerization*, G. Allan, J. C. Bevington, G. C. Eastmond, A. Ledwith, S. Russo, and P. Sigwald (Eds.), Pergamn, Oxford, 1989, p. 201.

42. R. J. Gaymans, J. Amirtharay, and H. Kamp, *J. Appl. Polym. Sci.*, **27**, 2513 (1982).

43. J. Zimmerman, *Polym. Lett.*, **2**, 955 (1964).

44. D. W. van Krevelen, *Properties of Polymers*, Elsevier, Amsterdam, 1990, p. 189.

45. Z. Tuzar, P. Kratochvíl, and M. Bohdanecký, *Adv. Polym. Sci.*, **30**, 117 (1979).

46. G. B. Taylor, *J. Am. Chem.Soc.*, **69**, 638 (1947).

47. I. E. Walz and G. B. Taylor, *Anal. Chem.*, **19**, 448 (1949).

48. A. Horbach, in *Kunststoff Handbuch*, Vol. 3/4: *Polyamide*, L. Bottenbruch and R. Binsack (Eds.), Hanser Verlag, Munich, 1998, p. 273.

49. G. Jacobi, H. Schuttenberg, and R. C. Schulz, *Makromol. Chem. Rappid Commun.*, **1**, 397 (1980).

50. W. Nierlinger, B. Brassat, and D. Neuray, *Angew. Makromol. Chem.*, **98**, 225 (1981).

51. G. E. Bush, C. E. Schwier, R. M. Lembeke, and S. W. Cook, U.S. Patent 6,169,162 (to Solutia Inc.), 1999.

52. A. M. Brearley, J. J. Lang, and E. K. A. Marchildon, U.S. Patent 5,674,974 (to Du Pont), 1974.

53. P. E. Beck and E. E. Margat, *Macromolecular Syntheses*, Vol. 6, C. G. Overberger (Ed.), Wiley, New York, 1977, p. 57.

54. W. R. Sorensen and T. W. Campbell, *Preparative Methods of Polymer Chemistry*, Wiley-Interscience, New York, 1961, p. 74.

55. F.-E. Baumann, K. Moser, U. Rohde-Liebenau, and F. G. Schmidt, in *Kunststoff Handbuch*, Vol. 3/4: *Polyamide*, L. Bottenbruch and R. Binsack (Eds.), Hanser Verlag, Munich, 1998, p. 769.

56. W. Nielinger, H. Brinkmeyer, R. Binsack, L. Bottenbruch, and H.-J. Füllmann, U.S. Patent 4,719,284 (to Bayer A.G.), 1988.

57. W. R. Sorensen and T. W. Campbell, *Preparative Methods of Polymer Chemistry*, Wiley-Interscience, New York, 1961, p. 86.

58. R. J. Gaymans, V. S. Venkatraman, and J. Schuijer, *J. Polym. Sci., Polym. Chem. Ed.*, **22**, 1373 (1984).

59. D. Braun, H. Cherdron, and H. Ritter, *Praktikum der Makromolekularen Stoffe*, Wiley-VCH, Weinheim, 1999, p. 244.

60. D. Braun, H. Cherdron, and H. Ritter, *Praktikum der Makromolekularen Stoffe*, Wiley-VCH, Weinheim, 1999, p. 172.

61. R. W. Wynn, S. A. Glickman, and M. E. Chiddix, *Macromolecular Syntheses*, Vol. 3, C. G. Overberger (Ed.), Wiley, New York, 1968, p. 106.

62. K. R. Meyer and H. Ohm, U.S. Patent 5,519,097 (to Huels), 1995.

63. W. R. Sorensen and T. W. Campbell, *Preparative Methods of Polymer Chemistry*, Wiley-Interscience, New York, 1961, p. 81.

64. W. Nielinger, W. Alewelt, and K. H. Hermann, U.S. Patent (to Bayer A.G.), 1994.

65. D. W. Fox and S. J. Shafer, U.S. Patent 4,567,349 (to GE Plastics), 1986.

66. V. E. Shashoua and W. M. Eareckson, *J. Polym. Sci.*, **40**, 343 (1959).

64. W. Nielinger, W. Alewelt, and K. H. Hermann, U.S. Patent (to Bayer A.G.), 1994.

65. D. W. Fox and S. J. Shafer, U.S. Patent 4,567,349 (to GE Plastics), 1986.

66. V. E. Shashoua and W. M. Eareckson, *J. Polym. Sci.*, **40**, 343 (1959).

67. R. J. Gaymans, *J. Polym. Sci., Polym. Chem. ed.*, **23**, 1599 (1985).

68. Y. Wei, X. Jia, D. Jin, and F. A. Davis, *Makromol. Rapid Commun.*, **17**, 897 (1996).

69. L. Vollbracht and T. J. Veerman, U.S. Patent 4,308,374 (to AKZO), 1981.

70. N. Yamazaki, F. Higashi, and J. Kawabata, *J. Polym. Sci. Polym. Chem. Ed.*, **12**, 2149 (1974).

71. S. Russo, A. Mariani, V. N. Ignatov, and I. I. Ponomarev, *Macromolecules*, **26**, 4984 (1993).

72. S. L. Kwolek, P. W. Morgan, I. R. Schaefgen, and L. W. Gulrich, *Macromolecules*, **10**, 1390 (1977).

73. K. Bouma, G. M. M. Groot, J. Feijen, and R. J. Gaymans, *Polymer*, **41**, 2727 (2000).

74. G. della Fortuna, E. Oberrauch, T. Salvatori, E. Sorta, and M. Bruzzone, *Polymer*, **18**, 269 (1977).

75. F. Cognigni and A. Mariano, U.S. Patent 4,614,815 (to Anic S.p.A.), 1986.

76. S. Mumcu, K. Burzin, R. Fieldmann, and R. Feinauer, *Angew. Makromol. Chem.*, **74**, 49 (1978).

4 Polyurethanes and Polyureas

Jeff Dodge
Bayer Corporation, Pittsburgh, Pennsylvania 15205

4.1 INTRODUCTION

Polyurethanes (PUs or PURs) are an extremely versatile class of polymers. A wide variety of raw materials coupled with adaptable synthetic techniques allows the polyurethane chemist to design useful materials for many applications. Table 4.1 lists commercially important usage categories for urethanes along with examples of manufacturing methods and physical properties. All of the items shown can be readily achieved by proper choice of starting materials, polymer design, processing conditions, and application technique. While polyurethanes certainly have limitations, it is arguable that no other class of polymers can match their collective versatility, usefulness, and performance.[1,2]

Urethane polymers are formed by reaction of polyisocyanates and polyols which create the urethane chemical linkage. The closely related polyureas are synthesized from polyisocyanates and polyamines, producing urea linkages (Scheme 4.1). Although many polyurethanes contain both urethane and urea groups and are therefore more properly termed polyurethane/ureas, it is common to refer to such polymers under the blanket term polyurethanes. The polyol and/or polyamine in these polymers often comprises the majority mass component, so the terms polyurethane, polyurethane/urea, and polyurea refer to the corresponding chemical group linkage and not the whole of the polymer backbone. The urethane and urea functional groups, however, do impart most of the important physical properties.

4.1.1 Historical Perspective

The history of polyurethanes begins with Otto Bayer[3] at Germany's I. G. Farbenindustrie (the predecessor company of Bayer AG[4]) in 1937, the year of the first disclosure of diisocyanate addition polymerization to form polyurethanes and polyureas. The main impetus for this work was the success of Wallace Carothers

Synthetic Methods in Step-Growth Polymers. Edited by Martin E. Rogers and Timothy E. Long
© 2003 John Wiley & Sons, Inc.   ISBN: 0-471-38769-X

TABLE 4.1 Polyurethanes: Applications, Properties, and Processing Methods

Applications	Physical States and Properties	Processing Methods
Foams	Thermosets	Open-cast molding
Elastomers	Thermoplastics	Fiber spinning
Coatings	Amorphous or	Blow molding
Adhesives/sealants/binders	microcrystalline	Injection molding
Encapsulants	Hard or soft	Extrusion/pultrusion
Elastomeric fibers	Transparent or opaque	Reaction injection
Films	High or low T_g	molding
Gels	Aromatic or aliphatic	Spraying/brushing/rolling
Composites	Hydrophilic or	Resin transfer molding
Microcellular elastomers	hydrophobic	Rotational casting
Rubbers/millable gums	100% Solids/solvent	Rotomolding
	borne/water borne	Thermoforming
	Resilient or energy	Compounding/vulcani-
	absorbing	zation
	Polyester/polyether/	Compression molding
	polyacrylate	Centrifugal molding
	High or low	
	molecular weight	

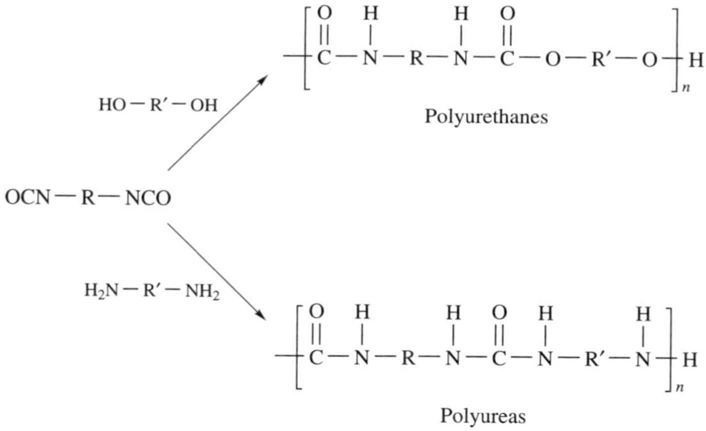

Scheme 4.1 Polyurethanes and polyureas from polyisocyanates.

at Du Pont in making polyamides and polyesters. Bayer and his colleagues were trying to find a route to similar materials without infringing on Du Pont's patents. The synthesis and basic reactions of isocyanates had been explored beginning in the nineteenth century by Wurtz and others.[2a] Bayer utilized the known addition reactions of the isocyanate group with alcohols and amines to form macromolecules from diisocyanates and either diols or diamines. This led first to the preparation of polyureas from hexamethylene diamine and hexamethylene diisocyanate (HDI; this and other isocyanates are shown in Scheme 4.2), but these polymers were intractable and proved to be unsuited for the preparation of fibers or thermoplastics. However, the reaction of HDI with glycols such as 1,4-butanediol led to polyurethanes which found limited applications under the tradenames Perlon U for fibers and Igamid U for thermoplastics. Throughout the 1940s, a whole new raw material product line was developed at I. G. Farben, including polyisocyanates under the tradename Desmodur and polyester polyols under the Desmophen tradename. Various products such as adhesives, rigid foams, elastomers, and coatings were also designed and tested on a developmental scale but were not sold outside the government.[5]

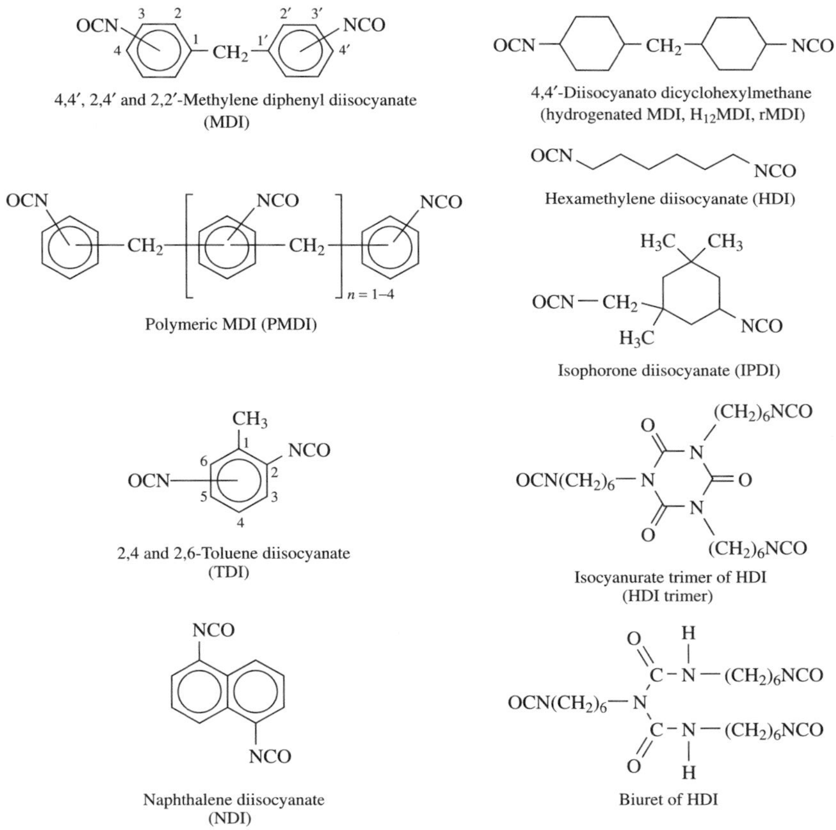

Scheme 4.2 Examples of commercially important isocyanates.

During this time, other companies in England and the United States quickly developed urethane products. Du Pont applied for patents on the subject as early as 1939, dealing mainly with elastomers and adhesives.[6] ICI worked in this area beginning in 1941 designing adhesives, elastomers, and coatings.[7] The company developed the Vulcaprene line of elastomers, which also found use as fabric coatings for barrage balloons due to their low hydrogen gas permeability. Raw materials began to be manufactured on a larger pilot plant scale at several companies, including I. G. Farben, ICI, Du Pont, Monsanto, National Aniline, and Wyandotte. Toluene diisocyanate (TDI) rapidly became the most common isocyanate, especially for flexible foams, which would become the largest urethane application. The most important production route to isocyanates was and still is the direct phosgenation of primary amines.

Until the early 1950s, polyurethanes were expensive materials for high-performance applications. At that time, large-scale commercial production of flexible foams occurred, with Bayer AG supplying raw materials and Hennecke AG the processing equipment. These slabstock foams were produced using TDI and polyester polyols and sold in the seating, bedding, and carpet industries. It is important to note that these raw material and machinery producers provided customers with the chemicals and know-how to manufacture the polymer product themselves. In other words, the customer carried out the actual urethane polymerization. It was unusual to provide reactive raw materials, such as isocyanates, to raw material customers for processing. Usually, the polymer manufacturer would produce or buy the chemicals, carry out the polymerization, and sell the product in the form of thermoplastic pellets, fibers, gum rubbers, or some other prereacted state.

The significance of the Bayer–Hennecke cooperation cannot be overemphasized because it was the development of entirely new processing methods and equipment that made large-scale commercialization of urethanes possible. Together, the two companies brought customers a complete package, including raw materials, production machinery, and technical expertise. An especially important milestone was the development of a high-pressure mixing technique by Weinbrenner and colleagues at Bayer in the 1940s, known today as *impingement* mixing.[5d] This allowed very rapid mixing of relatively viscous isocyanate and polyol streams in a compact mixhead. It was this invention, perhaps more than any other, which gave the Bayer–Hennecke team market dominance throughout the 1950s and 1960s.

By 1960, annual worldwide production of flexible urethane foams reached 100 million pounds. These polyester-based foams were hydrolytically sensitive and slowly disintegrated in air. It was not until the mid-1950s that urethane-grade polyether polyols became commercially available from Du Pont, Wyandotte, Dow, and Union Carbide. These were more hydrolytically stable, were less expensive than polyesters, and provided pathways to an expanded range of urethane materials. Many other products were also commercialized by the end of the 1950s, including applications in coatings, elastomers, thermoplastics, and noninsulating rigid foams.

Following the phenomenal success of flexible urethane foams, a broader range of isocyanates and polyols became available at reasonable prices, and polyurethanes garnered wider attention. With the advent of higher functionality reactants (such as polyethers and "polymeric" MDI), interest in rigid foam applications grew, especially closed-cell rigid foam for insulation. One important breakthrough was the conversion from water to chlorofluorocarbons (CFCs, such as CCl_3F and CCl_2F_2) as *blowing agents* in insulation foams. These *physical blowing agents* improved processing characteristics and insulative performance and lowered costs. They were widely used until the Montreal Protocol of 1987 and Clean Air Act of 1990 spurred development of non-CFC foam systems. Today, urethane foams are made using either hydrochlorofluorocarbons (HCFCs), hydrofluorocarbons (HFCs), hydrocarbons, or water as blowing agents. Another key development was the discovery of insulation foams formed from trimerization of isocyanates (see Section 4.3.1.5). These isocyanurate foams had reduced combustion rates and smoke formation compared to standard urethane foams because of the more thermally stable isocyanurate linkages. This was important because of increasingly stringent fire codes in the 1960s.

Elastomers were among the first urethane materials to be developed.[5b,c] At Bayer, a line of urethane rubber was developed in the 1940s under the Vulkollan tradename. These materials were based on polyester prepolymers made from naphthalene diisocyanate (NDI) and chain extended using short-chain glycols, diamines, or alkanolamines.[8] The early products were in the form of millable gums which could be processed and vulcanized like conventional rubber. Several other companies brought out similar products at this time, including ICI (Vulcaprene), Goodyear (Chemigum), and General Tire (Genthane). Later, in the 1950s, Du Pont introduced their Adiprene product line, the first millable urethane gums utilizing polyether polyols. The most significant development in the elastomer area occurred at Bayer with the development of liquid-castable systems.[9] These could be conveniently poured or pumped into molds as low-viscosity liquids and cured to shape, giving rise to the term *cast elastomers*. With the commercialization in 1960 of polymeric MDI (PMDI) by ICI in England and Carwin in the United States, the related *monomeric* MDI (4,4'-methylene diphenyl diisocyanate) became more widely available. Since then, MDI and PMDI have been used extensively in cast elastomer applications, except for Bayer's Vulkollan and a few TDI-based products. Polyurethane elastomers were initially slow to grow commercially, largely due to the success of synthetic rubbers. However, polyurethane elastomer use has dramatically increased and now they even dominate certain niche markets such as in-line skate wheels and high-performance bowling balls due to superior dynamic mechanical properties.

Thermoplastic polyurethanes (TPUs) made their debut in the early 1950s. B. F. Goodrich was the first successful commercial producer of TPUs with their Estane VC product line. While earlier TPUs had to be heat cured after initial molding, the Estane VC products, based on 4,4'-MDI, butylene adipate polyester polyols, and 1,4-butanediol, had such good physical properties they did not need the postcuring step. They were considered "virtually crosslinked" (hence the VC designation) due to the presence of hydrogen bonding and hard-block/soft-block phase separation (see Section 4.2.3). These products could be processed by standard thermoplastic methods, such as extrusion and injection molding. Today,

TPUs are more successful than ever and are used in a wide variety of applications, including automotive products, sports equipment, tool housings, and films.

Polyurethane coatings have proven to be highly suitable for a variety of substrates (such as metal, wood, and textiles) and applications (automotive, construction, flooring). They were among the first applications explored at I. G. Farben. For example, during World War II they were used to impregnate paper for gas-resistant suits and as protective coatings for aircraft.[5a] It was the need for coatings with improved light stability which drove most of the development work on aliphatic isocyanates. Also of importance was the development of aliphatic isocyanate adducts to reduce the vapor pressure of compounds like HDI. This led to products such as the biuret and isocyanurate trimer of HDI (Scheme 4.2). Coatings also spawned the first one-component formulations (Section 4.3.2.3), with the early uralkyd resins (or so-called urethane oils), and blocked isocyanates. Urethane coatings have generally followed industry trends with the advent of moisture-cured systems, powder coatings, and materials cured by ultraviolet or electron beam irradiation. Although many coating formulations were and still are solvent based, high solids based, or even 100% solids based, one of the most fundamental changes has been the more recent move toward water-based systems (Section 4.3.2.2). This trend has been driven by the need to reduce the emission of volatile organic compounds (VOCs), especially solvents.

4.1.2 Applications

An indication of the usefulness of polyurethanes can be found in worldwide consumption data; as of 2000, urethanes of all types amounted to 9.25 million tons.[10] The commercial success of urethane polymers can be attributed to the ability to conveniently manufacture and apply materials with specific combinations of physical properties. Furthermore, urethanes are often the best choice in demanding, high-performance applications. In this section, the applications listed in Table 4.1 are discussed and a few examples are given from each category.

Foams are the largest polyurethane application area and account for about 80% of sales (by weight) worldwide. The term *foam* covers a broad range of materials and physical properties, but there are three general categories: rigid, semirigid, and flexible.[11] Rigid foams are nonresilient, high compressive-strength materials used mainly for insulation and structural support. They can have low thermal conductivity when designed and processed properly. For example, refrigerators are made using rigid PU foam providing both insulation and structural support. Other uses include insulation foams for homes and buildings (the single largest urethane application), structural foams for lamination products and boat hulls, and energy-absorbing foams for packaging and automotive interiors. A rigid insulation foam formulation is given in Example 5, Section 4.4.

Flexible foams are used in applications where a high degree of resiliency is required with moderate load-bearing capacity. Essentially all foam seating is urethane based, including the furniture and automotive markets. Other examples are packaging, textiles, filters, sports equipment, and recreational items.

Semirigid foams are between rigid and flexible foams in terms of resiliency, energy absorption, and compression strength. These properties make them ideal for applications where a combination of flexibility and energy absorption is desired. A good example is padding for automotive instrument panels and steering wheels, where the flexibility provides tactility and the energy absorption characteristics offers a degree of occupant protection.

Coatings comprise the second largest polyurethane application area and have proven to be extremely effective for a wide variety of uses.[12] Automotive original equipment manufacturers (OEMs) as well as refinishers, for example, use urethane coatings for various types of body coatings, such as electrocoatings, primers, base (color) coatings, and clear/top coatings. Highly flexible coatings for plastics are used for fascias, bumpers, wheel covers, and body panels. Other uses include floors, furniture, sports equipment, textiles, leather, appliances, and heavy-duty industrial coatings. See Examples 6–9 in Section 4.4.

Polyurethanes also make excellent adhesives, sealants, encapsulants, and binders.[13] They have been successful in these areas due to their design flexibility in terms of physical properties and application techniques and their ability to adhere strongly to a wide variety of substrates. For example, automotive structural adhesives (Example 10 in Section 4.4) are used to bond sheet-molding compound (SMC) body panels to vehicle frames without any cleaning or priming pretreatment. Other examples are concrete sealants (Example 11), automotive windshield adhesives, roofing cements, and electrically and thermally conductive adhesives for electronic applications. Unmodified isocyanates are often used as moisture-cured binders for foundry cores, engineered wood products, and rebonded carpet underlayment. One-component, moisture-cured prepolymers and two-component polyurethane systems are used as binders for rubberized running tracks and playground surfaces. Urethanes are also used as encapsulants for carbonless copy paper ink and slow-release fertilizers. Electrical components such as airbag sensors and circuit boards can be protected using polyurethane potting compounds. As with coatings, polyurethane adhesives, sealants, encapsulants, and binders can be solvent based, 100% solids, or aqueous. They may also be one-component (moisture-cured, polyurethane dispersions, or blocked) or two-component systems (see Section 4.3.2.3).

Elastomers come next in order of quantity sold worldwide. Like the previous categories, elastomers cover an extremely broad range of material types and application areas. They include cast elastomers, spray and rotational casting elastomers, thermoplastic elastomers, microcellular elastomers, gels, and elastomeric fibers. Each type is covered separately.[13d, 14]

Cast elastomers are noncellular solids produced by pouring or injecting an isocyanate–polyol reaction mixture directly into a mold to form a finished part. For example, pipeline pigs and golf club grips can be made using urethane cast elastomers based on either MDI or TDI prepolymers (see Section 4.3.2.1). Other uses include industrial wheels and rollers, die-cut blankets, and table edgings. Closely related in terms of applications and physical properties are the polyurethane millable gums (urethane rubber). These materials are compounded, processed, and cured like natural rubber and are used in applications such as gear belts, seals, and bellows. All of these applications are very demanding, especially in terms of dynamic properties. Therefore, it is apparent that cast elastomers and millable rubbers are often used in high-end applications where performance requirements justify their relatively high cost as compared to competitive materials, such as rubbers and polyesters. See Examples 1 and 2 in Section 4.4.

Spray and rotational casting elastomer formulations resemble cast elastomers but are processed using different application techniques. Polyurethane and polyurea spray elastomers are handled using equipment similar to that used for spray-on coatings and are formulated to have nonsagging performance, enabling them to be applied thickly to vertical surfaces without running or dripping. They are also fast reacting, with "gel times" (Section 4.3.1.7) under half a minute. Examples include structural backings for acrylic spa shells, truck bed liners, and floor coverings. Rotational casting elastomers are also formulated to be nonsagging but are somewhat slower reacting due to the application technique. In this case, the elastomer is applied directly from a static mixer onto a rotating cylinder, roller, or drum mounted on a lathe. The mixhead moves in a controlled manner down the length of the part until coverage is complete. Multiple passes may be made, depending on the thickness desired for the covering. Uses include drum rollers used in paper and steel manufacturing and protective membranes for compressed gas cylinders. Spray and rotational casting elastomers differ from coatings in that they not only provide a protective and attractive covering but also are thicker and impart *elastomeric* properties to the substrate.

Thermoplastic polyurethanes[15] can be employed wherever standard thermoplastics are used, utilizing the same processing equipment. TPUs are high-molecular-weight, linear or lightly crosslinked polymers, but they differ from most other thermoplastics in that they have the added benefit of hydrogen bonding from the urethane linkages (see Sections 4.2.2 and 4.2.3). Hydrogen bonds act as physical crosslinks, but their thermal reversibility allows the material to be processed at elevated temperatures like a linear thermoplastic. Manufacturing via reaction extrusion is most common whereby a prepolymer is made in situ and then directly reacted with chain extender. The reaction mixture is then extruded into strands, cooled, pelletized, and sold to processors. Standard thermoplastic processing methods can be used such as extrusion, compression molding, calendering, blowing, and rotational molding (not to be confused with rotational casting). As with other urethane materials, TPUs are used in demanding applications requiring superior physical properties. Examples are ski boots, football bladders, films, automotive instrument and door panel skins, safety glass, soft spikes for golf shoes, and catheters. Example 13 in Section 4.4 is of a hand-formable TPU.

Microcellular elastomers[16] bridge the gap between solid elastomers and low-density foams. Although they may appear to be noncellular, these materials have a microscopic cell structure and so are by definition high-density foams, with densities between noncellular solids and standard foams. The most commercially important applications include shoe soles, carpet backing, chair armrests, nonpneumatic tires, and automotive body parts. Reaction injection molding (RIM) is one method for producing microcellular products whereby liquid raw materials are mixed under high pressure (impingement mixing) and injected into a high-tonnage mold. The reactants then cure and the part can be demolded shortly thereafter. Nucleation using dissolved nitrogen gas creates a microcellular structure to reduce material density, facilitate flow through the mold, and give high-quality surfaces. The primary advantages of RIM are high-volume manufacturing capability, lower capital investment and energy consumption compared to that of conventional injection molding, and superior end-product physical properties. This process is most often used in large-scale manufacturing of automotive and farm equipment parts, recreational products, and computer housings.

Shoe soles are made from a different kind of microcellular urethane which is lower in density compared to most RIM materials. Shoe sole products are quite flexible and elastic but have damping characteristics as well. They are one application of *integral skin* foams, where a tough, elastomer-like outer skin is designed to self-form on the part during the molding process. They may be manufactured using RIM but are mostly made using lower cost open-mold pouring techniques. Microcellularity is created using water as a *chemical blowing agent* to lower density (reduce part weight) and improve energy absorption characteristics. An auxiliary physical blowing agent (such as HFC-134a or pentane) is often added to further reduce density or to improve skin quality. The primary advantages of urethanes in this market are high production rates, design flexibility, and superior physical properties such as abrasion resistance, elasticity, and energy absorption. Urethanes are used in 6–8% of footwear products worldwide.

Polyurethane elastomeric fibers[17] are known generically in the United States as *spandex* (*elastane* in Europe). These were developed by Bayer under the tradename Dorlastan and by Du Pont as Lycra. These materials are highly elastic and can withstand elongation of several hundred percent before breaking. They are extruded into filaments for textiles and are commonly used in elastic fabrics, especially sportswear.

Urethane gels and ultrasoft elastomers are a more recent development.[18] They are made primarily by reacting high-molecular-weight polyether polyols with a stoichiometric deficiency of isocyanate. The low NCO-to-OH ratio allows for a wide latitude in hardness adjustment. These low-hardness elastomers are used for seating applications (such as gel bicycle seats), shoe inserts, and soft padding for orthopedic devices.

4.1.3 Research Trends and Critical Issues

Perhaps the most significant recent issue for the industry was the phase-out of CFCs as blowing agents resulting from the Montreal Protocol (1987) and Clean Air Act (1990).[19] These regulations mandate specific timelines for discontinuation and, since foams comprise the bulk of polyurethane applications, have prompted a worldwide search for alternatives. Hydrochlorofluorocarbons, HFCs, and hydrocarbons (HCs) are now the blowing agents of choice. Which one is used depends on many factors, including cost, performance, plant design, cost of retrofitting, and local and national regulations. Water serves well in many applications as a chemical blowing agent (it reacts with isocyanate to form gaseous CO_2; see Section 4.3.1.4), and its use continues to increase. However, for rigid insulation foams water is not preferred due to the relatively high thermal conductivity and permeativity of carbon dioxide. Hydrocarbons, including isomers of pentane and butane, are better blowing agents in insulation foams because they offer suitable boiling points, solubility characteristics, low permeation, low thermal conductivity compared to CO_2, and low cost. The largest hurdle for developers and processors has been the high flammability and associated hazards of these volatile hydrocarbons. The need for large-scale explosion-proof manufacturing equipment and facilities — not to mention new formulations, techniques, and products — can make plant conversion costs prohibitively high in many cases. Currently, most rigid insulation foam in Europe is made using hydrocarbons. In the United States, HCFC-141b is still being used but will not be manufactured for domestic use after 2002. Manufacturers will then have no alternative but to convert over to HFCs (e.g., HFC-245fa, HFC-134a) or hydrocarbons. Rigid lamination (structural) foam in the United States, however, is most often made using pentanes. Liquid carbon dioxide (*supercritical CO_2*) is also of great interest as a physical blowing agent, and its use is slowly increasing as technical hurdles are overcome. In addition to the disadvantages mentioned for water-blown insulation foams, liquid CO_2 also promotes open cell structure and so is used primarily in open-celled foams such as flexible foams and rigid lamination foams.

Another pressing issue over the last 15 years has been a reduction in the use of volatile organic compounds (VOCs).[20] This is important to the urethane industry because solvents have traditionally been used as diluents and carriers. Foams and elastomers have not been affected to a great extent because they do not rely as heavily on the use of solvents. For adhesives, binders, and coatings, however, a reduction in the use of organic solvents has required extensive research and development on waterborne and solvent-free systems (Section 4.3.2.2). Development of aqueous systems has been a daunting effort because of the water/isocyanate reaction, which has generally been avoided at all costs (except in water-blown foams where water is used as a chemical blowing agent). By about 1990, it had been found that certain isocyanates could indeed be dispersed in water and reacted preferentially with polyols in situ using proper catalysis and reaction conditions. In this case, water serves only as a carrier and viscosity reducer, allowing the isocyanate–polyol reaction mixture to be conveniently applied to a substrate, then evaporates to leave the cured product. This methodology is suitable for applications where the urethane is applied in thin cross section, enabling the water to evaporate quickly; otherwise, entrapped water can form bubbles, especially if it is present long enough to react with isocyanate. This has given rise to a whole new technology of two-component (2-K), waterborne polyurethane coatings (see Example 9 in Section 4.4), adhesives, and binders. The use of one component (1-K) *polyurethane dispersions* (PUDs) (Section 4.3.2.3 and Example 8), fully reacted urethane polymers dispersed in water, has also increased dramatically in the last decade. Although PUDs have been known since the 1960s, it was not until the crackdown on VOCs that waterborne technology gained wider attention. Another low-VOC alternative is high-solids coatings using low-viscosity reactive components (reactive diluents) to provide processibility without contributing to VOCs (see Section 4.2.3 and Example 6).

Medical applications for polyurethanes, especially thermoplastic and cast elastomers, is another area of current intensive interest.[21] Urethanes can be very biocompatible and suitable for blood contact and implantation applications. Of course, polyurethanes are used in many non–blood contact medical applications, such as for prostheses, gloves, limb casts, and wheelchair tires. However, the requirements are far more challenging when there is blood contact, especially in vivo. Polyurethanes can be designed to be biodurable, biocompatible, hemocompatible, and nonthrombogenic. Because of the ability to finely tune physical properties, urethanes can be designed for very exacting applications such as artificial veins and cardiac pacemaker lead insulators. Furthermore, target physical properties for elastomers can be achieved without introduction of plasticizers or fillers which can affect biocompatibility. Historically, the most challenging problem with implanted urethane elastomers has been the development of surface fissures and cracks. This environmental stress cracking is caused by biochemical reactions at the surface and can cause loss of mechanical properties and facilitate blood clotting. Other problems with implanted polyurethanes are calcification, hydrolysis, and hydration. The propensity for calcification depends on the specific urethane used but may be mitigated using surface treatments. Hydrolysis and hydration are both largely a result of the polyol used. Polyester polyols are more susceptible to hydrolysis reactions which can rapidly degrade properties. Hydration is common in urethanes, all of which will absorb some water, and can have a plasticizing effect on the material, making it softer. In some situations, this can be desirable, such as for catheters, where the part is hard and stiff for insertion but later softens for the patient's comfort. Recent improvements have been made by utilizing alternative, more robust polyols instead of the traditional polyethers and polyesters. Overall, the best results have been obtained with polycarbonate glycols, and products using them are still under development.

In vitro applications for blood contact are not as demanding on materials, and urethanes have made significant progress in this area. For example, hemodialysis cartridges have been manufactured using urethane encapsulants to hold the filtration filaments in place. Another area of recent activity is the development of urethane hydrogels for bandages. These gels are similar to those described in Section 4.1.2 but are based on hydrophilic poly(ethylene oxide) polyols and can be hydrated and swelled with water. They are used for wound dressings, especially severe, chronic wounds which have failed to heal, such as ulcers and bed sores. Their main advantage is that they are "skin friendly," keeping the wound moist and easing dressing removal. They also facilitate healing by being quite permeable to oxygen.

Recycling of urethane materials has been an important issue for the last decade or so.[22] This came as a result of increasing pressure for recycling in general or at least to demonstrate recyclability of products. An example is the automotive sector, where it is often necessary for suppliers to provide evidence that their materials can be recycled. Polyurethanes may arguably be recycled by more methods than any other polymer, although this is partially due to the large variety of product types which can be produced from them.

There are three basic categories for polyurethane recycling: mechanical recycling, chemical recycling, and energy recovery. Mechanical recycling (also known as material, physical, or particle recycling) is the oldest method for reuse of urethanes. For example, using flexible foam scrap for rebonded carpet underlayment has been practiced for decades. This is so successful that nearly one-quarter of the foam scrap used in the United States for this application is actually imported from Europe. Other examples of mechanical recycling include the use of powdered waste as fillers and absorbents, using granulated materials for compression molding of new products under heat and high pressure, and binding of particulated material for playground surfaces and park benches.

Chemical recycling involves degrading the urethane into either new polyurethane feedstocks (such as polyols and polyamines) or gases and oils for use in energy production or chemical manufacturing. There are several methods of chemical recycling: glycolysis, hydrolysis, hydrogenation, and pyrolysis. There are advantages and disadvantages for each, but the main issue for any recycling effort is that it must be done economically. Many of these methods require high temperature and pressure (especially pyrolysis and hydrogenation) and do not operate efficiently. To date, the most successful chemical recycling route is glycolysis, where granulated urethane scrap is heated in the presence of a glycol (such as diethylene glycol) to about 200°C. Under these conditions, the polymer undergoes glycolysis and reverts back to the original polyols and various side products. Careful prescreening of the raw polyurethane scrap can result in high-quality polyols which are nearly identical to the original. Several companies in Europe are currently using this method on a commercial scale.

The third general category of polyurethane recycling is energy recovery (incineration), where the scrap is burned, the energy utilized, and the waste captured and disposed of. Although there are many proven methods for recycling polyurethanes, with few exceptions they are just not cost competitive with the virgin materials. Progress is being made, however, and it is expected that more firms will commercialize related technology and products as they become more economically viable.

4.2 STRUCTURE–PROPERTY RELATIONSHIPS

4.2.1 General Considerations

An in-depth understanding of structure–property relationships is perhaps the most important concern for the urethane formulation chemist. Material design objectives often go far beyond physical property requirements and may also include considerations like processing characteristics (i.e., compatibility, reactivity, viscosity, ease of mixing), appearance (transparency, surface gloss, color), weathering (light stability, water resistance), and cost. A urethane product can be comprised of many ingredients and isocyanates, polyols, crosslinkers, chain extenders, fillers, catalysts, and other additives must each be carefully selected. In Section 4.2.2, the main raw material categories are discussed in terms of how molecular structure can influence reactivity, processing traits, mechanical properties, and physicochemical attributes. Later, in Section 4.2.3, the important subjects of backbone microstructure and polymer morphology are reviewed to show how they are influenced by the ingredients chosen and the synthetic technique used to make the material.

4.2.2 Role of Molecular Structure

There are several ways to categorize isocyanates, but the broadest delineation is aromatic versus aliphatic. Aromatic MDI, PMDI, and TDI (Scheme 4.2) constitute by far the largest worldwide volume of isocyanates manufactured. Two characteristics stand out with these compounds. First, their aromaticity causes materials based on them to absorb ultraviolet (UV) radiation. This triggers numerous oxidative side reactions, especially in the presence of atmospheric oxygen and water. These oxidation reactions form colored quinoidal and other delocalized moieties which cause a discoloration to yellow or brown, depending on extent of reaction. Discoloration is undesirable in most applications, such as in foams and elastomers, but usually does not affect bulk physical properties unless the extent of oxidation is extreme. In coating applications, this sensitivity to light is critical and can cause not only discoloration but also loss of surface gloss, crazing, and many other problems. Another reason light sensitivity is of broader importance in coatings and other "thin cross-sectional" applications is because UV radiation can penetrate a larger percentage of the material's thickness, affecting not only the surface, but the bulk properties of the material as well, causing embrittlement, cracking, and peeling. It is important to note that although aliphatic urethanes are dramatically less sensitive to light than aromatic formulations, they are still susceptible to UV-induced degradation and are extensively tested accordingly.

The second important characteristic of aromatic isocyanates is that they are much more reactive than aliphatics due to delocalization of electron density into the aromatic ring. Of the mesomer structures shown in Scheme 4.3, if the

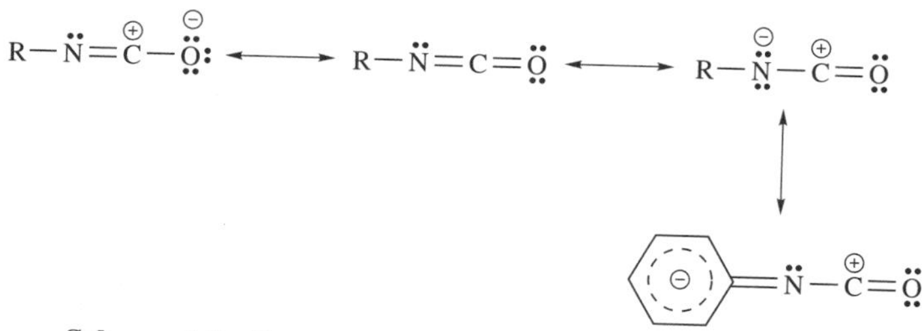

Scheme 4.3 Resonance structures of the isocyanate group.

isocyanate group is attached directly to an aromatic ring, the resonance effect places a higher average positive charge on the NCO carbon atom, making it more reactive toward nucleophiles such as alcohols and protic amines.

With few exceptions, the more expensive, less reactive aliphatic isocyanates are only used in situations where light stability is paramount. For coatings, films, and other materials which will not be exposed to light (especially sunlight), a less expensive aromatic isocyanate is more likely to be used.

Isocyanate functionality and isomer structure also influence physical properties and reactivity. For example, there are three major isomers for MDI (4,4', 2,4', and 2,2') and two for TDI (2,4 and 2,6). The most commercially important MDI product is the 4,4' isomer because its molecular linearity contributes to good polymer properties and it is more reactive compared to the relatively sterically hindered 2,4' and 2,2' isomers. However, higher 2,4' and 2,2' isomer content products may be desirable in certain situations, such as for their depressed freezing points, processing characteristics, and lower reactivity. Likewise, the two TDI isomers each have their own importance. The 2,4-TDI isomer is more reactive because the NCO group in the 4-position is relatively unencumbered. The 2,6-TDI isomer is less reactive, but it tends to give higher hardness flexible foams, allowing density to be reduced without sacrificing load-bearing capacity. There are also important differences in the way the two TDI isomers influence foam processing characteristics.[1a,11b]

Isocyanate functionality also plays a large role in formulation design. First and foremost, functionality of three or more contributes to crosslinking, leading to, for example, higher hardness, stiffness, and scratch resistance. Of course, too much crosslinking can be detrimental, causing embrittlement and reduced tear strength, for example. Isocyanate functionality is especially important in coating raw materials due to the relatively volatile aliphatic isocyanates commonly used, such as HDI (vapor pressure 0.011 mm Hg at 25°C). The 3-functional isocyanurate trimer of HDI (Scheme 4.2) not only gives good physical properties but also has a lower vapor pressure (5.2×10^{-9} mm Hg at 20°C), thereby reducing VOCs. The so-called PMDIs are manufactured mainly for their increased functionality and resultant physical property benefits, especially in one-shot elastomers (see Section 4.3.2) and rigid foams. However, PMDIs also have reduced freezing points, so they are convenient-to-use liquids at room temperature. The increased viscosity of the higher molecular weight PMDIs can often be problematic in terms of processing (e.g., ease of mixing, pumping, mold filling). However, this can sometimes be an advantage where a higher viscosity isocyanate is desired to facilitate mixing with the polyol, improve flow properties (reduce splashing), or increase dispersion stability in those rare cases where filler is added to the isocyanate.

There are many situations where isocyanates are derivatized before use, such as when MDI or TDI are reacted with high-molecular-weight, difunctional polyols to form low-NCO *prepolymers* used to make cast elastomers by the prepolymer method (Section 4.3.2.1). Because of their molecular weight, prepolymers can be of much higher viscosity than the starting isocyanates, and some are solid at room temperature. Prepolymers often require high processing temperatures (typically 80–100°C) to liquefy and/or reduce viscosity, so making products via this technique is also known as the *hot-cast* prepolymer method. There are also *quasi-prepolymers* in which fewer of the NCO groups are prereacted with polyol, and the viscosity is between that of "full" prepolymers and starting isocyanates. The various synthetic techniques all have important bearing on material types and properties and are discussed in more detail later in this section and in Section 4.3.2.

Another reason to derivatize raw isocyanates is to modify the characteristics of the compound itself. This is often done to lower the freezing point of the product, which makes it more user friendly to the processor. Thus, 4,4′-MDI freezes at 39°C so it is supplied as either a molten liquid or refrigerated solid (to reduce the rate of dimer formation; see Section 4.3.1.5). However, reacting this MDI isomer with a small amount of a low-molecular-weight glycol can reduce the melting point by several degrees. This allows the product to be shipped at more convenient temperatures and to be stored and used by the processor as a low-viscosity, pumpable liquid that is less susceptible to solidification. The penalty for this is higher viscosity and possibly a reduction in the ultimate physical properties which can be achieved due to reduced molecular symmetry and the incorporation of relatively flexible glycol units into the polymer backbone. Such products are known as *liquefied* (or *modified*) MDIs. Reacting an isocyanate with a water-soluble polyol can impart enough hydrophilicity so as to make it dispersible in water. This is one way to make two-component, waterborne coatings (see Section 4.3.2 and Example 9 in Section 4.4).

The next important category of urethane starting components is the polyol. The polyol makes many contributions to the finished polymer, including flexibility, softness, low-temperature properties, and processing characteristics. Scheme 4.4 shows representations of common polyol types, distinguished by their backbone structure. The most common urethane-grade polyols are the polyethers made from ethylene oxide (EO), propylene oxide (PO), and tetrahydrofuran, referred to as C_2, C_3, and C_4 ethers, respectively, from the number of carbon atoms in each repeat unit. The C_3 ethers are often tipped (or capped) with C_2–ether endgroups to provide the enhanced reactivity of unhindered, primary OH groups. Random and mixed-block copolymers of ethylene oxide and propylene oxide are used mainly in water-blown foams where they provide better control of miscibility between polyol, isocyanate, and water. They also can give better regulation of processing characteristics such as reactivity, demold, and flow properties.[11b]

Polyether polyols are predominant in urethanes because they are available in a wide range of molecular weight, viscosity, functionality, and composition. The polyethers can have very low glass transition temperatures (T_g) due to the flexible etheric backbone, and this benefits the low-temperature properties and flexibility of the resultant urethane. They also are generally less expensive than other polyol types. The C_3 ether polyols are the most common because they are much less hydrophilic/hygroscopic than the C_2 ethers, of which there are relatively few. The C_3 polyols also give quite good physical properties and are commonly used in almost all areas of urethane technology. However, when

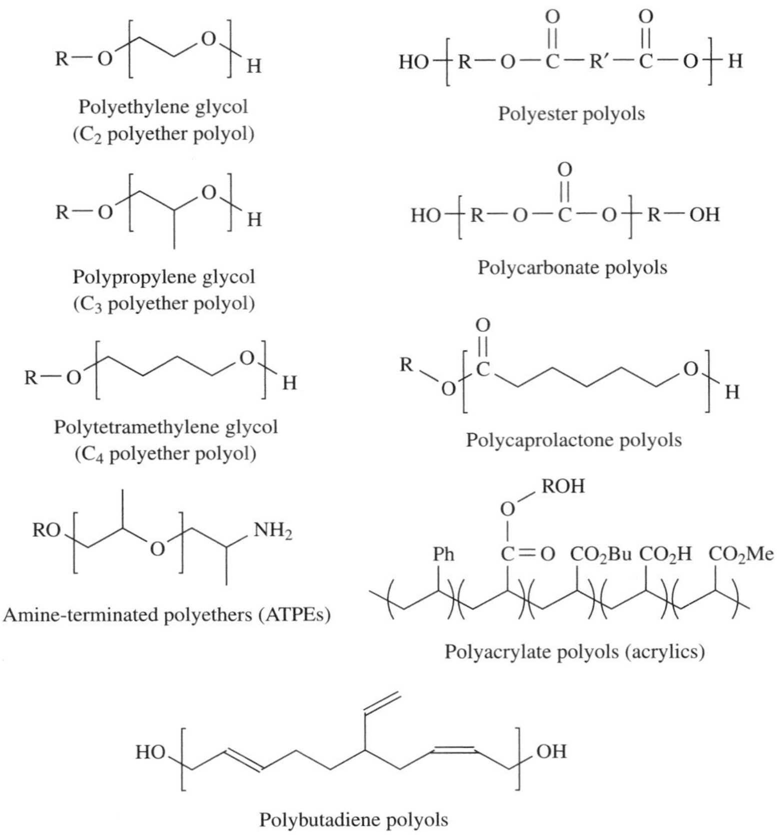

Polyethylene glycol
(C$_2$ polyether polyol)

Polyester polyols

Polypropylene glycol
(C$_3$ polyether polyol)

Polycarbonate polyols

Polytetramethylene glycol
(C$_4$ polyether polyol)

Polycaprolactone polyols

Amine-terminated polyethers (ATPEs)

Polyacrylate polyols (acrylics)

Polybutadiene polyols

Scheme 4.4   Common urethane-grade polyol types.

enhanced dynamic performance is required, the linear poly(tetramethylene oxide) polyols (C$_4$ polyethers) are often employed. Although the C$_4$ ethers tend to be more expensive, they contribute to enhanced phase separation and polymer morphology development (see Section 4.3.2) and so are used in demanding elastomer applications where higher prices are better tolerated. The C$_3$ and C$_4$ polyethers are known for their solvent and alkali resistance and hydrolytic stability. However, they are relatively sensitive to UV radiation and are not the best choice for outdoor ("weatherable") applications.

Polyesters are another important class of polyols. There are many polyester types used, so a generic structure is shown in Scheme 4.4. They are often based on adipic acid and either ethylene glycol (*ethylene adipates*) or 1,4-butanediol (*butylene adipates*). Polyesters, because of the polar carbonyl groups, contribute more to intermolecular forces, and physical properties such as tear and impact resistance are often improved by using them. They are also utilized for their solvent and acid resistance and light stability. Relatively poor hydrolytic stability is one of their disadvantages, and they are especially sensitive to alkali degradation. The polyesters tend to be more expensive than the C_3 and C_2 ethers, mainly because of higher raw material costs.

Acrylic polyols are commonly used in coatings formulations. They are made from acrylic and acrylate monomers in conjunction with styrene as a T_g modifier (and because it is less expensive). As shown in Scheme 4.4, these polyols have a carbon–carbon backbone with multiple side groups, some of them OH functional. These materials, therefore, are quite hydrophobic and typically have OH functionalities of 3–6. For coatings applications, these low-molecular-weight, high-functionality polyols provide high hardness materials with excellent weathering performance, light stability, chemical resistance, and durability. They are often used in demanding automotive coatings, including clear top coats.

Polybutadienes, polycaprolactones, polycarbonates, and amine-terminated polyethers (ATPEs) are shown in Scheme 4.4 as examples of other commercially available polyols. They are all specialty materials, used in situations where specific property profiles are required. For example, ATPEs are utilized in spray-applied elastomers where fast-reacting, high-molecular-weight polyamines give quick gel times and rapid viscosity buildup. Polycarbonates are used for implantation devices because polyurethanes based on them perform best in this very demanding environment. Polycaprolactones and polybutadienes may be chosen for applications which require exceptional light stability, hydrolysis resistance, and/or low-temperature flexibility.

Another family of polyols is the "filled" polyols.[11b] There are several types, but the *polymer polyols* are the most common. These are standard polyether polyols in which have been polymerized styrene, acrylonitrile, or a copolymer thereof. The resultant colloidal dispersions of micrometer-size particles are phase stable and usually contain 20–50% solids by weight. The primary application for these polyols is in flexible foams where the polymer filler serves to increase foam hardness and load-bearing capacity. Other filled polyol types that have been developed and used commercially (mainly to compete with the preeminent polymer polyols) include the polyurea-based PHD (*polyharnstoff dispersion*) polyols and the urethane-based PIPA (*polyisocyanate polyaddition*) polyols.

Chain extenders are low-molecular-weight diols or diamines used to increase urethane and/or urea content in the final polymer. They are also the sole curatives for low-NCO prepolymers (chain extension in the full prepolymer method; see above and Section 4.3.2). A few common chain extenders are shown in Scheme 4.5. Chain extender molecules can be relatively long or short, aromatic or aliphatic, hydrophilic or hydrophobic. Because they are low molecular weight and react with isocyanate, chain extenders become part of the so-called hard segment in the resultant polymer and can dramatically influence hardness, modulus, crystallinity, and so on (discussed below and in Section 4.2.3). They can also be utilized to modify processing characteristics, including gel time, viscosity buildup, and green strength (polymer integrity at demold, Section 4.4). For example, diamine chain extenders can be used as rheology modifiers because they react rapidly with isocyanate (before the polyol) to build molecular weight and viscosity. This is one way to make nonsagging adhesives and spray elastomers that can be applied to vertical surfaces without sagging, running, or dripping (see Example 10 in Section 4.4).

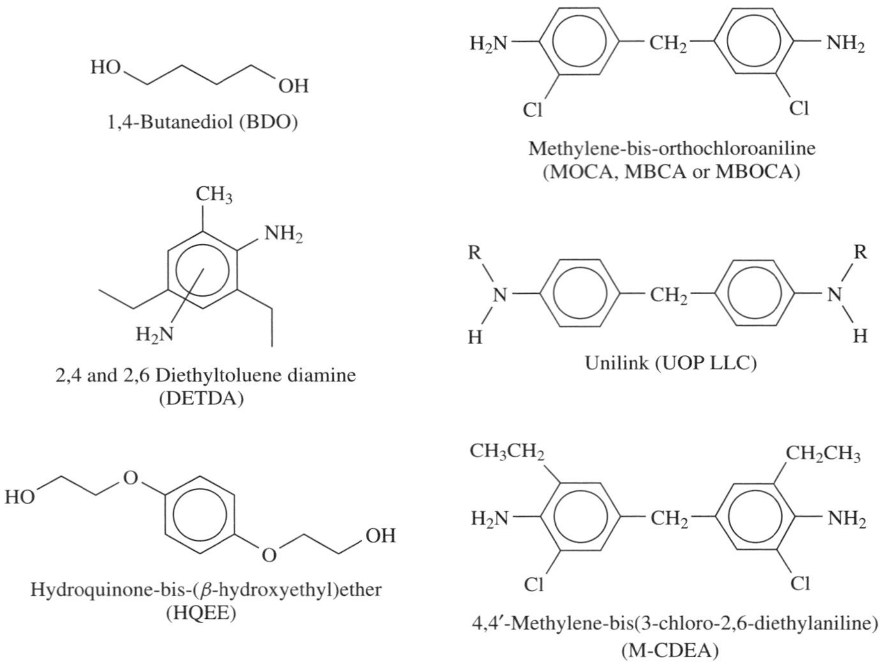

1,4-Butanediol (BDO)

Methylene-bis-orthochloroaniline
(MOCA, MBCA or MBOCA)

2,4 and 2,6 Diethyltoluene diamine
(DETDA)

Unilink (UOP LLC)

Hydroquinone-bis-(β-hydroxyethyl)ether
(HQEE)

4,4′-Methylene-bis(3-chloro-2,6-diethylaniline)
(M-CDEA)

Scheme 4.5 Examples of glycol and amine chain extenders. (Unilink is a registered trademark of UOP LLC.)

Crosslinking agents by definition have functionality $\geqslant 3$, and any such reactants in a formulation, including isocyanates and polyols, are crosslinkers. The term crosslinker is sometimes inappropriately used to denote what is really a two-functional chain extender. Examples of hydroxy-functional crosslinkers are shown in Scheme 4.6. The two most important features of a crosslinking molecule are functionality and molecular weight. Low-molecular-weight molecules are more effective at crosslinking the polymer matrix on a molar basis and are usually used at low concentrations. Higher molecular weight compounds, such as polyether triols, are often not thought of as crosslinkers per se, but their influences cannot be discounted because they do contribute to crosslink density and can increase resistance to swelling, improve tear strength, and decrease low-temperature flexibility. Trifunctional, low-molecular-weight compounds are most commonly used. Although 4-functional molecules are occasionally utilized as crosslinkers, higher functional materials (such as sucrose) are not because compounds with such high functionality will tend to become immobilized in the polymer network before all functional groups in the molecule can react.

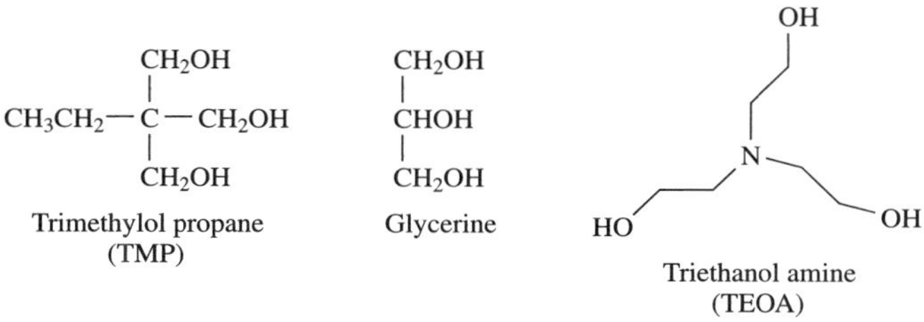
Trimethylol propane
(TMP)
Glycerine
Triethanol amine
(TEOA)

Scheme 4.6 Examples of crosslinkers.

The composition and structure of the raw materials presented in this section comprise only part of the molecular configuration of urethane polymers. How these compounds are covalently connected to form the skeletal backbone makes up the remaining portion of the polymer's primary structure. A schematic representation is shown in Scheme 4.7 of a linear urethane polymer chain made from high-molecular-weight polyol, isocyanate, and chain extender. The polyol forms the so-called *soft segment* of the backbone, while the isocyanate and chain extender form the *hard segment*. The high-molecular-weight, relatively nonpolar polyol soft segment and the short, polar isocyanate/chain extender hard segment of the polymer backbone are incompatible and tend to segregate into their own microdomains, called the *soft block* and *hard block*, respectively, as shown in Scheme 4.8. Thus, urethane elastomers are often referred to as *phase-separated* (or segregated) polymers. Another driving force for this segregation is intermolecular hydrogen bonding among the polymer hard segments. Scheme 4.8 depicts how this physical crosslinking might appear on a molecular level. The linkage sequence of the various components can be manipulated by the order of raw material addition during synthesis. This is especially true of fibers, thermoplastics, and cast elastomers where the synthetic method utilized can have important consequences on the properties of the polymer obtained. The two basic approaches for preparing such materials are known as the *one-shot* and prepolymer methods and are discussed below. More details regarding backbone microstructure and polymer morphology are presented in Section 4.2.3.

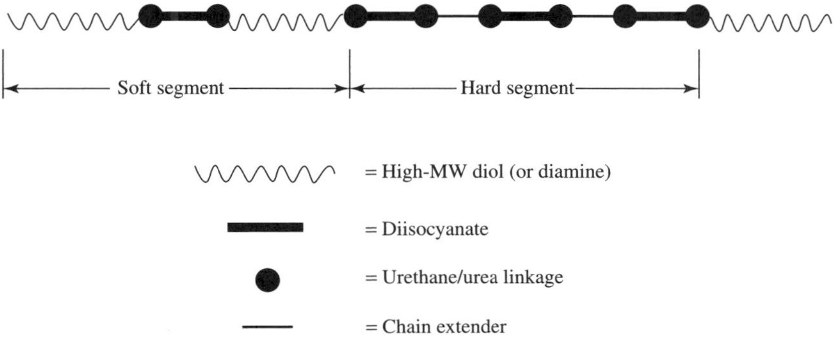
Soft segment
Hard segment
= High-MW diol (or diamine)
= Diisocyanate
= Urethane/urea linkage
= Chain extender

Scheme 4.7 Primary structure in polyurethanes: backbone microstructure.

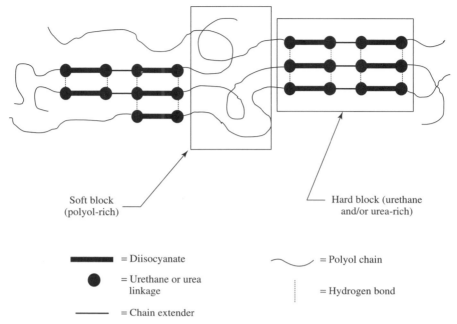

Soft block
(polyol-rich)

Hard block (urethane
and/or urea-rich)

▬▬▬▬ = Diisocyanate

● = Urethane or urea
linkage

──── = Chain extender

⁓⁓ = Polyol chain

⋮ = Hydrogen bond

Scheme 4.8 Secondary structure in polyurethanes: hard and soft block microdomains.

In the one-shot technique (Scheme 4.9a), isocyanate, polyols, chain extenders, and additives are all blended together at the same time and the material is poured into a mold (or processed in some other manner). The isocyanate, therefore, is free to react with any other compound in the system. The various molecules in a formulation, such as polyols and chain extenders, often have differing reactivities and some may react preferentially with the isocyanate. Since chain extenders are usually either glycols with 1° OH groups or amines, they are often the most reactive species in a formulation and will tend to react with the isocyanate before the high-molecular-weight polyols. This result is the formation of hard segments of random length: some long, some short. There is no control over hard-segment chain length and, as a result, the hard-block domain sizes are highly disperse as well.

The prepolymer method involves a different sequence of reactions. As a first step, a diisocyanate is reacted with a high-molecular-weight diol to form an NCO-terminated prepolymer of low NCO group concentration (Scheme 4.9b). The prepolymer is then reacted with a chain extender to yield the cured polymer. This second step is often referred to as chain extension because it links the high-molecular-weight prepolymer chains together. Since a prepolymer is synthesized in a separate reaction prior to forming the final polymer, this method is also called the *two-shot* technique. By creating the prepolymer, most of the diisocyanate molecules are bound to the end of a long polyol chain. Upon chain extension, the remaining NCO groups are reacted, creating uniform hard segments of one

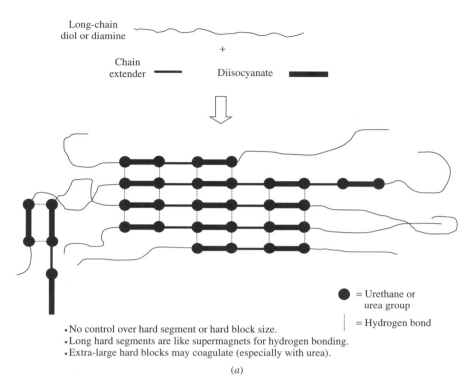

Long-chain
diol or diamine

+

Chain
extender   Diisocyanate

● = Urethane or
    urea group

┊ = Hydrogen bond

• No control over hard segment or hard block size.
• Long hard segments are like supermagnets for hydrogen bonding.
• Extra-large hard blocks may coagulate (especially with urea).

(*a*)

Scheme 4.9a   One-shot technique and molecular structure.

chain extender and two isocyanate units. This, in turn, tends to result in hard-block domains of relatively small, uniform size, as represented in Scheme 4.9b. Compared to the one-shot method, the hard segments are relatively short (almost exclusively composed of two diisocyanate and one chain extender molecule) and numerous. This molecular regularity results in better material physical properties. In fact, using the same reactants in the same molar ratios, radically different materials can be synthesized by the two techniques.

Prepolymer-based elastomers, fibers, and thermoplastics are known for excellent dynamic properties, cut resistance, and tear strength. One-shot elastomers, although not quite as good in terms of physical properties, do have the important advantage that they tend to be easier to work with because the high-molecular-weight, high-viscosity (or solid) prepolymers are avoided. These processing techniques are discussed in more detail in Section 4.3.2.1. How the hard and soft segments interact to influence physical properties is discussed in the next section.

4.2.3 Domain Morphology and Phase Separation

Considerations of the primary chemical structure (e.g., molecular composition, backbone microstructure, chain length, crosslinking) of urethane polymers are

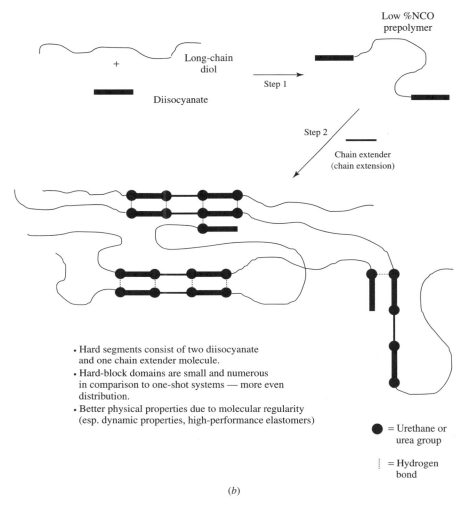

(b)

Scheme 4.9b   Prepolymer (two-shot) technique and molecular structure.

discussed in Section 4.2.2. In this section, hard-block/soft-block domain morphology (*secondary structure*) and phase segregation (*tertiary structure*) are presented with the objective of showing how they can be manipulated in designing materials. These topics are only important in segmented polymers such as fibers, thermoplastics, and cast elastomers. Monophase and phase-mixed materials are discussed later in this section.

The representation of hard-block domain structure shown in Scheme 4.8, implying rigid, crystallike molecular order, can be misleading because hard blocks are, at best, microcrystalline (as are soft blocks). Although microcrystallinity can be readily obtained, it requires careful selection of raw materials, processing parameters, and cure conditions. The hard-block domains are usually better represented as a conglomeration of noncrystalline, hydrogen-bonded regions of high urethane/urea concentration. They are more reminiscent of polar water droplets (the hard blocks) in nonpolar oil (the soft blocks). The illustration in Scheme 4.8 is meant only to portray the relative order and rigidity within the hard block compared to the relative disorder and flexibility within the soft block. This imparity in domain structure and physical characteristics can have important bearing on material properties and adds another level of sophistication to polyurethane material design.

The backbone hard and soft segments have different characteristics, and their composition and relative ratio are manipulated to achieve desired polymer properties. The hard segment of the molecule is relatively polar and rigid due to the urethane/urea groups and the short, inflexible combination of chain extender and diisocyanate. Raising the proportion of hard segments tends to increase the polymer's glass transition temperature, softening temperature, and solvent resistance.

If the material is prepared properly, the hard and soft segments of the polymer matrix will segregate into respective hard- and soft-block phases because of their immiscibility. In most materials the soft block forms the continuous phase because the polyol soft segments usually make up the majority of the formulation on a weight basis. Therefore, the relatively small hard blocks are dispersed throughout the soft-block phase and act as a kind of filler, or toughening agent. This is analogous to other biphase polymeric materials, such as acrylonitrile–butadiene–styrene (ABS) and high-impact polystyrene (HIPS). Hard-block concentration is increased by incorporating more hard segments into the polymer backbone and by proper synthetic procedure and cure/postcure conditions. The hard blocks are critically important for dynamic properties and serve to increase hardness, cut and abrasion resistance, tear strength, and impact resistance but will also decrease resiliency and elongation. This phase separation is the reason for the unbeatable physical properties of urethane elastomers and elastomeric fibers.

Hard-block structure and magnitude of phase segregation are influenced by molecular composition of the isocyanate and chain extender, extent of H bonding, soft-segment composition and compatibility with the hard segment, synthetic preparation method, and cure conditions. Starting with the isocyanate, it is apparent that 4,4'-MDI (Scheme 4.2) should be capable of forming good-quality hard blocks with a suitable chain extender because of its relative molecular linearity and symmetry. In fact, this compound is widely utilized as the preeminent isocyanate for high-performance, phase-separated elastomers. Other diisocyanates, such as TDI and IPDI, are not as effective at hard-block formation due to their nonlinear, bulky structures. Polymeric MDIs are also not a good choice because their contribution to crosslinking will tend to immobilize polymer chains and inhibit phase separation. The influence of chain extender structure follows trends similar to those of the isocyanate; hard-block formation and phase separation are facilitated by extenders that are linear and symmetrical. Good examples are 1,4-butanediol and hydroquinone-bis(β-hydroxyethyl)ether, both widely used in cast elastomers. These and other prevalent chain extenders are illustrated in Scheme 4.5.

Common wisdom states that H bonding is another important factor influencing extent and quality of hard-block formation. However, studies have shown that specially designed model polymers devoid of H-bonding sites are not only biphase but can have excellent physical properties as well.[2d,e] In fact, standard urethane and urea groups cause degradation reactions at relatively mild temperatures (ca. 150–200°C) and are the weak link that limits the high-temperature applicability of polyurethanes. The non-H-bonding polymers are stable up to 300°C. Such well-defined model systems are of great academic value and help us to better understand standard urethanes and ureas but have not found commercial application because their synthetic routes are too cumbersome. Nonetheless, the presence of hydrogen bonding and polar urethane and urea linkages are advantageous. Each urethane group has one N—H bond, but urea units contain two and therefore contribute twice the H-bonding power. Incorporation of urea groups into the polymer matrix can enhance many important properties such as hardness, tear strength, adhesion, and solvent resistance because the additional intermolecular forces hold the entire matrix together more effectively.

The soft-segment composition also strongly influences the degree of phase separation. Polyols with a functionality of 3 or greater contribute to crosslinking, which restricts molecular mobility and phase separation. Miscibility of the polyol with the isocyanate/chain extender hard segment is also important because polyol chains of relatively high polarity, such as polyesters, will tend to mix more with the polar hard segments. The carbonyl-containing polyester polyols will further boost phase mixing by H bonding with the hard segments. Therefore, for optimum hard-block/soft-block phase separation, relatively nonpolar, high-molecular-weight diols are used (MW ≈ 2000 g/m). High molecular weight is important because it will make the polyol even less compatible with the hard segments. Low-molecular-weight polyols, or even polyols with high polydispersity (such as polyesters), will tend to phase mix more. Polytetramethylene glycols (C_4 ethers) are especially effective because of their incompatibility with hard segments, even though they do have high polydispersity. Polypropylene polyols also work well and are very common in elastomer formulations.

For optimum two-phase morphology the method by which the material is prepared and the conditions under which it is cured are critical. The full prepolymer method using diisocyanate prepolymers and linear chain extenders is how phase-separated elastomers with superb dynamic properties are synthesized. High-temperature annealing (postcuring) provides the energy needed for the hard segments to realign and pack, breaking and remaking H bonds to increase molecular order. This process not only improves the "quality" of the hard blocks but also provides for maximum phase separation and physical property buildup.

Phase separation is usually only important in elastomers, thermoplastics, and fibers with excellent mechanical properties for applications like conveyor belts, tool housings, and spandex textiles. For many other material types, such as foams and coatings, phase separation is not an issue because these materials are usually prepared from polyols and polyisocyanates without the use of traditional chain extenders. By definition, polymer hard segments are formed from diisocyanates and chain extenders to form relatively rigid and polar linkages capable of intermolecular H bonding. If no chain extender is used, there are no hard segments and no phase separation into hard block domains. This is especially true when low-molecular-weight polyols of high functionality are used in conjunction with PMDI, such as for rigid foams. In these cases, chain extenders would serve no purpose because the material is highly crosslinked and there is no high-molecular-weight soft segment. Flexible foams are just the opposite: they are made from high-molecular-weight, branched polyols and either TDI or MDI, without any chain extender. In this case, the polymer matrix is composed mostly of soft segment with a little urethane holding it all together. These and other material types are nonsegmented, monophase polymers. Another example is the early Perlon U fibers made from HDI and 1,4-butanediol (discussed in Section 4.1.1), where the polymer backbone is composed entirely of hard segments, just isocyanate and chain extender.

Some coatings are designed to have high hard-segment content without phase separation. This is obviously important in clear coatings where extensive phase separation would cause the material to be opaque. High-functionality isocyanates and polyols, such as HDI trimer (Scheme 4.2) and acrylic resins (Scheme 4.4), are utilized to form tightly crosslinked networks for excellent physical properties such as scratch resistance, solvent resistance, and high hardness. A more recent development is the use of polyaspartic esters and aldimines as chain extenders (Scheme 4.10). These compounds are viscosity reducers, which lessen the need for volatile solvents, but they also react with isocyanates and become part of the polymer matrix, hence their description as *reactive diluents*. The hard segments they form are inhibited from hard-block formation and phase separation because of their bulk, nonlinearity, and relative compatibility with the soft block. In this way, optically clear coatings may be made with high urethane/urea content.

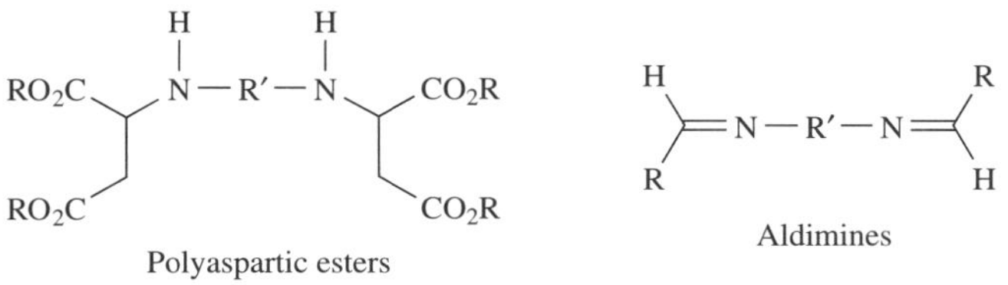

Polyaspartic esters

Aldimines

Scheme 4.10   Polyaspartic esters and aldimine reactive diluents.

Even with elastomers, it is often desirable to phase mix the material for certain combinations of physical properties. For example, elastomers with high hardness and flexural modulus are prepared from polyisocyanates, polyols, and polyamines of high functionality, creating a high crosslink density. Although polyol chains are present, if they are low molecular weight there is no real soft segment in the matrix. Such materials are a concentrated mixture of hard segments, hard blocks, low-molecular-weight polyol chains, and crosslinks.

4.3 SYNTHESIS AND MATERIAL CHARACTERIZATION

4.3.1 Chemistry and Catalysis

4.3.1.1 Preparation of Isocyanates

Isocyanates are manufactured on a large scale by direct, conventional phosgenation of primary amines.[1a,2a] In this process, generalized in Scheme 4.11, the amine (such as toluene diamine, TDA) is dissolved in an appropriate solvent and then treated with excess phosgene. This exothermic reaction leads to a slurry of partially-soluble carbamoyl chloride and is called *cold phosgenation* because no external heating is required. The HCl by-product coordinates with available amine to give an amine salt. In the second step, known as *hot phosgenation*, the reaction mixture is heated, decomposing the carbamoyl chloride and amine salt to generate isocyanate and free amine, respectively. A large excess of phosgene is necessary to minimize urea side-product formation. Solvent, phosgene, and HCl by-product are all carefully collected and recycled. There are several other synthetic routes to isocyanates, some of them phosgene free.[23]

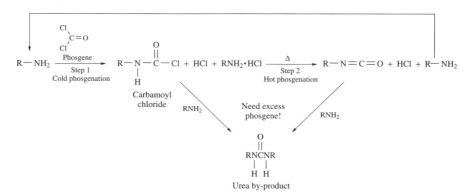

Scheme 4.11   Isocyanates via direct, conventional phosgenation of primary amines.

4.3.1.2 Synthesis of Polyols and Polyamines

Polyols and polyamines are the most important coreactants for isocyanates. As briefly outlined in Section 4.2.2, the two most common classes of urethane-grade polyols are the polyethers and polyesters. In this section their synthesis and structure are discussed. Other polyol types, such as acrylic resins and polycarbonates, are of more limited applicability and are not presented here.

Polyether polyols are synthesized by ring-opening polymerization of alkylene oxides, most commonly EO and PO. Due to their hygroscopicity, all-EO polyols (C_2 ethers) are rarely utilized. Propylene oxide polyols (C_3 ethers) are much more common, but because of the lower reactivity of their secondary (2°) hydroxyls, they are often tipped with ethylene oxide. The resultant primary (1°) OH endgroups are more reactive toward isocyanates, reducing the demand for catalysts and narrowing the reactivity gap between polyols and other reaction mixture ingredients, such as chain extenders and crosslinkers. The synthesis of an EO-capped, poly(propylene oxide) polyol is outlined in Scheme 4.12. The polymerization initiator is generated by partial KOH neutralization of the so-called *starter* (i.e., glycerine or propylene glycol), the OH functionality of which determines the functionality of the final product. Propylene oxide is oligomerized first, followed by ethylene oxide. It is difficult to cap all of the 2° alkoxylate groups because the 1° alkoxylates are so much more reactive toward the oxide. Reacidification followed by filtration and drying yields the finished product. Molecular weights in conventional polyols range up to about 6000 g/m. In this traditional synthetic route, a common side reaction is chain termination via proton abstraction. This limits molecular weight buildup, increases polydispersity, and generates allyl alkoxides which initiate polymerization to yield oligomeric "monols" with unsaturated endgroups. This is especially important in high-molecular-weight diols used in cast elastomers because the resultant monol species act as polymerization terminators which can prevent realization of ultimate physical properties. Furthermore, the double-bond functional groups can undergo unwanted reactions, including degradation by ozonolysis and photolysis. These facts have led to the development of new catalyst systems and processing technologies for the manufacture of "ultralow-monol" polyether polyols.[24]

Poly(tetramethylene oxide) polyols (see Scheme 4.4) are a special class of polyethers synthesized via acid-catalyzed ring-opening polymerization of tetrahydrofuran. Although less susceptible to side reactions, the synthesis of these C_4 ethers is less flexible in terms of product composition and structure. Thus, because of this synthetic route, only two-functional glycols are available and copolymers are not readily available. Molecular weights of commercial C_4 glycols range up to about 3000 g/m.

The ATPEs have a poly(propylene oxide) backbone with amine endgroups, as shown in Scheme 4.4. There are several routes to such materials,[25] but the only one in current commercial production is the direct amination of polyether polyols. A line of urethane-grade ATPEs is made by Huntsman under the tradename Jeffamine. Mono-, di-, and trifunctional products are available in molecular weights up to 5000 g/m.

Polyester polyols (Scheme 4.4) are prepared by condensation polymerization of dicarboxylic acids and diols. An excess of diol ensures OH functional product, minimizing the possibility of residual acid groups which react with isocyanates to generate CO_2 and act as inhibitors in catalyzed urethane reactions. The reactants are heated at 200–230°C under vacuum to remove the water by-product and drive the reaction to completion. The most common coreactants include adipic

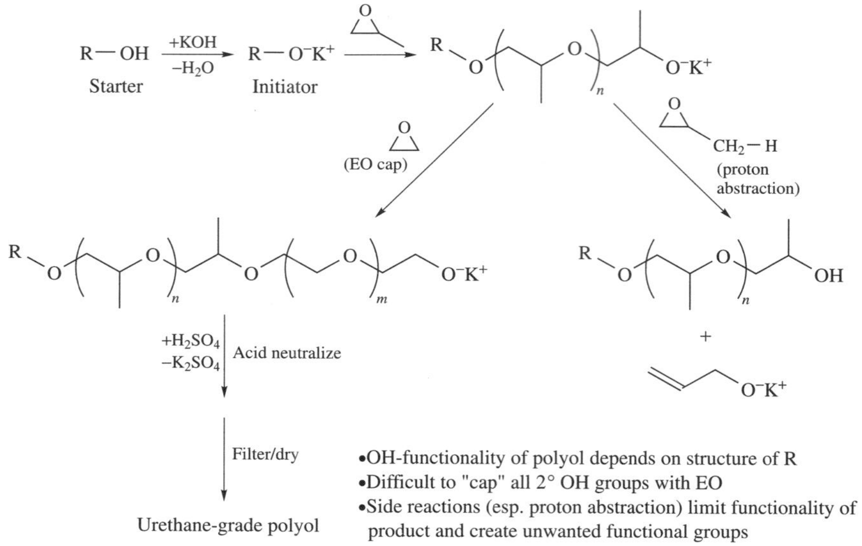

Scheme 4.12 Conventional synthesis of polyether polyols via base-catalyzed ring-opening polymerization of alkylene oxides.

acid, phthalic anhydride, ethylene glycol, butylene glycol, and hexane glycol. Functionality higher than 2 is achieved by adding a polyfunctional alcohol, such as trimethylol propane (TMP).

4.3.1.3   Reaction of Isocyanates with Alcohols and Amines

Polyurethane chemistry revolves around reactions of the isocyanate functional group. The influence of aromaticity and steric hindrance on isocyanate reactivity was briefly discussed in Section 4.2.2. The most important isocyanate reactions are those with hydroxyl- and amine-containing compounds to form urethanes and ureas, respectively (see Scheme 4.1). Primary, secondary, and tertiary alcohols exhibit decreasing reactivity, in that order, due to increasing steric hindrance. Tertiary alcohols can be several orders of magnitude slower than primary ones but are rarely employed. Not only are phenols slow reacting with isocyanates, but the urethane linkage obtained is thermally labile (see the discussion on blocked isocyanates in Section 4.3.2.3). Catalysts and/or elevated temperatures are usually required to drive the NCO–OH reactions to completion. The most common catalysts are Lewis acids and bases, especially Sn^{4+} and Sn^{2+} compounds, and tertiary amines. The isocyanate–hydroxyl reaction leads to solidification and development of polymeric properties, so compounds which strongly catalyze this are often referred to as *gellation catalysts* (discussed in more detail in Section 4.3.1.7).

Primary and secondary amines react readily with isocyanates without added catalyst due to their inherent nucleophilicity. As with hydroxyl compounds, aromatic amines are slower than aliphatics because of electron delocalization. Electron-withdrawing groups ortho or para to the aromatic amine moiety reduce nucleophilicity even further. Primary amines are generally more reactive than the more hindered secondary amines. These trends allow a high degree of control over the reactivity of amines toward isocyanates and can be utilized to design formulations with specific processing characteristics. An illustration of this is Example 10 in Section 4.4, which describes the use of an aromatic diamine as a reactive thickening agent.

4.3.1.4   Reaction of Isocyanates with Water and Carboxylic Acids

Isocyanates react with water to form the corresponding primary amine and carbon dioxide (Scheme 4.13). The amine formed will react with any available isocyanate to form a urea linkage. This reaction can serve many useful purposes. One example is to use water as a blowing agent in foam formulations. Such *water-blown foams* are becoming more common due to increasing pressure to reduce the use of traditional halogenated blowing agents (see Section 4.1.3). Water is often used as a coblowing agent in flexible foams because the urea groups it generates help to increase product hardness, resulting in higher load-bearing capacity. The isocyanate–water reaction is also taken advantage of in moisture-cured systems, including coatings, adhesives, sealants, binders, and encapsulants. In other ways this reaction can be a severe nuisance, such as in elastomers and coatings, where, for example, CO_2 bubbles can cause surface defects and degrade mechanical properties. The $NCO–H_2O$ reaction is catalyzed by many tertiary amines which are often referred to as *blowing catalysts*. Good examples are bis(dimethylaminoethyl)ether and N,N-dimorpholinodiethyl ether (DMDEE). See Section 4.3.1.7.

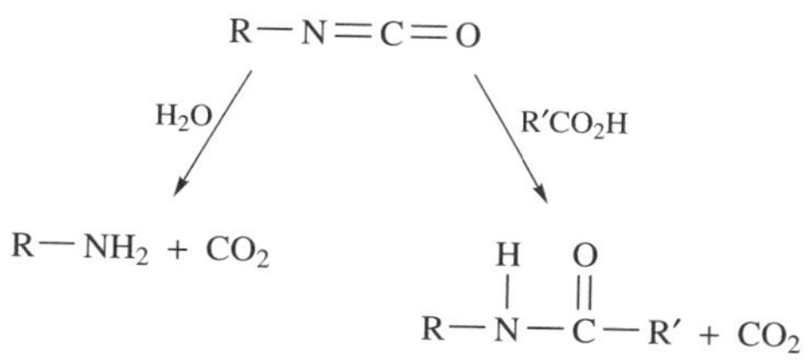

$$R-N=C=O$$

$$H_2O \diagup \qquad \diagdown R'CO_2H$$

$$R-NH_2 + CO_2 \qquad \qquad \underset{R-N-C-R' + CO_2}{\overset{H \quad O}{\underset{|}{} \overset{||}{}}}$$

Scheme 4.13   Reaction of isocyanates with water and carboxylic acids.

Isocyanates react with carboxylic acids to form amides, ureas, anhydrides, and carbon dioxide, depending on reaction conditions and the structure of the starting materials (Scheme 4.13). Aliphatic isocyanates more readily give amides. Aromatic isocyanates tend to react with carboxylic acids to first generate anhydrides and ureas, which at elevated temperatures (ca. 160°C) may further react to give amides. In practice, the isocyanate reaction with carboxylic acid is rarely utilized deliberately but can be an unwanted side reaction resulting from residual CO_2H functionality in polyester polyols.

4.3.1.5 NCO–NCO Reactions: Dimerization, Trimerization, and Carbodiimidization

Isocyanates also undergo dimerization and trimerization reactions which can be useful in many ways (Scheme 4.14). 4,4′-MDI will slowly dimerize to the uretdione upon standing in the solid or liquid state; to reduce the rate of dimer formation, it is stored either as a refrigerated solid ($<15°C$) or as a molten liquid just above its melting point of 39°C. In this case, formation of the insoluble dimer is an unwanted side reaction which creates turbidity and precipitation. Dimerization can be utilized, however, to modify physical properties, such as depressing the melting point or reducing the volatility of an isocyanate (see Section 4.2.2), or to make "blocked" isocyanates (Section 4.3.2.3). Dimerization is strongly catalyzed by trialkylphosphines.

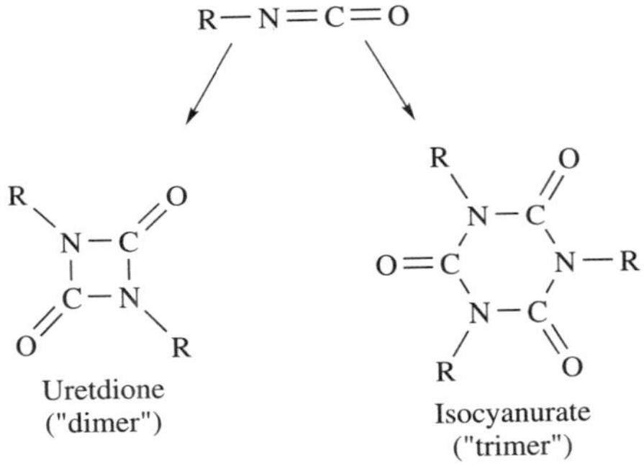

$$R-N=C=O$$

Uretdione ("dimer")	Isocyanurate ("trimer")

Scheme 4.14 Dimerization and trimerization of isocyanates.

Trimerization to isocyanurates (Scheme 4.14) is commonly used as a method for modifying the physical properties of both raw materials and polymeric products. For example, trimerization of aliphatic isocyanates is used to increase monomer functionality and reduce volatility (Section 4.2.2). This is especially important in raw materials for coatings applications where higher functionality is needed for crosslinking and decreased volatility is essential to reduce VOCs. Another application is rigid isocyanurate foams for insulation and structural support (Section 4.1.1) where trimerization is utilized to increase thermal stability and reduce combustibility and smoke formation. Effective trimer catalysts include potassium salts of carboxylic acids and quaternary ammonium salts for aliphatic isocyanates and Mannich bases for aromatic isocyanates.

Carbodiimide functionality can be produced by reacting isocyanates at elevated temperature with proper catalysis (Scheme 4.15). Although carbodiimides undergo a variety of reactions,[2a] most commonly as dehydrating agents, in the presence of excess isocyanate they will form uretone imines. This not only increases the average functionality of the isocyanate product but also lowers its freezing point. For example, a "liquefied" (or modified) version of 4,4′-MDI can be made by converting a small portion of the NCO groups to carbodiimide (which combines with excess NCO upon cooling to form uretone imine). As discussed previously in Section 4.2.2, such products are more convenient for processors because they freeze below room temperature and do not form dimer as readily as unmodified 4,4′-MDI. Phospholines and phospholine oxides are good catalysts for carbodiimidization.

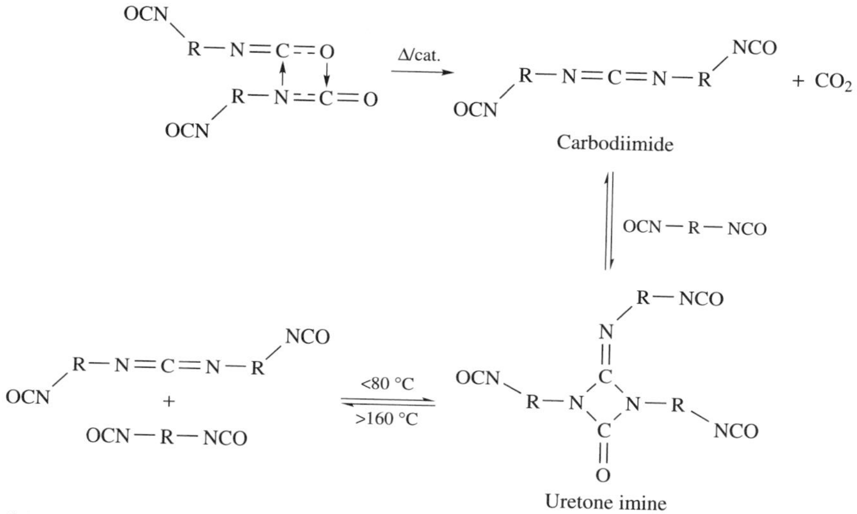

Scheme 4.15   Carbodiimidization of isocyanates and formation of uretone imines.

4.3.1.6   *Secondary Reactions: Allophanates and Biurets*

Urethane and urea linkages contain acidic protons capable of reacting with iso-cyanates at high temperatures. As shown in Scheme 4.16, urethanes react with isocyanates to form *allophanate* groups, while ureas give *biurets*. These reactions have become useful for modifying isocyanate raw materials. For example, allophanate modification is another route to liquefied 4,4′-MDI.[26] Similarly, HDI can be converted to biuret form (see biuret of HDI, Scheme 4.2), which increases functionality and reduces volatility.[27] Allophanate and biuret formation can also be an important consideration for thermal curing of elastomers. Polyurethanes and polyureas cured with an excess of isocyanate at high temperature ($>120°C$ for urethanes; $>100°C$ for the more nucleophilic ureas) for extended periods (12–18 h) can form allophanate and biuret crosslinks. Such reactions contribute to the buildup of physical properties, including hardness, modulus, and tear strength. Catalysts such as zinc acetylacetonate allow these secondary reactions to proceed at milder temperatures.

4.3.1.7   *Catalysts for Gellation and Blowing*

Appropriate catalysts for the reactions discussed in Sections 4.3.1.2–4.3.1.6 were briefly mentioned. Here, catalysts for the NCO–OH (gellation) and NCO–H_2O

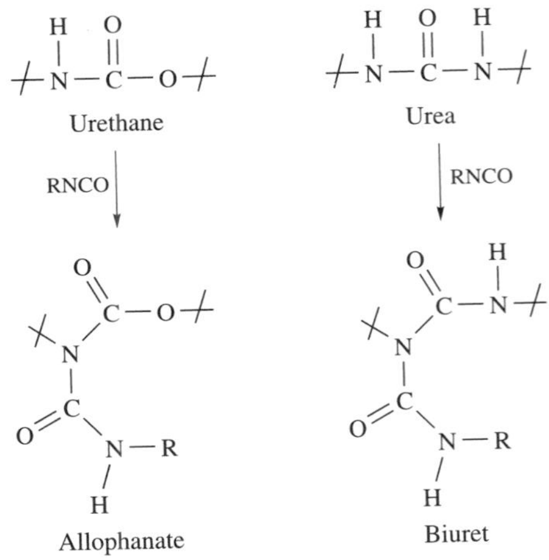

Scheme 4.16   Reactions of urethane and urea groups to form allophanates and biurets.

(blowing) reactions are discussed in more detail because they are the most important to the urethane chemist. Catalysts are utilized in the vast majority of urethane formulations, and their proper selection and use are critical to achieving desired physical properties and processing characteristics. Furthermore, because polyurethanes are amenable to so many material types and processing techniques, the demands placed on catalysts are much greater than with other polymers. Thus, the need to fine-tune reactivity profiles and physical properties has given rise to a large number of urethane catalyst products. Knowledge of the catalysts available, the reactions they promote, and the mechanisms by which they function is essential for successful system design. For these reasons and because there are few modern general reviews of polyurethane catalysis,[28a-c,f] a broad overview of the subject is presented here. The cited publications provide more detailed examinations.[2a,11a,b,28-37] Reference 11b contains an excellent — though dated — listing of common urethane catalysts, including tradenames, applications, and molecular structures.

The tertiary amines are the largest and most versatile class of urethane catalysts.[2a,28,29] They catalyze both gellation and blowing reactions and are used extensively in the industry. Two proposed mechanisms by which tertiary amines may catalyze these reactions are outlined in Scheme 4.17. In the first,[28a,f,j,29a,30c] the amine and alcohol interact via a hydrogen bond, weakening and lengthening the O—H bond, making the oxygen atom more nucleophilic, and enhancing the likelihood of attack across the C=N bond. The second mechanism[2a,28i,30c] involves activation of the isocyanate carbon atom by Lewis acid–base coordination with the amine, exposing it to nucleophilic attack. The O—H group then adds across the C=N bond to form the urethane linkage and regenerate the catalyst.

More important than the mechanism by which the tertiary amine catalysts function is how their molecular structure influences catalytic activity and selectivity

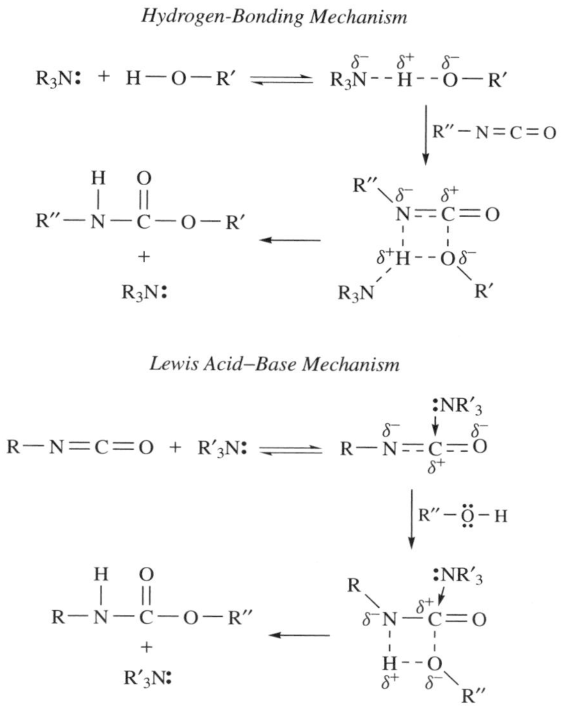

Hydrogen-Bonding Mechanism

$$R_3N: \; + \; H\!-\!O\!-\!R' \; \rightleftharpoons \; \overset{\delta^-}{R_3N}\text{-}\text{-}\overset{\delta^+}{H}\text{-}\text{-}\overset{\delta^-}{O}\!-\!R'$$

$$\downarrow R''\!-\!N\!=\!C\!=\!O$$

$$R''\!-\!\underset{\underset{H}{|}}{N}\!-\!\underset{\underset{||}{O}}{C}\!-\!O\!-\!R' \qquad \longleftarrow \qquad \underset{R_3N}{\overset{R''}{\diagdown}}\!\!\underset{\delta^+H\text{-}\text{-}O\delta^-}{\overset{\delta^-}{N}\!\!=\!\!\overset{\delta^+}{C}\!=\!O}$$

$$+ \qquad\qquad\qquad\qquad \diagup \qquad \diagdown R'$$

$$R_3N:$$

Lewis Acid–Base Mechanism

$$R\!-\!N\!=\!C\!=\!O \; + \; R'_3N: \; \rightleftharpoons \; R\!-\!\overset{\delta^-}{N}\!=\!\!\underset{\delta^+}{\overset{\overset{\displaystyle :NR'_3}{\downarrow}}{C}}\!\!=\!\!\overset{\delta^-}{O}$$

$$\downarrow R''\!-\!\overset{..}{\underset{..}{O}}\!-\!H$$

$$R\!-\!\underset{\underset{H}{|}}{N}\!-\!\underset{\underset{||}{O}}{C}\!-\!O\!-\!R'' \qquad \longleftarrow \qquad \underset{\overset{H\text{-}\text{-}O}{\delta^+ \;\; \delta^-}}{\overset{R}{\diagdown}\overset{\delta^-}{N}\!\!-\!\!\overset{:NR'_3}{\overset{\swarrow}{\underset{\delta^+}{C}}}\!\!=\!O}$$

$$+ \qquad\qquad\qquad\qquad\qquad\qquad\qquad \diagdown R''$$

$$R'_3N:$$

Scheme 4.17 Proposed mechanisms for amine catalyzation of the NCO/OH reaction.

of the gellation and blowing reactions. Their activity (or ability to promote a given reaction) depends on Lewis basicity and steric hindrance, with the latter being the overwhelming factor. For example, the protonated forms of N,N-dimethylcyclohexylamine (DMCHA) and N,N-diethylcyclohexylamine (DECHA, Scheme 4.18) have about the same pK_a values (10.1 and 10.0, respectively). However, the less sterically hindered methylated molecule is more effective in catalyzing the reaction of PhNCO and n-BuOH (6.0 for DMCHA versus 0.7 for DECHA, relative reaction rate). By contrast, protonated N-methyl piperidine and N-methyl morpholine have different acidities (pK_a of 10.08 vs. 7.41, respectively) but should have essentially the same steric accessibility about the nitrogen atom. The more effective catalyst for the PhNCO–2-ethyl hexanol reaction is N-methyl piperidine (6.0 vs. 1.0, relative rate for N-methyl piperidine and N-methyl morpholine, respectively) because its protonated form is a weaker acid (the neutral molecule is a stronger base).[29d]

The propensity of a given 3° amine to more strongly catalyze either gellation or blowing is referred to as *blow-to-gel selectivity*. This may be understood in terms of both the ability of the molecule to coordinate and hold water and reagent accessibility to the active nitrogen center. Five examples of amine catalysts are illustrated in Scheme 4.19 in order of selectivity, from high blowing selectivity

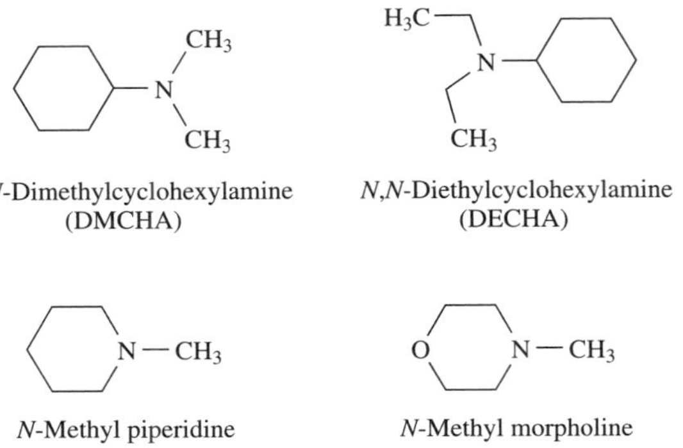

N,N-Dimethylcyclohexylamine   N,N-Diethylcyclohexylamine
(DMCHA)   (DECHA)

N-Methyl piperidine   N-Methyl morpholine

Scheme 4.18   Examples of tertiary amine catalysts.

to high gellation selectivity (top to bottom). At the top is the highly selective blowing catalyst N,N'-dimorpholinodiethyl ether (DMDEE). With two nitrogen and three oxygen atoms, this compound contains five potential hydrogen-bonding sites. Furthermore, multidentate chelation of H_2O is possible between the central etheric oxygen and the two morpholino nitrogen atoms, as illustrated in Scheme 4.19. Reaction between isocyanate and H_2O is more likely using this catalyst because of its coordination of water, holding it in close proximity to the catalytically active amine centers. The sterically hindered nitrogen atoms may also impede access of the relatively large isocyanate and alcohol reactants while the small H_2O molecule is comparatively unaffected, thus favoring the isocyanate–water reaction.

Bis(2-dimethylaminoethyl)ether (BDMAEE, Scheme 4.19) is another common catalyst with a strong preference for blowing,[29b] but its selectivity is not quite as high as that of DMDEE. Although BDMAEE effectively chelates water, it does not have as many H-bonding sites as DMDEE and so is not as selective. The NMe_2 nitrogen centers in BDMAEE are also less sterically hindered than the morpholino N atoms in DMDEE, allowing better access by isocyanate and alcohol.

Decreasing intermolecular forces between catalyst and water results in a higher preference for the NCO–OH gellation reaction. For example, N,N,N',N'',N''-pentamethyldiethylene triamine (PMDETA, Scheme 4.19) is structurally similar to BDMAEE but has a slightly lower preference for blowing. This is because the methyl group on the central nitrogen atom in PMDETA will tend to sterically hinder H bonding (especially when inversion is considered) whereas the 2-coordinate oxygen in BDMAEE is relatively unencumbered. Another example is N,N,N',N'',N''-pentamethyldipropylene triamine (PMDPTA, Scheme 4.19), which has a greater preference for gellation. In this molecule, the active amine centers are separated by propylene bridges as opposed to the ethylene units in PMDETA. The additional CH_2 group places the amine centers further apart, making chelation with water entropically less favorable.

Triethylene diamine (TEDA; also known as 1,4-diazabicyclo[2.2.2]octane, or DABCO) is a powerful catalyst with a high selectivity for gellation. One reason

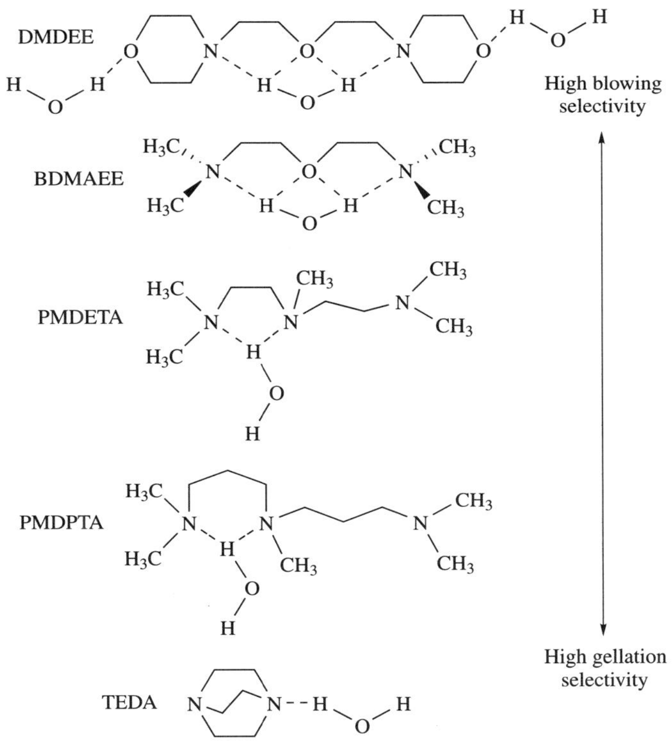

DMDEE

High blowing
selectivity

BDMAEE

PMDETA

PMDPTA

High gellation
selectivity

TEDA

Scheme 4.19 Water chelation to amine catalysts.

for this selectivity is the molecular structure does not allow multidentate chelation of water so there is relatively little enhancement of the proximity of water and isocyanate (Scheme 4.19). Another reason for the gellation preference is that the lone-pair electrons at nitrogen are relatively unencumbered and more basic because the three ethylene groups are held rigidly in place between the nitrogen atoms. This makes coordination of the larger isocyanate and alcohol reactants to the amine groups much easier, facilitating the NCO–OH reaction.

Organo–tin compounds are the next largest category of urethane catalysts.[28,30] They are utilized exclusively as gellation catalysts, although they can very weakly catalyze the NCO–H_2O reaction. Some examples of tin catalysts are shown in Scheme 4.20. There is no general consensus regarding exactly how these compounds promote the isocyanate–hydroxyl reaction, but they may function via an insertion mechanism, as illustrated in Scheme 4.21. This mechanism[30a] involves alcoholysis of the R_2SnX_2 compound, resulting in the formation of a tin alkoxide, $R_2Sn(X)OR'$, the actual catalytic species. The isocyanate then coordinates to the metal atom in $R_2Sn(X)OR'$ and inserts into the Sn–O bond (hence the term *insertion mechanism*). Displacement of the carbamate by incoming $R'OH$ forms the urethane group and regenerates the alkoxide. Another possible mechanism involves Lewis acid–base coordination of the isocyanate group to the tin atom (Scheme 4.21).[28g,i,j] This polarizes the NCO group, placing a higher average positive charge on the carbon atom and enhancing nucleophilic attack by alcohol.

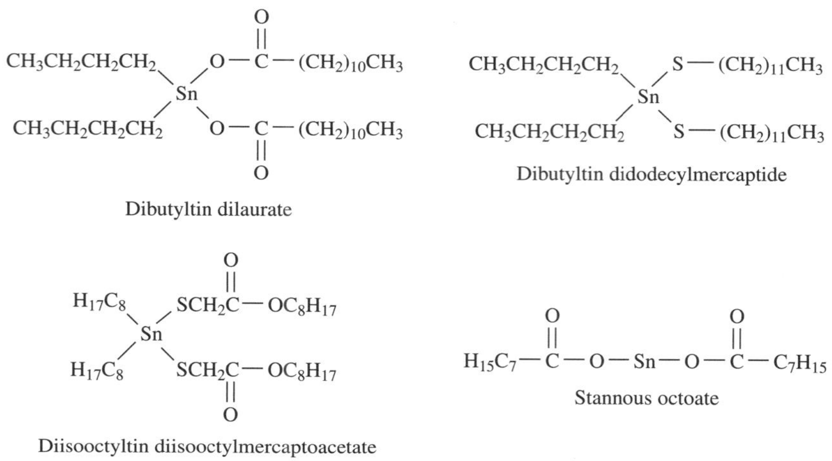

Scheme 4.20   Examples of organo-tin catalysts.

The catalytic effectiveness of the Sn^{4+} compounds depends on a balance of Lewis acidity and steric hindrance. Dibutyltin dilaurate (DBTDL, Scheme 4.20) is a very strong gellation catalyst and is widely used in the industry. One theory for its effectiveness is that the electron-withdrawing carboxylate groups enhance the Lewis acidity of the metal center while the relatively small butyl groups do not unduly hinder NCO access to the tin $5d$ orbitals. Furthermore, it may be that the structure of the molecule allows isocyanate access but limits coordination to the carbonyl groups of the formed urethane linkages, which would tend to remove it from the catalytic process and limit its effectiveness.[28g] By contrast, dioctyltin dilaurate is somewhat "slower" than DBTDL due to the larger, more sterically demanding octyl groups.

The tin mercaptides and tin mercaptoacetates are considered to be *delayed-action* catalysts, at least in comparison to the tin carboxylates. An example of each type is shown in Scheme 4.20: dibutyltin didodecylmercaptide and diisooctyltin diisooctylmercaptoacetate. Possible explanations for their delayed action are that they have lower Lewis acidity and that it is more difficult for the isocyanate group to insert into the stronger Sn–S bond. Whatever the mechanism, these catalysts provide delayed reactivity for lengthened *gel time* (also called *pot life* or *working time*). They are also more hydrolytically stable compared to the tin carboxylates and are utilized in systems where exposure to water is an issue, such as in waterborne coatings.

The Sn^{4+} compounds discussed above are the largest class of tin catalysts in terms of the number of commercial products available. Another category is the

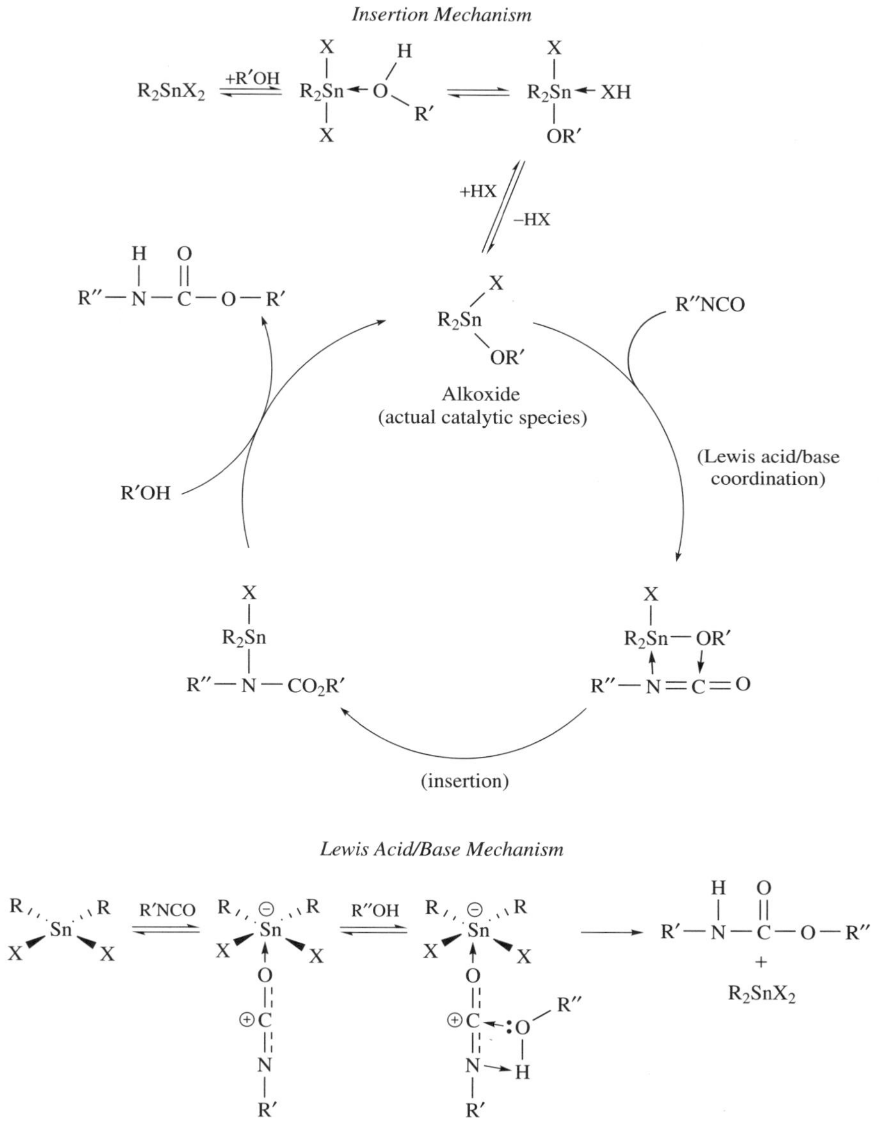

Insertion Mechanism

$$R_2SnX_2 \xrightleftharpoons{+R'OH} R_2Sn \overset{X}{\underset{X}{\leftarrow}} O \overset{H}{\underset{R'}{\diagup}} \rightleftharpoons R_2Sn \overset{X}{\leftarrow} XH \quad (OR')$$

$+HX$ $-HX$

$$R'' - \overset{H}{\underset{|}{N}} - \overset{O}{\underset{||}{C}} - O - R'$$

$$R_2Sn \overset{X}{\underset{OR'}{\diagdown}}$$

Alkoxide
(actual catalytic species)

R''NCO

(Lewis acid/base coordination)

R'OH

$$\overset{X}{\underset{|}{R_2Sn}} \quad R'' - N - CO_2R'$$

$$R_2Sn - OR' \quad R'' - N = C = O$$

(insertion)

Lewis Acid/Base Mechanism

$$\overset{R,,\quad R}{\underset{X\diagup Sn \diagdown X}{}} \xrightleftharpoons{R'NCO} \overset{R,,\ominus\ R}{\underset{X \diagup Sn \diagdown X}{}} \xrightleftharpoons{R''OH} \overset{R,,\ominus\ R}{\underset{X \diagup Sn \diagdown X}{}} \longrightarrow R' - \overset{H}{\underset{|}{N}} - \overset{O}{\underset{||}{C}} - O - R''$$

$$\underset{\oplus C}{O} \quad \underset{\oplus C - :O}{O} \qquad + \quad R_2SnX_2$$

$$N \qquad N \rightarrow H$$

$$R' \qquad R'$$

Scheme 4.21 Proposed mechanisms for organo-tin catalyzation of the NCO/OH reaction.

Sn^{2+} compounds, such as stannous octoate (Scheme 4.20), a gellation catalyst widely used in the slabstock foam industry. Its sensitivity to air and moisture is considered an advantage because the compound rapidly decomposes after foam formation, preventing thermocatalytic degradation such as during the curing stage where slabstock foams can generate high exotherm temperatures. Flexible slabstock foam formulations contain water in the polyol as a blowing agent, so stannous octoate must be added in a third stream at the point of mixing isocyanate and polyol. It is a cost-effective catalyst because of the molecule's high tin content.

There is a well-known synergism between tin compounds and tertiary amines whereby mixtures of the two can be more effective than either one alone.[28a,d,f,g,k] This is often taken advantage of to reduce catalyst usage, especially in slabstock foams where, for example, stannous octoate is used for gellation in combination with bis(2-dimethylaminoethyl)ether for blowing. A mechanism which may explain this synergism is shown in Scheme 4.22. In this scenario,[28a,g] the isocyanate coordinates to the tin catalyst while the amine hydrogen bonds with the alcohol, "activating" both isocyanate and alcohol and enhancing the likelihood of their reaction. This lends credence to the Lewis acid–base mechanism shown in Scheme 4.21 and the H-bonding mechanism in Scheme 4.17. A tin–amine coordination complex ($R_3N \rightarrow SnR_2X_2$) has also been proposed as an activated catalytic species.[31]

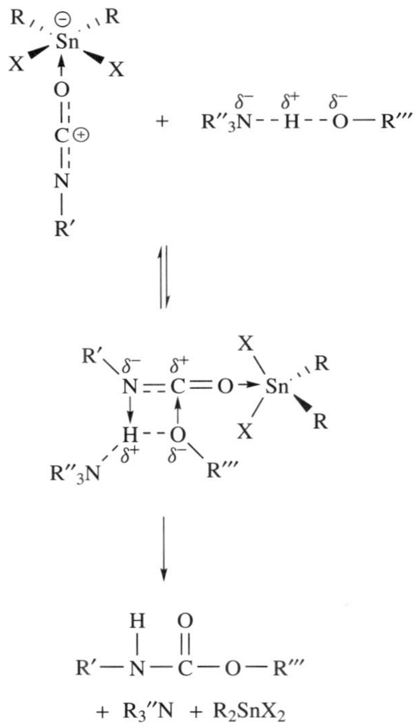

$$
\begin{array}{c}
R_{\prime\prime}\ \underset{\ominus}{Sn}\ \diagup\!\!\!\backslash R \\
X\ \ \uparrow\ \ X \\
O \\
\|\\
C\oplus \\
\|\\
N \\
|\\
R'
\end{array}
\qquad + \qquad
\overset{\delta^-}{R''_3N} - - \overset{\delta^+}{H} - - \overset{\delta^-}{O} - R'''
$$

$$
\begin{array}{c}
R'\underset{\delta^-}{\diagdown}\ \ \overset{X}{\underset{\delta^+}{N}}\!\!=\!\!C\!\!=\!\!O \rightarrow Sn\diagdown R \\
\downarrow\ \ \uparrow \\
\underset{R''_3N}{\overset{H--O}{\ \ \delta^+\ \delta^-}}\ \ \ \overset{X}{\diagup}\ R''' \\
\end{array}
$$

$$
\begin{array}{c}
\overset{H}{\underset{|}{\ \ }}\ \ \overset{O}{\underset{\|}{\ \ }} \\
R'-N-C-O-R''' \\
+\ R_3''N\ +\ R_2SnX_2
\end{array}
$$

Scheme 4.22   Proposed mechanism for the tin–amine catalytic synergism.

It is often desirable to retard the onset of gellation and blowing reactions in order to extend the working time available to process a given formulation, so as to fill a large mold or apply a coating. Then, at some appropriate point in the process, the catalyst (or catalyst mixture) becomes "active" and begins to strongly promote blowing, gellation, or both. This often results in a compromise between putting off chemical reactions in the early stages of a process (*front-end* reactions) yet vigorously enhancing curing reactions later on to provide rapid physical property build-up (*back-end* cure). Maximizing working time while minimizing time to final cure often requires the use of special, delayed-action catalysts.[28b,32] In the past, organomercury compounds (i.e., phenyl mercuric acetate) served as the industry benchmark for delayed-action gellation catalysts in the CASE (coatings, adhesives, sealants, and elastomers) area. These compounds provided long induction times followed by sharp back-end cure profiles. Their high toxicity, however, has caused them to be phased out, although they are still in limited use. Direct substitutes for mercury catalysts have been elusive. As mentioned above, the tin mercaptides and mercaptoacetates are considered delayed action, but only in comparison to the front-end tin carboxylates. Other delayed-action organometallic compounds available commercially include nickel acetylacetonate[32a] and bismuth and zinc carboxylates.[32c] There are many other reports of catalysts and methods for delaying system reactivity. Examples include dithiastannetanes,[33] encapsulated amines,[34] and complexation of organometallic compounds with amines,[35] pentanedione,[36] and mercaptans.[32a]

Perhaps the largest category of delayed-action catalysts is the *blocked amines*, formed by complexation of tertiary amines with acidic compounds such as formic acid, 2-ethylhexanoic acid, or phenol.[28b,32a,b] These ammonium salts are heat activated and dissociate to the acid and free amine upon exposure to elevated temperature. They were originally designed for the foam industry, where pouring and molding operations required longer processing times. However, blocked amines have found broad use in many nonfoam applications, including elastomers and adhesives. There are no apparent trends regarding activation temperature (or delaying effect) and the structure of the amine or the acid used.[32b]

The discussion above is largely geared toward catalysis of aromatic isocyanate reactions. There can be important differences in the way catalysts act on the various aromatic isocyanate types, especially between MDI and TDI. Some reports have described these differences,[28e,29b] but finding the right catalyst and concentration is at least partially trial and error. It is also important to note that aliphatic isocyanates have their own special catalytic requirements.[35,37] First, amine catalysts are generally not effective with aliphatic isocyanates, although they are occasionally used in special situations. Strong organometallic catalysts such as dibutyltin dilaurate, dimethyltin diacetate, and zinc octoate are commonly used in such systems because of the reduced reactivity of aliphatic isocyanates compared to the aromatics (see Section 4.2.2). Also, coating formulations are often aqueous, and this restricts the number and type of employable catalysts. For such waterborne systems, tin mercaptides are often used because of their good hydrolysis resistance.

Catalysts are almost always required in polyurethane formulations because the NCO–OH reaction is generally slow at ambient temperature. There are situations, however, where catalysts may not be necessary. For example, in all-urea systems additional catalyst is often not needed for the highly reactive 1° and 2° amine coreactants used and because curing is facilitated by the high exotherms these formulations tend to generate. This can be true in polyurea cast elastomers,[38a] spray elastomers,[38b,c] and even reaction injection-molding systems.[38d] Another situation where catalysts may not be necessary is in auto-catalytic formulations where the catalyst is built into the raw materials. The most common example is amine-started polyols (Section 4.3.1.2). Starters like triethanolamine and ethylenediamine create tertiary amine groups as an integral part of the polyol molecule, and these act as any other amine catalyst, promoting both gellation and blowing. This can reduce or eliminate the need for additional catalysts but can also make systems using amine-started polyols more sensitive to atmospheric moisture.

4.3.2 Synthetic Methods

Synthetic techniques and application methods are presented here in a general fashion, covering broad topics relevant to both the laboratory and manufacturing plant. Specific processes, such as reaction injection molding, rotational casting, and rubber compounding, are beyond the scope and purpose of this book and are amply reviewed elsewhere.[1a] In this section, considerations pertaining to the field of polyurethanes as a whole are discussed because they are important in selecting materials and designing systems. Specific examples can be found in Section 4.4.

4.3.2.1 One- and Two-Shot Techniques

The various ways of synthesizing and applying urethanes can generally be delin-eated into two categories: one- and two-shot methods. These were discussed at length in Sections 4.2.2 and 4.2.3 in terms of their effect on polymer microstruc-ture and properties of elastomers and fibers (see Schemes 4.9a and b). The most apparent difference between them is the sequence of raw material addition. The two-shot method involves prereaction of raw diisocyanate (such as MDI) and high-molecular-weight diol (MW $\approx 2000-4000$) to give an A-stage intermediate product. This product is called a prepolymer, and this technique is also known as the prepolymer method. The prepolymer contains relatively few unreacted NCO groups, usually from about 3% to about 10% by weight. Products of such low %NCO content are also called full prepolymers to distinguish them from quasi-prepolymers (ca. 10–20% NCO), modified isocyanates (ca. 20–28% NCO), and unmodified isocyanates (usually >28% NCO). Compared to other isocyanate products, prepolymers are usually higher in molecular weight and viscosity. Some prepolymers, especially those based on polyester polyols, are solid at room tem-perature. Because of this, prepolymers are typically heated during processing, usually at about 80°C. Heating is necessary not only to melt the prepolymer and/or reduce its viscosity but also to achieve desired physical properties in the final product. This is especially true of cast elastomers where tight control of reactant, mold, and cure temperatures can be vitally important. Another differ-ence between the one- and two-shot techniques is that full prepolymers are cured by chain extending with a low-molecular-weight diol or diamine (Section 4.2.2 and Scheme 4.5). Reaction ratios are usually about 10 : 1 by volume, prepolymer to chain extender, because there are only a few free NCO groups left to react in the prepolymer. Therefore, even small deviations in reactant ratio can cause large variations in the molar ratio of NCO–OH (reaction "index"), which can strongly influence material properties. Such high volume ratios require very accurate processing equipment.

In the one-shot technique isocyanate, polyol, and all other ingredients are mixed together in one step to form the final material or part. There is no intermediate prepolymer, although modified isocyanates are often used. Compared to prepolymer systems, one-shots are usually lower in viscosity, processed at lower temperatures, and reacted at lower volume ratios (often close to 1 : 1) and do not have such stringent curing and postcuring requirements. Also, because isocyanates of relatively high NCO content are used, the raw materials do not require the same exacting specifications as full prepolymers. These things make one-shot processes much more "user friendly" and are why the vast majority of urethane materials are made using this technique.

There is another method, called the quasi-prepolymer technique, which is similar to the full-prepolymer process but utilizes prereacted isocyanates in the 10–20% NCO range. This eases processing compared to full-prepolymer systems, requiring lower temperatures and volume ratios typically from 4 : 1 to nearly 1 : 1 (polyol to isocyanate).

The full-prepolymer, quasi-prepolymer, and one-shot techniques may appear to be quite similar, but they each have important bearing on equipment requirements and physical properties. The full-prepolymer method, for example, is the process of choice for making high-performance cast elastomers with superb dynamic properties, such as industrial bumpers and bushings. Quasi-prepolymers are used in less demanding applications like shoe soles, adhesives, and spray elastomers. One-shots make up the rest, including most foams and elastomers.

4.3.2.2 Solvent-borne, Water-borne, and 100% Solids Formulations

Almost all urethane materials are synthesized without the use of solvents or water as diluents or carriers and are referred to as being 100% solids. This is true of all foams and elastomers. There are many products, however, which do utilize solvents or water, and these are known as *solvent-borne* and *water-borne* systems, respectively. In the past, many coatings, adhesives, and binders were formulated using a solvent to reduce viscosity and/or ease application. However, the use of volatile solvents has been dramatically curtailed in favor of more environmentally friendly water (see Section 4.1.3), and now there are many aqueous coatings, adhesives, and associated raw materials. Hydrophilic raw materials capable of being dispersed in water are called *water reducible* (or *water dispersible*), meaning they are sufficiently hydrophilic so as to be readily emulsified in water to form stable colloidal dispersions.

Two important methods by which isocyanates, polyols, and polyurethanes are rendered water dispersible both involve incorporation of *internal emulsifiers*, which serve to suspend the molecule in the aqueous phase.[12c,f] *Ionic* internal emulsifiers include sodium sulfonate or carboxylic acid groups which become hydrophilic when neutralized with $N(CH_2CH_3)_3$. This is illustrated in Scheme 4.23 with dimethylol propionic acid (DMPA) as the emulsifying group in an NCO-terminated prepolymer. The CO_2H group in DMPA is sufficiently hindered that it does not react with isocyanate under typical reaction conditions. The carboxylic acid units are ionized when exposed to triethylamine, making the entire prepolymer dispersible in water. After reacting with a water-reducible diamine and application (of the coating, adhesive, or binder), the water and triethyl amine volatilize, allowing the polymer to cure and rendering it hydrophobic. This is an example of the prepolymer method used in an aqueous coating formulation.

A related technique involves incorporation of monofunctional poly(ethylene oxide) chains as nonionic, internal emulsifier groups. Even PMDI can be dispersed in water using this nonionic method (Scheme 4.24). High-molecular-weight (ca. 2000 g/m) monols are usually used which act as chain terminators and long, hydrophilic tails which function as an emulsifying agent.

4.3.2.3   One- and Two-Component Systems

Most polyurethane formulations are two-component systems, meaning they are made from two basic constituents which are mixed together and react to form the final polymer. The two components are an isocyanate (or isocyanate blend) and a polyol (or polyol blend). The isocyanate is often referred to as the "A side" and the polyol the "B side", although some use the opposite convention.

In special cases, urethanes can be formulated as one-component systems, where there is only one ready-to-use constituent which is cured by heating, exposure to atmospheric moisture, or some other method. These are also known as *one-pack* or *1-K* (after the German *komponente*) products. Materials amenable to 1-K technology are coatings, adhesives, sealants, and binders because these are usually applied in thin cross section, allowing a high surface area so that water, solvent, or volatile by-products can escape the polymer matrix upon curing. There are several types of one-component formulations, including moisture-curable isocyanates, blocked isocyanates, aqueous polyurethane dispersions, powder coatings, and urethane alkyds, oils, and lacquers.

Moisture-cured isocyanates are commonly used in coatings, adhesives, sealants, and binders. These rely on atmospheric moisture to react with free NCO groups, forming primary amines and, subsequently, urea linkages. The $NCO-H_2O$ reaction creates CO_2, so the application must allow for gas release. For thin coatings, adhesives, and binders, the CO_2 can escape without creating bubbles. Sealants, however, are usually applied in thicker cross section so gas liberation is more difficult. In the low-% NCO prepolymers (usually about 3% NCO) used in sealant formulations, the carbon dioxide may be absorbed by the polymer matrix and slowly volatilized. One trick of the trade is to add powdered calcium oxide (lime) to absorb the CO_2 and prevent the formation of bubbles. Moisture-cured coatings are based on either aliphatic or aromatic isocyanates. An example of a highly crosslinked

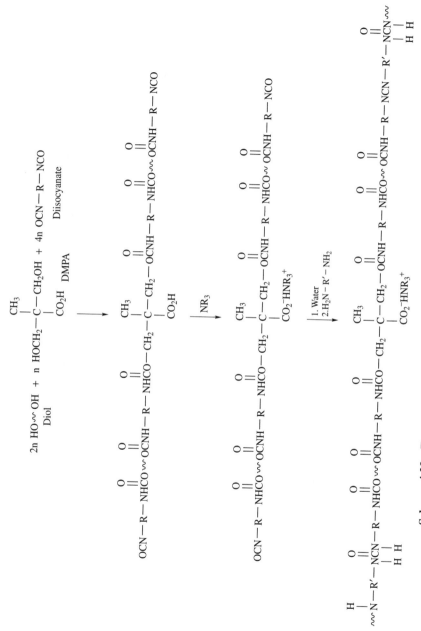

Scheme 4.23 Two-component water-borne polyurethane by the prepolymer method.

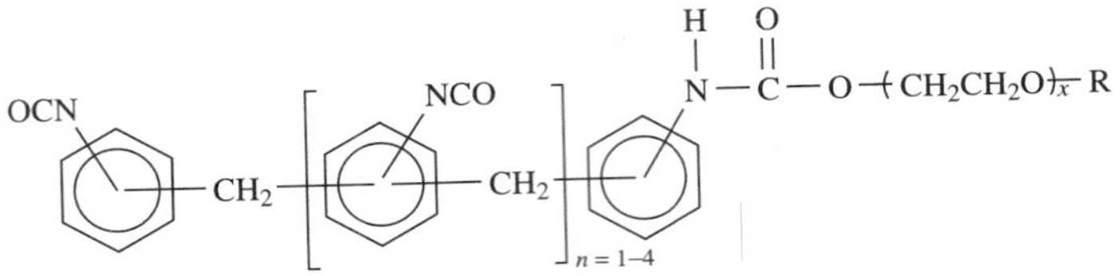

Scheme 4.24   Polymeric MDI with a poly(ethylene oxide) internal emulsifier.

clear coating ("clearcoat") based on a modified HDI is given in Example 7 in Section 4.4. Moisture-cured adhesives, sealants, and binders are most often made with aromatic isocyanates because the light stability of aliphatics is usually not required. Moisture-cured adhesives and sealants are typically designed with MDI-based polyether prepolymers of low %NCO (about 2–4% NCO). See Example 11 for a one-component, moisture-cured sealant. Binders are most often moisture cured. They are usually made from aromatic prepolymers, modified isocyanates, or even raw isocyanates. Example 12 is of a binder for wood using unmodified PMDI.

Blocked (or *capped*) isocyanates are stable and nonreactive at room temperature but are reactivated when heated to regenerate or release the free isocyanate.[39] They can be blended with polyol and other ingredients to make a single-component, ready-to-use system. One type of blocked isocyanate is prepared by reacting an isocyanate with a suitable monofunctional compound to form a thermally labile linkage. Common blocking agents include diethyl malonate, phenol, caprolactam, and methylethyl ketoxime. Deblocking (or splitting) temperatures range from about 120 to 200°C, depending on the isocyanate and blocking agent used. Blocked isocyanates can also be made by dimerization (Section 4.3.1.5) to uretdione linkages capable of dissociating at elevated temperature. The dimer of TDI (Desmodur TT, Bayer AG) is a solid at room temperature and the free NCO groups do not begin to react until the compound melts at 145°C. At about 160°C, the uretdione linkage "splits," regenerating free NCO and allowing the material to fully cure. Desmodur TT is most often used as a latent vulcanizing agent in compoundable polyurethane rubber formulations. Dimerization of aliphatic isocyanates is more complex, but there is much interest in such materials in the coatings area.[39c,d]

Aqueous PUDs are water-dispersed (see Section 4.3.2.2), OH-terminated prepolymers. They are "fully reacted," meaning there are no free NCO groups left in the prepolymer. Being stable dispersions in water, they may be used as coatings or adhesives which set upon drying and coagulation of the colloidal droplets (one-shot systems). PUDs may also be used in 2-K formulations and cured with a water-reducible isocyanate (two-shot process). Powder coatings are fine powders containing everything needed to make a urethane coating: blocked isocyanate, resinous polyol, pigments, and other additives. The powder is electrostatically sprayed onto a grounded substrate and then heated to melt the powder, deblock the isocyanate, and cure the coating. Urethane lacquers are similar to PUDs in that they are fully reacted, OH-functional prepolymers but are dissolved in a nonaqueous solvent. Stoving lacquers are blocked versions requiring relatively high baking temperatures to split the blocked isocyanate and cure the coating.

Urethane alkyds and urethane oils are oil and alkyd resin-modified polyurethanes dissolved in a volatile solvent. Upon application and solvent evaporation, the coating is crosslinked and cured via oxidation by atmospheric oxygen.

4.3.3 Analytical Techniques and Physical Testing

A detailed overview of the material characterization of polyurethanes and associated raw materials is not possible within the scope of this chapter. Comprehensive treatments of physical property and mechanical testing[1a,40] and chemical analyses[41] used in the industry are available elsewhere, and ref. 42 lists sources on analysis and testing of polymeric materials in general. The first four categories in the following discussion are compositional analysis, thermal characterization, mechanical properties, and application testing. The focus of these sections is on cured polymers, but a short discussion of the equally important subject of raw material analysis is given at the end. Each of these subjects is worthy of an entire volume, but they are briefly presented here in general terms to give some appreciation for the scope of their application to urethanes.

4.3.3.1 *Compositional Analysis*

The chemistry of polyurethanes is very well understood, so conformation of molecular composition and structure is usually not of importance. Nonetheless, it is occasionally desirable to determine the composition (formulation) of a material or to check a product's molecular weight, free isocyanate content, and so on. The first requirement is to determine if the polymer is a thermoplastic or thermoset (if not already known). A thermoplastic, or any linear uncrosslinked polymer, will usually be soluble in an appropriate solvent. If the polymer is dissoluble, it can be analyzed directly and fully characterized by standard methods. Insoluble thermosets, however, must be degraded and the solubilized by-products characterized individually in order to perform a complete formulation analysis.[43] It is often possible to quantitatively and/or qualitatively identify certain components without resorting to such degradation techniques. One example is analysis for free NCO by infrared spectroscopy.[41] However, full characterization of a thermoset polyurethane requires that the crosslinked polymer be carefully "disassembled" and the formulation deduced from the structure of each isolated component. Examples of useful analytical techniques include nuclear magnetic resonance, gel permeation chromatography, gas chromatography (often in combination with mass spectroscopy), liquid chromatography, and infrared spectroscopy.

4.3.3.2 *Thermal Analysis*

Thermal and thermomechanical analyses[44] are very important for determining the upper and lower usage temperature of polymeric materials as well as showing how they behave between those temperature extremes. An especially useful thermal technique for polyurethanes is dynamic mechanical analysis (DMA).[45] This is used to study dynamic viscoelastic properties and measures the ability to store and dissipate energy. It is also a sensitive method for measuring glass transition (T_g), melting transition (T_m), and softening (T_{soft}) temperatures. In biphase urethane polymers, the region between T_g and T_m or T_{soft} is the practical use temperature range, known as the "rubbery plateau" from the relatively constant modulus. Heat distortion temperature (HDT) is another technique for measuring softening point. The coefficient of linear thermal expansion (CLTE) is determined in applications where thermal expansion and contraction are important considerations. Thermogravimetric analysis (TGA) and differential scanning calorimetry (DSC) are also of occasional use, especially for elastomers.

4.3.3.3   Mechanical Properties

Determination of the mechanical properties of a cured polymer serves to characterize its macroscopic (bulk) features such as flexibility and hardness. Using standardized methods of the American Society for Testing and Materials (ASTM) and the International Standards Organization (ISO) allows direct comparison to other materials. The vast majority of polyurethane research and development is conducted in industry where mechanical properties are of vital importance because this information is used to design, evaluate, and market products. General test categories are presented here with a few illustrative examples.

Tensile strength is a common test for elastomers and coatings. This measures the force versus deflection (amount of elongation) to the break point of a small "dog bone"-shaped piece of the material. The resultant stress–strain curve provides much information, including yield point, elongation at yield, ultimate tensile strength, percent elongation at break, and stress at 100, 200, and/or 300% elongation. Stress–strain behavior depends on the nature of the material, such as whether it is hard or soft and the test rate.

Compression tests are routinely carried out on elastomers and foams of all kinds. In compression testing, a squeezing force is applied to the entire cross section of the sample. At a constant load application rate, the deformation and load are simultaneously recorded to generate a compression stress–strain curve which provides information regarding compression strength. The load removal curve indicates resiliency and energy absorption (hysteresis). Such data are especially important for flexible foams used in seating and bedding, and several specialized tests have been developed for these applications. Compression set measures how much a material permanently deforms under long-term compression load and is commonly used for flexible foams and elastomers.

Flexural modulus is the force required to deform a material in the elastic bending region. It is essentially a way to characterize stiffness. Urethane elastomers and rigid foams are usually tested in flexural mode via three-point bending and the flexural (or "flex") modulus is obtained from the initial, linear portion of the resultant stress–strain curve.

There are several ways to measure a material's resistance to tearing. In these tests, the applied force is not distributed over the entire specimen but is concentrated on a slit or notch and the tear strength is reported as the force required to propagate a tear from this point. For urethane elastomers and foams, the most common tear strength tests are Die C, Trouser and Split tear. These differ in the shape of the test sample, the way in which the samples are torn, and the results generated. The test chosen depends on the type of material; die C is used mainly for solid elastomers, while trouser and split tear are more often used for foams and microcellular materials.

Impact strength is a material's resistance to fracture when a sudden force is applied. This quantifies dynamic toughness and is also an indication of brittleness. The most common impact strength tests for elastomers are Izod, Dart, and Charpy. These differ in the size and shape of the sample, how it is held in place, and the mode of impact. There is a larger selection of impact tests for coatings. Since coatings are applied and tested on a substrate, these impact tests determine not only resistance to cracking or fracture but also delamination and chipping. Gardner impact is often used for evaluation of coatings whereby a coated metal panel is struck with the coating facing either toward (direct impact) or away from (reverse impact) the impact head. The gravelometer is a more specialized instrument for automotive coatings where an abrasive "gravel" is sprayed at the coating surface under air pressure as a way to simulate road conditions. The coating is then checked for chipping, marring, scratching, and loss of gloss.

There are several ways in which hardness is quantified, depending on the material and the physical property of importance. The most prevalent hardness measurement for urethane polymers is resistance to plastic deformation using a Shore hardness durometer (from Shore Manufacturing). A durometer measures the penetration of a spring-loaded steel probe into the surface of the material. The degree of penetration is calibrated to a built-in *Shore hardness* gauge. There are several Shore hardness scales, the most common for polyurethanes being the Shore A and Shore D scales for relatively soft and hard materials, respectively. Coatings, however, are too thin for the penetration method and other techniques must be used. A common hardness test for coatings is pencil hardness, where the squared (90°) tip of a lead pencil of a specific lead hardness is rubbed back and forth on the coating surface. The pencil hardness of the coating is the lead hardness required to dig into the coating and either gouge or delaminate it. The pendulum hardness of a coating is a little more quantitative whereby rounded, pendulum-mounted posts are rubbed back and forth on the coating surface under constant load. The pendulum hardness of the sample is given as the number of seconds it takes the pendulum arc to decrease from 6° to 3°.

Abrasion resistance is the ability of a surface to resist scratching and marring when contacted with an abrasive material. This is an important test for many applications of coatings, elastomers, and integral-skin foams (Section 4.1.2). The degree of abrasion is measured by loss of weight or by evidence of surface marring, such as loss of gloss. For elastomers, the most common test is *Taber abrasion*, where a sample is rotated a certain number of times under two abrasive spinning wheels at a constant load. Results are measured in terms of weight loss (ASTM D4060) or the percent decrease in light transmittance (ASTM D1044).[42c] Because surface quality and appearance are so important in coatings, several tests for abrasion resistance have been developed, the most common being falling sand, crock tester, and gravelometer (see above). As the name implies, in the falling-sand abrasion test a certain amount and type of gravity-fed sand is allowed to fall onto the surface of the coating. The crock tester has a weighted pad with an applied abrasive that is rubbed back and forth on the coating surface. These tests give an indication of a coating's resistance to marring and scratching.

Resiliency (or elasticity) is the tendency of a material to return to its original size and shape after the release of a short-term load applied in its elastic region. It is also a measure of the energy-absorbing characteristics of a material. For foams, the falling ball rebound and indentation force deflection (see below) tests are most common. Falling ball rebound simply measures the amount a standardized steel ball bounces when dropped from a specified height onto a sample and is usually used for flexible foams. Results are reported in the percentage of the original drop height that the ball bounces (percent rebound). Bashore "hardness" for solid elastomers is similar except it utilizes a rounded steel rod instead of a ball.

4.3.3.4 Application Testing

There is a seemingly infinite variety of tests for specific applications, and some even have alternate adaptations for particular industries or customers. This section contains a few examples of common application tests for urethane foams, coatings, adhesives, and elastomers that may be conducted in addition to those described above. Some of these are just specialized versions of tests discussed in the previous section but are presented here because they have been developed for specific applications. Others are more broadly useful and very common.

Urethane foams fall into three basic categories: rigid, semirigid, and flexible. A common evaluation test of rigid foams for insulation is determination of the so-called k-factor, or coefficient of thermal conductivity. This is a measure of the ability to transport heat from one side of a piece of the foam to the other. Lower k-factor foams are better insulators. Other important tests for rigid foams are dimensional stability (percent change in volume and linear dimensions after timed exposure to specific temperature and humidity) and moisture vapor transmission (weight uptake or loss after exposure to high or low humidity). Resiliency and hysteresis are often important properties for flexible foams, so a special type of compression test is used called indentation force deflection (IFD). In this test a small, circular indentation pad (or "foot") is depressed into the face of the foam and a stress–strain curve is generated. Of particular interest is the force required to reach certain deflections and the energy absorption (hysteresis) characteristics of the foam, which are determined by the difference between the loading and unloading portions of the IFD stress–strain curve. Semirigid foams are usually valued for their damping (energy-absorbing) properties, so, like flexible foams, compression stress–strain behavior is important, but in this case a high amount of hysteresis is generally desirable. For this, the foam's compression force deflection (CFD) is determined. CFD is similar to IFD except that the compression plate is large enough to compress the entire cross section of the foam sample.

For coatings, common evaluations are in regard to surface quality, chemical resistance, color, and weathering/light stability (discussed below). Gloss and distinctness of image are common surface measurements. Gloss is the percent of light reflected at a specific incident angle (usually 20° or 60°). Distinctness of image (DOI) is a relative evaluation of how perfectly a coating reflects images. There are many methods for determining chemical resistance, including various *spot* and *acid-etch* tests where a coating is placed in contact with solvents or acids under certain time–temperature conditions and then checked for blistering, softening, loss of gloss, and discoloration. A related test is the so-called MEK double-rub where a methyl ethyl ketone–soaked cloth is rubbed a certain number of times (one back and forth motion is one "double rub") across a coating surface or until break-through to the substrate. If the coating survives this treatment, the effect on its surface is then described. These chemical resistance tests are obviously very qualitative but are nonetheless important for product development.

Weathering tests are conducted on many types of materials to determine how well they stand up to the elements (air, water, sunlight, etc.). Weathering is very important for most coatings applications as well as any other materials subjected to outdoor exposure (such as elastomers in automotive applications). *Accelerated weathering* tests expose a material to intense UV radiation and may include simulated weather conditions, such as heat and/or moisture cycles. Exterior weathering involves long-term exposure, most commonly in Florida or Arizona. These may involve reflectors to intensify the sunlight and periodic spraying with water. For all these tests, the samples are checked for changes in color (yellowing) and surface quality (gloss and distinctness of image). Signs of chalking, a white, powdery formation from decomposition of the polymer surface, are also monitored.

Combustion testing[46] is critical in many applications such as seating, lamination, and insulation foams. There are many types of combustion tests, but most common for polyurethanes are UL (Underwriter's Laboratories) testing, the tunnel test, and the cone calorimeter. There are various types of UL tests, but they all involve exposing the polymer to a flame and noting how quickly it ignites, how long it continues to burn, whether there is drip formation, and so on. In the tunnel test, the sample is ignited inside a special tubular instrument (or burn tunnel) and the distance that it burns into the tube is measured, as is the density of smoke formation. The cone calorimeter involves exposing the sample surface to a powerful infrared heat lamp, and the resultant emissions are measured for smoke density, heat release, and the concentration of liberated CO and CO_2.

4.3.3.5 Analysis of Raw Materials

The discussion above is limited to finished polymers, but analysis of raw materials[40,41] is also very important. Regarding isocyanate-functional products, one common titration is for NCO content (%NCO by weight), which allows molar equivalency (or *equivalent weight*, molecular weight per functional group) to be calculated, which is helpful in determining stoichiometry because the starting materials used are frequently mixtures of 2-functional and higher molecules (e.g., PMDI). From %NCO content, average functionality can be calculated for polymeric MDIs. Other common tests for isocyanates include viscosity, specific gravity, acidity, color, and turbidity. For polyols, *OH number* (also OH# or OHN, milligrams of KOH required to neutralize 1 g of polyol) is measured via titration and serves the same purposes as %NCO in isocyanates. In addition to the other tests mentioned for isocyanates, of importance for polyols are acid number (Acid#, same units as OH#), water content, cloud point, and pH.

4.4 SYNTHETIC METHODS

This section contains detailed instructions for preparing several types of urethane materials. Most of the examples illustrate procedures used and materials made in the industrial synthetic urethane laboratory where the emphasis is usually on the bulk physical properties of the material. The examples are relatively simple formulations for instructional purposes only and should not be construed as being part of any product specifications or guarantees or representative of the best which urethanes have to offer. Most of the examples were not actually confirmed in the laboratory, but some were checked for accuracy and clarity by the original investigators. References are supplied where appropriate.

It is assumed that the user of this information is fully competent in synthetic laboratory methods and is aware of the hazards and safe handling of the chemicals and techniques utilized. The target audience is the college senior, first-year graduate student, or experienced industrial technician. Regarding the use of isocyanates, some are skin and respiratory tract sensitizers. Sensitization means you can become hypersensitive to even minute amounts of these substances. Such reactions are rare and happen only after overexposure. These hypersensitive reactions can be asthma-like, immediate or delayed, short or long lasting, and temporary of permanent. Isocyanates are also skin, eye, and respiratory tract irritants and can cause rashes, blisters, swelling, tearing, coughing, difficulty breathing, and so on. Acute overexposure can lead to permanent damage to skin, eyes, lungs, and so on. It is essential to review all relevant safety literature supplied with these products, such as material safety data sheets (MSDSs). Additional information is available from manufacturers. Users should never be exposed to any chemicals via contact or inhalation. Proper use of personal protective equipment (PPE; gloves, safety glasses, etc.) and laboratory equipment (hoods, fume ducts, safety shields, etc.) should prevent any such exposures.

The raw materials used are common and available from a variety of industrial sources and are always used as-received without further purification. All raw materials must be urethane grade, meaning mainly that the water content is less than about 0.05% by weight. If in doubt, water level should be measured.[41a,42] Bayer products are used where possible, but a detailed description of each compound is given so that substitutes from other manufacturers may be used. However, it cannot be overemphasized that extreme care must be taken when choosing substitutes because even small differences in these complex materials can cause marked discrepancies in results. Sources for common materials, such as 1,4-butanediol, are not specified as these are readily available from a variety of vendors. It should also be noted that deliberate modifications to any formulation can have dramatic effects on physical properties and processing characteristics and should only be done by an experienced chemist or for instructional purposes: Even a catalyst substitution can make your foam turn out more like a coating and your coating like a foam!

Most of the examples make use of common or relatively inexpensive laboratory equipment. There are many systems, such as RIM and spray elastomers, which are not suitable here because they require complex and expensive equipment. However, some of the procedures do require specialized equipment not commonly found in the synthetic laboratory. For example, preparation of high-quality coatings of uniform thickness will require draw-down bars. Foam processing will require high-speed mixer motors and high-shear stir blades. Availability of such specialized laboratory-scale equipment is assumed. In-depth involvement in urethane research, development, and application will require techniques and equipment for measuring NCO concentration.[41a,42] There are many manual titration methods for measuring %NCO, but it is most convenient to use an automatic titrator, such as those available from Brinkmann. Although the examples used herein utilize commercially available materials which require no further work-up, more involved research and development will require the preparation of new prepolymers and modified isocyanates, necessitating measurement of %NCO. The same is true of OH#, Acid#, viscosity, water content, and so on.

The raw materials used to make urethanes are usually complex mixtures of multifunctional compounds, so the concept of equivalent weight is used to determine reaction stoichiometry. Equivalent weight (or eq. wt.) is the average molecular weight divided by average functionality (MW_{ave}/f_{ave}). It has units of grams/equivalent. For isocyanates, this is calculated from the measured value for %NCO using the following equation: $(42/\%NCO)(100) = $ eq. wt. Similarly, for polyols, eq. wt. $= (56.1/OH\#)(1000)$. The number of equivalents equals grams of material divided by the equivalent weight. The ratio of NCO equivalents to OH equivalents determines the reaction stoichiometry and is called the reaction "index." Most often, a slight excess of isocyanate is used (index ≈ 1.05) to compensate for loss of NCO by reaction with water present in reactants, solvents, and/or air. Excess isocyanate also provides optimum physical property buildup in many cases. The ratio used may be above or below 1 for various reasons. As an example, to react 50 g of a polymeric MDI of 32.0% NCO at a 1.05 index with a polyol of OH# 250 requires:

$$4200 \div 32.0\% \text{ NCO} = 131.5 \text{ isocyanate eq. wt.}$$

$$50 \text{ g} \div 131.5 \text{ g/eq.} = 0.3802 \text{ eq. isocyanate}$$

$$0.3802 \text{ eq.} \times 1.05 \times (56,100/250) = 89.6 \text{ g polyol}$$

For polyol and isocyanate blends, the total equivalents of isocyanate divided by the total equivalents of OH equals the reaction index. For more on urethane calculations, see refs. 1d and 1k.

Exposure of raw materials to air should be limited as much as practicable, especially in humid climates. All materials should be stored in tightly capped jars or Nalgene bottles under a nitrogen blanket. Weighing and mixing may be conducted using disposable glass jars, plastic cups, or metal cans. Paper cups (including waxed paper) should never be used because the moisture they contain can react with the isocyanate, causing bubbles to form in the resultant polymer. For the same reason, metal stir blades should always be used, not wooden tongue depressors or sticks. All storage, mixing, and handling equipment should be clean and dry, but otherwise no special drying treatment is necessary.

A few key physical properties are listed for each product. They are given here as a way to quantitatively screen materials and are not intended to be a guarantee of results. The physical properties of the products can depend on many factors, including raw materials, processing techniques, and atmospheric conditions. Qualitative descriptions of the materials are also frequently supplied, and these should be closely observed. For example, if the material is described as a bubble-free solid and there are bubbles within your sample, something is wrong with either your processing technique or the raw materials. The physical properties listed will require access to testing equipment, some of which can be very expensive. Even simple hardness measurements for solid elastomers require a Shore durometer, which can cost several hundred dollars. Access to this type of analytical equipment is assumed.

4.4.1 Elastomers

An important prerequisite for the preparation of high-quality elastomers is thorough degassing of starting materials. Liquid components contain dissolved gases which must be removed prior to use or they will escape upon heating during casting and curing, forming bubbles which ruin the part. Degassing is accomplished by placing the liquid isocyanate, polyol, or polyol blend into a container (no more than about one-third full) and applying vacuum to the sample until the bubbling and frothing cease. This is most safely done in a vacuum oven under a pressure of at least 20–25 in. Hg. Most of the starting materials used here are quite viscous and can take up to an hour or two to fully degas. It may be helpful to warm the sample during this process to reduce viscosity and facilitate gas removal. If absolutely necessary, an antifoaming agent may be added to help break down the gas bubbles. For cast elastomers, degassing should be done immediately prior to use. Only extremely fast reacting elastomer systems, such as spray elastomers, can be used without degassing because they react and cure so quickly there is not enough time for bubbles to form in this way.

Elastomer samples are cast in molds, the size and shape of which depend on its purpose. Samples for physical properties can be produced using a custom-made "book" mold designed to create a thin sheet (0.1 in.) containing premolded test parts, such as those for die-C tear, flexural modulus, and so on. Alternatively, a flat plaque mold may be used to create a 6 × 6 × 0.1-in. sheet from which may be cut samples for testing. Thicker samples for hardness measurements may be produced from a "button" mold which creates a connected series of 1-in. circular buttons $\frac{1}{2}$ in. thick. These thick cross-sectional samples allow accurate hardness measurements. In lieu of a button mold, a thicker plaque may be made or a series of thin samples may be stacked and hardness measured from the top sample. Whichever methods are used, it is important to be consistent and only compare results obtained by identical techniques. When describing results, be sure to list the methods and samples used.

Example 1. One-Shot Cast Elastomer. *This is an example of a one-shot cast elastomer using a PMDI, two polyols, and a chain extender. It is a good formulation to begin with because it is simple and uses low-viscosity ingredients that can be blended and cast at room temperature.*

Charge Multranol 9139 [69.16 g, 0.0345 eq., an EO-tipped poly(propylene oxide) polyol of 3-functionality, OH# = 28, Bayer], Multranol 4012 [19.76 g, 0.1303 eq., an EO-tipped poly(propylene oxide) polyol of 3-functionality, OH# = 370, Bayer], 1,4-butanediol (9.87 g, 0.2193 eq.), RC 6080 (0.2 g, a solution of 25% triethylenediamine in 1,4-butanediol, Rhein Chemie), and Dabco T-12 (0.02 g, a solution of dibutyltin dilaurate in a glycol carrier, Air Products) into a suitable container and mix until a homogeneous blend is obtained. Thoroughly degas this blend in the manner described above. At the same time, degas about 100 g of Mondur MRS-4 (a PMDI of 32.5% NCO, eq. wt. = 129, viscosity at 25°C = 40 mPa·S, density at 25°C = 1.02 g/mL, Bayer). To the polyol blend, quickly add 54.0 g (0.4178 eq.) of the isocyanate. This may be done by either tare weighting the isocyanate or placing the polyol on a tared balance and adding the isocyanate to it. Either way, the isocyanate should be added as quickly as possible without splashing or incorporating bubbles to the mixture. Immediately upon adding the isocyanate, quickly mix the reaction mixture. A metal spatula may be used to mix by hand or a metal stir blade attached to a mechanical stirrer may be used. Whichever method is used, it is imperative to blend the reactants thoroughly, quickly, and without incorporating bubbles. Pour the mixture into a mold as described above and cure at 80°C for 1 h. Demold the part and condition at room temperature for at least one day.

Physical properties (ASTM methods in parentheses): Gel time = 2–5 min; hardness (D2240) = 49 Shore D (>90 Shore A); tensile strength (D412) = 2393 pounds per square inch (psi); ultimate elongation (D412) = 103%; die C tear (D624 − 91) = 341 foot-pounds per inch (lbf/in.) split tear (D3489 − 90) = 97 lbf/in.; T_g(DMA) = −50°C.

Example 2. Two-Shot Cast Elastomer (Prepolymer Method). *Synthesis of polyurethane elastomers via the prepolymer method is challenging and requires especial attention to details and techniques. This example is a relatively easy one, using a low-viscosity, 9.6% NCO MDI-based poly(tetramethylene ether) glycol prepolymer. It is adapted from the Bayer Product Information Bulletin for Baytec ME-090 prepolymer.*

Carefully degas Baytec ME-090 (100.0 g, 0.2286 eq.) in a vacuum oven while heating to 100°C. Degas and preheat a sample of 1,4-butanediol (9.8 g, 0.2178 eq.)

at the same time. Wet tare the 1,4-butanediol sample at temperature. Both samples must be evenly heated to 100°C. If necessary, the samples may be reheated in a conventional or microwave oven. Whichever heating method is used, care should be taken to avoid overheating and localized hot spots (stir occasionally). The sample can be stirred using the metal probe from a thermocouple, allowing concurrent monitoring of temperature. Hold the isocyanate sample under a motor-attached stir blade and raise the container so that the mixing blade is gently immersed into the liquid. Turn on the mixer motor and slowly stir the material. Avoid mixing bubbles into the isocyanate. If any bubbles appear, remove the sample and repeat the entire process, beginning with degassing. While holding the isocyanate container with one hand, quickly pour the wet-tared extender into it and then increase the motor speed as much as possible without creating a vortex or bubbles. When the sample is thoroughly mixed (about 15 s), turn off the mixer, quickly remove the sample from the stir blade, and pour the sample into a preheated (110°C) metal mold. The pot life of the reaction mixture is only 2–5 min, so all of this must be done quickly. Lower raw material and mold temperatures will extend gel time but will allow more time for phase separation, creating opaqueness in the sample. If cast at 100°C into a 110°C mold, a transparent, pale yellow sample should be obtained. After casting, the mold should immediately be placed into a convection oven at 110°C. If it is desired to use the mold again, the sample may be removed after 1 h and postcured at 110°C for an additional 16 h. This postcure step is important for two-shot cast elastomers and provides development of ultimate physical properties. Some materials may require additional conditioning at room temperature until physical properties (such as hardness) do not change further.

The material should have these properties: gel time = 2–5 min; hardness (D2240) = 95 Shore A; Bashore resilience (D2632) = 60%; tensile strength (D412) = 5400 psi; ultimate elongation (D412) = 500%; die C tear (D624) = 610 lbf/in.

Example 3. Urethane Hydrogel. *Hydrogels are prepared by swelling a crosslinked hydrophilic polymer in water. Polyurethane hydrogels are most commonly made using a poly(ethylene oxide) polyol for hydrophilicity in conjunction with a crosslinking agent which allows the material to swell in water but not dissolve. This example is from E. Haschke et al., Journal of Elastomers and Plastics, Vol. 26 (January 1994), p. 41. It is a simple formulation using $H_{12}MDI$, poly(ethylene oxide) polyol, and trimethylolpropane.*

A prepolymer is made first by charging Pluracol E2000 [1000.0 g, 1.0 eq., poly(ethylene oxide), 56 OH#, BASF] to a suitable container equipped with a mechanical stirrer and a nitrogen gas inlet. Flush the container with dry nitrogen and add Desmodur W (264.0 g, 2.0 eq., 4,4'-methylene-bis(cyclohexyl isocyanate), 31.8% NCO, Bayer). While maintaining a positive N_2 pressure on the reaction mixture, stir and heat at 80°C for 2 h. Cool the product to room temperature and check the NCO content (theory = 3.32%). It might be necessary to warm the highly viscous prepolymer to take samples for titration. To a portion of this prepolymer (250.0 g, 0.2 eq.), add Dabco T-12 (0.25 g, dibutyltin dilaurate,

Air Products) and heat to 80°C with stirring. Add trimethylol propane (8.94 g, 0.2 eq., TMP, Aldrich), stir for 1 min, and pour the reaction mixture into a pre-heated aluminum mold treated with mold release. Place the mold into an oven at 80°C for 6 h. Remove the sample and store in a desiccator.

From this material, samples are cut and swelled to constant weight in a buffered saline solution prepared from 8.43 g sodium chloride (NaCl), 9.26 g boric acid (H_3BO_3), 1.0 g sodium borate (Na_3BO_3), and 0.1 g of the disodium salt of the dihydrate of ethylenediaminetetraacetic acid $[Na_2 \cdot EDTA \cdot (H_2O)_2]$ in 1 L of distilled water.

For the swelled hydrogel: saline solution uptake (swollen weight minus dry weight/swollen weight × 100) = 70%; elongation (D412) = 25%; tensile strength (D412) = 2.0 MPa.

4.4.2 Foams

Example 4. Flexible Foam. *This is a conventional, water-blown, flexible, slab-stock foam formulation. It is a good formulation to begin making foams with because of its utilization of water as the blowing agent, which makes it relatively easy to process. It is also the only example here of a TDI-based polyurethane. This is a small, laboratory "cup foaming" procedure using common laboratory materials. The central piece of equipment is a variable-speed (1000–3000-rpm) mechanical stirrer with a metal mixing blade.*

Place Arcol Polyol F-3022 (100 g, 0.1 eq., 56 OH#, mixed PO/EO triol from Bayer) into a suitable container. To this add distilled water (3.3 g, 0.4125 eq.), Niax Silicone L-620 (0.5 g, a silicone surfactant from OSi Specialties), and Niax C-183 (0.12 g, an amine catalyst from OSi Specialties). Thoroughly blend this mixture without incorporating air bubbles. Then add Dabco T-9 (0.25 g, stannous octoate from Air Products) and mix again. The T-9 must be added last because it is quite water sensitive, so its exposure to the water-containing polyol blend should be kept to a minimum. To this polyol blend, quickly add Mondur TD-80 (42.6 g, 0.4868 eq., a mixture of 80% 2,4-TDI and 20% 2,6-TDI isomers from Bayer) and immediately stir at 3000 rpm for 5 s. Quickly pour the reaction mixture into a suitable container such as a 1-qt paper or plastic cup and allow the foam to "free-rise." The stir blade may be wiped or brushed clean.

For the reaction mixture: cream time (time from start of mixing to when there is a visual indication of initiation of foaming; usually evident as a change to a lighter color, hence the name "cream time"): 12 s; rise time (when the foam stops visibly "rising," or foaming): 110 s. For the finished foam: density = 1.8 pounds per cubic foot (pcf); 25% IFD = 40 lb; 25% CFD = 0.55 psi; ball rebound = 45%; tensile strength = 13 psi; elongation = 200%; tear strength = 1.8 pounds per linear inch (pli).

Example 5. Rigid Foam. *This is an example of an HCFC 141b–blown rigid, closed cell, insulation foam. Working with physical blowing agents such as 141b takes much practice and skill and should only be attempted after mastering the processing of water-blown foams such as that described in Example 4.*

Charge Multranol 4063 [36.4 g, 0.2985 eq., a poly(propylene oxide) polyol of 4-functionality, OH# = 460, Bayer] to a disposable 1-qt metal can. To this add Multranol 8114 [36.4 g, 0.2563 eq., a poly(propylene oxide) polyol of 4-functionality, OH# = 395, Bayer], Niax L-5440 (1.5 g, a silicone surfactant from OSi Specialties), Desmorapid PV (2.2 g, pentamethyl diethylene triamine, PMDETA, Rhein Chemie), Bulab 600 (1.1 g, tetramethyl ethylenediamine, TMEDA, Buckman Laboratories), and HCFC 141b (22.4 g, 1-fluoro-1,1-dichloro ethane, Atofina). Carefully note the weight of the filled container so that evaporative loss of HCFC 141b can be compensated for later. Mix the ingredients using a high-shear mixing blade and a high-speed mixer capable of maintaining a constant speed under load. This mixture should be transparent. For reproducibility, the temperature of the ingredients must be kept at a specific temperature, such as 20°C. The raw materials may be stored and mixed at a lower temperature (such as 15–17°C), so as to minimize blowing agent evaporation, and allowed to warm to 20°C prior to adding isocyanate. Care must also be taken that no resin remains on the stir blade and shaft. To minimize resin loss in this way, the stir blade can be removed from the resin while it is still rotating as long as the can is tall enough so that it will catch the resin slung off by centrifugal force. Weigh the container and add HCFC 141b to make up for lost weight, if necessary. In a separate container, wet tare Mondur MR (76.0 g, 0.5700 eq., a polymeric MDI, 31.5% NCO, Bayer) and add this quickly to the stirring polyol blend. Immediately start a timer and raise the stirrer speed to 1000 rpm. Mixing time should be about 5 s. After mixing is complete, quickly raise the rotating stir blade from the reaction mixture so as to force off the resin. Turn off the stirrer and quickly pour the contents into an open-top mold or other suitable container.

The following are noted: cream time: 6–7 s; string (or gel) time (when a small wooden stick or tongue depressor is dipped into the foaming reaction mixture until the time when thin threads of foam remain hanging from the stick as it is withdrawn): 29 s; tack-free time (when there is no detection of tackiness when a wooden stick or tongue depressor is lightly touched to the top foam surface): 35 s; free-rise core density: 1.77 pcf; k-factor: 0.125. This material should be nonfriable, meaning that it is not brittle and rubbing the hand across a freshly cut cross section of the foam should not result in the formation of any powdery dust.

4.4.3 Coatings

Example 6. High Solids, Two-Component, Clear, Aliphatic, Polyurea Coating. *This example of an aliphatic, 100% urea coating utilizes polyaspartic esters as reactive diluents (see Section 4.2.3). These compounds allow fine tuning of reactivity and physical properties and eliminate the need for viscosity-reducing solvents.*

Charge Desmophen NH 1420 (385.07 g, 1.38 eq., N,N'-[methylenebis(4,1-cyclohexanediyl]bis-tetraethyl ester, eq. wt. = 279, 100% solids, Bayer) to a suitable container. Add Desmophen XP-7068 (128.35 g, 0.4411 eq., N,N'-[methylenebis(2-methyl-4,1-cyclohexanediyl)]bis-tetraethyl ester, eq. wt. = 291,

100% solids, Bayer) and Byk 354 (13.69 g, polyacrylate flow and leveling aid, BYK-Chemie). Mix this resin solution thoroughly. If bubbles are incorporated during stirring, allow them to rise and break before using. To this blend, add Desmodur XP-7100N (382.21 g, 1.8655 eq., a low-viscosity modified HDI trimer, %NCO = 20.5, 100% solids, Bayer) and thoroughly mix without incorporating air bubbles. This system may be drawn into films using a wet film applicator or brushed/rolled onto substrates. For roller applications, addition of 0.1 to 2% w/w Airex 900 (a deaerator from Tego Chemie) will help to reduce bubble formation.

For the above polyol blend: viscosity (Brookfield, ASTM D-2196) = 1500 mPa · S at 23°C. For the reaction mixture: working (pot) life: 20 min; Gardner circular dry times [72°F, 54% relative humidity (RH)]; surface dry = 1.0 h, hard dry = 2.0 h, mar free = 3.5 h. For the finished coating: gloss (ASTM D-523) = 90+ at 60°; impact (ASTM D-2794) = 60 in.-lb direct, 10 in.-lb reverse; Tabor abrasion (ASTM D-4060, 1000 g load, 1000 cycles, CS-17 wheel) = 95.6 mg; pendulum hardness = 180 s; MEK double rubs (ASTM D4752-95, 50 double rubs) = softened.

Example 7. Solvent-Borne, One-Component, Moisture-Cure, Aliphatic Polyurea Coating. *This example is of a more traditional one-component coating formulation using a volatile solvent as a diluent. The "active ingredient" is a modified HDI, Desmodur N-75 BA from Bayer.*

Charge Desmodur N-75 BA (529.7 g, 2.0809 eq., a biuret of HDI, 75% solids containing n-butyl acetate, Bayer, %NCO = 16.5, eq. wt. = 255, viscosity = 90–210 mPa ·S) into a suitable container. Add Aromatic 100 [271.29 g, a naphtha (petroleum) aromatic solvent from Eastman Chemical, $C_8 – C_{10}$ aromatic hydrocarbon mixture, boiling point range = 159–170°C] and mix well using a mechanical stirrer. To this, add a 10% solution of CAB 381-0.1 (21.12 g, cellulose acetate butyrate, a flow aid from Eastman Chemical) in PM acetate (methoxypropanol acetate) and mix well. Next, add a 10% solution of Dabco T-12 (16.93 g, dibutyltin dilaurate, Air Products) in PM acetate. Where light stability is of concern, 1% by weight (based on resin solids) each of Tinuvin 292 and Tinuvin 1130 (UV stabilizers from Ciba-Geigy) can be added. After thoroughly mixing, a transparent, colorless solution should be obtained. This material will be stable for extended periods if it is protected from exposure to water or atmospheric moisture and from evaporation of the solvent. It may be drawn down into films on suitable substrates (such as glass plates) using a wet film applicator.

For the blend: viscosity = 40 mPa·S at 22°C, weight percent solids = 47.74, volume percent solids = 42.44, weight per gallon = 8.39 lb. Dry time (hours at 68% RH): set to touch = 0.3, surface dry = 1.6, hard dry = 3.3, mar free = 4.0. Film physical properties: impact = 140 in.-lb direct, 40 in.-lb reverse; Tabor abrasion (mg loss, CS-17 wheel, 1000 cycles, 1000 g load) = 8.4; pencil hardness = 2 H; spot tests (4 h): IPA, no effect; MEK, no effect; 50% NaOH, no effect; 37% HCl, lifted film; 100% acetic acid, failure.

Example 8. Water-Borne Polyurethane Dispersion (PUD). *This example illustrates the synthesis of a water-dispersible, OH-terminated polyurethane* *prepolymer. It utilizes dimethylolpropionic acid as a built-in dispersant when neutralized with triethylamine. It is, therefore, an anionic dispersion by definition. This example and the next were adapted from European Patent Application EP 0469389.*

Charge Desmophen E-609 (132.8 g, 0.1326 eq., a polyester diol from adipic acid and hexanediol, Bayer, OH# = 56, MW = 1000) to a 2-L, three-neck flask equipped with a mechanical stirrer, thermocouple, and N_2 inlet. Add neopentyl glycol (13.4 g, 0.2577 eq., 2,2-dimethyl-1,3-propanediol), DMPA (20.6 g, 0.3075 eq., α,α-dimethylolpropionic acid), polyether LB-25A (5.0 g, 0.0023 eq., a monofunctional alcohol based on EO and PO having an OH# of 26, Bayer AG), and N-methyl pyrrolidinone (NMP, 90.0 g). Blanket the flask with dry N_2 and heat to 70°C with continuous stirring. Add Desmodur W [127.6 g, 0.9661 eq., bis(4-isocyanatocyclohexyl)methane, %NCO = 31.8, Bayer], increase the reaction mixture temperature to 110°C, and stir for 1 h, or until the isocyanate content is 3.5%. Cool the resultant prepolymer to 70°C and add triethylamine (15.5 g, 0.1538 m, 1 m/m DMPA). Stir well for 15 min and add ethanolamine (17.0 g, 0.2787 eq., based on N−H) and N-methylpyrrolidinone (50.0 g). The reaction mixture should exotherm to about 90°C. Cool to 70°C and stir an additional hour or until found to be NCO free by IR analysis (loss of NCO band at 2270 cm^{-1}). Charge an additional 5.0 g (0.0023 eq.) of polyether LB-25A and stir for 30 min. To this, slowly add distilled water (391.3 g) at 50°C with vigorous mixing and stir for an additional hour.

For the resultant aqueous dispersion: pH = 9.3; percent solids = 35; functionality = 2 (OH); urethane−urea content = 13% (calculated as NH−C−O, MW 43).

Example 9.   Two-Component   Waterborne   Polyurethane   Coating. *This example utilizes the PUD from Example 8 in a two-component, waterborne coating formulation. An HDI-based, water-dispersible isocyanate is also prepared.*

A water-dispersible polyisocyanate is prepared as follows. Charge Desmodur N-3300 (900 g, 4.6154 eq., HDI trimer, Bayer, 100% solids, 195 eq. wt., 21.6% NCO) to a 2 L, three-neck flask equipped with a mechanical stirrer, a thermocouple, and a nitrogen inlet. To this, add Polyether LB-25A (100.0 g, 0.0463 eq.) and heat to 110°C with stirring for 1.5 h. For the isocyanate product: percent Solids = 100; viscosity = 3900 mPa·S (25°C, #4 spindle, 60 rpm); %NCO = 18.8.

Charge the OH functional urethane water dispersion prepared in Example 8 (200 g, 0.1603 eq. OH) to a suitable container. Vigorously stir the dispersion with a mechanical stirrer and slowly add the 18.8% NCO water-dispersible polyisocyanate described above (40.0 g, 0.1791 eq.). This will give an NCO−OH equivalent ratio (index) of 1.12. To this mixture, add Silwet L-77 (a flow aid from OSi Specialties) and thoroughly mix. Use a draw-down bar to make 6-mil (0.006-in.) wet-thickness films on glass plates and 5-mil (0.005-in.) films on Bonderite-treated steel panels (Bonderite is from Henkel). Condition the resultant films at 72°F and 50% RH for 2–3 weeks before evaluating. For the finished films: MEK double rubs = 200+; pendulum hardness = 127 s; Gardner impact (psi, ASTM D-3029-84, method G) = 160 direct, 160 reverse; tensile strength (ASTM D-638) = 4483 psi; elongation (ASTM D-638) = 25%.

4.4.4 Adhesives, Sealants, and Binders

Example 10. Two-Component, Nonsagging, Polyurea Structural Adhesive. *This example is an all-urea formulation: unusual for a 100% solids, hand-castable system. It is also a two-component, one-shot system designed to be nonsagging so that it may be applied to nonhorizontal surfaces without dripping or running. This and similar examples are described in U.S. Patent 5,654,085. See also Guether et al., presented at Polyurethanes Conference 2000, October 8–11, 2000 CRC Press: Boca Raton, FL, 2000.*

An amine-terminated polyether (ATPE) is prepared as follows. Charge poly(tetramethylene oxide) diol (PolyTHF 1000, BASF, 75.96 g, 0.0759 m) to a 500-mL three-neck round-bottom flask fitted with a thermocouple, a mechanical stirrer, and a vacuum port. Add tert-butylacetoacetate (24.04 g, 0.1582 m) and apply vacuum. Heat at 175°C for 4 h, Fourier transform infrared (FTIR) analysis should indicate complete loss of the polyol OH absorption at 3300 cm^{-1}. The room temperature viscosity of the product should be about 520 mPa·s. React this acetoacetylated product (85.5 g, 0.0649 m) with cyclohexylamine (14.5 g, 0.1465 m) at 110°C under vacuum for several hours. Cool the resultant cyclohexylaminocrotonate polyether product to room temperature (1790 mPa·s at room temperature).

Prepare an amine blend from the above cyclohexylaminocrotonate polyether (63.5 g, 0.0824 eq.), Ethacure 300 (4.6 g, 0.0429 eq., dimethylthiotoluene diamine isomer mixture, Albemarle), PACM-20 [1.5 g, 0.0143 eq., bis(4-aminocyclohexyl)methane, Air Products and Chemicals], Baylith L Powder (7.0 g, powdered sodium aluminosilicate, UOP), and Mistron Vapor (23.4 g, talc powder, 2.8 sp. gr., 1.7 μm avg. particle size, Luzenac America). Blend this mixture and degas to remove bubbles and dissolved air. Charge the mixture to a plastic bag, such as a plastic sandwich bag, by placing the bag in a can or cup such that the material is placed into one corner of the bag. Degas Mondur MR and place a sample (20.9 g, 0.1568 eq., 130 index, polymeric diphenylmethane diisocyanate, 31.5% NCO, Bayer) in the bag on top of the amine blend. Loosely twist off the corner of the bag such that there is little or no air entrapped in the portion containing the reaction mixture. The "pocket" containing the reaction mixture should still be loose enough, however, so that its contents may be kneaded and thoroughly mixed by hand. It is important to remove as much air as possible from the bag so as to avoid incorporating bubbles into the adhesive. Rapidly knead the bag until it is quite warm to the touch and the contents are highly viscous (approximately 1 min mix time; be careful not to break the bag!). The material must be of even consistency, with no signs of precipitation or gellation. While holding the bag and its contents with the corner pointing up, cut the tip off with scissors creating a hole roughly $\frac{1}{4}$ in. in diameter. Squeeze the reaction *mixture out of the bag onto either a vertical surface (to demonstrate its nonsag performance) or a plaque mold with a cover (for preparation of flat samples for physical testing) or test specimens for adhesion testing (see below). A bead of the material up to $\frac{1}{2}$ in. in diameter should not sag, run, or drip when placed on a vertical surface. Viscosity buildup can be adjusted as necessary by varying the ratio and concentration of PACM-20 and/or Ethacure 300.*

Samples prepared for testing should be conditioned at $72°F$ and 50% relative humidity for several days or until the hardness of the material plateaus. The ultimate hardness should be about 55 Shore D. The softening point should be about $168°C$ as determined by thermomechanical analysis (TMA). Adhesive performance can be demonstrated in a number of ways. Since this material was originally developed for automotive applications, bonding to sheet-molding compound (SMC) is described here using the lap shear test (SAE J1525). Two SMC plaques (Budd 971 from Budd Co., $4 \times 6 \times 0.125$ in.) are bonded together using metal spacers (metal beads or paper clips may be used) to ensure a consistent bond thickness of 0.030 in. A jig may be prepared for this procedure. Bond overlap of the plaques is 1 in., and the bonding surfaces are buffed clean with a cloth. The adhesive is applied and the parts mated in such a way that the bonding surfaces are completely covered and there are no entrapped bubbles. Wipe off excess adhesive, cure in a convection oven at $135°C$ for 1 h and then condition at room temperature for one day. Test coupons (4×1 in.) are cut from the cured plaque using a tile saw. Five, 1-in.-wide samples may be cut, discarding the outer $\frac{1}{2}$-in. edges of the 6-in.-wide plaque. The test samples are then pulled apart in shear mode using an Instron analyzer. The samples should have an average maximum load of 686 psi and 100% substrate failure.

Example 11. One-Component, Moisture-Cure Polyurethane Sealant. *This example is of a low-hardness, high-elongation, moisture-curable polyurethane sealant. The material is based on a low-%NCO prepolymer made from 4,4'-MDI and a low-unsaturation (low-monol-content) Acclaim polyol from Bayer. It is adapted from (a) J. Lear et al., Adhesives Age, February 1999, pp. 18–23; and (b) B. Lawrey, et al., presented at UTECH 2000, The Hague, The Netherlands, March 30, 2000, Crain Communications: London, 2000.*

Charge Acclaim Polyol 4200 (313 g, 0.1562 eq., a propylene oxide glycol, OH# = 28.1, f = 1.99) and Acclaim Polyol 6300 (313 g, 0.1043 eq., a propylene oxide triol, OH# = 18.7, f = 2.94) to a 2-L three-neck flask equipped with a mechanical stirrer, a thermocouple, and a nitrogen inlet/vacuum port. Flush the flask with nitrogen and then thoroughly mix the contents until a clear, homogenous blend is obtained. To this, stir in molten Mondur M (4,4'-MDI, ca. 45°C, 81.0 g, 0.6480 eq., 33.6% NCO, Bayer). Heat the reaction mixture to 80°C for 1 h or until the %NCO is 2.3. While stirring the prepolymer at 80°C, add CaCO₃ (313 g) and diisodecyl phthalate (DIDP) (391 g). After the mixture is thoroughly blended, apply vacuum and degas the material until no bubbles or froth is apparent at ⩽25 in. Hg. Cool to 50°C, add Dabco T-12 (0.07 g, 50 ppm, a solution of dibutyltin dilaurate in a glycol carrier, Air Products), and stir for an additional

20 min. Let the product cool to room temperature. The sealant can be stored at room temperature for extended periods if extreme care is taken to avoid exposure to the atmosphere. The material can be drawn down into films using a wet film applicator or it may be poured into syringes for preparation of thicker parts or application as a standard sealant. If these syringe samples are carefully capped and stored in a desiccator, the material will keep for extended periods. Sample size and shape will depend on their purpose. For hardness measurements, sample thickness should be at least $\frac{1}{2}$ in. to avoid read-through of the supporting surface. Thicker samples will require longer atmospheric cure time, but all samples prepared for testing should be conditioned for 14 days at 72°F and 50% RH. The product should have the following physical properties: hardness = 32 Shore A; tensile strength = 2.317 MPa; elongation at break = 377%; 100% modulus = 0.869 MPa; die C tear = 7.53 kN/m.

Example 12. Isocyanate Binder for Wood. *Polymeric MDIs are used as one-component binders to make high-performance "engineered" wood products such as oriented strand board, plywood, and particle board. This laboratory-scale example was adapted from U.S. Patent 6,214,265 and illustrates the preparation of a type of particle board using clean wood particles milled specifically for this purpose. This wood "furnish" (from Allegheny Particleboard, Kane, PA) must have a moisture content of <10% by weight. Moisture content can be checked by drying and checking for percent weight loss (103°C for 21 h, ASTM D1037; electronic moisture analyzers are also available). If necessary, moisture content is lowered by heating at 103°C and periodically checking the moisture content until it is <10%.*

Add wood furnish (384 g, moisture content 6.02%) to the bowl of a rotary blade paddle mixer (such as a Kitchen-Aid KSM90) and agitate at the lowest speed setting. Add Mondur 541 (7.39 g, 1.9% w/w, a polymeric diphenylmethane diisocyanate of 31.5% NCO, Bayer) dropwise over a 5-min period using a disposable syringe. Continue blending for an additional 10 min and then transfer the blend to an $8 \times 8 \times 2$-in. metal form at the bottom of which is a metal plate which fits inside. The resin-coated furnish is evenly spread inside the form and another metal plate is placed on top. All parts of the form and plates are presprayed with mold release. The completed form assembly is placed into a hydraulic press (such as a model PW-22 manufactured by Pasadena Hydraulics) with platens heated at 350°F. The furnish is then pressed between the two form plates to a thickness of $\frac{1}{2}$ in. Press controls are used to ensure consistency of board thickness. The assembly is heated for $4\frac{1}{2}$ min. before demolding the cured wood panel.

For the resultant $8 \times 8 \times \frac{1}{2}$-in. board: board density = 40.67 lb/ft³; internal bond strength (ASTM D1037) = 133 psi; thickness swell (ASTM D1037) = 18.8%.

4.4.5 Thermoplastics

Example 13. Thermoformable Polyurethane. *Traditional TPUs are processable only at temperatures above about 150°C, often much higher. The example* related here is unusual in that it is designed to be formable at only 75°C. It illustrates the principles of thermoplastic urethanes yet is easy to make and can be formed by hand in hot water. See U.S. Patent 5,656,713.

Charge Formrez 66-37 [106.78 g, 0.0704 eq., a poly(hexylene adipate) polyester polyol from Witco Chemical, OH# = 37, f = 2, viscosity = 2500 cps at 60°C) to a 500-mL three-neck round-bottom flask equipped with a mechanical stirrer, thermocouple, and vacuum port/nitrogen inlet. To this add Dabco T-12 (0.184 g, dibutyltin dilaurate in a glycol carrier, Air Products and Chemicals) and 1,6-hexanediol (2.36 g, 0.0400 eq.). Flush the flask with dry N_2 and heat the mixture to 75°C while stirring. Preheat Desmodur W [15.0 g, 0.1136 eq., bis(4-isocyanatocyclohexyl)methane, 31.8% NCO, Bayer] to 60°C and charge it directly to the stirred polyol solution. Apply vacuum to the reaction mixture while vigorously mixing for about 1 min. Release the vacuum pressure, and pour the contents of the flask into a mold (either closed or open). Cure for 16 h at 75°C. At the curing temperature, the material should be soft, transparent, and readily formable by hand. Upon cooling to room temperature, however, the sample should harden and turn white. The material can be reheated in an oven or by immersing in hot water. Physical properties are as follows: elongation = 657%; elongation set = 503%; tensile strength = 14.48 N/mm^2; hardness = 51 Shore D; flexural modulus = 365.5 N/mm^2.

ACKNOWLEDGMENTS

I thank the many Bayer colleagues whom I pestered with questions during preparation of this manuscript. I am especially grateful to the following industry veterans for reviewing whole sections of this work: Karl Haider, Walt Heckla, Neil Nodelman, Ashok Sarpeshkar, Bill Slack, and Ron Taylor. Thanks also to Peggy King for tracking down every reference I asked for.

REFERENCES

1. For general sources on polyurethanes see (a) G. Oertel (Ed.), *Polyurethane Handbook*, 2nd ed., Hanser/Gardner, Cincinnati, 1993; (b) J. K. Backus, C. D. Blue, P. M. Boyd, F. J. Cama, J. H. Chapman, J. L. Eakin, S. J. Harasin, E. R. McAfee, C. G. McCarty, N. H. Nodelman, J. N. Rieck, H. G. Schmelzer, and E. P. Squiller, in *Encyclopedia of Polymer Science and Engineering*, H. F. Mark, N. Bikales, C. G. Overberger, G. Menges, and J. I. Kroschwitz (Eds.), Vol. 13, 2nd ed., John Wiley & Sons, NY, 1988, pp. 243–303; (c) D. Dieterich and K. Uhlig, in *Ullman's Encyclopedia of Industrial Chemistry*, Vol. A21, 1992, pp. 665–716; (c) *60 Years of Polyurethanes, International Symposium and Exhibition*, January 15–16, 1998, Technomic, Lancaster, PA, 1998; (d) G. Woods, *ICI Polyurethanes Book*, 2nd ed., ICI Polyurethanes & Wiley & Sons, New York, 1990; (e) K. N. Edwards (Ed.), *Urethane Chemistry and Applications*, ACS Symposium Series 172, American Chemical Society, Washington, DC, 1981; (f) J. M. Buist and H. Gudgeon, *Advances in Polyurethane*

Technology, Wiley, 1968; (g) J. M. Buist (Ed.), *Developments in Polyurethane-1*, Applied Science, London, 1978; (h) K. Uhlig and A. Conrad, *Discovering Polyurethanes*, Hanser Gardner, Cincinnati, OH 1999; (i) *Polyurethanes Conference 2000: Defining the Future Through Technology*, CRC Press, Boca Raton, FL, 2000; (j) K. C. Frisch and D. Klempner, "Polyurethanes," in *Comprehensive Polymer Science*, Vol. 5, G. C. Eastman, A. Ledwith, S. Russo, and P. Sigwalt (Eds.), Pergamon, Oxford, England, 1989, Chapter 24; (k) M. Szycher, *Szycher's Handbook of Polyurethanes*, CRC Press, Boca Raton, FL, 1999.

2. For urethane chemistry and synthesis see (a) J. H. Saunders and K. C. Frisch, *Polyurethanes: Chemistry and Technology, Part I. Chemistry*, Wiley, New York, 1962; (b) Schmelzer, H. Georg, and P. H. Markusch, "Synthesis of Polyurethane Elastomers," Paper presented at American Chemical Society, Rubber Division 151st Spring Technical Meeting, May 6–9, 1997, Anaheim, CA, Paul Flory Colloquium on Polymer Synthesis, paper number VII; (c) C. D. Eisenbach and H. Nefzger, "Poly(urethanes) and Related Polymers," in *Handbook of Polymer Synthesis*, Part B. H. R. Kricheldorf (Ed.), Marcel Dekker, NY, 1992; (d) C. D. Eisenbach, K. Fischer, H. Hayen, H. Nefzger, A. Ribbe, and E. Stadler, "Polyurethane Elastomers, Segmented (Non-Hydrogen Bonding Systems)," in *Polymeric Materials Encyclopedia*, Vol. 9, J. C. Salamone (Ed.), CRC Press, Boca Raton, FL, 1996; p. 6957; (e) C. D. Eisenbach and H. Nefzger, "New Insights in the Structure and Properties of Segmented Polyurethane Elastomers from Non-Hydrogen-Bond Forming Model Systems," in *Contemporary Topics in Polymer Science*, Vol. 6: *Multiphase Macromolecular Systems*, Plenum, New York, 1989, pp. 339–361.

3. (a) O. Bayer, H. Rinke, W. Siefken, L. Ortner, and H. Schild, German Patent 728,981, December 7, 1942 (priority from November 13, 1937); (b) For Bayer's life and career see K. H. Buchel et al., *Chem. Berichte*, **120**, 21–35 (1987).

4. For the history of Bayer AG, see E. Verg, et al., *Meilenstiene*, Bayer AG, Corporate Communications, Geb. W4, D-51368, Leverkusen, Germany, 1988.

5. For more on the development history of polyurethanes at I. G. Farben and elsewhere, see (a) O. Bayer, *Angew. Chem.*, **A59**, 275 (1947); (b) O. Bayer, E. Mueller, S. Peterson, H. Piepenbrink and E. Windemuth, *Angew. Chem.*, **62**, 57 (1950); (c) O. Bayer, E. Mueller, S. Peterson, H. Piepenbrink, and E. Windemuth, *Rubber Chem. Technol.*, **23**, 812 (1950); (d) *Urethanes Technology*, Vol. 4, March and June issues (1987), Crain Communications, London; (e) K. C. Frisch, "Historical Developments of Polyurethanes," in *60 Years of Polyurethanes*, International Symposium and Exhibition, January 15–16, 1998, Technomic Publishing, Lancaster, PA, 1998.

6. W. E. Catlin, U.S. Patent 2,284,637; W. E. Hanford and D. F. Holmes, U.S. Patent 2,284,896 (both to Du Pont), June 2, 1942.

7. B. J. Habgood, D. A. Harper, and R. J. W. Reynolds, British Patent 580,524 (to Imperial Chemical Industries), issued 1946 (priority from 1941).

8. (a) O. Bayer, E. Mueller, S. Peterson, H. Piepenbrink, F. Schmidt and E. Weinbrenner, *Angew. Chem.*, **64**, 523 (1952); (b) O. Bayer, E. Mueller, S. Peterson, H. Piepenbrink, F. Schmidt, and E. Weinbrenner, *Rubber Chem. Technol.*, **26**, 493 (1953).

9. E. Mueller, *Rubber & Plastics Age*, **39**(3), 195 (1958).

10. Bayer Corporation internal statistics.

11. (a) G. Woods, *Flexible Polyurethane Foams: Chemistry and Technology*, Elsevier Applied Science, Oxford, 1982; (b) R. Herrington and K. Hock (Eds.), *Flexible*

Polyurethane Foams, 2nd ed., Dow Plastics, Midland, MI, 1997; (c) N. C. Hilyard and A. Cunningham, *Low Density Cellular Plastics*, Chapman and Hall, New York, 1994; (d) D. Klempner and K. C. Frisch, *Handbook of Polymer Foams and Foam Technology*, Hanser, Munich, 1991.

12. (a) Z. W. Wicks, F. N. Jones, and S. P. Pappas, *Organic Coatings: Science and Technology*, Vol. I, 2nd ed., Wiley, New York, 1999; (b) E. P. Squiller et al., "High Solids Polyurethane Coatings," in *Polymeric Materials Encyclopedia*, Vol. 5, J. C. Salamone, (Ed.), CRC Press, Boca Raton, FL, 1996, p. 3032; (c) P. B. Jacobs and P. C. Yu, *J. Coatings Technol.*, **65**, (822), (July 1993), p. 45; (d) T. A. Potter and J. L. Williams, *J. Coatings Technol.*, **59**, (749), (June, 1987); (e) H. G. Schmelzer, *Polyurethanes World Congress 1987: 50 Years of Polyurethanes, Proceedings of the FSK/SPI*, Lancaster, PA, Proceedings Technomic, 1987, p. 614 (f) J. W. Rosthauser and K. Nachtkamp, "Waterborne Polyurethanes," in *Advances in Urethane Science and Technology*, Vol. 10, K. C. Frisch and D. Klempner (Eds.), 1987, p. 121.

13. (a) I. Skeist, *Handbook of Adhesives*, 3rd ed., Van Nostrand, NY, 1990; (b) S. R. Hartshorn (Ed.), *Structural Adhesives: Chemistry and Technology*, Plenum, New York, 1986; (c) R. M. Evans, *Polyurethane Sealants: Technology and Applications*, Technomic, Lancaster, PA, 1993; (d) C. Hepburn, *Polyurethane Elastomers*, 2nd ed., Elsevier Applied Science, New York, 1992.

14. P. Wright and A. P. Cumming, *Solid Polyurethane Elastomers*, MacLaren: London, 1969.

15. G. Holden, N. R. Legge, R. P. Quirk, and H. E. Schroeder (Eds.), *Thermoplastic Elastomers*, 2nd ed., Hanser Gardner, Cincinnati, OH, 1996.

16. (a) D. Ho et al., PCT Int. Patent WO 97/44374 (to Arco Chemical), 1997; (b) U. E. Younes et al., PCT Int. Patent WO 0002940 (to Arco Chemical), 2000; (c) C. Macosko, *Fundamentals of Reaction Injection Molding*, Hanser Publishers, Munich, 1989; (d) S. H. Metzger, "Microcellular Elastomers," *Reaction Injection Molding*, W. E. Becker (Ed.), van Nostrand-Reinhold, New York, 1979, Chapter 2.

17. (a) J. C. Arthur, (Ed.), *Polymers for Fibers and Elastomers, ACS Symposium Series No. 260*, American Chemical Society, Washington, DC, 1984; (b) K. H. Wolf, M. Kausch, and H. Schroer, "Polyurethane Fibers," in *Ullmann's Encyclopedia of Industrial Chemistry*, M. Bohnet, C. J. Brinker, B. Cornils, T. J. Evans, H. Creim, L. L. Hegedus, J. Heitbaum, W. A. Herrmann, W. Keim, A. Kleemann, G. Kreysa, T. Laird, J. Loliger, R. O. McClellan, J. L. McGuire, J. W. Mitchell, A. Mitsutani, T. Onoda, L. Plass, G. Stephanopoulos, D. Werner, P. Woditsch, N. Yoda (Editorial Board). 1. Polymers and Plastics, Fibers, 2. Synthetic Organic, 3. Polyurethane Fibers, 6th ed., Wiley Interscience, NY, Lancaster, PA, 2001 Electronic Release.

18. (a) V. Kudela, in *Encyclopedia of Polymer Science and Engineering*, Vol. 7, 2nd ed., Wiley, New York, 1986, p. 703; (b) *Urethanes Technology*, December 1999/January 2000, Vol. **16**, No. 6, Crain Communications, London; (c) A. J. Quarfoot, et al., U.S. Patent 4,909,244, (to Kendall Co.), 1990.

19. (a) F. W. Lichtenberg (Ed.), *CFCs and the Polyurethane Industry*, Technomic, Lancaster, PA 1991; (b) *Urethanes Technology*, June/July 2000, Vol. 17, No. 3, p. 22, Crain Communications, London.

20. (a) *Chemical Market Reporter*, Vol. 258, No. 16, October 16, 2000, p. FR−11; (b) *Adhesives Age*, Vol. 43, No. 5, May 1, 2000, p. 15, (c) *Chemical Market Reporter*, October 12, 1998, p. FR−22.

21. (a) N. Lamba, K. Woodhouse, and S. Cooper, *Polyurethanes in Biomedical Applications*, CRC Press, Boca Raton, FL, 1998; (b) L. Pinchuk, J. *Biomater. Sci., Polym. Ed.*, **6**(3), 225–267 (1994).

22. (a) "Recycling of Polyurethanes," K. C. Frisch, D. Klempner, and G. Prentice, Eds., *Advances in Plastics Recycling Series*, Vol. 1, Technomic, Lancaster, PA, 1999; (b) D. Reed, "Recycling Polyurethanes," in *Urethanes Technology*, December 1998/January 1999, Vol. 15, No. 6, p. 26, Crain Communications, London, and other articles therein; London (c) W. F. Gum, in *Proceedings of the SPI-33rd Annual Polyurethane Technical/Marketing Conference*, Technomic, Lancaster, PA, 1990, p. 26.

23. (a) S. Ozaki, *Chem. Rev.*, **72**(5), p. 457 (1972); (b) H. Ulrich, *Chemistry and Technology of Isocyanates*, Wiley Ltd., Chichester, England, 1996.

24. (a) N. Barksby et al., "ACCLAIM Polyether Polyols for Cast Elastomers," paper presented at the Fall PMA Meeting, Pittsburgh, PA, 1995; (b) N. Barksby and G. L. Allen, "Low Monol Polyols and Their Effects in Urethane Systems," *Polyurethanes World Congress Proceedings*, CRC Press, Bocaraton, FL, 1993, pp. 445–50; (c) S. D. Seneker et al., in *Advances in Urethane Science and Technology*, D. Klempner and K. C. Frisch (Eds.), Rapra Technology, Shropshire, United Kingdom, 2001.

25. (a) R. L. Adkins and W. E. Slack, *Polym. Mater. Sci. Eng.*, **77**, 542 (1997); (b) Dobson, S. Froom, European Patent Application EP414425 (19,910,227); 1991; (c) H. G. Schmelzer et al., *J. Prakt. Chem./Chem.-Ztg.*, **336**(6), 483 (1994).

26. (a) P. H. Markusch, R. S. Pantone, R. G. Guether, W. E. Slack, U.S. Patent 6,242,556 (to Bayer Corp.), 2001; (b) J. Dormish, C. J. Lau, C. Kinney, and W. E. Slack, *Adhesives Age*, **43**(4), 33 (2000).

27. (a) W. E. Slack and R. L. Adkins, U.S. Patent 5,212,275 (to Miles) 1993; (b) W. E. Slack, U.S. Patent 4885353, (to Mobay) 1989; (c) H. Woynar, K. Konig, D. Odenthal, J. Pedain, German Offen. DE3700209 (to Bayer AG), 1988.

28. (a) R. Becker and L. Thiele, "Polyurethane Catalysis," in *Polymeric Materials Encyclopedia*, Vol. 9, J. C. Salamone (Ed.), CRC Press, Boca Raton, FL, 1996, p. 6940; (b) M. S. Vratsanos, "Polyurethane Catalysts," in *Polymeric Materials Encyclopedia*, Vol. 9, J. C. Salamone (Ed.), CRC Press, Bacon Raton, FL, 1996, p. 6947; (c) S. Hashimoto, *Int. Prog. Urethanes*, **3**, 43 (1981); (d) G. Burkhart et al., *J. Cell. Plast.*, **20**(1), 37 (1984); (e) *A Guide to Polyurethane Catalyst Choice*, Polyurethane Additives Technical Bulletin, Air Products Co. Pub. No. 140–9344, Allentown, PA, 1993; (f) L. Thiele and R. Becker, *Adv. Urethane Sci. Technol.*, **12**, 59 (1993); (g) J. W. Britain and P. G. Gemeinhardt, *J. Appl. Polym. Sci.*, **4**(11), 207–211 (1960); (h) S. L. Reegen and K. C. Frisch, *Adv. Urethane Sci. Technol.*, **1**, 1 (1971); (i) K. C. Frisch and L. P. Rumao, *J. Macromol. Sci. Rev. Macromol. Chem.*, **C5**(1), 103 (1970); (j) K. Wongkamolsesh and J. E. Kresta, *Polym. Mater. Sci. Eng.*, **49**, 1983.

29. (a) N. Malwitz and J. E. Kresta, in *Proceedings of the SPI/FSK Polyurethanes World Congress*, Technomic, Lancaster, PA, 1987, p. 826; (b) M. L. Listemann et al., in *Proceedings of the SPI/FSK Polyurethanes World Congress*, Technomic, Lancaster, PA, 1993, p. 595; (c) N. Malwitz, et al., in *Proceedings of the SPI-30th Annual Polyurethane Technical/Marketing Conference*, Technomic, Lancaster, PA, 1986, p. 338; (d) J. Burkus, *J. Org. Chem.*, **26**, 779 (1961).

30. (a) R. P. Houghton and A. W. Mulvaney, *J. Organometall. Chem.*, **518**, 21 (1996); (b) F. W. Van Der Weij, *J. Polym. Sci., Poly. Chem. Ed.*, **19**, 3063 (1981); (c) K. Wongkamolsesh, Ph.D. Thesis, University of Detroit, 1985; (d) K. Wongkamolsesh and J. E. Kresta, in *Reaction Injection Molding*, J. E. Kresta, Ed. ACS Symposium Series 270, American Chemical Society, Washington, DC, pp. 111–121, 1985.

31. I. S. Bechara, "The Mechanism of Tin-Amine Synergism in the Catalysis of Isocyanate Reaction with Alcohols," in *Urethane Chemistry and Applications*, ACS Symposium Series 172, K. N. Edwards, (Ed.), American Chemical Society, Washington, DC, 1981.

32. (a) K. C. Frisch, Jr., and P. T. Engen, "Novel Delayed Action Catalyst/Co-Catalyst System for C.A.S.E Applications," in *60 Years of Polyurethanes, International Symposium and Exhibition*, January 15–16, 1998, Technomic, Lancaster, PA; (b) K. Diblitz and C. Diblitz, in *Polyurethanes World Congress 1993*, October 10–13, 1993, p. 619; (c) A. R. Leckart and L. S. Slovin, Lancaster, PA, *Proceedings of the Polyurethanes World Congress 1987*; Vancouver, B.C., Canada, Oct. 10–13, 1993, Technomic Lancaster, PA, 1987, p. 351.

33. (a) J. W. Rosthauser et al., U.S. Patent 6,020,283 (to Bayer) 2000; (b)J. W. Rosthauser et al., U.S. Patent 6,140,381 (to Bayer), 2000.

34. S. P. Bitler et al., *Proceedings of the Polyurethanes World Congress*, The Netherlands, Sep. 29–Oct. 1, 1997. Technomic, Lancaster, PA, 1997, p. 338.

35. (a) A. C. Savoca and M. L. Listemann, U.S. Patent 5,212,306 (to Air Products and Chemicals), 1993; (b) L. A. Grier et al., U.S. Patent 5,491,174 (to Dow Chemical), 1996.

36. J. Florio, Paint & Coatings Industry, October 1997, **13**/ No. 10, p. 110.

37. (a) E. P. Squiller and J. Rosthauser, *Modern Paint and Coatings*, June 1987, p. 28; (b) E. P. Squiller and J. W. Rosthauser, *Polym. Mater. Sci. Eng.*, **55**, 640 (1987); (c) E. P. Squiller and J. W. Rosthauser, in *Proc. 14th Waterborne Higher Solids Coat. Symp.*, 1987, p. 460. (d) S. D. Seneker and T. A. Potter, *Coatings Techn.*, **63**, (793) 19 (1991).

38. (a) K. W. Haider, W. E. Slack, R. L. Adkins, J. W. Rosthauser, P. H. Markusch, U.S. Patent 5,510,445, (to Bayer), 1996; (b) D. J. Primeaux, K. C. Anglin, in *Proc. 34th Annu. Tech./Mark. Conf. (Polyurethanes 92)*, 1992, p. 598; (c) D. Primeaux, et al., *J. Elastomers Plast.*, **24**(4), 323 (1992); (d) D. D. Steppan, R. M. Mafoti, and W. E. Slack, Canadian Patent Application 2,084,826, Filed Dec. 8, 1992.

39. (a) D. A. Wicks and Z. W. Wicks, Jr., *Prog. Org. Coat.*, **36**, 148 (1999); (b) D. A. Wicks and Z. W. Wicks, *Prog. Org. Coat.*, **41**(1–3), 1–83 (2001); (c) F. Schmitt et al., *Proc. Int. Conference Organic Coatings: Proceedings of the 23rd Annual Waterborne, High Solids, Powder Coatings Symposium*, Athens, Greece, July 8, 1997, "Institute of Materials Science," New Paltz, NY, 1997 pp. 465–484; (d) P. Mischke, *Eur. Coat. J.*, (3), **3**, 81 (2000).

40. (a) *Test Methods for Polyurethane Raw Materials*, 3rd ed., Society of the Plastics Industry, Polyurethane Division, New York, 1996; (b) *Testing Standards for Polyurethane Products*, 5th ed., ASTM, Conshohocken, PA, 1996.

41. D. J. David and H. B. Staley, *Analytical Chemistry of the Polyurethanes*, Krieger, Huntington, NY, 1979.

42. (a) R. Brown (Ed.), *Handbook of Polymer Testing*, Marcel Dekker, New York, NY 1999; (b) V. Shah, *Handbook of Plastics Testing Technology*, 2nd ed., Wiley, New York, 1998; (c) S. B. Driscoll (Ed.), *The Basics of Testing Plastics: Mechanical Properties, Flame Exposure and General Guidelines*, ASTM Manual Series MNL35, ASTM, Conshohocken, PA., 1998.

43. (a) J. T. Mehl, R. Murgasova, X. Dong, D. M. Hercules, and H. Nefzger, *Anal. Chem.*, **72**(11), 2490 (2000); (b) D. M. Hercules et al., *Polym. Prepr. (Am. Chem. Soc., Div. Polym. Chem.)* **41**(1), 639 (2000); (c) U. Liman et al., German Offen. DE 4217524, (to Bayer AG), 1993; (d) J. K. Haken, *Prog. Org. Coat.*, **21**(2–3), 111 (1992); (e) K. Watanabe and T. Miyake, *Toyoda Gosei Giho*, **37**(1), 8 (1995).

44. E. A. Turi, *Thermal Characterization of Polymeric Materials*, Academic, New York, 1997, Chapter 5,

45. (a) S. M. Clift, *J. Elast. Plast.*, **23**, 66 (1991); (b) S. M. Clift, *Elastomerics*, March 1992, p. 16.

46. J. Troitzsch, *International Plastics Flammability Handbook: Principles, Regulations, Testing and Approval*, 2nd ed., Hanser, New York, 1990.

5 Polyimides and Other High-Temperature Polymers

B. Sillion
SCA, 69390 Vernaison, France

R. Mercier and D. Picq
LMOPS, 69390 Vernaison, France

5.1 INTRODUCTION

5.1.1 Historical Perspectives

The heterocyclic and aromatic polymers were introduced during the 1960s in order to match the thermal stability requirements of the aerospace industry. Aromatic and heterocyclic rings offer conjugated rigid structures with high glass transition and strong linkages allowing good resistance in harsh environments.

Two approaches were used for the development of these new polymers. An important effort involving new synthetic methods as well as specific solvents[1] was the key for the preparation of aromatic polyamides (Nomex and Kevlar for Du Pont) or aromatic polyester[2] exhibiting respectively lyotropic or thermotropic properties. At the end of the 1960s the relationship between structure and properties were well established, and different aromatic polyamides, polyesters, polysulfones, and polyetherketones, crystalline or amorphous, were commercially available. The concept for the second approach was more original: Basically the formation of the chain backbone was obtained by a reaction leading to the formation of a heterocycle according to the pioneering work of C. S. Marvel describing the polybenzimidazoles.[3] During the 1960s this polyheterocyclization concept was extremely used, and many polyheterocycles were published.[4,5] However, except for polyhydantoines, polyquinoleines, and polyquinoxalines, the properties and prices of these new materials did not adequately compete with the polyimides to justify commercialization.

The high-temperature glassy or crystalline transitions of the linear aromatic and heterocyclic polymers were an important drawback for the structural aerospace applications, which need a good flow for the adhesive or composite formulations.

Synthetic Methods in Step-Growth Polymers. Edited by Martin E. Rogers and Timothy E. Long
© 2003 John Wiley & Sons, Inc. ISBN: 0-471-38769-X

It is well known that by reducing the molecular weight of a polymer its T_g decreases and its solubility increases. At the end of the 1960s a new concept was introduced for the preparation of thermostable adhesives and matrices for composites. According to this concept, a low-molecular-weight oligomer (1000 < $\overline{M}_n$ < 5000 Daltons) was end capped by a maleimide or a nadimide group. These materials melt or flow below 200°C and the reactive groups lead to a crosslinked material above 300°C allowing interesting processing window (temperature interval between the flow temperature, glass temperature, and the polymerization exotherm). Although about 20 different end-capping groups have been published right now, only the maleimide and the nadimide end-capped oligomers are used in aerospace industries for the preparation of composite with carbon fibers.[6,7] Since the 1980s, the research is oriented toward the synthesis of tailored monomers or polymers for specific applications other than the ones in the aerospace domains. The thermal properties of an organic polymer as well as composites are governed by some thermophysical properties: specific heat (C_p), thermal conductivity (k), and thermal diffusivity (α), which are correlated according to the equation

$$\alpha = \frac{k}{\rho C_p}$$

where ρ is the specific gravity of the material.

Concerning the mechanical properties at high temperature, the glass transition temperature T_g and the crystalline transition temperature T_m are the relevant basic properties which indicate the limit for high-temperature service but also give information concerning the physical aging the higher T_g is, the better is the thermal aging.[8] The glass transition temperature (T_g) of a polymer has been correlated to the concept of free volume so that the relationship between T_g and the molecular mass M is expressed by

$$T_g = T_g^\infty - \frac{K}{M}$$

where T_g^∞ is the glass transition temperature for an infinite molecular weight, K is a constant, and M is the molecular weight. This relation is understandable if one considers that the fractional free volume due to the end chains decreases when the molecular weight increases. Considering that the possibility of motion is greater for the end chain compared with an internal part of the chain, a low-molecular-weight polymer will have a T_g lower than a corresponding high-molecular-weight polymer. The constant K is the contribution of the end groups:

$$K = \frac{2\rho N_A \theta}{\alpha_f}$$

where ρ is the polymer density, N_A is Avogradro's number, θ is the contribution of one chain end to the free volume, and α_f is the thermal expansion coefficient of the free volume. Some long branches on the polymer backbone will decrease the glass transition temperature by increasing the number of chain ends, while regular short branching along the chain increases both the stiffness and the glass transition temperature and decreases the mobility. Crosslinking reduces the specific volume so that the T_g rises as a consequence of the reduction of molecular motion.[9] For homocyclic and heterocyclic aromatic polymers, the structure–property relationships have been reviewed mainly in the polyimides series and some general features have been pointed out. In the case of aromatic substitution, meta versus para catenation decreases the T_g, and flexible bridging groups such as oxygen or methylene decrease the T_g more than highly dipolar groups such as carbonyl and sulfone. Bulky substituents or substituents restricting the free rotation between two phenyl rings also increase the T_g[10,11]. Aromatic and heterocyclic high-performance polymers are very often amorphous materials, and the variations of the properties are governed by glass transition temperature. The T_g of the semirigid aromatic and heterocyclic polymers can be predicted by calculation of a rigidity index for each group.[12]

Chemistry, reaction mechanisms, and properties have been extensively reviewed.[4,5,10–20] The present chapter deals with only one type of fully cyclized aromatic heterocyclic polymers: the high-molecular-weight linear polymer with a special emphasis on the synthesis and structure — property relationships for specific applications.

5.1.2 Trends in Polyimide Applications: Critical Issues

5.1.2.1 Aerospace

After the reduction of the military budget, the announcement that Boeing has postponed from 2010 to 2020 the launch of high-speed civil transport (HSCT), and the fact that no European program for HSCT is expected, the domain is gloomy and the industrial development of new high-temperature resins is questionable. However, recent improvements have been done in the field of thermosetting resins. Chemistry and properties of bismaleimide and bisnadimide are now well understood,[7] and the new phenyl-ethynyl-terminated resins are very promising, considering processability window and thermo-oxidative aging.[21,22] The subsonic airplane does not need such high-performance resins, but some technical requirements for structural foam (better mechanical behavior at very low temperature) are identified and could be solved with foam obtained with linear aromatic and heterocyclic polymers. Expanding processes for linear polymer foams are under investigations; for example, the condensation of polyarylether ketone with amino-terminated polyetherimide takes place during extrusion and the water evolution during the process is the blowing agent.[23]

5.1.2.2 Other Structural Applications

The automotive industry needs more and more organic material with improved thermomechanical properties up to 200–220°C, but the requirements for the processability and cost make the problem difficult to solve. The development of new products with moderate thermo-oxidative properties could probably be obtained by increasing the T_g of aliphatic condensation polymer with aromatic moieties. One possible way would be a random copolycondensation by reactive processing. Aliphatic aromatic copolyimides would be interesting candidates and a different synthetic way could be used: for example the copolymerization of one aliphatic diaminetetracarboxylic salt with an aromatic diaminetetracarboxylic salt.[24,25] On the other hand, the bulk transimidization is an interesting pathway to get copolyimides with tailored properties. The transimidization can take place by reaction of an aromatic imide with an aliphatic diamine. As showed in Fig. 5.1, the equilibrium between aliphatic and aromatic imide at 200°C is favorable to the formation of aliphatic imide.[26] However, good leaving groups such as N-ethoxycarbonyl imide or 2-pyridylimide allow a substitution with an aromatic diamine (Fig. 5.2). Recently a new bulk reaction of an aromatic dinitrile with a secondary aliphatic amide leading to aromatic aliphatic secondary amide with high yield has been

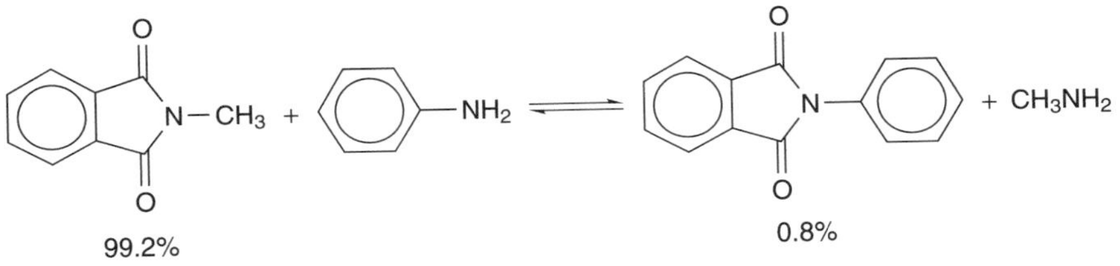

Figure 5.1 Aliphatic aromatic imide equilibrium at 200°C.

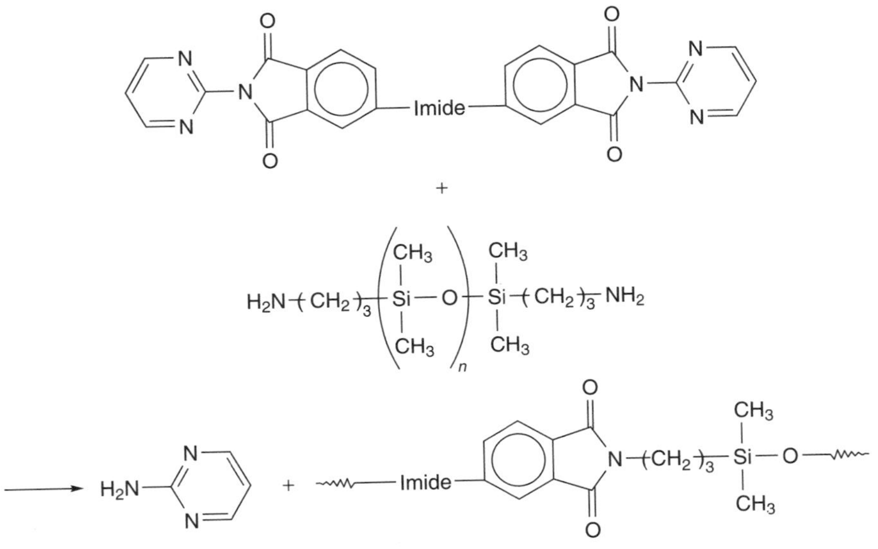

Figure 5.2 Polyimide syntheses by imide exchange with 2-aminopyrimidine as leaving group.

reported[27]; the results obtained with models are promising for an extension to the aliphatic polyamides.

5.1.2.3 Molecular Composites

The para catenation of the aromatic and heterocyclic structure leads to highly rigid rodlike macromolecules. When these macromolecules are soluble, they show a lyotropic behavior in solution (liquid crystal organization as a function of the concentration and orientation in the direction of the shear stress).[28] By spinning the solution, fibers with very high tenacity and high modulus are obtained. These polymers dispersed at the molecular level in a matrix greatly improve the mechanical properties.[28] Recently a molecular composite was obtained by mixing a rodlike precursor (obtained by reaction of biphenyl-tetracarboxylic dianhydride with *para*-phenylenediamine) with a crosslinkable acetylenic polyimide.[29] If a rodlike polyimide (or other heterocyclic polymer) could be dissolved at the molecular level without phase separation in a commercial matrix, the concept of molecular composite would find an important development.

5.1.2.4 Electrical Industries

There is an increasing demand for the insulation of wires working above 180°C. Nowadays three heterocyclic polymers [polyesterimide containing trishydroxyethyl isocyanurate (THEIC) in cresolic solution, polyimide in NMP, and polyhydantoines in cresol] are used.[30] For economical and ecological reasons, the trend is to replace the high-temperature technology for the solvent removal by a photochemical process. But as the most popular crosslinking agents are aliphatic (acrylic esters), the thermal stability will be decreased. Right now, the literature does not give any information about a breakthrough on the subject.

5.1.2.5 Electronic and Electro-Optic Applications

Since the end of the 1970s, the polyimides have been introduced for the production of electronic components mainly for the passivation. But more and more they are interesting for the integrated circuits and multichip modulus fabrications. Processability and dielectric and thermomechanical properties are the most attractive features of these materials for the electronic[31] and electro-optical applications.[32]

5.1.2.5.1 Alignment Coating for Liquid Crystal Devices (LCDs)

The top and bottom electrodes of the LCDs are covered by a blocking layer (protective coating) and by an alignment layer which plays two roles: the first one is the orientation of the liquid crystal (same orientation of the dipole) and the second is to create a small angle from the surface, when the liquid crystals are not addressed. The polyimides are deposited in thin coatings which are mechanically rubbed. The rubbing ensures the chain orientation of the polyimide responsible for the LCD organization. As the mechanism of the alignment coating is not yet well understood, much work is devoted to these phenomena either by physicians or by chemists:

- Distortions of the bond and out-of-plan rotation of polyimides have been suggested to explain the surface modification by rubbing.[33]
- The orientation of the polyimide coating by treatment with linearly polarized light is under investigation.[34]
- The synthesis of polyimides with the pending group (aliphatic or fluorinated substituent) and some relationships between the polymeric structure and the value of the tilt angle have been reported.[35]
- Improvement of the viewing angle by using rodlike polyimides — the in-plane and out-of-plane refractive indexes are different and the resulting negative birefringence increases the viewing angle.[36]

5.1.2.5.2   Dielectrics for Integrated Circuit and Multichip Modules

LOW DIELECTRIC CONSTANT   A very low dielectric constant cannot be obtained with mineral dielectrics. For different reasons, such as processing, mechanical properties, and adhesion on the substrate, polyimides are now accepted as dielectrics. The introduction of fluorine in polyimide structure reduces the dielectric constant mainly by modifying the free volume but also by changing the electronic mode of polarization.[37] However, the lowest dielectric constant obtained with fluorinated polyimide is about 2.5,[38] and in order to reduce the dielectric constant below 2, a new concept has been recently developed. As the air dielectric constant is 1, the concept is based on the generation of nanovoids in the polyimides. The synthesis of a block copolymer containing a high-T_g aromatic or heterocyclic segment such as polyquinoxaline or polyimide and a thermolabile segment (such as polypropyleneoxide, polymethylmethacrylate, and polystyrene) leads to a heterophased system after deposition on the support and solvent evaporation. By heating, the thermolabile block decomposes, and the decomposition products diffuse through the thermostable segments with formation of a nanofoam.[39] To be successful, the process needs an important difference between the decomposition temperature of the labile part and the T_g of the rigid segment in order to avoid coalescence. The synthesis of rigid polyimides containing a thermolabile function is another approach. tert-Butyl carbonate is easily introduced in a polymer containing a hydroxyl group by reaction of tert-butyl dicarbonate. The carbonate function decomposes at 180°C with CO_2 and isobutylene evolution.[40] The void generation allows the formation of dielectric films with dielectric constant ranging between 1 and 2 depending on the ratio of air to polymer.

PHOTOSENSITIVE POLYIMIDES   During the process in microelectronic, the dielectric has to be patterned in order to obtain the vias needed for the connection. The patterning is performed with photosensitive resins and masks. The process is time and money consuming, and the use of photosensitive polyimides could save three (wet etching) or five (dry etching) steps in the process.[31] The development of photosensitive polyimide is an important industrial challenge, and different chemical systems have been patented or published,[41] but some problems such as photosensitivity, contrast, control of the thickness, and dielectric constant have not yet definitely been solved.

5.1.2.6   Some Other Opto-Electronic Targets

5.1.2.6.1   New Linear Optical Devices

Some highly polarizable organic chromophores (containing an electron donor, a electron connective segment, and an electron acceptor group) organized in pure noncentrosymmetric lattices could exhibit very high electro-optic coefficients [100–1000 picometer/volt (pm/V) at 1.3 μm wavelength].[42] Does the introduction of such chromophores in a polymeric system compete with the properties offered by a pure lattice? The problem is to get and keep a cooperative effect of the chromophores linked to or dispersed in the matrix by orientation (poling). The chromophores can be oriented by photoinduced orientation (polarized light at room temperature) or by photo-assisted poling (polarized light plus electrical field) or thermal poling (electrical field at T_g).[43] Contradictory properties are needed for the thermosetting matrix or linear backbone: a flexible material to get a good orientation during the poling and a rigid system to maintain the orientation in a wide temperature range. On the other hand, the subglass transition allowing small molecular movements could also be responsible for partial relaxation of the orientation, and the flexibility of the connecting group between the chromophore and the backbone has to be considered. Interesting but not definitive results for the second-harmonic generation were obtained with chromophores linked to linear high-T_g polyimide or thermosetting like maleimide resins.[42,44]

5.1.2.6.2   Electroluminescent Devices

An electroluminescent device is based on two materials: an electron-transporting material and a hole-transporting material under an electrical field. The recombination of hole and electron gives the electroluminescent signal. Tailored aromatic diamines like 4,4''''-diaminoquinquephenyl[45a] or heterocyclic diamines like 2,7-(8)diaminothiantrene[45b] have been used to prepare polyimides which are good hole-transporting layers for blue-emitting layers in light-emitting diode (LED) devices. Other heterocyclic polymers such as polyquinoxalines[45c] exhibit electron-transporting properties.[45d] Recently a toluidine-based polyimide has been described as a hole-transporting material and gives an electroluminescent system (green 521–525 nm) with aluminum-8-hydroxyquinolate.[45e]

5.1.2.6.3   Optical Fiber Waveguide

The transport of an optical signal instead of an electrical one offers many advantages: no crosstalk, no electromagnetic interference, and the possibility of transporting more than one optical signal on the same guide. For opto-electronic applications, thermal stability of the waveguide is needed to withstand the processing temperature, and polyimides have been investigated. Fluorinated polyimides, which are more soluble than the nonfluorinated materials, can be processed as fully cyclized solution and give colorless films without flaws. As a consequence, the optical loss is low.[32] A recent example of fluorinated polyimide[46] with low optical loss (0.3 db/cm) and good thermal stability (1 h at 380°C or 200 h at 85°C) in humid atmosphere (80% H_2O) shows the interest of these materials.

5.1.2.7 Membrane Technologies

Gas separations by distillation are energy-consuming processes. The driving force for the gas permeation is only the pressure difference between two compartments separated by the membrane. The permeation is governed by two parameters — diffusion and solubility:

$$P = D \times S$$

When two gases A and B compete to cross a membrane, the selectivity is the ratio of permeability of A over the permeability of B:

$$\alpha = \frac{P_A}{P_B} = \frac{D_A}{D_B} \times \frac{S_A}{S_B}$$

It has been shown that the diffusion coefficient is the most important factor, and for two gases with different van der Waals volume, the selectivity is better when the membrane is a thermoplastic than when the membrane is an elastomer. Polyimides which have very high T_g has been extensively studied for O_2-N_2 separation but also for hydrocarbon–CO_2 or water separation and H_2–hydrocarbons.[47] The relationship between structure and property is always under investigation, and probably it would be interesting to pay more attention to film preparation and the resulting morphologies. The most important challenge for industrial development is to get high permeability with high selectivity. For that, the control of free volume by clever substitution of the backbone is probably a promising approach.

5.1.2.8 Proton Exchange Membrane Fuel Cells (PEMFCs)

A fuel cell is an electrochemical reactor with an anodic compartment for the fuel oxidation giving a proton and a cathodic compartment for the reaction of the proton with oxygen. Two scientific problems must be solved: finding a low-cost efficient catalyst and finding a membrane for the separation of anodic and cathodic compartments. The membrane is a polyelectrolyte allowing the transfer of hydrated proton but being barrier for the gases.

Fuel cell technology probably offers a new emerging area for polyheterocyclic polymers as membranes. Fuel cells are interesting in transport applications and are now being evaluated in Chicago in transit buses with a 275-hp engine working with three 13 kW Ballard fuel cell stacks.

The car market is demanding, but many questions are still without answer: What kind of fuel should be used: hydrogen? How can H_2 and methanol be stored? Should direct electrochemical conversion be used or reforming, giving H_2? Currently, a Mercedes has a 40-L fuel tank of methanol, allowing 250 miles.

Fuel cells are also used for stationary plants in hotels and hospitals, for example. The production of 5–10 kW residential stacks is expected in the United States.

What progress is needed?

- In the stack, the graphite plates for gas distribution are too expensive, and development of vinyl ester–graphite composites is expected.
- For membrane electrode assembly, the development of a Pt–Ru anode working in the presence of CO is a key point.
- A membrane to replace Nafion should be a goal.

Nafion is a good membrane for a fuel cell working with H_2 but does not work above 80°C or with methanol. An important work on sulfonated aromatic and heterocyclic systems is now involving many teams in North America, Europe, and Japan.[48,49]

5.1.2.9 *Miscellaneous Applications*

Polyimide can be used as host for conductive polymers: After diffusion of pyrrole and its dopant in a polyimide film, pyrrole polymerization is carried out on a metal electrode. The obtained films show an interesting conductivity (between 3.5×10^{-3} and 15 S/cm).[50] Recently, it has been shown that a coating obtained by blending a polyaniline with a polyimide improves corrosion resistance[51a] and a quinone containing polyimide has been prepared for the protection of iron against corrosion.[51b] Mica flakes coated with stannic oxide doped with antimony and titaniumdioxide have been added to a polyamic acid solution. The obtained polyimide film can give a laser receptive bar code label with good high-temperature property.[52] In the medical domain, some new polyimide fibers with good blood compatibility can be used as an intravascular oxygenator.[53]

5.2 STRUCTURE–PROPERTY RELATIONSHIPS

5.2.1 General Considerations

Polyimides and other aromatic heterocyclic polymers were investigated during the 1960s in order to obtain high-temperature resistant materials according to two basic concepts: formation of rigid polymers with low mobility obtained with the chemical structure prepared with the strongest covalent bonds. The consequences were a low solubility, rather high glass and crystalline transitions, and very high viscosity above the transitions. These properties must be considered as severe drawback for the processing of these materials. The structure–property relationships in aromatic and heterocyclic polymers have been reviewed by Harris and Lannier[13] and more recently by Saint Clair,[10] by our group,[14] and by de Abajo et al.,[20] and in this section we will discuss the relationships between structure and properties such as thermal transitions and optical and dielectric properties.

5.2.2 Role of Chemical Structure

5.2.2.1 *Glass Transition and Solubility*

Basically, the first approach to correlate the polyimide chain organization to the monomer structure was to take into consideration the electron affinity of the anhydride and the ionization potential of the diamine,[10] as shown in Fig. 5.3. The strongest interactions between the polymeric chain are expected when the polyimide is prepared with the dianhydride having the highest electron affinity and the diamine with the lowest ionization potential. The strongest interchain interaction leads to high T_g and low solubility.

Recently, Ronova and Pavlova[54] suggested a set of relations between the conformational rigidity and some physical properties of the aromatic polymers. The conformational rigidity is correlated to the Kuhn statistical segment A calculated by the Monte Carlo method:

$$A = \lim_{n \to \infty} \left(\frac{\langle R^2 \rangle}{n l_0} \right)$$

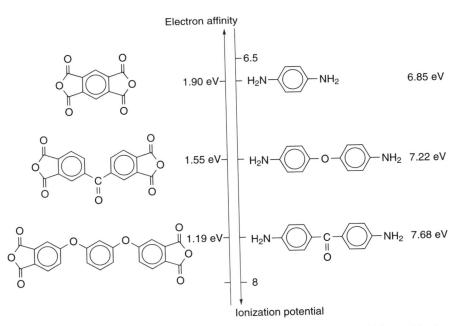

Figure 5.3   Reactivity scale as function of electron affinity for dianhydrides and ionization potential for diamines.

where $\langle R^2 \rangle / n l_0$ is the ratio of the average square end-to-end distance of a chain to its contour length, n is the number of repeating units, and l_0 is the contour length of the repeating unit. An increase in A means that the chain rigidity increases and the relationships between A and the softening temperature, friction coefficient, yield strength, fiber and film moduli, and more intriguingly the fire resistance and luminescent properties of the solutions have been discussed. The chains can interact according to different mechanisms depending on the chemical nature of the repeating units. For example, hydrogen bonding is responsible for the benzimidazole chain packing (Fig. 5.4). Thus, the substitution of the hydrogen atom by a phenyl group leads to a soluble benzimidazole with high T_g,[55] but charge transfers or electronic polarization would explain the polyimide interaction (Fig. 5.4).

Different methods have been studied to reduce the chain interactions or to promote a better interaction between a solvent and the polymeric chain:

1. Reactants with asymmetric structures (meta or ortho instead of para catenation).[10,14]
2. Reactants with flexible links (O, CO, SO_2, CH_2, etc.) which disrupt the conjugation and increase the chain flexibility.[10,14]

Many reactants like diamines and dianhydrides have been synthesized according to these previous concepts[10] and some of them are commercialized (see Section 5.3.3).

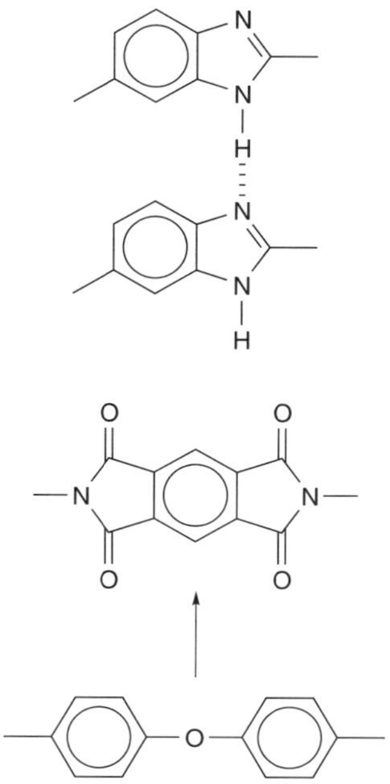

Figure 5.4 Hydrogen bonding between benzimidazole and charge transfer in polyimide.

3. *Introduction of polar groups in the repeating unit*: The replacement of the carbonyl group in the benzophenone tetracarboxylic anhydride by a benzhydrol group strongly enhances the solubility in aprotic polar solvents.[23,24]

4. *Introduction of bulky substituents*: This concept is very interesting because the bulky substituents in the chain increase the solubility by hindering the chain packing but do not affect the glass transition temperature. The phenyl group introduced in the 2- or 3-position on the quinoxaline ring allows the solubility of the polymers in metacresol–xylene mixtures.[58] This observation was probably the starting point for the synthesis of soluble high-T_g heterocyclic polymers which were developed by using cardo systems, *tert*-butyl substituents, methyl in the ortho position of the amino group, methyl or fluoromethyl group in the 2,2'-position of the biphenyl, 2,2'-isopropyl or 2,2'-trifluoroisopropyl groups, and adamantane (Fig. 5.5).

These general effects are summarized in Table 5.1. As can be seen in the table, unsymmetrical structure and flexible links adversely affect the glass transition. However, a combination of flexible and polar substituents can increase solubility without deleterious effect on T_g.

Two polymers have been prepared with the benzophenone tetracarboxylic dianhydride (BTDA) and either bis-1,3-(4-aminophenoxy)-benzene or bis-1,3-(4-aminophenoxy)-2-cyano-benzene (Fig. 5.6). Both polymers are soluble in NMP and their T_g values are respectively 191 and 243°C.[59]

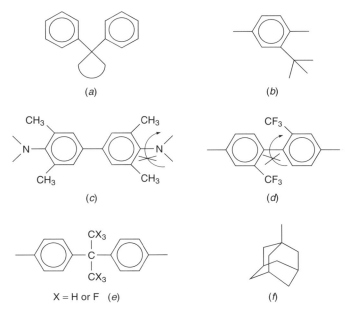

Figure 5.5 Bulky substituents used for preparation of soluble high-T_g polyimides.

TABLE 5.1   Effect of Chemical Structure on Solubility and Glass Transition

Structure Modification	Solubility	T_g	References
o, m versus *p* attachment	↗	↘	10, 14
Flexible links	↗	↘	10, 14
Polar groups	↗	↘	56, 57
Bulky substituents	↗	↗	14

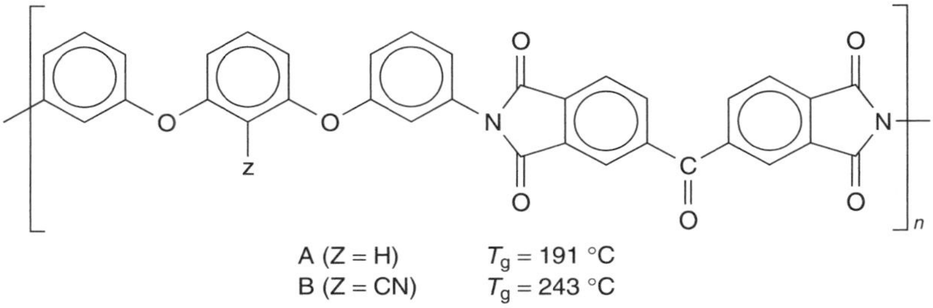

A (Z = H) T_g = 191 °C
B (Z = CN) T_g = 243 °C

Figure 5.6   Effect of polar substituents on glass transition.

5.2.2.2   *Optical Properties of Colorless Polyimides*

As for other polymeric materials, amorphous polyimides exhibit a good transparency. Colorless polyimides result in lowering the electronic conjugation or charge transfer, and it has been mentioned[60] that different fluorinated amines or anhydrides (Fig. 5.7) give polyimides with UV cutoff of about 315 nm. Controlling the refractive index of the colorless polyimides is an important requirement for the optical waveguide. It is performed by copolymerization of fluorinated dianhydride (6FDA) with a fluorinated diamine (2,2′-bis trifluoromethyl-4,4′-diaminobiphenyl, PFMB) and a nonfluorinated one (4,4′-diaminodiphenylether, ODA). The refractive index of the polymer goes up when the molar fraction of the nonfluorinated diamine increases. The possible drawback is a higher optical loss when the proportion of the nonfluorinated amine goes up. It has been shown that the replacement of the ODA by a chlorinated diamine (2,2′-dichloro-4,4′-diaminobiphenyl) can increase the refractive index without optical loss.[61]

Some diamines carrying very bulky substituents like cardo groups can give colorless polyimides. For example, the bis-9,9-(4-aminophenyl)fluorene (FDA) or brominated and acetylenic FDA derivatives react with 6FDA giving copolymer films[62] with low birefringence (low difference between in-plane and out-of-plane refraction index) (Fig. 5.8). A new cardo diamine 1,1-bis[4-(4-aminophenoxy)phenyl]cyclododecane (Fig. 5.8) reacts with different aromatic dianhydrides with formation of colorless polyimides.[63]

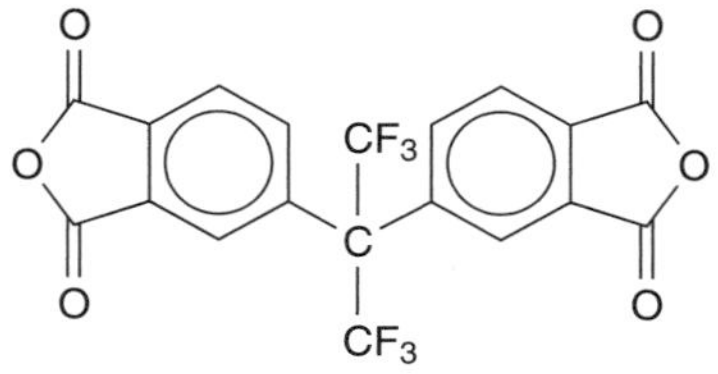

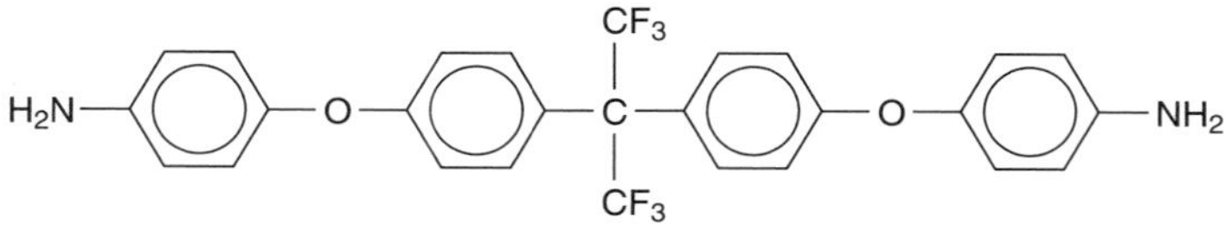

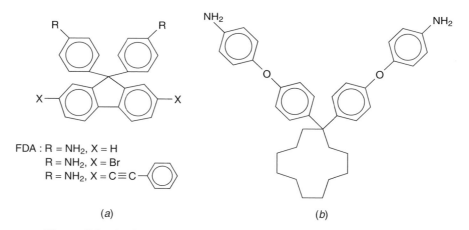

Figure 5.7   Fluorinated monomers used for preparation of colorless polyimides.

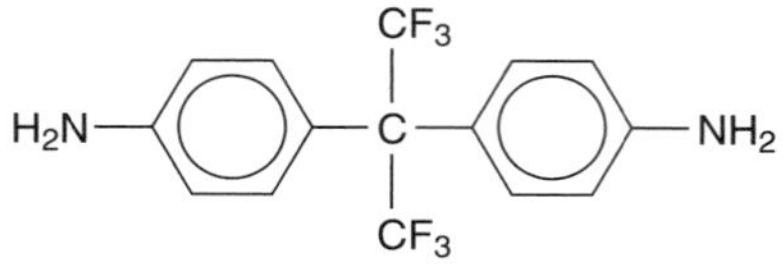

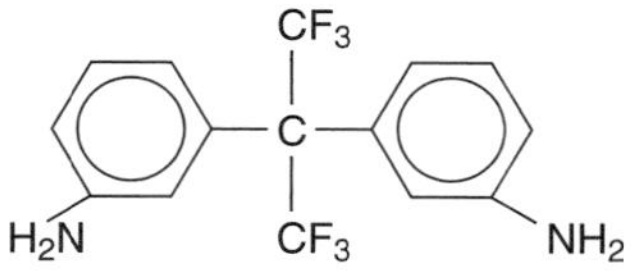

FDA : R = NH₂, X = H
R = NH₂, X = Br
R = NH₂, X = C≡C—

(a)

(b)

Figure 5.8   Cardo diamines used for preparation of colorless polyimides.

In addition to the research on fluorinated and cardo polyimides, an important work was devoted to the semiaromatic cycloaliphatic polyimides. Volksen points out the potential interest of these materials in electronic industry.[64] He reports that the simplest procedure to prepare these materials is to use a cycloaliphatic dianhydride and an aromatic diamine (Fig. 5.9) instead of an aliphatic diamine and an aromatic dianhydride, which leads to formation of gels.

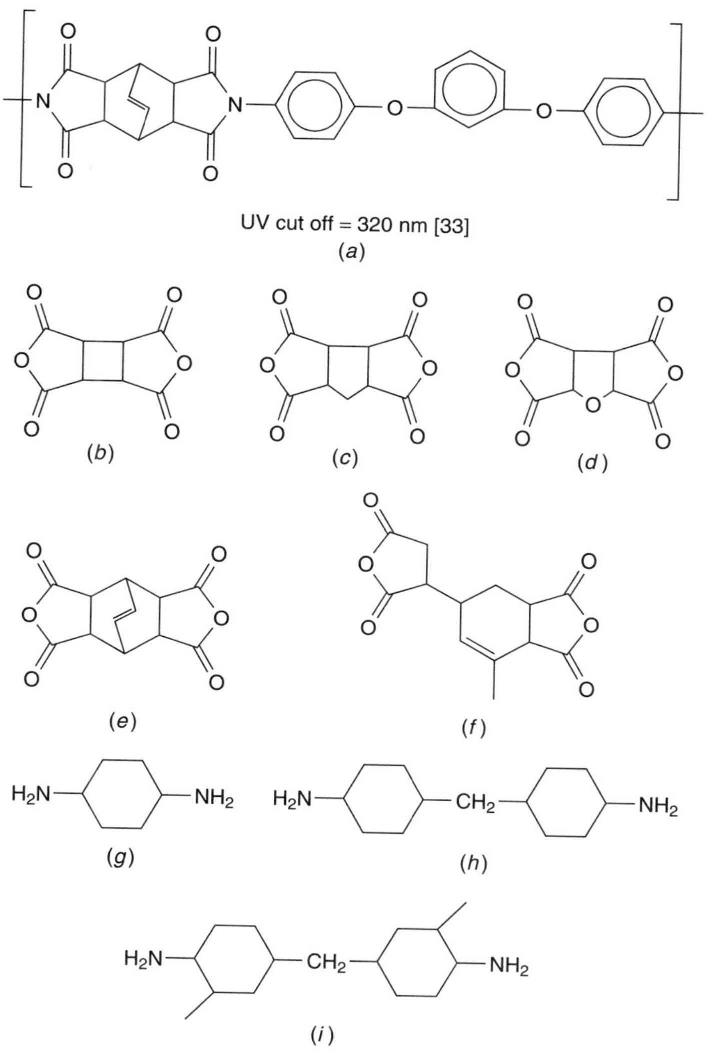

UV cut off = 320 nm [33]

(a)

(b)   (c)   (d)

(e)   (f)

(g)   (h)

(i)

Figure 5.9 (a) Colorless semialiphatic polyimide; (b–f) cycloaliphatic dianhydrides; (g–i) cycloaliphatic diamines.

Different dianhydrides have been synthesized or are commercially available, and some structures are shown in Fig. 5.9.[64–66] An improved method for preparation of cyclobutanetetracarboxylic dianhydride (CBDA) by photochemical dimerization of the maleic anhydride has been developed by Nissan.[67] The polyimide obtained by condensation of CBDA with oxydianiline gives a transparent and colorless material. The transmittance of 50-μm-thick film is 82% and the UV cutoff is 310 nm.

5.2.2.3   Dielectric Properties and Moisture Uptake

The chemical requirements leading to a low dielectric constant (below 3) and low moisture uptake are the same which were discussed in the previous section. Bulky substituents like fluoroalkyl, fluoroalkoxyl,[68,69] or cardo[70] groups allow the dielectric constant to drop to 2.6–2.7. The moisture uptake is also minimized for these polymers.[71] Similar results were observed with cycloaliphatic imides[64] but with a lower thermal stability.

5.2.2.4   Coefficient of Thermal Expansion (CTE)

When a polymeric material is in contact with an inorganic one (as in a composite or a thin dielectric coating in electronics), the behavior of the system depends on the CTE of the two materials when a thermal shock occurs. Figure 5.10 shows the CTE of inorganic and organic materials. The CTE of metals and ceramics lies between 2×10^{-5} and 1×10^{-6} K^{-1}, compared with amorphous polyimides in the range of $5.0–8.0 \times 10^{-5}$ K^{-1}. Any mismatch between organic and inorganic CTEs will create a mechanical stress during thermal shock. More flexible is the polymeric backbone, the higher is the CTE (the structure relationships are well documented[14]). In order to decrease the CTE, dianhydrides and diamines without flexible links and with para catenation have been extensively used. Concerning BPDA, two possible conformations trans and cis have been discussed.[69] Figure 5.11 shows some aromatic diamines with fluoro substituents giving soluble polyimides with low CTE.[72] Some rigid heterocyclic diamines with linear orientation of the diamino groups have been synthesized in order to prepare

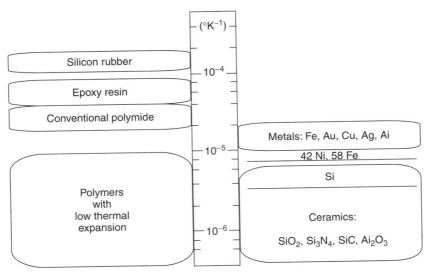

Figure 5.10 Chart showing coefficient of thermal expansion of organic and inorganic materials.

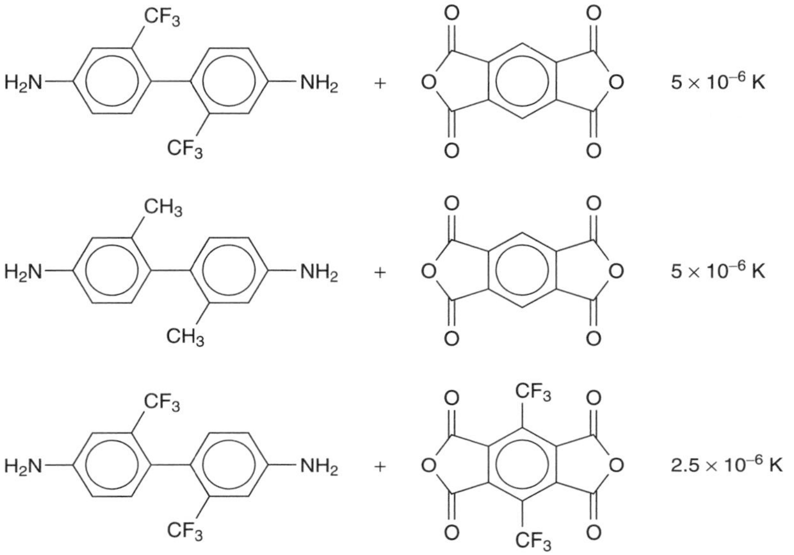

Figure 5.11 Monomers for preparation of rodlike polyimides with low coefficient of thermal expansion.

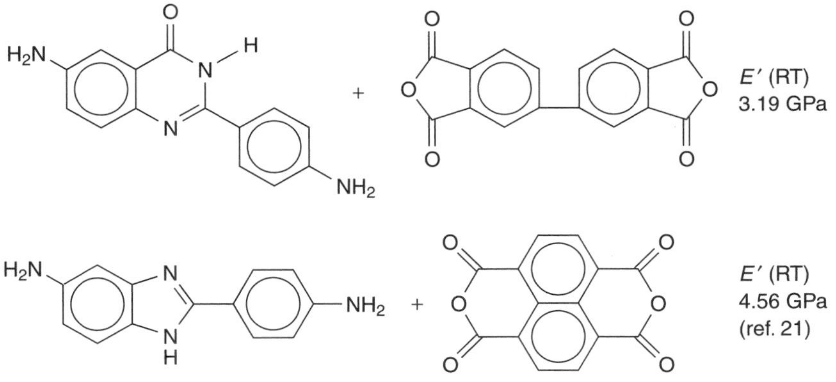

Figure 5.12 Rigid heterocyclic diamines and dianhydrides for preparation of high-moduli polyimides.

rodlike polyimides. 2-(4-Aminophenyl)-6-amino-4-(3H)quinazolone (APQ) and 2-(4-aminophenyl)-6-amino benzimidazole (APB) (Fig. 5.12) were respectively prepared by Yoshida and Hay[73] and Ponomarev et al.[74] The condensation of these diamines with dianhydrides gives rodlike polymers with high moduli. The chemistry and properties of the rodlike aromatic and heterocyclic structures have been reviewed by Arnold and Arnold.[75]

5.2.3 Role of Microstructure and Architecture

5.2.3.1 Aromatic and Heterocyclic Block Copolymers

Many copolymers containing a polyarylether segment and a heterocyclic connection, such as quinoxaline, imidazole, oxadiazole, benzoxazole, or benzimidazole, have been prepared either by nucleophilic displacement on the fluorophenyl heterocycle (the withdrawing effect of the heterocycle group enhances the nucleophilic substitution of the fluorine atom) by an α, ω polyarylether bisphenate (Fig. 5.13) or by the reaction of a dihydroxy heterocycle with an α, ω dihalogeno polyphenylether. The heterocycle introduction increases the T_g and, in some cases, a very high crystalline transition was observed.[76] However, phase segregation with two different glass transitions, which is usual for incompatible blocks or blends, was not observed, probably because the sequences were too short.

Recently an alternate and a block aliphatic and aromatic oxadiazole have been compared. The copolymers were obtained by reaction of dodecanedioic acid, (50%) isophthalic acid (50%), and a hydrazine derivative in phosphorus pentoxide and methanesulfonic acid (Fig. 5.14). When the reaction is carried out with isophthalic dihydrazide, an alternate copolymer is obtained, but by using hydrazine sulfate and the mixture of diacids, the more reactive and more soluble aliphatic diacid first reacts and a block copolymer is obtained. The alternate copolymer exhibits a crystalline melting point at 100°C, close to the melting point of the corresponding aliphatic homopolyoxadiazole. The block copolymer shows a crystalline transition at 50°C due to the melting of the aliphatic block and a glass transition at 240°C due to the isophthalic oxadiazole block. Heated and stretched at 150°C, then cooled down under stress, the block copolymer exhibits a temporary stable shape without stress at room temperature. The initial shape is restored by heating without stress at 150°C. Such a shape memory behavior would probably be observed with other heterocyclic aliphatic block copolymers with a large rubbery domain.[77]

Block and random naphthalenic copolyimides obtained with 4,4'-diaminobiphenyl-2,2'-disulfonic acid (BDSA) and nonsulfonated aromatic diamine (Fig. 5.15) were compared from the point of view of moisture uptake and conductivity. The copolymers were prepared in meta-cresol solution. The block copolymer was obtained in two steps. First was the condensation of BDSA with the naphthalene tetracarboxylic dianhydride (NDTA) in molar ratio BDSA–NDTA calculated for the control of sequence length. Then the nonsulfonated diamine and NTDA were added to the cresolic solution of the diamino telechelic sulfonated oligomer. Moisture uptake and conductivity were higher in the case of the block copolymers, and these phenomena were attributed to a phase separation in aqueous medium between ionic domains and hydrophobic segments. Using small-angle neutron scattering (SANS) experiments, a broad scattering maximum observed in the case of the block copolymer (and not observed with the random one) confirms that hypothesis.[78]

From a mechanistic point of view, the polyimide synthesis via the polyamic acid is not a convenient method of preparing the block copolyimide for two main

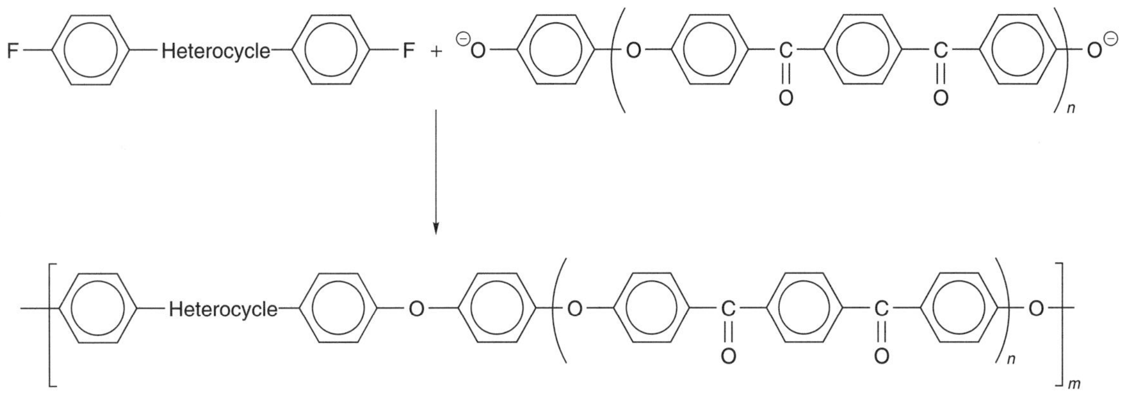

Heterocycle = quinoxaline, imidazole, oxadiazole, benzoxazole, benzimidazole

Figure 5.13 Heterocyclic aromatic block copolymers obtained by aromatic nucleophilic substitution.

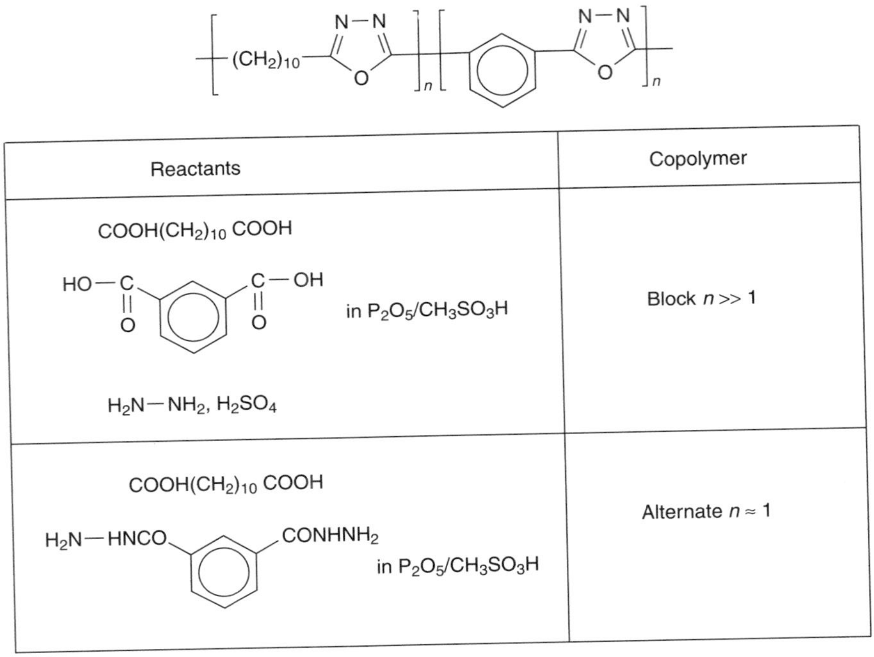

Figure 5.14   Aromatic aliphatic copolyoxadiazoles.

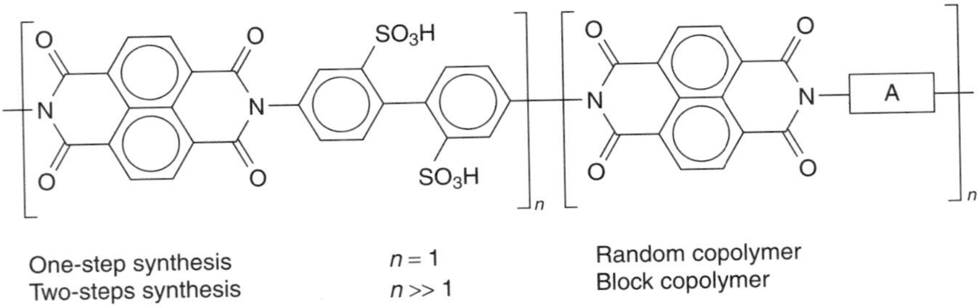

One-step synthesis	$n = 1$	Random copolymer
Two-steps synthesis	$n \gg 1$	Block copolymer

Figure 5.15   Naphthalenic copolyimides with sulfonated block.

reasons: The formation of the amic acid is reversible and the transamidification of the polyamic acid is favored by the neighboring participation of the adjacent carboxylic group.[79]

5.2.3.2 Hyperbranched Structures

In 1941, Flory[80] introduced the concept of hyperbranched macromolecules by polymerization of a trifunctional AB$_2$ monomer (Fig. 5.16). From a chemical point of view, perfectly defined structures were obtained when the monomer reactions were performed step by step: At each step, a polyfunctional molecule

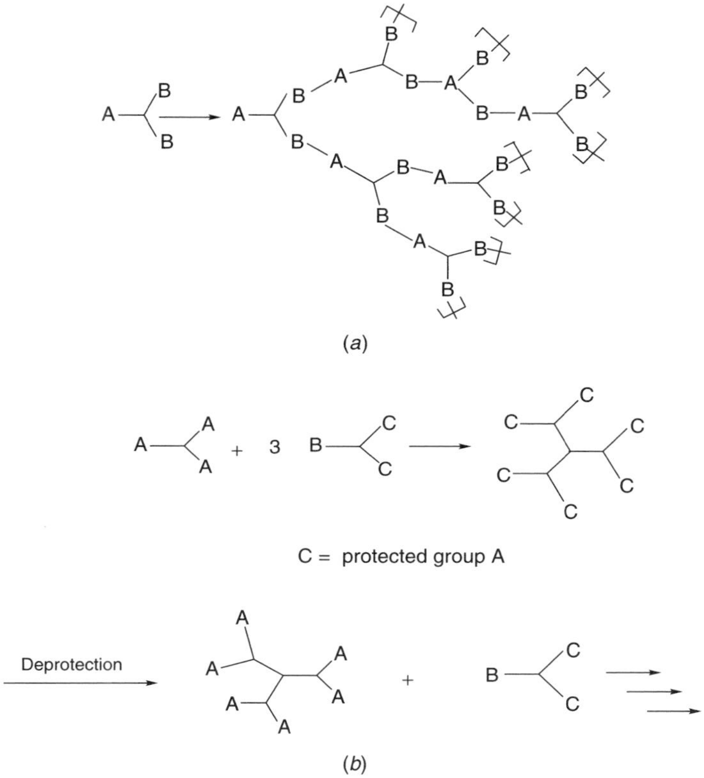

(a)

C = protected group A

(b)

Figure 5.16 Formation of (a) hyperbranched and (b) dendritic macromolecules.

AB_2 reacts selectively with only one antagonist function of a second poly-functional molecule, the other ones being protected[81] (Fig. 5.16). The perfect hyperbranched molecules obtained according to that step-by-step process are called dendrimers. The degree of branching characterizes the structure of a hyper-branched polymer and has been defined by Hawker et al.[82] as

$$DB = \frac{D + T}{D + T + L}$$

where D, T, and L are the number of dendritic, terminal, and linear units in the polymers, which are determinated by ^1H and ^{13}C nuclear magnetic resonance (NMR) spectroscopy by comparison with the model structure for each polymer unit.

In 1996, Hawker and Frechet[83] discussed a comparison between linear hyperbranched and dendritic macromolecules (Fig. 5.17) obtained with the same monomeric structure, 3,5-dihydroxybenzoic. The thermal properties (glass transition and thermal decomposition) were not affected by the architecture.

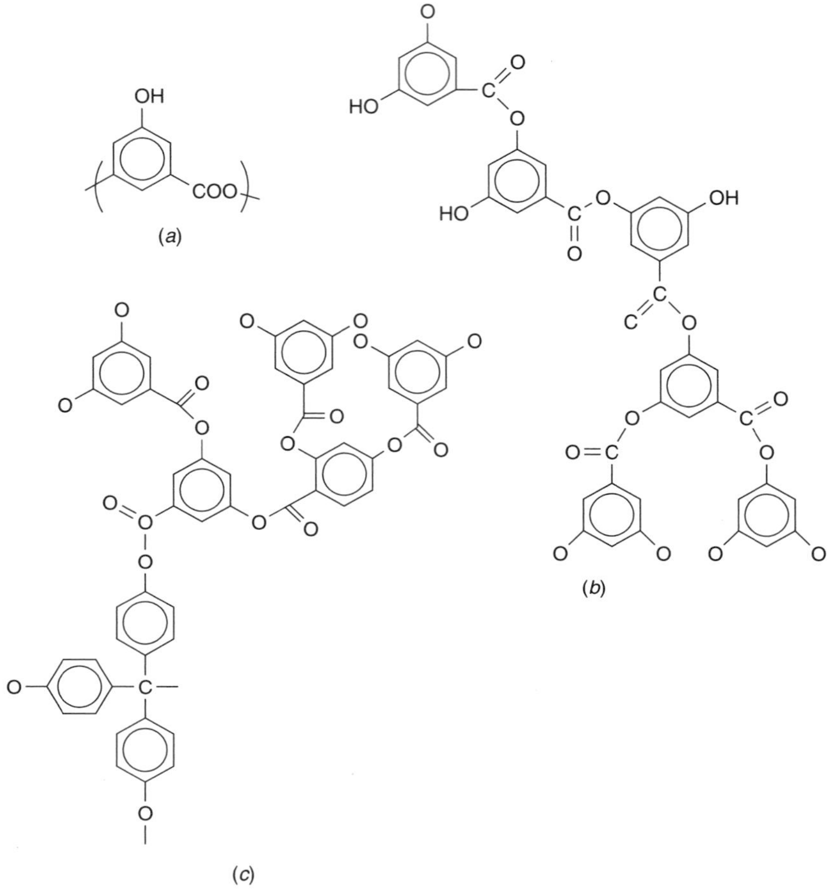

Figure 5.17   (a) Linear, (b) hyperbranched, and (c) dendritic aromatic polymers.

However, dendrimeric and hyperbranched polyesters are more soluble than the linear ones (respectively 1.05, 0.70, and 0.02 g/mL in acetone). The solution behavior has been investigated, and in the case of aromatic hyperbranched polyesters,[84] a very low α-value of the Mark–Houvink–Sakurada equation ($\eta = K M^{\alpha}$) and low intrinsic viscosity were observed. Frechet presented a description of the intrinsic viscosity as a function of the molar mass[85] for different architectures: The hyperbranched macromolecules show a nonlinear variation for low molecular weight and a bell-shaped curve is observed in the case of dendrimers (Fig. 5.18).

A review of hyperbranched $A_x B$ monomers gives a general view of the properties of these materials[86]:

- Glass transition is strongly affected by the chemical nature of the endgroups.

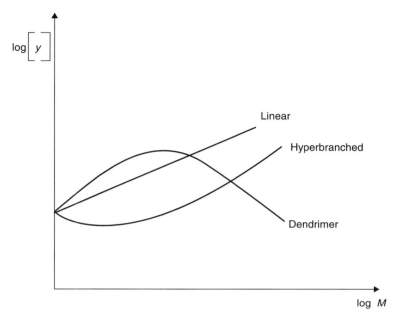

Figure 5.18   Variation of intrinsic viscosity as function of molar mass for linear, hyperbranched, and dendritic macromolecules.

- In the molten state, a Newtonian behavior was observed, a consequence of lack of entanglements. The melt behavior is also dependent on the structure of the endgroups.
- Melt viscosity is lower than the one of corresponding linear polymers, but polar endgroups can increase the viscosity by several orders of magnitude.

However, Sendijarevic and Mac Hugh[87] pointed out that hyperbranched polyetherimides (PEIs)[88] exhibit a rheological behavior depending on the molar mass. They concluded that the polymers are characterized by a more open structure with entanglement leading to a viscoelastic behavior when the molecular weight M_n reached 20,000. The strategy of synthesis of hyperbranched polyimides will be discussed in Section 5.4.

5.3 OVERVIEW OF CHEMISTRY AND ANALYTICAL TECHNIQUES

5.3.1 Chemistry and Catalysis

5.3.1.1 Catalytic Systems

This section presents catalytic systems that have been studied for either the synthesis of monomers or polymerization. In addition, some new catalysts, working in smooth conditions, which could be used for high-performance polymers synthesis are discussed.

5.3.1.1.1   Phase Transfer Catalysts

By exchange or complexation with the counterion of a nucleophilic reagent, such as a phenate, a phase transfer catalyst enhances the nucleophilic character of the reagent and makes it soluble in less polar medium. The main phase transfer agents are crown ethers or ammonium salts. The classical ammonium salts (tetrabutylammonium salts) do not withstand high temperature due to β-hydrogen elimination. Brunelle and Singleton[89] found that some β-*tert*-alkyl ammonium salts are stable up to 300°C. This type of phase transfer agent was used by Hoffman et al.[90] to activate the reaction of some bisphenates with the 4,4'-(bis-4-chlorobenzoyl) benzene, and high-molecular-weight polyetherketones were obtained.

5.3.1.1.2   Lanthanide Catalysts

Samarium iodide has been used for the reductive coupling of ether acid chloride, giving α-diketone, or aldehyde with formation of α-diols. Similarly, chlorosilane gives disilane. These reactions take place in smooth conditions (room temperature) with high yield.[91] The trifluoromethane salts of several lanthanides behave like Lewis acids and are stable in aqueous media.[92] In these conditions their catalytic effect on the Diels-Alder reactions in the presence of water, the aldol reaction, or Michael addition was demonstrated. An interesting example of the condensation of benzaldehyde with indole is shown (Fig. 5.19) and probably could be used in polycondensation. Lanthanide triflates[93] have been used for condensation of aliphatic or aromatic nitriles with α,ω-diamino ethane or propane, leading to the corresponding cyclic amidines (Fig. 5.19).

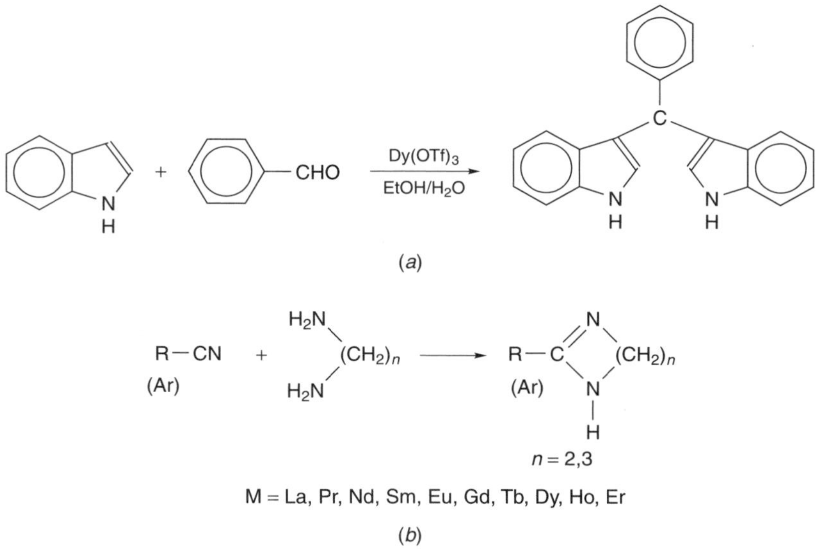

(a)

$n = 2,3$

M = La, Pr, Nd, Sm, Eu, Gd, Tb, Dy, Ho, Er

(b)

Figure 5.19   Lanthanide trifluoromethanesulfonic salts as catalysts.

5.3.1.1.3 Catalyst for Aromatic Coupling

Aromatic coupling has been used for the preparation of functionalized monomers and also for polymerization. With Ni (obtained by reduction of $NiCl_2$ by Zn), triphenylphosphine, and 2,2-bipyridine, Colon and Kwiatkowski[94] obtained a high-molecular-weight polyethersulfone in DMAC by poly coupling of 4,4′-(bis-4-chlorophenoxy)diphenylsulfone. The presence of $ZnCl_2$ in the polymer could have a deleterious effect on the aging behavior, and another nickel-based catalytic system was published by Yamamoto et al.[95] The reduction of nickel acetate was performed by the sodium hydride. With this catalyst the coupling of 2,5-dialkoxy-1,4-dichlorobenzene takes place in tetrahydrofuran (THF). No chlorine was observed in the polymer and the rather low molecular weight obtained ($10^3 - 10^4$ Da) is probably the consequence of a competitive reduction of the chlorine by the catalytic system. The coupling is usually performed with a halogen derivative: however Percec et al.[96a] showed that the coupling could occur with a triflate instead of a halogen. The reaction was performed with a benzene ring carrying nitrile or ester in the ortho, meta, or para position (Fig. 5.20). The very high yield makes this method very interesting for the preparation of monomers and polymers.[96b]

The Suzuki reaction[97] allows the coupling of two aromatic rings by reaction of an arylboronic compound with a iodo or bromo aryl derivative. The tetrakis (triphenylphosphine) Pd is the catalyst working in the basic medium. This reaction was recently used[98] in aqueous media for the preparation of different isomers of diphenyldicarboxylic acids (Fig. 5.21) but also for the synthesis of soluble rodlike polyimides[99] by coupling the 3,6-diphenyl-N,N'-bis(4-bromo-2,5-didodecylphenylene)pyromellitic diimide with 2,5-didodecylbenzene-1,4-diboronic acid (Fig. 5.21). The reaction was carried out in a heterogeneous system (water and toluene). The molar mass ($\overline{M_w} = 72,300$ Da) was measured by membrane osmometry in o-dichlorobenzene.

5.3.1.1.4 Other Metal-Catalyzed Polymerizations

A direct route to obtain polyimides and polybenzoxazoles without diacid has been explored by using direct carbonylation of an aromatic diiodo compound.[100] The reactions were performed in polar aprotic solvent. The cheaper chlorine derivative can be used instead of the iodo one when it is located on a strongly

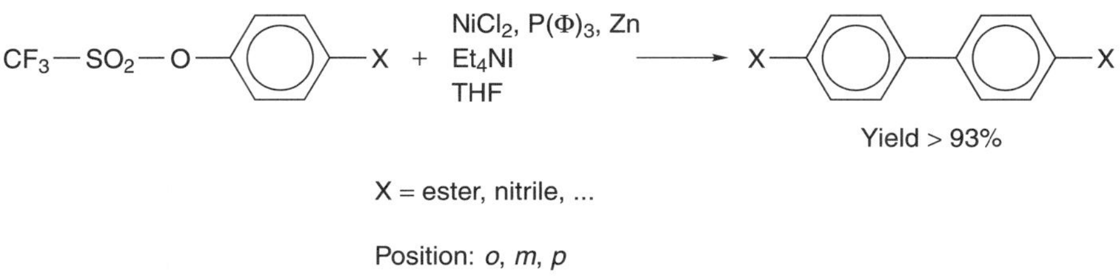

$$CF_3-SO_2-O-\langle\bigcirc\rangle-X \ + \ \begin{matrix} NiCl_2, P(\Phi)_3, Zn \\ Et_4NI \\ THF \end{matrix} \ \longrightarrow \ X-\langle\bigcirc\rangle-\langle\bigcirc\rangle-X$$

Yield > 93%

X = ester, nitrile, ...

Position: o, m, p

Figure 5.20 Aromatic coupling with trifluoromethanesulfonate as leaving group.

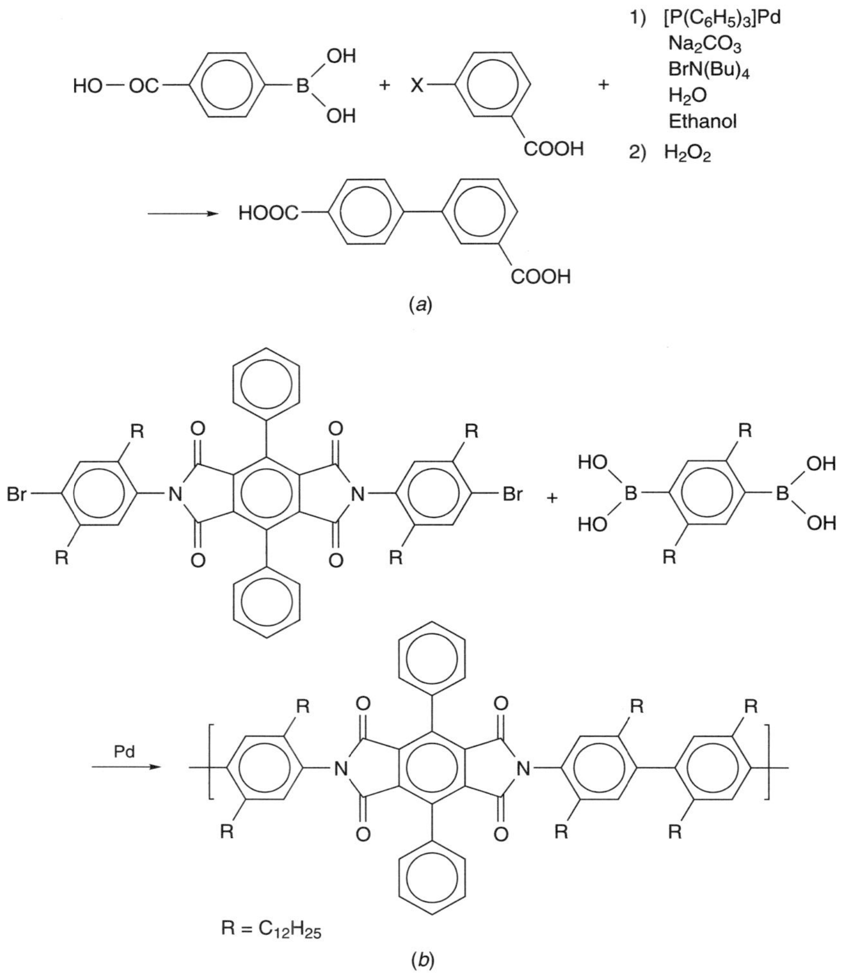

Figure 5.21 (*a*) Monomers and (*b*) polymers syntheses by Suzuki reaction.

electron withdrawing ring. For example, 4,3′-bis(4-chlorophthalimido)diphenyl ether reacts with 4,4′-(4-aminophenoxy)biphenyl at 90 psig CO in the presence of PdCl$_2$(PO$_3$)$_2$ in NMP at 90–100°C. A polyamide imide ($\overline{M_n}$ = 38,600) was obtained (Fig. 5.22). A similar reaction starting from the 4,4′-diiododiphenyl ether and a bis-*ortho*-aminophenol leads to a polyamidophenol which is cyclized by thermal treatment at 250–300°C (Fig. 5.23) in polybenzoxazole. An original synthesis of polyquinoxaline-2,3-diyls has been reported. The monomer was 1,2-diisocyanobenzene, and polymerization had been initiated by a methyl Pd(II) complex. The polymerization proceeds with step-growth insertion of the isocyano group on carbon palladium bonds[101] (Fig. 5.24).

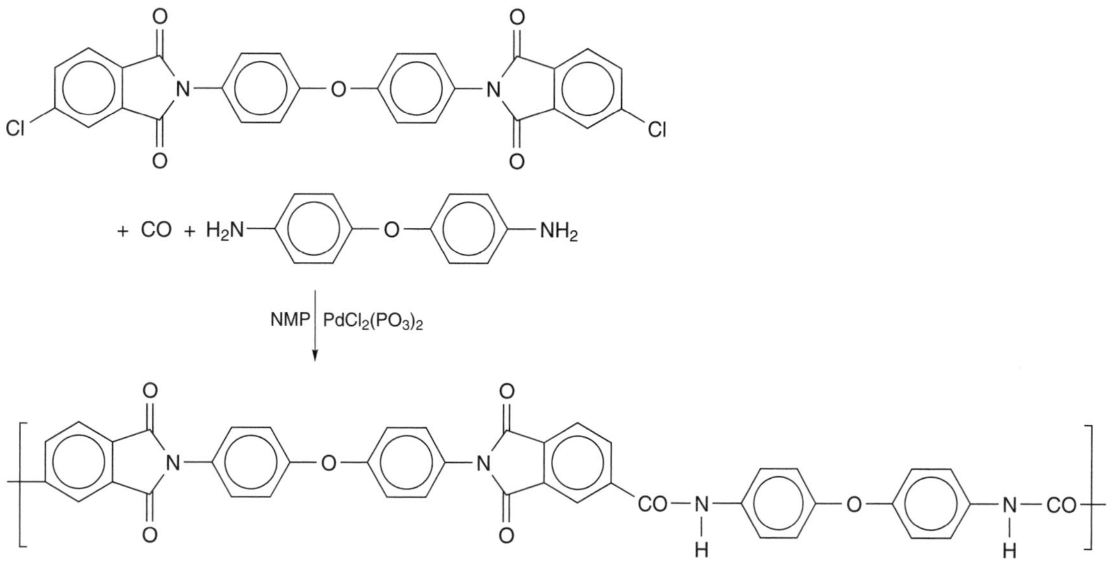

Figure 5.22 Synthesis of polyamide-imide by catalytic carbonylation of bis-4-chlorophthalimide in presence of aromatic diamine.

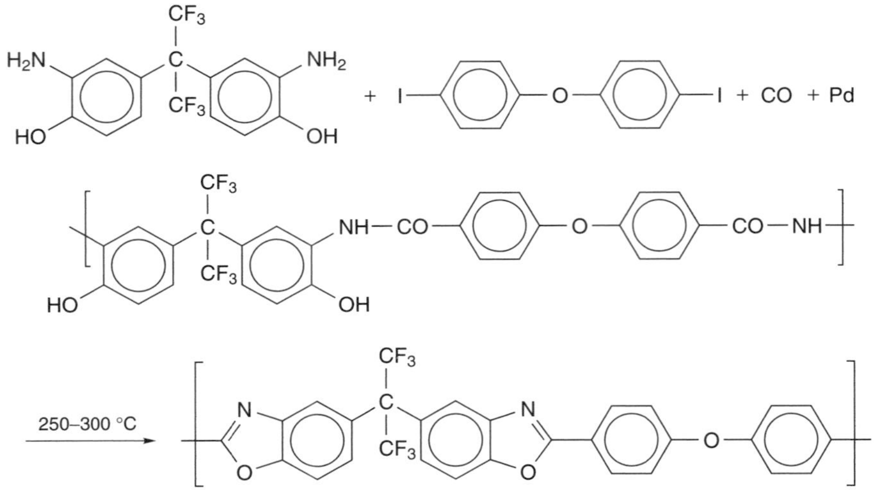

Figure 5.23   Synthesis of polybenzoxazoles by catalytic carbonylation of 4,4′-diiododiphenylether.

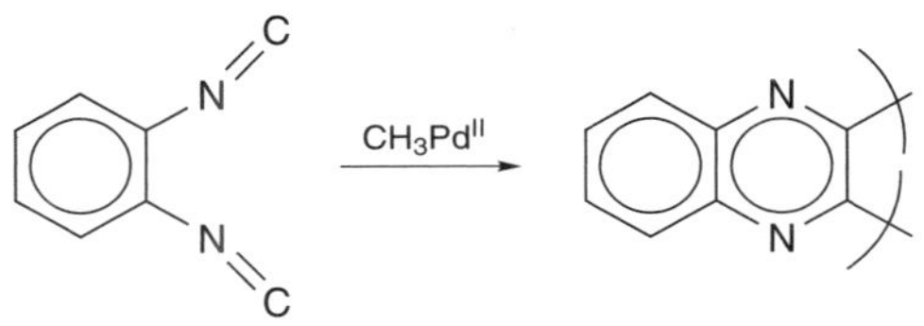

Figure 5.24   Synthesis of polyquinoxaline-2,3-diyl.

5.3.2 Specific Imide Chemistry

5.3.2.1 Condensation in Smooth Conditions

Photosensitive functions are in many cases also heat sensitive, so the preparation of photosensitive polyimides needs smooth conditions for the condensations and imidization reactions. Some chemical reactants, which can be used for polyamide preparation, have been patented for the synthesis of polyimides and polyimide precursors. For example, chemical imidization takes place at room temperature by using phosphonic derivative of a thiabenzothiazoline.[102] A mixture of N-hydroxybenzotriazole and dicyclohexylcarbodiimide allows the room temperature condensation of diacid di(photosensitive) ester with a diamine.[103] Dimethyl-2-chloro-imidazolinium chloride (Fig. 5.25) has been patented for the cyclization of a maleamic acid in toluene at 90°C.[104] The chemistry of imidazolide has been recently investigated for the synthesis of polyimide precursor.[105] As shown in Fig. 5.26, a secondary amine reacts with a dianhydride giving *meta-* and *para*-diamide diacid. The carbonyldiimidazole

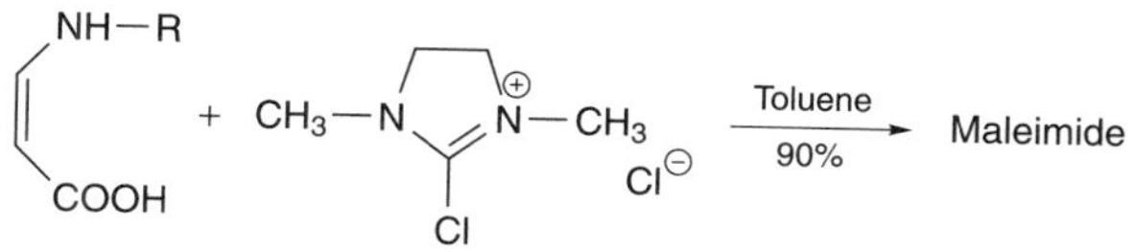

Figure 5.25 Cyclization of maleamic acid with imidazolinium salt.

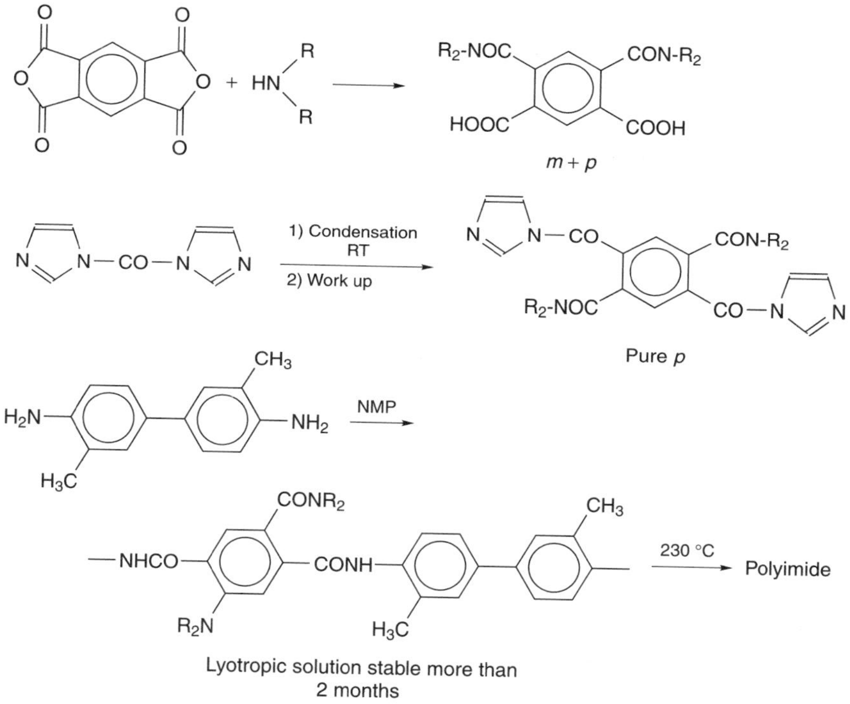

Figure 5.26 Synthesis of rodlike polyimide using imidazolide chemistry.

gives smoothly at room temperature the bisimidazolide. The para isomer condensation with a 4,4′-diaminodiphenyl monomer gives a lyotropic solution of polyimide precursor. Imidazole blocked isocyanates react with a dianhydride in rather smooth conditions. Both monomers dissolved in N,N'-dimethylpropylene urea containing 4-dimethylaminopyridine were heated at 120°C and a fully cyclized polyimide was obtained (Fig. 5.27).[106]

5.3.2.2 Solid-State Condensation

Some examples concerning the polyimide synthesis by condensation of the crystalline salts obtained with the aliphatic or aromatic diamines and aromatic tetraacids will be discussed in Section 5.4.1.

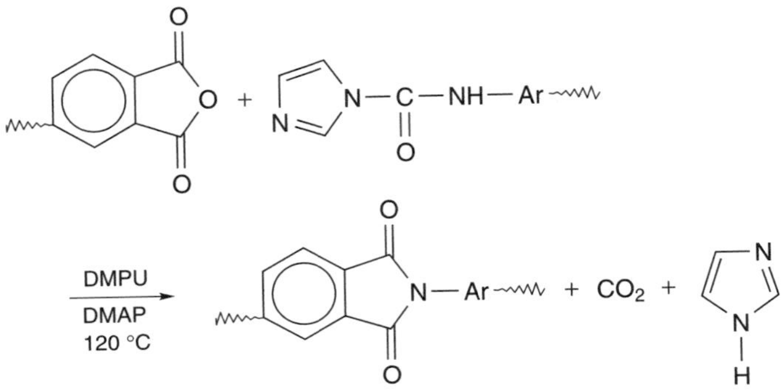

Figure 5.27   Polyimide synthesis with imidazole blocked diisocyanate and dianhydride.

5.3.2.3 Polyimide Preparation in New Solvent

Diacetone alcohol has been recently claimed as a solvent for polyimide preparation. According to the patent,[107] ODA was dissolved in the solvent and then PMDA added to the solution. The patent claims good storage stability and easy removal of the solvent (boiling point: 166°C).

5.3.2.4 Experimental Processes

General procedures[108] of both main solution preparations of polyimides are given below: These procedures are representative and can be applied to all types of dianhydrides.

1. *Synthesis in NMP.*   The dianhydride (1.35 mmol) is added to a stirred solution of 1.35 mmol of the diamine in NMP (solid content 15% w/w) under N_2 at room temperature. After 24 h, 2.97 mmol of pyridine and 2.97 mmol of acetic anhydride are added. After 24 h, the solution is diluted with 8 mL of NMP and then added to 600 mL of ethanol. The polyimide is collected by filtration, washed with ethanol and ether and then dried at 125°C under reduced pressure for 24 h.

2. *Synthesis in m-cresol.*   The dianhydride (1.35 mmol) is added to a stirred solution of 1.35 mmol of the diamine in the appropriate amount of *m*-cresol containing 6 drops of isoquinoline under N_2 at room temperature. After 3 h, it is heated to reflux (ca. 200°C) and maintained at that temperature for 3 h. During this time, the water of imidization is distilled from the reaction mixture along with 1 to 3 mL of *m*-cresol. The *m*-cresol is continually replaced to keep the total volume of the solution constant. After the solution is allowed to cool to room temperature, it is diluted with 20 mL of *m*-cresol and then slowly added to 1 L of vigorously stirred 95% ethanol. The precipitated polymer is collected by filtration, washed with ethanol and ether and dried under reduced pressure at 125°C for 24 h.

This one step method has been used for the syntheses of rodlike polyimides, for example with 3,3′,4,4′-biphenyltetracarboxylic dianhydride (BPDA) and 2,2′-bis(trifluoromethyl)-4,4′-diaminobiphenyl (PFMB).[109]

5.3.3 Monomer Syntheses

The development of polyimides either matching different processability requirements or exhibiting new, specific properties is the driving force for the research in new monomers. For structural applications, easy processability, good mechanical properties, and good aging behavior are needed, but the low cost is an important constraint. Considering the functional applications (electronic, liquid crystal devices, membrane for gas separations, fuel cell membranes), the cost is somewhat bit less important, and in some cases (e.g., nonlinear optical devices), preparation of the monomer carrying the functional group will be a multistep synthesis. For industrial reasons, the modification of the diamines is more investigated than the modification of dianhydrides.

5.3.3.1 *Diamines and Dianhydrides for Processable Polyimides*

The relationships between structures and properties are well documented, and the different approaches to get processable polyimides have been discussed in Section 5.2.2.

5.3.3.1.1 *Diamines: General and Experimental Methods*

A variety of aromatic diamines have been used for the synthesis of polyimides. It would be too time consuming to report all the pathways described in literature, so here only the most common ones are considered. The biggest part of these different methods leads to nitro compounds generally reduced using H_2 or hydrazine monohydrate with catalytic amounts of Pd–C in refluxed ethanol or dioxane. The numbers in parentheses refer to Fig. 5.28:

(1) Reaction of bisphenol with chloronitroaromatic compounds was generally performed in dimethylformamide (DMF) or dimethyl sulfoxide (DMSO) at reflux using K_2CO_3 as a base.[108,109] It is possible to achieve this condensation in Ullmann's conditions by using a cuprous chloride or iodide–pyridine system as a catalyst; when this reaction is performed with deactivated aromatic compounds, it gives too poor yields[110]; ultrasounds can dramatically improve yields without solvent.[111]

(2) Fluorenone is treated by excess of aniline in the presence of aniline hydrochloride.[112]

(3) Halogenonitroaromatic compound is heated to reflux of DMF with excess of dendritic copper powder.[113,114] It is possible to use Ni or Zn.

(4) Reduction of the nitroaromatic with $NaBH_4$ in DMSO or Zn in NaOH/EtOH leads to the hydrazo derivative and treatment with aqueous HCl gives benzidine rearrangement.[115]

$$HO{-}Ar{-}OH + Cl{-}Ar{-}NO_2 \longrightarrow O_2N{-}Ar{-}O{-}Ar{-}O{-}Ar{-}NO_2 \quad (1)$$

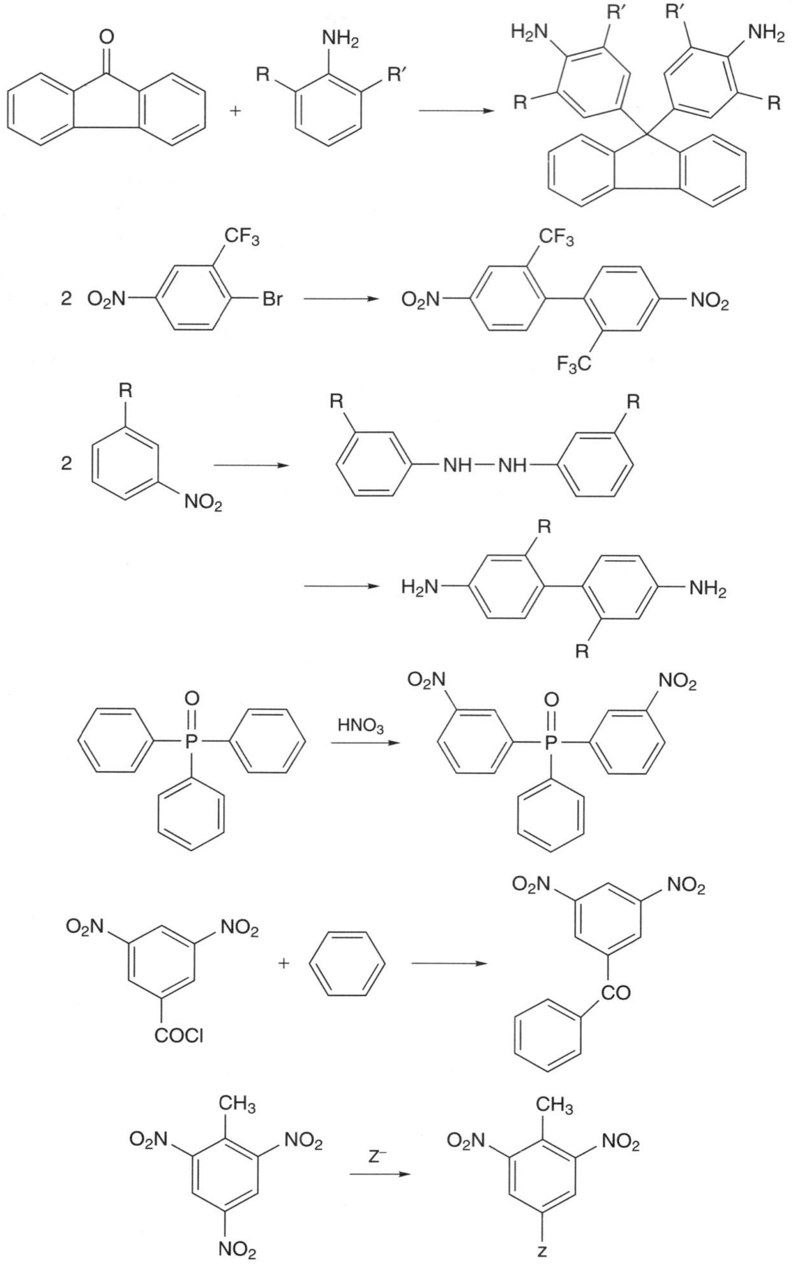

(2)

(3)

(4)

(5)

(6)

(7)

Figure 5.28 General scheme for syntheses of aromatic diamines or dinitro precursors.

(5) Triphenylphosphine oxide is dissolved in cooled H_2SO_4 and fuming nitric acid is added.[116] Dinitro derivative is the major compound (80%) with mononitro and trinitro as by-products (about 10% each).

(6) It is possible to apply one of the oldest reaction, Friedel–Crafts, for the synthesis of diamines.

(7) TNT-based condensation monomers. The synthesis of these materials is presented humorously by Russian authors a as consequence of the end of the Cold War. The synthesis, based on a very easy nucleophilic displacement of an activated nitro group, offers many possibilities for the synthesis of diamines bearing a functional group.[117]

5.3.3.1.2 Functional Diamines

Some recently published diamines could find interesting applications. For example, the structure of the diaminoquinone-diimine[118] is very close to the one of the polyaniline. A rigid para five-ring pyridine salt diamine reacts with the dianhydride leading to a lyotropic solution. After cyclization, the polyimide exhibits reversible color upon irradiation.[119] A furane-containing diamine shows a blue photoluminescence at 442 nm with a high quantum yield (0.92). The corresponding polyimide with pyromellitic dianhydride shows also a blue photoluminescence at 419 and 436 nm.[120]

5.3.3.1.3 Dianhydrides: General and Experimental Methods

There are only two inexpensive dianhydrides: 3,3′,4,4′-benzophenonetetracarboxylic dianhydride (BTDA) and pyromellitic dianhydride (PMDA). Some other commercially available dianhydrides were used, like 3,3′,4,4′-biphenyltetracarboxylic dianhydride (BPDA), 2,2′-bis(3,4-dicarboxyphenyl) hexafluoropropane dianhydride (6FDA), oxydiphthalic anhydride (ODPA), and 3,3′,4,4′-diphenylsulfonetetracarboxylic dianhydride (DSDA) (Fig. 5.29). Many other dianhydrides are reported in the literature (Fig. 5.30):

(8) A patent claims the semihydrolysis of PMDA for the synthesis of alternative polyimides with different diamines.[121]

(9) and (10) Many dianhydrides are prepared by a Diels-Alder reaction.[122,123]

(11) An elegant synthesis of a dianhydride with pendant p-nitrophenyl group has been reported by a Spanish group.[124]

Alicyclic dianhydrides are interesting for electronic applications. The polyimides obtained from them are colorless with high transparency in the visible range, exhibit low birefringence,[125] and have a low dielectric constant.[126] The reactivity of the polycyclic aliphatic dianhydride has been investigated. For example, bicyclo-[2,2,2]-oct-7-ene tetracarboxylic dianhydride reacts quickly with an aromatic amine because the bicyclo-imide is less strained than the corresponding dianhydride.[127]

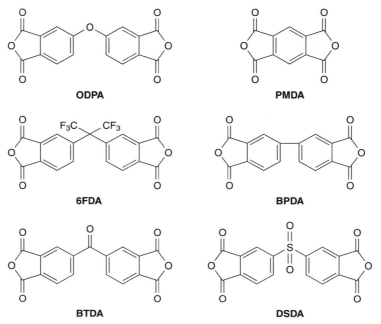

Figure 5.29 Commercially available dianhydrides.

5.3.4 Analytical Methods

5.3.4.1 Differential Scanning Calorimetry

Measurement of glass transition temperature is usually done either by the calorimetric method (DSC, which "sees" the variation of C_p) or by the mechanical method [mainly dynamical mechanical analysis (DMA), which can measure the loss tangent or the loss modulus E'' or elastic modulus decrease onset]. The determination of T_g by DSC is a controversial subject: Some authors use the midpoint of the heat capacity variation and others use the extrapolated onset temperature. A comparison of both DSC and DMA methods for the T_g determination has been reported. In the case of polyimide the difference was between 2 and 4°C.[128] The determination of glass temperature for a polyamic acid cannot be done by classical DSC because the heat flow due to the imidization is important, so the variation of the specific heat flow cannot be observed. An oscillating differential scanning calorimeter allows a distinction between the two phenomena.[129] The T_g of several polyamic acid films containing different percentage of NMP were obtained. Then a calculation based on the Gordon–Taylor equation gives the T_g of the solvent free polyamic acid:

PMDA–ODA: polyamic acid $T_g = 207°C$.
DPDA–PDA: polyamic acid $T_g = 166°C$.
BTDA–PDA: polyamic acid $T_g = 213°C$.

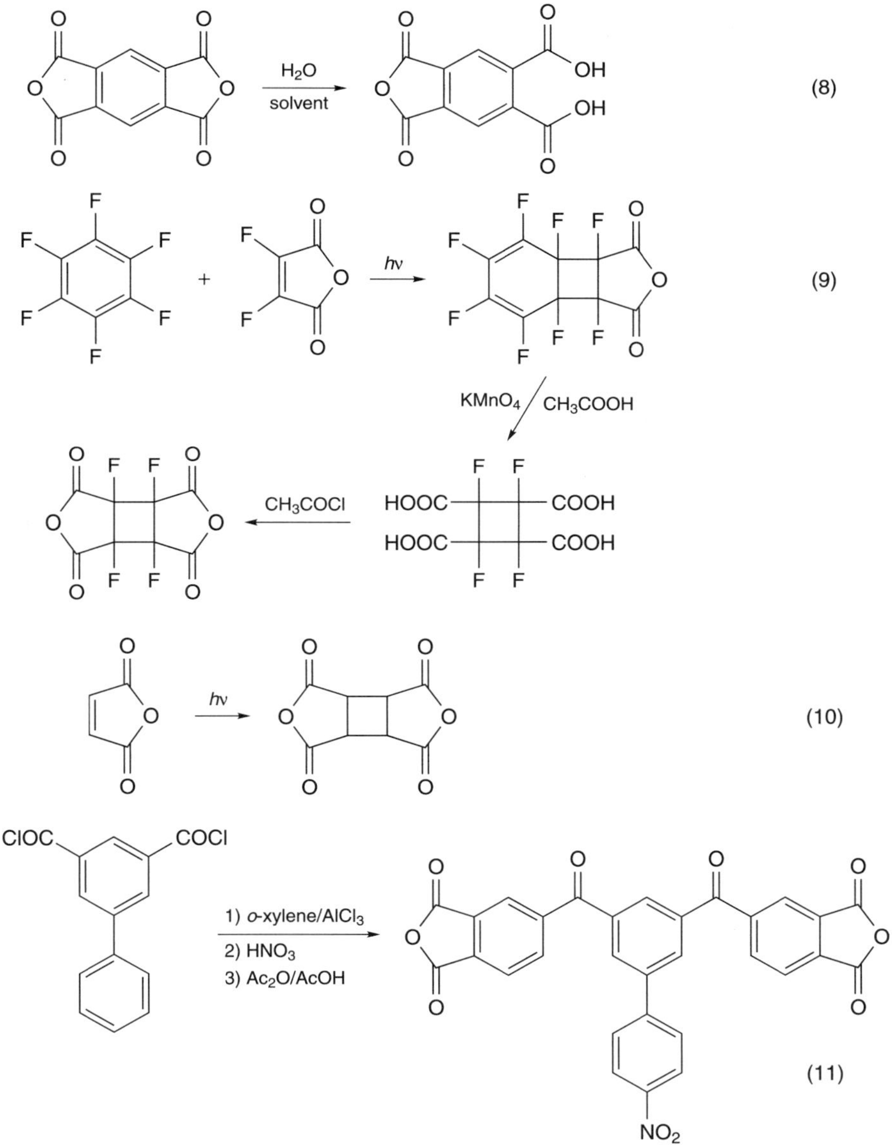

Figure 5.30 Syntheses of some specific dianhydrides.

A report[130] of DSC measurements on polybenzimidazole fibers describes important differences for the glass transition temperature depending on the mechanical treatment of the fiber. An as-spun fiber exhibits a T_g at 387°C instead of 401°C for a drawn fiber free to shrink or 435°C for a drawn fiber with fixed length.

5.3.4.2   Infrared Spectroscopy

Fourier transform infrared (FTIR) spectroscopy is the most popular method for determining the imidization process in the solid state and identifying specific substituents on the macromolecular backbone (e.g., CN, SO_3H, CO, SO_2).[131] A method for calculating the thermal imidization extent based on FTIR data has been reported by Pride.[132] Raman spectroscopy was used on the model study of PMDA–ODA condensation, and the possible formation of an imine bond by reaction of an amino group with an imide carboxyle was evidenced.[133]

5.3.4.3   NMR Spectroscopy

Proton and carbon NMR are extensively used for the characterization of monomers and soluble polyimides and other heterocyclic polymers.[117,119,125–127] Studies concerning the polyamic acids are limited due to the complex spectra of these materials (uncertainties of peak identification) and to the presence of solvents. However, amic acid reverse reaction and redistribution (Fig. 5.31) were studied using this technique.[26] The thermal curing of PMDA–ODA polyamic acid and the hydrolysis of the corresponding polyimide have been investigated by N-15 solid-state NMR with cross-polarization magic-angle spinning (CPMAS). This method furnishes indications concerning the nature of the nitrogen (imide, isoimide, amide, amine) in the macromolecule.[134] Fluorine-19 NMR can be used in any solvent with no spectral interference. The method is very sensitive because fluorine chemical shift is strongly dependent on the chemical environment. It can be used for mechanistic studies concerning chemistry of the amic acids in NMP solutions.[127,135]

5.3.4.4   Positron Annihilation Lifetime Spectroscopy (PALS)

The sizes and concentration of the free-volume cells in a polyimide film can be measured by PALS. The positrons injected into polymeric material combine with electrons to form positroniums. The lifetime (nanoseconds) of the trapped positronium in the film is related to the free-volume radius (few angstroms) and the free-volume fraction in the polyimide can be calculated.[136] This technique allows a calculation of the dielectric constant in good agreement with the experimental value.[137] An interesting correlation was found between the lifetime of the positronium and the diffusion coefficient of gas in polyimide.[138,139] High permeabilities are associated with high intensities and long lifetime for positron annihilation.

5.4   SYNTHETIC METHODS

5.4.1   Polyimides

5.4.1.1   Basic Principles for Synthesis of AABB Polyimides

Polyimides syntheses starting from dianhydrides (or derivatives such as bis acid-esters) and diamines (or derivatives like silyl- substituted diamine or isocyanate)

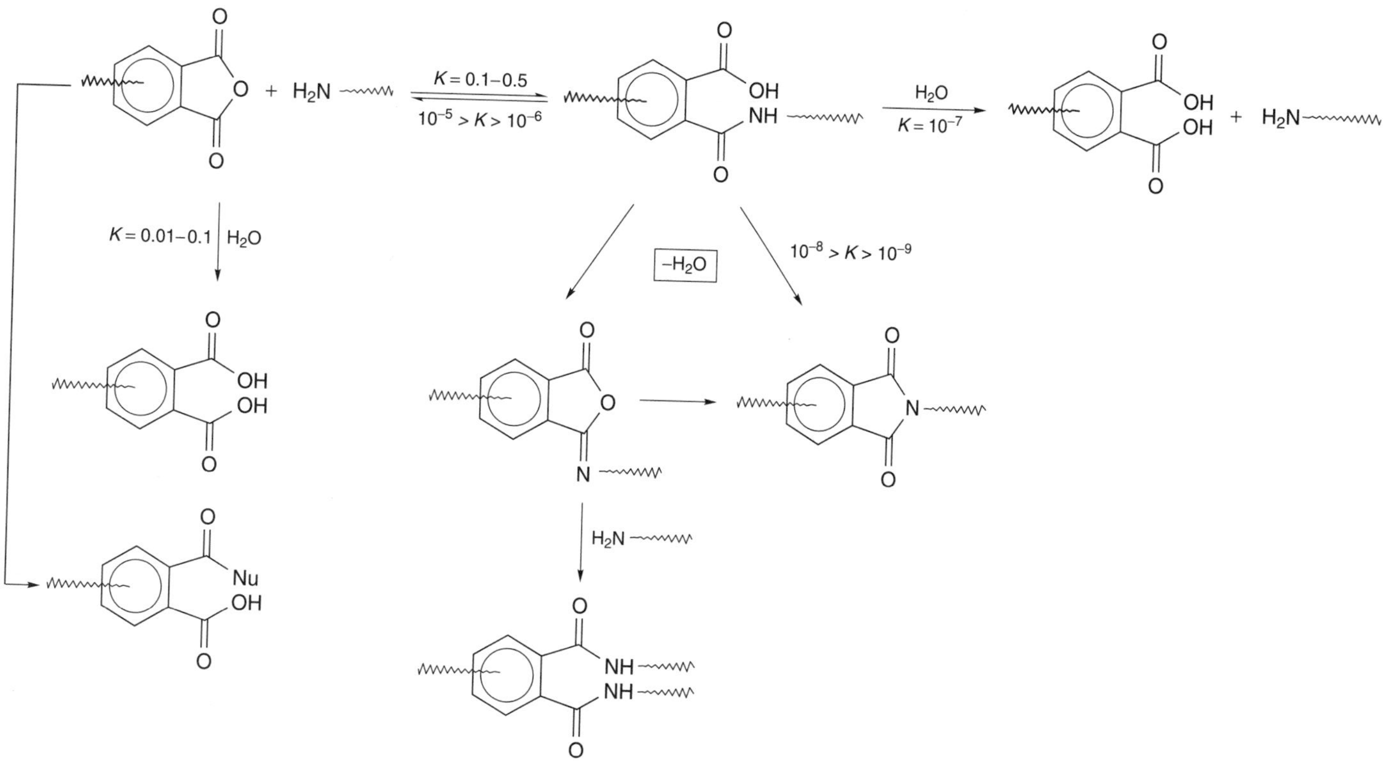

Figure 5.31 Amic acid formation and imide cyclization.

have been extensively reviewed by Takekoshi.[26] We will give here only the general features of the two most important methods used for industrial production.

5.4.1.1.1 Low-Temperature Condensation in Aprotic Solvent

N-methyl pyrolidinone is used in most cases. Figure 5.31 summarizes the main reaction which can take place during the process and the corresponding rate constant. The formation of diamide has also been evidenced.[140] The reactivity is governed by the electron affinity of the anhydride and the ionization potential or basicity of the diamine (see Section 5.2.2.1). When a diacid with a low electron affinity reacts with a weak nucleophilic diamine, a low-molecular-weight is obtained, because the reverse reaction is not negligible compared with the forward reaction.

Although the condensation of a dianhydride with a diamine is catalyzed by organic acids, in aprotic solvent, the formation of the amic acid is not autocatalytic because the amic acid is in a strong and exothermic interaction with the solvent. Among the side reactions, the anhydride hydrolysis is important in the case of the most reactive dianhydride such as PMDA, but the water can also hydrolyze the amic acid (Fig. 5.31). The viscosity of the solution reaches a maximum and decreases as a function of time. When the reaction starts, the large observed polydispersity of the molar mass with very high $\overline{M_w}$ is responsible for the high solution viscosity. The reverse reaction leads to an equilibrium and the $\overline{M_w}/\overline{M_n}$ ratio decreases.[79] When the formation of polyimide occurs by thermal imidization, a phase separation takes place before the total cyclization of the chain, and after solvent removal, the cyclization rate is governed by the T_g of the polyimide. The chemical imidization is performed by reaction with acetic (or other) anhydride added to the solution. Tertiary alkyl amines or aromatic pyridine derivatives catalyze the reaction. The problem of the cyclization in heterogeneous medium still remains except for soluble polyimide.

5.4.1.1.2 High-Temperature Condensation in Cresolic Medium

This method is being increasingly used for the preparation of soluble polyimides, and the rate is faster than in aprotic solvents. This method allows for preparation of block copolyimides.[78]

5.4.1.2 Trends in Synthetic Methods

5.4.1.2.1 Transimidization

The competition at 200°C between an aliphatic and an aromatic amine toward the formation of an imide is a very selective reaction (Fig. 5.1) for the formation of an aliphatic imide.[141] This reaction suggests that the reactive processing in extruder, for example, could be used to transform a melt-processable polyimide with an oligomer end capped with an aliphatic amine. In order to get a perfectly alternate block polyimide–block siloxane, Rogers et al. used low-temperature transimidization.[142] An oligomeric aromatic imide was end capped with 2-aminopyrimidine, which is a nice leaving group when the aminopropyl-terminated polydimethylsiloxane reacts at 110°C in chlorobenzene (Fig. 5.2).

5.4.1.2.2 Polyimides by Vapor-Phase Deposition

An IBM team[143] described an apparatus for the vapor-phase polymerization of PMDA and ODA. The condensation in the vapor phase is avoided, and the polycondensation takes place on the solid surface giving a polyamic acid. The thickness reproducibility is about 2% for a thickness from 10 nm to 20 μm. Imidization is performed by thermal treatment at a temperature above 175°C. The dielectric constant of the polyimide film (2.91 ± 0.03) is lower than the one obtained by classical deposition of polyamic acid in NMP solution (3.5). The birefringence (in-plane refractive index minus out-of-plane refractive index) is also lower than the birefringence of the spin-coated polyimide. More recently the tensile strength of PMDA–ODA polyimide film obtained by vapor-phase deposition has been measured. Values ranging between 85 and 100 MPa are interesting but lower than the ones obtained with commercial films (Kapton: 230 MPa; Upilex: 400 MPa).[144] The preparation of PMDA–ODA polyimide shells was performed by vapor-phase polymerization of the monomers on to spherical mandrels of poly-α-methyl styrene. By heating at 300°C, the mandrel is decomposed into α-methyl styrene which permeated through the polyimide shell.[145] The vapor-phase preparation of polyimide has been reviewed until 1995.[146]

5.4.1.2.3 Polyimide Synthesis without Solvent via a Monomeric Salt

The first patent of Edwards and Robinson[147] claims the condensations of pyromellitic acid and aliphatic diamine salt to prepare polyimide. Recently, that approach has been revisited, and biphenyl tetracarboxylic and pyromellitic acids give a salt monomer by reaction with 1 mol of an aliphatic diamine (octamethylene diamine and dodecamethylene diamine). The salts were polymerized under 250 MPa at 250°C for 5 h in closed reaction vessels (Fig. 5.32) giving crystalline polymers.[148] By reaction of pyromellitic tetraacid with oxydianiline, it has been possible to isolate a monomeric salt. It was polymerized under 30 MPa giving a PMDA–ODA polyimide with water elimination.

5.4.1.2.4 Synthesis or Processing of Polyimide in Water

For ecological and economical reasons, water is the most interesting solvent for the chemical industry. Although it is well known that imide formation gives 1 mol of water per imide ring and that the intermediate amic acid is moisture sensitive, the synthesis of polyimide has been observed in water medium probably by formation of a salt, as evidenced by Imai et al.[149] The formation of polyimide from polyamic acid trialkylamine salt has been investigated for two main reasons. First, the formation of a salt may increase the hydrolytic stability by preventing the reverse reaction (formation of amine and anhydride). The second advantage would be an enhanced imidization rate as previously mentioned by Kreuz et al.[150] The polyamic salts were prepared by dissolution of the polyamic acid in water or

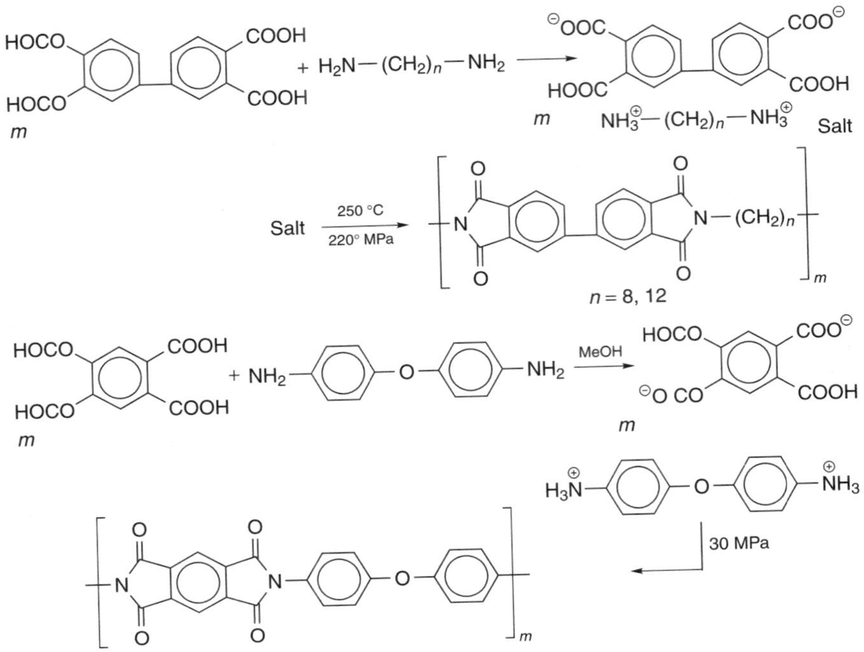

Figure 5.32   Solid-state syntheses of polyimides.

methanol and a tertiary amine was added with an excess of 10%. Thermal imidization was monitored, detecting imidization products by gas chromatography/mass spectrometry (GC/MS) analysis. The cyclization would proceed according to a nucleophilic substitution to form the polyamic ester (Fig. 5.33).

5.4.1.2.5 Crystalline Polyimide Powders

The thermal imidization of a polyamic acid film (PMDA–ODA or BPDA–ODA) obtained by casting an NMP solution leads to an amorphous polyimide. Two different teams have shown that a polyamic acid solutions in NMP heated at 200°C for a short time (20 min) gives polyimide particles fully cyclized and highly crystalline, as shown by X-ray diffraction and solid ^{13}C NMR spectroscopy.[151,152] The chemical imidization of the same solution gives only amorphous particles. The difference between the cyclization of a solution and a casted film in the same solvent is intriguing. In the case of the solution, the temperature and the heating time are lower than in the case of the casted film; as a consequence, a less organized structure would be expected for the particle.

5.4.1.3 AB Polyimides

The first AB polyimide was observed 92 years ago when reduction of 4-nitrophthalic anhydride did not give the expected aminophthalic anhydride.[153]

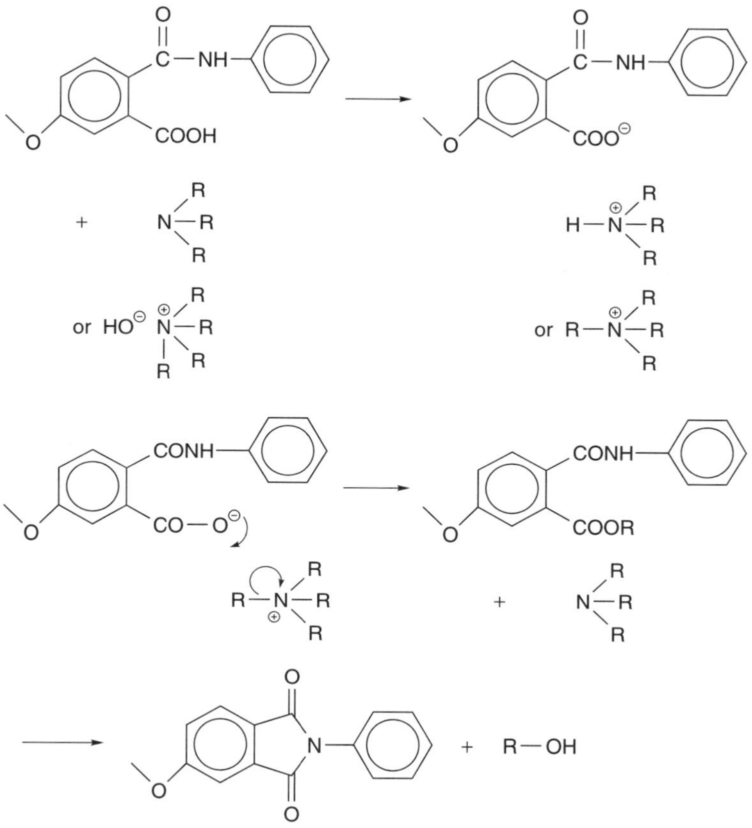

Figure 5.33 Polyimide cyclization via amic acid ammonium salt.

In fact, the 4- or 3-aminophthalic anhydride (4-APA and 3-APA) gives only intractable oligomers as pointed out later.[154] Interesting are the attempts of vapor-phase deposition of 4-APA, 3-APA, and 4-amino naphthalic anhydride (4-ANA). The sublimation of 3-APA gives only a low yield of polyimide. With both 4-APA and 4-ANA no sublimation took place and oligomeric products were formed before melting.[154]

Tractable polymers can be prepared when amino and anhydride functions are not located on the same aromatic ring, and different strategies were employed to obtain soluble polymer. AB benzhydrol imide was prepared by polycondensation of 4-(3-amino-1-hydroxymethylene) phthalic acid monomethyl ester in NMP. The polymer soluble in NMP has been used as adhesive and coating.[56] A second approach was based on an ether imide structure. AB aminophenylether phthalic acids (Fig. 5.34) were prepared by a multistep synthesis from bisphenols.[155] The products are stable as hydrochloride, and the polycondensation takes place by activation with triphenylphosphite. The polymers are soluble in an aprotic polar

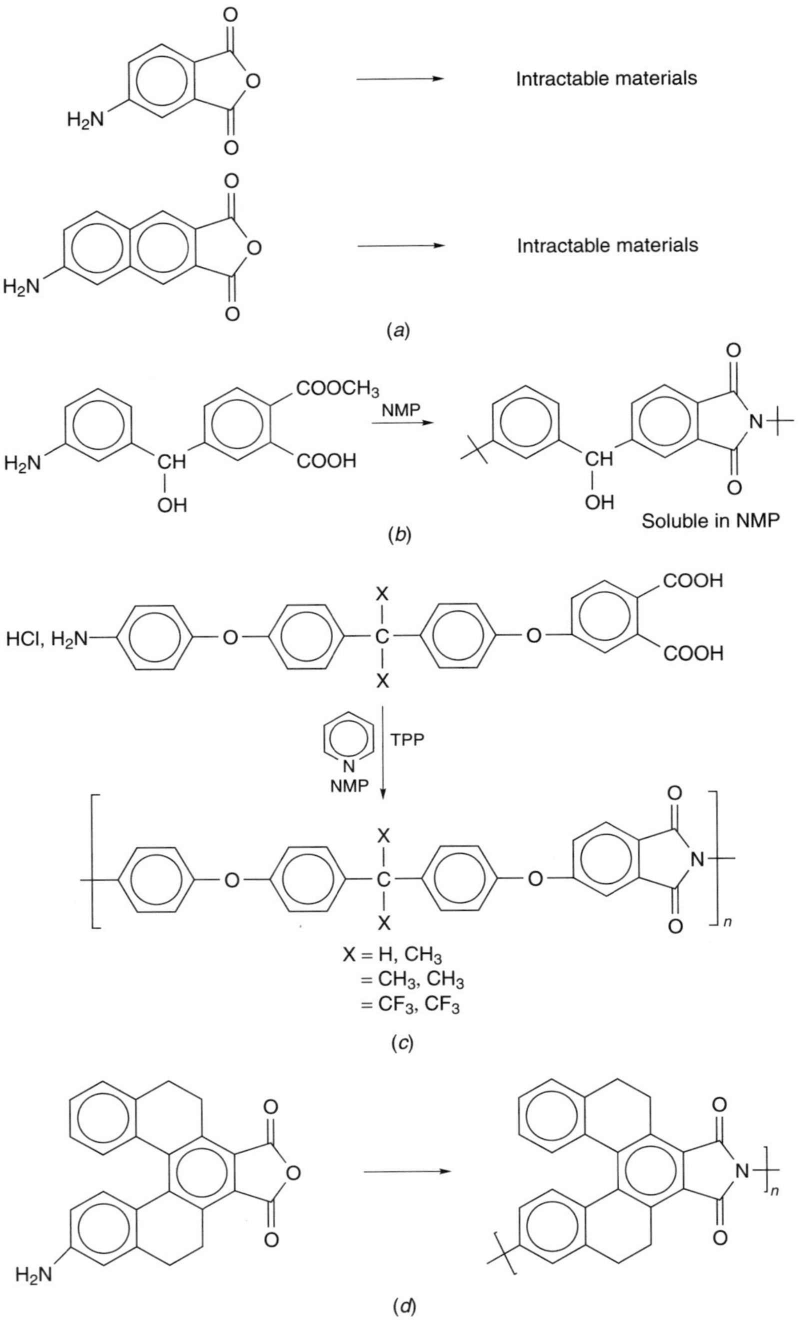

Figure 5.34 AB polyimide syntheses.

solvent such as DMAc, NMP, and *m*-cresol. The utilization of bulky structure was the third approach. The 3-amino-5,6,9,10-tetrahydro[5]helicene-7,8-dicarboxylic anhydride has been isolated in stable pure form. The polymerization takes place by heating at 200°C in cresol, giving a high-molecular-weight product.[156]

5.4.1.4 Hyperbranched Polyimides

The literature devoted to hyperbranched polyimides is not yet very abundant compared with other dendritic macromolecules. The one-pot polycondensation of an AB_x monomer leads to systems with a degree of branching dependent on polymerization degree[86] but always less than 100% due to partial formation of linear units. Considering an AB_2 monomer, a degree of branching of 100% can be reached if, after the reaction of A with one B, the reactivity of a second B is enhanced by a kind of anchimeric assistance according to the scheme in Fig. 5.35. B* will react more easily than the other B with A and the yield for each step will be close to 100%. A maleimido azine is basically an AB_2 monomer in which the maleimide is the A group and the azine function is a BB system because two maleimides are needed for one azine, as shown in Fig. 5.35.[157] An AB_2 monomer based on 3,5-dihydroxyphenyl-4-fluoro-phthalimide was prepared by reaction with tertiobutyldimethylchlorosilane[158]

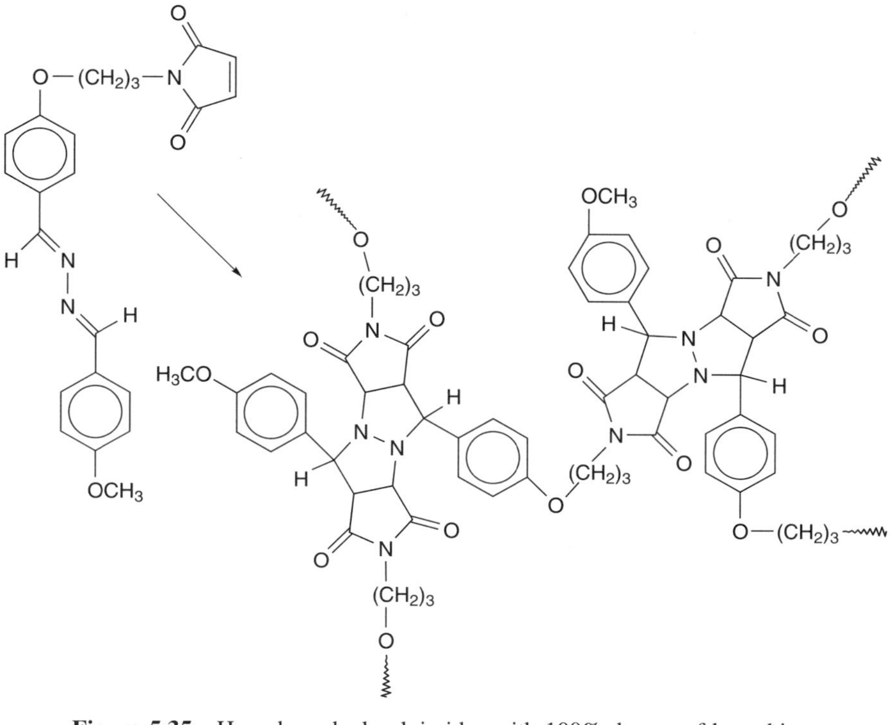

Figure 5.35 Hyperbranched polyimides with 100% degree of branching.

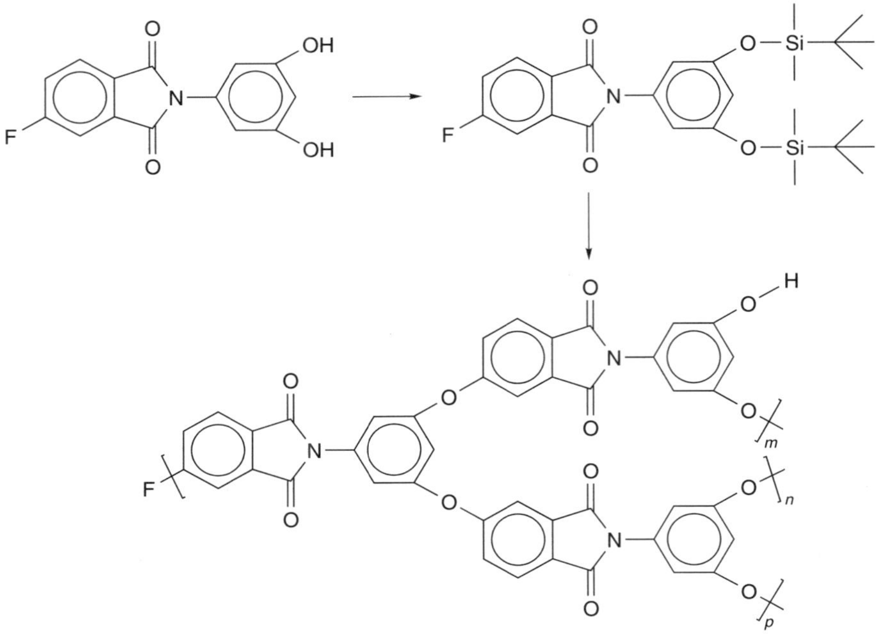

Figure 5.36 Hyperbranched polyimides via nucleophilic substitution.

(Fig. 5.36). The polycondensation took place by heating with CsF as a catalyst at 240°C in diphenylsulfone. The degree of branching of 65% has been determinated by proton NMR comparing with model. This hyperbranched polyetherimide is soluble in many organic solvents except paraffins, methanol, ethanol, and water.

Another synthesis of AB_2 hyperbranched polyimide based on the more classical reaction of an acid ester group (part A) with two amino groups (part B) was recently described.[88] The polycondensation was a two-step synthesis. First, a hyperbranched polyamide ester was prepared by polycondensation in the presence of 2,3-dihydro-2-thioxo-3-benzoxazolyl phosphonic acid diphenyl ester (DBOP). Second, a chemical cyclization was performed with acetic anhydride and pyridine, and the unreacted amino group undergoes an acetylation (Fig. 5.37). The degree of branching of 48% was calculated with proton NMR spectroscopy after identification of the signals corresponding to the dendritic, linear, and terminal aromatic protons with three model compounds. This type of hyperbranched polymer exhibits a good solubility in NMP and the DSC trace shows a glass transition temperature at 193°C.

An unusual method has been used to prepare a hyperbranched polyimide starting from two monomers: a difunctional A_2 and a trifunctional B_3. The gel formation can be avoided with careful control of the polycondensation conditions (molar ratio, order of the monomer addition, and low concentration). The A_2 and B_3 monomers were respectively 6FDA and tris(4-aminophenyl)

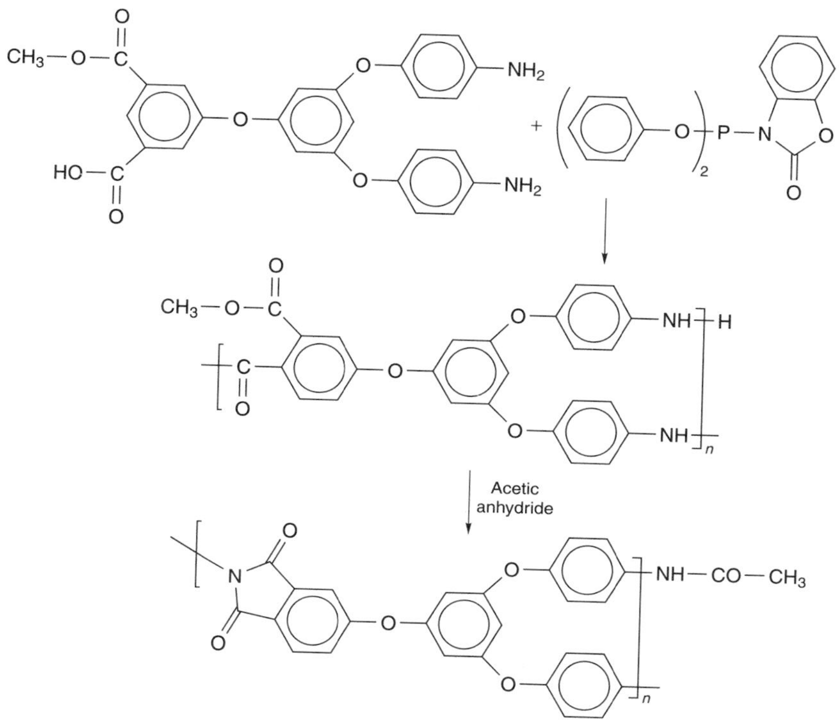

Figure 5.37 Hyperbranched polyimides in smooth conditions.

amine (TAPA). The molar ratio of 6FDA–TAPA was 1/1 when the dianhydride was added to a solution of TAPA and an amino-terminated hyperbranched poly-imide was obtained. When the TAPA was added to a dianhydride solution, a 2/1 6FDA–TAPA molar ratio is needed in order to get an anhydride-terminated polymer.[159] The degree of branching estimated by ^{1}H NMR determination was in the range of 70%. The glass transition temperature of the amino-terminated hyperbranched polyimide was a little bit higher than the one of the anhydride-terminated polymer (339 vs. 320°C).

5.4.2 Polyquinoxaline (PQ) and Polyphenylquinoxaline (PPQ)

The synthesis and properties of PQ and PPQ have been previously reviewed,[17,160] and this section will discuss mainly the recent trends in synthesis.

5.4.2.1 Synthesis by Formation of Heterocyclic Ring: Main Features

The basic reaction is the condensation of a bisorthodiamine with a bisethane-dione (Fig. 5.38). The first papers described the unsubstituted quinoxalines,[161] but the phenylquinoxalines described later[162] are more stable against oxydative

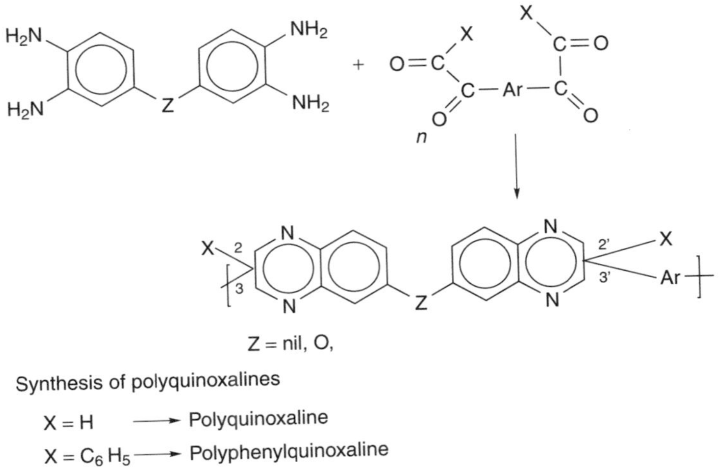

Z = nil, O,

Synthesis of polyquinoxalines

X = H $\longrightarrow$ Polyquinoxaline

X = C$_6$H$_5$ $\longrightarrow$ Polyphenylquinoxaline

Figure 5.38   Polyquinoxaline syntheses.

degradation and more soluble. The synthesis can occur in many solvents. However, *meta*-cresol is the best one for both kinetic and solubility aspects; PPQs are amorphous materials probably because three types of linkages (22', 23', 33') have been evidenced (Fig. 5.38) by ^{13}C NMR spectroscopy.[17] The relationship between structure and properties discussed in Section 5.2 has been observed for the polyquinoxalines. It can be pointed out that the effect of the chain termination on the stability against oxidation was clearly identified in the PPQ series. The stability decreases according to the following order: phenyl and capped PPQ > ketone-terminated PPQ > amine-terminated PPQ.[163]

5.4.2.2   Synthesis by Nucleophilic Displacement

The S$_N$Ar reaction takes place when a phenoxide reacts with an activated aromatic halogen or nitro group (Fig. 5.39). Takekoshi showed that an imide group can play the role of a withdrawing group and activate the nucleophilic displacement of a halogene or nitro group located on the 4-carbon.[164] Labadie et al.[165] investigated the withdrawing behavior of other heterocyclic groups such as quinoxaline, benzimidazole, benzoxazole, and benzothiazole. These heterocycles activate the fluoro (and in some cases the nitro) displacement from aromatic ring. The polycondensation is carried out in the presence of potassium carbonate in an anhydrous NMP–toluene mixture when the activating group is the phenylquinoxaline and in the NMP in the case of benzazole activation (Fig. 5.39). The bisphenates coming from bisphenol-A and bisphenol-6F were used, and the obtained polyarylether phenylquinoxalines showed T_g ranging between 230 and 270°C and ductile mechanical properties (elongation on the order of 20%). The synthesis by nucleophilic displacement does not avoid the handling of the aromatic tetraamines (carcinogenic products) for the synthesis

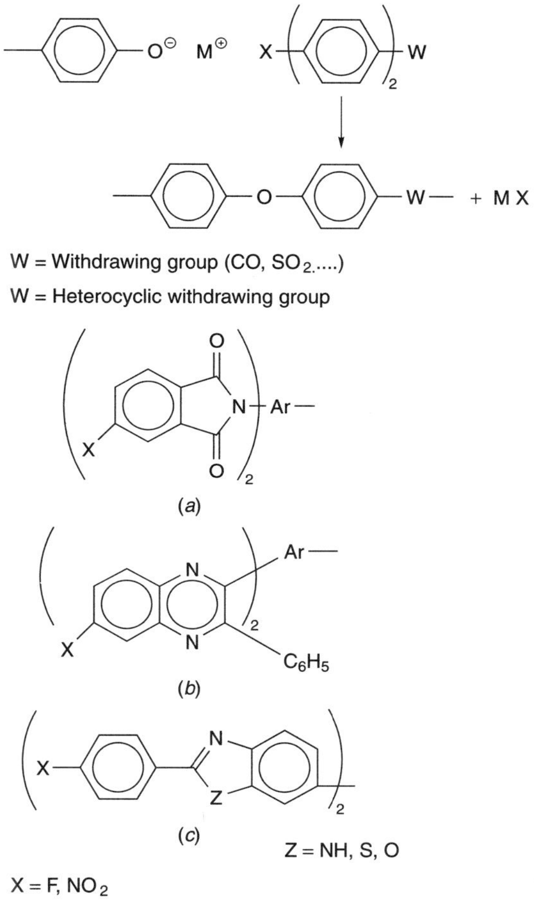

W = Withdrawing group (CO, SO$_2$....)
W = Heterocyclic withdrawing group

(a)

(b)

(c)   Z = NH, S, O

X = F, NO$_2$

Figure 5.39   Heterocyclic ring as activating group for nucleophilic substitutions.

of the bisfluoromonomers. The preparation of AB monomer is probably a safer approach. Recently Baek and Harris[166] prepared a new AB polyphenylquinoxaline precursor mixture of 3-(4-hydroxyphenyl)-2-phenyl-6-fluoro quinoxaline and isomer with the hydroxyphenyl group located on the 2-carbon. The synthesis of the monomer is shown for the last step in Fig. 5.40 and the polymerization is carried out in NMP–toluene in the presence of potassium carbonate. The viscosities of the AB PPQs obtained by this method ranged from 1.11 to 1.29 dL/g. The properties of these materials have been investigated as film adhesive and composite. The film had a tensile strength of 114 MPa, Young's modulus of 3.7 MPa, and elongation at break of 93%. The pristine resin exhibits a fracture energy (G_{1c}) of 2170 J/m^2 and a titanium–titanium lap shear strength of 31.4 MPa, but the high-temperature properties were limited by the T_g (257°C).[167]

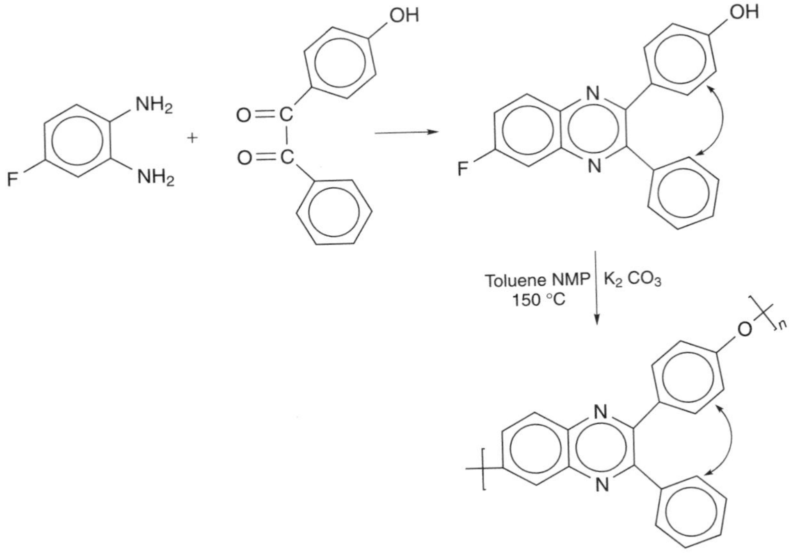

Figure 5.40 AB polyquinoxaline via nucleophilic substitutions.

5.4.2.3 Synthesis by Dehalogenative Organometallic Polycondensation

Various 1,5-dibromoquinoxalines have been polymerized by organometallic dehalogenation (Fig. 5.41). The reaction takes place in DMF with (1,5-cyclooctadiene) Ni(0) in the presence of 2,2'-bipyridine at 60°C for 48 h.[168,169] Highly conjugated acenaphthene quinoxalines were prepared by this procedure and exhibit photoluminescence peaks at 400 and 514 nm.[170]

5.4.2.4 Syntheses by Vapor Deposition Polymerization

1,1'-Phenylene-bis(2-phenylethanedione) and 3,3'-diaminobenzidine were evaporated in a vacuum chamber on a substrate. Proton NMR and GPC were used to monitor the ratio of evaporation rates on the stoichiometry. After deposition, the film was baked at 300°C and the formation of the polyquinoxaline was followed by the increase of the absorption at 390 nm.[171] An initial molecular weight ($\overline{M_n}$) of about 60,000 Da was obtained and this increased by baking at 300°C. The film was used as an electron transport layer on top of the poly-1,4-phenylenevinylene hole-transporting layer in organic light-emitting diodes (OLEDs).

5.4.2.5 Hyperbranched Polyphenylquinoxalines

The synthesis of polyether quinoxaline by nucleophilic displacement discussed in Section 5.4.2.2 has also been used for the preparation of hyperbranched polyquinoxaline based on AB$_2$ or A$_2$B monomers. In the AB$_2$ monomer the focal point is a single fluoro group (Fig. 5.42)[172]; in the A$_2$B the focal point is a single

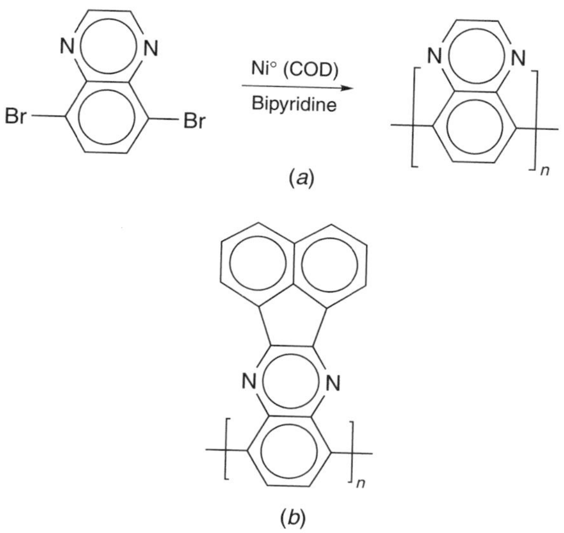

(a)

(b)

Figure 5.41   Polyquinoxaline syntheses by catalyzed dehalogenative polycondensation.

phenolic group.[173] The polycondensations were carried out in NMP, NMP/CHP (N-cyclohexylpyrrolidone), or DMPU (N,N'-dimethylpropylene urea). It has been pointed out[172] that MALDI-TOF and ^{1}H NMR analyses showed total fluorine elimination. Il was concluded that an intramolecular ring closure took place during the polycondensation.

5.4.3 Polybenzimidazoles

The chemistry and properties of polybenzimidazole (PBI) have been recently reviewed by Chung,[174] and the paper discusses the two main syntheses of PBI (Fig. 5.43): condensation of the 3,3',4,4'-tetraaminobiphenyl ether with the diphenyl ester of an aromatic acid (in two steps) or with an aromatic diacid in the presence of phosphorous catalyst.[175] An amazing new synthetic route to polybenzimidazole has been published by a Japanese team.[176] One mole of 3,3',4,4'-tetraaminobiphenyl reacts with 2 mol of ferrocene dimethanol in the presence of $RuCl_2 P(C_6H_5)_3$ as a catalyst. The polycondensation gives a high-molecular-weight polybenzimidazole ($\overline{M}_n = 2.1 \times 10^4$ Da). The reaction of only 1 mol of ferrocenediol gives only a low yield of the polymer. An interesting chemoselective sequence of reactions takes place: one alcoholic function reacts with the tetraamine giving a benzimidazole end capped with a ferrocenyl alcohol. The second step is the formation of a polyether (Fig. 5.44). During the 1960s and 1970s, the polybenzimidazoles were mainly developed for composite, adhesive, and fiber applications, but now, new research areas are opened in the field of membranes for fuel cells[177,178] or supported catalysts.[179]

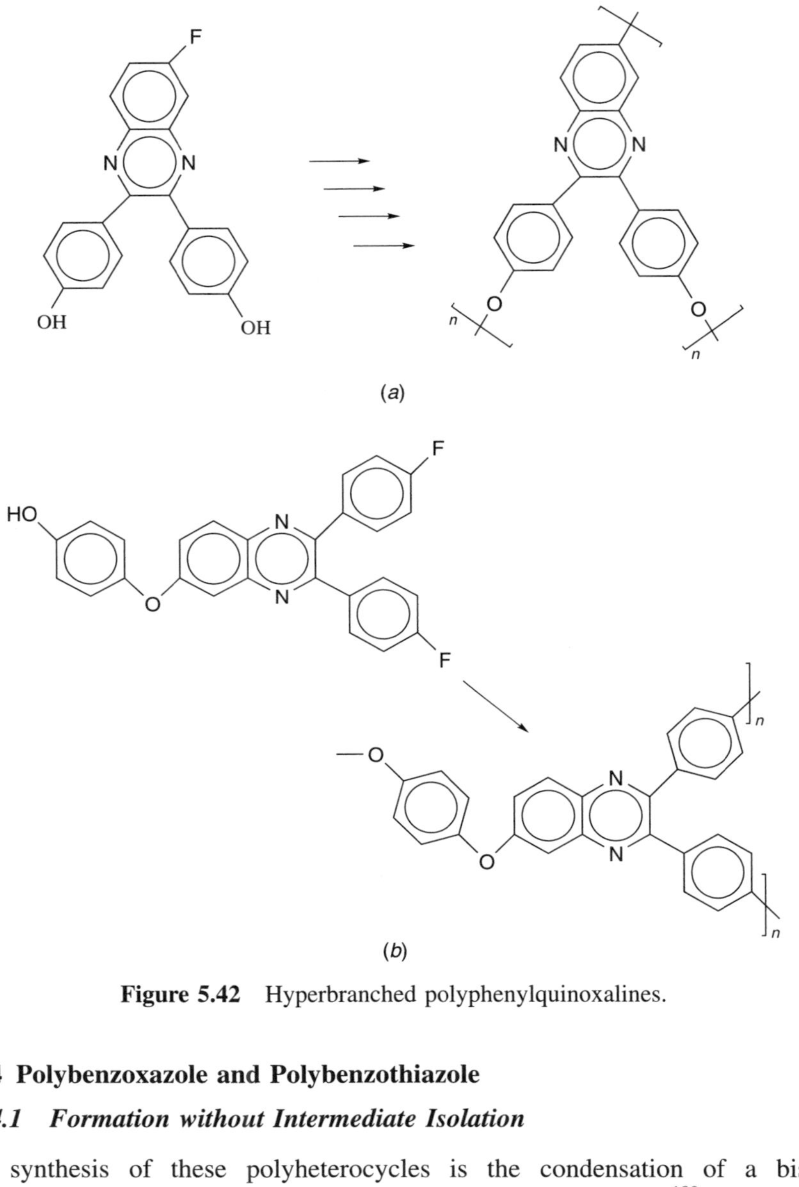

(a)

(b)

Figure 5.42 Hyperbranched polyphenylquinoxalines.

5.4.4 Polybenzoxazole and Polybenzothiazole

5.4.4.1 Formation without Intermediate Isolation

The synthesis of these polyheterocycles is the condensation of a bis-*o*-amino phenol or bis-*o*-aminothiol with a diacid diphenylester[180] (Fig. 5.45). Rigid rodlike structures were obtained by the condensation of 2,5-diamino-1,4-benzene dithiol dihydrochloride[181] and 4,6-diamino-1,3-benzenediol[182] with terephthalic acid in polyphosphoric acid. The reaction of 5,5'-dicarbonyl-2,2'-bipyridine dichloride with the same monomer gives high-molecular-weight polymer exhibiting also lyotropic behavior.[183] These polymers displayed electrical conductivity of 10^{-10}–10^{-8} S/cm, which was

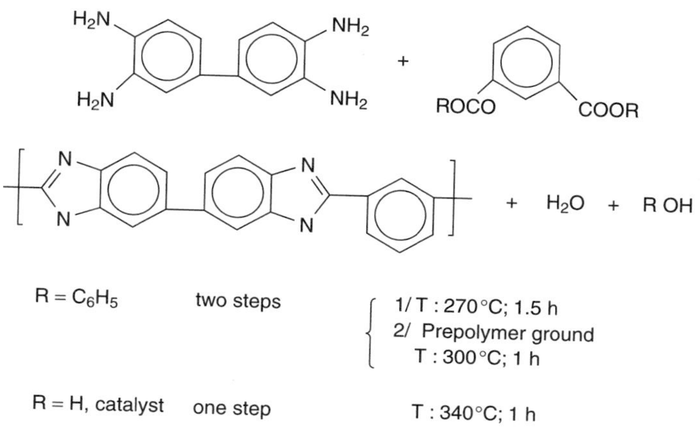

Figure 5.43   Polybenzimidazole syntheses.

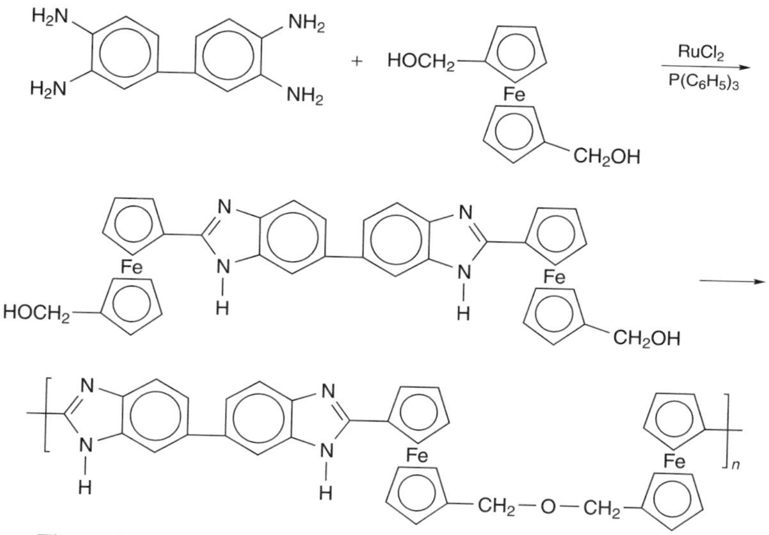

Figure 5.44   Synthesis of ferrocene-containing polybenzimidazole.

increased by introducing Ag^+ ion ($10^{-5}-10^{-4}$ S/cm) or Ag^0 (25–42 S/cm). Amorphous polybenzoxazoles have been prepared by condensation of aromatic diacid containing arylether groups with commercial 3,3'-dihydroxy-4,4'-(hexafluoroisopropylidene)bis-2-aminophenol. The condensation was carried out in a mixture of phosphorus pentoxide–methanesulfonic acid (PPMA) in sulfolane or trimethylsilylpolyphosphate (PPSE) in dichlorobenzene (Fig. 5.46). The glass transition temperatures of these polymers were detected in a temperature range of 236–270°C, but only the polymers containing the 6F group were soluble in organic solvents.[184]

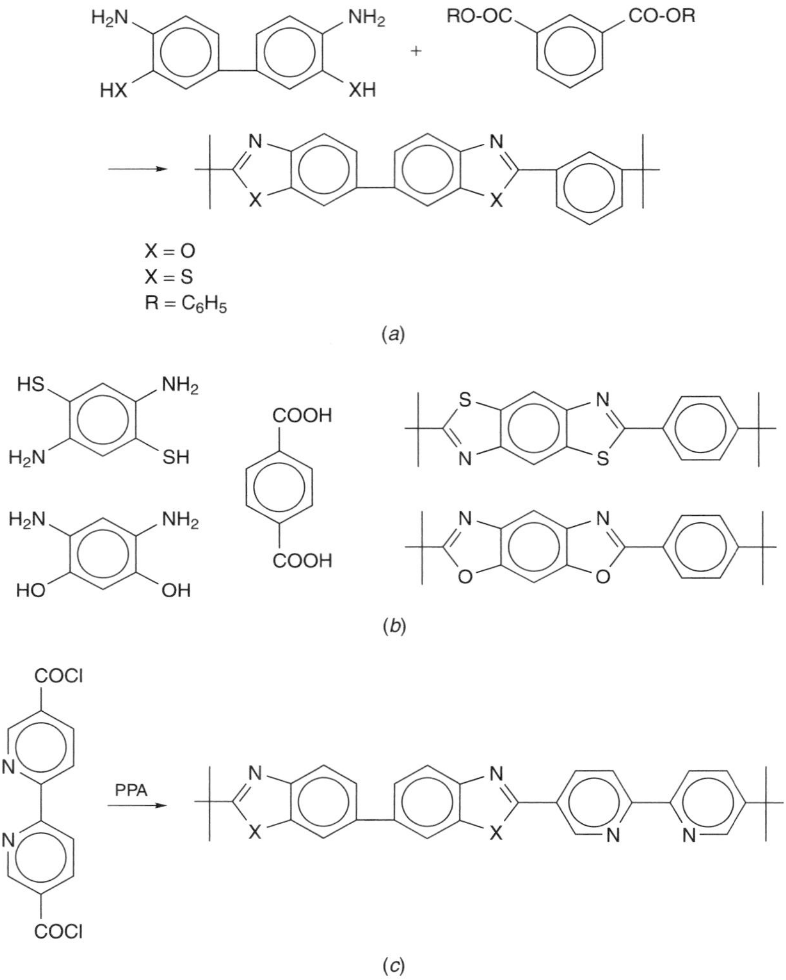

X = O
X = S
R = C₆H₅

(a)

(b)

(c)

Figure 5.45   Polybenzoxazole and polybenzothiazole syntheses.

5.4.4.2   Synthesis with Isolation of an Intermediate

The low condensation of diacyl chloride and bis-*ortho*-aminophenol gave a polyhydroxyamide that was submitted to a thermal dehydration.[185–187] The two functions of an aminophenol could be acylated, and the question of chemoselective acylation has been discussed.[188] The problem is not O-acylation versus N-acylation because the N-acylated product is the more stable isomer, but the question is how to control N-monoacylation versus diacylation that could reduce the molecular weight. It has been shown that the condensation in the presence of inorganic salt (LiCl) in NMP gives selectively the N-monoacylated

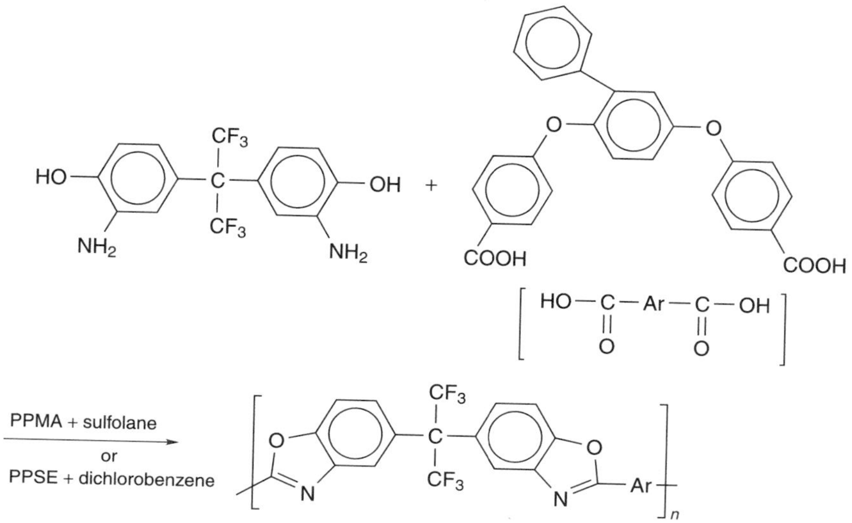

Figure 5.46 Amorphous soluble fluorinated polybenzoxazole.

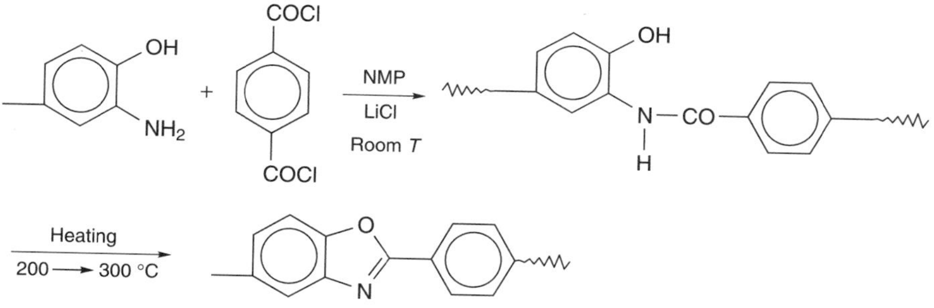

Figure 5.47 Synthesis of polybenzoxazole with isolation of polyamide phenol intermediate.

product (Fig. 5.47). The thermal cyclodehydration of the polyhydroxyamide precursor to polybenzoxazole has been compared with the cyclodehydration of the polyamic acid precursor to polyimide. The comparison has been done by heating precursor film; cyclization onset ranges from 255–309°C for the benzoxazole precursor compared to 175–234°C for the imide precursor.[189] An interesting modification of a polyhydroxyamide was performed with ditertiobutylcarbonate in γ-butyrolactone. The modified polymer became insoluble in aqueous bases. After formulation with a photoacid generator (PAG), the UV irradiation generated an acid that cleaved the *t*-boc protecting group leading to a polymer soluble in aqueous basic medium. Heat treatment of protected and unprotected precursor gives the polybenzoxazole (Fig. 5.48).[190] Such a chemical approach can be used

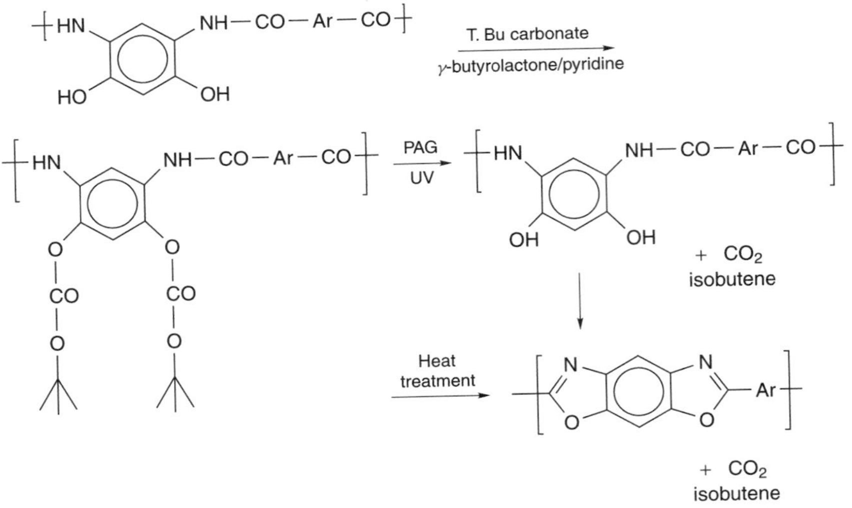

Figure 5.48   Polybenzoxazole synthesis via protected intermediate.

for the development of positive photosensitive thermostable material in the semi-conductor industry. Hydroxy-containing polyimides were prepared by reaction of dianhydride and bis-*o*hydroxyamine. These polyimides with a pendant hydroxyl group ortho to the imide nitrogen were found to undergo thermal conversion at a temperature up to 500°C. That conversion leads to a polyimide with carbon dioxide evolution[191] (Fig. 5.49).

5.4.4.3 Properties

The hydrolytic stability of the benzobisoxazole and benzobisthiazole has been investigated. Either in acidic conditions at 100°C for 21 days or in sodium hydroxyde solution, the model compounds and a poly(*p*-phenylene) benzobisox-azole did not show any decomposition. In very severe conditions such as in methanesulfonic acid containing water medium, generation of carboxylic acid was observed.[192] As a conclusion, PBO and PBT are probably more stable in hydrolytic media than polyimides. Concerning the mechanical properties, the highly oriented fibers exhibit fascinating tensile strength and tensile modulus

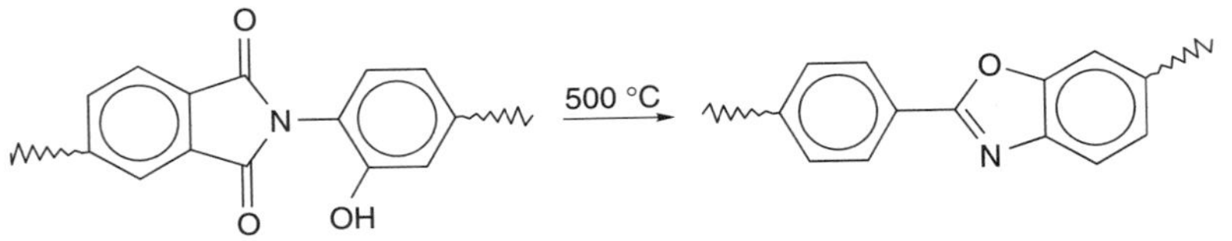

Figure 5.49   Polybenzoxazole synthesis via hydroxyphenylimide.

properties. However, improvement is needed in compressive strength. A recent review discussing the possible improvement of compressive strength by lateral interaction between the chains has been published.[193] This paper also deals with the synthesis of the monomer 1-5-dihydroxy-2,4-diaminobenzene (DADHB) either from 1,2,3-trichlorobenzene or from resorcinol. The formation of PBO in a polyphosphoric acid–P_2O_5 mixture is discussed and a step growth based on the formation of oligomeric PBO end capped with DADHB is proposed. Different approaches based on lateral substitution of the PBO backbone are presented in order to increase the interaction between the rodlike chains (chemical interaction with crosslinking or physical entanglements have been studied). The conclusions are a bit pessimistic concerning improvement in the compressive properties of the fibers. However, film properties were improved by introducing a sol–gel glass or a thermosetting matrix.

REFERENCES

1. P. W. Morgan, *Condensation Polymers: By Interfacial and Solution Methods*, Interscience, New York, 1965.
2. J. Economy, R. S. Storm, V. I. Matkovich, S. G. Cottis, and B. E. Novak, *J. Polym. Sci., Polym. Chem. Ed.*, **14**, 2207 (1976).
3. H. Vogel and C. S. Marvel, *J. Polym. Sci.*, **50**, 511 (1961).
4. R. J. Cotter and M. Matzner, *Ring Forming Polymerization, Part B2: Heterocyclic Rings*, Academic, New York, 1972.
5. P. E. Cassidy, *Thermally Stable Polymers, Syntheses and Properties*, Marcel Dekker, New York, 1980.
6. H. D. Stenzenberger, *Adv. Polym. Sci.*, **117**, 163 (1994).
7. P. Mison and B. Sillion, *Adv. Polym. Sci.*, **140**, 137 (1999).
8. M. R. Tant, H. L. N. MacManus, and M. E. Rogers, *ACS Symp.*, **603**, 1 (1995).
9. R. Young, *Introduction to Polymeric Materials*, Chapman and Hall, London, 1981, p. 204.
10. T. L. Saint Clair, in *Polyimides*, D. Wilson, H. D. Stenzenberger, and P. M. Hergenrother (Eds.), Blackie, London and Glasgow, 1990, p. 55.
11. C. E. Sroog, *J. Polym. Sci., Macromol. Rev.*, **11**, 161 (1976).
12. C. J. Lee, *J. Macromol. Sci., Rev. Macromol. Chem. Phys.*, **C29**, 431 (1989).
13. F. W. Harris and L. H. Lannier, *Structure-Solubility Relationships in Polymers*, Academic, New York, 1987.
14. B. Sillion and G. Rabilloud, *New Methods in Polymer Synthesis*, Vol. 2, J. R. Ebdon and G. C. Eastmond (Eds.), Blackie, London, 1995 p. 236.
15. M. I. Bessonov, M. M. Koton, V. V. Kudryavtsev, and L. A. Laius, *Polyimides— Thermally Stable Polymers*, Plenum, New York, 1987.
16. M. I. Bessonov and V. A. Zubkov, *Polyamic Acids Polyimides*, CRC, Boca Raton, FL, 1993.
17. B. Sillion, *Comprehensive Polymer Science*, Vol. 5, Pergamon, London, 1989.

18. D. Wilson, H. D. Stenzenberger, and P. M. Hergenrother, *Polyimides*, Chapman and Hall, New York, 1989.

19. M. K. Ghosh and K. L. Mittal (Eds.), *Polyimides Fundamental and Applications*, Marcel Dekker, New York, 1996.

20. J. H. de Abajo and J. G. de la Campa, *Adv. Polym. Sci.*, **140**, 23 (1999).

21. P. M. Hergenrother, *SAMPE J.*, **35**, 30 (1999).

22. X. Fang, D. F. Rogers, D. A. Scola, and P. M. Stevens, *J. Polym. Sci., Part A: Polym. Chem.*, **36**, 461 (1998).

23. D. K. Brandom, J. P. Desoura, D. J. Baird, and G. L. Wilkes, *J. Appl. Polym. Sci.*, **66**, 1543 (1997).

24. Y. Imai, *Adv. Polym. Sci.*, **140**, 1 (1999).

25. Y. Imai, T. Fueki, T. Inoue, and M. A. Kakimoto, *J. Polym. Sci. Part A: Polym. Chem.*, **36**, 1341 (1998).

26. T. Takekoshi, in *Polyimides Fundamental and Applications*, M. K. Ghosh and K. L. Mittal (Eds.), Marcel Dekker, New York, 1996, p. 7.

27. D. Picq, L. Ben Mahdi, K. Soulard, and B. Sillion, *Polymer*, **40**, 5219 (1999).

28. F. E. Arnold, Jr. and F. E. Arnold, *Adv. Polym. Sci.*, **117**, 257 (1994).

29. T. Takeichi, K. Nakajima, M. Zuo, and R. Yokota, *High Perform. Polym.*, **10**, 111 (1998).

30. K. W. Lienert, in *Progress in Polyimides*, Vol. 2, H. R. Kricheldorf (Ed.), Springer-Verlag, Berlin and Heidelberg, 1999, p. 45.

31. J. H. Lai, *Polymers for Electronic Applications*, CRC Press, Boca Raton, FL, 1989, p. 34.

32. C. Feger and H. Franke, in *Polyimides Fundamental and Applications*, M. K. Ghosh and K. L. Mittal (Eds.), Marcel Dekker, New York, 1996, p. 759.

33. G. D. Hietpas, J. M. Sands, and D. L. Allara, *Macromolecules*, **31**, 3374 (1998).

34. Y. Wang, C. Xu, A. Kanazawa, T. Shiono, T. Ikedo, Y. Matsuki, and Y. Takeushi, *J. Appl. Phys.*, **84**, 181 (1998).

35. K. W. Lee, S. A. Lien, S. H. Paek, and C. Durning, *Ann. Tech. Conf. Soc. Plast. Eng.*, **2**, 2623 (1995).

36. B. Li, T. He, and M. Ding, *Thin Solid Films*, **320**, 280 (1998).

37. G. Hougham, G. Tesoro, A. Viehbeck, and J. D. Chapple-Sokol, *Macromolecules*, **27**, 5964 (1994).

38. B. C. Auman, *Mat. Res. Soc. Proc.*, **381**, 19 (1995).

39. K. Carter, *Mat. Res. Soc. Proc.*, **476**, 87 (1997).

40. B. Sillion, G. Rabilloud, J. Garapon, O. Gain, and J. Vallet, *Mat. Res. Soc. Proc.*, **381**, 93 (1995).

41. K. Horie and T. Yamashita, *Photosensitive Polyimides*, Technomic, Lancaster, PA, 1995.

42. L. R. Dalton, in *Electrical and Optical Polymer Systems*, D. L. Wise, G. E. Wnek, D. J. Transtolo, T. M. Cooper, and J. D. Gresser (Eds.), Marcel Dekker, New York, 1998, p. 609.

43. Z. Sekkat, A. Knoessen, V. Y. Lee, and R. D. Miller, *J. Polym. Sci., Part B: Polym. Phys.*, **36**, 1669 (1998).

44. R. Meyrueix and J. C. Dubois, *Matériaux Polymères, enjeux et perspectives*, Masson, Paris, 1995, p. 204.

45. (a) I. K. Spiliopoulos and J. A. Mikroyannidis, *Macromolecules*, **31**, 515 (1998); (b) S. Janietz, A. Wedel, R. Friedrich, and S. Anlauf, *Polym. Prepr.*, **40**, 1219 (1999); (c) D. O. Briens, A. Bleyer, D. D. C. Bradley, and S. Meng, *Synth. Meth.*, **76**, 105 (1996); (d) Y. Cui, X. Zhang, and S. A. Jenekhe, *Macromolecules*, **32**, 3824 (1999); (e) Y. F. Wang, T. M. Chen, K. Okada, M. Uekawa, and T. Nakaya, *Macromol. Chem. Phys.*, **199**, 1263 (1998).

46. J. Kobayashi, T. Matsuura, S. Sazaki, and T. Maruno, *Appl. Opt.*, **37**, 1032 (1998).

47. M. Langsam, in *Polyimides Fundamental and Applications*, M. K. Ghosh and K. L. Mittal (Eds.), Marcel Dekker, New York, 1996, p. 697.

48. Anonymous, *Metal. Rev.*, **43**, 14 (1999).

49. O. Savogado and P. R. Roberge (Eds.), *Proceedings of the Second International Symposium on New Material for Fuel Cell and Modern Battery Systems*, Ecole Polytechnique de Montréal, 1997.

50. F. Selampinar, U. Akbulut, T. Yilmaz, A. Gungor, and L. Toppare, *J. Polym. Sci., Part A: Polym. Chem.*, **35**, 3009 (1997).

51. (a) T. P. Mc Andrew, S. A. Miller, A. G. Gilicinski, and L. M. Robeson, *ACS Symp. Ser.*, **689**, 396 (1998); (b) M. Han and D. E. Nikles, *Polym. Prep.*, **40**, 849 (1999).

52. Anonymous, *Res. Discl.*, **409**, 542 (1998).

53. H. Kawakame, Y. Mori, J. Takagi, S. Nagaoka, T. Kanamori, T. Shimbo, and S. Kubota, *ASAIO J.*, **43**, M490 (1997); *Chem. Abstr.*, **128**, 26896 (1998).

54. I. A. Ronova and S. S. A. Pavlova, *High Perform. Polym.*, **10**, 309 (1998).

55. F. W. Harris, B. H. Ahn, and S. Z. D. Chang, *Polymer*, **34**, 3083 (1993).

56. J. Malinge, J. Garapon, and B. Sillion, *Br. Polym. J.*, **20**, 431 (1988).

57. D. J. Liaw, B.-Y. Liaw, L.-J. Li, B. Sillion, R. Mercier, and R. Thiria, *Chem. Mat.*, **10**, 734 (1998).

58. P. M. Hergenrother and H. H. Levine, *J. Polym. Sci. A1*, **5**, 1453 (1967).

59. (a) T. Pascal, B. Sillion, F. Grosjean, M. F. Grenier-Loustalot, and P. Grenier, *High Perform. Polym.*, **2**, 95 (1990); (b) D. Ayala, A. E. Lozano, J. de Abajo, and J. G. de la Campa, *J. Polym. Sci., Part A: Polym. Chem.*, **37**, 3377 (1999).

60. A. K. Saint Clair and W. S. Slemp, *SAMPE J.*, **21**, 28 (1985).

61. K. Han, H.-J. Lee, and T. H. Rhee, *J. Appl. Polym. Sci.*, **74**, 107 (1999).

62. B. Li, L. A. Prexta, S. Zhihao Shen, Z. D. Cheng, and F. Harris, *Polym. Prep.*, **41**, 105 (2000).

63. D. J. Liaw and B.-Y. Liaw, *Polymer*, **40**, 3183 (1999).

64. W. Volksen, H. J. Cha, M. I. Sanchez, and D. Y. Yoon, *React. Funct. Polym.*, **30**, 61 (1996).

65. F. Turpin, P. Mison, and B. Sillion, *Polyimides Trends in Materials and Applications*, C. Feger, M. M. Khojasteh, and S. E. Molis (Eds.), Hopewell, New York, 1996, p. 169.

66. T. Matsumoto and T. Kurosaki, *React. Funct. Polym.*, **30**, 55 (1996).

67. H. Suzuki, T. Abe, K. Takaishi, M. Narita, and I. Hamada, *J. Polym. Sci., Part A: Polym. Chem.*, **38**, 108 (2000).

68. T. Ichino, S. Suzuki, T. Matsuura, and S. Nishi, *J. Polym. Sci., Part A: Polym. Chem.*, **28**, 323 (1990).

69. B. C. Auman, T. L. Myers, and D. P. Higley, *J. Polym. Sci., Part A: Polym. Chem.*, **35**, 2441 (1997).

70. G. Boiteux, J. M. Oraison, G. Seytre, M. Senneron, and B. Sillion, *Electr. Opt. Acoust. Prop. Polym. Int. Conf.*, **2**, P18/1 (1990).

71. A. K. Saint Clair, T. L. Saint Clair, and W. P. Winfree, *Polym. Mat. Sci. Eng.*, **59**, 28 (1988).

72. T. Matsuura, M. Ishizawa, Y. Hasuda, and S. Nishi, *Macromolecules*, **25**, 3540 (1992).

73. S. Yoshida and A. S. Hay, *Macromolecules*, **30**, 3979 (1997).

74. I. I. Ponomarev, O. G. Nikol'skii, J. A. Volkova, and A. V. Zakharov, *Visohomol. Soed.*, **A36**, 1429 (1994).

75. F. E. Arnold, Jr. and F. E. Arnold, *Adv. Polym. Sci.*, **117**, 45 (1994).

76. P. M. Hergenrother, in *Polyimides and Other High Temperature Polymers*, M. Abadie and B. Sillion (Eds.), Elsevier, Amsterdam, 1991, p. 1.

77. T. Briffaud, J. Garapon, and B. Sillion, *High Performance Polym.*, **13**, S197 (2001).

78. C. Genies, R. Mercier, B. Sillion, N. Cornet, G. Gebel, and M. Pineri, *Polymer*, **42**, 359 (2001).

79. W. Wolksen and P. M. Cotts, in *Polyimides*, K. L. Mittal (Ed.), Plenum, New York, 1984 p. 163.

80. P. J. Flory, *J. Am. Chem. Soc.*, **63**, 3083 (1941).

81. (a) D. A. Tomalia, A. N. Naylor, and W. A. Goddard III, *Angew. Chem. Int. Ed. Engl.*, **29**, 138 (1990); (b) C. J. Hawker and J. M. J. Frechet, *J. Am. Chem. Soc.*, **112**, 7638 (1990).

82. C. J. Hawker, R. Lee, and J. M. J. Frechet, *J. Am. Chem. Soc.*, **113**, 4583 (1991).

83. C. J. Hawker and J. M. J. Frechet, *ACS. Symp. Ser.*, **624**, 132 (1996).

84. S. R. Turner, B. I. Voit, and T. H. Mourey, *Macromolecules*, **27**, 1611 (1994).

85. J. M. J. Frechet, *Science*, **263**, 1710 (1994).

86. A. Hult, J. Mats, and E. Malmström, *Adv. Polym. Sci.*, **143**, 1 (1999).

87. I. Sendijarevic and J. Mac Hugh, *Macromolecules*, **33**, 590 (2000).

88. D. Scott Thomson, L. J. Markoski, and J. S. Moore, *Macromolecules*, **32**, 4764 (1999).

89. D. J. Brunelle and D. A. Singleton, *Tetrahed. Lett.*, **25**, 3383 (1994).

90. U. Hoffman, F. Helmer Metzman, M. Klappen, and K. Mullen, *Macromolecules*, **27**, 3575 (1994).

91. Y. S. Vydgoskii, *Macromol. Symp.*, **128**, 71 (1998).

92. W. Xie, Y. Jin, and P. G. Wang, *Chemtech*, **29**, 23 (1999).

93. J. H. Forsberg, V. T. Spaziano, T. M. Balasubramanian, G. K. Liu, S. A. Kinsley, C. A. Duckworth, J. J. Poteruca, P. S. Brown, and J. L. Miller, *J. Org. Chem.*, **52**, 1017 (1987).

94. I. Colon and G. T. Kwiatkowski, *J. Polym. Sci., Polym. Chem. Ed.*, **28**, 367 (1990).

95. T. Yamamoto, N. Hayashida, and T. Maruyama, *Macromol. Chem. Phys.*, **198**, 341 (1997).

96. (a) V. Percec, J. Y. Bae, M. Zhao, and D. H. Hill, *J. Org. Chem.*, **60**, 176 (1995); (b) V. Percec, M. Zhao, J. Y. Bae, and D. H. Hill, *Macromolecules*, **29**, 3727 (1996).

97. N. Miyaura and A. Suzuki, *Chem. Rev.*, **95**, 2457 (1995).

98. D. Brown, R. S. Clary, C. D. Lee, W. G. Monroe, D. A. Vaughn, R. T. Ragheb, J. W. Erter III, and D. A. Shiraldi, *Polym. Prepr.*, **41**, 123 (2000).

99. L. Smitz, M. Rebahn, and M. Ballauff, *Polymer*, **34**, 646 (1993).

100. R. J. Perry, R. Turner, R. W. Blevins, and B. D. Wilson, *Polym. Prepr.*, **35**, 706 (1994).

101. Y. Yto, E. Ihara, T. Uesaka, and M. Murakami, *Macromolecules*, **25**, 6711 (1992).

102. S. L. C. Hsu, A. Naiini, W. D. Weber, and A. J. Blakeney, U.S. Patent 5,789,524, 1998.

103. N. Takeda, H. Makabe, T. Hirano, and T. Takeda, Japanese Patent 09329893, 1997; *Chem. Abstr.*, **128**, 147497 (1998).

104. I. Ikeda, W. Yamashita, and S. Tamai, Japanese Patent 10175952, 1998; *Chem. Abstr.*, **129**, 81662 (1998).

105. J. P. Marasco, J. Garapon, and B. Sillion, *Mater. Res. Soc. Symp. Proc.*, **476**, 249 (1997).

106. J. C. Jung and S. B. Park, *Polym. Bull.*, **35**, 423 (1995).

107. K. Seto, M. Okamoto, I. Tomiaka, and Y. Echigo, Japanese Patent 09263698, 1997; *Chem. Abstr.*, **127**, 332324 (1997).

108. D. J. Liaw and B. Liaw, *Macromolecules*, **32**, 7248 (1999).

109. S. Tamai, A. Yamaguchi, and M. Ohta, *Polymer*, **37**, 3683 (1996).

110. T. Pascal, R. Mercier, and B. Sillion, *Polymer*, **30**, 739 (1989).

111. K. Smith and D. Jones, *J. Chem. Soc. Perkin Trans I*, 407 (1997).

112. B. Li, L. A. Prexta, Z. Shen, S. Z. D. Cheng, and F. W. Harris, *Polym. Prep.*, **41**, 105 (2000).

113. T. Matsuura, Y. Hasuda, S. Nishi, and N. Yamada, *Macromolecules*, **24**, 5001 (1991).

114. B. C. Auman, T. L. Myers, and D. P. Higley, *J. Polym. Sci., Polym. Chem.*, **35**, 2441 (1997).

115. A. E. Feiring, B. C. Auman, and E. R. Wonchoba, *Macromolecules*, **26**, 2779 (1993).

116. M. F. Martinez-Nuñez, V. N. Sekharipuram, and J. E. McGrath, *Polym. Prep.*, **35**, 709 (1994).

117. A. L. Rusanov, L. G. Konomarova, M. D. Dutov, I. A. Vatsadze, and O. V. Serushkina, *Macromol. Symp.*, **122**, 123 (1997).

118. K. Sripadaj, N. Gupta, and I. K. Varma, *Proc. IUPAC Int. Symp. Polym. Sci. Technol.*, **1**, 134 (1998).

119. X. Sum, Y.-K. Yang, and F. Lu, *Macromolecules*, **31**, 4291 (1998).

120. S. M. Pyo, S. I. Kim, T. J. Shin, H. K. Park, M. Ree, K. H. Park, and J. S. Kang, *Macromolecules*, **31**, 4777 (1998).

121. M. Haubs, P. Foley, and D. L. Cangiano, U.S. Patent 5,710,241, (1998).

122. K. Y. Song and J. A. Moore, *Polym. Prep.*, **41**, 192 (2000).

123. H. Suzuki, T. Abe, K. Takaishi, M. Narita, and F. Hamada, *J. Polym. Sci., Polym. Chem.*, **38**, 108 (2000).

124. D. Ayala, A. E. Lozano, J. De Abajo, and J. De La Campa, *J. Polym. Sci., Polym. Chem.*, **37**, 3377 (1999).

125. T. Matsumoto and C. Feger, *J. Photo Polymer Sci. Techn.*, **11**, 231 (1998).

126. W. Volksen, H. J. Cha, M. I. Sanchez, and D. Y. Yoon, *React. Funct. Polym.*, **30**, 61 (1996).

127. C. Tamagna, P. Mison, T. Pascal, R. Petiaud, and B. Sillion, *Polymer*, **40**, 5523 (1999).

128. R. G. Ferrillo and P. J. Achorn, *J. Appl. Polym. Sci.*, **64**, 191 (1997).

129. S. I. Kim, S. M. Pyo, and M. Ree, *Macromolecules*, **30**, 7890 (1997).

130. J. D. Menczel, *J. Therm. Anal. Calor.*, **59**, 1023 (2000).

131. M. W. Snider, in *Polyimides: Material Chemistry and Characterization*, C. Feger, M. M. Khojastch, and J. E. Mc Grath, (Eds.), Elsevier, Amsterdam, 1989 p. 363.

132. C. A. Pride, *J. Polym. Sci., Part A: Polym. Chem.*, **27**, 711 (1989).

133. A. K. Saini and H. H. Paterson, *Polym. Prepr.*, **32**, 301 (1991).

134. C. D. Smith, R. Mercier, H. Waton, and B. Sillion, in *Fluoropolymers 2: Properties*, G. Hougham (Ed.), Plenum, New York, 1999, pp. 371–399.

135. P. D. Murphy, R. A. Di Pietro, C. J. Lund, and W. D. Weber, *Macromolecules*, **27**, 279 (1994).

136. Y. C. Jean, H. Cao, X. Hong, Y. Gu, C. M. Huang, C. Y. Chung, and J. Liu, *PMSE*, **75**, 86 (1996).

137. J. J. Singh, *Nucl. Instr. Meth. Phys. Res.*, **B79**, 349 (1993).

138. G. C. Eastmond, J. H. Daly, A. S. McKinnon, and R. A. Pethrick, *Polymer*, **40**, 3605 (1999).

139. K. Tanaka, T. Kawai, H. Kita, K. I. Okamoto, and Y. Ito, *Macromolecules*, **33**, 5513 (2000).

140. R. W. Snider, B. Thomson, B. Bartges, D. Czerniawski, and P. C. Painter, *Macromolecules*, **22**, 4166 (1989).

141. T. Takekoshi, in *Polyimides Fundamental and Applications*, M. K. Ghosh and K. L. Mittal (Eds.), Marcel Dekker, New York, 1996, p. 38.

142. M. E. Rogers, T. E. Glass, S. J. Mecham, D. Rodrigues, G. L. Wilkes, and J. E. McGrath, *J. Polym. Sci., Part A: Polym. Chem.*, **32**, 2663 (1994).

143. J. R. Salem, F. O. Sequeda, J. Duran, W. Y. Lee, and R. M. Yang, *J. Vac. Sci. Technol.*, **A4**, 369 (1986).

144. C. Chad Roberts, S. A. Letts, M. D. Saculla, E. J. Hsieh, and R. C. Cook, *Fusion Technol.*, **35**, 138 (1999).

145. E. L. Alfonso, F. Y. Tsai, S. H. Chen, R. Q. Gram, and D. R. Harding, *Fusion Technol.*, **35**, 131 (1999).

146. T. Strunskus and M. Grunze, in *Polyimides Fundamental and Applications*, M. K. Ghosh and K. L. Mittal (Eds.), Marcel Dekker, New York, 1996, p. 187.

147. W. M. Edwards and I. M. Robinson, U.S. Patent 2,710,853, 1955.

148. T. Inoue, Y. Kumagai, M. Kakimoto, Y. Imai, and J. Watanabe, *Macromolecules*, **30**, 1921 (1997).

149. Y. Imai, T. Fueki, T. Inoue, and M. Kakimoto, *J. Polym. Sci., Part A: Polym. Chem.*, **36**, 2663 (1998).

150. J. A. Kreuz, A. L. Endrey, F. P. Gay, and C. E. Sroog, *J. Polym. Sci., Part A1*, **4**, 2607 (1966).

151. Y. Nagata, Y. Ohnishi, and T. Kajiyama, *Polym. J.*, **28**, 980 (1996).

152. F. Basset, A. Lefrant, T. Pascal, B. Gallot, and B. Sillion, *Polym. Adv. Technol.*, **9**, 202 (1997).

153. M. T. Bogert and R. R. Renshaw, *J. Amer. Chem. Soc.*, **30**, 1135 (1908).

154. J. Coudane and M. Vert, *Eur. Polym. J.*, **23**, 501 (1987).

155. J. K. Im and J. C. Jung, *Polym Bull.*, **41**, 409 (1998).

156. Z. Y. Wang, Y. Qi, T. P. Bender, and J. P. Gao, *Macromolecules*, **30**, 764 (1997).

157. G. Maier, C. Zech, B. Voit, and H. Komber, *Macromol. Chem. Phys.*, **199**, 2655 (1998).

158. K. Yamanaka, M. Jikei, and M. A. Kakimoto, *Macromolecules*, **33**, 1111 (2000).

159. J. Fang, H. Kita, and K. I. Okamoto, *Macromolecules*, **33**, 4639 (2000).

160. P. M. Hergenrother, in *Encyclopedia of Polymer Science*, Vol. 13, 2nd ed., H. F. Mark, N. M. Bikales, C. G. Overberger, and G. Menges (Eds.), Wiley, New York, 1988, p. 55.

161. (a) G. De Gaudemaris and B. Sillion, *J. Polym. Sci.: Part B*, **2**, 203 (1964); (b) J. K. Stille and J. R. Williamson, *J. Polym. Sci.: Part B*, **2**, 209 (1964).

162. P. M. Hergenrother and H. H. Levine, *J. Polym. Sci.: Part A1*, **5**, 1453 (1967).

163. J. M. Angl, *J. Polym. Sci.: Part A1*, **10**, 2403 (1972).

164. T. Takekoshi, in *Polyimides Fundamental and Applications*, M. K. Ghosh and K. L. Mittal (Eds.), Marcel Dekker, New York, 1996, p. 39.

165. J. W. Labadie and J. L. Hedrick, *Makromol. Chem., Macromol. Symp.*, **54/55**, 313 (1992).

166. J. B. Baek and F. W. Harris, *Polym. Prepr.*, **41**, 886 (2000).

167. B. S. Kim, J. E. Korleski, Y. Zhang, D. J. Klein, and F. W. Harris, *Polymer*, **40**, 4553 (1999).

168. T. Kambara and T. Yamamoto, *Chem. Lett.*, 1459 (1993).

169. T. Yamamoto, K. Sugiyama, T. Kushida, T. Inoue, and T. Kambara, *J. Am. Chem. Soc.*, **118**, 3930 (1996).

170. I. Nurulla, A. Yamaguchi, and T. Yamamoto, *Polym. Bull.*, **44**, 231 (2000).

171. M. Jandke, K. Kreger, and P. Strohriegl, *Synth. Met.*, **111/112**, 221 (2000).

172. S. Srinivasan, R. Twieg, J. L. Hedrick, and C. J. Hawker, *Macromolecules*, **29**, 8543 (1996).

173. J. B. Baek and F. W. Harris, *Polym. Prepr.*, **41**, 157 (2000).

174. T. S. Chung, *J. Macromol. Sci., Rev. Macromol. Chem. Phys.*, **C 37**, 277 (1997).

175. E. W. Choe, *J. Appl. Polym. Sci.*, **53**, 497 (1994).

176. I. Yamaguchi, K. Osakada, T. Yamamoto, and M. Katada, *Bull. Chem. Soc. Jpn.*, **72**, 2557 (1999).

177. P. Staiati, M. Minutoli, and S. Hocevar, *J. Power Sources*, **90**, 231 (1999).

178. M. Kawahara, J. Morita, M. Rikukawa, K. Sanui, and N. Ogata, *Electrochem. Acta*, **45**, 1395 (2000).

179. G. Olason and D. C. Sherrington, *React. Funct. Polym.*, **42**, 163 (1999).

180. J. F. Wolfe, in *Encyclopedia of Polymer Science*, Vol. 11, 2nd ed., H. F. Mark, N. M. Bikales, C. G. Overberger, and G. Menges (Eds.), Wiley, New York, 1988, p. 601.

181. J. F. Wolfe, B. H. Loo, and F. E. Arnold, *Macromolecules*, **14**, 915 (1981).

182. J. F. Wolfe and F. E. Arnold, *Macromolecules*, **14**, 909 (1981).

183. L. S. Tan, J. L. Burkett, S. R. Simko, and M. D. Alexander, *Makromol. Rapid Commun.*, **20**, 16 (1999).

184. G. Maglio, R. Palumbo, and M. Tortora, *J. Polym. Sci., Part A: Polym. Chem.*, **38**, 1172 (2000).

185. Y. Maruyama, Y. Oishi, M. Kakimoto, and Y. Imai, *Macromolecules*, **21**, 2305 (1998).

186. R. A. Johnson and L. J. Mathias, *J. Polym. Sci., Part A: Polym. Chem.*, **33**, 1901 (1995).

187. S. H. Hsiao and L. R. Dai, *J. Polym. Sci., Part A: Polym. Chem.*, **37**, 2129 (1999).

188. H. Seino, K. Igushi, O. Haba, Y. Oba, and M. Ueda, *Polym. J.*, **31**, 822 (1999).

189. J. H. Chang, K. M. Park, and I. C. Lee, *Polym. Bull.*, **44**, 63 (2000).

190. T. K. Kim, K. Y. Choi, K. S. Lee, D. W. Park, and M. Y. Jin, *Polym. Bull.*, **44**, 55 (2000).

191. G. L. Tullos and L. J. Mathias, *Polymer*, **40**, 3463 (1999).

192. Y. H. So, S. J. Martin, K. Owen, P. B. Smith, and C. Karas, *J. Polym. Sci., Part A: Polym. Chem.*, **37**, 2637 (1999).

193. Y. H. So, *Progr. Polym. Sci.*, **25**, 137 (2000).

6 Synthesis of Poly(arylene ether)s

Sheng Wang and J. E. McGrath

Department of Chemistry, Virginia Polytechnic Institute and State University, Blacksburg, Virginia

6.1 INTRODUCTION

Poly(arylene ether)s are a family of high-performance engineering thermoplastic materials with high glass transition temperature, high thermal stability, good mechanical properties, and excellent resistance to hydrolysis and oxidation.[1-13] Their outstanding properties have resulted in several important commercial sulfone- or ketone-containing products, such as bisphenol A polysulfone, Udel (Solvay Advanced Polymers); poly(ether sulfone), Victrex (ICI); poly(arylene ether ether ketone) (PEEK), Victrex PEEK (ICI); poly(arylene ether ketone) (PEK), Stilan (Raychem Corporation); and poly(arylene ether ketone ether ketone ketone) (PEKEKK), Ultrapek (BASF), shown in Scheme 6.1. Several are no longer commercially available.

A generalized structure of the poly(arylene ether)s can be represented as in Scheme 6.2.

Poly(arylene ether)s have been widely used as engineering thermoplastics due to their good mechanical properties and thermal stability, which depend on their unique structures. In the polysulfone main chain, the bond energy of an aromatic carbon–oxygen ether linkage (84.0 kcal/mol) is slightly higher than that of the carbon–carbon counterpart (83.1 kcal/mol). Aromatic ether linkages (Ar–O–Ar) are stabilized through resonance, as is illustrated in Scheme 6.3 for diphenyl ether.[14]

The aryl C–O–C linkage has a lower rotation barrier, lower excluded volume, and decreased van der Waals interaction forces compared to the C–C bond. Therefore, the backbone containing C–O–C linkage is highly flexible. In addition, the low barrier to rotation about the aromatic ether bond provides a mechanism for energy dispersion which is believed to be the principal reason for the toughness or impact resistance observed for these materials.[15-17]

Robeson et al. studied the secondary loss transitions of a series of poly(arylene ether)s using a torsion pendulum.[15] They found that the secondary loss transitions are closely related to the segmental motion of the aryl ether bonds. The secondary

Synthetic Methods in Step-Growth Polymers. Edited by Martin E. Rogers and Timothy E. Long
© 2003 John Wiley & Sons, Inc. ISBN: 0-471-38769-X

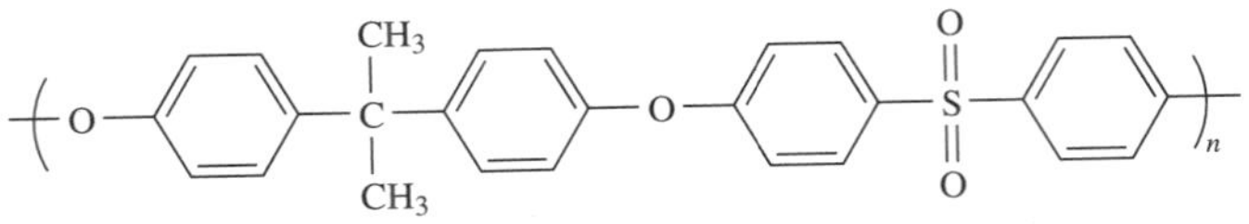

Udel®

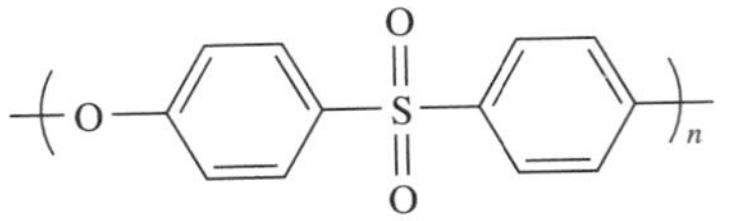

Victrex®

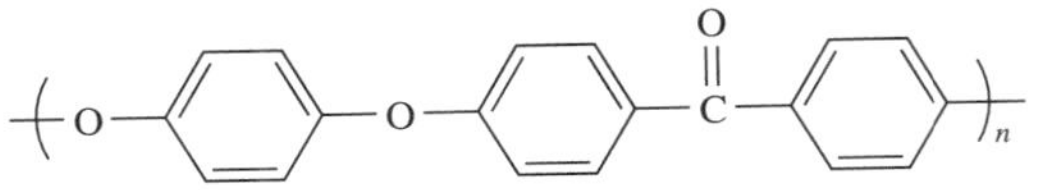

Victrex PEEK®

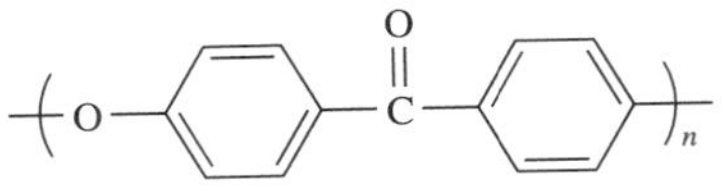

Stilan®

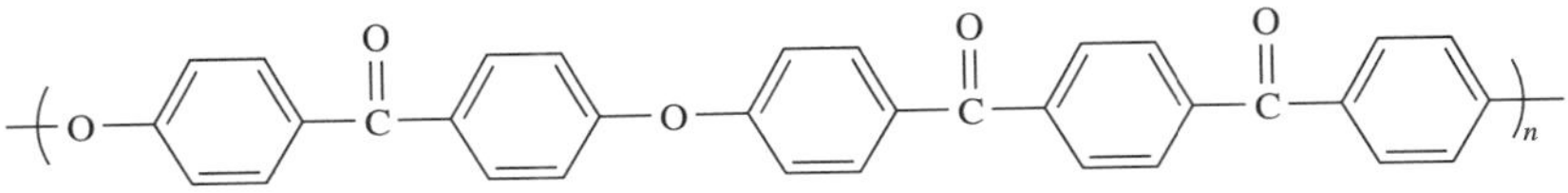

Ultrapek®

Scheme 6.1 Examples of commercial poly(arylene ether)s.

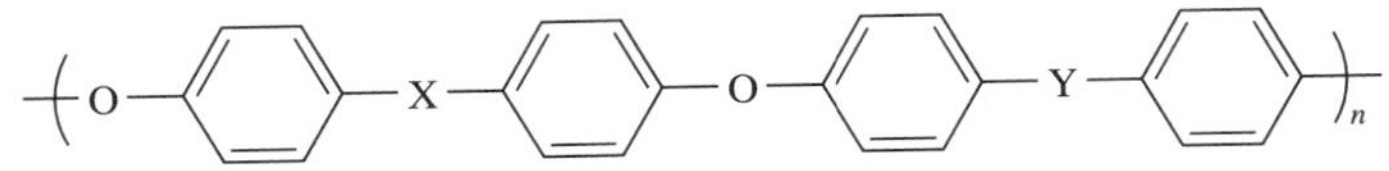

Scheme 6.2 Generic representation of poly(arylene ether) backbone structure.

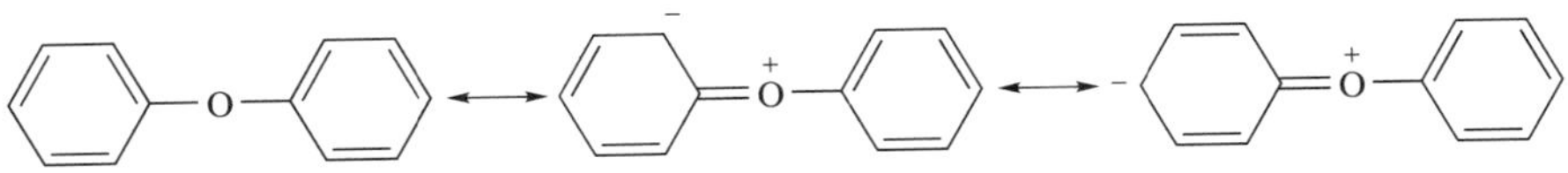

Scheme 6.3 Resonance stabilization in aryl ethers.

loss transitions due to the interactions between the sulfonyl groups and water molecules are not dependent on the structure of the polymers. Dumais et al. studied the β-relaxation in poly(arylene ether sulfone)s by deuterium nuclear magnetic resonance (NMR).[17] The results suggested that the primary mode of motion of the aromatic rings in these polymers is 180° phenyl ring flips with a broad distribution of characteristic frequencies (ca. $10^2 - 10^7$ s^{-1}). Addition of antiplasticizers decreases the phenyl ring-flipping rate.

A wide variety of high-performance polymeric materials have been synthesized by incorporating thermally stable moieties such as sulfone, ketone, or aryl or alkyl phosphine oxide in addition to the ether linkage in poly(arylene ether)s.

6.2 GENERAL APPROACHES FOR THE SYNTHESIS OF POLY(ARYLENE ETHER)S

The general approaches for the synthesis of poly(arylene ether)s include electrophilic aromatic substitution, nucleophilic aromatic substitution, and metal-catalyzed coupling reactions. Poly(arylene ether sulfone)s and poly(arylene ether ketone)s have quite similar structures and properties, and the synthesis approaches are quite similar in many respects. However, most of the poly(arylene ether sulfone)s are amorphous while some of the poly(arylene ether)s are semicrystalline, which requires different reaction conditions and approaches to the synthesis of these two polymer families in many cases. In the following sections, the methods for the synthesis of these two families will be reviewed.

6.2.1 Electrophilic Aromatic Substitution

6.2.1.1 Mechanism for Friedel–Crafts Sulfonylation and Acylation

The sulfonylation mechanism involves a two-stage reaction, as shown in Scheme 6.4. The sulfonylium cation ArSO$_2$$^+$ attacks the carbon on the aromatic nucleus to generate an intermediate complex, which subsequently decomposes to afford the final product by eliminating a proton. It is postulated that the effective sulfonylating agent is the sulfonylium salt generated by action of the Lewis-acid catalyst on the sulfonyl halide.[18]

The synthesis of sulfones included electrophilic substitution, specifically, Friedel–Crafts catalysis by AlCl$_3$, FeCl$_3$, SbCl$_5$, AlBr$_3$ and BF$_3$, and so on, which are efficient catalysts for the sulfonylation by arenesulfonyl halides.[19-22]

Similarly, the acylation mechanism also involves a two-stage reaction.[23] Polyphosphoric acid, AlCl$_3$, HF/BF$_3$, and CF$_3$SO$_3$H. are typical catalysts.

6.2.1.2 Friedel–Crafts Sulfonylation and Polysulfonylation

Earlier synthesis of polysulfones included electrophilic aromatic substitution. Traditionally, this reaction requires high catalyst concentrations. However, polysulfonylation to achieve high molecular weights can utilize elevated temperatures and low concentration of the Lewis acids. For example, arenesulfonyl chlorides react smoothly in the molten state at $120-140°$C with aromatics in the presence of $1-5$ mol % of FeCl$_3$ [or SbCl$_5$, InCl$_3$, Fe(II) or Fe(III) acetylacetonate (acetylacetone is 2,4-pentanedione), or BiCl$_3$], giving, in a few hours, the corresponding sulfones in high yields.[24,25]

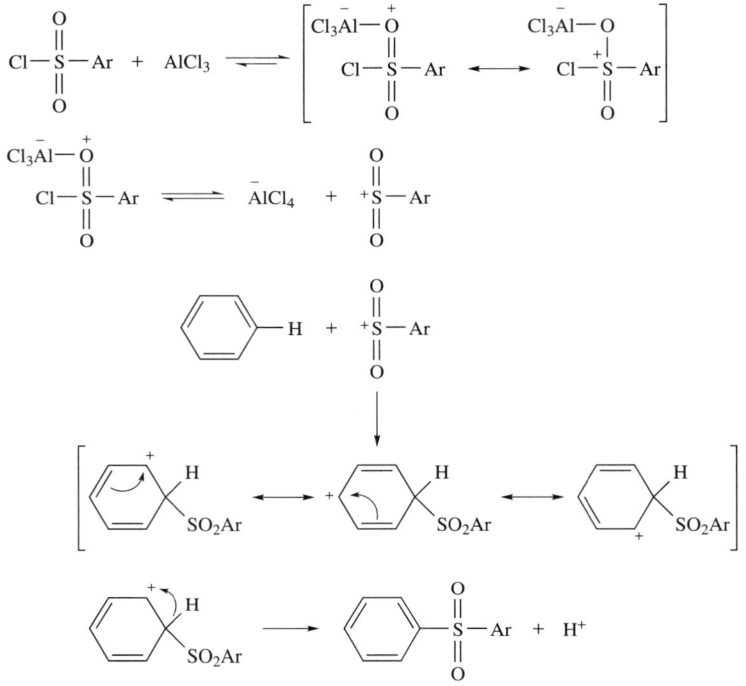

Scheme 6.4   Sulfonylation mechanism.

Poly(arylene ether sulfone)s can be synthesized by two different polysulfonylation reaction routes: condensation of AA and BB monomers or self-condensation of AB monomer.[26,27] Scheme 6.5 shows two approaches.

In the AA–BB type of sulfonylation, two or more activated aromatic hydrogen atoms are commonly present in the reacting molecules. Therefore, this polycondensation process may result in different repeating units. Structural irregularities

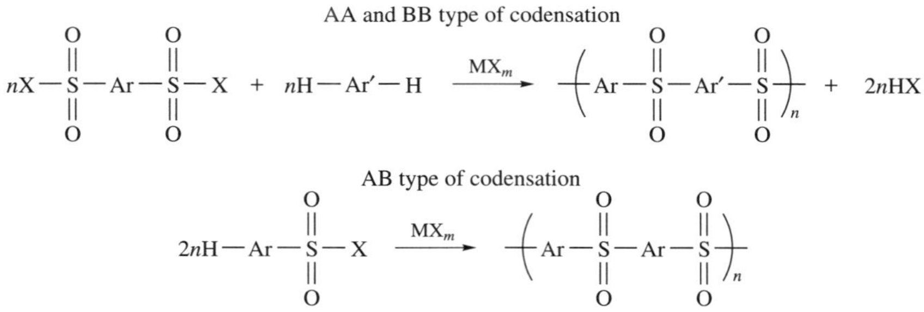

Scheme 6.5   Friedel–Crafts sulfonylation of AA–BB and AB types of polycondensations.

can occur due to possible sulfonylation at various ring positions, disulfonylations on one aromatic nucleus, as well as side reactions directly promoted by the Friedel–Crafts catalyst.[28–30] When high amounts of catalyst are used, more than the theoretical amount of HCl is evolved, probably because of the attack of anhydrous metal halides on the aromatic nuclei, a known reaction for FeCl$_3$ and AlCl$_3$[26,27]. Indeed, polymers containing structures in Scheme 6.6 were proposed.[28,29]

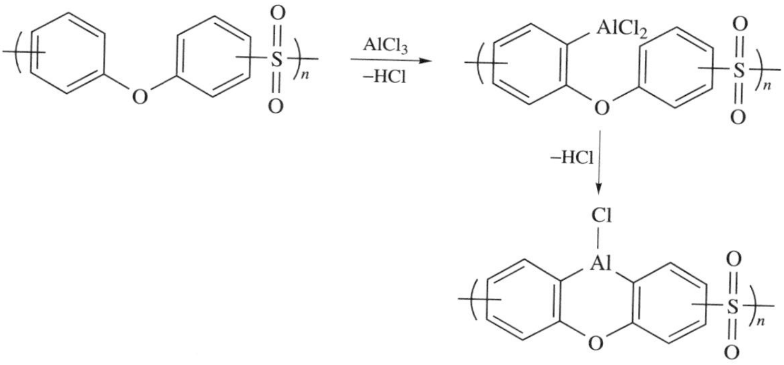

Scheme 6.6 Side reactions of polysulfonation by high catalyst concentration.[29]

Generally, higher catalyst concentration leads to side reactions. An efficient approach to minimize the side reactions is to use the minimum amounts of Friedel–Crafts catalyst (e.g., FeCl$_3$, 0.1–4 wt %). The reaction can be performed either in bulk or in solution using, for example, nitrobenzene, dimethyl sulfone, or chlorinated biphenyls as the reaction media.

Bulk polysulfonylation was initially performed in the molten state, and the resulting solid low-molecular-weight polymer can be ground, followed by powder sintering at high temperature to produce high molecular weight. The final reaction temperatures were from 150 to 320°C under nitrogen, and vacuum eventually was applied to remove HCl.[25,31] Since sulfonyl chlorides decompose above 250°C via a radical mechanism, the best reaction temperature range was from 230 to 250°C.[32] However, the polymers prepared by this approach were insoluble because of crosslinking side reactions.

In contrast to bulk polymerization, solution polymerization provided soluble polymers with high molecular weights using low FeCl$_3$ concentration at 120–140°C.[31] A major disadvantage of the above approaches is that all the metal–halide catalysts need to be removed, since the catalyst residue will deteriorate the thermal stability and electrical and other properties.

Electron-donating groups strongly activate sulfonylation at ortho or para positions, with para sulfonylation more favored due to less steric hindrance. Variables, such as highly reactive monomers, elevated reaction temperatures, and high amounts of catalyst, promote high conversion but may also result in structural irregularity and even crosslinking.

In the self-polycondensation of monosulfonyl chlorides, A–B systems can demonstrate advantages over A–A, B–B systems. The former stoichiometry is not affected by preferential vaporization. Therefore, self-polycondensation yields polymers with high molecular weight and regular structures. Since the sulfonyl group is a strong deactivating group for further sulfonylation, usually high reaction temperature is needed.

Sulfonic acid catalysts can also be used to prepare poly(arylene ether sulfone)s, such as $(CF_3CO)_2O$, polyphosphoric acid (PPA),[26] $MeSO_3H–P_2O_5$ mixture,[33] and CF_3SO_3H.[34]

6.2.1.3 *Friedel–Crafts Acrylation Polymerization*

Similar to the synthesis of polysulfones, poly(arylene ether ketone)s can also be prepared by using either AA and BB monomers or an AB monomer.

The first poly(arylene ether ketone)s were prepared by Bonner at DuPont in 1962.[35] The wholly aromatic poly(ether ketone ketone)s (PEKK) were prepared by Friedel–Crafts acylation. Isophthaloyl chloride or terephthaloyl chloride was condensed with diphenyl ether using nitrobenzene as solvent and $AlCl_3$ as a catalyst, as shown in Scheme 6.7. However, only low-molecular-weight polymers were obtained due to the insolubility of the resulting polymers. Later, Goodman et al. at ICI reported the successful synthesis of poly(ether ketone) by the self-condensation of *p*-phenoxybenzoyl chloride using methylene chloride as the reaction medium under similar reaction conditions.[36] This polymer had a relatively high molecular weight, as indicated by reduced viscosity values.

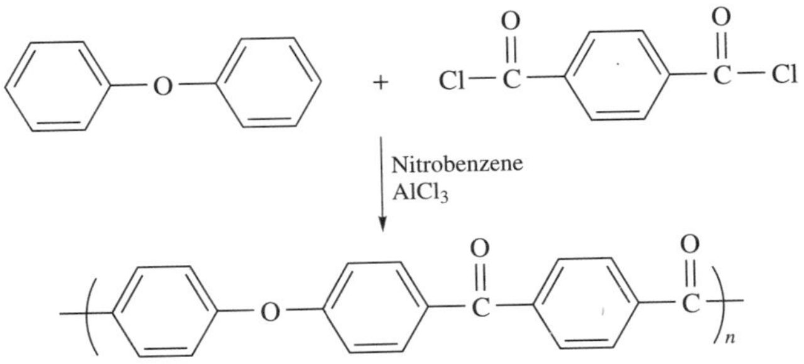

Scheme 6.7   Synthesis of poly(ether ketone ketone) via aromatic electrophilic substitution catalyzed by $AlCl_3$.

The $AlCl_3$-catalyzed polycondensation of diphenyl ether with a mixture of terephthaloyl chloride and isophthaloyl chloride is a relatively inexpensive route to poly(ether ketone)s. The polymerizations were carried out in chlorinated solvents such as 1,2-dichloroethane, dichloromethane, or *o*-chlorobenzene in the presence of an excess of aluminum chloride.[37] Some problems limited the application of AlCl$_3$. The insolubility of the polymers in these solvents limits affording high molecular weight. Introducing some groups such as *tert*-butyl significantly increased the solubility and thus increased the molecular weight.[38] Posttreatment is generally required to improve the melt stability. It also increases structural complexity. For example, in some cases, ortho acylation and alkylation of aromatic polymer chains were also observed.[39] In addition, some inexpensive diacid chlorides, such as terephthaloyl chloride and isophthaloyl chloride, are less reactive than 4,4'-oxybisbenzoyl chloride and thus are more difficult to be activated by AlCl$_3$. The resulting carbocationic intermediates are more reactive and less selective to aromatic sites, which lead to increased ortho acylation during polymerization.

Jansons et al. reported that cleaner and higher molecular weight polymers were prepared at or below room temperature in the presence of a large excess of AlCl$_3$ complexed with a Lewis base, such as *N*, *N*-dimethylformamide (DMF), LiCl, dimethyl sulfone, pyridine or *N*-methylpyridinium chloride.[40,41] The Lewis base can increase the solubility of AlCl$_3$ in a nonbasic solvent, such as CH$_2$Cl$_2$ or 1,2-dichloroethane, by complexation. Furthermore, the polar complex solutions increase the solubility of polymer complexes that lead to high-molecular-weight polymers. Most of the work in this area was submitted for patents. One example in literature publications was reported by Fukawa and Tanabe.[42] *p*-Phenoxybenzoyl chloride was polymerized in the presence of excess AlCl$_3$ and phenyl ether capping agent in a solution of DMF in methylene chloride to give PEK with high inherent viscosity values. Another example is polymerization of isophthaloyl chloride and diphenyl ether.[43] The polymerization was controlled by catalyst quantity, reaction temperature, monomer ratio, and polymer concentration.

An alternative approach to the preparation of PEKK is by condensation of isophthaloyl chloride and/or terephthaloyl chloride with isolated intermediates such as 4,4'-diphenoxyterephthaloylphenone and 4,4'-diphenoxyisophthaloylphenone. This method affords cleaner polymers, especially in the presence of AlCl$_3$ complexed with a basic coagent. A copolymer was prepared by Gay and Brunette at DuPont from the isolated intermediates by reacting 1,3-bis(*p*-phenoxybenzoyl)-benzene with terephthaloyl chloride, and 1,4-bis(*p*-phenoxybenzoyl)benzene with isophthaloyl chloride (Scheme 6.8).[44] This perfectly alternating copolymer of meta and ortho phenylene group had a high T_g (166°C) and low T_m (332°C), which was considered desirable for good processibility and high heat deformation resistance.

Polyphosphoric acid can be used as a solvent to produce high-molecular-weight poly(ether ketone ketone)s, because polyphosphoric acid dissolves the polymers, probably by protonation of the carbonyl groups. Iwakura et al. reported the synthesis of PEK with inherent viscosity of 0.5 dL/g by self-condensation of *p*-phenoxybenzoic acid using polyphosphoric acid as both the solvent and catalyst.[45] Devaux et al. demonstrated that diphenyl ether could be used to condense with diacids or diacid chlorides in the presence of AlCl$_3$ or SnCl$_4$ to produce high-molecular-weight poly(ether ketone)s.[46]

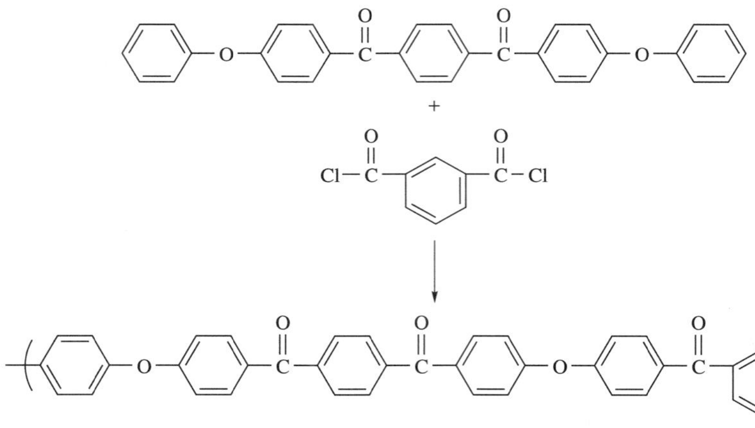

Scheme 6.8   Synthesis of alternating PEKK copolymer.

Marks developed a more effective solvent–catalyst (HF–BF$_3$) system to synthesize high-molecular-weight PEK.[47] Under well-controlled reaction conditions, high-molecular-weight PEK with an inherent viscosity of 1.33 dL/g was prepared in the HF–BF$_3$ solvent system. Large excesses of solvent and catalyst have to be used to obtain high-molecular-weight polymers. Rose introduced another strong acid, trifluoromethanesulfonic acid, to catalyze the condensation of carboxylic acid with activated phenyl ether, and high-molecular-weight polymers were synthesized using this solvent system (Scheme 6.9).[48] Replacement of CF$_3$SO$_3$H with the less expensive CH$_3$SO$_3$H failed to yield high-molecular-weight semicrystalline poly(arylene ether ketone) (PAEKs). Neither of these two strong-acid systems are very attractive for industrial applications due to the harsh reaction conditions and high cost.

6.2.2 Nucleophilic Aromatic Substitution Reaction Polymerization

6.2.2.1 Mechanism for Nucleophilic Aromatic Substitution

The most practical method for the preparation of poly(arylene ether)s employs nucleophilic aromatic substitution (S$_N$Ar). Although nucleophilic substitution can occur via four principal mechanisms,[49] the most important mechanism utilized for the synthesis of poly(arylene ether)s has been S$_N$Ar, in which activating groups are present on the aromatic ring (Scheme 6.10).

In the first step, the nucleophile attacks the carbon atom of the activated C–X bond, resulting in a resonance-stabilized intermediate, often termed a Meisenheimer complex.[50] The second step involves the departure of the leaving group. The first step is generally the rate-determining step. The proposed mechanism can be supported by the isolation of Meisenheimer salts. Investigating the effects of the leaving group on the reaction rate provided further evidence for this mechanism. In spite of the strength of the carbon–fluorine bond, the order of

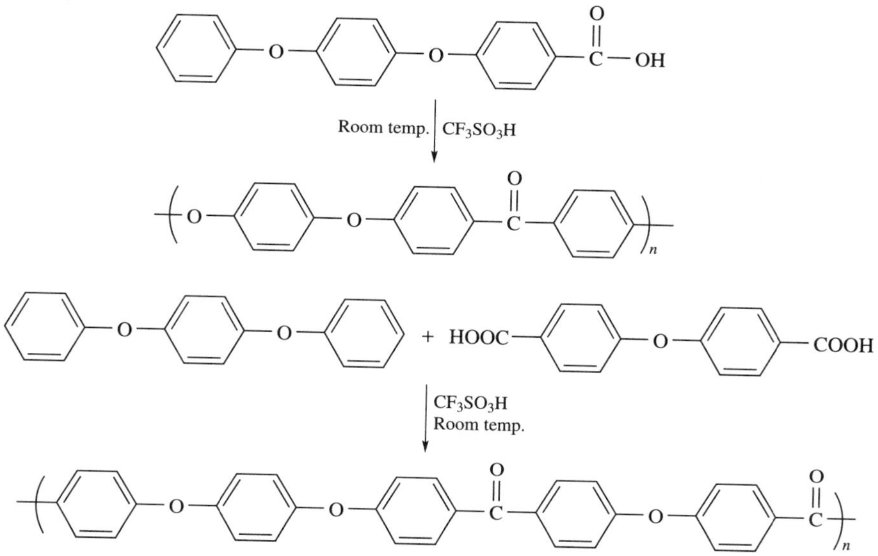

Scheme 6.9 Synthesis of poly(arylene ether ether ketone) (PEEK) and poly(ether ether ketone ether ketone (PEEKEK) in triflic acid.

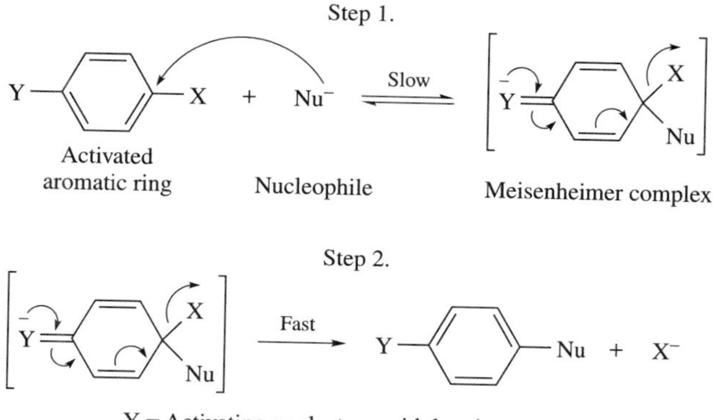

Step 1.

Step 2.

Y = Activating or electron-withdrawing group
X = Leaving group
Nu = Attacking nucleophilic anion

Scheme 6.10 Nucleophilic aromatic substitution mechanism.

reactivity for the halogens was found to be F >> Cl > Br > I, suggesting the rate-determining step does not involve the departure of the leaving group. Therefore, the observed order of reactivity can be explained by the better stabilization of the Meisenheimer intermediate by the highly electronegative fluorine through inductive electron-withdrawing effects. Furthermore, the carbon directly attached to the fluorine will be more electrophilic and, consequently, more susceptible to nucleophilic attack.

Many factors, such as the strength of the activating group, the nucleophilicity of the attacking nucleophile, the electronegativity of the leaving group, the reaction conditions, and the nature of the solvent, determine the kinetics of the S_NAr reaction. Some major activating groups, such as

$$-N\underset{O}{\overset{O}{\diagdown}} \qquad -\overset{\overset{O}{\|}}{\underset{\underset{O}{\|}}{S}}- \qquad -\overset{\overset{O}{\|}}{C}- \qquad -\overset{\overset{O}{\|}}{\underset{R}{P}}- \qquad -C\equiv N$$

in the ortho and para positions of the leaving groups accelerate the nucleophilic substitution since they strongly stabilize Meisenheimer intermediate. On the contrary, electron-donating groups such as amine or methoxy decrease the stability of Meisenheimer intermediate, thus hindering the substitution.

The reaction rate increases with increasing strength of the nucleophile. The overall approximate order of nucleophilicity is $ArS^- > RO^- > ArO^- > OH^- > ArNH_2 > NH_3 > I^- > Br^- > Cl^- > H_2O > ROH^{51}$.

Since the decomposition of the Meisenheimer intermediate is not the rate-determining step, the trends for the leaving groups are quite different from common nucleophilic substitution. The leaving group trends for S_NAr substitutions were reported to have the following order: $F^- > NO_2^- >^- SOPh > Cl^- > Br^- \sim I^- >^- OAr >^- OR >^- SR^{52,53}$.

6.2.2.2   Poly(arylene ether) Synthesis Via S_NAr Reaction

Compared with the aromatic electrophilic substitution approach, the S_NAr approach general requires higher reaction temperatures. The polymers generally have well-defined structures. Therefore, it is more facile to control the structures of the products. In addition, it is more tolerable to some reactive functional groups, which makes it possible to synthesize reactive-group end-capped prepolymers and functional copolymers using functional monomers.

6.2.2.3   Poly(arylene ether sulfone) Synthesis

The S_NAr reactions were first successfully used in the synthesis of high-molecular-weight poly(arylene ether)s by Johnson et al.[4,5] This reaction represents a good example for poly(ether sulfone)s in general, either in laboratory- or industrial-scale preparations. In this procedure, the bisphenol A and sodium hydroxide with an exact mole ratio of 1 : 2 were dissolved into dimethyl sulfoxide (DMSO)–chlorobenzene. The bisphenol A was converted into disodium bisphenolate A, and water was removed by azeotropic distillation. After the formation of the anhydrous disodium bisphenolate A, an equal molar amount of 4,4′-dichlorodiphenyl sulfone (DCDPS) was added in chlorobenzene under anhydrous conditions and the temperature was increased to 160°C for over 1 h to afford high molecular weights. Chlorobenzene is added to the viscous polymer solution during cooling, since it is a solvent at room temperature and facilitates precipitation of the salts (Scheme 6.11).

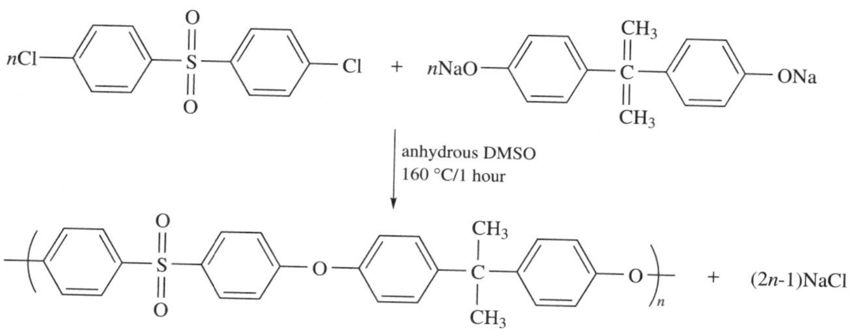

Scheme 6.11   Typical nucleophilic synthesis of bisphenol A polysulfone.[4,5]

As mentioned before, 4,4′-difluorodiphenyl sulfone (DFDPS) is much more reactive, but its application is limited because of economics. It can be prepared by the melt reaction of KF with DCDPS. The mechanism and kinetics of polyetherification have been investigated in detail for the polymerization of AA–BB monomers[5,54,55] and AB monomers.[56,57] The reactivity of the activated halides can be estimated by measuring ^{1}H, ^{13}C, and ^{19}F NMR chemical shifts. The nucleophilic substitution reaction only proceeds in the halides with ortho or para positions substituted by a strong electron-withdrawing group. Therefore, ^{1}H NMR chemical shift data from the protons ortho or para to the electron-withdrawing group can be used to determine the reactivity of the monomer indirectly.[58] Carbon-13 and ^{19}F NMR can be used to probe the chemical shift at the actual site of nucleophilic reaction. In general, lower chemical shifts correlate with lower monomer reactivity. Carter reported that a compound might be appropriate for nucleophilic displacement if the ^{13}C chemical shift of an activated fluoride ranges from 164.5 to 166.2 ppm in CDCl$_3$[59].

Preparing bisphenolates by using a strong base is often feasible, and the phenolate readily reacts with activated dihalides (DCDPS or DFDPS) at high temperature to yield high-molecular-weight in a short period of time.

However, many salts such as the hydroquinone or biphenol salt are so insoluble that they do not work well by this procedure. Furthermore, a stoichiometric amount of base used for the reaction is critical to obtain high-molecular-weight polymers. Moreover, the strong base may undesirably hydrolyze the dihalides to afford deactivated diphenolates, which upset the stoichiometry. Clendining et al. reported that potassium carbonate or bicarbonate could be used in these reactions instead of corresponding hydroxides.[60] McGrath and co-workers were the first to systematically study the use of the weak base K$_2$CO$_3$ instead of a strong base to obtain phenolate salts.[8,61,62] Potassium carbonate was found to be better than

Na_2CO_3 due to its relative stronger basicity and higher solubility in the reaction medium. The proposed mechanism for this method is shown in Scheme 6.12. This approach makes it fairly easy to control the molecular weights and the endgroups. An additional advantage is that the precise amount of weak base is not extremely critical for achieving high molecular weight, as long as it is in excess. The reaction rates of these polymerizations are generally slower than the strong-base method.

Aprotic polar solvents have to be used for several reasons. They are often good solvents for both monomers (including phenolates) and amorphous polymers. In addition, they can also stabilize the Meisenheimer intermediates. Common aprotic polar solvents, such as DMSO, N,N-dimethyl acetamide (DMAc), DMF, N-methyl pyrrolidone (NMP), and cyclohexylpyrrolidone (CHP) can be used. Under some circumstances, very high reaction temperature and boiling point solvents such as sulfolane and diphenyl sulfone (DPS) have to be used due to the poor reactivity of the monomers or poor solubility of the resulting, possibly semicrystalline polymers, as in the PEEK systems.

The choice of solvent is not trivial and, generally, the reaction medium must be a good solvent for both monomers and polymer product. In addition, to obtain high molecular weight, water needs to be removed from the system to avoid hydrolyzing the activated substrate, since hydrolysis reduces the reaction rate and upsets the stoichiometry of the monomers.[61-63]

The reaction generates easily oxidizable alkali phenates. Thus, the polymerization must be conducted in an inert atmosphere. The sodium and potassium salts of the bisphenols are widely used because they are more soluble in polar aprotic solvents.

Other salts, especially fluoride salts, (e.g., KF) can be used to perform nucleophilic substitution. As is well known, halides, and particularly the fluoride anions, are rather powerful Lewis bases and can exert a catalytic effect on aromatic nucleophilic substitutions in dipolar aprotic solvents. Phenols can be alkylated in the presence of KF (or CsF) absorbed on Celite[64,65] or Et_4NF.[66] Taking advantage of this reaction, halophenols and dihalides with bisphenols have been successfully polymerized in sulfolane at 220–280°C by using KF as the base.

While most of the polymers have been synthesized by solution polymerization through nucleophilic substitution, Kricheldorf and Bier showed that melt or solvent free polymerization can be achieved by reacting bis(trimethylsilyl ether)s of various bisphenols with bis(4-fluorophenyl) sulfone (Scheme 6.13).[67] This idea came from well-known reactions on trimethylsilyl ethers, esters, etc., that were performed with fluoride-catalyzed desilylation in the presence of KF, CsF, or $(C_4H_9)_4NF$.[68,69] Silylation followed by distillation is an efficient method for purifying bisphenols, aminophenols, and diamines. The product can be directly processed without further purification since the highly volatile fluorotrimethylsilane is the only by-product and the catalytic amount of CsF does not deteriorate the mechanical properties of the final products. In addition, the removal of the volatile fluorotrimethylsilane by-product further drives the reaction to yield high molecular weight. No halogen-substituted diphenols can be used, since an exchange reaction may occur.[70]

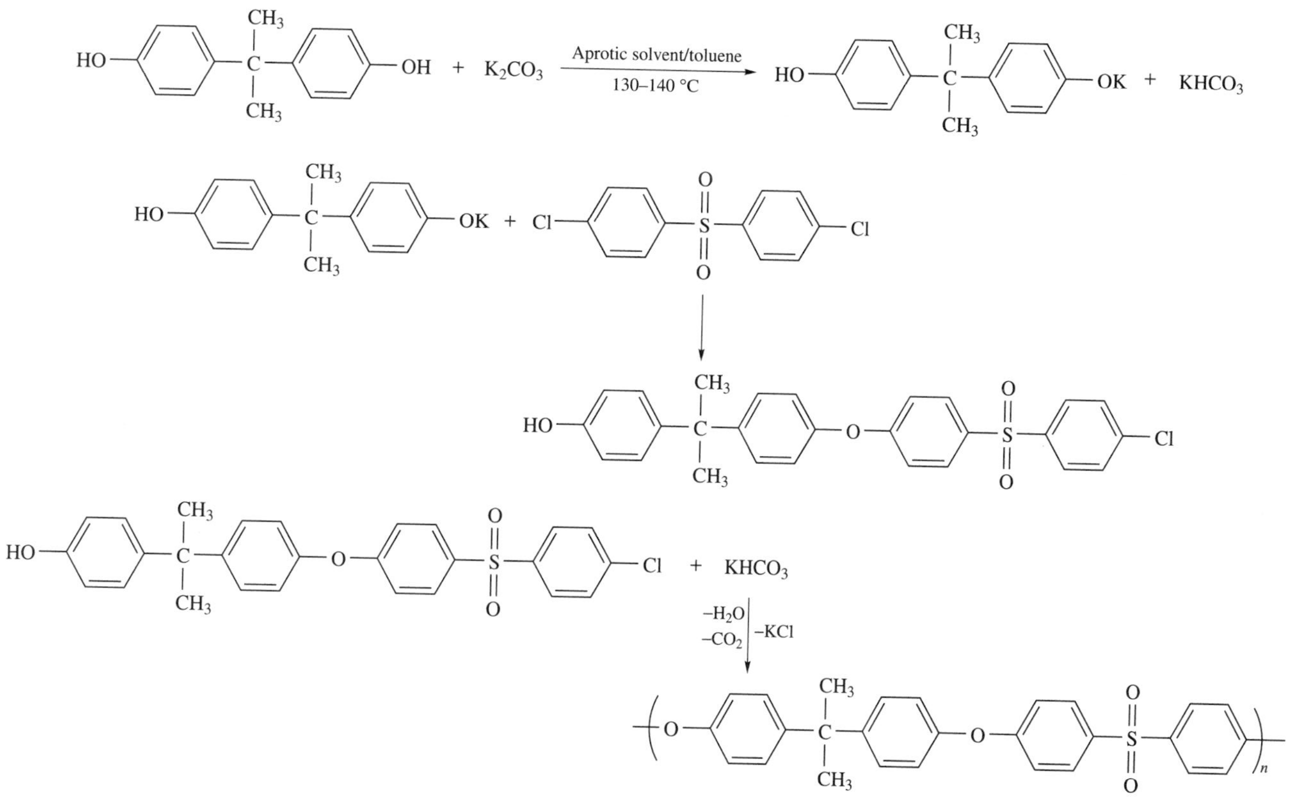

Scheme 6.12 Proposed mechanism for the synthesis of poly(arylene ether sulfone) *via* the potassium carbonate process.[8]

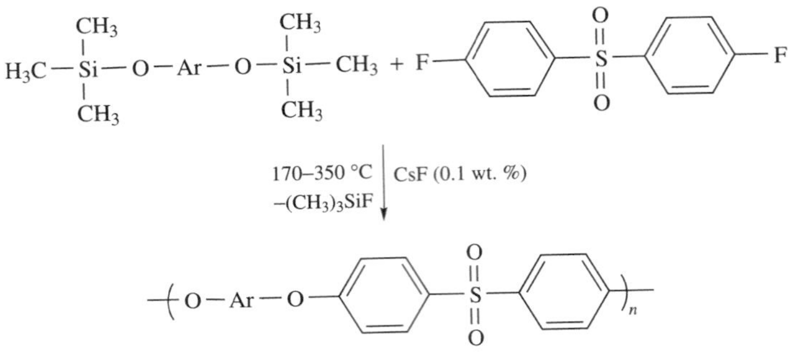

$$CH_3 \quad\quad CH_3$$
$$H_3C-\overset{|}{\underset{|}{Si}}-O-Ar-O-\overset{|}{\underset{|}{Si}}-CH_3 + F-\!\!\!\diamond\!\!\!-\overset{O}{\underset{O}{\overset{\|}{S}}}-\!\!\!\diamond\!\!\!-F$$
$$CH_3 \quad\quad CH_3$$

170–350 °C | CsF (0.1 wt. %)
$-(CH_3)_3SiF$ ↓

$$-\!\!\!\big(O-Ar-O-\!\!\!\diamond\!\!\!-\overset{O}{\underset{O}{\overset{\|}{S}}}-\!\!\!\diamond\!\!\!\big)\!_n$$

Scheme 6.13 Synthesis of poly(arylene ether sulfone)s via silyl ether displacement.

A proposed mechanism for silyl ether displacement is shown in Scheme 6.14. In the first step, the fluoride anion converts the trimethyl siloxy group into a phenolate salt. In the following step, the phenolate anion attacks the activated fluoro monomer to generate an ether bond. The amount of catalyst required is about 0.1–0.3 mol %. Catalyst type and concentration are crucial for this reaction.

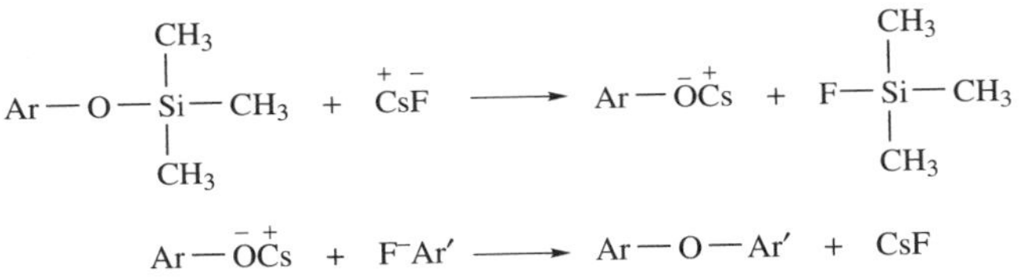

$$CH_3 \quad\quad\quad\quad\quad\quad\quad\quad CH_3$$
$$Ar-O-\overset{|}{\underset{|}{Si}}-CH_3 + \overset{+}{C}s\overset{-}{F} \longrightarrow Ar-\overset{-}{O}\overset{+}{C}s + F-\overset{|}{\underset{|}{Si}}-CH_3$$
$$CH_3 \quad\quad\quad\quad\quad\quad\quad\quad CH_3$$

$$Ar-\overset{-}{O}\overset{+}{C}s + F^-Ar' \longrightarrow Ar-O-Ar' + CsF$$

Scheme 6.14 Reaction mechanism of silyl ether displacement.

Unfortunately, the method is only suitable for fluorinated systems such as DFDPS. Using chloro monomers generally affords low molecular weight, because a weak base like KF or CsF is needed and DCDPS is not reactive enough under these reaction conditions. However, the activated dichloro compounds can be successfully polymerized in NMP in the presence of equimolar amounts of K_2CO_3[71].

6.2.2.4 Poly(arylene ether ketone) Synthesis

6.2.2.4.1 Common One-Step Process

The nucleophilic aromatic substitution reaction for the synthesis of poly(arylene ether ketone)s is similar to that of polysulfone, involving aromatic dihalides and aromatic diphenolates. Since carbonyl is a weaker electron-withdrawing group than sulfonyl, in most cases, difluorides need to be used to afford high-molecular-weight polymers. Typically potassium carbonate is used as a base to avoid the side reaction of alkali metal hydroxides[72] and dipolar aprotic solvents such as DMSO, DMAc, and NMP are utilized.

Johnson et al. reported the first attempt to synthesize PEEK by polycondensation of bisphenolate with activated dihalides using DMSO as a solvent and NaOH as a base.[5] High-molecular-weight polymers were difficult to obtain due to the crystallization and solubility of the polymers in DMSO. Attwood et al.[73] and Rose and Staniland[74] used diphenyl sulfone as the reaction medium, and the polymerization was carried out near the melting point of the polymer to maintain solubility. Carefully optimal reaction conditions are required to afford high-molecular-weight polymers, because side reactions became significant at high temperatures. Only expensive difluoro monomers can be use for this approach. Less reactive dichloro monomers fail to produce high-molecular-weight polymers, partially because of the side reactions such as single electron transfer reaction.[75,76]

Attempts to replace expensive fluoro monomers such as 4,4'-dichlorobenzophenone (DFBP) with activated chloro monomers have been reported. Fukawa et al. reported preparing high-molecular-weight PEK with 1.15 dL/g inherent viscosity using 4,4'-dichlorobenzophenone (DCBP).[77] The polymerization was carried out in diphenyl sulfone at high temperatures (2 h at 280°C, 1 h at 300°C, and then 1 h at 320°C) using DCBP to react with sodium carbonate in the presence of SiO_2-CuCl_2 catalyst. Silicon dioxide was removed by washing the product with 4% aqueous NaOH solution at 60°C.

Dichloro monomers can also be polymerized with bisphenols in the presence of fluorides as promoting agents.[78] The fluoride ions promote the displacement of the chloride sites to form more reactive fluoride sites, which react with phenolate anion to form high-molecular-weight polymers. Adding 5–10 mol% phase transfer catalysts such as N-alkyl-4-(dialkylamino)pyridium chlorides significantly increased the nucleophilicity and solubility of phenoxide anion and thus shortened the reaction time to one fifth of the uncatalyzed reaction to achieve the same molecular weight.[79]

Kricheldorf et al. used trimethylsilylated phenols to react with activated difluoro monomers, such as DFBP, in the presence of catalytical amounts of CsF to prepare poly(ether ketone)s.[80,81] The polymers were prepared in the melt without solvent and can be directly processed.

Carbonyl groups on the side chains also activate the dihalides. These particular dihalides were employed to prepare poly(arylene ether)s containing pendent benzoyl groups.[82] The polymers may be used as a positive photoresist for UV irradiation.

6.2.2.4.2 Soluble Precursor Approaches

Several soluble precursor approaches were reported to produce high-molecular-weight soluble precursors, which could be later reacted to remove the solubilizing groups. The introduction of bulky substitutions can reduce crystallinities or even suppress crystallization of a polymer, thereby improving its solubility substantially. Thus, the incorporation of removable substituents provides a strategy for preparing high-molecular-weight polymers that otherwise are not easily accessible. The pendent soluble moieties can be incorporated onto either the diphenols or the activated dihalides.

Mohanty et al. were the first to introduce pendent *t*-butyl groups in the polymer backbones. The resulting material was quite soluble in aprotic dipolar solvents.[83] The PEEK precursors were prepared under a mild reaction condition at 170°C. The polymer precursor can be converted to PEEK in the presence of Lewis acid catalyst $AlCl_3$ via a retro Friedel–Crafts alkylation. Approximately 50% of the *tert*-butyl substitutes were removed due to the insolubility of the product in the solvent used. Later, Risse et al. showed that complete cleavage of *tert*-butyl substitutes could be achieved using a strong Lewis acid CF_3SO_3H as both the catalyst and the reaction medium (Scheme 6.15).[84]

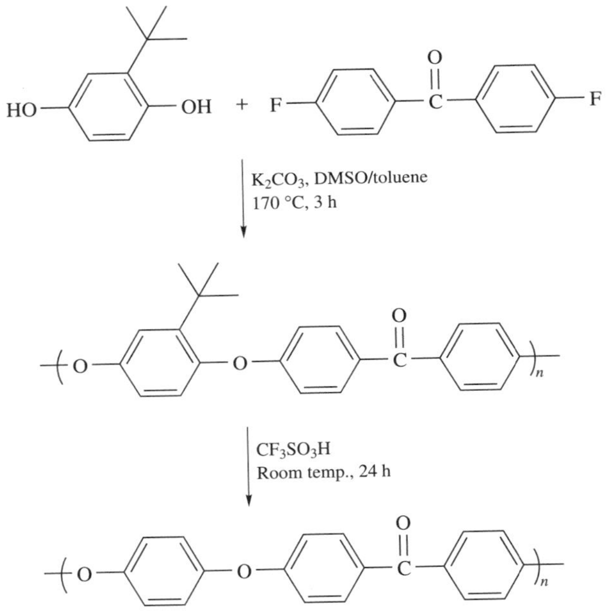

Scheme 6.15   Synthesis of high-molecular-weight PEEK via *t*-butyl containing a soluble PEK precursor.

Kelsey et al. reported that the cyclic ketal of 4,4'-dihydroxybenzophenone (DHBP) can polymerize with 4,4'-difluorobenzophenone in DMAc at 150°C (Scheme 6.16).[85] The polymerization afforded soluble amorphous polyketal that was quantitatively converted to PEK. Because of relatively lower reaction temperature, the PEK had minimal defect structures and thus possesses higher crystallinity and higher T_g, and has better physical properties than its counterpart made under higher temperatures.

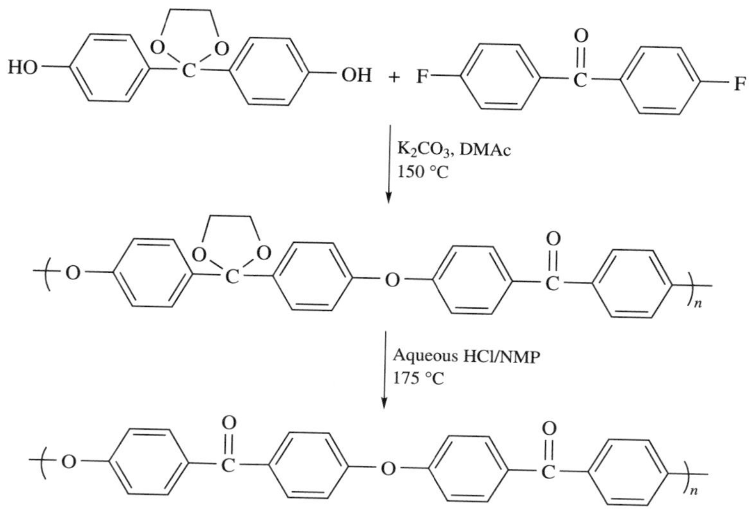

Scheme 6.16 Synthesis of high-molecular-weight PEEK via a ketal containing PEK precursor.

DFBP can be converted into difluorophenylketimine, which was used to prepare ketimine-modified PEEK at 180°C in NMP (Scheme 6.17).[86–89] The soluble

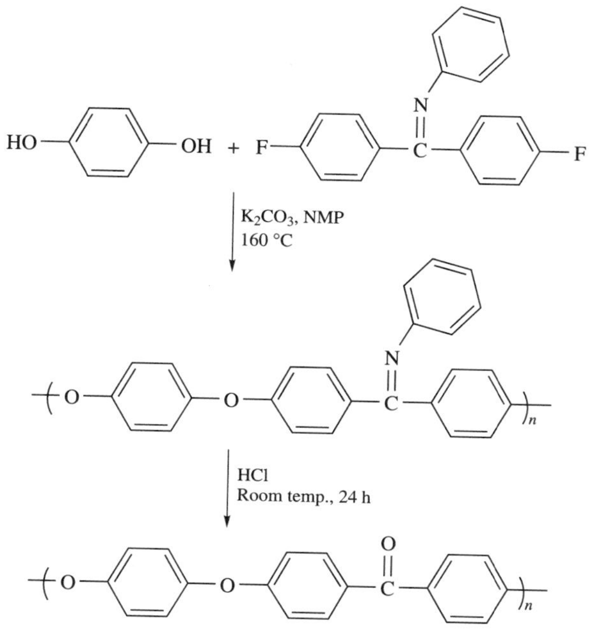

Scheme 6.17 Synthesis of poly(ether ketone) via poly(ether phenyl ketimine).

prepolymer can be hydrolyzed with dilute acid to yield very fine powders of PEEK. The biphenol-based ketimine-modified poly(ether ether ketone)s also have been prepared via the same approach.[90,91]

Compared with the *tert*-butyl substitutes PEEK, carbonyl-functionalized poly-ketones can be converted to the target polyketones at relatively mild conditions.

Pandya et al. developed a new approach to the synthesis of polyketone based on the nucleophilic substitution of bis(*a*-aminonitrile) derivatives with activated aromatic dihalides.[92] Bis(*a*-aminonitrile)s were prepared from aromatic dihalides in high yield by Strecker synthesis.[90] The conjugated bases of the *α*-aminonitriles are selective nucleophiles and thus can react with activated aromatic dihalides to form a carbon–carbon bond. Polymerization of isophthalaminonitrile with DFDPS in anhydrous DMF at room temperature using NaH as the base yields a soluble high-molecular-weight polyaminonitrile. Acid hydrolysis of this polyaminonitrile afforded the corresponding poly(ketone ketone sulfone). This method was applied to the synthesis of a series of highly ketone containing polyketones with or without ether linkages (Scheme 6.18).[90] These polymers were crystalline and have very high T_g and T_m ($>400°C$).

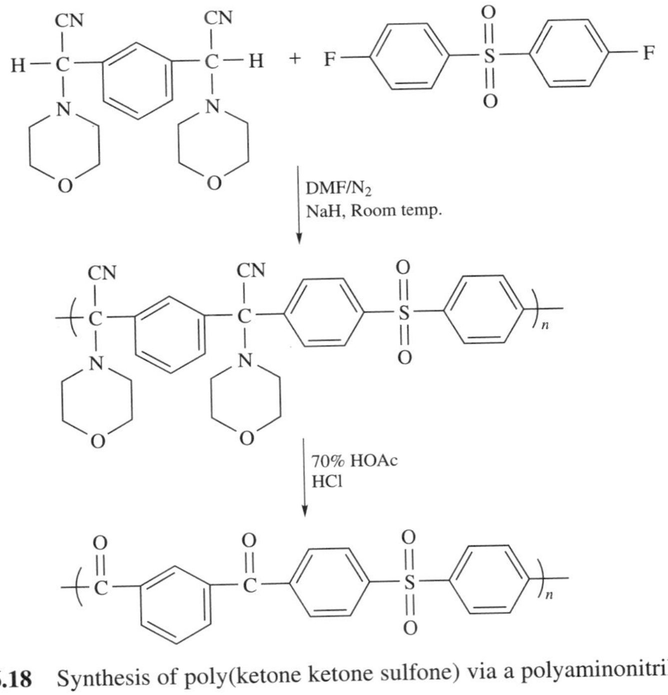

Scheme 6.18   Synthesis of poly(ketone ketone sulfone) via a polyaminonitrile precursor.

6.2.2.4.3   Ring-Opening Routes

Colquhoun at ICI first reported the synthesis of macrocyclic monomers containing ether and ketone linkages through nickel-catalyzed coupling of aryl dihalides under *pseudo* high dilution conditions.[93] Later, Chen et al. prepared poly(ether ether ketone) single-size macrocycles and macrocyclic oligomer mixtures via both nucleophilic aromatic substitution and Friedel–Crafts acylation approaches.[94–96] The ring-opening polymerization is initiated by a catalytic amount of nucleophiles, such as CsF or potassium salts of phenoxides. Low viscosities of the oligomers ensure good processibility. However, control of ring-opening polymerization is difficult and, complete conversion of the macromonomers is very difficult to achieve. Under some circumstances, melting points of the cyclics or cyclic mixtures are too high for practical ring-opening polymerization. One way to circumvent this problem may be achieved by preparing low-melting-point or even amorphous oligomers. Qi et al. recently reported new cyclic oligomers prepared from hydroquinone and N-phenyl(4,4′-difluorodiphenyl) ketimine.[97]

6.2.2.5 Poly(arylene ether phosphine oxide)s

Poly(arylene ether phosphine oxide)s are of special interest due to their fire resistance, optical properties,[11] and miscibility with other polymers.[98] Poly(arylene ether phosphine oxide) (PEPO) was first reported by Hashimoto et al.[99] However, only low-molecular-weight PEPOs were obtained by reacting bis(4-chlorophenyl)phenyl phosphine oxide (BCPPO) with bisphenols in various aprotic dipolar solvents utilizing sodium hydroxide as the base, because the electron-withdrawing ability of the phosphonyl group is not as strong as sulfonyl. Later, Smith et al. prepared high-molecular-weight PEPOs using bis(4-fluorophenyl)phenyl phosphine oxide (BFPPO) instead of BCPPO.[100] Riley et al. extended this work, and many structural units, such as carbonyl and sulfonyl, were successfully incorporated into the polymer main chain.[11] These noncoplanar polymers are amorphous with high T_gs. They are also thermally stable with very high decomposition temperatures. Cone calorimetry measurements showed their much lower heat release rate compared to polysulfones with similar structures, indicating their potential application as fire-retardant materials. They are optically clear and colorless with much higher refractive index than polycarbonate ($n = 1.58$). In addition, they also exhibited highly radiation-resistant character.[101] Very recently, McGrath and co-workers successfully synthesized a bisphenol-A-based PEPO by melt polymerization[102] via a silyl ether displacement. This approach does not require the removal of catalyst and avoids using solvent as the reaction medium. Therefore, the resulting polymer can be directly melt processed.

Phosphine-oxide-containing polymers can be chemically modified by reducing the phosphonyl group to phosphine using phenylsilane as the reducing agent.[103] The phosphine group can react with halo compounds and phosphonium salt can be generated.[104] This kind of polyelectrolyte was employed to prepare nonlinear optical (NLO) materials.[11,105]

New poly(ether ketone)s derived from bis[4-(4-fluorobenzoyl)phenyl]phenylphosphine oxide was reported by Fitch et al.[106] These polymers are thermally stable and have high T_g values (about 200°C).

6.2.2.6 Poly(arylene ether)s Synthesized via other Activated Dihalides

Although sulfonyl, carbonyl, and phosphonyl groups are the most common activating groups, other groups and compounds containing some electron-withdrawing groups, such as heterocyclic rings,[107] amide,[108] 2,6-dihalobenzonitrile,[109] 4,4'-difluoroazobenzene,[110] 4,4'-difluoroazoxybenzene,[111] and fluoro-substituted diarylacetylenes,[112] have been employed for the nucleophilic aromatic substitution. Polymers containing stilbene moieties,[113] chalcone group,[114] and dicyanoethylene group[115] were synthesized, and all of these materials can be crosslinked by heating.

The imide groups can also activate the aromatic halo and nitro groups on the bishaloimides,[116] or bisnitroimides.[117] Indeed, this approach has been commercialized to produce a bisphenol-A-based poly(ether imide) with the tradename Ultem (General Electric). A common feature of these groups is the stabilization of the negative charges developed at the 2- or 4-position of the aryl moiety in the transition state of the nucleophilic halogen displacement reaction through a Meisenheimer complex.[118] Labadie and Hedrick reported the perfluoroalkyl groups, either pendent or on the main chain, activate fluoro or nitro displacement by phenoxides.[119] Since electron-withdrawing perfluoroalkyl groups cannot participate in resonance stabilization, the activation by this group is expected due to the stabilization of the negative charges at the 2- or 4-position by hyperconjugation and by the negative inductive $(-I)$ effect. The steric congestion due to a bulky trifluoromethyl group may also facilitate the formation of a stable Meisenheimer complex with the release of steric strain.[120]

The *para*-fluorine atoms on highly fluorinated aromatic compounds such as hexafluorobenzene or decafluorobiphenyl are activated and thus can go through aromatic nucleophilic substitution with bisphenols in an aprotic solvent at low temperatures ($<80°C$).[121–123]

6.2.3 Other Approaches for the Synthesis of Poly(arylene ethers)s

In addition to nucleophilic and electrophilic substitution reactions, other reactions have also been used to prepare poly(arylene ether)s, especially those with special structures which otherwise could not be prepared. The following paragraph briefly reviews these reactions.

6.2.3.1 The Ullman Reaction

The Ullman reaction has long been known as a method for the synthesis of aromatic ethers by the reaction of a phenol with an aromatic halide in the presence of a copper compound as a catalyst. It is a variation on the nucleophilic substitution reaction since a phenolic salt reacts with the halide. Nonactivated aromatic halides can be used in the synthesis of poly(arylene ether)s, thus providing a way of obtaining structures not available by the conventional nucleophilic route. The ease of halogen displacement was found to be the reverse of that observed for activated nucleophilic substitution reaction, that is, I > Br > Cl ≫ F. The polymerizations are conducted in benzophenone with a cuprous chloride–pyridine complex as a catalyst. Bromine compounds are the favored reactants.[5a,124–127] Poly(arylene ether)s have been prepared by Ullman coupling of bisphenols and dibromoarylenes in the presence of a copper catalyst.[125] Poor reproducibility, the need for brominated monomers, and the difficulty of removing copper salts are the major disadvantages of this reaction.

6.2.3.2   *Metal Coupling Reactions*

A relatively new approach for preparing poly(arylene ether sulfone)s involves the formation of an aromatic carbon–carbon bond by coupling halogenated aromatic compounds with Ni^0[128]. Thus, polymers with biphenyl units can be made without starting with a biphenyl compound. Both unactivated and activated dihalides could be used. Unlike the Ullman reaction, the nickel coupling reaction works well with chloro compounds and the reaction conditions are much milder. The reaction is carried out in an aprotic solvent at $60-80°C$[129–132]. A zero-valent nickel–triphenylphosphine complex was used as the catalyst prepared from nickel chloride and zinc metal. This reaction was also used to polymerize a Schiff base of 4,4'-dichlorobenzophenone to yield a soluble polyketimine in NMP. The polyketimine was readily hydrolyzed to afford crystalline ketone.[133] Poly(4,4'-diphenylphenylphosphine oxide) (PAPO) was synthesized by nickel-catalyzed coupling polymerization. Ghassemi and McGrath reported preparing PAPO using bis(4-chlorophenyl)phenyl phosphine oxide catalyzed by a mixture of $NiCl_2$, 2,2'-bipyridine, zinc powder, and triphenyl phosphine in a nitrogen atmosphere.[103] The polymer developed a deep red color once the phosphine oxide groups in the polymer were reduced to phosphine using phenylsilane, possibly indicating conjugated behavior.

Other coupling reactions were also employed to prepare poly(arylene ether)s. Polymerization of bis(aryloxy) monomers was demonstrated to occur in the presence of an Fe(III) chloride catalyst via a cation radical mechanism (Scholl reaction).[134] This reaction also involves carbon–carbon bond formation and has been used to prepare soluble poly(ether sulfone)s, poly(ether ketone)s, and aromatic polyethers.

The Suzuki reaction was also used to prepare the polyketone since this particular reaction tolerates the subsequent step (Scheme 6.19).[135] Palladium-catalyzed cross-coupling of aromatic diacid chlorides and bis(trimethylstannane) monomers was utilized to prepare poly(arylene ether ketone)s.[136]

6.3   CONTROL OF MOLECULAR WEIGHT AND/OR ENDGROUPS

Control of molecular weight is very important, because the molecular weight needs to be high enough to afford polymers with good mechanical properties and yet low enough to have good processability. Generally, by upsetting the mole ratio of the two monomers, one can easily control the molecular weight. In some cases, the reactive endgroups may significantly deteriorate the polymer stability. One way to minimize the problem is to use a monofunctional monomer as a comonomer to end cap the polymer chain. Thus, it is known that melt stable end-capped bisphenol A based polysulfones can be prepared by bubbling methyl chloride into phenolate-reactive systems.

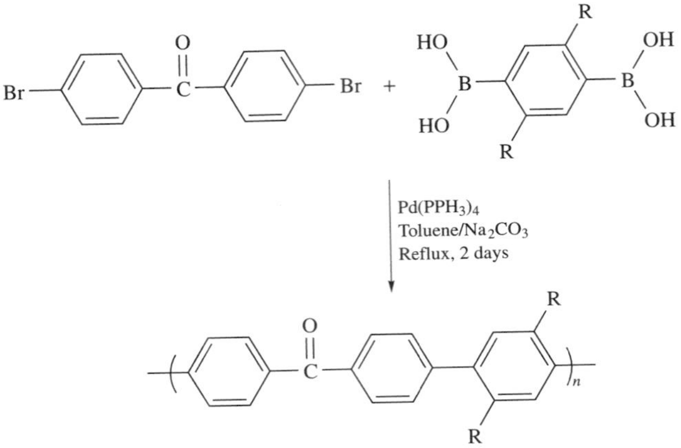

Scheme 6.19   Synthesis of polyketone via Suzuki reaction.

End capping the polymer chain with functional groups is also a very useful approach to prepare reactive oligomers for generating some block copolymers or modifying polymer networks. Polysulfones, poly(arylene ether ketone)s, and phosphorus-containing poly(arylene ether)s were prepared with different terminal functional groups, such as carboxyl,[137–140] amine,[141–145] hydroxyl,[146,147] phenylethynyl,[148] and cyanate[149] chain ends. Amino-terminated poly(arylene ether sulfone)s or poly(arylene ether ketone)s were synthesized and used to modify epoxies.[150–153]

Mandal and Hay reported poly(sulfone)s end capped with metal containing phthalocyanines (PcM).[154] The polymers were soluble in common organic solvents. The T_g was increased probably due to the stacking of metallophthalocyanine (PcM) moieties. In addition, annealing the polymers afforded a melting endotherm due to crystallization. Again, this may be due to the stacking of PcM rings.

Noell et al. reported the preparation of silica–poly(ether ether ketone) hybrid materials with improved physical properties.[155] An amine-end-capped poly(ether ether ketone) was used to react with isocyanatopropyltriethoxysilane in tetrahydrofuran (THF). The triethylsilane-end-capped poly(ether ether ketone) was mixed with tetraethoxysilane (TEOS) in THF. Quantitative amounts of water were introduced into the system, and the mixture was refluxed at 80°C. The entire reaction mixture was allowed to further react in Teflon molds. Tough transparent materials were obtained by this approach.

6.4 CONTROL OF TOPOLOGIES

6.4.1 Hyperbranched Polymers

Hyperbranched polymers generally have very low melt and intrinsic viscosities. The large number of chain-end functional groups present in hyperbranched macromolecules have also been shown to dramatically affect physical properties such as the glass transition temperature, solubility, and chemical reactivity. The degree of branching (DB) would be 0 and 1 for a perfect linear molecule and for a perfect dendritic molecule, respectively.[156] In all the cases, DB is between 0 and 100%.

Miller et al. were the first to report the synthesis of hyperbranched poly(arylene ether ketone)s.[157] They used 3,5-bis(4-fluorobenzoyl)phenol as the starting material, as shown in Scheme 6.20.

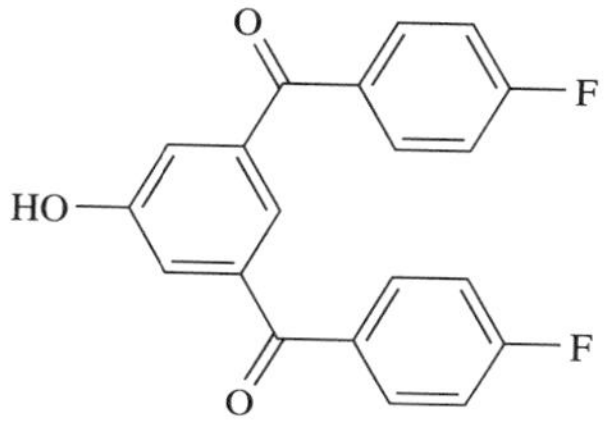

Scheme 6.20   Structure of 3,5-bis(4-fluorobenzoyl)phenol.

Later, Hawker and Chu prepared two different monomers, 3,5,-difluoro-4′-hydroxybenzophenone and 3,5-dihydroxy-4′-fluorobenzophenone (Scheme 6.21).[158]

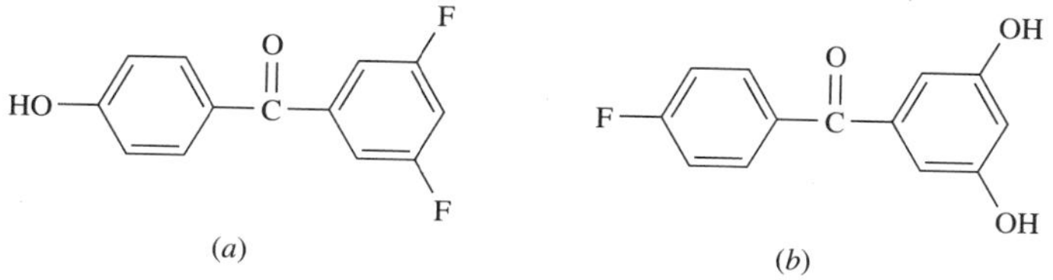

(a)   (b)

Scheme 6.21   Structures of (a) 3,5-difluoro-4′-hydroxybenzophenone and (b) 3,5-dihydroxy-4′-fluorobenzophenone.

Polymerization of these two monomers afforded the same internal linkages but with different terminal groups. High-molecular-weight polymers were obtained when the polymerization was at 200°C in an NMP–toluene solvent mixture using sodium carbonate as a base. Higher molecular weight polymer ($M_n = 95$ kg/mol) was obtained from B than from A ($M_n = 20$ kg/mol) due to the different reactivities of these two monomers. The degree of branching was also different for these two hyperbranched poly(ether ether ketone) analogues. They ranged from 14% for the phenolic-terminated polymer to 49% for the fluoro-terminated polymer. Kwak and Lee synthesized poly(ether ketone) with controlled degree of branching using monomer A by introducing some trifluoromonomer as the core.[159] The properties of the hyperbranched polymer have been reported to be largely dependent on the nature of the chain endgroups. Morikawa modified the monomer Miller et al. used by introducing a various number of phenylene units in the backbone.[160]

The polymers can either be amorphous or semicrystalline, with increased glass transition temperatures by increasing the phenyl units in the backbone.

Aromatic electronic substitution approach can also be used to prepare the branched poly(ether ketone)s. Shu et al. prepared two monomers: 3,5-diphenoxybenzoic acid (A)[161] and 5-phenoxyisophthalic acid (B) (Scheme 6.22).[162] These two monomers were polymerized in phosphorus pentoxide–methanesulfonic acid via an electrophilic aromatic substitution. The reaction temperature for monomer A was 90°C. Because monomer B was only partially soluble at 100°C, reaction temperature higher than 120°C was used. The degree of branching for these polymers was about 55%. The phenoxy groups at the chain ends of the hyperbranched poly(ether ketone) are highly reactive for further electrophilic aromatic substitution and readily react with various carboxylic acids. The carboxylic acid groups at the chain ends were readily accessible to reagent in solution and were converted to a variety of functional groups. Physical properties, such as the glass transition temperature and the solubility of the hyperbranched poly(ether ketone)s, depended heavily on the nature of the chain ends. An ammonium salt form of carboxylic-terminated poly(ether ketone) was soluble in water and was demonstrated to form unimolecular micelle.

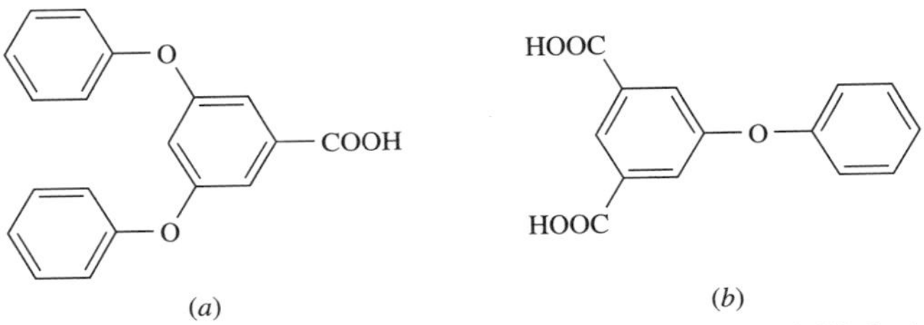

(a)   (b)

Scheme 6.22   Structures of (a) 3,5-diphenoxybenzoic acid (A) and (b) 5-phenoxyisophthalic acid (B).

Recently, Fossum et al. prepared several phosphine-oxide-containing monomers (Scheme 6.23).[163] These monomers were used to prepare hyperbranched polymers in a typical aromatic nucleophilic substitution. However, only oligomers with M_n lower than 2500 g/mol were obtained. These results did not surprise us, since our previous work demonstrated that the *para*-hydroxyl group of the phosphonyl group is not very reactive and would require higher reaction temperatures.[11]

6.4.2 Dendritic Poly(arylene ether)s

Dendrimers have structures similar to that of hyperbranched polymer and can be taken as the "perfectly" branched polymer with monodispersity. However, they need to be prepared by a multistep procedure. Therefore, very little work has been done on dendritic poly(arylene ether)s. Morikawa et al. prepared a series of monomers with a various number of phenylene units.[164,165] These monomer were used to prepare poly(ether ketone) dendrons with graded structures (Scheme 6.24).

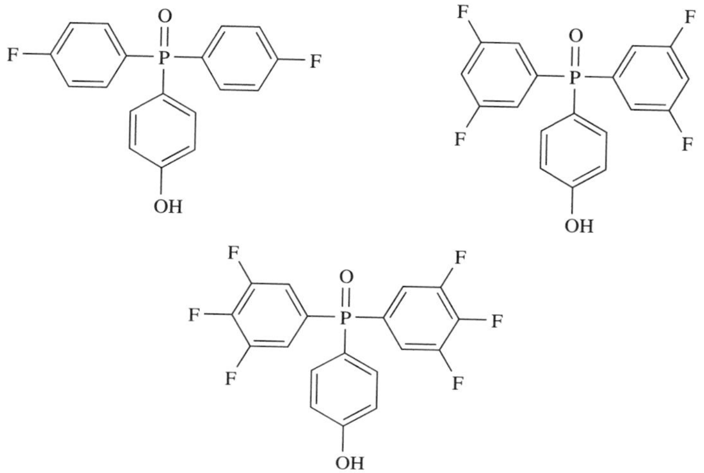

Scheme 6.23   Structures of AB_x triarylphosphine oxides.

6.5 MODIFICATION OF POLY(ARYLENE ETHER)S

One objective of modifying poly(arylene ether)s was to alter their chemical nature to some degree without sacrificing their excellent physical and other properties. Functionalized polysulfones may find many applications as membrane materials. These materials may be used as gas separation membranes or membranes for water desalination. In addition, functionalized polymers can be further modified to generate graft copolymers or used to modify networks such as epoxies. Recent advances in fuel cells require higher performance proton exchange membranes, which, in turn, have motivated research on sulfonated high-performance polymers. Generally, there are two methods to functionalize polymers. One is to chemically modify the preformed polymer. Another is to use a functionalized monomer to prepare the polymer via direct copolymerization.

6.5.1 Modification of Polymers

Introducing amino groups onto the phenylene rings comprising poly(phenylene ether sulfone) chains was reported by Conningham et al. by carrying out nitration–reduction procedures on preformed PES.[166] However, side reactions can occur in nitration–reduction procedures due to the attack of nitronium ion at carbon-to-sulfur chain linkages, as previously suggested by Schofield.[167]

Quentin may have been the first to sulfonate (arylene ether sulfone).[168] In this patent, it was demonstrated that the bisphenol A polysulfone could be sulfonated by chlorosulfonic acid to produce a sulfonated poly(arylene ether sulfone), which was used for desalination via reverse osmosis. However, the chlorosulfonic acid may be capable of cleaving the bisphenol A polysulfone partially at the isopropylidene link or it might induce branching and crosslinking reactions by

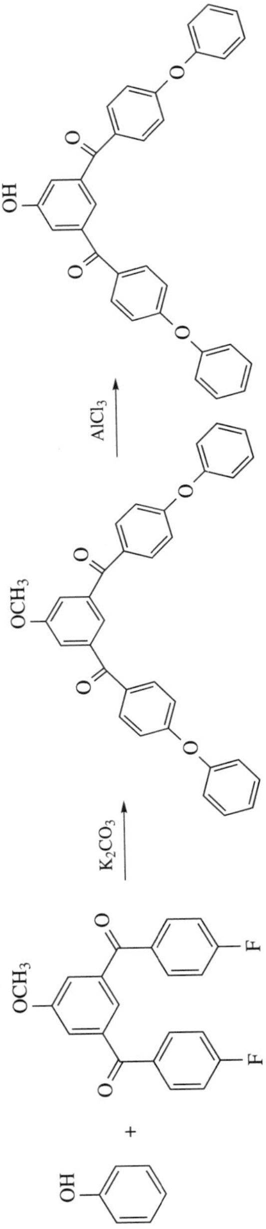
OH
+
OCH3
F
F
O
O
K2CO3
AlCl3

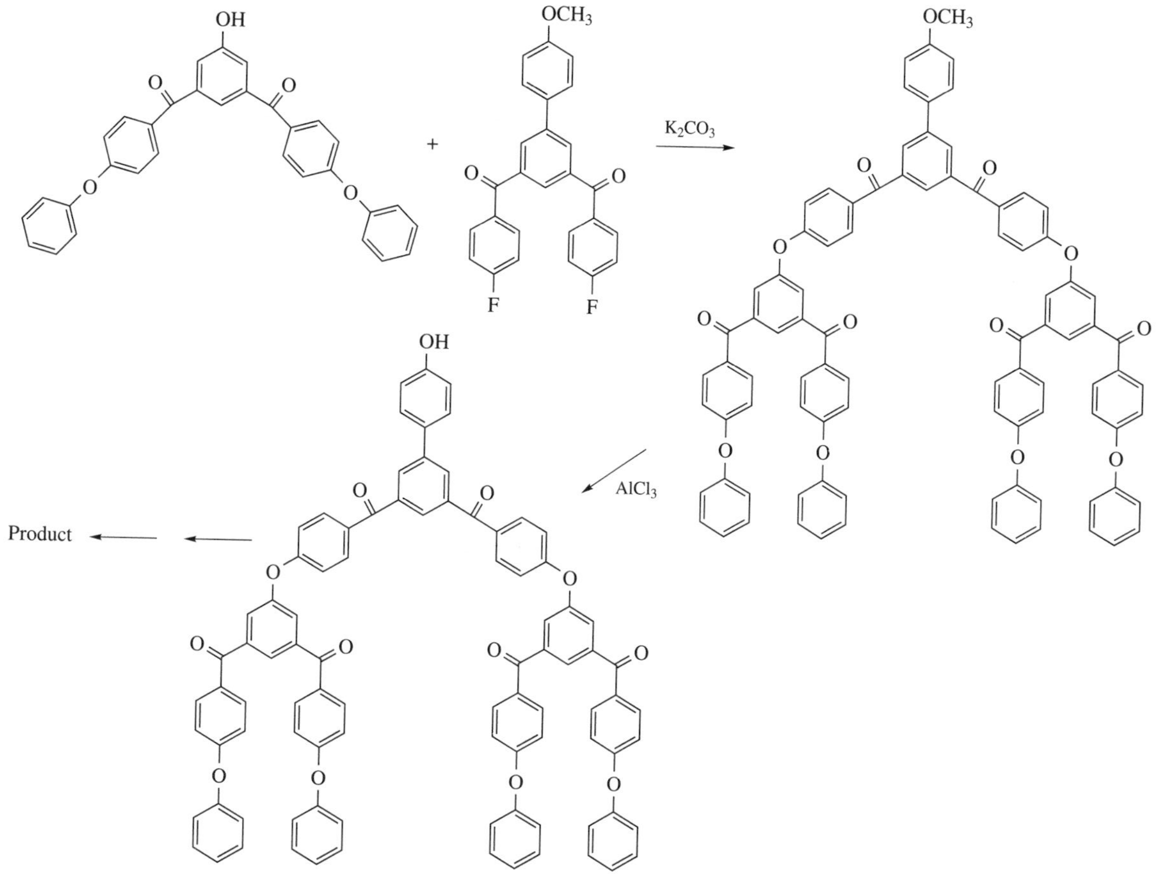

Scheme 6.24 Synthesis of poly(ether ketone) dendrons with graded structures

converting the intermediate sulfonic acid group into a partially branched or crosslinked sulfone unit. An alternative route was employed to sulfonate bisphenol A polysulfone. In this approach, a complex of SO_3 and triethyl phosphate (TEP) with a mole ratio of 2 : 1 was used to sulfonate the polymer at room temperature.[169,170] This mild sulfonation treatment could minimize or even eliminate possible side reactions.

Chloromethylation was also carried out to modify the polysulfone for preparing membrane materials.[171]

Recently, Guiver et al. reported a number of derivatives of polysulfone and poly(aryl sulfone).[172–188] Polysulfones were activated either on the ortho-sulfone sites or the ortho-ether sites by direct lithiation or bromination–lithiation. The lithiated intermediates were claimed to be quantitatively converted to azides by treatment with tosyl azides. Azides are thermally and photochemically labile groups capable of being transformed readily into a number of other useful derivatives.

Poly(arylene ether ketone)s can also be modified by introducing the functional groups using similar approaches to polysulfones. For example, poly(arylene ether ketone)s were sulfonated.[189] In addition, o-dibenzoylbenzene moieties in the poly(arylene ether)s can be transformed to heterocycles by cyclization with small molecules. These polymers can react with hydrazine monohydrate in the presence of a mild acid in chlorobenzene or with benzylamine in a basic medium.[190] Another example of the use of the o-benzyl cyclization strategy is the intramolecular ring closure of poly(arylene ketone)s containing 2,2′-dibenzoylbiphenyl units to form poly(arylene ether phenanthrenes).[191]

Modifying polymers is very convenient and quite economic. However, control of sulfonation is very difficult since the sulfonation occurs on the activated ring, which may deteriorate thermal stability of the polymer. The extent of sulfonation is also hard to control and side reactions may occur.

6.5.2 Modification of Monomers

Modification of monomers is fundamentally different than postreactions in that it can allow better control of the molecular structure. Both diphenols and dihalides can be modified to incorporate functional groups or new monomers containing functional groups can be synthesized with similar structures as their counterparts.

6.5.2.1 Modification of Diphenols

Mathias et al. synthesized poly(ether sulfone)s and poly(ether ether ketone)s with pendent adamantane groups.[192] Incorporating adamantane into a polymer as a pendent group has been demonstrated to significantly increase the glass transition temperature.

Herbert and Hay reported a bisphenolic monomer, 3,8-bis(4-hydroxyphenyl)-N-phenyl-1,2-naphthalimide (Table 6.1), as well as its corresponding poly(N-phenyl imido aryl ether sulfone) via transimidization reactions with hydrazine monohydrate, aliphatic amines, and an amino acid.[193] These polysulfones with

TABLE 6.1   Diphenol Monomers Modified with Functional Groups

Diphenol Monomers	References
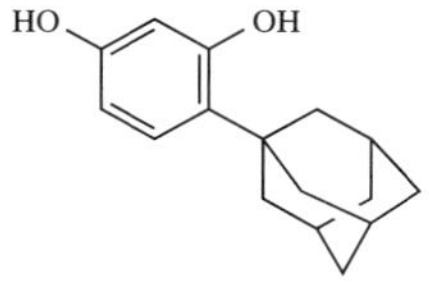 4-(1-Adamentyl)-1,3-benzenediol	170
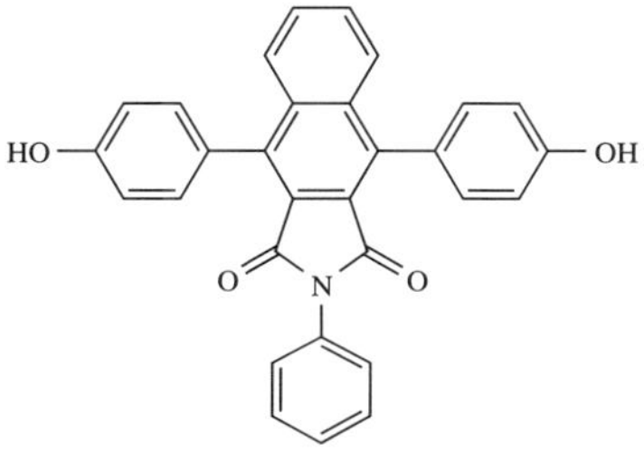 3,8-Bis(4-hydroxyphenyl)-*N*-phenyl-1,2-naphthalimide	171
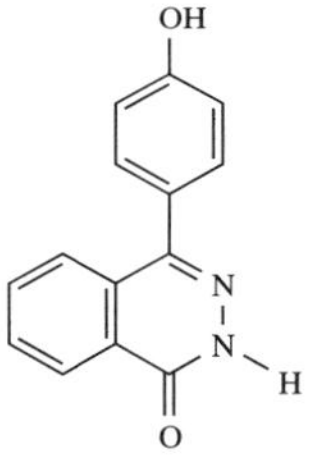 Phthalazinone	172, 173
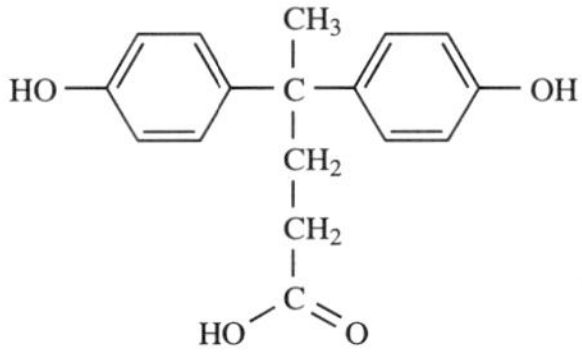 4,4′-Bis(4-hydroxyphenyl)pentanoic acid (BHPA)	174–177
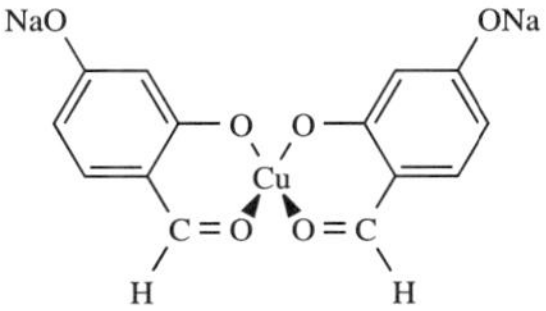 Copper–chelate monomer	178, 179
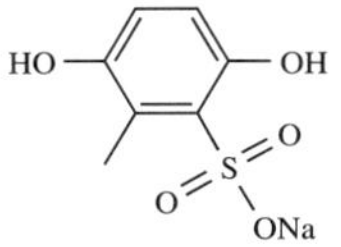 Sulfonated hydroquinone	180

reactive functional groups can be used to prepare graft copolymers or polymer networks by thermal curing.

Phenolphthalein reacts with hydroxylamine to give 4-aminophenol and 2(4-hydroxybenzoyl)benzoic acid in high yield. 2-(4-Hydroxybenzoyl)benzoic acid reacts with hydrazine to yield phthalazinone.[194] This monomer behaves like a bisphenol and reacts with activated difluorides (Scheme 6.25).[195] The resulting poly(arylene ether)s have high glass transition temperatures (over 265°C).

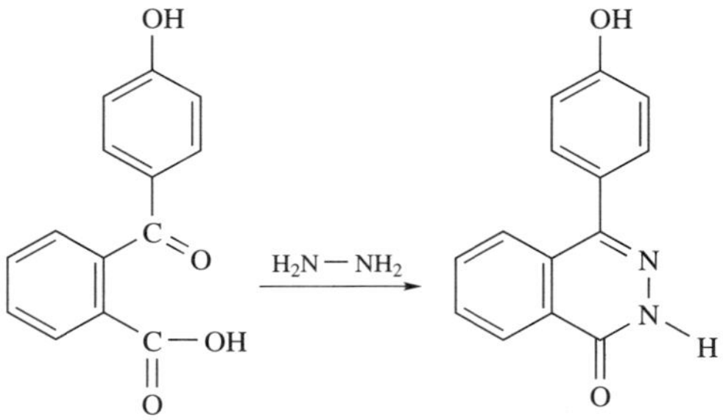

Scheme 6.25   Synthesis of phthalazinone.

Using 4,4′-bis(4-hydroxyphenyl)pentanoic acid (BHPA)[196,197] as comonomer, some polysulfones with pendent carboxylic groups were successfully synthesized.[198,199] Table 6.1 shows the structure of BHPA. The functional groups can be used for the preparation of graft copolymers.

A new monomer containing Cu(II) was also used to synthesize the electrically conductive polysulfone (Table 6.1).[200,201]

Sulfonated hydroquinone was used to prepare functional poly(arylene ether ketone)s, which may be used a gas separation membranes.[202]

6.5.2.2 Modification of Activated Dihalides

A few functional dihalides have been prepared. Some of these monomers are tabulated in Table 6.2. The monomers could be nitrated and then reduced to amine-functionalized monomers (Scheme 6.26).[203] This approach was used to nitrate DCDPS or bis-4-fluorophenyl phenyl phosphine oxide.[204] These monomers were used to copolymerize with some other activated dihalides as the comonomers.

Robeson and Matzner were the first to report the synthesis of the sulfonation of DCDPS.[205] This work makes it possible to synthesize sulfonated poly(arylene ether sulfone) with well-controlled structures. Ueda et al. used this monomer (Scheme 6.27) as a comonomer of DCDPS to react with bisphenol A and high-molecular-weight bisphenol-A-based copolymers with up to 30 mol% sulfonation achieved.[206] Biphenol-based copolymers with up to 100 mol% sulfonation were recently reported by Wang et al.[207]

Wang et al. also successfully sulfonated DFDPS (Scheme 6.28).[208] Since the DFDPS is more reactive than DCDPS, it is expected that the sulfonated DFDPS

TABLE 6.2   Dihalide Monomers Modified with Functional Groups

Monomer	Reference
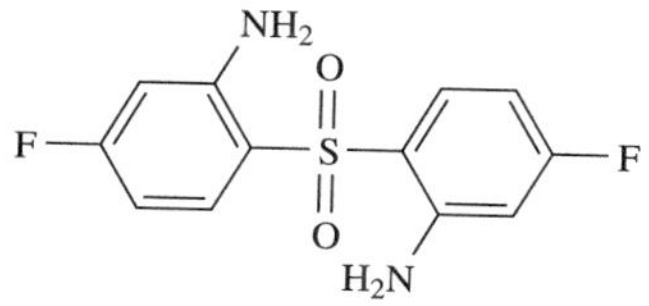 Amine functional DFDPS	181
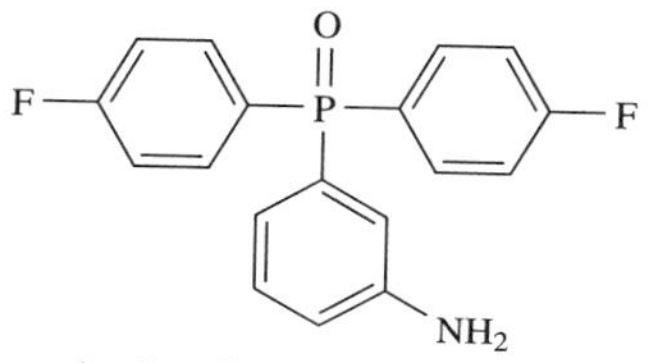 Amine functional BFPPO	182
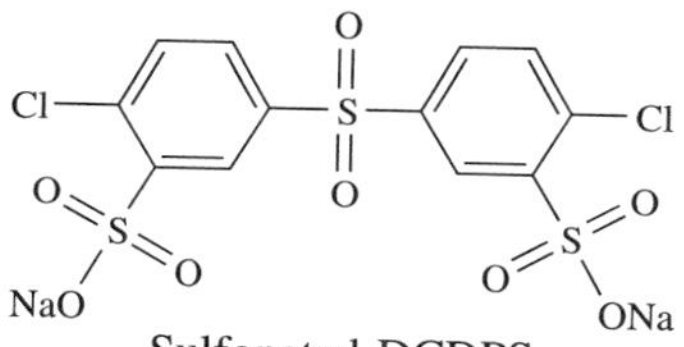 Sulfonated DCDPS	183–185
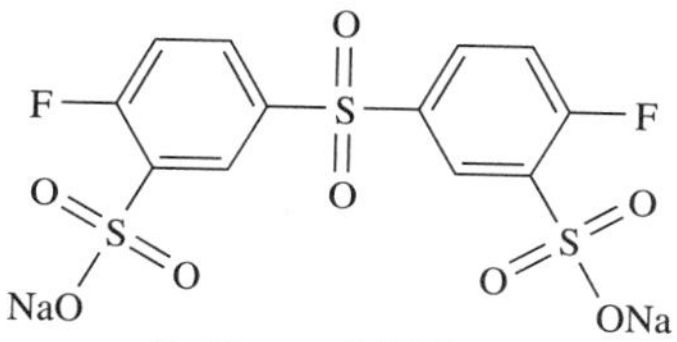 Sulfonated DFDPS	186
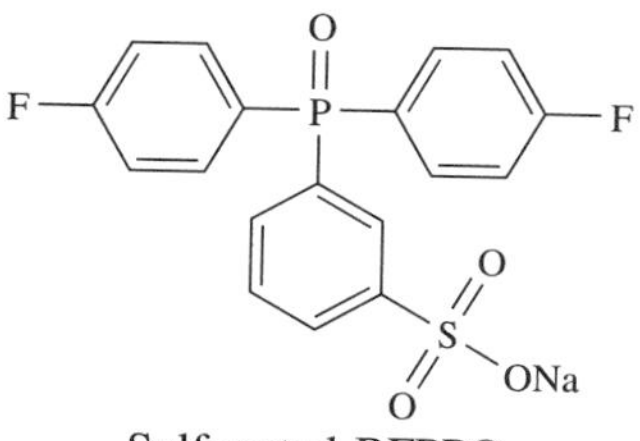 Sulfonated BFPPO	187

is more reactive than sulfonated DCDPS. Therefore, the sulfonated DFDPS can react with deactivated diphenols, such as bisphenol S and bis(4-hydroxyphenyl phenyl phosphine oxide).

Shohba et al. reported a sulfonated BFPPO monomer.[209] The sulfonated BFPPO was used as a comonomer of BFPPO to react with biphenol. Fairly high molecular

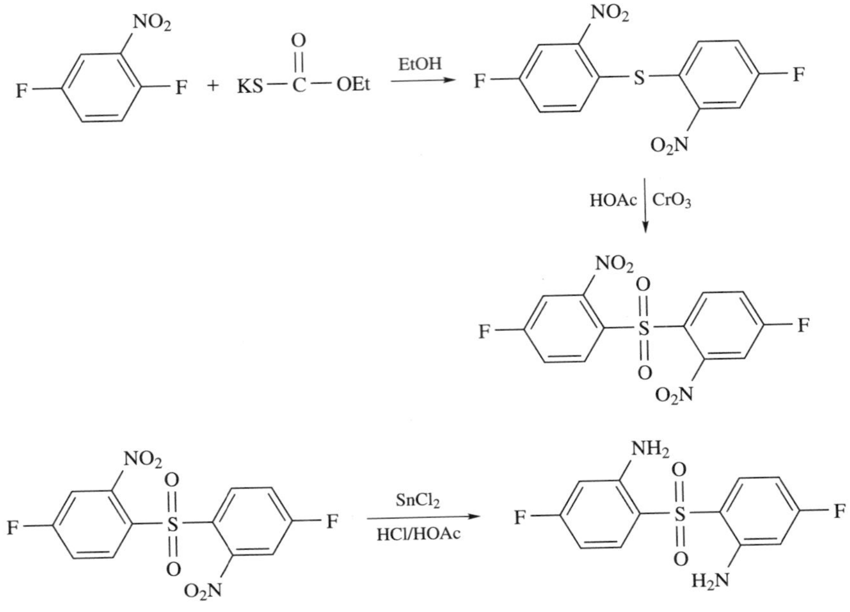

Scheme 6.26   Synthetic route for the synthesis of amino DFDPS.

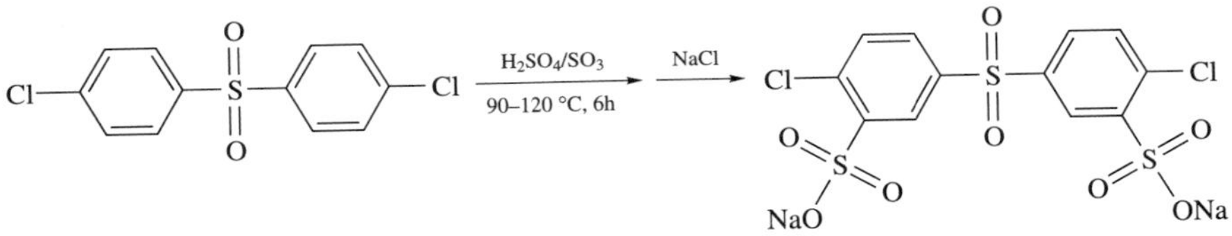

Scheme 6.27   Synthesis route for disulfonated DCDPS.

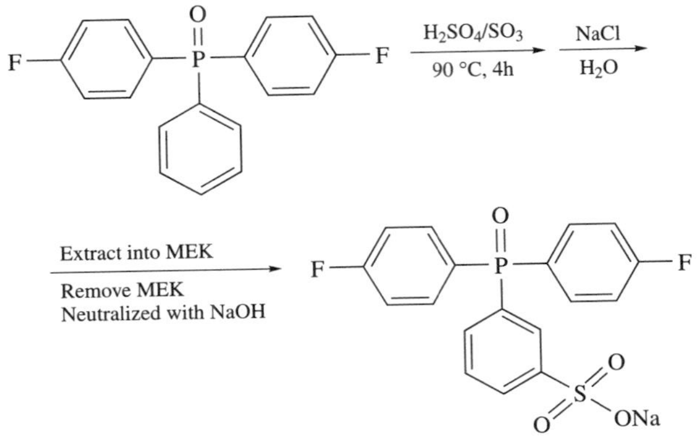

Scheme 6.28   Synthesis route for sulfonated BFPPO.

weights were obtained and the mole percent of sulfonated monomer was up to 100 mol %. The polymers afforded transparent films.

6.6 BLOCK AND GRAFT COPOLYMERS

Poly(arylene ether sulfone)s and poly(arylene ether ketone) have been employed to prepare block and graft copolymers. Generally, the block copolymers can be prepared by reacting functional-group-terminated oligomers with other functional oligomers and monomers.

Poly(arylene ether)s are high-performance polymers. Introducing soft segments generates a series of block copolymers. In addition, the soft segments can be either hydrophobic or hydrophilic. The first perfectly alternating polysulfone-*block*-polydimethylsiloxane was reported by Noshay et al.[210] They used a hydroxyl-terminated bisphenol A polysulfone to react with a dimethylamine-terminated polydimethylsiloxane in chlorobenzene at 132°C. The copolymers are surprisingly quite hydrolytically stable, and varying the block lengths can control the morphologies. Hedrick et al. reported the synthesis of poly(arylene ether ketone)s and polydimethylsiloxane segmented copolymer.[211] The poly(ether ether ketone)-*block*-polydimethylsiloxane copolymers have some special properties. Structure–property studies showed the properties of the block copolymers are highly related to the structures.[212] Perfectly alternating poly(arylene ether phenyl phosphine oxide)-*block*-polydimethylsiloxane copolymers were recently reported by Polk and co-workers.[213] Due to the polarity of the phosphonyl group, these series of block copolymers are expected to have some novel properties.

Polystyrene-*block*-polysulfone-*block*-polystyrene and poly(butyl acrylate)-*block*-polysulfone-*block*-poly(butyl acrylate) triblock copolymers were prepared using a macroinitiator.[214] The hydroxyl-terminated polysulfone was allowed to react with 2-bromopropionyl bromide, an atomic transfer radical polymerization (ATRP) initiator, in the presence of pyridine. The modified macroinitiator could initiate the styrene polymerization under controlled conditions.

Incorporating a hydrophilic block into polysulfones may improve certain properties. Zhao et al. reported a A–B block copolymer based on poly(tetramethylene oxide) (PTMO) and polysulfone (PSF).[215] They first prepared pyridine-terminated PSF oligomers of controlled molecular weights via nucleophilic aromatic substitution reactions. PTMO oligomers were prepared by ring-opening polymerization processes using trifluoromethanesulfonic anhydride as the initiator and were terminated with the PSF oligomers, resulting in quaternary nitrogen linkages along the backbone. The PSF-*b*-PTMO block copolymers were also prepared via a melt transetherification procedure.[216]

Preparation of poly(ethylene oxide) (PEO) and poly(arylene ether) based hydrophilic–hydrophobic block copolymer is of special interest because PEO has been proven to be particularly reliable and versatile for the surface modification of biomaterials. The first poly(ethylene oxide)-*block*-polysulfone (PEO-*b*-PSF) copolymers were reported by Aksenov et al.[217] They employed diisocyanate chemistry to link hydroxy-terminated sulfone oligomers and poly(ethylene glycol). Recently, Ting and Hancock prepared PEO-*b*-PSF-*b*-PEO by a traditional nucleophilic substitution method using monohydroxyl PEO and bisphenol A as the comonomers to react with DCDPS.[218] Hancock and Fagen reported an alternative approach.[219] They separately prepared aryl-chloride-terminated PSF oligomer and PEO oligomer. These two polymers condensed with each other via a typical nucleophilic aromatic substitution. With regard to the similar microporous hollow-fiber membranes, a PEO-*b*-PSF membrane had a dynamic water contact angle of 33° compared to 111° for a polysulfone membrane, demonstrating highly improved hydrophilicity of block copolymer, which suggests that they might be unique materials for medical devices.

Synthesis of block copolymers containing two high-performance block polymers is of interest, because this approach may generate new polymeric materials with novel morphologies and properties. McGrath et al. prepared polycarbonate–polysulfone block copolymer using hydroxyl-terminated polysulfone as the prepolymer.[220] Lambert et al. reported segmented copolymers based on poly(arylene terephthalate) or polyoxygenzoates and poly(arylene ether sulfone)s.[221] Wu et al. developed a method to prepare poly(arylene ether ketone sulfone) copolymers.[222] In this approach, diphenols and DFBP were polymerized in the presence of hydroxyl-terminated polysulfone. A segmented block copolymer was obtained. The copolymer has increased T_g while not sacrificing crystalline melting behavior relative to the PEEK.

Wu et al. synthesized PES with sodium phenate endgroups and reacted this with fluorine-ended PEEK.[223] However, PES[224] and PEK[225,226] polymers are known to undergo transetherifications rapidly at high temperatures in the presence of nucleophilic phenolates. Therefore, any synthesis involving a nucleophilic substitution reaction can lead either to a molecular weight decrease and/or to molecular rearrangement (scrambling) of the structure units.

Bourgeois et al. reported a new approach[227] wherein poly(arylene ether sulfone) and poly(arylene ether ketone) were functionalized with anhydride and amine chain ends, respectively, to afford imide-linked copolymers.

Hedrick et al. reported imide aryl ether ketone segmented block copolymers.[228] The block copolymers were prepared via a two-step process. Both a bisphenol-A-based amorphous block and a semicrystalline block were prepared from a soluble and amorphous ketimine precursor. The blocks of poly(arylene ether ether ketone) oligomers with M_n range of 6000–12,000 g/mol were coreacted with 4,4′-oxydianiline (ODA) and pyromellitic dianhydride (PMDA) diethyl ester diacyl chloride in NMP in the presence of N-methylmorphiline. Clear films with high moduli by solution casting and followed by curing were obtained. Multiphase morphologies were observed in both cases.

Very few graft copolymers based on poly(arylene ether)s have been synthesized, probably because of their chemical inertness. Klapper et al. reported grafting the polystyrene or polyisoprene onto the poly(ether ether ketone ketone) (PEEKK) by anionic deactivation.[229] The carbonyl groups on the backbone can be attacked by the polystyrene monoanion or polyisoprene anion (M_n about 3000). Due to the steric hindrance only about 30% of the carbonyl groups can be reacted.

Control of the degree of grafting could be achieved should the molar ratio of poly(styrenelithium) or poly(isoprenelithium) versus PEEKK be kept below the maximum grafting efficiency. Grafting polystyrene onto the backbone did not change its T_g whereas grafting polyisoprene onto the backbone did increase its T_g, indicating grafting makes the polyisoprene chain more rigid. This approach demonstrated that combining the extraordinary properties of a high-performance polymer with those of standard polymers can be achieved by grafting polystyrene and polyisoprene onto a poly(ether ether ketone ketone) backbone.

Le Guen et al. modified an amine-functionalized poly(arylene ether ketone) by condensation with carboxy-terminated oligobenzamides as well as benzoyl chloride.[230] Grafting oligobenzamides increased the T_g of the polymer control. However, grafting the benzoyl group decreased T_g. Young's modulus increases regularly with the length of the oligobenzamide side chain introduced. These results indicated that hydrogen bonding of the side chains plays a crucial role in the improvement of the polymer properties.

6.7 MISCELLANEOUS POLY(ARYLENE ETHER)S, POLY(ARYLENE THIOETHER)S, AND RELATED POLYMERS

6.7.1 Fluorinated Polymers

Fluorinated poly(arylene ether)s are of special interest because of their low surface energy, remarkably low water absorption, and low dielectric constants. The bulk$-CF_3$ group also serves to increase the free volume of the polymer, thereby improving various properties of polymers, including gas permeabilities and electrical insulating properties. The 6F group in the polymer backbone enhances polymer solubility (commonly referred to as the "fluorine effect") without forfeiture of the thermal stability. It also increases the glass transition temperature with concomitant decrease of crystallinity.

Stamatoff and Wittmann reported a synthesis of a 2-(4-phenoxyphenyl)hexafluoroisopropanol in the presence of HF and an organic solvent via a Friedel–Crafts reaction, as shown in Scheme 6.29.[231] The resulting polymer could be compression molded at 330–350°C. It also exhibited excellent thermostability and mechanical properties.

Scheme 6.29 Synthesis of poly(hexafluorobisphenol A).

Later, Farnham and Johnson reported the synthesis of higher molecular weight, thermoplastic poly(arylene ether)s with good thermal, oxidative, chemical stability, and physical properties by reacting (at 120–260°C) a phenolate metal salt with a dibromide in the presence of a cuprous salt or cuprous salt complex in an inert diluent (Scheme 6.30).[232]

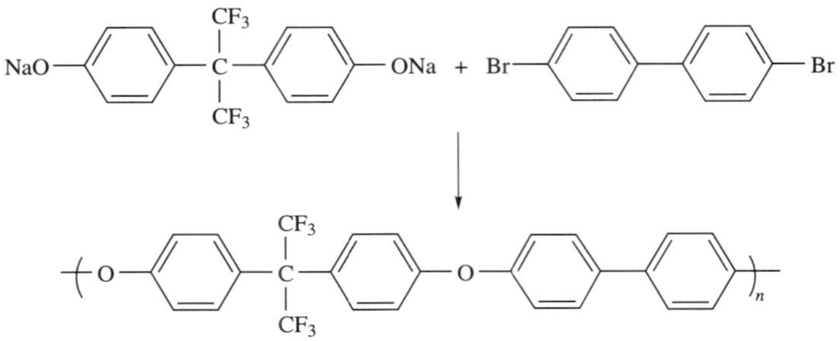

Scheme 6.30   Synthesis of HFBPA-based poly(arylene ether)s.

Hexa-fluorobisphenol A (HFBPA) based polysulfone and poly(arylene ether phosphine oxide) were prepared by nucleophilic aromatic substitution similar to that of bisphenol-A-based polysulfone and poly(arylene ether phosphine oxide).[11]

Poly(arylene ether)s containing 6F and ketone groups were synthesized via nucleophilic aromatic substitution by bisphenol AF with different dihalides containing ketone groups.[233-236] Mercer et al. reported a highly fluorinated poly(arylene ether ether ketone) by reacting decafluorobenzophenone with diphenols via solution condensation polymerization.[237] The results suggested that the polar groups needed to be avoided to obtain low-k materials. As a result, the fluorinated polymers have enhanced solubility in common organic solvents, high thermal stability (485–553°C), T_gs of 180°C, and low dielectric constants (2.4 at 10 GHz), while a nonfluorinated poly(ether ketone) analog had a T_g of 143°C and dielectric constant of 2.85. A systematic study on the dielectric constant and moisture absorption in fluorinated polyethers, poly(ether sulfone)s, and poly(ether ketone)s showed that dielectric constants were in a range of 2.72–3.03 and moisture absorption 0.78–1.21%, which was lower for poly(arylene ether)s without any sulfone, ketone, or ether polarizable groups. The conclusion is that polymers prepared without polarizable groups, with a high level of fluorine and symmetrical substitution, would have very low dielectric constants.[238]

Further lowering the dielectric constants has been achieved by preparing highly fluorinated polyethers without any sulfone, ketone, or other polarizable groups.[239-241] Typically, the *para*-fluorine atoms on highly fluorinated aromatic compounds, such as hexafluorobenzene and decafluorobiphenyl, are activated and thus can go through aromatic nucleophilic substitution with HFBPA under typical reaction conditions (Scheme 6.31).[217]

Recently, Banerjee et al. prepared a series of difluoro monomers with pendent trifluoromethyl groups using a Pd(0)-catalyzed cross-coupling reaction (Scheme 6.32).[242,243] These monomers were converted to poly(arylene ether)s by nucleophilic displacement of the halogen atoms on the benzene ring with several

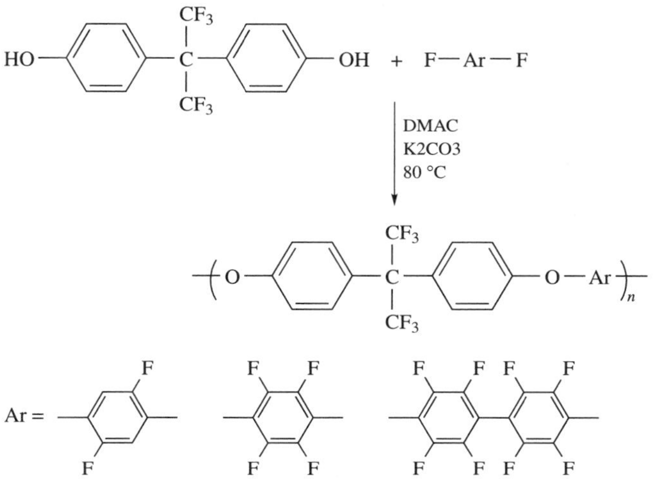

Scheme 6.31 Synthesis of highly fluorinated polyethers.[217]

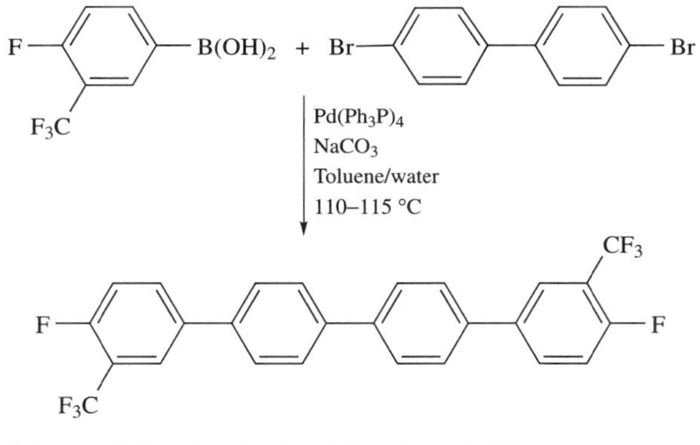

Scheme 6.32 Synthesis of fluorinated difluoro monomer.

bisphenols. Most of these polymers are thermally stable with T_g up to 234°C. They are quite soluble in common organic solvents. The tensile strength can be up to 115 MPa. The heterocyclic ring containing monomers were also prepared.[244] However, only the oxazole monomer afforded high molecular weight.

6.7.2 Poly(arylene thioether)s

Polysulfonation of self-polycondensation of 4-(phenylthio)benzenesulfonyl chloride was also used to prepare poly(arylene sulfide sulfone)s.[245,246] Condensation of diphenyl sulfide with 4,4′-oxydibenzenesulfonic acid or 4,4′-thiodibenzenesulfonic acid[247] or by polyetherification of polycondensation of DCDPS with 4,4′-dihydroxydiphenol sulfide occurred.[5]

Condensation of an activated aromatic halide with either an alkali metal thiophenoxide or an alkali metal sulfide is performed under the same experimental condensations as used for polyetherification (Scheme 6.33).[248]

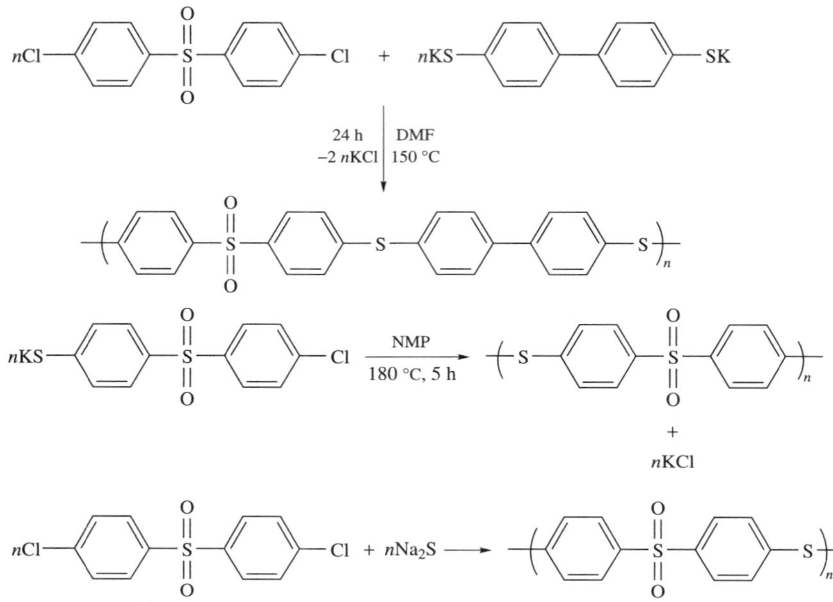

Scheme 6.33   Some examples of synthesis of poly(arylene thioether sulfone)s.

Bis(haloaryl) sulfones can indeed be polycondensed with dialkali metal bisthiophenoxides,[249–251] and alkali metal (haloarylsulfonyl)thiophenoxides can be self-polycondensed.[252] In some cases, the polymerization of (haloarylsulfononyl)thiophenol can be promoted by KF.[253,254] Bis(haloaryl) sulfones may also be polycondensed with alkali metal sulfides in the presence of CH_3COOLi or heterocyclic amines,[255] or additionally, copolycondensed with dihalobenzenes.[256] The reaction must be optimized by controlling the reaction temperature, time, and amounts of water to obtain high-molecular-weight poly(sulfide sulfone)s.[257]

Recently, Liu et al. used DCDPS to prepare thiobisphenol S (TBPS) by hydrolysis.[258] DTPS can be used to react with DCDPS in a dipolar aprotic solvent with K_2CO_3 as the weak base at 165°C and high molecular weight can be obtained due to higher reactivity of TBPS compared with its BPS counterpart.

REFERENCES

1. R. J. Cotter, *Engineering Plastics: Handbook of Polyarylethers*, Gordon and Breach, Basel, Switzerland, (1995).

2. K. V. Gatham and S. Turner, *Polymer*, **15**, 665 (1974).

3. D. C. Clagett, in *Encyclopedia of Polymer Science and Engineering*, Vol. 6, H. F. Mark, N. M. Bikales, C. G. Overberger, and G., Menges (Eds.), Wiley, New York, 1986, p 94.

4. R. N. Johnson, in *Encyclopedia of Polymer Science and Technology*, N. M. Bikales, (Ed.), Wiley, New York, 1969.

5. (a) R. N. Johnson, A. G. Farnham, R. A. Clendinning, W. F. Hale, and C. L. Merriam, *J. Polym. Sci., Polym. Chem. Ed.*, **5**, 2375 (1967); (b) A. G. Farnham, L. M. Robeson, and J. E. McGrath, *J. Appl. Polym. Sci., Appl. Poly. Symp.*, **26**, 373 (1975).

6. T. E. Attwood, P. C. Dawson, J. L. Freeman, L. R. J. Hoy, J. B. Rose, and P. A. Staniland, *Polymer*, **22**, 1096 (1981).

7. O. B. Searly and R. H. Pfeifer, *Polym. Eng. Sci.*, **25**, 474 (1985).

8. (a) R. Viswanathan, B. C. Johnson, and J. E. McGrath, *Polymer*, **25**, 1927 (1984); (b) R. Viswanathan, Ph.D. Thesis, Virginia Polytechnic Institute and State University, Blacksburg, VA, (1981).

9. P. M. Hergenrother, B. J. Jensen, and S. J. Havens, *Polymer*, **29**, 358 (1988).

10. J. L. Hedrick and J. W. Labadie, *Step-Growth Polymers for High-Performance Materials: New Synthetic Methods*, ACS Symposium Series 624, American Chemical Society, Washington, DC, 1996.

11. D. J. Riley, A. Gungor, S. Srinivasan, M. Sankarapandian, C. Tchatchoua, M. W. Muggli, T. C. Ward, J. E. McGrath, and T. Kashiwagi, *Polym. Eng. Sci.*, **37**, 1501 (1997).

12. V. L. Rao, *J. Macromol. Sci., Rev. Macromol. Chem. Phys.*, **C39**, 655 (1999).

13. F. Theil, *Angew. Chem. Int. Ed.*, **38**, 2345 (1999).

14. S. Patai, *The Chemistry of the Ether Linkage*, Interscience, London, 1967.

15. L. M. Robeson, A. G. Farnham, and J. E. McGrath, in *Molecular Basis of Transitions and Relaxations*, R. F. Boyer and D. J. Meyer (Eds.), Gordon, New York, 1978; pp. 405–425.

16. A. F. Yee and S. A. Smith, *Macromolecules*, **14**, 54 (1981).

17. J. J. Dumais, A. L. Cholli, L. W. Jelinski, J. L. Hedrick, and J. E. McGrath, *Macromolecules*, **19**, 1884 (1986).

18. G. A. Olah, *Friedel–Crafts Chemistry*, Wiley, New York, 1973, p. 488.

19. C. M. Suter, *The Organic Chemistry of Sulfur*, Wiley, New York, 1944, p. 673.

20. G. A. Olah, *Friedel–Crafts Chemistry*, Wiley, New York, 1973, p. 122.

21. S. F. Fox and K. G. Neil (ICI Ltd.), Austrion Patent 242187; 1962; *Chem. Abstr.*, **63**, 420g (1965).

22. J. Huisman (A. G. Farbenind I. G.), German Patent 701954; **1941**; *Chem. Abstr.*, **36**, 987 (1942).

23. J. March, *Advanced Organic Chemistry: Reactions, Mechanisms, and Structure*, 4th ed., Wiley, New York, 1992, p. 539.

24. M. E. Jones (ICI Ltd.), British Patent 979111, 1965; *Chem. Abstr.*, **62**, 9065h (1965).

25. B. E. Jennings, M. E. B. Jones, and J. B. Rose, *J. Polym. Sci., Part C*, **16**, 715 (1967).

26. S. M. Cohen and R. H. Young, *J. Polym. Sci., Polym. Chem. Ed.*, **4**, 722 (1966).

27. M. E. A. Cudby, R. G. Feasey, S. Gaskin, M. E. B. Jones, and J. B. Rose, *J. Polym. Sci., Part C.*, **22**, 747 (1969).

28. P. Kovacic and C. Wu, *J. Org. Chem.*, **26**, 759, 762 (1961).

29. C. A. Thomas, *Anhydrous Aluminum Chloride in Organic Chemistry*, Reinhold, New York, 1961, p. 648.

30. S. M. Cohen and R. H. Young, *J. Polym. Sci., Polym. Chem. Ed.*, **4**, 722 (1966).

31. M. E. A. Cudby, R. G. Feasey, B. E. Jennings, M. E. B. Jones, and J. B. Rose, *Polymer*, **6**, 589 (1965).

32. P. J. Bain, E. J. Blackman, W. Cummings, S. A. Hughes, E. R. Lynch, E. B. McCall, and R.J. Roberts, *J. Proc. Chem. Soc.* London, 86 (1962).

33. M. Ueda (Idemitsu Kosan Co.), Japanese Patent 85228541, 1985; *Chem. Abstr.*, **104**, 187100 (1986).

34. J. B. Rose (ICI plc), European Patent Application 49,070, 1982; *Chem. Abstr.*, **97**, 6999 (1982).

35. W. H. Bonner, U.S. Patent 3,065,205, 1962.

36. I. Goodman, J. E. McIntyre, and W. Russell, British Patent 971,227, 1964; *Chem. Abstr.*, **61**, 14805b (1964).

37. E. Brugel, U.S. Patent 4,987,171 (to DuPont), 1991.

38. M. G. Zolotukhin, J. D. Abajo, J. C. Alvarez, J. G. D. L. Campa, and D. R. Rueda, *J. Polym. Sci., Polym. Chem. Ed.*, **36**, 1251 (1998).

39. R. Clendinning, D. Kelsey, J. Botkin, P. Winslow, M. Yousefi, R. Cotter, R. Matzner, and G. Kwiatowski, *Macromolecules*, **26**, 2361 (1993).

40. V. Jansons and H. C. Gors, *PCT Int. Appl. WO*, 84 03,891 (1984).

41. V. Jansons, H. C. Gors, S. Moore, R. H. Reamey, and P. Becker (Raychem Corp.), U.S. Patent, 4,698,393, 1987.

42. I. Fukawa and T. Tanabe, *J. Polym. Sci., Polym. Chem. Ed.*, **31**, 535 (1993).

43. Y. Sakaguchi, M. Tokai, and Y. Kato, *Polymer*, **7**, 1512 (1993).

44. F. Gay and C. Brunette, U.S. Patent 4,816,556, 1989.

45. Y. Iwakura, K. Uno, and T. Takiguchi, *J. Polym. Sci., Part A-1* **6**, 3345 (1968).

46. J. Devaux, D. Delimoy, D. Daost, R. Legras, J. P. Mercier, C. Straszielle, and E. Nield, *Polymer*, **26**, 1994 (1985).

47. B. M. Marks, U.S. Patent 3,441,538, 1964.

48. J. B. Rose, European Patent, **63**, 874, 1983; *Chem. Abstr.*, **98**, 180081 (1983).

49. J. March, *Advanced Organic Chemistry: Reactions, Mechanisms, and Structure*, 4th ed., Wiley, 1994, p. 641.

50. J. Meisenheimer, *Liebigs Ann. Chem.*, **323**, 205 (1902).

51. J. F. Bunnett and R. E. Zahler, *Chem. Rev.*, **49**, 273 (1951).

52. J. A. Miller, *Aromatic Nucleophilic Substitution*, Elsevier, London, 1968, p. 61.

53. J. R. Beck, *Tetrahedron*, **34**, 2057 (1978).

54. S. R. Shultze and A. L. Baron, *Adv. Chem. Ser.*, **96**, 692 (1969).

55. T. E. Attwood, A. B. Newton, and J. B. Rose, *Br. Polym. J.*, **4**, 391 (1972).

56. T. E. Attwood, D. A. Barr, T. Kin, A. B. Newton, and J. B. Rose, *Polymer*, **4**, 391 (1972).

57. A. B. Newton and J. B. Rose, *Polymer*, **13**, 465 (1972).

58. J. L. Hedrick and J. W. Labadie, *Macromolecules*, **23**, 1561 (1990).

59. K. R. Carter, *Polym. Mater. Sci. Eng.*, **69**, 432 (1993).

60. R. A. Clendining, A. G. Farnham, N. L. Zutty, and D. C. Priest, Canadian Patent 847 963, Union Carbide Corp., 1970.

61. J. L. Hedrick, D. K. Mohanty, B. C. Johnson, R. Viswanathan, J. A. Hinkley, and J. E. McGrath, *J. Polym. Sci. Polym. Chem.*, **23**, 287 (1986).

62. J. L. Hedrick, J. J. Dumais, L. W. Jelinski, R. A. Patsiga, and J. E. McGrath, *J. Polym. Sci. Polym. Chem.*, **25**, 2289 (1987).

63. (a) R. N. Johnson and A. G. Farnham, *J. Polym. Sci., Polym. Chem.*, **5**, 2415 (1967); (b) G. Conio, E. Bianchi, A. Ciferri, and A. Tealdi, *Macromolecules*, **14**, 1084 (1981).

64. T. Ando, J. Yamawaki, and T. Kawate, *Chem. Lett.*, **45** (1979).

65. T. Ando, J. Yamawaki, T. Kawate, S. Sumi, and T. Hanfusda, *Bull. Chem. Soc. Jpn.*, **55**, 2504 (1982).

66. J. M. Miller, K. H. So, and J. H. Clark, *Can. J. Chem.*, **57**, 1887 (1979).

67. H. R. Kricheldorf and G. Bier, *J. Polym. Sci., Polym. Chem. Ed.*, **21**, 2283 (1983).

68. E. J. Corey and B. B. Snider, *J. Am. Chem. Soc.*, **94**, 2549 (1972).

69. M. Fieser, *Reagents for Organic Synthesis*, Vol. 10, Wiley, New York, 1982, pp. 81, 325, 378.

70. H. R. Kricheldorf, U. Delius, and K. U. Tonnes, *New Polym. Mater.*, **1**, 127 (1988).

71. H. R. Kricheldorf and P. Jahnke, *Makromol. Chem.*, **191**, 2027 (1990).

72. D. K. Mohanty, Y. Sachdeva, J. L. Hedrick, J. F. Wolfe, and J. E. McGrath, *Polym. Prepr., ACS*, **25**, 19 (1984).

73. T. E. Attwood, P. C. Dawson, J. L. Freeman, L. R. Hoy, J. B. Rose, and P. A. Staniland, *Polymer*, **20**, 1204 (1981).

74. J. B. Rose and P. A. Staniland, U.S. Patent 4,320,224, (1982).

75. V. Percec, R. S. Clough, M. Grigors, P. L. Rinaldi, and V. E. Litman, *Macromolecules*, **26**, 3650 (1993).

76. V. Percec and R. S. Clough, *Macromolecules*, **27**, 1535 (1994).

77. I. Fukawa, T. Tanabe, and T. Dozono, *Macromolecules*, **24**, 3838 (1991).

78. S. Ebata and Y. Higuchi, European Patent EP 88 11407 (to Mitsubishi Gas Chem. Co.), 1988.

79. U. Hoffmann, F. Helmer-Metzmann, M. Klapper, and K. Mullen, *Macromolecules*, **27**, 3575 (1994).

80. H. R. Kricheldorf and B. Bier, *Polymer*, **25**, 1151 (1984).

81. H. R. Kricheldorf and U. Delius, *Macromolecules*, **22**, 517 (1989).

82. K. Konno, M. Ueda, P. E. Cassidy, and J. W. Fitch, *J. Macrom. Sci., Pure Appl. Chem.*, **A34**, 929 (1997).

83. D. K. Mohanty, T. S. Lin, T. C. Ward, and J. E. McGrath, *Int. SAMPE Symp. Exp.*, **31**, 945 (1986).

84. W. Risse and D. Y. Sogah, *Macromolecules*, **23**, 4029 (1990).

85. D. R. Kelsey, L. M. Robeson, and R. A. Clendinning, *Macromolecules*, **20**, 1204 (1987).

86. D. K. Mohanty, R. C. Lowery, G. D. Lyle, and J. E. McGrath, *Int. SAMPE Symp. Exp.*, **32**, 408 (1987).

87. B. E. Lindfors, R. S. Mani, J. E. McGrath, and D. K. Mohanty, *Makromol. Chem., Rapid. Commun.*, **12**, 337 (1991).

88. J. Roovers, J. D. Cooney, and P. M. Toporowski, *Macromolecules*, **23**, 1611 (1990).

89. A. E. Brink, S. Gutzeit, T. Lin, H. Marand, K. Lyon, T. Hua, R. Davis, and J. S. Riffle, *Polymer*, **34**, 825 (1993).

90. A. Strecker, *Ann. Chem. Pharm.*, **75**, 27 (1850).

91. (a) J. Yang, C. S. Tyberg, and H. W. Gibson, *Macromolecules*, **32**, 8259 (1999); (b) J. Yang and H. W. Gibson, *Macromolecules*, **32**, 8740 (1999).

92. A. Pandya, J. Yang, and H. W. Gibson, *Macromolecules*, **27**, 1367 (1994).

93. H. M. Colquhoun, C. C. Dudman, M. Thomas, C. A. O'Mahoney, and D. J. Williams, *J. Chem. Soc., Chem. Commun.*, **336** (1990).

94. M. Chen, F. Fronczek, and H. W. Gibson, *Macromol. Chem. Phys.*, **197**, 4069 (1996).

95. M. Chen and H. W. Gibson, *Macromolecules*, **29**, 5502 (1996).

96. M. Chen and H. W. Gibson, *Macromolecules*, **30**, 2516 (1997).

97. Y. H. Qi, T. L. Chen, and J. P. Xu, *Polym. Bull.*, **42**, 245 (1999).

98. (a) S. Srinivasan, L. Kagumba, D. J. Riley, and J. E. McGrath, *Macromol. Symp.*, **122**, 95 (1997); (b) S. Wang, Q. Ji, C. N. Tchatchoua, A. R. Shultz, and J. E. McGrath, *J. Polym. Sci., Polym. Phys. Ed.*, 37, 1849 (1999); (c) S. Wang, J. Wang, Q. Ji, A. R. Shultz, T. C. Ward, and J. E. McGrath, *J. Polym. Sci., Polym. Phys. Ed.*, **38**, 2409 (2000).

99. S. Hashimoto, I. Furukawa, and K. Ueyama, *J. Macromol. Sci. Chem.*, **A11**, 2167 (1977).

100. C. D. Smith, H. Grubbs, H. F. Webster, A. Gungor, J. P. Wightman, and J. E. McGrath, *High Perform. Polym.*, **4**, 211 (1991).

101. J. L. Hopewell, D. J. T. Hill, J. H. O'Donnell, P. J. Pomery, J. E. McGrath, D. B. Priddy Jr., and C. D. Smith, *Polym. Degrad. Stab.*, **45**, 293 (1994).

102. (a) S. J. Mecham, M. A. Hickner, M. Sankarapandian, L. M. Grieco, and J. E. McGrath, *Int. SAMPE Symp.*, **44**, 16 (1999); (b) J. E. McGrath, S. J. Mecham, M. A. Hickner, S. Wang, H. B. Shoba, Y. Oishi, and M. Sankarapandian, *Polym. Prepr., ACS*, **40**(2), 1291 (1999).

103. H. Ghassemi and J. E. McGrath, *Polymer*, **38**, 3139 (1997).

104. H. Ghassemi, D. J. Riley, M. Curtis, E. Bonaplata, and J. E. McGrath, *Appl. Organomet. Chem.*, **12**, 781 (1998).

105. D. J. Riley, Ph.D. Thesis Virginia, Tech, Blacksburg, VA, 1997.

106. J. W. Fitch, V. S. Reddy, P. W. Youngman, G. A. Wohlfahrt, and P. E. Cassidy, *Polymer*, **41**, 2301 (2000).

107. (a) J. L. Hedrick and J. W. Labadie, *Macromolecules*, **21**, 1883 (1988); (b) J. L. Hedrick and J. W. Labadie, *Macromolecules*, **25**, 2021 (1992); (c) F. E. Arnold and R. L. van Deusen, *Macromolecules*, **2**, 497 (1969); (d) H. R. Kricheldorf, G. Schwarz, and J. Erxleben, *Makromol. Chem.*, **189**, 2255 (1989); (e) C. Blaise, A. Bouanane, A. Brembilla, P. Lochon, and J. Neel, *J. Polym. Sci., Polym. Symp.*, **52**, 137 (1975); (f) R. F. Kovar and F. E. Arnold, *J. Polym. Sci., Polym. Chem. Ed.*, **14**, 2807 (1976); (g) J. L. Hedrick, *Macromolecules*, **24**, 6361 (1991); (h) H. R. Kricheldorf, *Macromol. Chem., Macromol. Symp.*, **54/55**, 365 (1992); (i) Y. Saegusa, T. Iwasaki, and S. Nakamura, *J. Polym. Sci., Polym. Chem. Ed.*, **32**, 249 (1994); (j) J. L. Hedrick and J. W. Labadie, *Macromolecules*, **28**, 4342 (1995); (k) J. L. Hedrick, H. Jonsson, and K. R. Carter, *Macromolecules*, **28**, 4342

(1995); (l) J. G. Hilborn, J. W. Labadie, and J. L. Hedrick, *Macromolecules*, **23**, 2854 (1990).

108. (a) J. L. Hedrick, *Macromolecules*, **24**, 812 (1991); (b) M. Lucas and J. L. Hedrick, *Polym. Bull.*, **8**, 129 (1992); (c) M. Lucas and J. L. Hedrick, *J. Polym. Sci., Polym. Chem. Ed.*, **31**, 2179 (1993).

109. S. Matsuo, T. Murakami, and R. Takasawa, *J. Polym. Sci., Polym. Chem. Ed.*, **31**, 3439 (1993).

110. A. Bhatnagar, J. Mueller, D. J. Osborn, T. L. Martin, J. Wirtz, and D. K. Mohanty, *Polymer*, **36**, 3019 (1995).

111. A. Bhatnagar, A. Schroder, and D. K. Mohanty, *Polymer*, **38**, 239 (1997).

112. M. Strukelj, M. Paventi, and A. S. Hay, *Macromolecules*, **26**, 1777 (1993).

113. C. Gao and A. S. Hay, *J. Polym. Sci., Polym. Chem. Ed.*, **33**, 2347 (1995).

114. C. Gao and A. S. Hay, *J. Macromol. Sci.*, **A33**, 157 (1996).

115. K. A. Yeomans and A. S. Hay, *Polym. Prepr., ACS*, **69**, 240 (1993).

116. (a) J. G. Wirth and D. R. Health, U.S. Patent 3,786,364 (to General Electric), 1974; (b) F. J. Williams, U.S. Patent 3,847,869 (to General Electric), 1974.

117. (a) J. G. Wirth and D. R. Health, U.S. Patent 3,838,097 (to General Electric), 1974; (b) T. Takekoshi, J. G. Wirth, D. R. Health, J. E. Kochanowski, J. S. Manello, and M. J. Webber, *J. Polym. Sci., Polym. Chem. Ed.*, **18**, 3069 (1980).

118. S. Maiti and B. Mandal, *Prog. Polym. Sci.*, **12**, 111 (1986).

119. J. W. Labadie and J. L. Hedrick, *Polym. Prepr., ACS*, **31**(1), 344 (1990).

120. S. S. K. Park and S. Y. Kim, *Macromolecules*, **31**, 3385 (1998).

121. J. A. Irvin, C. J. Neef, K. M. Kane, P. E. Cassidy, G. Tullos, and A. K. St. Clair, *J. Polym. Sci., Polym. Chem. Ed.*, **30**, 1675 (1992).

122. F. W. Mercer, T. D. Goodman, A. N. K. Lau, L. P. Vo, and R. C. Sovish, U.S. Patent, 5,114,780, 1992.

123. F. W. Mercer, D. Duff, J. Wojtowicz, and T. D. Goodman, *Polym. Mater. Sci. Eng.*, **66**, 198 (1992).

124. A. G. Farnham and R. N. Johnson, U.S. Patent 3,332909, 1967.

125. (a) M. J. Jurek and J. E. McGrath, *Polym. Prepr., ACS*, **28**(1), 180 (1987); (b) M. J. Jurek, Ph.D. Thesis, Virginia Tech, Blacksburg, VA, 1987.

126. W. Burgoyne and L. M. Robeson, U.S. Patent 5,658,994, 1997.

127. R. N. Vrtis, K. A. Heap, W. F. Burgoyne, and L. M. Robeson, *Mater. Res. Soc. Symp. Proc.*, **443** (Low-Dielectric Constant Materials II), 171–176 (1997).

128. I. Colon and G. T. Kwiatokowski, *J. Polym. Sci., Polym. Chem.*, **28**, 367 (1990).

129. I. Colon, L. M. Maresca, and G. T. Kwiatkowski, U.S. Patent 4,263,466, 1991.

130. P. A. Havelka and V. V. Sheares, *Macromolecules*, **32**, 6418 (1999).

131. P. D. Bloom and V. V. Sheares, *J. Polym. Sci: Part A: Polym. Chem.*, **39**, 3505–3512 (2001).

132. G. T. Kwiatkowski, M. Matzner, and I. Colon, *J. Macromol. Sci., Pure Appl. Chem.*, **A34**, 1945 (1997).

133. R. W. Phillips, V. V. Sheares, E. T. Samulski, and J. M. DeSimone, *Macromolecules*, **27**, 2354 (1994).

134. V. Percec, J. H. Wang, and Y. Oishi, *J. Polym. Sci., Polym. Chem. Ed.*, **29**, 949 (1991).

135. M. Rehahn, A.-D. Schlüter, G. Wegner, *Makromol. Chem., Rapid Commun.*, **11**, 535 (1990).

136. G. A. Deeter and J. S. Moore, *Macromolecules*, **26**, 2535 (1993).

137. W. Waehamad, K. L. Cooper, and J. E. McGrath, *Polym. Prepr.*, **30**(2), 441 (1989).

138. K. L. Cooper, W. Waehamad, H. Huang, D. Chen, G. L. Wilkes, and J. E. McGrath, *Polym. Prepr., ACS*, **464** (1989).

139. Ph.D. Dissertations of (a) J. M. Lambert, 1986; (b) K. L. Cooper, 1991; (c) W. Waehamad, 1991, Virginia Tech, Blacksburg, VA.

140. I.-Y. Wan, R. Srinivasan, and J. E. McGrath, *Polym. Prepr.*, **33**(2), 223 (1992).

141. G. D. Lyle, M. J. Jurek, D. K. Mohanty, S. D. Wu, J. C. Hedrick, and J. E. McGrath, *Polym. Prepr., ACS*, **28**(1), 77 (1987).

142. J. L. Hedrick, I. Yilgor, M. Jurek, J. C. Hedrick, G. L. Wilkes, and J. E. McGrath, *Polymer*, **32**(11), 2011 (1991).

143. J. C. Hedrick, N. M. Patel, and J. E. McGrath, "Toughening of Epoxy Resin Networks with Functionalized Engineering Thermoplastics," in *Rubber Toughened Plastics*, K. Riew (Ed.), American Chemical Society, Washington, DC, 1993.

144. T. H. Yoon, D. B. Priddy, Jr., G. D. Lyle, and J. E. McGrath, *Macromol. Symp.*, **98**, 673–686 (1995).

145. (a) M. J. Jurek and J. E. McGrath, *Int. SAMPE Symp.*, **31**, 913 (1986); (b) *Polymer*, **30**, 1552 (1989).

146. J. L. Hedrick, I. Yilgor, G. L. Wilkes, and J. E. McGrath, *Polym. Bull.*, **13**, 201–208 (1985).

147. (a) G. D. Lyle, J. S. Senger, D. H. Chen, S. Kilic, S. D. Wu, D. K. Mohanty, and J. E. McGrath, *Polymer*, **30**, 978 (1989); (b) S. A. Srinivasan and J. E. McGrath, *Polym.*, **39**(12), 2415

148. A. Ayamba, S. Mecham, and J. E. McGrath, *Polymer*, **41**, 5109 (2000).

149. S. A. Srinivasan and J. E. McGrath, *J. Appl. Polym. Sci.*, **39**(12), 2415 (1998).

150. J. A. Cecere, J. L. Hedrick, and J. E. McGrath, *Int. SAMPE Symp.*, **31**, 580 (1986).

151. D. K. Mohanty, R. C. Lowery, G. D. Lyle, and J. E. McGrath, *Int. SAMPE Symp.*, **32**, 408 (1987).

152. D. K. Mohanty, J. S. Senger, C. D. Smith, and J. E. McGrath, *Int. SAMPE Symp.*, **33**, 970 (1988).

153. G. S. Bennett and J. Farris, *J. Polym. Sci., Polym. Chem. Ed.*, **32**, 73 (1994).

154. H. Mandal and A. S. Hay, *J. Macromol. Sci., Pure Appl. Chem.*, **A35**, 1979 (1998).

155. J. L. W. Noell, G. L. Wilkes, D. K. Mohanty, and J. E. McGrath, *J. Appl. Polym. Sci.*, **40**, 1177 (1990).

156. C. J., Hawker, R. Lee and J. M. J. Fréchet, *J. Am. Chem. Soc.*, **113**, 4583 (1991).

157. T. M. Miller, T. X. Neenan, E. W. Kwock, and S. M. Stein, *J. Am. Chem. Soc.*, **115**, 356 (1993).

158. C. J. Hawker and F. Chu, *Macromolecules*, **29**, 4370 (1996).

159. S.-Y. Kwak and H. Y. Lee, *Polym. Prepr., ACS*, **41**(1), 333 (2000).

160. A. Morikawa, *Macromolecules*, **31**, 5999 (1998).

161. C.-F. Shu, C.-M. Leu, and F.-Y. Huang, *Polymer*, **40**, 6591 (1999).

162. C.-F. Shu and C.-M. Leu, *Macromolecules*, **32**, 100 (1999).

163. E. Fossum, M. Grapenthien, B. Kopan, and E. Olson, *Polym. Prepr., ACS*, **41**(1), 201 (2000).

164. A. Morikawa, M. Kakimoto, and Y. Imai, *Macromolecules*, **25**, 3247 (1992).

165. A. Morikawa and K. Ono, *Macromolecules*, **32**, 1062 (1999).

166. P. Conningham, R. J. Roach, J. B. Rose, and P. T. McGrail, *Polymer*, **33**, 3951 (1992).

167. K. Schofield, *Aromatic Nitration*, Cambridge University Press, Cambridge, 1980, p. 209.

168. J. P. Quentin, U.S. Patent 3,709,841, 1973.

169. A. Noshay and L. M. Robeson, *J. Appl. Polym. Sci.*, **20**, 1885 (1976).

170. B. C. Johnson, I. Yilgor, C. Tran, M. Iqbal, J. P. Wightman, D. R. Lloyd, and J. E. McGrath, *J. Polym. Sci., Polym. Chem. Ed.*, **22**, 721 (1984).

171. W. H. Daly, *J. Macromol. Sci., Chem.*, **A22**, 713 (1985).

172. M. D. Guiver and G. P. Robertson, *Macromolecules*, **28**(1), 294–301 (1995).

173. M. D. Guiver, G. P. Robertson, and S. Foley, *Macromolecules*, **28**(23), 7612–7621 (1995).

174. M. D. Guiver and O. Kutowy, U.S. Patent 4,996,271, 1991.

175. M. D. Guiver, Y. Dai, G. P. Robertson, K. J. Lee, J. Y. Jho, and Y. S. Kang, *Polym. Mater. Sci., Eng.*, **85**, 94 (2001).

176. I. W. Kim, K. J. Lee, J. Y. Jho, H. C. Park, J. Won, Y. S. Kang, M. D. Guiver, G. P. Robertson, and Y. Dai, *Macromolecules*, **34**(9), 2908–2913 (2001).

177. M. D. Guiver, H. Zhang, G. P. Robertson, and Y. Dai, *J. Polym. Sci., Part A: Polym. Chem.*, **39**(5), 675–682 (2001).

178. Y. Dai, X. Jian, X. Liu, and M. D. Guiver, *J. Appl. Polym. Sci.*, **79**(9), 1685–1692 (2000).

179. D. Nguyen, J. S. Kim, M. D. Guiver, and A. Eisenberg, *J. Polym. Sci., Part B: Polym. Phys.*, **37**(22), 3226–3232 (1999).

180. M. Yoshikawa, K. Tsubouchi, M. D. Guiver, and G. P. Robertson, *J. Appl. Polym. Sci.*, **74**(2), 407–412 (1999).

181. D. Mockel, E. Staude, and M. D. Guiver, *J. Membr. Sci.*, **158**(1–2), 63–75 (1999).

182. M. D. Guiver, G. P. Robertson, and C. M. Tam, *Polym. Mater. Sci. Eng.*, **77**, 347–348 (1997).

183. M. Kan, M. D. Guiver, G. P. Robertson, S. N. Willie, and R. E. Sturgeon, *React. Funct. Polym.*, **31**(3), 207–218 (1996).

184. M. D. Guiver and C. M. Tam, Invited Lect., IUPAC 34th Int. Symp. Macromol, 1992.

185. M. D. Guiver, A. Y. Tremblay, and C. M. Tam, National Research Council (Canada), NRCC Report No. 29895, Ottawa, Ontario, 1988.

186. M. D. Guiver, J. W. Apsimon, and O. Kutowy, *J. Polym. Sci., Part C; Polym. Lett.*, **26**(2), 123–127 (1988).

187. K. Miyatake and A. S. Hay, *J. Polym. Sci., Part A: Polym. Chem.*, **39**(11), 1854–1859 (2001).

188. M. D. Guiver, P. Black, C. M. Tam, and Y. Deslandes, *J. Appl. Polym. Sci.*, **48**(9), 1597–606 (1993).

189. (a) J. Lee and C. S. Marvel, *J. Polym. Sci., Polym. Chem. Ed.*, **22**, 295 (1984); (b) M. I. Litter and C. S. Marvel, *J. Polym. Sci., Polym. Chem. Ed.*, **23**, 2205 (1985).

190. (a) R. Singh and A. S. Hay, *Macromolecules*, **25**, 1017 (1992); (b) *Macromolecules*, **25**, 1025 (1992); (c) *Macromolecules*, **25**, 1033 (1992).

191. (a) Z. Y. Wang and C. Zhang, *Macromolecules*, **25**, 5851 (1992); (b) *Macromolecules*, **26**, 3330 (1993); (c) *Macromolecules*, **27**, 4415 (1994).

192. L. J. Mathias, M. L. Charles, and K. N. Wiegel, *Macromolecules*, **30**, 5970 (1997).

193. C. G. Herbert and A. S. Hay, *J. Polym. Sci., Polym. Chem. Ed.*, **35**, 1095 (1997).

194. (a) P. Friedlander, *Chem. Ber.*, **26**, 172 (1893); (b) H. Lund, *Acta Chem. Scand.*, **14**, 359 (1960).

195. N. Berard, M. Paventi, K. P. Chan, and A. S. Hay, *Makromol. Chem., Macromol. Symp.*, **77**, 379 (1994).

196. A. R. Bader and A. D. Kontowicz, *J. Am. Chem. Soc.*, **76**, 4465 (1954).

197. J. A. Mikroyannidis, *Eur. Polym. J.*, **21**, 1031 (1985).

198. I. C. H. M. Esser and I. Parsons, *Polymer*, **34**, 2836 (1993).

199. T. Koch and H. Ritter, *Macromol. Chem. Phys.*, **195**, 1709 (1994).

200. E. Butuc, V. Cozan, I. Giurgiu, I. Mihalache, Y. Ni, and M. Ding, *J. Macromol. Sci., Pure Appl. Chem.*, **A31**, 219 (1994).

201. M. Rusu, A. Airinei, E. Butuc, G. G. Rusus, C. Baban, and Rusu, *J. Macromol. Sci., Phys.*, **B37**, 73 (1998).

202. F. Wang, T. Chen, and J. Xu, *Macromol. Chem. Phys.*, **199**, 1421 (1998).

203. F. A. Bottino, A. Mamo, A. Recca, J. Brady, A. C. Street, and P. T. McGrail, *Polymer*, **34**, 2901 (1993).

204. (a) S. J. Park, G. D. Lyle, R. Mercier, and J. E. McGrath, *Polymer*, **34**, 885 (1993); (b) S. J. Park, Ph.D. Thesis, Virgnia Tech, Blacksburg, VA, 1992.

205. L. .M. Robeson and M. Matzner, U.S. Patent 4,380,598, 1983.

206. M. Ueda, H. Toyota, T. Ouchi, J. Sugiyama, K. Yonetake, T. Masuko, and T. Teramoto, *J. Polym. Sci., Polym. Chem. Ed.*, **31**, 853 (1993).

207. (a) F. Wang, Q. Ji, W. Harrison, J. Mecham, R. Formato, R. Kovar, P. Osenar, and J. E. McGrath, *Polym. Prepr., ACS*, **41**(1), 237 (2000); (b) F. Wang, M. Hickner, Y. S. Kim, J. E. McGrath, and T. Zawodzinski, *J. Memb. Sci.*, **197**, 231 (2002).

208. F. Wang, J. Mecham, W. Harrison, and J. E. McGrath, *Polym. Prepr., ACS*, **41**(2), 1401 (2000); *J. Poly Sci.*, in press (2003).

209. H. K. Shobha, G. R. Smalley, M. Sankarapnadian, and J. E. McGrath, *Polym. Prepr., ACS*, **41**(1), 180 (2000).

210. A. Noshay, M. Matzner, and T. C. Williams, *Ind. Eng. Chem. Prod. Res. Develop.*, **12**, 268 (1973).

211. J. C. Hedrick, C. A. Arnold, M. A. Zumbrum, T. C. Ward, and J. E. McGrath, *Int. SAMPE Symp.*, **35**(1), 82 (1990).

212. B. Risch, D. E. Rodrigues, K. Lyon, J. E. McGrath, and G. L. Wilkes, *Polymer*, **37**, 1229 (1996).

213. W. D. Polk, S. Wang, Y. Kim, M. Sankarapandian, T. E. Glass, and J. E. McGrath, *Polym. Prepr., ACS*, **41**(2), 1385 (2000); W. D. Polk, Ph.D. Thesis, Virginia Tech, Blacksburg, VA, May 2002.

214. S. G. Gaynor and K. Matyjaszewski, *Macromolecules*, **30**, 4241 (1997).

215. L. Y. Zhao, R. S. Mani, T. L. Martin, J. Mueller, and D. K. Mohanty, *J. Mater. Chem.*, **4**, 623 (1994).

216. D. Pospiech, K. Eckstein, L. Haussler, H. Komber, D. Jehnichen, K. Grundke, and F. Simon, *Macromol. Chem. Phys.*, **200**, 1311 (1999).

217. A. I. Aksenov, I. P. Storozhuk, T. S. Aksenova, and V. V. Korshak, *Vysokomol. Soedin., Ser. A*, **26**(1), 97 (1984).

218. Y.-P. R. Ting and L. F. Hancock, *Macromolecules*, **29**, 7619 (1996).

219. L. F. Hancock, S. M. Fagen, and M. S., Ziolo, *Biomaterials*, **21**, 725 (2000).

220. J. E. McGrath, T. C. Ward, E. Shchori, and A. Wnuk, *J. Polym. Eng. Sci.*, **17**, 47 (1977).

221. J. M. Lambert, B. E. McGrath, G. L. Wilkes, and J. E. McGrath, *Polym. Mater. Sci. Eng.*, **55**, 1 (1986).

222. S. D. Wu, J. L. Hedrick, D. K. Mohanty, B. K. Carter, G. L. Wilkes, and J. E. McGrath, *Int. SAMPE Symp.*, **31**, 933 (1986).

223. Z. Wu, Y. Zheng, H. Yan, T. Nakamura, T. Nozawa, and R. Yosomiya, *Angew. Makromol. Chem.*, **173**, 163 (1989).

224. A. B. Newton and J. B. Rose, *Polymer*, **13**, 465 (1972).

225. I. Fukawa and T. Tanabe, *J. Polym. Sci., Polym. Chem. Ed.*, **31**, 535 (1993).

226. I. Fukawa, T. Tanabe, and H. Hachiya, *Polym. J.*, **24**, 173 (1992).

227. Y. Bourgeois, Y. Charlier, J. Devaux, and R. Legras, *Polymer*, **37**, 5503 (1996).

228. J. L. Hedrick, W. Volksen, and D. K. Mohanty, *J. Polym. Sci., Polym. Chem. Ed.*, **30**, 2085 (1992).

229. M. Klapper, T. Wehrmeister, and K. Müllen, *Macromolecules*, **29**, 5805 (1996).

230. A. Le Guen, M. Klapper, and K. Müllen, *Macromolecules*, **31**, 6565 (1998).

231. G. S. Stamatoff and G. Wittmann, French Patent 1,394,897 (E. I. Du Pont de Nemours & Co.), 1965; *Chem. Abstr.*, **63**, 18297C (1965).

232. A. G. Farnham and R. N. Johnson, U.S. Patent 3,332,909 (Union Carbide Corp.), 1967; *Chem. Abstr.*, **68**, 69543a (1968).

233. G. L. Tullos, P. E. Cassidy, and A. K. St. Clair, *Int. SAMPE Electron. Conf.*, **3**, 219 (1989).

234. G. L. Tullos, P. E. Cassidy, and A. K. St. Clair, *Macromolecules*, **24**, 6059 (1991).

235. P. E. Cassidy, G. L. Tullos, and A. K. St. Clair, U.S. Patent Appl. 248,000, 1989.

236. F. Yamato, S. Nogami, Japanese Patent 01,169,456, 1989; *Chem. Abstr.*, **112**, 149022x (1990).

237. F. W. Mercer, M. M. Fone, V. N. Reddy, and A. A. Goodwin, *Polymer*, **38**, 1989 (1997).

238. F. W. Mercer and T. D. Goodman, in *Proceedings International Electronics Packing Conference* (*International Electronics Packaging Society*), Wheaton, IL, 1990, p. 1042.

239. J. A. Irvin, C. J. Neef, K. M. Kane, P. E. Cassidy, G. Tullos, and A. K. St. Clair, *J. Polym. Sci., Polym. Chem. Ed.*, **30**, 1675 (1992).

240. F. W. Mercer, T. D. Goodman, A. N. K. Lau, L. P. Vo, and R. C. Sovish, U.S. Patent 5,114,780, 1992.

241. F. W. Mercer, D. Duff, J. Wojtowicz, and T. D. Goodman, *Polym. Mater. Sci. Eng.*, **66**, 198 (1992).

242. S. Banerjee, G. Maier, and M. Burger, *Macromolecules*, **32**, 4279 (1999).

243. S. Banerjee and G. Maier, *Chem. Mater.*, **11**, 2179 (1999).

244. G. Maier, R. Hecht, O. Nuyken, B. Helmreich, and K. Burger, *Macromolecules*, **26**, 2583 (1993).

245. M. E. A. Cudby, R. G. Feasey, B. E. Jennings, M. E. B. Jones, and J. B. Jones, *Polymer*, **5**, 589 (1965).

246. ICI Ltd., Belgium Patent 639,634, 1964; *Chem. Astr.*, **63**, 700f (1965).

247. S. M. Cohen and R. H. Young, Jr. (Monsanto Co.), U.S. Patent 3,418,277, 1968; *Chem. Abstr.*, 1969, **70**, 48045 (1969).

248. S. R. Sandler and W. Karo, *Polymer Syntheses*, Vol. 3, Academics, New York, 1980, p. 98.

249. A. Kreuchunas, U.S. Patent 2,822,351, 1958; *Chem. Abstr.*, **52**, 7778b (1958).

250. A. L. Baron and D. R. Blank, *Makromol. Chem.*, **140**, 83 (1970).

251. R. Gabler, German Patent 2,009,323 (Inventa, A.G.), 1970; *Chem. Abstr.*, **52**, 7778b (1958).

252. R. G. Feasey, U.S. Patent 3,819,582 (ICI Ltd.), 1974; *Chem. Abstr.*, **82**, 157207 (1975).

253. R. G. Feasey and J. B. Rose, German Patent 2,156,343 (ICI Ltd.), 1972; *Chem. Abstr.*, **77**, 75774 (1972).

254. M. S. Fortuin, British Patent 1,369,217 (ICI Ltd.), 1974; *Chem. Abstr.*, **82**, 98741 (1975).

255. R. W. Campbell, Phillips Petroleum: (a) German Patent 2,726,862, 1977 (*Chem. Abstr.*, **88**, 137178, (1978); (b) U.S. Patent 4,070,349, 1978 (*Chem. Abstr.*, **89**, 25071 1978); (c) U.S. Patent 4,125,535, 1978 (*Chem. Abstr.*, **90**, 55528 (1979).

256. K. Idel, E. Ostlinning, W. Koch, and W. Heitz, German Patent 3312254 (Bayer A. G.), 1984; *Chem. Abstr.*, **102**, 25584 (1985).

257. Y. Liu, A. Bhatnagar, Q. Ji, J. S. Riffle, J. E. McGrath, J. F., Geibel, and T. Kashiwagi, *Polymer*, **41**, 5137 (2000).

258. Y. N. Liu, Ph.D. Thesis, Virginia Tech, Blacksburg, VA, 1998.

7   Chemistry and Properties of Phenolic Resins and Networks

S. Lin-Gibson

Polymers Division, NIST, Gaithersburg, Maryland 20899-8543

J. S. Riffle

Department of Chemistry, Virginia Polytechnic Institute and State University, Blacksburg, Virginia 24061

7.1 INTRODUCTION

Phenolic resins comprise a large family of oligomers and polymers (Table 7.1), which are various products of phenols reacted with formaldehyde. They are versatile synthetic materials with a large range of commercial applications. Plywood adhesives account for nearly half of all phenolic applications while wood-binding and insulation materials also make up a significant portion.[1] Other uses for phenolics include coatings, adhesives, binders for abrasives, automotive and electrical components, electronic packaging, and matrices for composites.

   Phenolic oligomers are prepared by reacting phenol or substituted phenols with formaldehyde or other aldehydes. Depending on the reaction conditions (e.g., pH) and the ratio of phenol to formaldehyde, two types of phenolic resins are obtained. Novolacs are derived from an excess of phenol under neutral to acidic conditions, while reactions under basic conditions using an excess of formaldehyde result in resoles.

   Phenolic resins were discovered by Baeyer in 1872 through acid-catalyzed reactions of phenols and acetaldehyde. Kleeberg found in 1891 that resinous products could also be formed by reacting phenol with formaldehyde. But it was Baekeland who was granted patents in 1909 describing both base-catalyzed resoles (known as Bakelite resins) and acid-catalyzed novolac products.[2]

   This chapter emphasizes the recent mechanistic and kinetic findings on phenolic oligomer syntheses and network formation. The synthesis and characterization of both novolac- and resole-type phenolic resins and their resulting networks are described. Three types of networks, novolac–hexamethylenetetramine (HMTA),

Synthetic Methods in Step-Growth Polymers.   Edited by Martin E. Rogers and Timothy E. Long
© 2003 John Wiley & Sons, Inc.   ISBN: 0-471-38769-X

TABLE 7.1   U.S. Phenolic Production (in millions of pounds on a gross weight basis)[a]

1998	1997	% Change
3940	3734	5.5

[a]From Society of Plastic Industries Facts and Figures, SPI, Washington, DC, 1999.

novolac–epoxies, and thermally cured resoles will be primarily discussed. Other phenolic-based networks include benzoxazines and cyanate esters. Since phenolic materials possess excellent flame retardance, a discussion of the thermal and thermo-oxidative degradation pathways will be included. Detailed information on the chemistry, applications, and processing of phenolic materials can be found in a number of references.[2–6]

7.2 MATERIALS FOR THE SYNTHESIS OF NOVOLAC AND RESOLE PHENOLIC OLIGOMERS

7.2.1 Phenols

The most common precursor to phenolic resins is phenol. More than 95% of phenol is produced via the cumene process developed by Hock and Lang (Fig. 7.1). Cumene is obtained from the reaction of propylene and benzene through acid-catalyzed alkylation. Oxidation of cumene in air gives rise to cumene hydroperoxide, which decomposes rapidly at elevated temperatures under acidic conditions to form phenol and acetone. A small amount of phenol is also derived from coal.

Substituted phenols such as cresols, *p-tert*-butylphenol, *p*-phenylphenol, resorcinol, and cardanol (derived from cashew nut shells) have also been used as precursors for phenolic resins. Alkylphenols with at least three carbons in the substituent lead to more hydrophobic phenolic resins that are compatible with many oils, natural resins, and rubbers.[7] Such alkylphenolic resins are used as modifying and crosslinking agents for oil varnishes, as coatings and printing inks, and as antioxidants and stabilizers. Bisphenol-A (2,2-*p*-hydroxyphenylpropane),

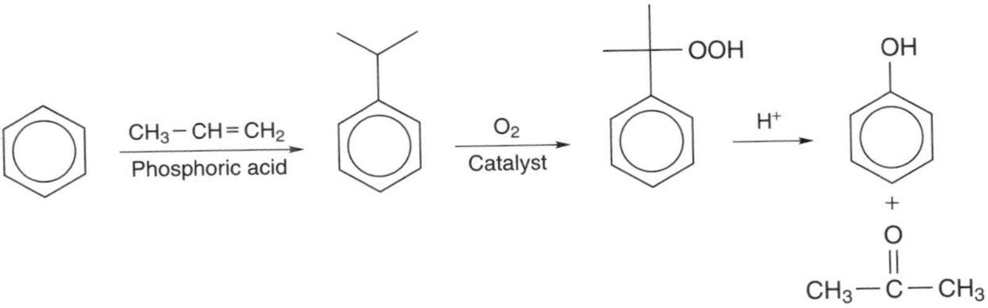

Figure 7.1   Preparation of phenol monomer.

a precursor to a number of phenolic resins, is the reaction product of phenol and acetone under acidic conditions.

An additional activating hydroxyl group on the phenolic ring allows resorcinol to react rapidly with formaldehyde even in the absence of catalysts.[8] This provides a method for room temperature cure of resorcinol–formaldehyde resins or mixed phenol–formaldehyde/resorcinol–formaldehyde resins. Trihydric phenols have not achieved commercial importance, probably due to their higher costs.

7.2.2 Formaldehyde and Formaldehyde Sources

Formaldehyde, produced by dehydrogenation of methanol, is used almost exclusively in the synthesis of phenolic resins (Fig. 7.2). Iron oxide, molybdenum oxide, or silver catalysts are typically used for preparing formaldehyde. Air is a safe source of oxygen for this oxidation process.

Since formaldehyde is a colorless pungent irritating gas, it is generally marketed as a mixture of oligomers of polymethylene glycols either in aqueous solutions (formalin) or in more concentrated solid forms (paraformaldehyde). The concentration of formalin ranges between about 37 and 50 wt %. A 40 wt % aqueous formalin solution at 35°C typically consists of methylene glycols with 1–10 repeat units. The molar concentration of methylene glycol with one repeat unit ($HO-CH_2-OH$) is highest and the concentrations decrease with increasing numbers of repeat units.[9] Paraformaldehyde, a white solid, contains mostly polymethylene glycols with 10–100 repeat units. It is prepared by distilling aqueous formaldehyde solutions and generally contains 1–7 wt % water.

Methanol, the starting reagent for producing formaldehyde, stabilizes the formalin solution by forming acetal endgroups and is usually present in at least small amounts (Fig. 7.3). Methanol may also be formed by disproportionation during storage. The presence of methanol reduces the rate of phenol–formaldehyde reaction but does not affect the activation energies.[10] It is generally removed by stripping at the end of the reaction.

Water is necessary for decomposing paraformaldehyde to formaldehyde (Fig. 7.4). However, water can serve as an ion sink and water–phenol mixtures

$$CH_3-OH \ + \ 1/2 \ O_2 \ \xrightarrow{\text{Catalyst}} \ H-\overset{\overset{\displaystyle O}{\|}}{C}-H \ + \ H_2O$$

Figure 7.2   Synthesis of formaldehyde.

$$CH_3-OH \ + HO-(CH_2-O)_n H \ \rightleftharpoons \ CH_3-O-(CH_2-O)_n H \ + \ H_2O$$

Figure 7.3   Formation of hemiformals.

$$HO-(CH_2-O)_n H \ + \ H_2O \ \longrightarrow \ HO-(CH_2-O)_{n-1} H \ + \ HO-CH-OH$$

Figure 7.4   Depolymerization of aqueous polyoxymethylene glycol.

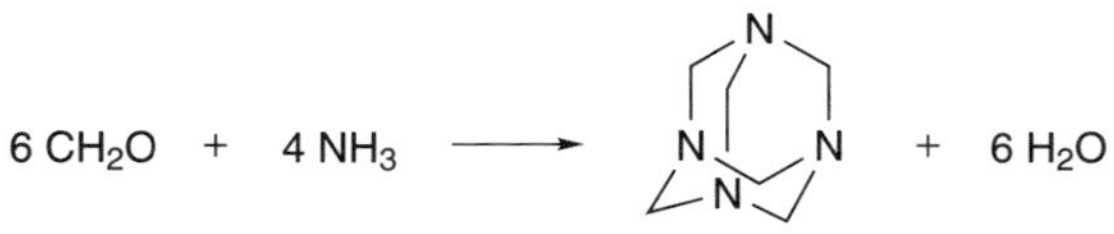

$$6\ CH_2O\ +\ 4\ NH_3\ \longrightarrow\ \text{[structure]}\ +\ 6\ H_2O$$

Figure 7.5   Synthesis of HMTA.

phase separate as the water concentration increases. Therefore, large amounts of water reduce the rate of reaction between phenol and formaldehyde.[11]

Used for crosslinking novolacs or catalyzing resole syntheses, HMTA is prepared by reacting formaldehyde with ammonia (Fig. 7.5). The reaction is reversible at high temperatures, especially above 250°C. HMTA can also be hydrolyzed in the presence of water.

7.3 NOVOLAC RESINS

The most common precursors for preparing novolac oligomers and resins are phenol, formaldehyde sources, and, to a lesser extent, cresols. Three reactive sites for electrophilic aromatic substitution are available on phenol which give rise to three types of linkages between aromatic rings, that is, ortho–ortho, ortho–para, and para–para. The complexity of the isomers leads to amorphous materials. For a novolac chain with 10 phenol groups, 13,203 isomers[12] can statistically form, making the separation of pure phenolic compounds from novolacs nearly impossible.

7.3.1 Synthesis of Novolac Resins

Novolacs are prepared with an excess of phenol over formaldehyde under acidic conditions (Fig. 7.6). A methylene glycol is protonated by an acid from the reaction medium, which then releases water to form a hydroxymethylene cation (step 1 in Fig. 7.6). This ion hydroxyalkylates a phenol via electrophilic aromatic substitution. The rate-determining step of the sequence occurs in step 2 where a pair of electrons from the phenol ring attacks the electrophile forming a carbocation intermediate. The methylol group of the hydroxymethylated phenol is unstable in the presence of acid and loses water readily to form a benzylic carbonium ion (step 3). This ion then reacts with another phenol to form a methylene bridge in another electrophilic aromatic substitution. This major process repeats until the formaldehyde is exhausted.

The reaction between phenol and formaldehyde is exothermic. Therefore, the temperature must be controlled to prevent the buildup of heat, particularly during the early stages of reaction.[4] When formalin is used, water provides a medium for heat dissipation.

Typical formaldehyde-to-phenol ratios in novolac syntheses range from about 0.7 to 0.85 to maintain oligomers with sufficiently low molecular weights

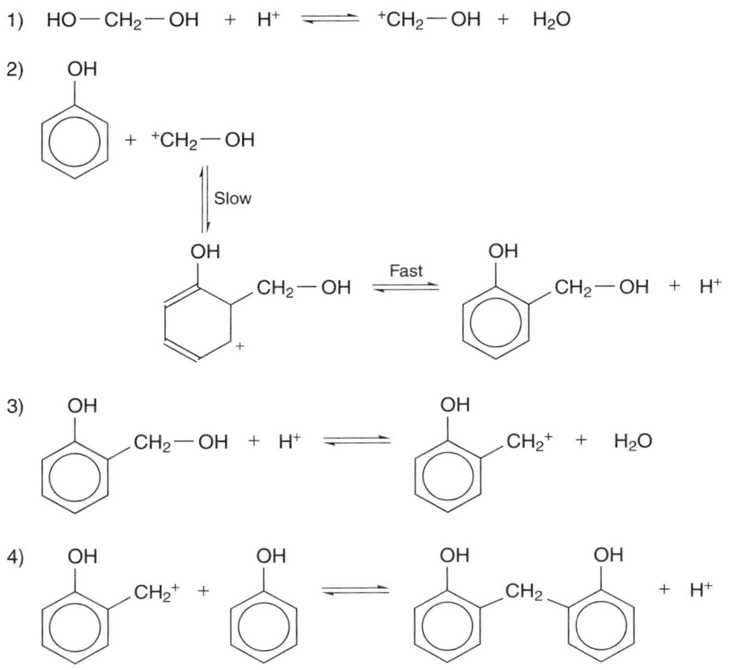

Figure 7.6 Mechanism of novolac synthesis via electrophilic aromatic substitution.

and reasonable melt viscosities. This is especially important since phenol is trifunctional and a gel fraction begins to form as conversion increases. As a result, the number-average molecular weights (M_n) of novolac resins are generally below 1000 g/mol.

The acidic catalysts used for these reactions include formic acid, HX (X = F, Cl, Br), oxalic acid, phosphoric acid, sulfuric acid, sulfamic acid, and p-toluenesulfonic acid.[4] Oxalic acid is preferred since resins with low color can be obtained. Oxalic acid also decomposes at high temperatures (>180°C) to CO_2, CO, and water, which facilitates the removal of this catalyst thermally. Typically, 1–6 wt % catalyst is used. Hydrochloric acid results in corrosive materials and reportedly releases carcinogenic chloromethyl ether by-products during resin synthesis.[2]

Approximately 4–6 wt % phenol can typically be recovered following novolac reactions. Free phenol can be removed by washing with water repeatedly. The recovered phenolic components may contain 1,3-benzodioxane, probably derived from benzyl hemiformals (Fig. 7.7).[2]

7.3.2 "High-Ortho" Novolac Resins

High-ortho novolacs (Fig. 7.8) are sometimes more desirable since they cure more rapidly with HMTA. A number of oxides, hydroxides, or organic salts of electropositive metals increase the reactivity of the ortho position during

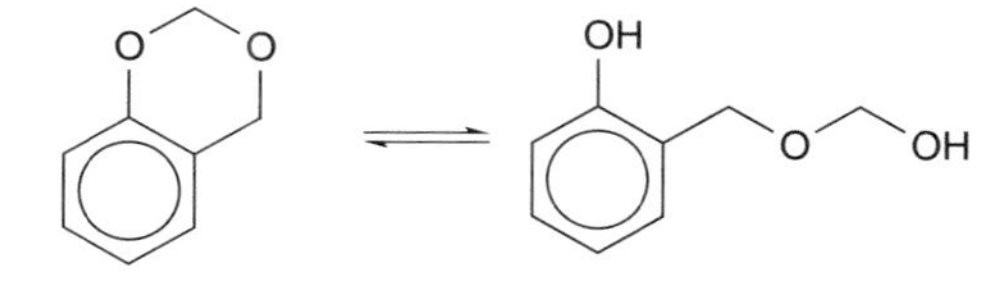

1,3-Benzodioxane   Benzyl hemiformal

Figure 7.7 By-products of novolac synthesis.

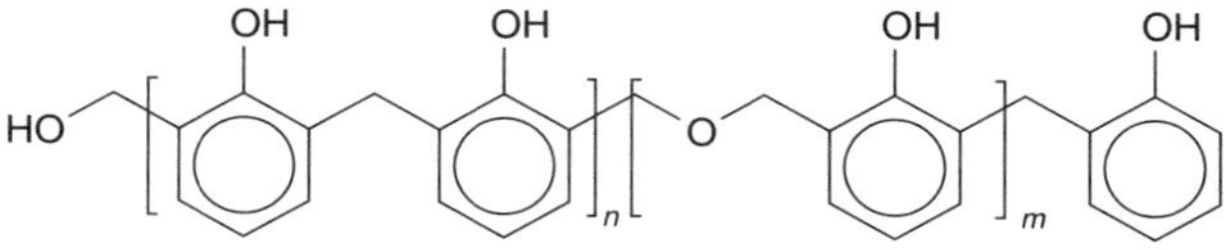

Figure 7.8 High ortho novolacs.

oligomer formation.[13] These high ortho novolacs are typically formed at pH 4–6 as opposed to the more common strongly acidic conditions.

Metal hydroxides of first- and second-group elements can enhance ortho substitution, the degree of which depends on the strength of metal-chelating effects linking the phenolic oxygen with the formaldehyde as it approaches the ortho position. Transition metal ions of elements such as Fe, Cu, Cr, Ni, Co, Mn, and Zn as well as boric acid also direct ortho substitutions via chelating effects (Fig. 7.9).

Phenol–formaldehyde reactions catalyzed by zinc acetate as opposed to strong acids have been investigated, but this results in lower yields and requires longer reaction times. The reported ortho–ortho content yield was as high as 97%. Several divalent metal species such as Ca, Ba, Sr, Mg, Zn, Co, and Pb combined with an organic acid (such as sulfonic and/or fluoroboric acid) improved the reaction efficiencies.[14] The importance of an acid catalyst was attributed to facilitated decomposition of any dibenzyl ether groups formed in the process. It was also found that reaction rates could be accelerated with continuous azeotropic removal of water.

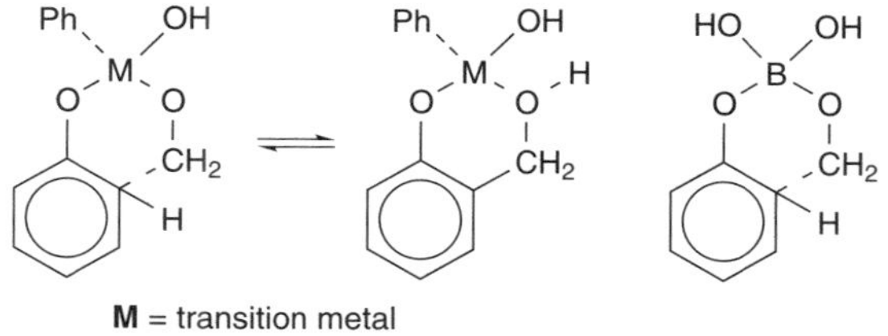

M = transition metal

Figure 7.9 Proposed chelate structures in the synthesis of high ortho novolac oligomers.

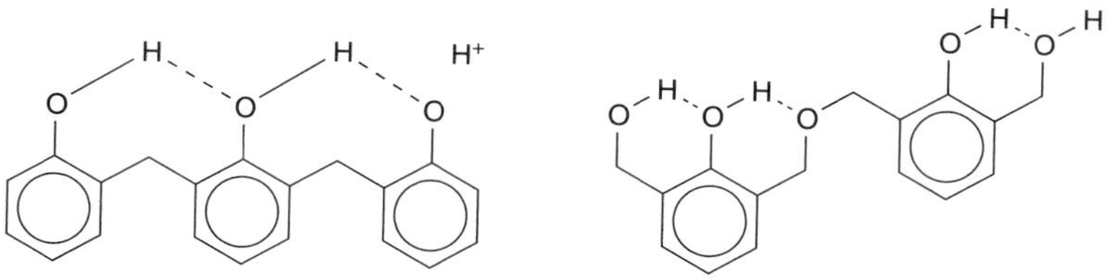

Figure 7.10   Intramolecular hydrogen bonding of high ortho novolacs.

An interesting aspect of high-ortho novolac oligomers is their so-called 'hyper-acidity'. The enhanced acidity of high-ortho novolac resins, intermediate between phenols and carboxylic acids, has been attributed to increased dissociation of the phenol protons due to strong intramolecular hydrogen bonding (Fig. 7.10). These materials are also reported to form strong complexes with di- and trivalent metals and nonmetals.[2]

7.3.3 Model Phenolic Oligomer Synthesis

Linear novolac oligomers containing only ortho linkages were prepared using bromomagnesium salts under dry conditions.[15] The bromomagnesium salt of phenol coordinates with the incoming formaldehyde (Fig. 7.11a) or quinone methide (Fig. 7.11b) directing the reaction onto only ortho positions.

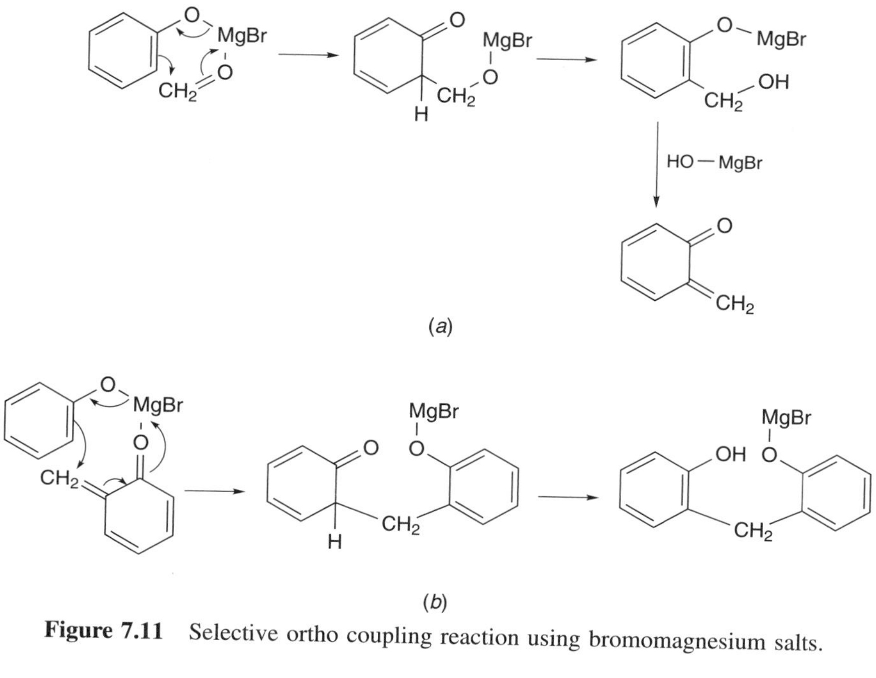

(a)

(b)

Figure 7.11   Selective ortho coupling reaction using bromomagnesium salts.

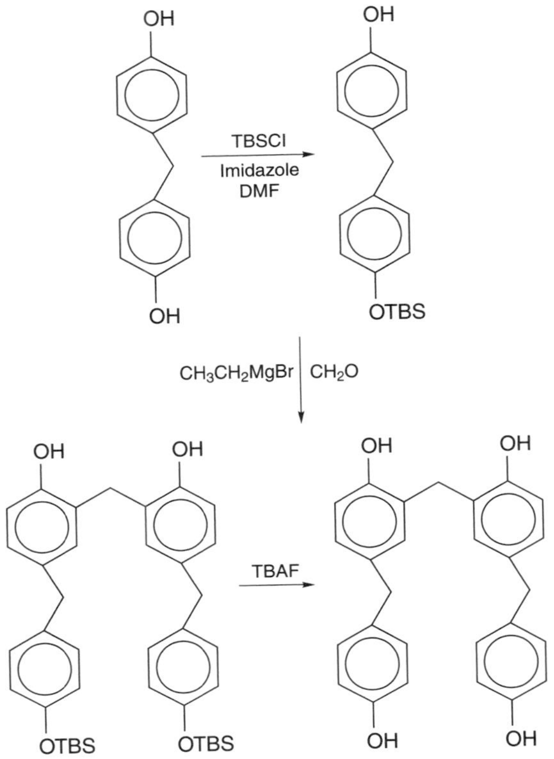

Figure 7.12 Synthesis of model phenolic compound.

de Bruyn et al.[16] prepared low molecular weight model novolac compounds comprised of four to eight phenolic units utilizing the bromomagnesium salt methodology (Fig. 7.12). A para–para-linked dimer was used as the starting material where *tert*-butyldimethylsilyl chloride (TBSCl) was reacted with one phenol on a dimer to deactivate its ring against electrophilic reaction with formaldehyde. Selective ortho coupling formed bridges between the remaining phenol rings; then the *tert*-butyldimethylsilyl protecting groups were removed with fluoride ion. These compounds and all ortho-linked model compounds prepared using bromomagnesium salts were subsequently used as molecular weight standards for calibrating gel permeation chromatography and to study model reactions with HMTA (see Section 7.3.8).

7.3.4 Controlled Molecular Weight Cresol Novolac Oligomers

Although molecular weights in typical phenolic novolac syntheses are intentionally limited, these procedures lack molecular weight control. The reactions are generally terminated prior to full conversion after a certain reaction time or

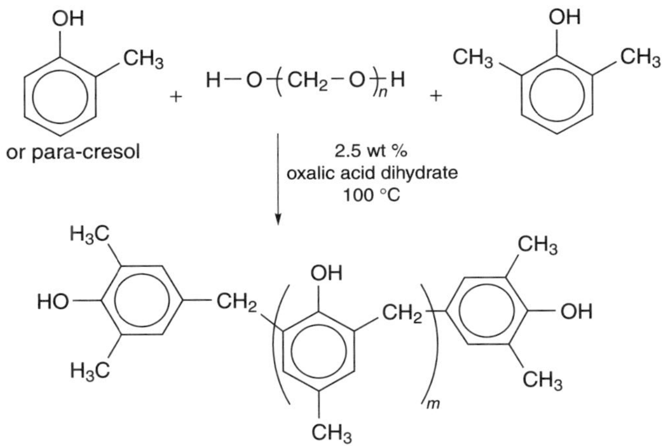

Figure 7.13 Synthesis of 2,6-dimethylphenol end-capped cresol novolac resin.

once a specified viscosity is reached. The synthesis of linear controlled molecular weight cresol novolac resins has been investigated by strategically controlling the stoichiometric ratio of cresol to 2,6-dimethylphenol (as the end-capping reagent) as calculated via the Carothers equation (Fig. 7.13) and reacting the materials until full conversion is achieved.[17] Targeted molecular weights could be obtained by using a slight excess of formaldehyde relative to the calculated values. The reasons for needing an excess of formaldehyde were attributed to some reagent loss during the early exothermic reaction stages and also to formation of a small amount of dimethylene ether linkages between cresol rings (thus using 2 mol of formaldehyde in one link as opposed to 1 mol).

The molecular weights were analyzed via ^{13}C nuclear magnetic resonance (NMR) by ratioing the methyl peaks on endgroups versus the methyl peaks within the repeat units. The molecular weights for these novolac resins made with ortho- or para-cresol showed good agreement between experimental results and the targeted values (Table 7.2).

TABLE 7.2 Calculated M_n versus Targeted M_n

Target M_n (g/mol)	N_{AA} (cresol) (mol)	N_{ZA}: (2,6-DMP)[a] (mols)	Ortho Series M_n (g/mol)	Para Series M_n (g/mol)
500	1	0.984	490	510
1000	1	0.323	930	1010
1500	1	0.193	1380	1460
2000	1	0.138	2250	2150

[a]2,6-Dimethylphenol.

7.3.5 Reaction Conditions and Copolymer Effects

Alkyl-substituted phenols have different reactivities than phenol toward reaction with formaldehyde. Relative reactivities determined by monitoring the disappearance of formaldehyde in phenol–paraformaldehyde reactions (Table 7.3) show that, under basic conditions, meta-cresol reacts with formaldehyde approximately three times faster than phenol while ortho- and para-cresols react at approximately one-third the rate of phenol.[18] Similar trends were observed for the reactivities of acid-catalyzed phenolic monomers with formaldehyde.

One comparison study of oxalic-acid-catalyzed reactions involving ortho- and meta-cresol mixtures demonstrated that meta-cresol was preferentially incorporated into the oligomers during the early stages of reaction.[19] Given the same reaction conditions and time, higher o-cresol compositions (of the mixtures) resulted in decreased overall yields since there was insufficient time for o-cresol to fully react. Consequently, the molecular weights and glass transition temperatures (T_g) were also lower in these partially reacted materials. As expected, the molecular weight increased if a larger amount of catalyst was used or if more time was allowed for reaction. Increased catalyst concentrations also broadened molecular weight distributions.

Bogan et al. conducted similar studies in which meta- and/or para-cresols were reacted with formaldehyde at 99°C for 3 h using oxalic acid dihydrate as the catalyst to form novolac-type structures.[20] Using a relative reactivity of 0.09 ± 0.03 for p-cresol with formaldehyde versus m-cresol with formaldehyde, a statistical model was employed to predict the amounts of unreacted cresols during the reactions, branching density, and m/p-cresol copolymer compositions. Good agreement was found between the predictions and experimental results. Since p-cresol reacted much slower that m-cresol, it was to a first approximation considered an unreactive diluent. When meta- and para-cresol mixtures were reacted, oligomers consisting of mostly m-cresol formed first; then when the m-cresol content was depleted, p-cresol incorporation was observed (mostly at the

TABLE 7.3   Relative Reaction Rates of Various Phenols with Formaldehyde under Basic Conditions[a]

Compound	Relative Reactivity
2,6-Xylenol	0.16
ortho-Cresol	0.26
para-Cresol	0.35
2,5-Xylenol	0.71
3,4-Xylenol	0.83
Phenol	1.00
2,3,5-Trimethylphenol	1.49
meta-Cresol	2.88
3,5-Xylenol	7.75

[a]From ref. 3.

chain ends). Full conversions were not achieved in these investigations, probably due to insufficient reaction times for *p*-cresol to react completely.

Linear novolac resins prepared by reacting para-alkylphenols with paraformaldehyde are of interest for adhesive tackifiers. As expected for step-growth polymerization, the molecular weights and viscosities of such oligomers prepared in one exemplary study increased as the ratio of formaldehyde to para-nonylphenol was increased from 0.32 to 1.00.[21] As is usually the case, however, these reactions were not carried out to full conversion, and the measured M_n of an oligomer prepared with an equimolar phenol-to-formaldehyde ratio was 1400 g/mol. Plots of apparent shear viscosity versus shear rate of these *p*-nonylphenol novolac resins showed non-Newtonian rheological behavior.

Reaction media play an important role in *m*-cresol–paraformaldehyde reactions.[22] Higher molecular weight resins, especially those formed from near-equimolar *m*-cresol–formaldehyde ratios, can be obtained by introducing a water miscible solvent such as ethanol, methanol, or dioxane to the reaction. Small amounts of solvent (0.5 mol solvent/mol cresol) increased reaction rates by reducing the viscosity and improving homogeneity. Further increases in solvent, however, diluted the reagent concentrations to an extent that decreased the rates of reaction.

7.3.6 Molecular Weight and Molecular Weight Distribution Calculations

The molecular weights and molecular weight distributions (MWD) of phenolic oligomers have been evaluated using gel permeation chromatography (GPC),[23,24] NMR spectroscopy,[25] vapor pressure osmometry (VPO),[26] intrinsic viscosity,[27] and more recently matrix-assisted laser desorption/ionization time-of-flight mass spectrometry (MALDI–TOF MS).[28]

The most widely used molecular weight characterization method has been GPC, which separates compounds based on hydrodynamic volume. State-of-the-art GPC instruments are equipped with a concentration detector (e.g., differential refractometer, UV, and/or IR) in combination with viscosity or light scattering. A viscosity detector provides in-line solution viscosity data at each elution volume, which in combination with a concentration measurement can be converted to specific viscosity. Since the polymer concentration at each elution volume is quite dilute, the specific viscosity is considered a reasonable approximation for the dilute solution's intrinsic viscosity. The plot of $\log[\eta]M$ versus elution volume (where $[\eta]$ is the intrinsic viscosity) provides a universal calibration curve from which absolute molecular weights of a variety of polymers can be obtained. Unfortunately, many reported analyses for phenolic oligomers and resins are simply based on polystyrene standards and only provide relative molecular weights instead of absolute numbers.

Dargaville et al.[29] and Yoshikawa et al.[23] recognized the difficulties in obtaining accurate GPC molecular weights of phenolic resins due to large amounts of isomers and their associated differences in hydrodynamic sizes. These workers generated GPC calibration curves using a series of low-molecular-weight model novolac compounds: (1) linear compounds with only ortho–ortho methylene linkages, (2) compounds with ortho–ortho methylene-linked backbones and where each unit had a pendent para–para methylene-linked unit, and (3) compounds with ortho–ortho methylene-linked backbones and where each unit had a pendent para-ortho methylene-linked unit.[29] For a given molecular weight, the hydrodynamic volume of oligomers with only the ortho–ortho methylene links was smaller than the others. It was reasoned that the reduced hydrodynamic volume was caused by "extra" intramolecular hydrogen bonding in high-ortho novolacs, which was a similar argument to that suggested previously by Yoshikawa et al.[23] Based on the GPC calibration curves of the model compounds and their known chemical structures, simulated calibration curves were generated for idealized 100% ortho-para methylene-linked oligomers and for 100% para–para-linked oligomers.

GPC chromatograms for a series of commercial novolacs, including resins with statistical distributions of ortho and para linkages and high ortho novolac resins were measured. Carbon-13 NMR provided the relative compositions of ortho–ortho-, ortho–para-, and para–para-linked methylene groups. Molecular weights from GPC were calculated by considering the fractions of each type of linkage multiplied by the molecular weights calculated from each of the three ortho–ortho (experimental), ortho–para (simulated), and ortho–ortho (simulated) GPC calibration curves. Good agreement was found between the resin molecular weights measured from ^{1}H NMR and the interpolated GPC numbers for oligomers up to an average of four to five units per chain, whereas more deviation was observed for higher molecular weights. This was attributed to complicated intramolecular hydrogen bonding in the higher molecular weight materials. Another factor may be that branching becomes significant in the higher molecular weight materials and the hydrodynamic volume effects of architecture are also complicated.

Proton NMR integrations of methylene and aromatic regions can be used to calculate the number-average molecular weights of novolac resins[29]:

$$\frac{[CH_2]}{[Ar]} = \frac{2n - 2}{3n + 2} \tag{7.1}$$

where $[CH_2]/[Ar]$ is the ratio of methylene protons to aromatic protons and n is the number of phenolic units. The method is quite accurate for novolacs with less than eight repeat units.

Solution ^{13}C NMR has been used extensively to examine the chemical structures of phenolic resins.[23,30] By ratioing the integration of peaks, degree of polymerization, number-average molecular weights, degrees of branching, numbers of free ortho and para positions, and isomer distributions have been evaluated. A typical ^{13}C NMR spectrum of a novolac resin shows three regions (Table 7.4): The methylene linkages resonate between 30 and 40 ppm; the peaks between 146 and 157 ppm are due to hydroxyl-substituted aromatic carbons; and peaks between 113 and 135 ppm represent the remainder of the aromatic carbons.

TABLE 7.4   Peak Assignments for ^{13}C NMR Chemical Shifts of Phenolic Resins^a

Chemical Shift Region (ppm)	Assignment
150–156	Hydroxyl-substituted phenolic carbons
127–135	Other phenolic carbons
121	Para-unsubstituted phenolic carbons
116	Ortho-unsubstituted phenolic carbons
85.9	$HO-CH_2-O-CH_2-OH$
81.4	$HO-CH_2-OH$
71.1	Para-linked dimethylene ether
68.2	Ortho-linked dimethylene ether
40.8	Para–para methylene linkages
35.5	Para–ortho methylene linkages
31.5	Ortho–ortho methylene linkages

^{a}From refs. 31 and 41.

The number of remaining ortho reactive sites versus the number of para reactive sites can also be calculated using ^{13}C NMR (Table 7.4). Since the rates of novolac cure reactions differ with the amount of ortho versus para reactive sites available, it is of great interest to calculate these parameters.

Degrees of polymerization can be calculated from quantitative ^{13}C NMR data by considering the number of substituted (reacted) relative to unsubstituted (not yet reacted) ortho and para phenolic carbons where $[S]$ is the sum of substituted ortho and para carbons and $[S] + [U]$ is the total ortho and para carbons. The fraction of reacted ortho and para sites is denoted by f_s [Eq. (7.2)]. Thus, the number-average number of phenol units per chain (x) can be calculated using Eq. (7.3). This leads to a simple calculation of $M_n = x \times 106 - 14$:

$$f_s = \frac{[S]}{[S] + [U]} \tag{7.2}$$

$$x = \frac{1}{(1 - 1.5\ f_s)} \tag{7.3}$$

Fourier transform infrared (FTIR) and FT Raman spectroscopy have been used to characterize phenolic compounds. The lack of hydroxyl interference is a major advantage of using FT Raman spectroscopy as opposed to FTIR to characterize phenolic compounds. Two regions of interest in Raman spectra are between 2800 and 4000 cm^{-1}, where phenyl C–H stretching and methylene bridges are observed, and between 400 and 1800 cm^{-1}.[31] For a high-ortho novolac resin, the phenyl C–H stretch and methylene bridge appear at 3060 and 2940 cm^{-1}, respectively. In the fingerprint region, the main bands are 1430–1470 cm^{-1} representative of methylene linkages and 600–950 cm^{-1} for out-of-plane phenyl

C$-$H bonds. Phenol, mono-ortho, and di- and tri-substituted phenolic rings can be monitored between 814$-$831, 753$-$794, 820$-$855, and 912$-$917 cm^{-1}, respectively. Para-substituted phenolic rings also absorb in the 820$-$855-cm^{-1} region.

Mandal and Hay[28] used MALDI$-$TOF mass spectrometry to determine the absolute molecular masses and endgroups of 4-phenylphenol novolac resins prepared in xylene or chlorobenzene. Peaks with a mass difference of 44 (the molecular weight of a xylene endgroup) suggested that reactions conducted in xylene included some incorporation of xylene onto the chain ends when a strong acid such as sulfuric acid was used to catalyze the reaction. By contrast, no xylene was reacted into the chain when a milder acid catalyst such as oxalic acid was used. No chlorobenzene was incorporated regardless of the catalyst used.

7.3.7 Hydrogen Bonding

The abundant hydroxyl groups on phenolic resins cause these materials to form strong intra- and intermolecular hydrogen bonds. Intramolecular hydrogen bonding of phenolic resins gives rise to their hyperacidity while intermolecular hydrogen bonding facilitates miscibility with a number of materials containing electron donors such as carbonyl, amide, hydroxyl, ether, and ester groups. Miscible polymer blends of novolac resins include those with some polyamides,[32] poly(ethylene oxide),[33] poly(hydroxyether)s,[34] poly(vinyl alcohol),[35] and poly(decamethylene adipate) and other poly(adipate ester)s.[36] The specific strength of hydrogen bonding is a function of the groups involved; for example, hydroxyl$-$hydroxyl interactions are stronger than hydroxyl$-$ether interactions.[37]

The effects of intermolecular hydrogen bonding on neat novolac resins with compounds containing hydrogen acceptors (e.g., 1,4-diazabicyclo[2,2,2]octane (DABCO) and hexamethylene tetramine) were also investigated.[38] Glass transition temperatures of neat resins and blends were measured using differential scanning calorimetry (DSC) to assess the degrees of hydrogen bonding. Hydrogen-bonding interactions of novolac resins with electron donor sites such as oxygen, nitrogen, or chlorine atoms resulted in increased T_g.

The propensity for dry novolac resins to absorb water at room temperature under 100% humidity is another indication that strong hydrogen bonds form. Approximately 15 wt % water is absorbed by the novolac after 4 d, which corresponds to one water molecule per hydroxyl group.[38]

Dielectric measurements were used to evaluate the degrees of inter- and intramolecular hydrogen bonding in novolac resins.[39] The frequency dependence of complex permittivity (ε^*) within a relaxation region can be described with a Havriliak and Negami function (HN function):

$$\varepsilon^* = \varepsilon_\infty + \frac{\varepsilon_S - \varepsilon_\infty}{[1 + (i\omega\tau_0)^\beta]^\gamma} \tag{7.4}$$

where ε_S and ε_∞ are the relaxed and unrelaxed dielectric constants, ω is the angular frequency, τ_0 is the relaxation time, and β and γ are fitting parameters. The complex permittivity is comprised of permittivity (ε') and dielectric loss

(ε''). Fitting parameters in the HN function are related to shape parameters m and n, which describe the limiting behavior of dielectric loss (ε'') at low and high frequencies, respectively. Intermolecular (characterized by m) and intramolecular (characterized by n) hydrogen bonding can be correlated with m and n values which range from 0 to 1 (where lower values correspond to stronger hydrogen bonding). For one novolac resin examined ($M_n = 1526$ determined via GPC using polystyrene standards, MWD = 2.6, $T_g = 57°C$), m was 0.52 and n was 0.2. These results were considered indicative of strong intramolecular hydrogen bonding within the novolac structures.

7.3.8 Novolac Crosslinking with HMTA

The most common crosslinking agent for novolac resins is HMTA which provides a source of formaldehyde. Novolac resins prepared from a phenol–formaldehyde (F/P) ratio of 1/0.8 can be cured with 8–15 wt % HMTA, although it has been reported that 9–10 wt % results in networks with the best overall performance.[3]

7.3.8.1 Initial Reactions of Novolacs with HMTA

The initial cure reactions of a novolac with HMTA were studied by heating the reactants at 90°C for 6 h, then raising the temperature incrementally to a maximum of 205°C. This led to mostly hydroxybenzylamine and benzoxazine intermediates (Fig. 7.14).[40,41] Hydroxybenzylamines form via repeated electrophilic aromatic substitutions of the active phenolic ring carbons on the methylenes of the HMTA (and derivatives of HMTA). Since novolac resins form strong intermolecular hydrogen bonds with electron donors, a plausible mechanism for the initial reaction between novolac and HMTA involves hydrogen bonding between phenolic hydroxyl groups and an HMTA nitrogen (Fig. 7.14). Such hydrogen bonding can lead to a proton transfer from which a phenolate ion is generated. A negatively charged ortho or para carbon can attack the methylene carbon next to the positively charged nitrogen on HTMA, which results in cleavage of a C–N bond. Benzoxazines form by nucleophilic attack of the phenolic oxygen on HMTA–phenolic intermediates (Fig. 7.14). Upon further reaction, methylene linkages form as the major product of both types of intermediates through various thermal decomposition pathways.

Since a small amount of water is always present in novolac resins, it has also been suggested that some decomposition of HMTA proceeds by hydrolysis, leading to the elimination of formaldehyde and amino–methylol compounds (Fig. 7.15).[42] Phenols can react with the formaldehyde elimination product to extend the novolac chain or form methylene-bridged crosslinks. Alternatively, phenol can react with amino–methylol intermediates in combination with formaldehyde to produce ortho- or para-hydroxybenzylamines (i.e., Mannich-type reactions).

Reaction pathways involved in the curing of novolacs with HMTA have been extensively investigated by Solomon and co-workers.[43–50] In a series of model studies where 2,6-xylenol and/or 2,4-xylenol was reacted with HMTA, these workers found that the types of linkages formed were affected by the initial *tris*-Hydroxybenzylamine   Benzoxazine

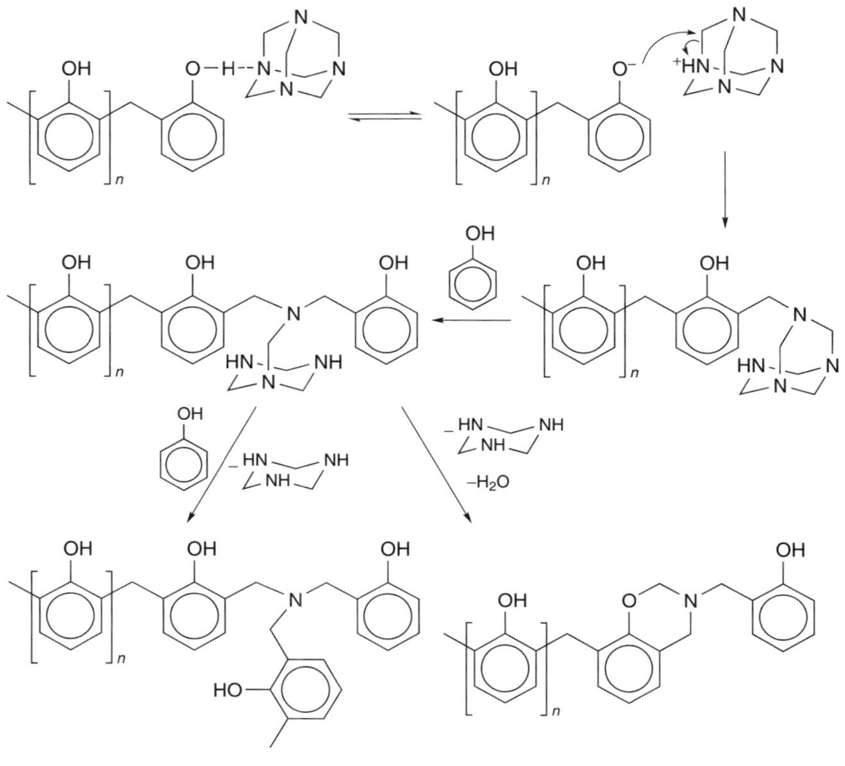

Figure 7.14 Initial reaction of novolac and HMTA via hydrogen-bonding mechanism.

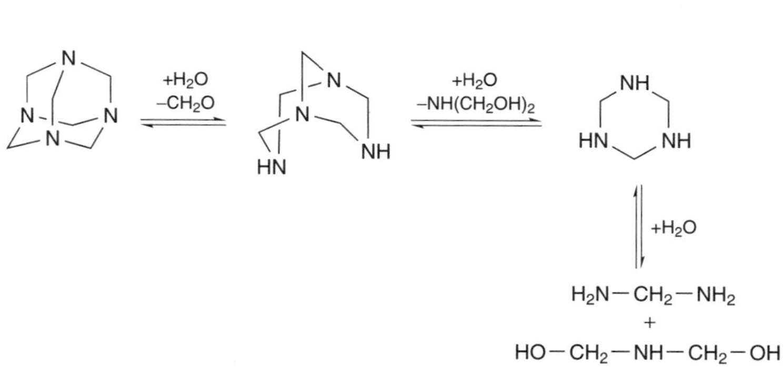

Figure 7.15 Decomposition of HMTA.

chemical structure of the novolac, that is, amount of ortho versus para reactive positions, amount of HMTA, and pH. Reaction intermediates for the cures were identified, mostly via FTIR, ^{13}C NMR, and ^{15}N NMR.

As previously described, the main intermediates generated from the initial reaction between ortho reactive sites on novolac resins and HMTA are hydroxy-benzylamines and benzoxazines.[44] Triazines, diamines, and, in the presence of trace amounts of water, benzyl alcohols and ethers also form (Fig. 7.16). Similar intermediates, with the exception of benzoxazines, are also observed when para sites react with HMTA.

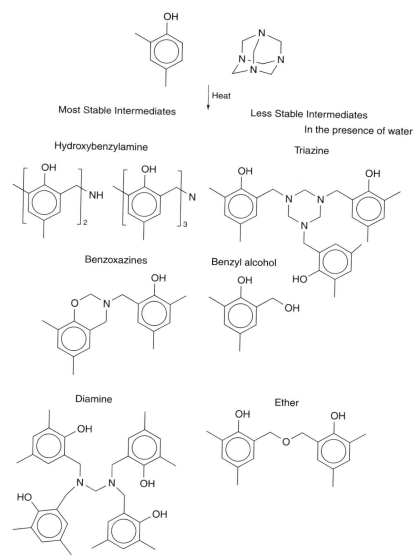

Figure 7.16 Possible reaction intermediates for the reaction of 2,4-xylenol with HTMA.

Thermolysis rates to form methylene linkages depend on the stabilities of hydroxybenzylamine and benzoxazine intermediates. Comparatively, ortho-linked hydroxybenzylamine intermediates are more stable than para-linked structures because six-membered rings can form between the nitrogen and phenolic hydroxyl groups via intramolecular hydrogen bonding. For the same reason, benzoxazines are the most stable intermediates and decompose only at higher temperatures (185°C).[45] If a high-ortho novolac resin is cured with HMTA, the reaction occurs at lower temperatures due to formation of relatively unstable para-linked intermediates and the amount of side products is low. If, however, a typical novolac is used, the reaction temperature must be higher to decompose the more thermally stable ortho intermediates, and the amount of nitrogen-containing side products is significantly higher.[46,47]

If only ortho sites are available for reaction, the amount of hydroxybenzylamine versus benzoxazine generated is largely dependent on the novolac–HMTA ratio. Hydroxybenzylamine is favored when the HMTA content is low whereas more benzoxazine is formed at higher HMTA concentrations. This is expected since only one HMTA carbon is needed per reactive ortho position in the formation of hydroxybenzylamine, but the formation of benzoxazine requires three HMTA carbons per two reactive ortho positions. The HMTA concentration therefore is one key in determining the structure of the resulting networks. Lower HMTA contents leading to more hydroxybenzylamine intermediates means that lower temperatures can be used for decomposition into methylene bridges and correspondingly lower levels of side products form under such conditions.

7.3.8.2   *Hydroxybenzylamine and Benzoxazine Decompositions in Novolac–HMTA Cures*

THERMAL DECOMPOSITION OF HYDROXYBENZYLAMINES   Depending on the concentration of HMTA and mobility of the system, hydroxybenzylamine and benzoxazine intermediates react by a number of pathways to form crosslinked novolac networks. Trishydroxybenzylamines eliminate benzoquinone methide between 90 and 120°C to form bishydroxybenzylamines, which decompose to methylene linkages with the elimination of $CH_2=NH$ at higher temperatures (Fig. 7.17).[46]

THERMAL DECOMPOSITION OF BENZOXAZINES   Thermal decomposition of benzoxazines does not occur substantially until the temperature reaches ~160°C. This begins with proton transfer from a phenolic hydroxyl group to a nitrogen. Cleavage of the C–O bond with water generates a tertiary hydroxymethylamine, which can eliminate formaldehyde, then $CH_2=NH$, to form methylene linkages (Fig. 7.18a). Alternatively, C–N bond cleavage in the benzoxazine leads to the elimination of a benzoquinone methide, which can react with phenols to primarily yield the product methylene-bridged species (Fig. 7.18b).[45] Further decomposition of benzoxazines can also lead to a variety of side products in small amounts.

REACTIONS OF BENZOXAZINES WITH PHENOLS   In the presence of 2,4-xylenol, benzoxazine intermediates react at lower temperatures (~90°C) to form hydroxybenzylamines (Fig. 7.19), which can then decompose to ortho–ortho methylene

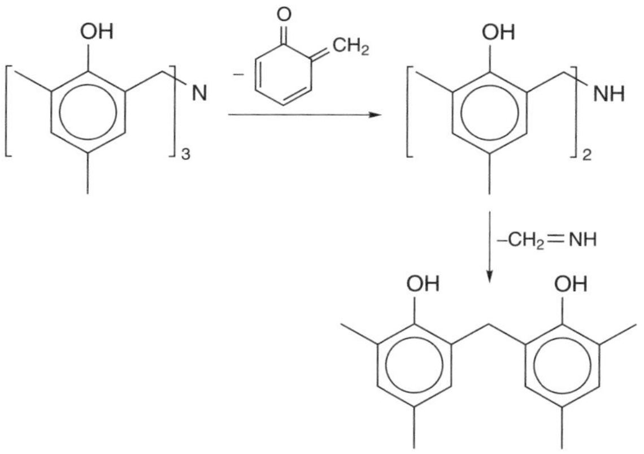

Figure 7.17   Thermal decomposition of hydroxybenzylamine.

linkages (as described in Fig. 7.17).[46] The reaction between benzoxazine and free ortho reactive positions on 2,4-xylenol occurs via electrophilic aromatic substitution facilitated by hydrogen bonding between benzoxazine oxygen and phenolic hydroxyl groups (Fig. 7.19).

The reaction of benzoxazine in the presence of 2,6-xylenol does not occur until $\sim 135^\circ$C, presumably because the hydrogen-bonded intermediate depicted for the 2,4-xylenol reaction (Fig. 7.19) cannot occur. All three types of linkages are obtained in this case. Para–para methylene-linked 2,6-xylenol dimers, obtained from the reaction of 2,6-xylenol with formaldehyde, formed in the decomposition of the benzoxazine (or with other by-products of that process) dominate. Possible side products from benzoxazine decomposition include formaldehyde and $CH_2=NH$, either of which may provide the source of methylene linkages. The amount of ortho–para linkages formed by reaction of 2,6-xylenol with benzoxazine is low. Ortho–ortho methylene-linked products presumably form by a decomposition pathway from benzoxazine (as in Fig. 7.18).

HMTA CROSSLINKING REACTIONS OF NOVOLACS CONTAINING BOTH ORTHO AND PARA REACTIVE SITES   When both ortho and para positions on novolac materials are available for reaction with HMTA, ortho–ortho, ortho–para, and para–para methylene linkages form through several pathways. This section will address crosslinking reaction pathways in which components that have been eliminated as "by-products" reenter the reactions. In particular, reactions of quinone methides, formaldehyde, and imine will be discussed. We will also describe exchange reactions between hydroxybenzylamine intermediates with phenolic methylol derivatives which lead to methylene-bridged final products. Exchange reactions between two different hydroxybenzylamine intermediates, which lead to primarily ortho–ortho-linked products, are also important.

In one model reaction where tris(para-hydroxybenzyl)amine was heated to 205°C in the presence of 2,4-xylenol (1 : 1 ratio), the ortho–ortho, ortho–para,

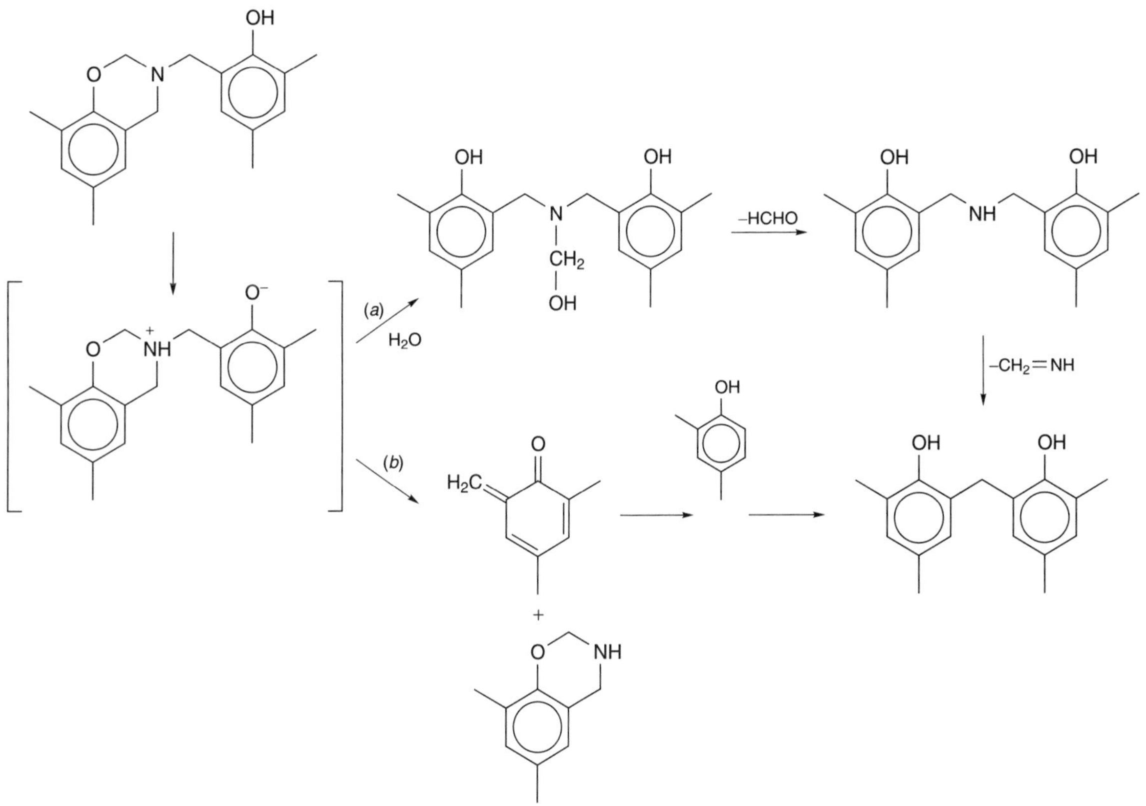

Figure 7.18 Thermal decomposition of benzoxazine.

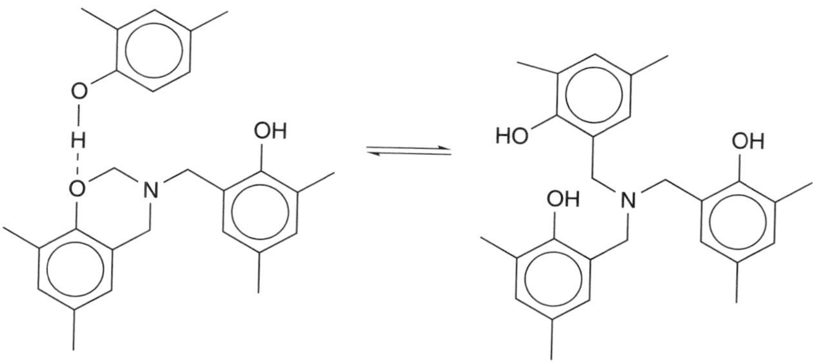

Figure 7.19   Reaction of benzoxazines and 2,4-xylenol.

and para–para methylene bridge ratios in the products were found to be 44, 14, and 38%, respectively (Fig. 7.20).[48] This model study demonstrated the importance of benzoquinone methide intermediates in the formation of various products in the novolac–HMTA curing reaction. Formaldehyde, $CH_2=NH$, and water liberated during the cure reaction also affect the reaction pathways (pathways 3, 4, and 5). Approximately 4% of 1,2-bis(para-hydroxyphenyl)ethane was also observed, presumably formed through dimerization of two quinone methides.

Para–para methylene linkages appeared first via hydroxybenzylamine decomposition at lower temperatures (pathway 1 in Fig. 7.20). Ortho–para methylene linkages also formed at the lower reaction temperatures (pathway 2). Since the only source of an ortho methylene-linked phenol product was the 2,4-xylenol starting material, these mixed products must have formed by the reaction of 2,4-xylenol with either a para-hydroxybenzylamine or with a quinone methide eliminated in pathway 1. Ortho–para methylene linkages also formed at higher reaction temperatures, which were attributed to exchange reactions between a methylol derivative of 2,4-xylenol and a hydroxybenzylamine (pathway 3). Ortho–ortho methylene linkages formed only at higher temperatures via hydroxybenzylamine exchange and methylol dimerization reactions described in pathways 4 and 5. The reactions depicted in pathway 4 involved sequential exchanges between para- and ortho-substituted intermediates through nucleophilic substitutions on hydroxybenzylamines. Since the amount of ortho–para-linked products was low, it was suggested that the major product of pathway 4 was the ortho–ortho linkage. This is reasonable since the equilibrium of these exchange reactions lies toward ortho-hydroxybenzylamines where hydrogen bonding provides stability. These more thermally stable hydroxybenzylamines then decompose at higher temperatures to form ortho–ortho linkages.

Small amounts of various phenolic side products that incorporate groups such as imines, amides, ethers, and ethanes into the networks also form. A number of these side products undergo further reactions which eventually lead to methylene linkages. Some side products generally remain in the networks even after heating at 205°C.

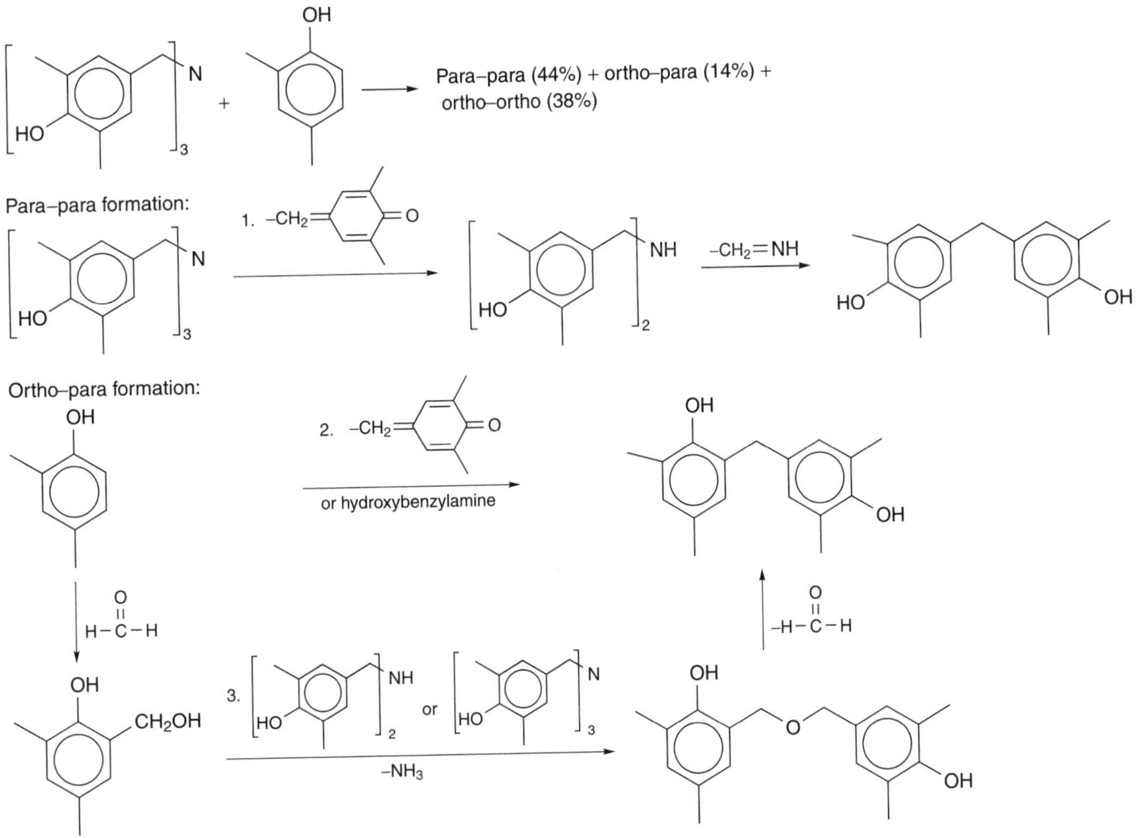

Para–para (44%) + ortho–para (14%) + ortho–ortho (38%)
Para–para formation:
1. –CH2=
–CH2=NH
Ortho–para formation:
2. –CH2=
or hydroxybenzylamine
3.
or
–NH3
CH2OH

Ortho–ortho formation:

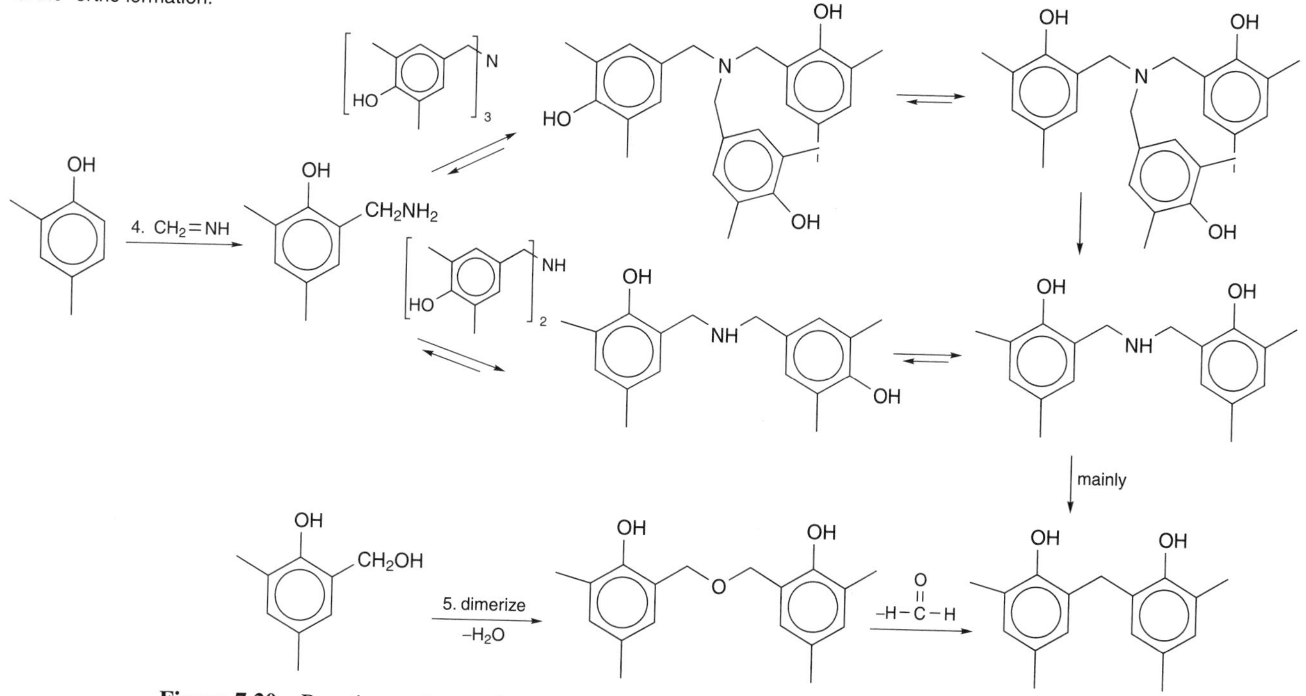

Figure 7.20 Reaction pathways for formation of ortho–ortho, ortho–para, and para–para through the reaction of *para*-trishydroxybenzylamine and 2,4-xylenol.

397

Lim et al. also investigated HMTA–phenolic reactions with somewhat larger model compounds (e.g., two- and four-ring compounds) and established that similar reaction pathways to those described previously occurred.[50] For these model compounds (as opposed to one-ring model compounds), which are more representative of typical oligomeric systems, increased molecular weight favored the formation of hydroxybenzylamines but not benzoxazines. This was suggested to be a steric effect.

Other crosslinking agents that provide sources of formaldehyde for methylene linkages include paraformaldehyde and trioxane, but these have only achieved limited importance. Quantitative ^{13}C solid-state NMR and FT Raman spectroscopy were used to monitor the cure reactions of a high-ortho novolac resin using paraformaldehyde under different conditions.[31] The weight percent paraformaldehyde needed to achieve the maximum crosslinking (1.5 mol formaldehyde/mol phenol) for the particular novolac examined ($M_n = 430$ g/mol determined via ^{13}C NMR) was calculated to be 17.76 wt %. Eleven weight percent formaldehyde was used in these studies so that phenol sites were in excess. The degree of conversion was assessed by comparing the formaldehyde-to-phenol ratio in the polymer to 1.18. As expected, higher temperatures and/or pressures lead to higher reaction conversions. However, none of these reaction conversions reached 100%, and this was attributed to a lack of mobility.

7.4 RESOLE RESINS AND NETWORKS

7.4.1 Resole Resin Syntheses

Resoles are prepared under alkaline conditions using an excess of formaldehyde over phenol (1 : 1 to 3 : 1) at typical temperatures of 60–80°C. The basic catalysts commonly used are NaOH, Na$_2$CO$_3$, KOH, K$_2$CO$_3$, Ba(OH)$_2$, R$_4$NOH, NH$_3$, RNH$_2$, and R$_2$NH.[4] In aqueous solutions, ammonia and HMTA are easily hydrolyzed to amines and also catalyze resole syntheses. Typical resole resins comprise a mixture of monomers, dimers, trimers, and small amounts of higher molecular weight oligomers with multiple methylol functional groups.

Resole syntheses entail substitution of formaldehyde (or formaldehyde derivatives) on phenolic ortho and para positions followed by methylol condensation reactions which form dimers and oligomers. Under basic conditions, phenolate rings are the reactive species for electrophilic aromatic substitution reactions. A simplified mechanism is generally used to depict the formaldehyde substitution on the phenol rings (Fig. 7.21). It should be noted that this mechanism does not account for pH effects, the type of catalyst, or the formation of hemiformals. Mixtures of mono-, di-, and trihydroxymethyl-substituted phenols are produced.

Phenol reacts with formaldehyde in either the ortho or the para position to form monohydroxymethyl-substituted phenols, which further react with formaldehyde to form di- and trihydroxymethyl-substituted phenols (Fig. 7.22).

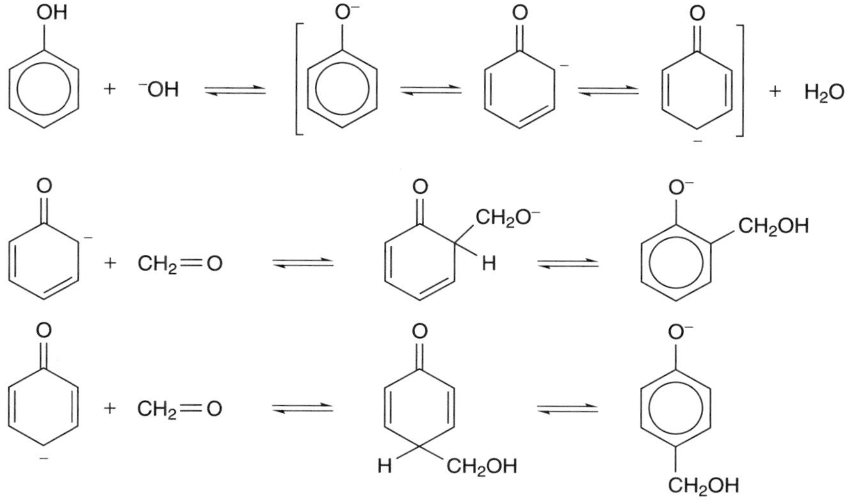

Figure 7.21   Mechanism of resole synthesis.

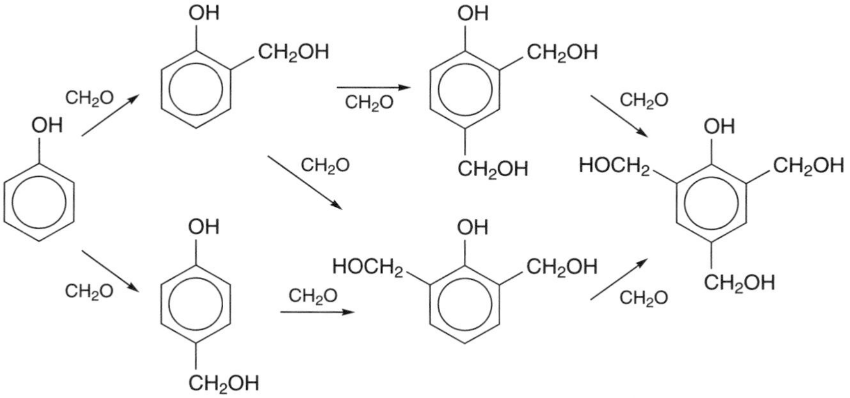

Figure 7.22   Reaction pathways for phenol–formaldehyde reactions under alkaline conditions.

Condensation reactions between two hydroxymethyl substituents eliminate water to form ether linkages (Fig. 7.23*a*) or eliminate both water and formaldehyde to form methylene linkages (Fig. 7.23*b*). Ether formation is favored under neutral or acidic conditions and up to ~130°C above which formaldehyde departs and methylene linkages are generated. The methylene linkage formation reaction, which eliminates water and formaldehyde, is more prevalent under basic conditions. Condensation reactions between hydroxymethyl groups and reactive

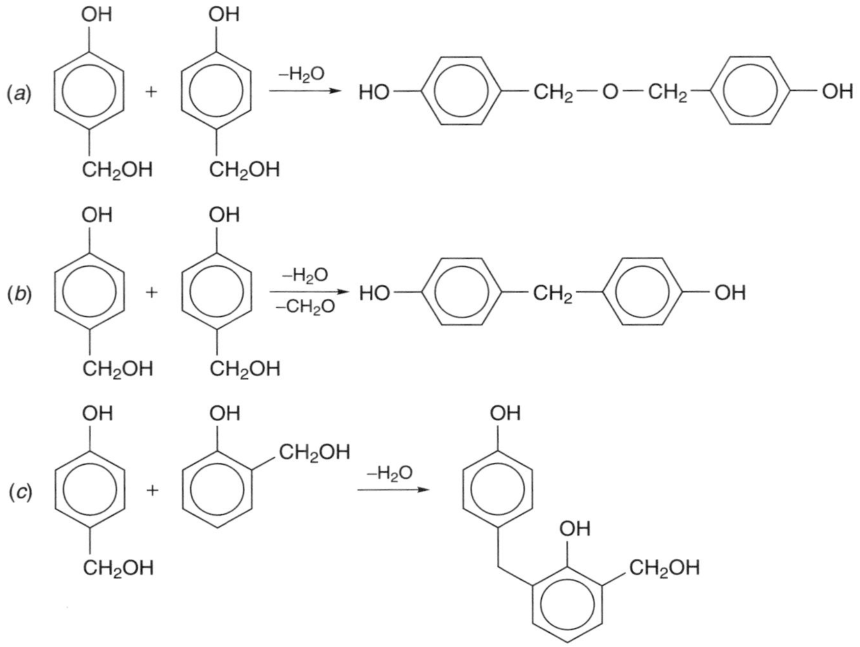

Figure 7.23   Condensation of hydroxymethyl groups.

ortho or para ring positions also lead to methylene bridges between phenolic rings (Fig. 7.23c). Relative reactivities of hydroxymethyl-substituted phenols with formaldehyde and with other hydroxymethyl-substituted phenols appear to be strongly dependent on interactions between ortho-methylol groups and the phenolic hydroxyl position. Hydroxymethyl condensation reactions under basic conditions strongly favor the formation of para–para and ortho–para methylene linkages.

Quinone methides are the key intermediates in both resole resin syntheses and crosslinking reactions. They form by the dehydration of hydroxymethylphenols or dimethylether linkages (Fig. 7.24). Resonance forms for quinone methides include both quinoid and benzoid structures (Fig. 7.25). The oligomerization or crosslinking reaction proceeds by nucleophilic attack on the quinone methide carbon.

The ortho-quinone methides are difficult to isolate due to their high reactivity, which leads to rapid Diels-Alder dimerization or trimerization (Fig. 7.26). At 150°C, a partial retro-Diels-Alder reaction of the trimer can occur to form ortho-quinone methide and bis(2-hydroxy-3,5-dimethylphenyl) ethane (dimer).[51]

Base-catalyzed phenol–formaldehyde reactions exhibit second-order kinetics [Eq. (5)]. Several alkylphenols such as cresols also follow this rate equation:

$$\text{Rate} = k[\text{phenolate}][\text{formaldehyde}] \tag{7.5}$$

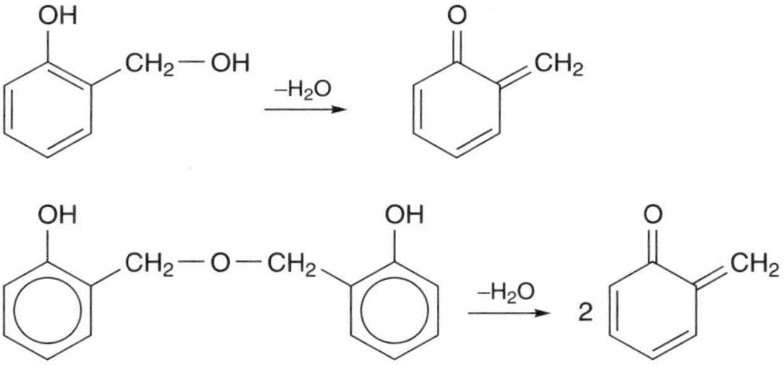

Figure 7.24 Dehydration of methylols or benzylic ethers to form quinone methides.

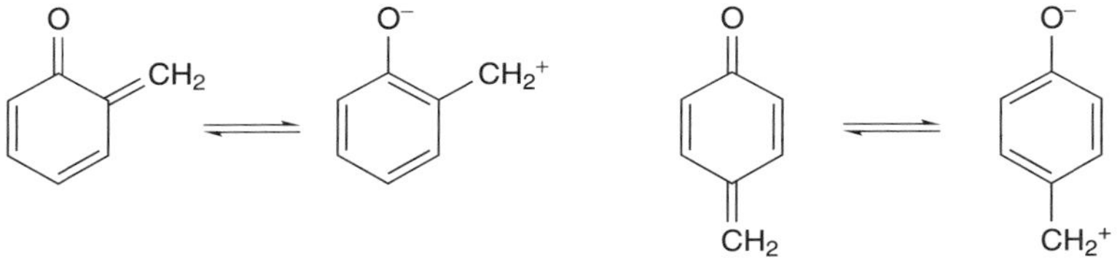

Figure 7.25 Resonance of quinone methides.

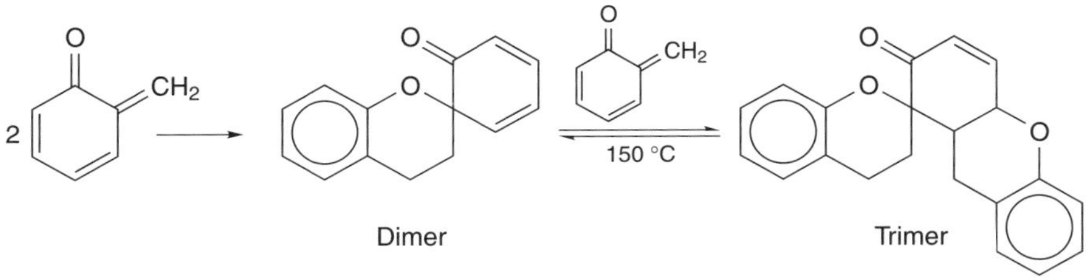

Figure 7.26 Dimer and trimer structures of ortho-quinone methides.

The rate constants for various hydroxymethylation steps (Fig. 7.22) have been evaluated by several groups (Table 7.5)[52–54] and more recently by Grenier-Loustalot et al.[56]

Some consensus observations for reactions conducted at 30°C indicate that the para reactive site on phenol is slightly more reactive than ortho reactive sites due to higher electron density on the para position. In addition, ortho-hydroxymethyl substituents significantly activate the rings toward further electrophilic addition of formaldehyde. This is especially pronounced for 2,6-dihydroxymethylphenols. The ortho-hydroxymethyl substituents are proposed to stabilize the quinoid resonance form via hydrogen bonding between the phenolic hydroxyl and ortho-hydroxymethyl groups in basic aqueous media (Fig. 7.27). This intramolecular stabilization activates the para position by intensifying electron density on the

TABLE 7.5 Relative Positional Reaction Rates in Base-Catalyzed Phenol–Formaldehyde Reaction

	Relative Reaction Rates		
	Reference 52	Reference 53	Reference 54
Phenol → 2-hydroxymethylphenol	1.00	1.00	1.00
Phenol → 4-hydroxymethylphenol	1.18	1.09	1.46
2-Hydroxymethylphenol → 2,6-dihydroxymethylphenol	1.66	1.98	1.75
2-Hydroxymethylphenol → 2,4-dihydroxymethylphenol	1.39	1.80	3.05
4-Hydroxymethylphenol → 2,4-dihydroxymethylphenol	0.71	0.79	0.85
2,4-Dihydroxymethylphenol → 2,4,6-trihydroxymethylphenol	1.73	1.67	2.04
2,6-Dihydroxymethylphenol → 2,4,6-trihydroxymethylphenol	7.94	3.33	4.36

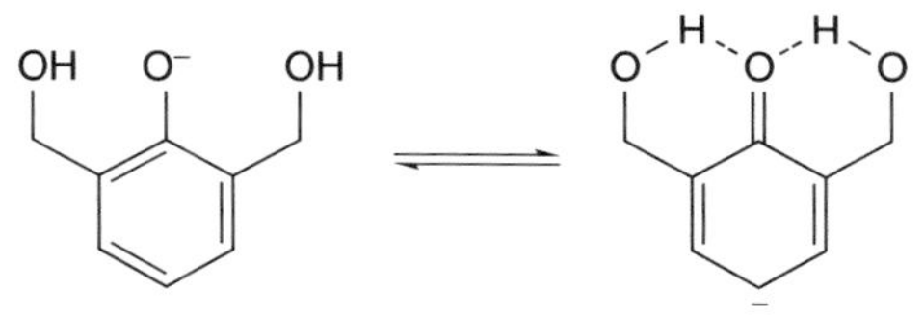

Figure 7.27 Quinoid resonance forms activating the para ring position.

para carbon. This reasoning, however, does not explain the reduced reactivity reported for 2,4-dihydroxymethylphenol.

More recently, the reaction advancement of resole syntheses (pH = 8 and 60°C) was monitored using high-performance liquid chromatography (HPLC), ^{13}C NMR, and chemical assays.[55,56] The disappearance of phenol and the appearances of various hydroxymethyl-substituted phenolic monomers and dimers have been measured. By assessing the residual monomer as a function of reaction time, this work also demonstrated the unusually high reactivity of 2,6-dihydroxymethylphenol. The rate constants for phenolic monomers toward formaldehyde substitution have been measured (Table 7.6).

As the reactions proceed, the disappearance of phenol is delayed due to competition for reaction with formaldehyde between phenol and the faster reacting hydroxymethyl-substituted phenols. Competition also exists between formaldehyde substitution reactions and condensation reactions between rings. Condensation reactions between two ortho-hydroxymethyl substituents are the least favorable condensation pathway. Depending on the reaction conditions, substitutions occur

TABLE 7.6   Second-Order Rate Constants for Reaction of Phenolic Monomers with Formaldehyde[a]

Compound	k (mol$^{-1} \cdot$h^{-1}) $\times 10^2$
Phenol	5.1
2-Hydroxymethylphenol	9.9
4-Hydroxymethylphenol	10.7
2,4-Dihydroxymethylphenol	8.6
2,6-Dihydroxymethylphenol	13.0

[a]From Ref. 56.

predominately in the earlier stages of reaction, and condensations become the major reactions in later stages.[55]

As described previously, condensation reactions of hydroxymethyl substituents strongly favor the formation of para–para and ortho–para linkages.[57–59] Various hydroxymethyl-substituted phenolic monomers were heated in the absence of formaldehyde (60°C, pH = 8.0) to investigate condensation reactions under typical resole synthesis conditions but without formaldehyde substitution.[58] Only methylene linkages were observed under the basic experimental conditions. Highly substituted dimers were predominant in the product mixture since monomers with more hydroxymethyl substituents had higher probabilities for condensation. The ortho-hydroxymethyl groups only condensed with substituents in the para position, and therefore no ortho–ortho methylene linkages were observed. The para-hydroxymethyl substituents, on the other hand, reacted with either ortho- or para-hydroxymethyl substituents or reactive ring positions, but preferentially with para-hydroxymethyl groups. Carbon-13 and ^{1}H NMR monitoring condensation reactions of resole resins comprised of two to five phenolic units showed that, with the exception of one trimer containing a dimethylene ether linkage, only para–para and ortho–para methylene linkages formed.

Upon further reaction, especially at higher temperatures (70–100°C), hydroxymethylated compounds reacted to form almost exclusively para–para and ortho–para methylene linkages. Since the key intermediates for the condensation of hydroxymethylphenols are quinone methides, the formation of para–para and ortho–para methylene linkages is attributed to the exclusive formation of para-quinone methide intermediates (Fig. 7.28).[5] This is attributed to intramolecular hydrogen bonding of both *ortho*-hydroxymethyl substituents with the quinone methide oxygen, which lead to stable para-quinone methide structures. The para-quinone methide intermediates then react with ortho or para reactive positions to form ortho–para and para–para methylene linkages or the quinone methide reacts with hydroxymethyl groups to form ethers, which further advance to methylene linkages.

The mechanisms for model condensation reactions of para-hydroxymethyl-substituted phenol (and therefore para-quinone methide) with reactive ortho positions are described in Fig. 7.29. The phenolate derivatives react with para-quinone

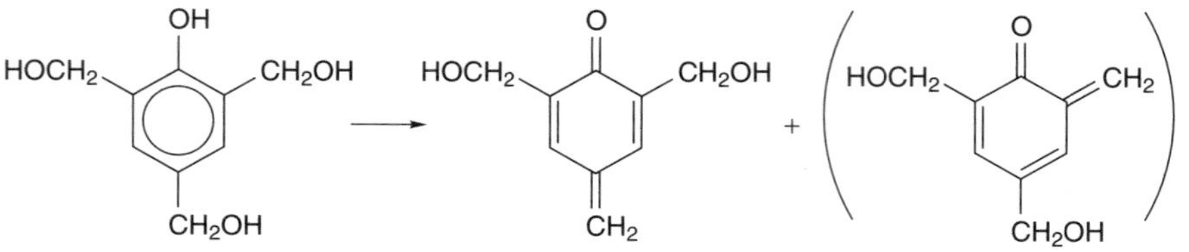

Figure 7.28 Preferential formation of para-quinone methides.

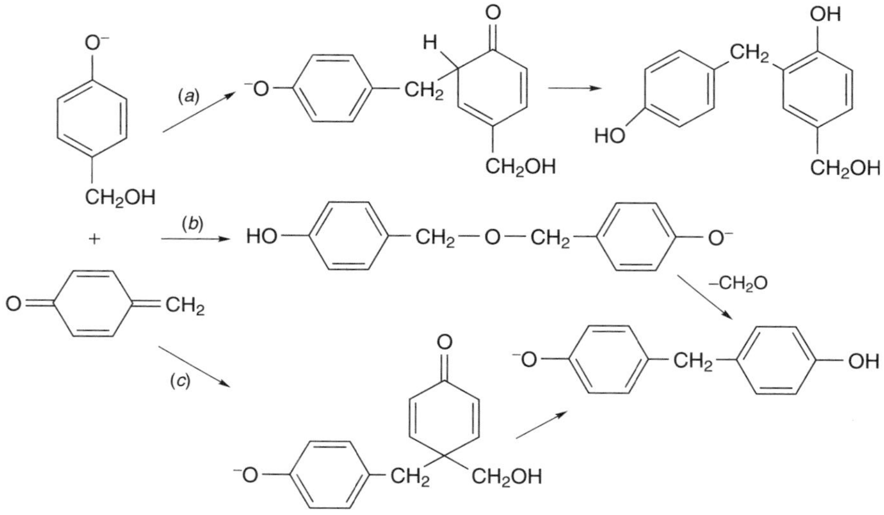

Figure 7.29 Reactions of a quinone methide with a hydroxymethyl-substituted phenolate.

methide via a Michael-type addition to form methylene linkages (Fig. 7.29*a*). Hydroxyl groups on methylol can also attack methide carbons to form dibenzyl ether linkages which subsequently eliminate formaldehyde to form methylene links (Fig. 7.29*b*). An *ipso* substitution in which a nucleophilic ring carbon having a hydroxymethyl substituent attacks a quinone methide has also been postulated to generate methylene linkages (Fig. 7.29*c*).

The reaction conditions, formaldehyde-to-phenol ratios, and concentration and type of catalyst govern the mechanisms and kinetics of resole syntheses. Higher formaldehyde-to-phenol ratios accelerate the reaction rates. This is to be expected since phenol–formaldehyde reactions follow second-order kinetics. Increased hydroxymethyl substitution on phenols due to higher formaldehyde compositions also leads to more condensation products.[55]

The amount of catalyst and pH of the reaction determine the extent of phenolate formation. Phenol–formaldehyde mixtures (F/P = 1.5, 60°C) did not react at pH = 5.5 and reaction rate increased as the pH was increased to about 9.25.[55]

There was a linear relationship between the rate constant and the [NaOH]–[phenol] ratio (pH between 5.5 and 9.25). It was suggested that a limiting pH of approximately 9 exists, above which an increase in pH does not enhance the rate of reaction due to saturation of phenolate anions. Considerable Canizarro side reactions occurred on formaldehyde at pH > 10.[55,60]

The type of catalyst influences the rate and reaction mechanism. Reactions catalyzed with both monovalent and divalent metal hydroxides, KOH, NaOH, LiOH and Ba(OH)$_2$, Ca(OH)$_2$, and Mg(OH)$_2$, showed that both valence and ionic radius of hydrated cations affect the formation rate and final concentrations of various reaction intermediates and products.[61] For the same valence, a linear relationship was observed between the formaldehyde disappearance rate and ionic radius of hydrated cations where larger cation radii gave rise to higher rate constants. In addition, irrespective of the ionic radii, divalent cations lead to faster formaldehyde disappearance rates than monovalent cations. For the proposed mechanism where an intermediate chelate participates in the reaction (Fig. 7.30), an increase in positive charge density in smaller cations was suggested to improve the stability of the chelate complex and, therefore, decrease the rate of the reaction. The radii and valence also affect the formation and disappearance of various hydroxymethylated phenolic compounds which dictate the composition of final products.

Tetraalkylammonium hydroxides have slightly lower catalytic activities than NaOH in resole syntheses. Increased alkyl length on tetraalkylammonium ions (larger ionic radii) decreased the catalytic activity. Contrary to the chelating effect, the reduced activity observed with tetraalkylammonium hydroxides was attributed to the screening effects of alkyl groups. Water solubility was limited to resole resins prepared with tetramethylammonium hydroxide and tetraethylammonium hydroxide. These catalysts also give rise to resins with longer gelation times.

Resole syntheses catalyzed with various amounts of triethylamine (pH adjusted to 8 using NaOH) and at various pH (8.0, 8.23, and 8.36) were monitored.[62] As

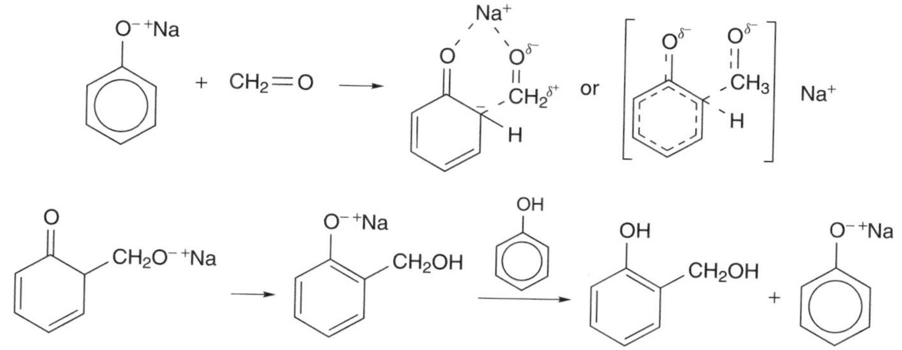

Figure 7.30 Mechanism of phenol and formaldehyde reaction using base catalyst involving the formation of chelate.

expected, shorter condensation times, faster reaction rates, and higher advancement in polymerizations were reached with increased catalyst concentrations. The pH, on the other hand, did not affect these parameters significantly. The reaction mechanisms differed when NaOH was used to adjust the pH since the hydroxide formed phenolate ions which favored para addition reactions. In the absence of NaOH, free phenolic hydroxyl groups formed complexes with triethylamine to promote ortho substitution.

7.4.2 Crosslinking Reactions of Resole Resins

Resole resins are generally crosslinked under neutral conditions between 130 and 200°C or in the presence of an acid catalyst such as hydrochloric acid, phosphoric acid, p-toluenesulfonic acid, and phenolsulfonic acid under ambient conditions.[3] The mechanisms for crosslinking under acidic conditions are similar to acid-catalyzed novolac formation. Quinone methides are the key reaction intermediates. Further condensation reactions in resole resin syntheses under basic conditions at elevated temperatures also lead to crosslinking.

The self-condensation of ortho-hydroxymethyl substituents and the condensation between this substituent with ortho or para reactive sites were investigated under neutral conditions.[51] 2-Hydroxymethyl-4,6-dimethylphenol was reacted (1) alone, (2) in the presence of 2,4-xylenol, and (3) in the presence of 2,6-xylenol. The rates of methylene versus dimethylether formation between rings at 120°C were monitored as a function of time and the percent yields after 5 h were recorded (Table 7.7). The ether linkage was more prevalent in the self-condensation of 2-hydroxymethyl-4,6-dimethylphenol. Possibly the 5% methylene-bridged product formed via the *ipso* substitution of an ortho-quinone methide electrophile onto the methylene position of another ring. Essentially no differences in product composition were observed between the 2-hydroxymethyl-4,6-dimethylphenol self-condensation and reaction of this compound in the presence of 2,6-xylenol. The formation of methylene linkages proceeded much more favorably in the presence of 2,4-xylenol. Moreover, increases in 2,4-xylenol concentrations further increased the methylene linkage yield. This suggests vacant ortho

TABLE 7.7 Percent Yield of Methylene and Ether Linkages of 2-Hydroxylmethyl-4,6-Dimethylphenol Self-reaction, 1 : 1 with 2,4-Xylenol, and 1 : 1 with 2,6-Xylenol

	Percent Yield	
	Methylene	Ether
Self-reaction	5	80
With 2,4-xylenol	38	65
With 2,6-xylenol	5	80

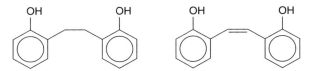

Figure 7.31 Ethane and ethene linkages derived from quinone methide structures.

positions are significantly more reactive than para reactive sites in reactions with ortho-quinone methide. These model reactions provide further evidence supporting quinone methides as the key reactive intermediates.

In addition to methylene and dimethylether linkages, cured networks contain ethane and ethene linkages (Fig. 7.31). These side products are proposed to form through quinone methide intermediates.

Crosslinking resoles in the presence of sodium carbonate or potassium carbonate lead to preferential formation of ortho–ortho methylene linkages.[63] Resole networks crosslinked under basic conditions showed that crosslink density depends on the degree of hydroxymethyl substitution, which is affected by the formaldehyde-to-phenol ratio, the reaction time, and the type and concentration of catalyst (uncatalyzed, with 2% NaOH, with 5% NaOH).[64] As expected, NaOH accelerated the rates of both hydroxymethyl substitution and methylene ether formation. Significant rate increases were observed for ortho substitutions as the amount of NaOH increased. The para substitution, which does not occur in the absence of the catalyst, formed only in small amounts in the presence of NaOH.

7.4.3 Resole Characterization

A number of analytical techniques such as FTIR spectroscopy,[65,66] ^{13}C NMR,[67,68] solid-state ^{13}C NMR,[69] GPC or size exclusion chromatography (SEC),[67-72] HPLC,[73] mass spectrometric analysis,[74] differential scanning calorimetry (DSC),[67,75,76] and dynamic mechanical analysis (DMA)[77,78] have been utilized to characterize resole syntheses and crosslinking reactions. Packed-column supercritical fluid chromatography with a negative-ion atmospheric pressure chemical ionization mass spectrometric detector has also been used to separate and characterize resoles resins.[79] This section provides some examples of how these techniques are used in practical applications.

Using FTIR spectroscopy, resole resin formation and cure reactions can be examined (Table 7.8). FTIR can be used to monitor the appearance and disappearance of hydroxymethyl groups and/or methylene ether linkages, ortho reactive groups, and para reactive groups for resole resin syntheses. Other useful information deduced from FTIR are the type of hydrogen bonding, that is, intra- versus intermolecular, the amount of free phenol present in the product, and the formaldehyde–phenol molar ratio. FTIR bands and patterns for various mono-, di-, and trisubstituted phenols have been identified using a series of model compounds.

TABLE 7.8   FTIR Absorption Band Assignment of Resole Resins

Wave Number (cm^{-1})	Assignment	Nature
3350	$v(CH)$	Phenolic and methylol
3060	$v(CH)$	Aromatic
3020	$v(CH)$	Aromatic
2930	$v_{ip}(CH_2)$	Aliphatic
2860	$v_{op}(CH_2)$	Aliphatic
1610	$v(C=C)$	Benzene ring
1500	$v(C=C)$	Benzene ring
1470	$d(CH_2)$	Aliphatic
1450	$v(C=C)$	Benzene ring
1370	$d_{ip}(OH)$	Phenolic
1240	$v_{ip}(C-O)$	Phenolic
1160	$d_{ip}(CH)$	Aromatic
1100	$d_{ip}(CH)$	Aromatic
1010	$v(C-O)$	Methylol
880	$d_{op}(CH)$	Isolated H
820	$d_{op}(CH)$	Adjacent 2H, para substituted
790	$d_{op}(CH)$	Adjacent 3H
760	$d_{op}(CH)$	Adjacent 4H, ortho substituted
690	$d_{op}(CH)$	Adjacent 5H, phenol

[a]From ref. 65. Abbreviations: v = stretching, d = deformation, ip = in plane, op = out of plane.

The kinetics of resole cure reactions monitored via FTIR suggest that a diffusion mechanism dominates below 140°C. The cure above 140°C exhibits a homogeneous first-order reaction rate. The activation energy of the cure reaction was ~49.6 kJ/mole.[66]

Carbon-13 NMR has proven to be an extremely powerful technique for both monitoring the phenolic resin synthesis and determining the product compositions and structures. Insoluble resole networks can be examined using solid-state ^{13}C NMR which characterizes substitutions on ortho and para positions, the formation and disappearance of hydroxymethyl groups, and the formation of para–para methylene linkages. Analyses using ^{13}C NMR have shown good agreement with those obtained from FTIR.[64]

Various ionization methods were used to bombard phenol–formaldehyde oligomers in mass spectroscopic analysis. The molecular weights of resole resins were calculated using field desorption mass spectroscopy of acetyl-derivatized samples.[74] Phenol acetylation was used to enable quantitative characterization of all molecular fractions by increasing the molecular weights in increments of 42.

Dynamic DSC scans of resole resins show two distinguishable reaction peaks, which correspond to formaldehyde addition and the formation of ether and methylene bridges characterized by different activation energies. Kinetic parameters calculated using a regression analysis show good agreement with experimental values.[75]

DMA was used to determine the cure times and the onset of vitrification in resole cure reactions.[77] The time at which two tangents to the storage modulus curve intersect (near the final storage modulus plateau) was suggested to correspond to the cure times. In addition, the time to reach the peak of the $\tan\delta$ curve was suggested to correspond to the vitrification point. As expected, higher cure temperatures reduced the cure times. DMA was also used to measure the degree of cure achieved by resole resins subsequent to their exposure to combinations of reaction time, temperature, and humidity.[78] The ultimate moduli increased with longer reaction times and lower initial moisture contents. The area under the $\tan\delta$ curves during isothermal experiments was suggested to be inversely proportional to the degree of cure developed in samples prior to the measurement.

A NaOH-catalyzed resole resin, acetylated or treated with an ion exchange resin (neutralized and free of sodium), was analyzed using GPC in tetrahydrofuran (THF) solvent.[80] The molecular weight of the ion-exchange-treated resin, calculated by GPC using polystyrene standards, was significantly lower than that estimated for the acetylated resin. The molecular weight for the ion-exchange-treated resin calculated by ^{1}H NMR and vapor pressure osmometry (VPO) agreed with the results from GPC. The higher molecular weight observed for the acetylated resin was attributed to higher hydrodynamic volume and/or intermolecular association in acetylated samples.

7.4.4 Resole Network Properties

Voids in resole networks detract from the mechanical properties. Irrespective of the curing conditions, all resole networks contain a significant amount of voids due to volatiles released during the cure reactions. The catalyst concentration in resole crosslinking reactions can lead to different pore microstructures, which influence the mechanical properties.[81] Resole networks cured using p-toluenesulfonic acid between 40 and 80°C showed that increased catalyst concentrations led to reduced average void diameters. Higher cure temperatures also resulted in reduced void diameters, although the effect was not as substantial. The same study showed that while the catalyst concentration did not affect the network T_gs, higher cure and postcure temperatures increased T_gs and reduced fracture strains.

In another exemplary study, optical microscopy revealed that the void content of resole networks ranged from 0.13 to 0.21.[82] Resole networks prepared from different F/P molar ratios showed comparable void size distributions. A bimodal distribution was observed for all networks, which was attributed to thermodynamic phase separation of reaction volatiles (free phenol, formaldehyde, and water) and reaction kinetics.

The glass transition temperatures were determined from the peaks of tan δ curves measured using dynamic mechanical analysis for a series of resole networks (prepared with F/P molar ratios ranging from 1 to 2.5).[82] Networks obtained from resoles with high F/P molar ratios (>1.2) had fairly consistent T_gs (between 240 and 260°C). The lowest T_g (190°C) was observed for networks prepared with low hydroxymethyl-substituted resoles (F/P = 1), and this was attributed to the low network crosslink densities at this ratio. The highest T_g occurred at F/P = 1.2 (~280°C), but it superimposed the degradation temperatures. The width of the tan δ curves was used to assess the distributions of chain length as well as the crosslink densities. Networks cured with resoles having F/P ratios of 1.3 and 1.4 exhibited the highest T_gs and therefore the highest crosslink densities.

7.4.5 Modified Phenol–Formaldehyde Resins

Phenol, formaldehyde, and urea have been copolymerized to achieve resins and subsequent networks with improved flame retardance and lower cost relative to phenol–formaldehyde analogues. The condensation of a phenolic methylol group with urea (Fig. 7.32) is believed to be the primary reaction under the weakly acidic conditions normally used.

Resins were prepared by cocondensing low-molecular-weight hydroxymethyl-substituted resoles with urea. The rate of the urea–methylol reaction was greatly enhanced by increased acidity in the reaction medium.[83] The para-methylol groups reacted faster than the ortho-methylol groups (presumably due to hydrogen-bonded ortho-methylol groups with phenol). The extent of urea incorporation depended on the F/P ratio and the resole–urea concentration. Increased urea incorporation ensued at higher urea concentrations and/or in the presence of highly hydroxymethylated resole resins (prepared from larger F/P ratios). Since the reactions were conducted under acidic conditions, the condensation of methylol and urea competes with methylol self-condensation. Increased urea compositions suppress the self-condensation reactions.

The curing process of trihydroxymethylphenol reacted with urea was monitored using torsional braid analysis.[84] Curing proceeded in two stages where the first stage occurred at lower temperatures and was attributed to the reaction of para-hydroxymethyl and urea groups. The second stage was due to the higher temperature reaction of ortho-hydroxymethyl and urea groups.

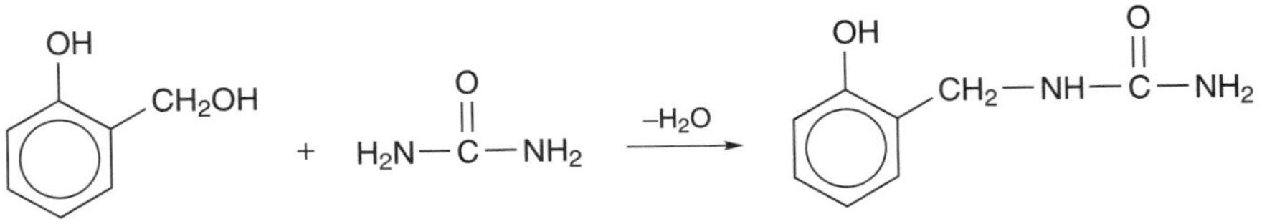

Figure 7.32   Reaction of hydroxymethylphenol and urea.

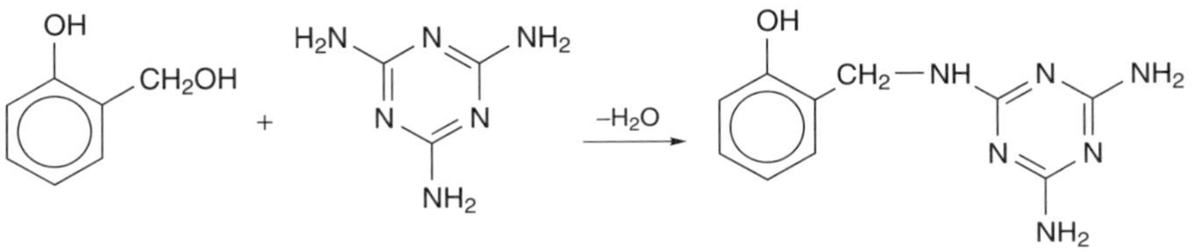

Figure 7.33 Reaction of hydroxymethylphenol and melamine.

Condensation reactions of hydroxymethyl groups on phenolic resoles and amines on melamine take place between pH 5 and 6 (Fig. 7.33). Only self-condensations of hydroxymethyl substituents occur under strongly acidic or basic conditions.

7.5 EPOXY–PHENOLIC NETWORKS

Void-free phenolic networks can be prepared by crosslinking novolacs with epoxies instead of HMTA. A variety of difunctional and multifunctional epoxy reagents can be used to generate networks with excellent dielectric properties.[2] One example of epoxy reagents used in this manner is the epoxidized novolac (Fig. 7.34) derived from the reaction of novolac oligomers with an excess of epichlorohydrin.

7.5.1 Mechanism of Epoxy–Phenolic Reaction

The reactions between phenolic hydroxyl groups and epoxides have been cat-alyzed by a variety of acid and base catalysts, group 5a compounds, and quater-nary ammonium complexes,[85] although they are typically catalyzed by tertiary amines or phosphines, with triphenylphosphine being the most commonly used reagent. The reaction mechanism (Fig. 7.35) involves triphenylphosphine attack-ing an epoxide, which results in ring opening and produces a zwitterion. Rapid proton transfer occurs from the phenolic hydroxyl group to the zwitterion to form a phenoxide anion and a secondary alcohol. The phenoxide anion subsequently reacts with either an electrophilic carbon next to the phosphorus regenerating the

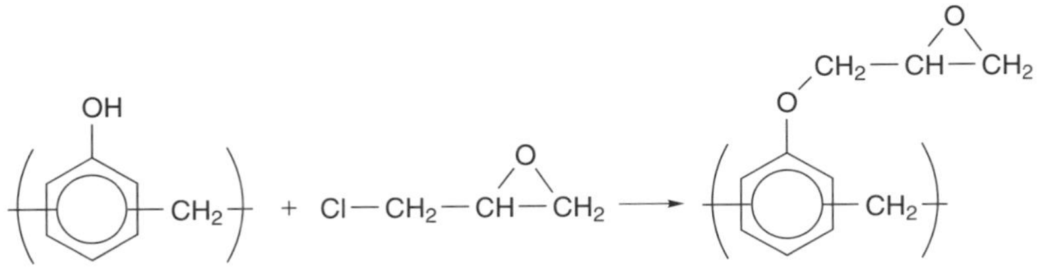

Figure 7.34 Reaction of phenol and epichlorohydrin to form epoxidized novolacs.

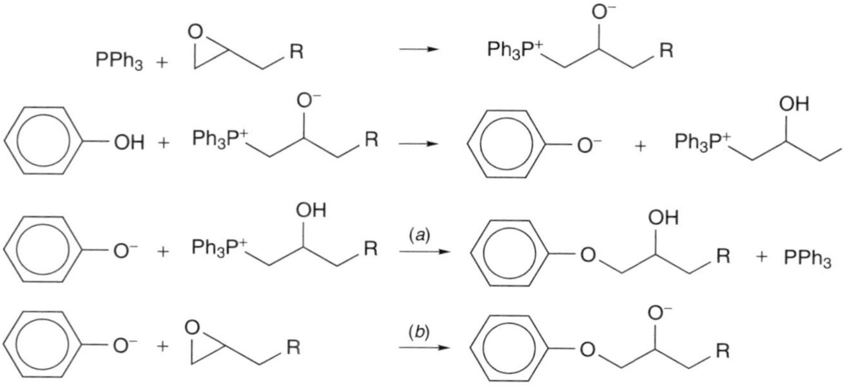

Figure 7.35 Mechanism for triphenylphosphine-catalyzed phenol–epoxy reaction.

triphenylphosphine (Fig. 7.35a)[86] or it can ring-open an epoxy followed by proton transfer from another phenol to regenerate the phenoxide anion (Fig. 7.35b). The phenolate anion is the reactive species for the crosslinking reaction.

Melt reaction mechanisms of tertiary aliphatic amine catalyzed phenolic–epoxy reactions were proposed to begin with a trialkylamine abstracting a phenolic hydroxyl proton to form an ion pair (Fig. 7.36). The ion pair was suggested to complex with an epoxy ring, which then dissociated to form a β-hydroxyether and a regenerated trialkylamine.[87]

Side reactions involving branching through a secondary hydroxyl group can also occur. The extent of these side reactions should decrease as the ratio of epoxy to phenol decreases since phenolate anions are significantly more nucleophilic than aliphatic hydroxyl groups.

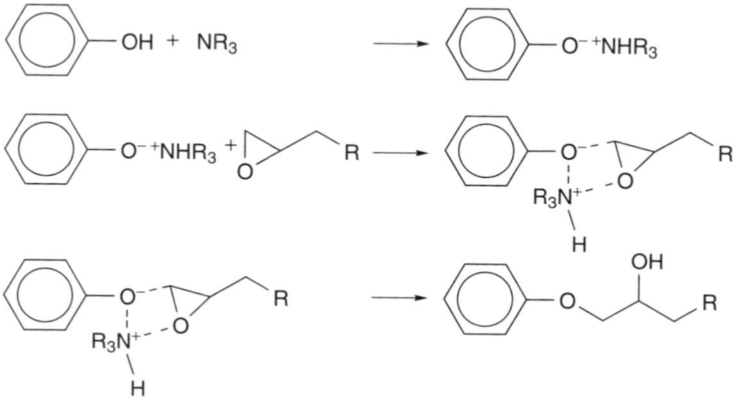

Figure 7.36 Proposed mechanism for tertiary amine–catalyzed phenol–epoxy reaction.

7.5.2 Epoxy–Phenolic Reaction Kinetics

A review of epoxy–novolac reaction mechanisms and kinetics is provided by Biernath et al.[85] Depending on the structures of the novolac and the epoxy, reactions have been reported to proceed through an nth-order mechanism or an autocatalytic mechanism.[88–92]

Biernath et al. concluded that phenolic novolac and epoxidized cresol novolac cure reactions using triphenylphosphine as the catalyst had a short initiation period wherein the concentration of phenolate ion increased, followed by a (steady-state) propagation regime where the number of reactive phenolate species was constant.[85] The epoxy ring opening was reportedly first order in the "steady-state" regime.

7.5.3 Epoxy–Phenol Network Properties

Void-free phenolic–epoxy networks prepared from an excess of phenolic novolac resins and various diepoxides have been investigated by Tyberg et al. (Fig. 7.37).[93–95] The novolacs and diepoxides were cured at approximately 200°C in the presence of triphenylphosphine and other phosphine derivatives. Network densities were controlled by stoichiometric offsets between phenol and

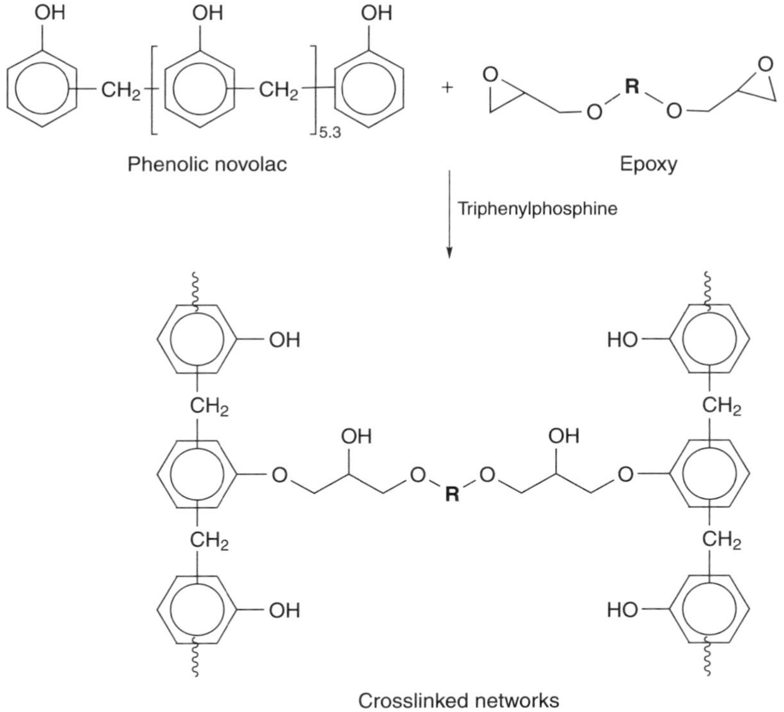

Figure 7.37 Network formation of phenolic novolac and epoxy.

epoxide groups. These networks contained high phenolic concentrations (up to ~80 wt %) to retain the high flame retardance of the phenolic materials while the mechanical properties were tailored by controlling the crosslink densities and molecular structures.

Network structure–property relationships and flame properties were determined for a novolac cured with various diepoxides (Fig. 7.38) at defined compositions. The fracture toughness of the networks, determined by the plane strain–stress intensity factors (K_{IC}), increased with increased stoichiometric offset to a maximum of approximately 1.0 MPa·m$^{1/2}$ for the 5 phenol/1 epoxy equivalence ratio (Table 7.9). All novolac–epoxy networks were significantly tougher than a thermally cured resole network (0.16 MPa·m$^{1/2}$). Further stoichiometric offset (to 7/1) reduced the fracture toughness. This was undoubtedly related to the increase in dangling ends and unconnected phenolic chains at these very high phenol-to-epoxy ratios. Likewise, as expected, glass transition temperatures of fully cured networks decreased as the distances between crosslinks (M_c) increased.

The flame retardance was measured using a cone calorimeter with a heat flux of 50 kW/m^2 and 20.95 mol % O_2 content (atmospheric oxygen). All phenolic novolac–epoxy networks with relatively high novolac compositions showed much lower peak heat release rates (PHRRs) than a typical amine-cured epoxy network (bisphenol-A epoxy stoichiometrically cured with p,p'-diaminodiphenylsulfone) (Table 7.10). Brominated epoxy reagents were also investigated since halogenated materials are well known for their roles in promoting flame retardance. Networks cured with the brominated diepoxide showed the lowest peak heat release rates, but the char yields of these networks were lower and the smoke toxicity (CO yield/CO$_2$ yield) was increased. Incorporating siloxane moieties into networks reduced the peak heat release rates and smoke toxicities compared to the novolac networks cured with bisphenol-A diepoxides.

A biphenyl diglycidyl ether based epoxy resin was crosslinked with amine-curing agents (4,4'-diaminodiphenylmethane and aniline novolac) and phenol-curing agents (phenol novolac and catechol novolac), and the thermomechanical

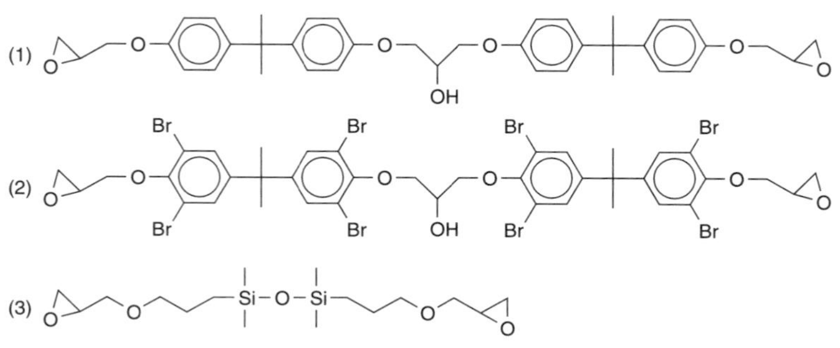

Figure 7.38 Epoxy structures: (1) bisphenol-A-based epoxy, (2) brominated bisphenol-A-based epoxy, and (3) siloxane epoxy.

TABLE 7.9 T_g and K_{IC} of Phenolic Novolac–Epoxy Networks

Epoxy	Phenol–Epoxy (wt/wt)	Phenol–Epoxy (mol/mol)	T_g	K_{IC} (MPa·m$^{1/2}$)	M_c (g/mol)
Bisphenol-A epoxy cured with 4,4′-DDS	—	—	127	0.62	—
Phenolic control (thermally cured resole)	—	—	—	0.16	—
Bisphenol-A	80/20	7 : 1	114	070	4539
	—	5 : 1	110	1.02	—
	65/35	3 : 1	127	0.85	1413
	50/50	2 : 1	151	0.64	643
Brominated bisphenol-A	65/35	5.8 : 1	130	0.74	3511
	50/50	3.1 : 1	148	0.84	1554
Disiloxane epoxy	80/20	7.2 : 1	96	0.62	4051
	65/35	3 : 1	87	0.77	1030

TABLE 7.10 Flame Retardance of Networks Prepared from Phenolic Novolac Crosslinked with Various Epoxies

Epoxy	Phenolic–Epoxy (wt/wt)	PHRR (KW/m^2)	Char Yield (%)	Smoke Toxicity^a ($\times 10^{-3}$)
Bisphenol-A epoxy cured with 4,4′-DDS	—	1230	5	44
Phenolic control (thermally cured resole)	—	116	63	—
Bisphenol-A	80/20	260	33	27
	65/35	360	29	34
	50/50	380	23	36
Brominated bisphenol-A	65/35	165	8	189
	50/50	158	9	175
Disiloxane epoxy	80/20	226	35	15
	65/35	325	24	27

^{a}CO/CO$_2$ release.

properties were investigated.[96] The presence of a distinct T_g when phenols were used to cure biphenyl-based diepoxide depended on the phenolic structure. Whereas a distinct T_g was evident in the phenolic-novolac-cured systems, no definite T_gs were observed when catechol novolac was used. The higher moduli shown by the catechol-novolac-cured networks were attributed to the orientation of mesogenic biphenyl groups which suppressed micro-Brownian chain motions.

Network properties and microscopic structures of various epoxy resins cross-linked by phenolic novolacs were investigated by Suzuki et al.[97] Positron anni-hilation spectroscopy (PAS) was utilized to characterize intermolecular spacing of networks and the results were compared to bulk polymer properties. The life-times (τ_3) and intensities (I_3) of the active species (positronium ions) correspond to volume and number of "holes" which constitute the free volume in the network. Networks cured with flexible epoxies had more holes throughout the temperature range, and the space increased with temperature increases. Glass transition tem-peratures and thermal expansion coefficients (α) were calculated from plots of τ_3 versus temperature. The T_gs and thermal expansion coefficients obtained from PAS were lower than those obtained from thermomechanical analysis. These dif-ferences were attributed to micro-Brownian motions determined by PAS versus macroscopic polymer properties determined by thermomechanical analysis.

7.6 BENZOXAZINES

Benzoxazines are heterocyclic compounds obtained from reaction of phenols, primary amines, and formaldehyde.[98,99] As described previously, they are key reaction intermediates in the HMTA–novolac cure reaction.[40,43] Crosslinking benzoxazine monomers at high temperatures gives rise to void-free networks with high T_gs, excellent heat resistance, good flame retardance, and low smoke toxicity.[100] As in HMTA-cured novolac networks, further structural rearrange-ment may occur at higher temperatures.

A difunctional bisphenol-A-based benzoxazine has been synthesized and char-acterized by GPC and ^{1}H NMR (Fig. 7.39). A small of amount of dimers and oligomers also formed. Thermal crosslinking of bisphenol-A benzoxazine con-taining dimers and oligomers resulted in networks with relatively high T_gs. Dynamic mechanical analysis of the network showed a peak of tan δ at approxi-mately 185°C.

The kinetics of bisphenol-A benzoxazine crosslinking reactions was studied using DSC.[100] The activation energy, estimated from plots of conversion as a function of time for different isothermal cure temperatures, was between 102 and 116 kJ/mole. Phenolic compounds with free ortho positions were suggested to initiate the benzoxazine reaction (Fig. 7.40).[101] Fast reactions between benzox-azines and free ortho phenolic positions, which formed hydroxybenzylamines, were facilitated by hydrogen bonding between the phenol hydroxyl and benzox-azine oxygen (as shown in Fig. 7.19). Subsequent thermal decompositions of these less stable hydroxybenzylamines led to more rapid thermal crosslinking (as described for the HMTA–novolac cure).

The reaction of bisphenol-A benzoxazine under strong and weak acidic con-ditions was also investigated.[102] The proposed mechanism for the benzoxazine ring-opening reaction in the presence of a weak acid involves an initial tau-tomerization between the benzoxazine ring and chain forms. In an electrophilic

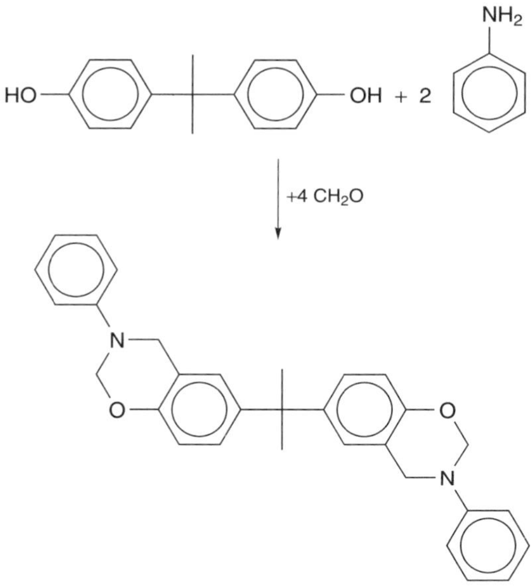

Figure 7.39   Synthesis of bisphenol-A-based benzoxazines.

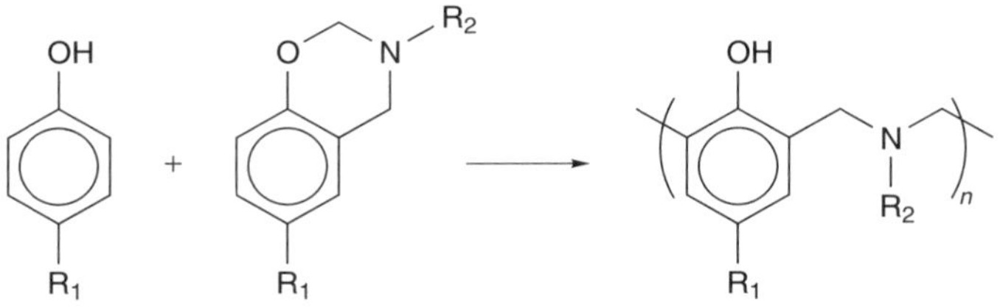

Figure 7.40   Reaction of benzoxazines with free ortho positions on phenolic compounds.

aromatic substitution reaction between a phenolic ring position and the chain tautomer, an iminium ion was suggested to follow. Strongly acidic conditions, high temperatures, and the presence of water lead to various side reactions, including benzoxazine hydrolysis in a reverse Mannich reaction. Side reactions could also terminate the reaction or lead to crosslinking.

The oxazine ring in benzoxazine assumes a distorted semichair conformation.[103] The ring strain and the strong basicity of the nitrogen and oxygen allow benzoxazines to undergo cationic ring-opening reactions. A number of catalysts and/or initiators such as PCl_5, PCl_3, $POCl_3$, $TiCl_4$, $AlCl_3$, and MeOTf are effective in promoting benzoxazine polymerization at moderate temperatures (20–50°C).[104] Dynamic DSC studies revealed multiple exotherms in polymerization of benzoxazine, indicating a complex reaction mechanism.

7.7 PHENOLIC CYANATE RESINS

Novolac hydroxyl groups reacted with cyanogen bromide under basic conditions to produce cyanate ester resins (Fig. 7.41).[105,106] Cyanate esters can thermally crosslink to form void-free networks, wherein at least some triazine rings form. The resultant networks possess high T_gs, high char yields at 900°C, and high decomposition temperatures.[105]

Novolac resins containing cardanol moieties have also been converted to cyanate ester resins.[107] The thermal stability and char yields, however, were reduced when cardanol was incorporated into the networks.

7.8 THERMAL AND THERMO-OXIDATIVE DEGRADATION

Phenolic networks are well known for their excellent thermal and thermo-oxidative stabilities. The mechanisms for high-temperature phenolic degradation include dehydration, thermal crosslinking, and oxidation, which eventually lead to char.

Thermal degradation below 300°C in inert atmospheres produces only small amounts of gaseous products. These are mostly unreacted monomers or water, which are by-products eliminated from condensation reactions between hydroxymethyl groups and reactive ortho or para positions on phenolic rings. A small

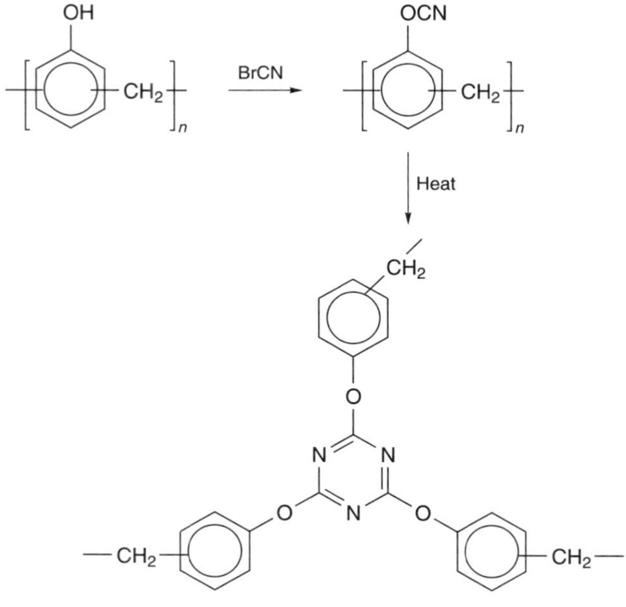

Figure 7.41 Synthesis of phenolic triazine resins.

amount of oxidation may occur in air as some carbonyl peaks have been observed using ^{13}C NMR.[108]

Degradation in inert atmospheres between 300 and 600°C results in porous materials. Little shrinkage has been observed in this temperature range. Water, carbon monoxide, carbon dioxide, formaldehyde, methane, phenol, cresols, and xylenols are released. According to various thermogravimetric analyses, the weight loss rate reaches a maximum during this temperature range. The elimination of water at this stage may also be caused by the phenolic hydroxyl condensations which give rise to biphenyl ether linkages (Fig. 7.42).

Morterra and Low[109,110] proposed that thermal crosslinking may occur between 300°C and 500°C where phenolic hydroxyl groups react with methylene linkages to eliminate water (Fig. 7.43). Evidence for this mechanism is provided by IR spectra which show decreased OH stretches and bending absorptions as well as increased complexity of the aliphatic CH stretch patterns in this temperature range.

At elevated temperatures, methylene carbons cleave from aromatic rings to form radicals (Fig. 7.44). Further fragmentation decomposes xylenol to cresols and methane (Fig. 7.44*a*). Alternatively, auto-oxidation occurs (Fig. 7.44*b*). Aldehydes and ketones are intermediates before decarboxylation or decarbonylation takes place to generate cresols and carbon dioxide. These oxidative reactions are possible even in inert atmospheres due to the presence of hydroxyl radicals and water.[5]

Oxidative degradation begins at lower temperatures in air (<300°C) and oxidation occurs most readily on benzylic methylene carbons since they are the most

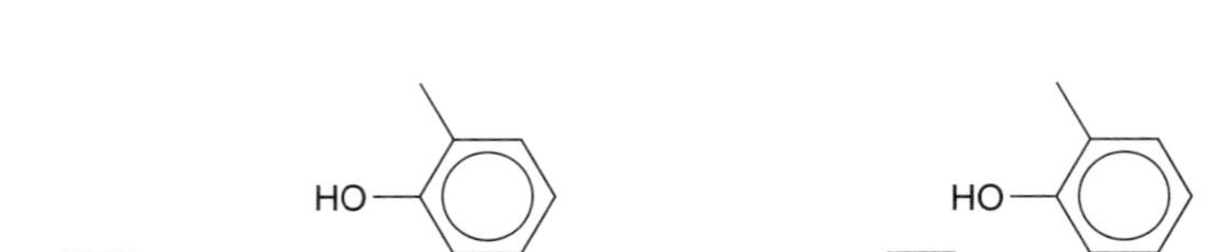

Figure 7.42   Dehydration of hydroxyl groups.

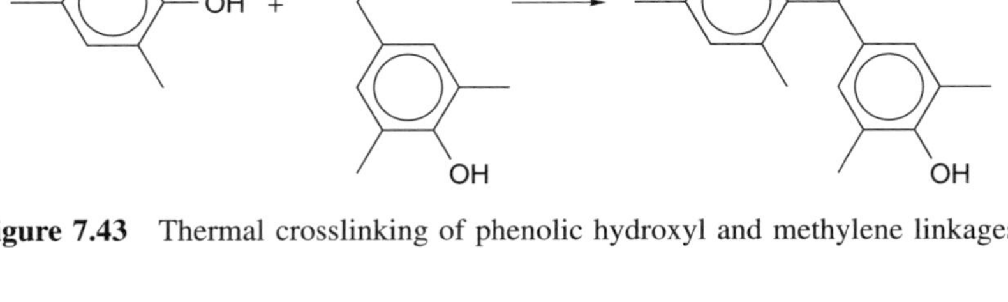

Figure 7.43   Thermal crosslinking of phenolic hydroxyl and methylene linkages.

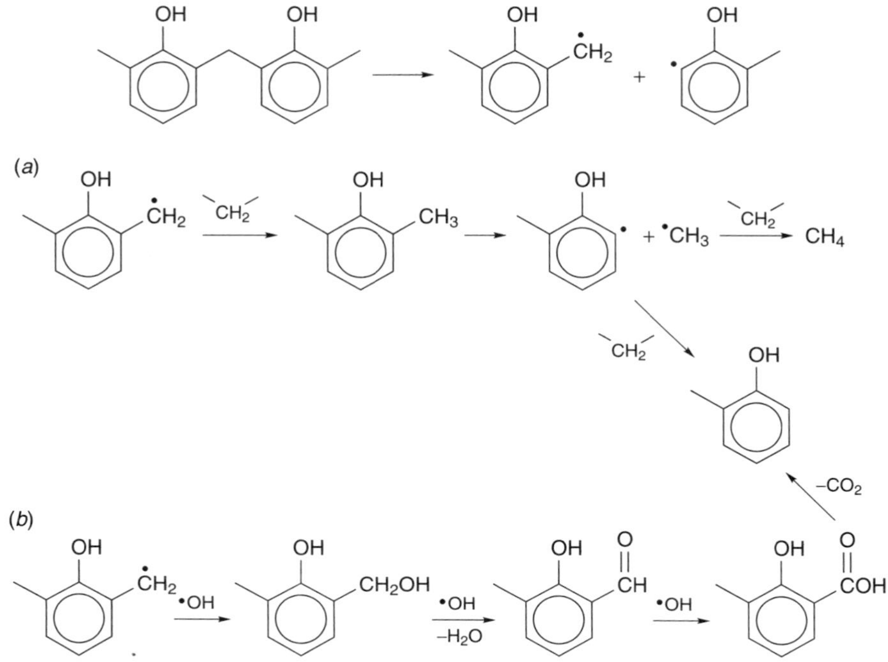

Figure 7.44   Thermal bond rupture: (*a*) fragmentation reaction. (*b*) oxidation degradation.

vulnerable sites. This leads to dihydroxybenzohydrole and dihydroxybenzophenone derivatives (Fig. 7.45). Dihydroxybenzophenone may cleave and further oxidize to carboxylic acid before decomposing to form cresols, xylenols, CO, and CO_2.

Hydroxyl elimination is necessary for the formation of benzaldehyde and benzoic acid derivatives and, ultimately, benzene and toluene (Fig. 7.46).[2] It is proposed that a cleavage between the hydroxyl group and aromatic ring leads to benzenoid species which undergo further cleavage coupled with oxidation to give various decomposition products.

Oxidative branching and crosslinking are the prevalent degradation pathways in air (Fig. 7.47). Phenoxy radicals are formed via hydrogen abstraction. These relatively stable intermediates can couple with each other and, depending on carbon–carbon or carbon–oxygen coupling, form ether linkages or ketones. Diphenolquinones derived from carbon–carbon dimerization further oxidize.[2]

Upon further heating above 600°C, the density increases as shrinkage occurs at a high rate. High-temperature degradation monitored via gas chromatography indicated that the formation of carbon char parallels carbon monoxide evolution. Along with the same by-products released at lower temperatures, pyrolysis gas chromatography/mass spectrometry (GC/MS) identified the formation of other

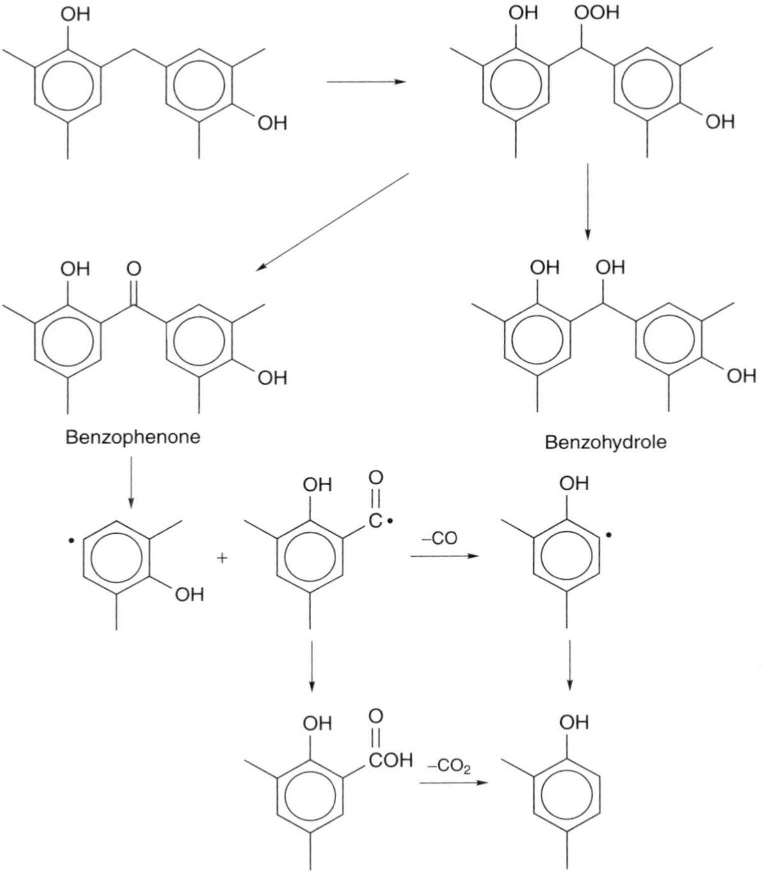

Figure 7.45 Oxidation degradation on methylene carbon.

low-volatility compounds, including naphthalene, methylnaphthalenes, biphenyl, dibenzofuran, fluorene, phenanthrene, and anthracene. These products may result from condensation of hydroxyl groups of adjacent ortho–ortho-linked phenolic rings[111] (Fig. 7.48).

The reaction sequence for char formation was also proposed to occur through the initial formation of quinone-type linkages, which lead to polycyclic products[5] (Fig. 7.49). The quinone functionalities were confirmed through IR studies. NMR spectra showed that at 400°C almost all methylene linkages and carbonyl groups shift or disappear. Mostly aromatic species are present. This is consistent with the proposed char-forming mechanism but is also compatible with the formation of biphenyls through direct elimination of CO.[108]

Conley showed that the primary degradation route for resole networks is oxidation regardless of the atmosphere.[6] By contrast, Morterra and Low found that

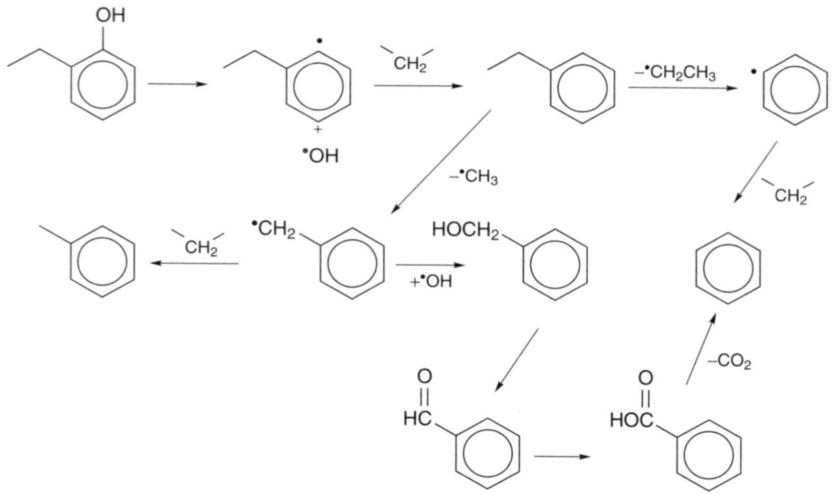

Figure 7.46   Formation of benzenoid species.

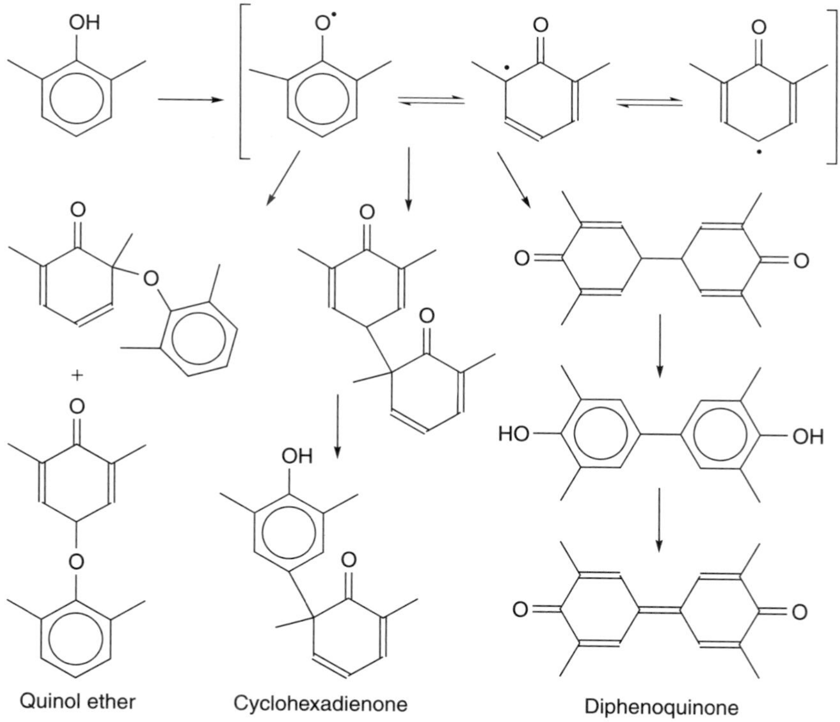

Quinol ether   Cyclohexadienone   Diphenoquinone

Figure 7.47   Decomposition via phenoxy radical pathways.

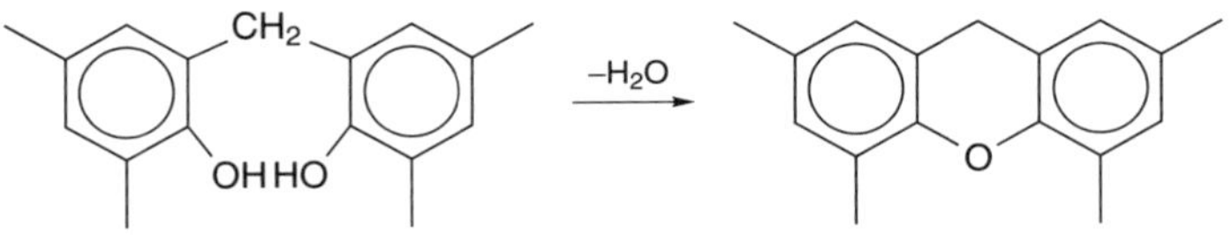

Figure 7.48   Condensation of ortho-hydroxyl groups.

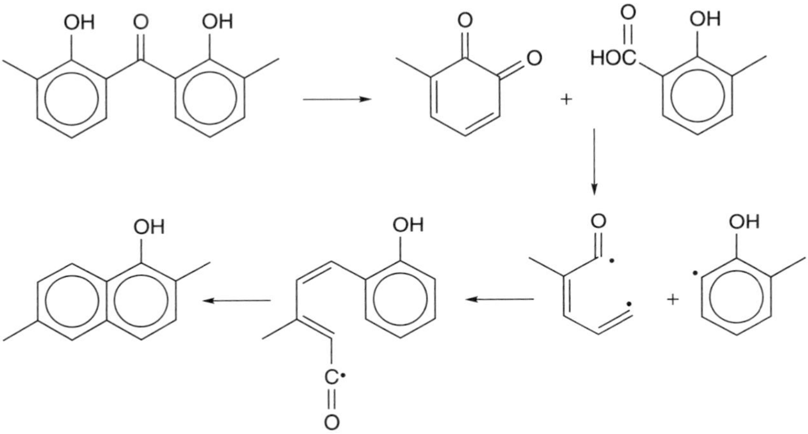

Figure 7.49   Char formation.

auto-oxidation was not the major degradation pathway for novolac resins.[109] The degradation is proposed to occur in two stages depending on the temperature. Fragmentation of the polymer chain, beginning at approximately 350°C, does not affect the polymer integrity. Above 500°C the network collapses as polyaromatic domains form.

Novolac network degradation mechanisms vary from those of resole networks due to differences in crosslinking methods. Nitrogen-containing linkages must also be considered when HMTA (or other crosslinking agent) was used to cure novolac networks. For example, tribenzylamines, formed in HMTA-cured novolac networks, decompose to cresols and azomethines (Fig. 7.50).

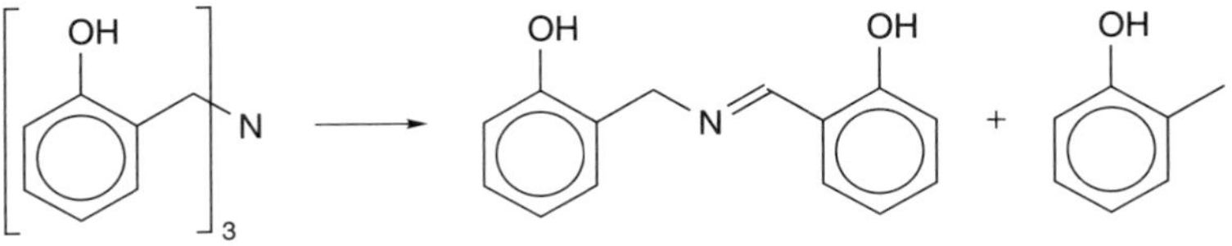

Figure 7.50   Decomposition of tribenzylamine.

Jha et al.[112] studied novolacs and networks which had been slowly heated in an inert atmosphere. Between 100 and 350°C, 1–2 wt % of water and phenol evolved. Above that temperature, small quantities of CO, CO_2, CH_4, and low-molecular-weight aromatics evolved, leaving behind 75–85% by weight of polymer. The degradation of these preexposed networks were then studied at temperatures up to ~700°C in both air and nitrogen. The rate of weight loss in air suggested that degradation under these conditions is a multistage process. Evidence suggested that this multistage thermo-oxidative process was due to two types of linking units present, which pyrolyzed at 375–400 and 500–550°C, respectively. By contrast, the degradation process in nitrogen was a single-stage process. Both processes formed a char representing 90–95% by weight of the preexposed materials. As expected, lower weight loss occurred with increased levels of the HMTA crosslinking reagent introduced into the network.

Oxygen indices of a series of phenolic networks (i.e., heat-cured resoles and novolacs cured with formaldehyde, trioxane, or terephthaloyl chloride) were investigated.[113] It has been previously shown that oxygen indices (OIs) correlate linearly with char yields at 800°C in nitrogen except for cases involving gas-phase retardants such as halogenated aromatic materials. The meta- and para-cresol formaldehyde networks exhibited slightly lower OI values than unsubstituted phenolics, presumably due to the presence of flammable methyl substituents. The halogenated phenolics showed higher OI values but lower char yields since halogenated materials undergo a gas-phase retardation of burning and the heavy halogen substituents convert to gaseous products. The crosslinking agent also has an affect on OI. Networks with methylene linkages derived from formaldehyde crosslinking reagents gave rise to higher OI values than those with ester linkages derived from terephthaloyl chloride. The formaldehyde-cured novolacs also showed higher OI than trioxane-cured materials since less stable ether linkages were formed in the trioxane crosslinked systems.

The effects of novolac structures and levels of crosslink densities on thermal stabilities were also investigated.[114] Three types of novolacs were prepared by reacting phenol with formaldehyde, meta-cresol with formaldehyde, and para-cresol with formaldehyde in a formaldehyde–phenol molar ratio of 0.95. Lightly crosslinked networks were obtained when the trifunctional phenol or *meta*-cresol was used. As expected, para-cresol novolac was linear. The thermal degradation behavior for the lightly crosslinked networks, the para-cresol novolac, as well as typical novolac resins were examined. Under inert atmosphere, thermogravimetric results revealed that the crosslinked samples had lower decomposition temperatures but the char yields were significantly higher. The greater stabilities exhibited at higher temperatures (500°C +) for the crosslinked materials were attributed to lesser fragmentation and, therefore, lower volatiles. In air, the crosslinked materials were less stable throughout the whole temperature range with complete degradation occurring 50°C below the low-molecular-weight resins. It was suggested that the crosslinked materials may be less susceptible to oxidation and, therefore, unable to form more thermally stable intermediates. Comparatively, the phenol–formaldehyde resin showed the highest thermal and thermal oxidative stability, followed by the meta-cresol–formaldehyde resin. The para-cresol–formaldehyde showed the least stability. Since the reactivity of starting phenolic monomer was different, novolac oligomers prepared via the same approach resulted in different structures. The results obtained in this study therefore could not be compared quantitatively.

A low-molecular-weight para-cresol novolac resin ($M_n \sim 560$) showed a high crystalline ratio according to X-ray diffraction.[115] No weight loss was observed below $400°C$. However, the morphology changed to a semicrystalline state after repeat heating–cooling cycles.

REFERENCES

1. *Society of Plastic Industries Facts and Figures*, SPI, Washington, DC, 1994.

2. A. Knop and L. A. Pilato, *Phenolic Resins — Chemistry, Applications and Performance*, Springer-Verlag, Berlin, 1985.

3. A. Knop and W. Scheib, *Chemistry and Application of Phenolic Resins*, Springer-Verlag, New York, 1979.

4. S. R. Sandler and W. Karo, *Polymer Synthesis*, 2nd ed., Vol. 2, Academic Boston, 1992.

5. P. W. Kopf, in *Encyclopedia of Chemical Technology*, 4th ed., Vol. 18, J. I. Kroschulitz (Ed.), Wiley, New York, 1996, pp. 603–644.

6. R. T. Conley, *Thermal Stability of Polymers*, Marcel Dekker, New York, 1970, pp. 459–496.

7. K. Hultzsch, *Am. Chem. Soc. Div. Organic Coating Plastic Chem.*, **26**(1), 121–128 (1966).

8. A. J. Norton, U.S. Patent 2,385,370, 1947; P. H. Phodes, U.S. Patent 2,385,372, 1946.

9. H Diehm and A. Hit, "Formaldehyde," in *Ullmanns Encyclopadie der Techn. Chem.*, 4th ed. Vol. 11, Verlag Chemie, Weinheim, 1976.

10. C. M. Chen and S. E. Chen, *Forest Prod. J.*, **38**(5), 49–52 (1988).

11. A. J. Rojas and R. J. J. Williams, *J. Appl. Polym. Sci.*, **23**, 2083–2088 (1979).

12. N. J. L. Megson, *Chem.-Ztg.*, **96**(1–2), 15–19 (1972).

13. H. L. Bender, A. G. Farnharn, and J. W. Guyer, U.S. Patents 2,464,207 and 2,475, 587, 1949.

14. W. Aubertson, U.S. Patent 4,113,700 (to Monsanto Co.), September 12, 1978.

15. G. Casnati. A. Pochini, G. Sartori, and R. Ungaro, *Pure Appl. Chem.*, **55**(11), 1677–1688 (1983).

16. P. J. de Bruyn, A. S. L. Lim, M. G. Looney, and D. H. Solomon, *Tetrahedron Lett.*, **35**(26), 4627–4630 (1994).

17. S. Lin, M. Rutnakornpituk, C. S. Tyberg, J. S. Riffle, and U. Sorathia, *Bridging the Centuries with SAMPE's Materials and Process Technology*, **45**(2), 1730–1743 (2000).

18. M. M. Sprung, *J. Appl. Polym. Sci.*, **63**(2), 334–343 (1941).

19. S. Miloshev, P. Novakov, V. I. Dimitrov, and I. Gitsov, *Chemtronics*, **4**, 251–253 (1989).

20. L. E. Bogan, Jr., "Understanding the Novolac Synthesis Reaction," in *Frontiers of Polymers and Advanced Materials*, P. N. Prasad (Ed.), Plenum, New York, 1994, pp. 311–318.

21. C. N. Cascaval, D. Rosu, and F. Mustata, *Eur. Polym. J.*, **30**(3), 329–333 (1994).

22. S. Miloshev, P. Novakov, V. I. Dimitrov, and I. Gitsov, *Polymer*, **32**(16), 3067–3070 (1991).

23. T. Yoshikawa, K. Kimura, and S. Fujimura, *Jo. Appl. Polym. Sci.*, **15**, 2513–2520 (1971).

24. T. A. Yamagishi, M. Nomoto, S. Ito, S. Ishida, and Y. Nakamoto, *Polym. Bull.*, **32**, 501–507 (1994).

25. L. E. Bogan, Jr., *Macromolecules*, **24**, 4807–4812 (1991).

26. M. G. Kim, W. L. Nieh, T. Sellers, Jr., W. W. Wilson, and J. W. Mays, *Ind. Eng. Chem. Res.*, **31**(3), 973–979 (1992).

27. F. L. Tobiason, C. Chandler, and F. E. Schwarz, *Macromolecules*, **5**(3), 321–325 (1972).

28. H. Mandal and A. S. Hay, *Polymer*, **38**(26), 6267–6271 (1997).

29. T. R. Dargaville, F. N. Guerzoni, M. G. Looney, D. A. Shipp, D. H. Solomon, and X. Zhang, *J. Polym. Sci. Pt. A: Polym. Chem.*, **35**(8), 1399–1407 (1997).

30. R. A. Pethrick and B. Thomson, *Br. Polym. J.*, **18**(3), 171–180 (1986).

31. B. Ottenbourgs, P. Adriaensens, R. Carleer, D. Vanderzande, and J. Gelan, *Polymer*, **39**(22), 5293–5300 (1998).

32. F. Y. Wang, C. C. M. Ma, and H. D. Wu, *J. Appl. Polym. Sci.*, **74**, 2283–2289 (1999).

33. P. P. Chu, H. D. Wu, and C. T. Lee, *J. Polym. Sci. Pt .B: Polym. Phys.*, **36**(10), 1647–1655 (1998).

34. H. D. Wu, C. C. M. Ma, and P. P. Chu, *Polymer*, **38**(21), 5419–5429 (1997).

35. H. D. Wu, P. P. Chu, C. C. M. Ma, and F. C. Chang, *Macromolecules*, **32**(9), 3097–3105 (1999).

36. H. D. Wu, C. C. M. Ma, P. P. Chu, H. T. Tseng, and C. T. Lee, *Polymer*, **39**(13), 2856–2865 (1998).

37. C. C. M. Ma, H. D. Wu, and C. T. Lee, *J. Polym. Sci. Pt. B: Polym. Phys.*, **36**(10), 1721–1729 (1998).

38. Z. Katovic and M. Stefanic, *Ind. Eng. Chem. Product Res. Devel.*, **24**, 179–185 (1985).

39. C. Holland, W. Stark, and G. Hinrichsen, *Acta Polym.*, **46**, 64–67 (1995).

40. P. W. Kopf and E. R. Wagner, *J. Polym. Sci. Polym. Chem. Ed.*, **11**(5), 939–960 (1973).

41. S. A. Sojka, R. A. Wolfe, and G. D. Guenther, *Macromolecules*, **14**, 1539–1543 (1981).

42. Y. Ogata and A. Kawasaki, "Equilibrium Additions to Carbonyl Compounds," in *The Chemistry of the Carbonyl Group*, J. Zabicky (Ed.), Vol. 2, Interscience, London, 1970.

43. T. R. Dargaville, P. J. De Bruyn, A. S. C. Lim, M. G. Looney, A. C. Potter, and D. H. Solomon, *J. Polym. Sci. Pt. A*, **35**, 1389–1398 (1997).

44. X. Zhang, M. G. Looney, D. H. Solomon, and A. K. Whittaker, *Polymer*, **38**(23), 5835–5948 (1997).

45. X. Zhang, A. C. Potter, and D. H. Solomon, *Polymer*, **39**(2), 399–404 (1998).

46. X. Zhang and D. H. Solomon, *Polymer*, **39**(2), 405–412 (1998).

47. X. Zhang, A. C. Potter, and D. H. Solomon, *Polymer*, **39**(10), 1957–1966 (1998).

48. X. Zhang, A. C. Potter, and D. H. Solomon, *Polymer*, **39**(10), 1967–1975 (1998).

49. X. Zhang and D. H. Solomon, *Polymer*, **39**(24), 6153–6162 (1998).

50. A. S. C. Lim, D. H. Solomon, and X. Zhang, *J. Polym. Sci. Pt. A: Polym. Chem.*, **37**, 1347–1355 (1999).

51. K. Lenghaus, G. G. Qiao, and D. H. Solomon, *Polymer*, **41**, 1973–1979 (2000).

52. J. H. Freemann and C. W. Lewis, *J. Am. Chem. Soc.*, **76**(8), 2080–2087 (1954).

53. A. A. Zsavitsas and R. D. Beaulieu, *J. Polym. Sci. AI*, **6**, 2451 (1969).

54. K. C. Eapen and L. M. Yeddanapalli, *Makromolek. Chem.*, **119**, 4 (1968).

55. M. F. Grenier-Loustalot, S. Larroque, P. Grenier, J. Leca, and K. Bedel, *Polymer*, **35**(14), 3046–3054 (1994).

56. M. F. Grenier-Loustalot, S. Larroque, P. Grenier, and D. Bedel, *Polymer*, **37**(6), 939–953 (1996).

57. B. Mechin, D. Hanton, J. Le Goff, and J. P. Tanneur, *Eur. Polym. J.*, **22**(2), 115–124 (1986).

58. M. F. Grenier-Loustalot, S. Larroque, P. Grenier, and D. Bedel, *Polymer*, **37**(6), 955–964 (1996).

59. L. Prokai, *J. Chromatogr.*, **333**(1), 91–98 (1985).

60. R. A. Haupt, and T. Sellers, *Ind. Eng. Chem. Res.*, **33**(3), 693–697 (1994).

61. M. F. Grenier-Loustalot, S. Larroque, D. Grande, P. Grenier, and D. Bedel, *Polymer*, **37**(8), 1363–1369 (1996).

62. G. Astarloa-Aierbe, J. M. Echeverria, A. Vazquez, and I. Mondragon, *Polymer*, **41**, 3311–3315 (2000).

63. B. D. Park and B. Riedl, *J. Appl. Polym. Sci.*, **77**(6), 1284–1293 (2000).

64. M. Grenier-Loustalot, S. Larroque, and P. Grenier, *Polymer*, **37**(4), 639–650 (1996).

65. T. Holopainen, L. Alvila, J. Rainio, and T. T. Pakkanen, *J. Appl. Polym. Sci.*, **69**(11), 2175–2185 (1998).

66. G. Carotenuto and L. Nicolais, *J. Appl. Polym. Sci.*, **74**(11), 2703–2715 (1999).

67. T. Holopainen, L. Alvila, J. Rainio, and T. T. Pakkanen, *J. Appl. Polym. Sci.*, **66**(6), 1183–1193 (1997).

68. M. G. Kim, L. W. Amos, and E. E. Barnes, *Ind. Eng. Chem. Res.*, **29**(10), 2032–2037 (1990).

69. P. Luukko, L. Alvila, T. Holopainen, J. Rainio, and T. T. Pakkanen, *J. Appl. Polym. Sci.*, **69**, 1805–1812 (1998).

70. T. Sellers and M. L. Prewitt, *J. Chromatogr.*, **513**, 271–278 (1990).

71. G. Gobec, M. Dunky, T. Zich, and K. Lederer, *Ang. Makromolek. Chem.*, **251**, 171–179 (1997).

72. M. G. Kim, W. L. Nieh, T. Sellers, W. W. Wilson, and J. W. Mays, *Ind. Eng. Chem. Res.*, **31**(3), 973–979 (1992).

73. G. Astarloa-Aierbe, J. M. Echeverria, J. L. Egiburu, M. Ormaetxea, and I. Mondradon, *Polymer*, **39**(14), 3147–3153 (1998).

74. L. Prokai and W. J. Simonsick, *Macromolecules*, **25**(24), 6532–6539 (1992).

75. J. M. Kenny, G. Pisaniello, F. Farina, and S. Puzziello, *Thermochim. Acta*, **269**, 201–211 (1995).

76. W. W. Focke, M. S. Smit, A. T. Tolmay, L. S. Vandervalt, and W. L. Vanwyk, *Polym. Eng. Sci.*, **31**(23), 1665–1669 (1991).

77. M. G. Kim, W. L. S. Nich, and R. M. Meacham, *Ind. Eng. Chem. Res.*, **30**(4), 798–803 (1991).

78. R. A. Follensbee, J. A. Koutsky, A. W. Christiansen, G. E. Myers, and R. L. Geimer, *J. Appl. Polym. Sci.*, **47**(8), 1481–1496 (1993).

79. M. J. Carrott and G. Davidson, *Analyst*, **124**(7), 993–997 (1999).

80. Y. Yazaki, P. J. Collins, M. J. Reilly, S. D. Terrill, and T. Nikpour, *Holzforschung*, **48**(1), 41–48 (1994).

81. J. Wolfrum and G. W. Ehrenstein, *J. Appl. Polym. Sci.*, **74**, 3173–3185 (1999).

82. L. B. Manfredi, O. de la Osa, N. Galego Fernandez, and A. Vazquez, *Polymer*, **40**, 3867–3875 (1999).

83. B. Tomita and C. Hse, *Mokuzai Gakkaishi*, **39**(11), 1276–1284 (1993).

84. B. Tomita, M. Ohyama, A. Itoh, K. Doi, and C. Hse, *Mokuzai Gakkaishi*, **40**(2), 170–175 (1994).

85. R. W. Biernath and D. S. Soane, "Cure Kinetics of Epoxy Cresol Novolac Encapsulant for Microelectronic Packaging," in *Contemporary Topic in Polymer Science. Advances in New Material.* Vol. 7, J. S. Salamone and J. S. Riffle (Eds.), Plenum, New York, 1992, pp. 103–160.

86. W. A. Romanchick, J. E. Sohn, and J. F. Geibel "Synthesis, Morphology, and Thermal Stability of Elastomer-Modified Epoxy Resin," in *ACS Symposium Series 221 — Epoxy Resin Chemistry II*, R. S. Bauer (ed.), American Chemical Society, Washington DC, 1982, pp. 85–118.

87. D. Gagnebien, P. J. Madec, and E. Marechal, *Eur. Polym. J.*, **21**(3), 273–283 (1985).

88. W. G. Kim, J. Y. Lee, and K. Y. Park, *J. Polym. Sci. Pt. A: Polym. Chem.*, **31**, 633–639 (1993).

89. M. F. Sorokin and L. G. Shode, *Zhurnal Organicheskoi Khimii*, **2**(8), 1463–1468 (1966).

90. S. Han, H. G. Yoon, K. S. Suh, W. G. Kim, and T. J. Moon, *J. Polym. Sci. Pt. A: Polym. Chem.*, **37**, 713–720 (1999).

91. C. S. Chern and G. W. Poehlein, *Polym. Eng. Sci.*, **27**(11), 788–795 (1987).

92. X. W. Luo, Z. H. Ping, J. P. Ding, Y. D. Ding, and S. J. Li, *Pure Appl. Chem.*, **A34**(11), 2279–2291 (1997).

93. C. S. Tyberg, M. Sankarapandian, K. Bears, P. Shih, A. C. Loos, D. Dillard, J. E. McGrath, and J. S. Riffle, *Construct. Build. Mater.*, **13**, 343–353 (1999).

94. C. S. Tyberg, K. Bergeron, M. Sankarapandian, P. Shih, A. C. Loos, D. A. Dillard, J. E. McGrath, and J. S. Riffle, *Polymer*, **41**(13), 5053–5062 (2000).

95. C. S. Tyberg, P. Shih, K N. E. Verghese, A. C. Loos, J. J. Lesko, and J. S. Riffle, *Polymer*, **41**, 9033–9048 (2000).

96. M. Ochi, N. Tsuyuno, K. Sakaga, Y. Nakanishi, and Y. Murata, *J. Appl. Polym. Sci.*, **56**, 1161–1167 (1995).

97. T. Suzuki, Y. Oki, M. Numajiri, T. Miura, K. Kondo, Y. Shiomi, and Y. Ito, *J. Appl. Polym. Sci.*, **49**, 1921–1929 (1993).

98. W. J. Burke, E. L. M. Glennie, and C. Weatherbee, *J. Organic Chem.*, **29**, 909 (1964).

99. X. Ning and H. Ishida, *J. Polym. Sci. Pt. A: Polym. Chem.*, **32**, 1121–1129 (1994).

100. H. Ishida and Y. Rodriguez, *Polymer*, **36**(16), 3151–3158 (1995).

101. G. Riess, J. M. Schwob, G. Guth, M. Roche, and B. Lande, "Ring Opening Polymerization of Benzoxazines — A New Route to Phenolic Resins," in *Advances in Polymer Synthesis*, B. M. Culbertson and J. E. McGrath (Eds.), Plenum, New York, 1985, pp 27–50.

102. J. Dunkers and H. Ishida, *J. Polym. Sci. Pt. A: Polym. Chem.*, **37**(13), 1913–1921 (1999).

103. H. Ishida and D. J. Allen, *J. Polym. Sci. Polym. Phys. Ed.*, **34**(6), 1019–1031 (1996).

104. Y. X. Wang and H. Ishida, *Polymer*, **40**(16), 4563–4570 (1999).

105. S. Das and D. C. Prevorsek, U.S. Patent 4,831,086 (to Allied-Signal, Inc.), May 16, 1989.

106. S. Das, *Abstr. Pap. Am. Chem. Soc. — PMSE*, **203**, 259 (1992).

107. C. P. R. Nair, R. L. Bindu, and V. C. Joseph, *J. Polym. Sci. Pt. A: Polym. Chem.*, **33** (4), 621–627 (1995).

108. C. A. Fyfe, M. S. McKinnon, A. Rudin, and W. J. Tchir, *Macromolecules*, **16**, 1216–1219 (1983).

109. C. Morterra and M. I. D. Low, *Carbon*, **23**(5), 525–530 (1985).

110. C. Morterra and M. I. D. Low, *Langmuir*, **1**, 320–326 (1985).

111. J. Hetper and M. Sobera, *J. Chromatogr. A*, **833**, 277–281 (1999).

112. V. Jha, A. Banthia, and A. Paul, *J. Thermal Anal.*, **35**, 1229–1235 (1989).

113. Y. Zaks, J. Lo, D. Raucher, and E.M. Pearce, *J. Appl. Polym. Sci.*, **27**, 913–930 (1982).

114. L. Costa, L. Rossi di Montelera, G. Camino, E. D. Weil, and E. M. Pearce, *Polym. Degrad. Stabil.*, **56**, 27–35 (1997).

115. A. Hamou, C. Devallencourt, F. Burel, J. M. Saiter, and M. Belbachir, *J. Thermal Anal.*, **52**, 697–703 (1998).

7.9 APPENDIX

The procedures which follow were adapted from *Polymer Synthesis*, 2nd ed., Vol. II, by S. R. Sandler and W. Karo, Academic, New York, 1992.

7.9.1 Preparations of Phenol–Formaldehyde Resole Resins

7.9.1.1 Method A

Eight hundred grams (8.5 mol) of phenol, 80.0 g water, 940.5 g 37 wt % aqueous formaldehyde (11.6 mol), and 40 g of barium hydroxide octahydrate were charged to a resin kettle. The reaction mixture was maintained at 70°C for 2 h with stirring; then sufficient oxalic acid was added to bring the pH to 6–7. The resinous material was partially condensed by removing water at 30–50 mm Hg at a temperature no higher that 70°C. Samples (1–2 mL) were withdrawn every 15 min to check the extent of condensation. The endpoint was taken when a sample of resin placed on a hot plate at 160°C gelled in less than 10 s or when the cooled resin was brittle and "nontacky" at room temperature. The resin at this point was termed an A-stage resole. Further heating converts this into a B-stage and finally to a C-stage resin.

7.9.1.2 Method B

Forty-seven grams (0.5 mol) of phenol, 80 mL of 37 wt % aqueous formaldehyde (1.0 mol), and 100 mol of 4 N NaOH were charged to a flask equipped with a reflux condenser and mechanical stirrer. The reaction mixture was stirred at room temperature for 16 h, then heated on a steam bath for 1 h. The mixture was cooled and the pH adjusted to 7.0. The aqueous layer was decanted from the viscous brown liquid product, the wet organic phase was taken up in 500 mL of acetone and dried over anhydrous $MgSO_4$, then over molecular sieves. The dried acetone product solution was filtered and evaporated to yield a water-free light brown syrup.

7.9.2 Preparation of Phenol–Formaldehyde Novolac Resin

Twenty-four hundred grams (25.6 mol) of phenol, 1645 g of 37 wt % formaldehyde (20.3 mol), and 30 g (0.33 mol) of oxalic acid were charged to a 5-L, three-necked resin kettle. The mixture was stirred and refluxed until the distillate was free of formaldehyde (1–3 h); then water was distilled from the mixture until the resin temperature reached 154°C. The viscosity of the resin at this point at 150°C was such that 105 s was required for flow in an inclined plate test. The pressure was slowly reduced while a slow current of nitrogen was bubbled through the resin, and the mixture was heated to 175°C at 6 mm Hg. Approximately 6 wt % phenol was recovered; then the resin was poured into an aluminum dish and cooled. The resin had a melt viscosity of 510 s at 150°C.

8 Nontraditional Step-Growth Polymerization: ADMET

A. Cameron Church, Jason A. Smith, James H. Pawlow, and Kenneth B. Wagener

Department of Chemistry, George and Josephine Butler Polymer Research Laboratory, University of Florida Gainesville, Florida 32611

8.1 INTRODUCTION

Olefin metathesis, an expression coined by Calderon in 1967,[1] has been accurately described in Ivin and Mol's seminal text *Olefin Metathesis and Metathesis Polymerization* as "the (apparent) interchange of carbon atoms between a pair of double bonds" (ref. 2, p. 1). This remarkable conversion can be divided into three types of reactions, as illustrated in Fig. 8.1. These reactions have been used extensively in the synthesis of a broad range of both macromolecules and small molecules[3]; this chapter focuses on acyclic diene metathesis (ADMET) polymerization as a versatile route for the production of a wide range of functionalized polymers.

8.1.1 Historical Perspective

The history of metathesis chemistry dates back over 50 years[4] to the development of Ziegler–Natta-type transition metal catalysts for olefin polymerization in the late 1940s.[5] While olefin exchange reactions were observed by several research groups using such catalysts, the exact nature of the active catalyst species and mechanism were not understood at the time. Most early metathesis catalyst systems, similar to either Phillips-type (supported metal oxides at high temperatures) or Ziegler–Natta-type (metal halide–alkyl aluminum mixtures at low temperatures) catalysts for the polymerization of α-olefins and ethylene involves the in situ formation of the catalyst. Such nebulous, multicomponent systems are complicated and, as a result, difficult to study and understand. Only after detailed labeling studies by Natta and co-workers was it conclusively proven that, in metathesis, the olefin double bond is cleaved and reassembled by the active

Synthetic Methods in Step-Growth Polymers.   Edited by Martin E. Rogers and Timothy E. Long
© 2003 John Wiley & Sons, Inc.   ISBN: 0-471-38769-X

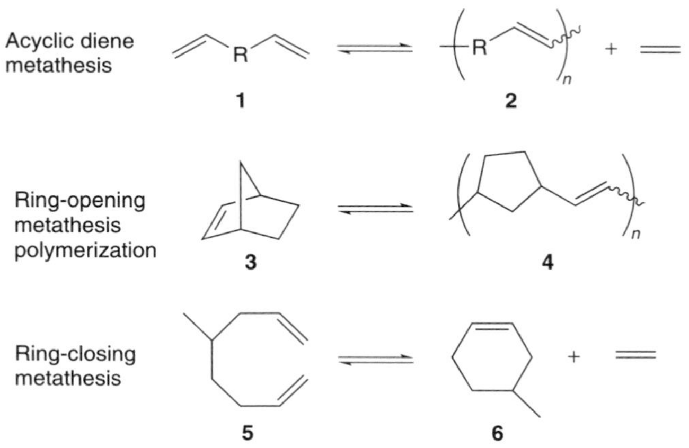

Figure 8.1   Three types of metathesis reactions.

catalyst.[6] The mechanism of such cleavage remained unknown until Herrison and Chauvin solved it in 1970.[1e] They proposed that the metathesis reaction involves a metallacyclobutane ring intermediate, which is formed by the interaction of a metal carbene with an olefin. This is the key step in metathesis chemistry and is common to all metathesis reactions. Although this intermediate was questioned initially, the scientific community now universally accepts its presence in the mechanism.

8.1.2 Metathesis Catalysts

The discovery and preparation of organometallic carbene complexes in 1964 not only provided a vital clue for the determination of the metathesis mechanism, but also has subsequently led to the deliberate design of improved, later generation catalysts. We present here a brief overview of several catalytic systems used for metathesis polymerization, in particular ADMET polymerization. The most important catalysts (determined by usage and numbers of publications) are based on 10 elements in the transition metal series of the periodic table: Mo, W, Ru, Re, V, Zr, Cr, Co, Rh, and Tc. It is uncommon to observe metathesis activity with non–transition metal elements. It was not until 1963 that Natta et al.[6] were able to prepare high-molecular-weight polymers via ring opening metathesis polymerization (ROMP) of the strained cyclic olefin cyclopentene (**7**), producing poly(1-pentenylene) (**8**) (Fig. 8.2).

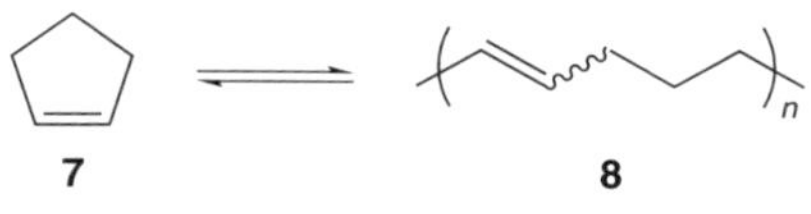

Figure 8.2   Natta et al.'s ROMP polymerization of cyclopentene.

Even more challenging was the synthesis of step-growth polymers using acyclic dienes, since high conversions (>99%) are mandatory for high-molecular-weight polymers to be produced.[3] Initially, attempts to prepare such polymers failed, and research into this area became quite limited. It was not until 1990, following the discovery of active, acid-free, single-site tungsten-based alkylidenes by Schrock's group,[7] that Wagener and co-workers were able to successfully perform the ADMET reaction using 1,9-decadiene (**9**) as the monomer, producing poly(1-octenylene) (**10**) (Fig. 8.3).[8] We now know that three catalyst systems are viable in ADMET polymerization: "classical" catalysts (**11**),[2] Schrock-type alkylidenes (**14**),[7] and Grubbs-type carbenes (**12, 13**).[9,10] Each catalyst type has advantages and limitations, and the choice of catalyst is dependent on the nature of the diene to be polymerized (e.g., monomer functionalities present and physical properties of the polymer formed). Classical catalysts (**11**) are, in reality, ill-defined catalytic mixtures containing two or more components in which the active species (alkylidene) is generated in situ.[2] Typically, a classical catalyst is composed of an early transition metal halide such as WCl_6 reacted with an alkylating agent or activator, such as Bu_4Sn or $EtAlCl_2$. Such catalysts are rather effective under certain conditions, such as high temperatures (100°C), and proceed with reasonable rates. A practical disadvantage to such catalytic mixtures is their ill-defined nature; the active species often cannot be isolated or in some cases identified even by spectroscopic means. Consequently, catalyst concentration cannot be easily quantified and monitored in the course of a reaction.

These limitations were overcome with the introduction of the well-defined, single-component tungsten and molybdenum (**14**) alkylidenes in 1990. (Fig. 8.4).[7] Schrock's discovery revolutionized the metathesis field and vastly increased the utility of this reaction. The Schrock alkylidenes are particularly reactive species, have no side reactions, and are quite effective as polymerization catalysts for both ROMP and ADMET. Due to the oxophilicity of molybdenum, these alkylidenes are moisture and air sensitive, so all reactions using these catalysts must be performed under anaerobic conditions, requiring Schlenk and/or glovebox techniques.

Grubbs-type carbenes (**12, 13**) based on the late transition metal ruthenium are more tolerant to air, moisture, and most functional groups.[9] The original Grubbs carbene (**12**) is slower compared to Schrock's [Mo] catalyst (**14**). A significant improvement to [Ru] carbene catalysts was discovered in 1999 by Herrmann, with the introduction of the *N*-heterocyclic carbene ligand.[11] Similar catalysts using this type of ligand, such as **13**,[10] have enabled the [Ru] catalysts to match or

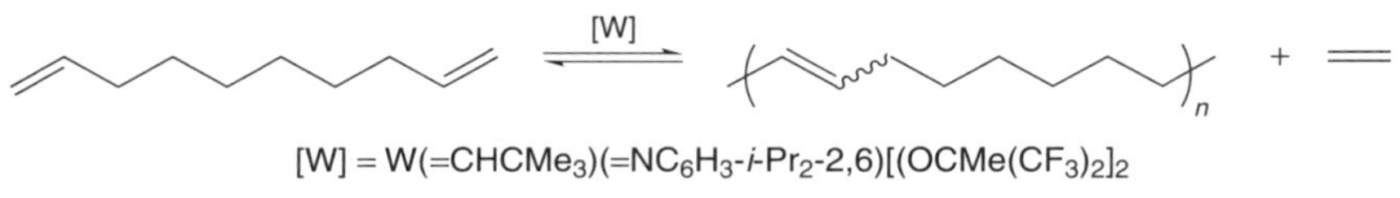

$[W] = W(=CHCMe_3)(=NC_6H_3\text{-}i\text{-}Pr_2\text{-}2,6)[(OCMe(CF_3)_2]_2$

9   **10**

Figure 8.3 Wagener's ADMET of 1,9-decadiene using Schrock's [*W*] catalyst.

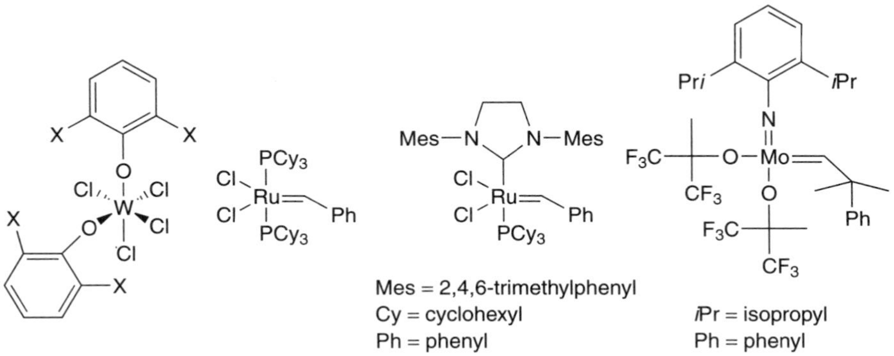

Classical	First-generation Grubbs benzylidene	Second-generation Grubbs benzylidene	Schrock alkylidene
11	**12**	**13**	**14**

Figure 8.4 Metathesis catalyst systems.

exceed the reaction rate of the Schrock alkylidenes, at least in several ADMET and ROMP reactions. This enhancement, when combined with the functional group tolerance of these carbenes, has led to its increased use in organic synthesis, particularly in ring-closing metathesis (RCM).[10]

8.1.3 Requirements for ADMET Polymerization

ADMET of α,ω-dienes has been a focus of research in the Wagener laboratories for many years now, where we have studied this chemistry to explore its viability in synthesizing polymers possessing both precisely designed microstructures as well as a variety of functionalities. The requirements for this reaction, such as steric and electronic factors, functionalities allowed, appropriate choice of catalyst, and necessary length or structure of the diene, have been examined.[3,12–14] A detailed discussion will be presented later in this chapter with a brief synopsis of general rules for successful ADMET polymerization presented here.

Like most polycondensation reactions, ADMET is most effective under bulk (neat) reaction conditions with the monomer itself acting as both reactant and solvent. Doing so maximizes the monomer concentration and promotes a shift in the reaction equilibrium toward unsaturated polymer formation. Similar to polycondensation reactions, ADMET is typically performed under reduced pressure ($<10^{-2}$ mmHg) to remove the ethylene by-product. This irreversibly shifts the monomer/polymer equilibrium toward polymer formation, promoting faster monomer conversion and raising the molecular weight of the product. ADMET is a true step-growth polymerization reaction exhibiting typical kinetic behavior and molecular weight distribution ($M_w/M_n \cong 2.0$) commonly observed for this class of polymerization.[12] In order to suppress monomer cyclization via the RCM route (cyclization is common in polycondensation chemistry), the α,ω-diene usually consists of a chain of 10 atoms or greater to favor ADMET polymerization.

Shorter chain dienes have an increased propensity to form stable five-, six-, and seven-membered rings. This thermodynamically controlled phenomenon is known as the Thorpe–Ingold effect.[15] Since ADMET polymerization is performed over extended time periods under equilibrium conditions, it is ultimately thermodynamics rather than kinetics that determine the choice between a selected diene monomer undergoing either polycondensation or cyclization.

8.1.4 Applications

ADMET is quite possibly the most flexible transition-metal-catalyzed polymerization route known to date. With the introduction of new, functionality-tolerant robust catalysts, the primary limitation of this chemistry involves the synthesis and cost of the diene monomer that is used. ADMET gives the chemist a powerful tool for the synthesis of polymers not easily accessible via other means, and in this chapter, we designate the key elements of ADMET. We detail the synthetic techniques required to perform this reaction and discuss the wide range of properties observed from the variety of polymers that can be synthesized. For example, branched and functionalized polymers produced by this route provide excellent models (after quantitative hydrogenation) for the study of many large-volume commercial copolymers, and the synthesis of reactive carbosilane polymers provides a flexible route to solvent-resistant elastomers with variable properties. Telechelic oligomers can also be made which offer an excellent means for polymer modification or incorporation into block copolymers. All of these examples illustrate the versatility of ADMET.

8.2 OVERVIEW OF CHEMISTRY AND ANALYTICAL TECHNIQUES

8.2.1 Chemistry and Catalysis

8.2.1.1 Brief History

As mentioned earlier, the chemistry and mechanism surrounding the olefin metathesis reaction (Fig. 8.1) have been a point of intense interest since its discovery over 50 years ago.[1,2] In the late 1980s, research by the Wagener group demonstrated approaches to meet the stringent requirements of step polymerization.[13] Defining the opportunity has been the direct result of revolutionary developments of single-site metathesis catalysts described earlier, and subsequently a series of high-molecular-weight polymers possessing various functionalities have been made by ADMET (Fig. 8.5).[14]

At this point it is appropriate to discuss the mechanism for ADMET, because ADMET polymerization is more involved than its chain polymerization counterpart — ROMP. Figure 8.6 illustrates the accepted mechanistic pathway which leads to productive metathesis polymerization, as first described by Wagener et al.[14a] A general model reaction between an α,ω-diene with a metal alkylidene

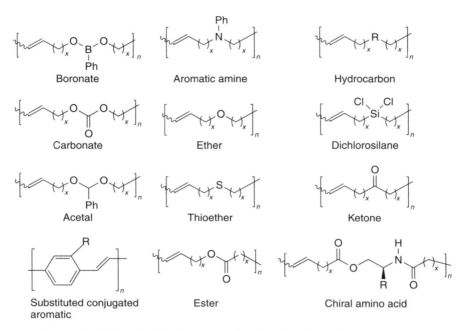

Boronate	Aromatic amine	Hydrocarbon
Carbonate	Ether	Dichlorosilane
Acetal	Thioether	Ketone
Substituted conjugated aromatic	Ester	Chiral amino acid

Figure 8.5 Sample of polymers produced by ADMET in Wagener group.

initiator is presented here for simplicity, and upon examination of this polymerization cycle, several interesting features become apparent. Since each and every reaction step in this cycle is in equilibrium, the ultimate driving force of the polymerization is the removal of a small olefinic species, typically ethylene, in the case of terminal dienes. Therefore, ADMET is an equilibrium, step-growth condensation-type polymerization (generally known as polycondensation) that displays typical molecular weight distributions (~2.0) and kinetics for this type of polymerization.[12]

The distinguishing features of this reaction mechanism are the following:

1. Ethylene is produced as the small molecule, which upon removal drives the reaction forward.
2. The ADMET cycle involves the formation of two metallacyclobutane intermediates [**D, F**], whereas the ROMP mechanism contains only one.
3. The methylidene complex [**E**] is the true catalyst, which as far as we can tell has the shortest half-life, and its reaction with an olefin produces ethylene as a by-product.
4. A second metallacyclobutane [**F**] is formed via the reassociation of terminal olefin from discrete oligomers (or monomer) with the active methylidene, produced in [**E**] (see above).

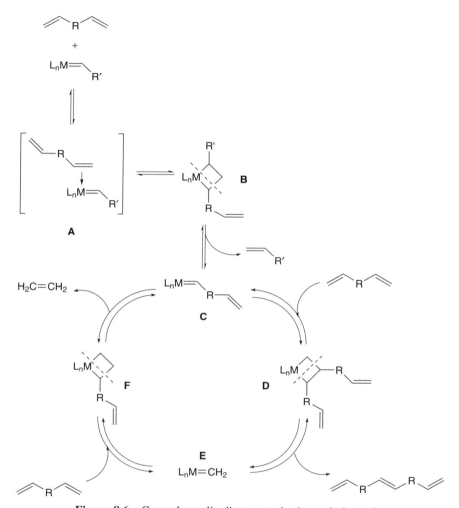

Figure 8.6 General acyclic diene metathesis catalytic cycle.

5. Upon collapse of this metallacyclobutane [**F**], ethylene is driven off and an intermediate alkylidene is formed which bears the growing polymer chain [**C**].

6. The cycle is then repeated numerous times, generating a high molecular weight polymer.

As is the case for other polycondensation reactions, internal interchange reactions are possible for ADMET, similar to that of polyesters and polyamides.[16] Interchange reactions involve a catalyst molecule on a polymer chain end reacting with an internal double bond in another polymer chain. The result is two new polymer chains; however, no change in the molecular weight distribution is observed from that expected for a random polymerization. The distribution observed is the Flory, or most probable, distribution.

Another factor in step-growth polymerizations is cyclization versus linear polymerization.[15,16] Since ADMET is a step-growth polymerization, most reactions are carried out in the bulk using high concentrations of the reactant in order to suppress most cyclic formation. A small percentage of cyclic species is always present but is dependent upon thermodynamic factors, typical of any polycondensation reaction.

8.2.1.2 Importance of Catalyst Selection

The obviation of side reactions is essential to the success of ADMET, and this can be realized if the proper catalyst is chosen. Catalyst choice must avoid the possibility of cation formation,[13] vinyl addition, and/or formation of multiple catalytic species, all of which are detrimental to clean metathesis chemistry. Over the past 10 years, our group has utilized a variety of different catalysts, several of which are illustrated in Fig. 8.4.

The two most commonly used single-site catalysts for ADMET today are (1) Schrock's alkylidene catalysts of the type $M(CHR')(NAr')(OR)_2$ where $M = W$ or Mo, $Ar' = 2, 6\text{-}C_6H_3\text{-}i\text{-}Pr_2$, $R' = CMe_2Ph$, and $R = CMe(CF_3)_2$ (**14**)[7] and (2) Grubbs' ruthenium-based catalyst, $RuCl_2(=CHPh)(PCy_3)_2$ (**12**) where Cy = cyclohexyl.[9] While both catalysts meet the requirements to be successful in ADMET, they are markedly different in their reactivity and in the results each can produce.

Kinetic studies using 1,9-decadiene and 1,5-hexadiene in comparison with catalyst **14** and catalyst **12** demonstrate an order-of-magnitude difference in their rates of polymerization, with **14** being the faster of the two.[12] Further, this study shows that different products are produced when the two catalysts are reacted with 1,5-hexadiene. Catalyst **14** generates principally linear polymer with the small amount of cyclics normally observed in step condensation chemistry, while **12** produces only small amounts of linear oligomers with the major product being cyclics such as 1,5-cyclooctadiene.[12] Catalyst **12**, a late transition metal benzylidene (carbene), has vastly different steric and electronic factors compared to catalyst **14**, an early transition metal alkylidene. Since the results were observed after extended reaction time periods and no catalyst quenching or kinetic product isolation was performed, this anomaly is attributed to mechanistic differences between these two catalysts under identical reaction conditions.

Another difference between these catalysts is found in their functional group tolerance. Catalysts such as **12** are more robust to most functionalities (except sulfur and phosphorus), moisture, oxygen, and impurities, enabling them to easily polymerize dienes containing functional groups such as esters, alcohols, and ketones.[9] On the other hand, catalyst **14** is more tolerant of sulfur-based functionalities.[7] The researcher must choose the appropriate catalyst by considering the chemical interactions between monomer and catalyst as well as the reaction conditions needed.

8.2.2 Experimental Methods

The successful polymerization of α,ω-dienes via ADMET continually produces a small molecule, typically ethylene, and the removal of this small molecule drives the reaction. When Schrock's [W] and [Mo] alkylidenes (**14**) are used, care has to be taken in maintaining an inert atmosphere devoid of both moisture and air in order to avoid decomposition of the catalyst. For this reason, Schlenk line techniques such as those used to handle Ziegler–Natta or metallocene catalysts and high purity monomers are important.

8.2.2.1 Discussion of General Glassware

A discussion of this polymerization method would not be complete without mention of the development of specialized glassware utilized over the years. It has evolved from very elaborate, sophisticated, and specially designed glass-ware to fairly simple setups. Initially, elaborate break-seal technology was used to complete the entire polymerization process,[14a] similar to anionic polymer-ization methodology.[17] Break-seal techniques were employed to fully under-stand many monomer structure–reactivity relationships; these techniques are no longer needed.

Today, the glassware required consists of either a round-bottom flask or a Schlenk tube serving as the reaction chamber. This chamber is equipped with a magnetic stirbar and a Teflon high-vacuum valve (or glass stopcock) which provides for easy vacuum control after attachment to a vacuum line (Fig. 8.7).

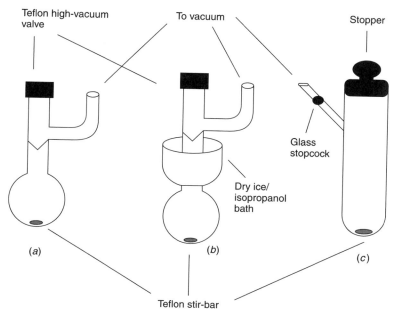

Figure 8.7 Second generation of ADMET polymerization glassware.

Figure 8.7b illustrates the vessel used when the monomer is very volatile (high vapor pressure). This design prevents low-boiling monomer from being lost under reduced pressure conditions due to the presence of a cold finger/trap, which usually contains a cooling mixture of dry ice–isopropanol ($-78°C$). This setup also can be used when a monomer is a solid, and solvent must be added in small quantities in order to facilitate the polymerization. In these cases, care must be taken to utilize *many intermittent vacuum cycles* to ensure the formation of a viscous polymer product before full vacuum ($<10^{-2}$ mm Hg) is employed. Figures 8.7a, c illustrate the glassware used when the monomer remains a stirrable, nonvolatile liquid under normal vacuum conditions at room temperature.

8.2.2.2 Purification of Monomers, Reagents, and Solvents

Care must be taken in the preparation of any reagent or monomer that comes in contact with the catalyst. All reagents and chemicals that are purchased or that are synthesized must be of high purity ($>99\%$ is preferred). Typical methods of monomer purification that have been used include recrystallization; simple, vigreux, or spinning band distillation; flash chromatography; and high-performance liquid chromatography (HPLC).[18] If catalyst **14** is used, then care must be taken to ensure absolute dryness and an oxygen-free atmosphere. Monomers used for these reactions usually are further purified and degassed by three freeze–pump–thaw cycles prior to polymerization. The N-heterocyclic-carbene-ligated ruthenium catalyst (**13**) requires substantially less stringent conditions.[10]

8.2.2.3 General Metathesis Polymerization Conditions

Usually ADMET polymerizations are conducted in the bulk state (neat) to maximize the molar concentration of the olefin, and so the examples discussed in this chapter describe bulk polymerization conditions.

For a typical ADMET polymerization, all monomers, reagents, and solvents are purified as described earlier and degassed. The glassware is dried in an oven for at least 4 h and taken into an argon (or nitrogen) glovebox. The monomer is placed in a tared reaction vessel, and then the appropriate catalyst is weighed out. Typical monomer-to-catalyst ratios are on the order of 500–1000 : 1 (**14**), 100–500 : 1 (**12**), and 100–500 : 1 (**13**). The catalyst then is added to the monomer in order to initiate the polymerization reaction, which is characterized by visible bubbling (evolution of ethylene). The reaction vessel is sealed off to the atmosphere, removed from the glovebox, and immediately connected to an evacuated vacuum line. Intermittent vacuum cycles are carried out until the viscosity increases enough to hinder or completely stop stirring.

The temperature of the oil bath in the initial stages and throughout the course of the polymerization is monomer and catalyst dependent. As a general rule, low-boiling, volatile monomers are started at room temperature ($20–25°C$), whereas higher boiling substrates may be started at $30–40°C$. Polymerizations using catalyst **14** should be started at lower temperatures ($20–30°C$) compared to reactions with **12** (30–40°C). This is primarily due to the possibility of decomposition when using Schrock's catalyst at higher temperatures, but it also takes into account the differences in relative reactivity of the two catalysts (rate of polymerization: Schrock's [Mo] > Grubbs' [Ru]). The best laboratory-scale ADMET reaction is a slow and controlled one, progressively building up the molecular weight of the product. Rapid reactions splatter monomer inside the flask, which can lead to decreased conversion.

After several intermittent vacuum cycles, the reaction mixture becomes viscous (bubbling of ethylene is less vigorous) and full, continuous vacuum ($<10^{-2}$ mm Hg) is then applied. The reaction temperature is slowly ramped to ultimately reach temperatures of 50–55°C when using **14**, 60–70°C for **12**, and 70–90°C for **13**. Once the evolution of ethylene is complete, the reaction vessel is cooled to room temperature and the catalyst is quenched by exposure to the atmosphere, addition of a terminating agent (benzaldehyde or ethyl vinyl ether), or the introduction of nonpurified, laboratory-grade solvent (usually benzene, chloroform, toluene, etc.). Polymers are isolated from the dissolved solution by precipitation into a nonsolvent such as methanol, isopropanol, or hexanes. Successive reprecipitations may be required to reduce the level of catalyst residue to acceptable levels. Alternative methods have been developed for Grubbs' catalyst removal using chelating phosphines.[19] Provided below are two examples of actual ADMET polymerizations using either catalyst as a reference (Fig. 8.8).

8.2.2.4 ADMET Polymerization of 1,9-Decadiene (9)

The following steps are performed in an argon atmosphere glovebox. In a 50-mL round-bottom flask equipped with a Teflon magnetic stirbar, 2.03 g (14.7 mmol) of previously distilled and degassed 1,9-decadiene (**9**) (Aldrich) and 27.6 mg (3.61×10^{-2} mmol) of Schrock's catalyst (**14**) (400 : 1) are combined. In a matter of seconds, the catalyst is dissolved into the monomer and a vigorous evolution of ethylene is observed. The flask is sealed with a Teflon vacuum adapter and removed from the box. The polymerization vessel is immediately connected to the vacuum line, placed into an oil bath, and stirred at 30°C. Intermittent vacuum

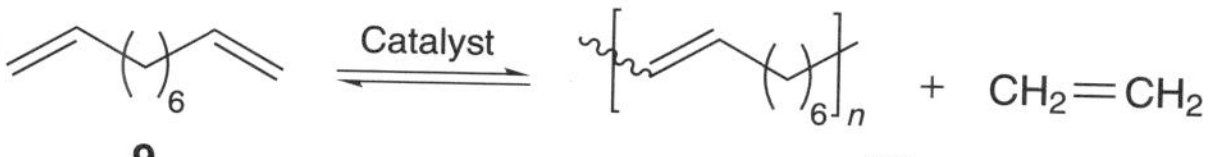

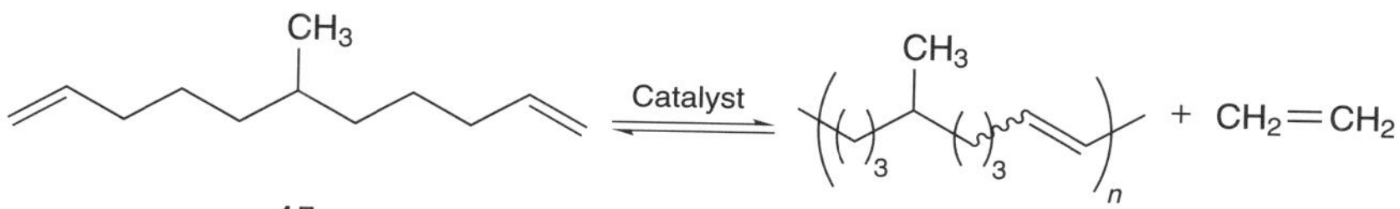

Figure 8.8 ADMET polymerization of (a) 1,9-decadiene and (b) 6-methyl-1,10-undecadiene.

cycles are performed until the evolution of ethylene slows and the increased viscosity of the polymer solution makes stirring very difficult. At this point, 1,9-decadiene is exposed to full vacuum ($<10^{-2}$ mmHg) after 1.5 h, and the temperature is increased slowly up to 50°C. After 3–5 days total reaction time, the polymerization is cooled to room temperature and quenched by exposure to air. The poly(l-octenylene) (**10**) is isolated in high yield (95%) by dissolving in toluene and precipitating into methanol. The dissolved polymer may be purified by flash chromatography through a short silica gel column in order to remove a large portion of decomposed catalyst residue. However, trace amounts of residue could still be present even after this procedure, which could lead to slight discoloration of the polymer. After this procedure, the polymer is subsequently characterized using typical analytical techniques (see Section 8.2.3).

8.2.2.5   ADMET Polymerization of 6-Methyl-1,10-Undecadiene (15)

In an argon atmosphere glovebox, 1.40 g (8.40 mmol) of distilled and degassed 6-methyl-1,10-undecadiene (**15**) and 28.0 mg (3.41×10^{-2} mmol) of Grubbs' catalyst (**12**) (250 : 1) are combined in a 50-mL round-bottom flask equipped with a Teflon magnetic stirbar.[20] In a matter of seconds, the catalyst is dissolved into the monomer and slow evolution of ethylene is observed. The flask is sealed to the atmosphere, removed from the box, immediately connected to the vacuum line, and placed in an oil bath with stirring at 40°C. Intermittent vacuum cycles are applied as described previously, and 6-methyl-1,10-undecadiene is exposed to full vacuum ($<10^{-2}$ mm Hg) after 4.5 h. The reaction is continued at these conditions for 24 h. The polymerization temperature is then slowly increased to 60°C (over a period of 1–2 days) to facilitate stirring and reacted until stirring ceases. After 3–5 days, the polymerization is cooled to room temperature and quenched by exposure to air. The polymer (**16**) is subsequently purified and characterized as described below.

8.2.3 Characterization of ADMET Polymers

ADMET polymers are easily characterized using common analysis techniques, including nuclear magnetic resonance (^{1}H and ^{13}C NMR), infrared (IR) spectra, elemental analysis, gel permeation chromatography (GPC), vapor pressure osmometry (VPO), membrane osmometry (MO), thermal gravimetric analysis (TGA), and differential scanning calorimetry (DSC). The preparation of poly(l-octenylene) (**10**) via the metathesis of 1,9-decadiene (**9**) is an excellent model polymerization to study ADMET, since the monomer is readily available and the polymer is well known.[21] The NMR characterization data (Fig. 8.9) for the hydrogenated versions of poly(l-octenylene) illustrate the clean and selective nature of ADMET.

These spectra not only confirm the primary structure of the repeat unit of the polymer but also strongly suggest that no side reactions are detectable within the limitations of the instrument. In the ^{13}C NMR spectrum (*vide infra*) all resonances can be unequivocally assigned, demonstrating the clean nature of the

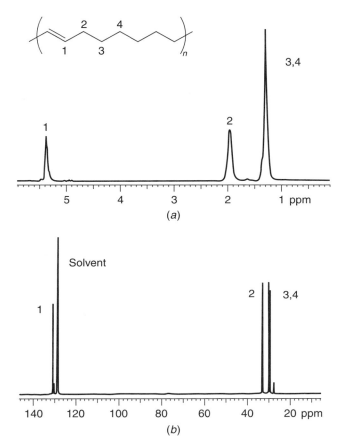

Figure 8.9 (*a*) 200-MHz ^{1}H NMR of linear poly(1-octenylene) produced by ADMET polymerization of 1,9-decadiene. (*b*) 50-MHz ^{13}C NMR of linear poly(octenamer) produced by ADMET polymerization of 1,9-decadiene. [Reproduced with permission from *Macromolecules*, **24**, 2649–2657 (1991). Copyright 1991, American Chemical Society.]

ADMET reaction. The ^{13}C NMR spectrum also allows the scientist to distinguish between cis and trans internal sp^2 carbons as well as the allylic carbons, which are adjacent to the internal vinyl position. Using quantitative ^{13}C NMR analysis, the integration of the peak intensities between the allylic carbon resonances and those of the internal vinyl carbons gives the percentage of trans/cis stereochemistry that is present for the polymer.[22] Empirically, the ratio of trans to cis linkages in ADMET polymers has typically been found to be 80 : 20. Elemental analysis results of polymers produced via ADMET demonstrate excellent agreement between experimental and theoretical values.

Membrane osmometry, vapor pressure osmometry, gel permeation chromatography, light scattering, and intrinsic viscosity measurements have been used to

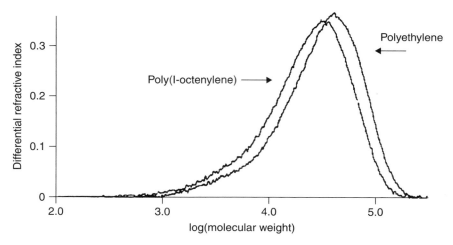

Figure 8.10   GPC analysis of poly(l-octenylene) (**10**) and hydrogenated poly(octenamer) in trichlorobenzene at 135°C. [Reproduced with permission from *Macromol. Chem., Rapid Commun.* **14**, 657–662 (1993). Copyright 1993 Wiley-VCH.]

determine the molecular weight for a series of linear poly(l-octenylene) samples made via ADMET and have been examined elsewhere.[14a] For simplicity, only the example of GPC analysis is given here (Fig. 8.10). This GPC trace shows that hydrogenation does little to effect the molecular weight distribution of poly(l-octenylene). Further, it is important to note that the polydispersity index (M_w/M_n) for polymers produced by ADMET typically approaches 2.0 (within the normal range for step condensation chemistry), and characteristic molecular weights M_n for ADMET polymers are in the range of 10,000–70,000.

Thermogravimetric analysis is used to determine the stability of a compound to decomposition upon heating, and DSC is employed to measure glass transitions (amorphous domain), melting/recrystallization behavior (polymers with crystalline domains), and the effects of thermal history or annealing in a polymer sample. Figure 8.11 illustrates the dramatic effect that hydrogenation has on a sample — the melting point is increased from 69 to 134°C, a typical value for high-density polyethylene. This is due to an increase in the crystallinity or order in the hydrogenated version (linear polyethylene) compared to its unsaturated parent. Additionally, the melting transitions for both the unsaturated and saturated versions of the ADMET polymer of 1,9-decadiene (**10**) are rather sharp and clean compared to polymers of a similar backbone in the literature, which usually show broad melting curves that extend over a wide temperature range. The sharpness of the melting transition is another characteristic that points to the clean microstructures of polymers made using ADMET. In the next section we examine the effect this precise control has on the structure–property relationships. In particular, we will focus on polymers that model branching effects in polyethylene.

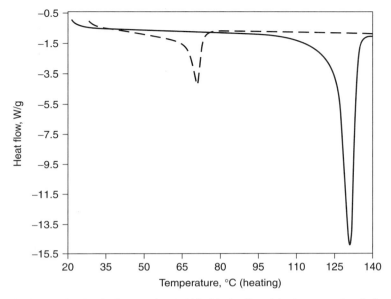

Figure 8.11 DSC of poly(l-octenylene) (**10**) [dashed] and hydrogenated poly(l-octenylene) (solid line). [Reproduced with permission from *Macromol. Chem. Rapid Commun.* **14**, 657–662 (1993). Copyright 1993 Wiley-VCH.]

8.3 STRUCTURE–PROPERTY RELATIONSHIPS

8.3.1 General Considerations

What can ADMET offer in terms of tailoring the properties of a given polymer? The answer lies in the clean chemistry of metathesis. If a metathesis active α,ω-diene can be synthesized, then a known polymer can be produced. Few other polymerization techniques are so versatile, yet so precise. In recent years, our group has focused attention toward modeling polymers and copolymers made from ethylene; in particular, we have been examining the effect of precise placement of alkyl and polar branches sequentially along the backbone of polyethylene.

8.3.2 Influence of Precise Branching on Polymer Microstructure

Polyethylene's simplicity of structure has made it one of the most thoroughly studied polymeric materials. With an estimated demand of close to 109 billion pounds in 2000 of the homopolymer and various copolymers of polyethylene,[24] it is by far the world's highest volume synthetic macromolecule. Therefore, it is still pertinent to study its structure–property relationships, thermal behavior, morphology, and effects of adding branches and functional groups to the polymer backbone.

Polyethylene is usually thought of as a linear, straight-chain macromolecule. In reality, it is a hydrocarbon backbone with variable branching. Purely linear polyethylene has been produced by both the catalytic decomposition of diazomethane[25] and ADMET of 1,9-decadiene followed by hydrogenation.[23] However, these are not the methods of choice to produce polyethylene industrially. Polyethylene is commonly produced via chain propagation chemistry using free-radical initiation,[26] Ziegler–Natta catalysis,[27] metallocene catalysis,[28] and, most recently, late transition metal single-site catalysts.[29] Inevitably, chain transfer occurs in all of these systems causing varying degrees of random branching. This creates polyethylenes with ill-defined, randomly branched microstructures, which have been used to an advantage industrially, creating a wider materials response.

What is so important about branching in polyethylene, and why would scientists like to study it or, better yet, control it? The extent of branching content in polyethylene is dependent on many factors, such as the method of polymerization and reaction temperature. The overall branch density may approach 15–30 branches per 500 mer units. Numerous studies using chain propagation chemistry have focused on delineating the effect of these irregularities in polyethylene random copolymers[30] with both alkyl[31] and polar pendant groups.[32] The studies have produced a consensus that the melting point is decreased with increasing frequency and steric bulk of the imperfections. Branching sequence distributions also play a major role in the final material properties of the polymer. Copolymerization of ethylene with different comonomers (usually α-olefins) is the best way to examine this phenomenon; however, due to differing reactivity ratios between monomers, it is not trivial to prepare polymers with precise placement of desired irregularity/branch point. In order to make a valid study of the exact impact an individual branch has on polymeric materials response, it is necessary to find a way to precisely place a pendant group into a set sequence length. Consequently, strategically designed ethylene-based copolymers with highly defined microstructures should prove extremely valuable as material models.

Ethylene copolymers with strict mer sequences (pendant group on every fifth carbon) have previously been prepared using an indirect method, accomplished by a perfectly alternating copolymerization of butadiene with vinyl monomers, followed by hydrogenation.[33] In order to further the understanding of the impact of precise pendant group placement, we began to synthesize a number of similar ethylene copolymers. Using ADMET, model branched polyethylenes with highly defined microstructures were synthesized. Ethylene–functional olefin copolymers with several commercially relevant substituents were also synthesized. This success has allowed us to more precisely evaluate the change in materials response caused by alkyl or polar functionalities.

By designing the repeat unit into the parent diene (containing either an alkyl branch or functionality), only a single type of repeat unit is formed upon polymerization, giving pure polymer microstructures. To date, perfectly controlled ADMET ethylene copolymers have included ethylene–CO,[34] ethylene–vinyl alcohol,[35] ethylene–vinyl acetate,[36] and ethylene–propylene.[20] Figure 8.12

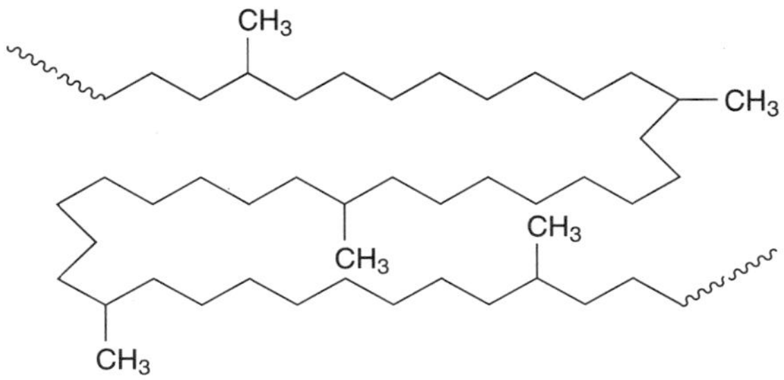

17

Precise methyl branch placement on every
9th carbon on the polythylene backbone

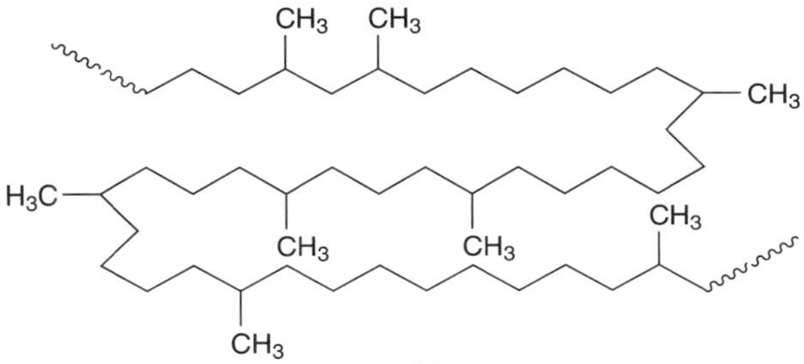

18

Random methyl branch placement on
the polyethylene backbone

Figure 8.12   Precise vs. random methyl placement in polyethylene. [Reproduced with permission from *Macromolecules*, **33**, 3781–3794 (2000). Copyright 2000, American Chemical Society.]

illustrates a possible pictorial difference in the structure of an ethylene–propylene copolymer produced via chain propagation (**18**) versus that of ADMET step condensation chemistry (**17**). The exact synthetic methodology to make the monomers is provided in the references contained within this text. Performing ADMET polymerization of the appropriate symmetrical α,ω-diene monomer with a central pendant group yields unsaturated precursor model polymers. Subsequent hydrogenation eliminates the last variable of the repeating structure, cis/trans distribution of olefinic linkages, thereby perfecting the mer sequence between branches in the polymer microstructure.

8.3.3 Thermal Behavior of Precisely Placed Branches in ADMET Polymer Microstructures

The precise placement of methyl branches on every 9th, 11th, 15th, 19th, and 21st carbon atom has had a significant impact on the thermal properties of these model polymers. Table 8.1 gives a numerical indication of how different the

TABLE 8.1   Thermal Properties of Selected Polyethylenes[a]

Polyethylene Type	$\overline{M}_n$ (10³ g/mol)	Methyl Branches per 1000 Carbon Atoms	T_m (°C)	Δh_m (J/g)
Theoretical PE[b]	24–114	0	141.5–146.5	293
ADMET PE[c]	2–15	0	131–134	204–252
Metallocene PE[d]	30–1500	0.9–1.2	137–140	—
HDPE[e]	50–250	1–6	133–138	219–245
Brookhart PE[f]	14–65	1.2–74	97–132	—
Ethylene/propylene copolymers[g]	20–70	2–100	80–133	—
LDPE[e]	20–100	30–60	105–115	95–141
Methyl branched (MB) ADMET PE[h]	8–72	48–111	−14–62	28–103

[a]From ref. 20. HDPE, LDPE = high- and low-density polyethylene.
[b]Hoffman's[37] equilibrium values derived from an infinitely long PE chain.
[c]PE Produced by ADMET polycondensation.[23]
[d]See ref. 38 for reviews on metallocene PE catalysis.
[e]From ref. 39.
[f]Brookhart's[29i] new late transition metal systems using Pd and Ni.
[g]From ref. 31a.
[h]Work in this study PE model polymers made by ADMET with precise placement of methyl branches along the backbone.

thermal properties for a number of polyethylenes can be. Why the differences? The polymerization mechanism and conditions by which polyethylenes are made will determine the amount and sometimes the identity of the branches that are present. The polyethylenes listed here are given in descending order from theoretical polyethylene, with an infinitely long chain and no branches, to those with varying degrees of branching. Contained herein are linear and branched ADMET polyethylenes. What is interesting to notice here is the impact the frequency of branch content has on the final melting point and heat of fusion of the final polymer. The theoretical melting point for ideal linear polyethylene is 145.5°C. Apparently, when the branch content is increased, the melting point and heat of fusion are subsequently decreased. All of the polyethylenes illustrated, except for those produced via ADMET, are produced by chain propagation techniques and have been studied extensively, with the exception of the polyethylenes synthesized by Brookhart et al.[29c,g–j] The most important result to note for these chain-made polymers is that as the methyl branch content increases, the melting temperature is lowered, and the shape of the endotherm is broadened, often to a point where no distinct melt can be observed.

This thermal behavior is completely opposite from the model polyethylenes that we have synthesized via ADMET. All the endotherms for the methyl-branched ADMET polyethylenes are sharp in comparison with their chain-made counterparts. Surprisingly lower melting temperature ranges for approximately the same quantity of branching as that for the ethylene–propylene copolymers and low-density polyethylene (LDPE) are also evident from this work. This may be attributed to the model methyl-branched ADMET polyethylenes having drastically different, controlled microstructures, containing precisely placed methyl branches when compared to the chain propagation alternative—randomly placed branches with differing branch identity.

DSC has been employed to examine the effect that exact placement of methyl branches has on the thermal properties for a set of model ADMET polyethylenes. For example, random copolymers of ethylene and propylene made using Ziegler–Natta catalysis[31a] show endotherms with a very broad shouldered DSC curve with indistinguishable melting points when the percentage of propylene is greater than 15 mol %.[31a,c] However, the precisely placed methyl branches in the ethylene–propylene copolymer made via ADMET produce both sharp melting endotherms and recrystallization exotherms. This is evident in Fig. 8.13, which illustrates a polymer possessing a methyl branch on every 21st carbon along the backbone of polyethylene (48 methyls/1000 carbon atoms). All the methyl-branched ADMET polyethylene model polymers exhibit similarly well-defined exotherms and endotherm DSC traces.

As shown earlier, the melting point for perfectly linear ADMET polyethylene (Table 8.1) approaches the range of melting point values exhibited by industrially produced, highly linear polyethylenes.[40] However, for ADMET polyethylene with a methyl branch placed on every ninth carbon, the T_m is lowered significantly. The melting point of $-14°C$ and enthalpy of 28 J/g is 150°C and approximately

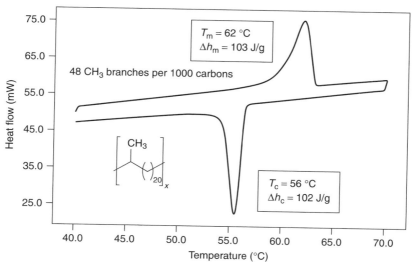

Figure 8.13 DSC data for model polyethylene with a methyl group on every 21st carbon. [Reproduced with permission from *Macromolecules*, **33**, 3781–3794, 2000. Copyright 2000, American Chemical Society.]

one order of magnitude lower than that of linear ADMET polyethylene. A distinct, sharp melting point for all of the methyl-branched polymers is found, particularly for the sample with a methyl branch placed on every ninth carbon (111 methyls/1000 carbon atoms). Further, model studies using chain techniques have shown that a completely amorphous polymer is formed when the methyl branch frequency is approximately 150/1000 carbon units. However, placing the branch in a specific sequence length along the backbone of polyethylene creates a semicrystalline order that has not been observed in the chain-made branched polyethylenes. There is no doubt that these materials have a well-defined, sharp melt and that there is also some type of secondary structure ordering in the material. The preliminary DSC data are just the first indication of how special these materials may be for the purposes of modeling.

8.4 SYNTHETIC METHODS: SILICON-CONTAINING POLYMERS, FUNCTIONALIZED POLYOLEFINS, AND TELECHELICS

8.4.1 Polycarbosilanes and Siloxanes

Polycarbosilanes are an interesting class of polymers from a materials standpoint; however, they do not occur naturally. A variety of synthetic methods can be employed to produce such polymers, but this part of the chapter will focus on using ADMET as a viable route to synthesize polycarbosilanes, siloxanes, and other silicon-containing polymers.

The first type of polycarbosilane synthesized by using ADMET methodology was a poly[carbo(dimethyl)silane].[14c] Linear poly(carbosilanes) are an important class of silicon-containing polymers due to their thermal, electronic, and optical properties.[41] They are also ceramic precursors to silicon carbide after pyrolysis. ADMET opens up a new route to synthesize poly(carbosilanes), one that avoids many of the limitations found in earlier synthetic methods.[41]

Initial attempts to condense divinyldimethylsilane were unsuccessful due to steric interactions between the methyl groups on the silicon atom adjacent to the double bond and the catalyst, which does not allow for the formation of the critical metallacyclobutane intermediate.[7b,14c] However, copolymers with 1,9-decadiene are possible. Methylene spacer units between the double-bond site and the silyl group are necessary in order to condense monomer to polymer. Consequently, diallyldimethylsilane (**19**), bis(allyldimethylsilyl)ethane (**21a**), and bis(allyldimethylsilyl)benzene (**21b**) were successfully polymerized using ADMET (Fig. 8.14).[14c] All three of these monomers can be polymerized using [W] catalyst (**23**), producing clean microstructures with a small amount of cyclic byproducts caused by catalyst backbiting. Table 8.2 provides molecular weight data (M_n), and differential scanning calorimetry illustrates low glass transition temperatures (T_g), with no melting temperatures (T_m), indicative of amorphous polymers.

Organofunctionalized polysiloxanes possess a unique combination of properties, including hydrophobicity, low-temperature flexibility, and thermal stability.[41]

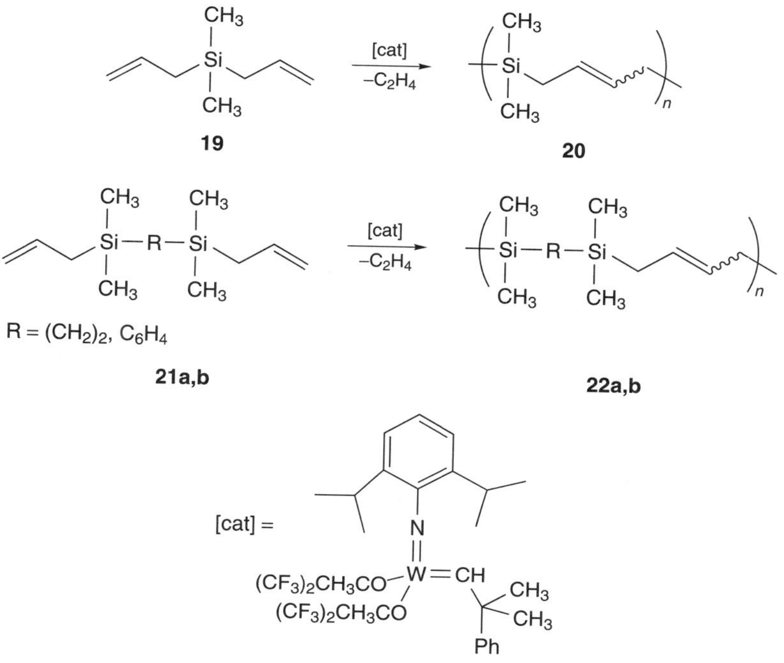

Figure 8.14 Classes of poly[carbo(dimethyl)silanes] produced by ADMET.

TABLE 8.2   Characterization Data for Poly[carbo(dimethyl)silane]s

Polymer	GPC $(M_n)^a$	M_w/M_n	T_g (°C)	TGA (°C)^b
20	12,000	1.94	−67	330
22a	50,000	1.10	−59	440,^{c} 230^d
22b	1,500	1.34	−46	

^{a}GPC M_n relative to polystyrene.
^{b}Temperature where 10% weight loss occurs.
^{c}In nitrogen.
^{d}In air.

Therefore, the incorporation of siloxane linkages into polymer chains as well as copolymers of siloxane and polyolefin blocks is of interest. Regularly alternating carbon and siloxane block copolymers combine the characteristics of the hydrocarbon segment, increasing the mechanical properties of the polymer while still taking advantage of the properties of the siloxane segment.[42] Several main-chain unsaturated poly(carbosiloxanes) have been synthesized in our group, most containing dimethyl-substituted silyl groups.[43]

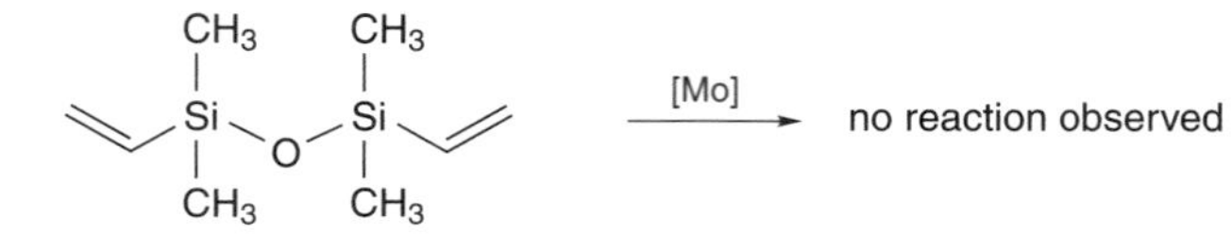

Figure 8.15 Two carbosiloxadienes which fail to produce linear polymer.

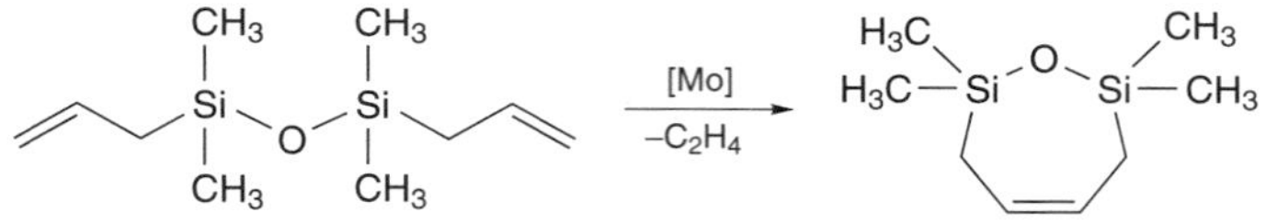

Monomers **24** and **25** behave differently when exposed to catalyst **14**, shown in Fig. 8.15. Divinyltetramethyldisiloxane **24** is found to be metathesis inactive due to similar steric inhibitions experienced with divinyldimethylsilane. Monomer **25** is synthesized with one additional methylene spacer unit between the silicon atom and the olefin moiety, which then is reacted with Schrock's [Mo] catalyst. Here, metathesis occurs quite readily, exclusively forming a seven-membered cyclic molecule (**26**) instead of polymer. The formation of the cyclic product can be explained by the Thorpe–Ingold effect.[15]

Cyclization is prevented by extending the siloxane linkage or by lengthening the carbon tether (methylene spacers between the siloxane and the olefin). Both routes are utilized in order to produce methyl-substituted polycarbosiloxanes. Monomer **27**, possessing a trisiloxane linkage, is synthesized by reacting a dichlorosiloxane with allylmagnesium bromide (Fig. 8.16).[43] Increasing the number of methylene spacers between the silicon atom and the olefin is accomplished by hydrosilation of 1,4-pentadiene using Karstedt's catalyst, tris(divinyltetramethyldisiloxane)diplatinum(0), producing monomer **28**.[44] Monomer **29** is

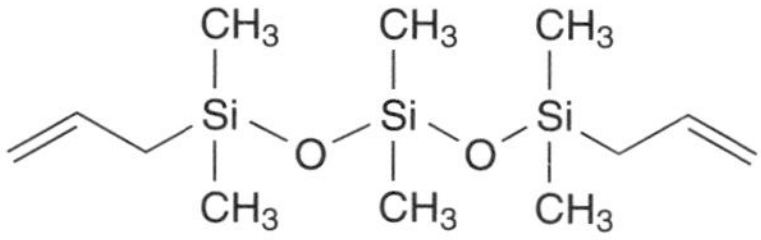

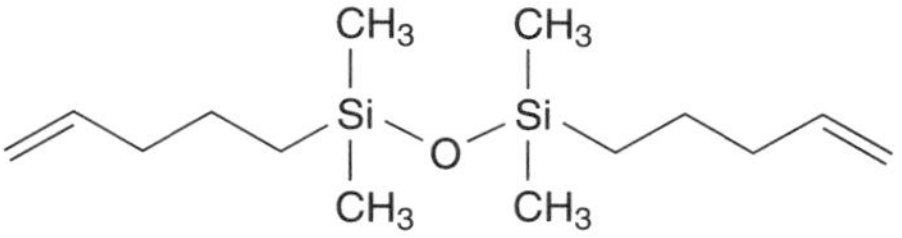

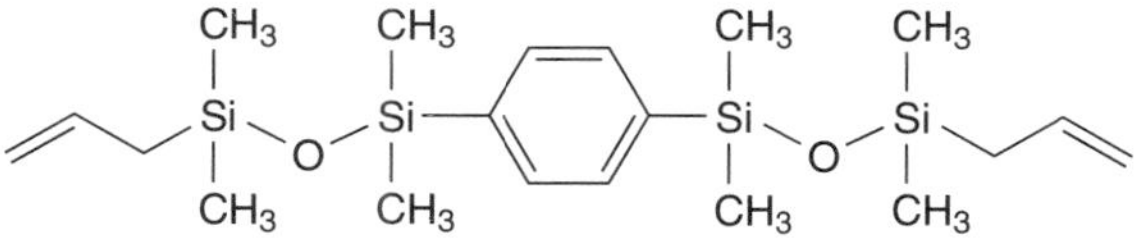

Figure 8.16 Three types of methyl-substituted carbosiloxadienes.

synthesized by the condensation of bis(hydroxysilyl)benzene with allylchlorodimethylsilane. All of these monomers are successfully converted from monomer to polymer using **14**, clearly demonstrating the compatibility of this catalyst with the siloxane functionality.

Monomers **27** and **28** are viscous liquids at room temperature, thus allowing for the formation of reasonably high molecular weight materials during polymerization in the bulk (Table 8.3). Monomer **28**, a solid at room temperature, yields the lowest degree of polymerization of the three monomers studied. Low glass transition temperatures and high thermal stability are observed for all the poly(carbosiloxanes), which is expected for polymers containing the highly flexible and stable siloxane linkage.

Another example of the flexibility of ADMET is the demonstration of successful polymerization of α,ω-telechelic diene carbosilane macromonomers.[45] The synthesis of macromonomer **30** is achieved using catalyst **23** and copolymerized with a rigid small-molecule diene, 4,4'-di-trans-1-propenylbiphenyl (Fig. 8.17).

TABLE 8.3 Characterization Data for Poly(carbosiloxane)s Produced by ADMET

Polymer	GPC $(M_n)^a$	$M_w/M_n{}^b$	T_g (°C)	TGA (°C)^{c,d}
Monomer **27**	15,000	1.7	−97	398,^{c} 284^d
Monomer **28**	32,000	1.7	−93	427,^{c} 282^d
Monomer **29**	1,400	1.2		

^{a}Relative to polybutadiene standards.
^{b}Reaction mixture.
^{c}In N$_2$.
^{d}In air.

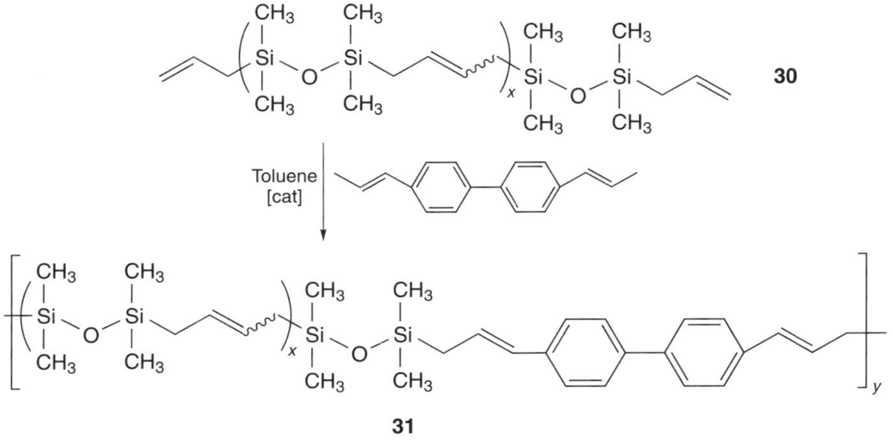

Figure 8.17 Copolymerization of telechelic oligomer **30** with a rigid aromatic diene.

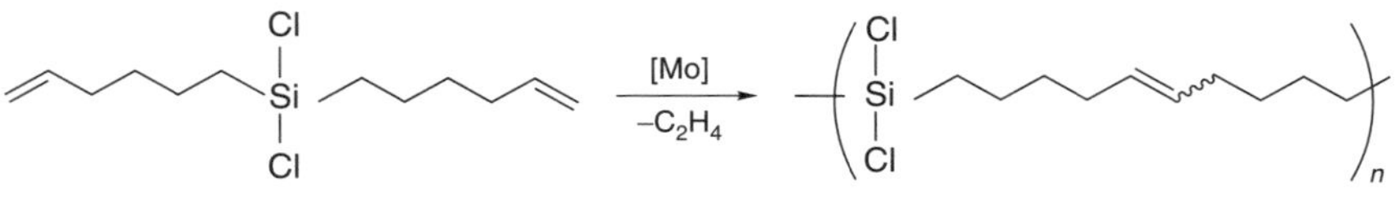

Figure 8.18   Synthesis of poly(di-5-hexenyldichlorosilane) (33) via ADMET.

The molecular weight of **30** is purposely stopped with $M_n = 7650$. The rationale for producing an oligomer before copolymerization occurs is to prevent irregularity in the overall polymer structure. This can be caused by unintended reactions such as self-metathesis of the biphenyl diene. Since this self-metathesis product is an insoluble solid, copolymerization with another diene containing siloxane linkages, such as **30**, is advantageous because it allows for higher molecular weight materials to be attained as well as imparting excellent solubility and improved physical properties to the polymer produced.

The designed structural regularity of this copolymer is one reason why we studied this system. Under normal conditions, when two monomers are mixed and reacted in the presence of an ADMET catalyst, random copolymers are produced with the final composition based upon the feed ratios of the two monomers. However, in this case, controlled addition of the aromatic diene with telechelic oligomer **30** produces a high-molecular-weight material with a designed number of aromatic units in the polymer backbone. The controlled addition of 4,4′-di-trans-1-propenylbiphenyl is achieved by simply adding a toluene solution of the biphenyldiene to a solution of **30** in toluene and catalyst **23** under inert conditions. About 11 mol % cyclic products are produced by catalyst backbiting after dissolving the telechelic oligomer in solvent. The cyclization does not increase significantly during the copolymerization, and the resulting copolymer contains 13 mol % of the biphenyldiene subunit in the polymer backbone. Integration of the aromatic resonances in both ^{1}H and ^{13}C NMR confirms that the degree of incorporation of the biphenylene unit is quantitative. No evidence of biphenylene self-metathesis is observed spectroscopically. Further support of copolymer formation is displayed by GPC analysis, which shows a unimodal peak with $M_n = 27,000$ (relative to polybutadiene standards) and $M_w/M_n = 1.9$, typical for polycondensation reactions.

Another class of silicon-containing polymers that have great potential to be extremely useful precursor materials are poly(chlorocarbosilanes).[14f,46] Poly(chlorocarbosilanes) are not useful without modification because of the rapid hydrolysis of Si−Cl bonds, forming HCl and an insoluble crosslinked polymer network. However, nucleophilic substitution of these Si−Cl bonds with various reagents produces materials with a broad range of properties that are determined by the nature of the nucleophile used.[47] Poly(chlorocarbosilanes) can be easily synthesized by ADMET (Fig. 8.18) without any detrimental side reactions, since the Si−Cl bond is inert to both catalysts **12** and **14**. Early studies produced a polymer with $M_n = 3000$.[14f]

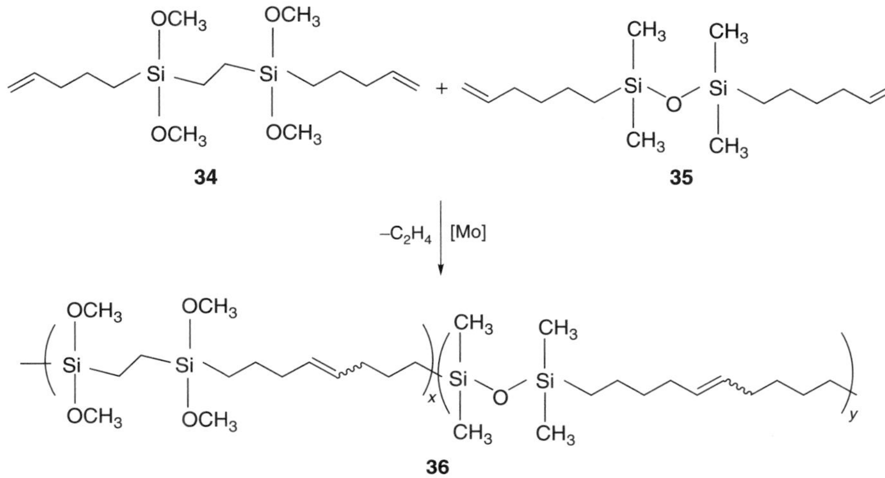

Figure 8.19   Copolymerization of monomers **34** and **35**.

Another system under investigation in our laboratory involves the use of silicon alkoxide groups in ADMET polymers.[14j,47] In particular, methoxy-functionalized silane polymers are chosen because the hydrolytic reactivity of the Si—OCH$_3$ bond is kinetically slower than the Si—Cl bond.[14j] We refer to these as latent-reactive polymers, since slow crosslinking occurs after polymerization. As an example of this concept, the copolymerization of **34** and **35** is performed by ADMET, giving a random distribution of the two monomers along the back-bone (Fig. 8.19). Like the Si—Cl bond, the Si—OCH$_3$ bond is also inert to the metathesis catalysts **12**, **13**, and **14**, and the resulting ADMET copolymer **36** is a linear, thermoplastic material.[14j] Each monomer in the copolymer imparts its distinct properties, contributing to the overall material characteristics of the resulting copolymer. The "soft" siloxane unit, consisting of unreactive methyl groups attached to the silicon atom, imparts flexibility and low T_g, whereas the "hard" carbosilane segment possesses four reactive silicon–methoxy moieties per repeat unit, providing strength to the material. The presence of these groups on the polymer allow for slow crosslinking over a period of time when introduced to atmospheric moisture, making a strong thermoset network material. As a result, the copolymer of both dienes ($M_n = 40{,}000$) has high thermal stability, yet a low glass transition temperature. GPC analysis of this copolymer confirms its copoly-mer nature, displaying a unimodal distribution and no evidence of any bimodal characteristics, thus eliminating the possibility of dual homopolymerization.

The slower, more controlled hydrolysis of silicon–methoxy bonds enables the synthesis of a material that is processable and extremely tough, yet flexible, with the degree of flexibility a direct result of the molar feed ratio of the two monomers since both are incorporated equally in the ADMET copolymerization. Therefore, a series of materials can be made where the properties are adjusted by varying the percentage of crosslinkable bonds present. We are continuing our study of polymers with sites of latent reactivity as well as examining improved methods of controlling the crosslinking reaction.

8.4.2 Telechelics from Polybutadiene

The breakdown and recycling of polymers through depolymerization mechanisms have been a focus of research for many scientists. Several different methods have been successfully used to achieve oligomeric products, and we have found that ADMET is useful for the depolymerization of unsaturated polymers. The simplest way to produce oligomers by this method is by the addition of ethylene to a solution of polymer and metathesis catalyst. Since ADMET is an equilibrium reaction, adding ethylene pushes the equilibrium toward depolymerization to monomer. We have also studied methodologies to produce exact-mass telechelic oligomers by using metathesis. The first example is the reaction of a monoolefin with 1,4-polybutadiene (**38**).[48] Allyltrimethylsilane (**37**) (10 equivalents per repeat unit of 1,4-polybutadiene) is added to a solution of polymer in the presence of catalyst **14** or **23**, evolving ethylene gas as the silane self-metathesizes. The ethylene is removed in vacuo and toluene is then added to the reaction flask. All volatiles (including excess silane olefin) are removed from the reaction, and exact-mass α,ω-bis(trimethylsilyl)-1,4-polybutadiene is produced (Fig. 8.20).

This strategy is used for the synthesis of three different exact-mass telechelic oligomers. GPC, NMR, and GC/MS evidence indicates that clean depolymerization chemistry occurs for all three samples. Poly(1,4-butadiene) (**38**) is broken down into oligomeric units with two, three, and four repeat units using catalyst **23**. Catalyst **14** is more efficient and produces even lower molecular weight oligomers, primarily one and two repeat units. When allylchlorodimethylsilane is used instead of ethylene with **14**, telechelic dimers are the only product.

Other groups such as esters, silylethers, and imides are also successfully incorporated through ADMET depolymerization with **14** (Fig. 8.21).[49] For an ester functionality, at least two methylene spacer units must be present between the olefin site and the functional group in order to achieve depolymerization. This is due to the "negative neighboring group effect," a deactivation of the catalyst by coordination of the functionality heteroatoms to the catalyst.[50] By physically

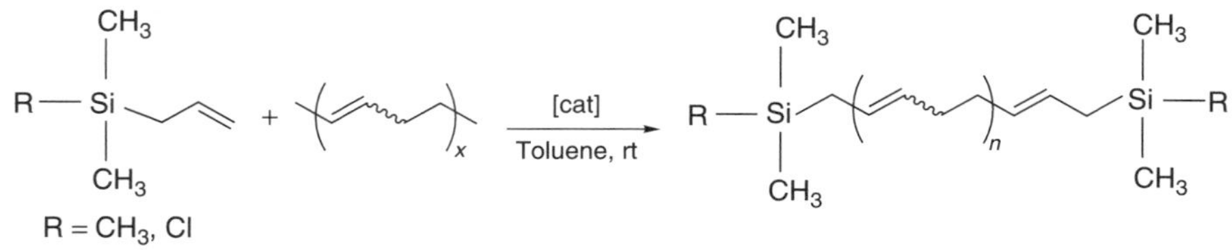

Figure 8.20 Metathesis depolymerization produces exact-mass telechelic oligomers.

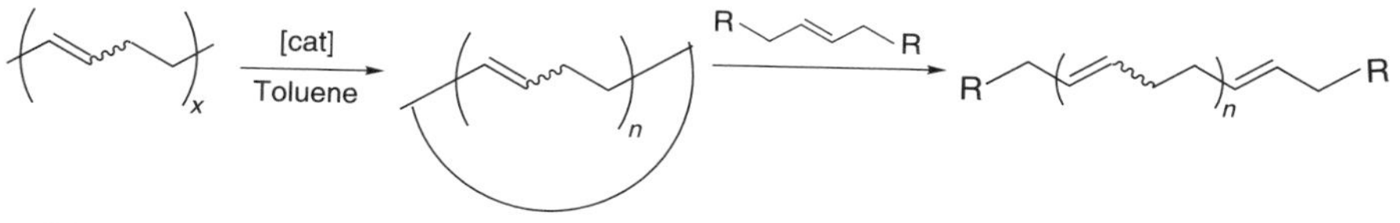

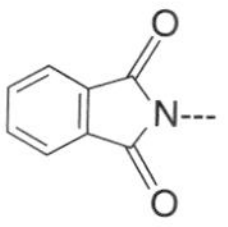

Figure 8.21 Synthesis of various difunctional telechelic oligomers via ADMET depolymerization.

separating the ester from the olefin site by using spacer groups, depolymerization is able to occur, producing ester-difunctionalized telechelic butadiene oligomers.

Diol-functionalized telechelic polymers have been desired for the synthesis of polyurethanes; however, utilizing alcohol-functionalized α-olefins degrades both **14** and **23**. Consequently, in order for alcohols to be useful in metathesis depolymerization, the functionality must be protected and the oxygen atom must not be β to the olefin or only cyclic species will be formed. Protection is accomplished using a t-butyldimethylsiloxy group, and once protected, successful depolymerization to telechelics occurs readily.

Imide-terminated telechelics are also synthesized by metathesis depolymerization, and it is found that phthalimide-substituted olefins allow for productive depolymerization when only one methylene spacer separates the nitrogen atom and the olefin (Fig. 8.21). This combination of steric hindrance around the nitrogen lone pair and decreased electron donation from resonance prevents the negative neighboring group effect. However, secondary acyclic amines are unable to produce telechelics through metathesis depolymerization because of unfavorable catalyst–amine interactions.

In order to prove that intramolecular cyclization occurs before telechelic oligomer formation, an experiment similar to previous work by Calderon[1f] is performed using **14** in place of Calderon's classical catalyst system. Macrocyclic species are formed when a toluene solution of polybutadiene is exposed to this catalyst, supported by both NMR and GC data. The vinylic resonances are clearly shifted upfield from polybutadiene. GC analysis shows macrocyclic trimers and tetramer regioisomers.

The most recent work in our group on metathesis depolymerization has focused on the solvent-free depolymerization of 1,4-polybutadiene.[51] In our previous work using catalyst **14**, a solvent is required in order to bring the catalyst and polymer into the same phase for depolymerization to take place. However, we have found that catalyst **12** can effectively depolymerize 1,4-polybutadiene with no solvent or any chain transfer agent required. In one experiment, **12** (0.25 mol %) is sprinkled onto cis-1,4-polybutadiene at room temperature under argon. Within 1 h, the polymer surface began to liquefy, producing a viscous liquid after 2–3 h. GPC analysis shows a decrease in molecular weight of two orders of magnitude.

Because of the absence of chain limiter, the catalyst itself may initially act as the chain limiter (Fig. 8.22). The catalyst reacts with the olefinic regions of the polymer backbone and causes chain scission to occur, forming two new chains. The reactive carbene which is produced then moves from chain to chain, forming two new chains with each scission until the most probable molecular weight distribution is reached ($M_w/M_n \cong 2$), producing linear chains end capped with [Ru] catalyst residues.

If the catalyst is the only chain limiter present, then the molecular weight should be proportional to the repeat unit–catalyst molar ratio. However, when the catalyst is reduced 10-fold from 0.25 to 0.025 mol %, the resulting molecular weight of the depolymerization product is the same. Therefore, another chain-limiting species must be present in the system. These results are due to 1,2-vinyl linkages present in all commercial 1,4-polybutadiene.[52] The 1,2-linkages act as "intramolecular chain limiters" through ring-closing metathesis, which then act as a chain limiter by cross-metathesis of the polymer backbone with the vinyl group (Fig. 8.23). Direct chain scission occurs when the vinyl group of the triad reacts with an olefin on the polymer backbone. The final result is a mixture of oligomeric 1,4-polybutadiene end capped with vinyl (**46, 50**), cyclohexenyl (**51**), and cyclopentenyl (**47**) groups. Since no external chain limiters are present in this depolymerization process, the position of the depolymerization equilibrium is determined by three metathesis processes: conversion of polymer to cyclics, incorporation of the catalyst as a chain limiter, and conversion of 1,4–1,2–1,4 triads as chain limiters.

Catalyst **14** has also been used, and we find that solvent-free depolymerization does not occur to a large extent.[51,53] For the first 30 min, the solution begins to liquefy but then becomes waxy. This result occurs because **14** produces a higher trans content of chains, which are more crystalline, in the early stages of the reaction. Hence, the mobility of the chains is hindered[54] and depolymerization is inhibited because of the lack of chain-to-chain migration.

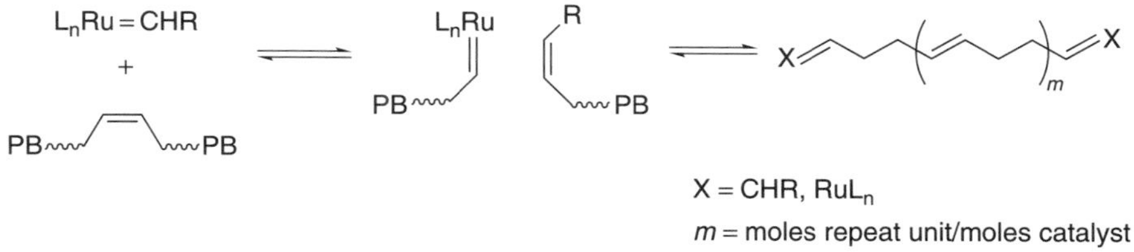

$X = CHR, RuL_n$

m = moles repeat unit/moles catalyst

42   **43**   **44**

Figure 8.22   Catalyst is functioning as a chain limiter.

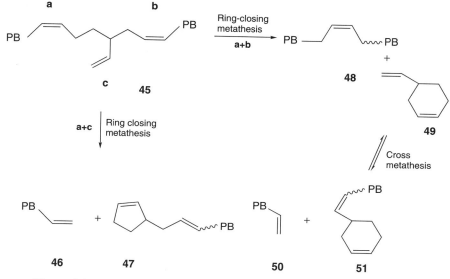

Figure 8.23   Depolymerization using 1,4–1,2–1,4 triads as chain limiters.

8.4.3 Functionalized Polyolefins

Functionalized polyethylene would be of great industrial importance, and if synthetic methods to control the microstructure of functionalized polymers using transition-metal-based catalysis are developed, it would significantly broaden the utility and range of properties of this class of polymers. Recent progress in the field of late transition metal chemistry, such as Brookhart's use of nickel-based diimine catalysts, has enabled the copolymerization of ethylene with functional α-olefins.[29] However, these systems incorporate functionalized olefins randomly and with limited quantity (mol percent) into the polymer backbone.

Since the reactivity ratios of ethylene–polar monomer pairs are quite different, the preparation of copolymers with precisely the same comonomer composition can be a challenging endeavor. Earlier in this chapter, we described the synthesis and characterization of precisely placed methyl groups on a polyethylene

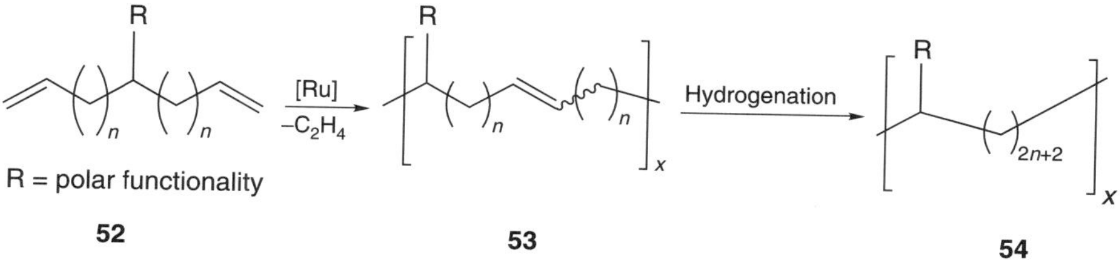

Figure 8.24   Using ADMET/hydrogenation to produce polyethylene with polar functionalities.

backbone.[20] An extension of this work is discussed here, where ADMET is used to prepare a number of linear polymers with precisely placed functional groups that have commercial relevance. The goal is to study the effect of the functionality (position and number) on a saturated backbone, in essence preparing copolymers of polyethylene with functionalized α-olefins. ADMET polymerization is performed on a series of symmetrical α,ω-dienes containing a central pendant functionality (**52**).[34,36,55] Subsequent hydrogenation of the unsaturated sites on the backbone converts these polymers to "perfect" copolymers of polyethylene (**54**) (Fig. 8.24). We refer to these models as ethylene–polar monomer copolymers (EPMs); however, one feature that distinguishes these models from perfect EPM models is the number of spacers between functionality.[55] Perfect EPM models have functional groups that are separated by an odd number of methylene spacer units, whereas our models have an even number of spacer units between functionalities.

Both ester and carbonyl groups have been incorporated into the polymer backbone in order to study how the lamellar crystal structure is affected by regular polymer defects.[34,36] Placement of a carbonyl functionality into a polyethylene-like backbone (**57**) mimics an ethylene–CO copolymer.[34] While regularly alternating ethylene–CO copolymers do exist, there is no published method to date besides ADMET hydrogenation that produces a copolymer of high molecular weight which controls the number of ethylene units between the carbonyl functionality. ADMET polymerization of monomer **55**, followed by hydrogenation, produces a model polymer **57** with a $T_m = 134°C$, having a lower molecular weight and broader molecular weight distribution than usual for ADMET polymers. This is because of the higher melting temperatures/intermolecular interactions of the monomer (Fig. 8.25). A small amount of toluene is added to facilitate the reaction, thus lowering polymer molecular weight by competition between macrocycle formation and polymerization.

Other commercially relevant monomers have also been modeled in this study, including acrylates, styrene, and vinyl chloride.[55] Symmetrical α,ω-dienes substituted with the appropriate pendant functional group are polymerized via ADMET and utilized to model ethylene–styrene, ethylene–vinyl chloride, and ethylene–methyl acrylate copolymers. Since these models have "perfect" microstructure repeat units, they are a useful tool to study the effects of the functionality on the physical properties of these industrially important materials. The polymers produced have molecular weights in the range of 20,000–60,000, well within the range necessary to possess similar properties to commercial high-molecular-weight material.

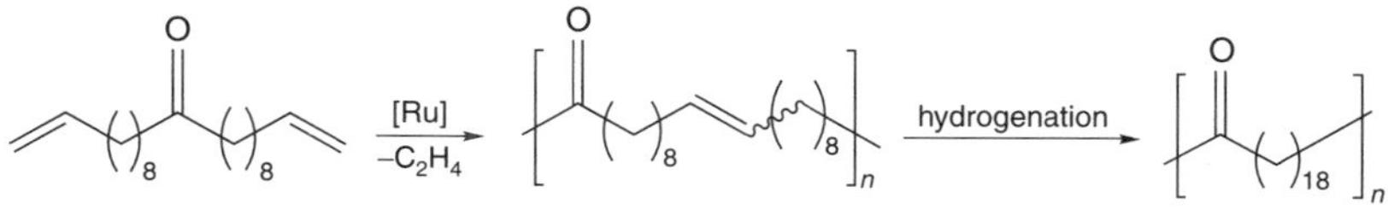

Figure 8.25   Ethylene–CO copolymer synthesized by ADMET.

8.5 CONCLUSIONS

Acyclic diene metathesis (ADMET), a rare example of step-growth polymerization catalyzed by a transition metal complex, has proven to be quite a versatile reaction. As the catalog of organometallic M=C catalysts for metathesis has grown, so has the flexibility of this reaction. Originally thought to be an impractical reaction scheme using early metal halide–Lewis acid catalyst systems, ADMET has benefited tremendously from the advent of breakthroughs in single site, well-defined carbene catalysts from the Schrock and Grubbs laboratories. These discoveries have allowed ADMET polymerization to be remarkably tolerant to variations in both structure and atom composition in the monomer. We have incorporated heteroatoms and a variety of functionalities as well as prepared telechelic and branched polymers via this route. The key advantage of ADMET is the complete lack of side reactions, producing a polymer with a predictable, designed architecture. As a result, we have been able to synthesize polymers with precise microstructures and produce "perfect" models for industrially relevant copolymers, work that is quite challenging to accomplish by other means. Research involving ADMET continues as we utilize this chemistry in the investigation of ADMET segmented and comb copolymers, reactive polymers, biopolymers, and polymer crystallization behavior.

ACKNOWLEDGMENTS

The authors would like to thank the Army Research Office and the National Science Foundation for financial support of this research.

REFERENCES

1. (a) N. Calderon, E. A. Ofstead, and W. A. Judy, *J. Polym. Sci. P. A*, **5**, 2209 (1967); (b) N. Calderon, H. Y. Chen, and K. W. Scott, *Tetrahedr. Lett.*, **8**, 3327 (1967); (c) N. Calderon, E. A. Ofstead, J. P. Ward, W. A. Judy, and K. W. Scott, *J. Am. Chem. Soc.*, **90**, 4133 (1968); (d) K. W. Scott, N. Calderon, E. A. Ofstead, W. A. Judy, and J. P. Ward, *Am. Chem. Soc. Adv. Chem. Ser.*, **91**, 399 (1969); (e) J. L. Herrison and Y. Chauvin, *Makromol. Chem.*, **141**, 161 (1970); (f) N. Calderon, E. A. Ofstead, and W. A. Judy, *Angew. Chem. Int. Ed.*, **15**, 401 (1976).

2. K. J. Ivin and J. C. Mol, *Olefin Metathesis and Metathesis Polymerization*, Academic, San Diego, 1997.

3. General metathesis reviews: A. Furstner, *Angew. Chem. Int. Ed.*, **39**, 3012 (2000). **RCM**: (a) R. H. Grubbs, S. J. Miller, and G. C. Fu, *Acc. Chem. Res.*, **28**, 446 (1995); (b) A. Furstner, *Topic Catal.*, **4**, 285 (1997); (c) M. Schuster and S. Blechert, *Angew. Chem. Int. Ed.*, **36**, 2037 (1997); (d) R. H. Grubbs and S. Chang, *Tetrahedron*, **54**, 4413 (1998). **ROMP**: (a) R. R. Schrock, *Acc. Chem. Res.*, **23**, 158 (1990); (b) R. R. Schrock, "Ring Opening Metathesis Polymerization,"

in *Ring Opening Polymerization*, D. J. Bruneile (Ed.), Hanser, Munich, 1993, p. 129; (c) R. H. Grubbs and E. Khosravi, "Ring Opening Metathesis Polymerization (ROMP) and Related Processes," in *Synthesis of Polymers*, A. D. Schluter (Ed.), Materials Science and Technology Series, Wiley, Weinheim, 1999, p. 65. **ADMET**: (a) D. Tindall, J. H. Pawlow, and K. B. Wagener, "Recent Advances in ADMET Chemistry," in *Topics in Organometallic Chemistry: Alkene Metathesis in Organic Synthesis*, A. Furstner (Ed.), Springer, Berlin, 1998, p. 184; (b) T. A. Davidson and K. B. Wagener, "Acyclic Diene Metathesis (ADMET) Polymerization," in *Synthesis of Polymers*, A. D. Schluter (Ed.), Materials Science and Technology Series, Wiley, Weinheim, 1999.

4. H. S. Eleuterio, U.S. Patent 3,074,918, 1963.

5. R. H. Crabtree, *The Organometallic Chemistry of the Transition Metals*, 3rd ed., Wiley, New York, 2001.

6. (a) G. Natta, G. Dall'Asta, and G. Mazzanti, *Angew. Chem., Int. Ed.*, **3**, 723 (1964); (b) P. Günther, F. Haas, G. Marwede, K. Nützel, W. Oberkirch, G. Pampus, N. Schön, and J. Witte, *Die Angew. Makro. Chem.*, **14**, 87 (1970); (c) G. Dall'Asta, *Makromol. Chem.*, **154**, 1 (1972); (d) G. Dall'Asta, *Rubber Chem. Tech.*, **47**, 510 (1974).

7. (a) P. R. Sharp, S. J. Holmes, and R. R. Schrock, *J. Am. Chem. Soc.*, **103**, 965 (1981); (b) R. R. Schrock, *Science*, **219**, 13 (1983); (c) R. R. Schrock, R. T. DePue, J. Feldman, C. J. Schaverien, J. C. Dewan, and A. H. Liu, *J. Am. Chem. Soc.*, **110**, 1423 (1988); (d) R. R. Schrock, J. S. Murdzek, G. C. Bazan, J. Robbins, M. DiMare, and M. O'Regan, *J. Am. Chem. Soc.*, **112**, 3875 (1990); (e) R. R. Schrock, R. T. DePue, J. Feldman, K. B. Yap, D. C. Yang, W. M. Davis, L. Park, M. DiMare, M. Schoefield, J. Anhaus, E. Walborsky, E. Evitt, C. Kruger, and P. Betz, *Organometallics*, **9**, 2262 (1990); (f) J. Feldman and R. R. Schrock, *Prog. Inorg. Chem.*, **39**, 1 (1991).

8. (a) M. Lindmark-Hamburg and K. B. Wagener, *Macromolecules*, **20**, 2949 (1987); (b) K. B. Wagener, J. G. Nel, J. Konzelman, and J. M. Boncella, *Macromolecules*, **23**, 5155 (1990).

9. (a) S. T. Nguyen, R. H. Grubbs, and J. W. Ziller, *J. Am. Chem. Soc.*, **115**, 9858 (1993); (b) P. Schwab, M. B. France, J. W. Ziller, and R. H. Grubbs, *Angew. Chem. Int. Ed.*, **34**, 2039 (1995); (c) P. Schwab, R. H. Grubbs, and J. W. Ziller, *J. Am. Chem. Soc.*, **118**, 100 (1996); (d) R. H. Grubbs, M. J. Marsella, and H. D. Maynard, *Angew. Chem. Int. Ed.*, **36**, 1101 (1997); (e) E. L. Dias, S. T. Nguyen, and R. H. Grubbs, *J. Am. Chem. Soc.*, **119**, 3887 (1997); (f) E. L. Dias and R. H. Grubbs, *Organometallics*, **17**, 2758 (1998); (g) T. M. Trnka and R. H. Grubbs, *Acc. Chem. Res.*, **34**, 18 (2001).

10. (a) M. Scholl, S. Ding, C. W. Lee, and R. H. Grubbs, *Organic Lett.*, **1**, 953 (1999); (b) M. Scholl, T. M. Trnka, J. P. Morgan, and R. H. Grubbs, *Tetrahedr. Lett.*, **40**, 2247 (1999); (c) A. K. Chatterjee and R. H. Grubbs, *Organic Lett.*, **1**, 1751 (1999); (d) M. S. Sanford, L. M. Henling, M. W. Day, and R. H. Grubbs, *Angew. Chem. Int. Ed.*, **39**, 3451 (2000); (e) C. W. Bielawski and R. H. Grubbs, *Angew. Chem. Int. Ed.*, **39**, 2903 (2000).

11. (a) W. A. Herrmann and C. Kocher, *Angew. Chem. Int. Ed.*, **36**, 2162 (1997); (b) T. Weskamp, W. C. Schattenmann, M. Spiegler, and W. A. Herrmann, *Angew. Chem. Int. Ed.*, **37**, 2490 (1998); (c) T. Weskamp, F. J. Kohl, W. Hieringer, D. Gleich, and W. A. Herrmann, *Angew. Chem. Int. Ed.*, **38**, 2416 (1999); (d) J. Huang, E. D. Stevens, S. P. Nolan, and J. L. Peterson, *J. Am. Chem. Soc.*, **121**, 2674 (1999).

12. K. B. Wagener, K. Brzezinska, J. D. Anderson, T. R. Younkin, K. Steppe, and W. DeBoer, *Macromolecules*, **30**, 7363 (1997).

13. K. B. Wagener, J. M. Boncella, J. G. Nel, R. P. Duttweiler, and M. A. Hillmyer, *Makromol. Chem.*, **191**, 365 (1990).

14. (a) K. B. Wagener, J. M. Boncella, and J. G. Nel, *Macromolecules*, **24**, 2649 (1991); (b) K. B. Wagener and K. Brzezinska, *Macromolecules*, **24**, 5273 (1991); (c) K. B. Wagener and ·D. W. Smith, Jr., *Macromolecules*, **24**, 6073 (1991); (d) K. B. Wagener, J. T. Patton, and J. M. Boncella, *Macromolecules*, **25**, 5273 (1992); (e) K. B. Wagener and J. T. Patton, *Macromolecules*, **26**, 249 (1993); (f) S. K. Cummings, D. W. Smith, Jr., and K. B. Wagener, *Macromol. Rapid Comm.*, **16**(5), 347 (1995); (g) J. D. Portmess and K. B. Wagener, *J. Polym. Sci., Part A*, **34**(7), 1353 (1996); (h) P. S. Wolfe and K. B. Wagener, *Macromol. Rapid Comm.*, **19**(6), 305 (1998); (i) P. S. Wolfe and K. B. Wagener, *Macromolecules*, **32**, 7961 (1999); (j) K. R. Brzezinska, R. Schitter, and K. B. Wagener, *J. Polym. Sci., Part A*, **38**, 1544 (2000).

15. (a) C. K. Ingold, *J. Chem. Soc.*, **119**, 305 (1921); (b) J. F. Thorpe and S. S. Deshapande, *J. Chem. Soc.*, **121**, 1430 (1922); (c) E. L. Eliel, *Stereochemistry of Carbon Compounds*, McGraw-Hill, New York, 1962, p. 197.

16. G. Odian, *Principles of Polymerization*, 3rd ed., Wiley, New York, 1991, pp. 74–77, 87.

17. N. Hadjichristidis, H. Iatrou, S. Pispas, and M. Pitsikalis, *J. Polym. Sci., Part A*, **38**, 3211 (2000).

18. (a) D. F. Shriver and M. A. Drezdzon, *The Manipulation of Air-Sensitive Compounds*, 2nd ed., Wiley, New York, 1986; (b) D. D. Perrin and W. L. F. Armarego, *Purification of Laboratory Chemicals*, 3rd ed., Pergamon, New York, 1988; (c) B. S. Furniss, A. J. Hannaford, P. W. G. Smith, and A. R. Tatchell, *Vogel's: Textbook of Practical Organic Chemistry*, 5th ed., Longman Group and Wiley and Sons, New York, 1989.

19. H. D. Maynard and R. H. Grubbs, *Tetrahedr. Lett.*, **40**, 4137 (1999).

20. J. A. Smith, K. R. Brzezinska, D. J. Valenti, and K. B. Wagener, *Macromolecules*, **33**, 3781 (2000).

21. (a) G. Natta, G. Dall'Asta, I. W. Bassi, and G. Carella, *Die Angew. Makromol. Chem.*, **91**, 87 (1966); (b) K. W. Scott, N. Calderon, E. A. Ofstead, W. A. Judy, and J. P. Ward, *Rubber Chem. Technol.*, **44**, 1341 (1971); (c) V. A. Kormer, I. A. Polcbaera, and T. L. Yufa, *J. Polym. Sci., Polym. Chem. Ed.*, **10**, 251 (1972); (d) W. Glenz, W. Holtrup, F. W. Kupper, and H. H. Meyer, *Die Angew. Makromol. Chem.*, **37**, 97 (1974); (e) L. Porri, P. Piversi, A. Lucherini, and R. Rossi, *Makromol. Chem.*, **176**, 3121 (1975); (f) T. J. Katz, S. J. Lee, and N. Acton, *Tetrahedr. Lett.*, **17**, 4247 (1976); (g) H. Sato, K. Okimoto, and Y. Tanaka, *J. Macromol. Sci.*, **A11**, 767 (1977); (h) A. J. Syatkowsky, T. T. Denisova, N. A. Buzina, and B. O. Babitsky, *Polymer*, **21**, 1112 (1980); (i) J. Finter, G. Wegner, E. J. Nagel, and R. W. Lenz, *Makromol. Chem.*, **181**, 1619 (1980); (j) H. Hocker, W. Reimann, L. Reif, and K. Riebel, *J. Mol. Cat.*, **8**, 191 (1980).

22. (a) T. J. Katz, S. J. Lee, and N. Acton, *Tetrahedr. Lett.*, 4247 (1976); (b) T. J. Katz, N. Acton, *Tetrahedr. Lett.*, **17**, 4251 (1976).

23. J. E. O'Gara, K. B. Wagener, and S. F. Hahn, *Makromol. Chem. Rapid Commun.*, **14**(10), 657 (1993).

24. M. S. Reisch, *Chem. Eng. News*, **75**, 14 (1997).

25. (a) H. Meerwein, *Angew. Chem.*, **60**, 78 (1948); (b) G. D. Buckley, L. H. Cross, and N. H. Ray, *J. Chem. Soc.*, **151**, 2714 (1950); (c) S. W. Kantor and R. C. Osthoff, *J. Am. Chem. Soc.*, **75**, 931 (1953); (d) L. Mandelkern, M. Hellmann, D. W. Brown, D. E. Roberts, and F. A. Quinn, Jr., *J. Am. Chem. Soc.*, **75**, 4093 (1953); (e) H. von Pechmann, *Ber. Dtsch. Chem. Ges.*, **31**, 2640 (1989).

26. E. W. Fawcett, R. Q. Gibson, M. H. Perrin, J. G. Patton, and E. G. Williams, British Patent 2,816,883 (to Imperial Chemical Industries), September 7, 1937.

27. (a) K. Ziegler, *Kunstoffe*, **45**, 506 (1955); (b) K. Ziegler, Belgium Patent 533,326, May 5, 1955.

28. (a) H. Sinn and W. Kaminsky, *Ziegler-Natta Catalysts. Advances in Organometallic Chemistry*, Academic, London, 1980, pp. 99–149; (b) D. E. James, *Ethylene Polymers: Encyclopedia of Polymer Science and Engineering*, 2nd ed., Wiley-Interscience, New York, 1986, p. 329.

29. (a) W. Keim, F. H. Kowaldt, R. Goddard, and C. Krüger, *Angew. Chem. Int. Ed.*, **17**, 466 (1978); (b) M. Peuckert and W. Keim, *Organometallics*, **2**, 594 (1983); (c) G. F. Schmidt and M. Brookhart, *J. Am. Chem. Soc.*, **107**, 1443 (1985); (d) V. M. Möhring and G. Fink, *Angew. Chem. Int. Ed.*, **24**, 1001 (1985); (e) U. Klabunde and S. D. Ittel, *J. Mol. Cat.*, **41**, 123 (1987); (f) G. Wilke, *Angew. Chem. Int. Ed.*, **27**, 185 (1988); (g) M. Brookhart, A. F. Volpe, Jr., D. M. Lincoln, I. T. Horvath, and J. M. Millar, *J. Am. Chem. Soc.*, **112**, 5634 (1990); (h) F. Rix and M. Brookhart, *J. Am. Chem. Soc.*, **117**, 1137 (1995); (i) L. K. Johnson, C. M. Killian, and M. Brookhart, *J. Am. Chem. Soc.*, **117**, 6414 (1995); (j) L. K. Johnson, C. S. Killian, S. D. Author, J. Feldman, E. F. McCord, S. J. McLain, K. A. Kreutzer, M. A. Bennett, E. B. Coughlin, S. D. Ittel, A. Parthasarathy, D. J. Tempel, and M. S. Brookhart, International Patent Application WO96/23010, 1996. (k) D. P. Long and P. A. Bianconi, *J. Am. Chem. Soc.*, **118**, 12453 (1996); (l) J. Feldman, S. J. McLain, A. Parthasarathy, W. J. Marshall, J. C. Calabrese, and S. D. Arthur, *Organometallics*, **16**, 1514 (1997); (m) J. S. Kim, J. H. Pawlow, L. M. Wojcinski II, S. Murtuza, S. Kacker, and A. Sen, *J. Am. Chem. Soc.*, **120**, 1932 (1998); (n) B. L. Small, M. Brookhart, and A. M. A. Bennett, *J. Am. Chem. Soc.*, **120**, 4049 (1998); (o) T. R. Younkin, E. F. Connor, J. I. Henderson, S. K. Friedrich, R. H. Grubbs, and D. A. Bansleben, *Science*, **287**, 460 (2000).

30. B. Wunderlich, *Macromolecular Physics*, Vol. 3, Academic, New York, 1980, pp. 275–278.

31. (a) B. Wunderlich and D. Poland, *J. Polym. Sci., Part A*, **1**, 357 (1963); (b) F. Gutzler and G. Wegner, *Colloid Polym. Sci.*, **258**, 776 (1980); (c) F. M. Mirabella, Jr., and E. A. Ford, *J. Polym. Sci., Part B: Polym. Phys.*, **25**, 777 (1987); (d) R. G. Alamo and L. Mandelkern, *Macromolecules*, **22**, 1273 (1989); (e) M. D. Failla, J. C. Lucas, and L. Mandelkern, *Macromolecules*, **27**, 1334 (1994); (f) W. S. Lambert and P. J. Phillips, *Polymer*, **37**, 3585 (1996).

32. (a) G. Wegner and F. Gutzler, *Coll. Polym. Sci.*, **258**, 776 (1980); (b) T. N. Bowmer and A. E. Tonelli, *Polymer*, **26**, 1195 (1985); (c) T. N. Bowmer and A. E. Tonelli, *Macromolecules*, **19**, 498 (1986); (d) A. E. Tonelli and M. Valenciano, *Macromolecules*, **19**, 2643 (1986); (e) D. E. Buerger and R. H. Boyd, *Macromolecules*, **22**, 2694 (1989); (f) D. E. Buerger and R. H. Boyd, *Macromolecules*, **22**, 2699 (1989); (g) M. A. Gomez, A. E. Tonelli, A. J. Lovinger, F. C. Schilling, M. H. Cozine, and D. D. Davis, *Macromolecules*, **22**, 4441 (1989); (h) G. D. Smith and R. H. Boyd,

Macromolecules, **24**, 2725 (1991); (i) G. D. Smith and R. H. Boyd, *Macromolecules*, **24**, 2731 (1991); (j) G. D. Smith, F. Liu, R. W. Devereaux, and R. H. Boyd, *Macro-molecules*, **25**, 703 (1992); (k) G. D. Smith and R. H. Boyd, *Macromolecules*, **25**, 1326 (1992); (l) N. Pourahmady and P. I. Bak, *J. Macromol. Sci.-Pure Appl. Chem.*, **A29**(11), 959 (1992); (m) F. Chowdhury, J. A. Haigh, L. Mandelkern, and R. G. Alamo, *Polym. Bull.*, **41**, 463 (1998); (n) S. Bistac, P. Kunemann, and J. Schultz, *Polymer*, **39**, 4875 (1998).

33. (a) K. Yakota and T. Hirabayashi, *Macromolecules*, **14**, 1613 (1981); (b) K. Yakota, T. Kouga, and T. Hirabayashi, *Polym. J.*, **15**, 349 (1983); (c) T. Hirabayashi, K. Yamauchi, and K. Yakota, *Macromolecules*, **24**, 4543 (1991); (d) K. Yakota, M. Miwa, T. Hirabayashi, and Y. Inai, *Macromolecules*, **25**, 5821 (1992); (e) W. Gerum, G. W. H. Höhne, W. Wilke, M. Arnold, and T. Wegner, *Macromol. Chem. Phys.*, **196**, 3797 (1995).

34. M. D. Watson and K. B. Wagener, *Macromolecules*, **33**, 3196 (2000).

35. D. J. Valenti, K. B. Wagener, and S. F. Hahn, *Macromolecules*, **30**, 6688 (1997).

36. M. D. Watson and K. B. Wagener, *Macromolecules*, **33**, 5411 (2000).

37. J. D. Hoffman, *Polymer*, **24**, 3 (1983).

38. (a) U. Zucchini and G. Cecchin, *Adv. Poly. Sci.*, **51**, 101 (1983); (b) W. Kaminsky, *Macromol. Chem. Phys.*, **197**, 3907 (1996).

39. R. P. Quirk and M. A. A. Alsamarrale, in *Polymer Handbook*, 3rd ed., J. Brandup and E. H. Immergut (Eds.), Wiley, New York, 1989, pp. 15–26.

40. (a) M. Gopalan and L. Mandelkern, *J. Phys. Chem.*, **71**, 3833 (1967); (b) L. Mandel-kern, A. Prasad, R. G. Alamo, and G. M. Stack, *Macromolecules*, **23**, 3696 (1990).

41. (a) J. E. McGrath and L. Yligor, *Adv. Polym. Sci.*, **86** (1988); (b) M. Zeldin, K. J. Wynne, and H. R. Allcock (Eds.), *Inorganic and Organometallic Polymers: Macro-molecules Containing Silicon, Phosphorous, and Other Inorganic Elements*, American Chemical Society, Washington, DC, 1988; (c) J. M. Zeigler and F. W. G. Fearon (Eds.), *Silicon-Based Polymer Science*, American Chemical Society, Washington, DC, 1990; (d) M. A. Brook, *Silicon in Organic, Organometallic, and Polymer Chemistry*, Wiley, New York, 2000.

42. (a) M. Grassies and I. G. Macfarlane, *Eur. Polym. J.*, **14**, 875 (1978); (b) P. R. Dvor-nic and R. W. Lenz, *High Temperature Siloxane Elastomers*, Huthig and Verlag, New York, 1990.

43. D. W. Smith, Jr. and K. B. Wagener, *Macromolecules*, **26**, 1633 (1993).

44. (a) B. D. Karstedt, U.S. Patent 3 775 452, 1973; (b) G. Chandra, P. Y. Lo, P. B. Hitchcock, and M. F. Lappert, *Organometallics*, **6**, 191 (1987); (c) P. B. Hitchcock, M. F. Lappert, and N. J. W. Warhurst, *Angew. Chem., Int. Ed.*, **30**, 438 (1991).

45. D. W. Smith, Jr. and K. B. Wagener, *Macromolecules*, **26**, 3533 (1993).

46. S. Cummings, E. Ginsburg, R. Miller, J. Portness, D. W. Smith, Jr., and K. Wagener, in *Step Growth Polymers for High Performance Materials: New Synthetic Methods*, J. L. Hedrick and J. W. Labadie (Eds.), ACS Symposium Series No. 624, American Chemical Society, Washington, DC, 1996, p. 113.

47. (a) J. D. Anderson, J. P. Portmess, S. K. Cummings, and K. B. Wagener, *Polym. Prepr. (Am. Chem. Soc. Div. Polym. Chem.)*, **36**(2), 162 (1995); (b) A. C. Church, J. H. Pawlow, and K. B. Wagener, *Polym. Prepr. (Am. Chem. Soc. Div. Polym. Chem.)*, **40**(1), 129 (1999); (c) A. C. Church, J. H. Pawlow, and K. B. Wagener, *Polym. Prepr. (Am. Chem. Soc. Div. Polym. Chem.)*, **42**(1), 235 (2001);

(d) A. C. Church, J. H. Pawlow, and K. B. Wagener, *Macromolecules*, **35**, 5746 (2002).

48. J. C. Marmo and K. B. Wagener, *Macromolecules*, **26**, 2137 (1993).

49. J. C. Marmo and K. B. Wagener, *Macromolecules*, **28**, 2602 (1995).

50. (a) K. B. Wagener and J. Patton, *Macromolecules*, **26**, 249 (1993); (b) J. E. O'Gara, J. D. Portmess, and K. B. Wagener, *Macromolecules*, **26**, 2837 (1993).

51. M. D. Watson and K. B. Wagener, *Macromolecules*, **33**, 1494 (2000).

52. H. L. Hsieh, N. Roderick, and P. Quirk, *Anionic Polymerization: Principles and Practical Applications*, Marcel Dekker, New York, 1996.

53. K. B. Wagener and J. C. Marmo, *Macromol. Rapid Commun.*, **16**, 557 (1995).

54. J. S. Vrentas and J. L. Duda, in *Encyclopedia of Polymer Science and Engineering*, Vol. 15, 2nd ed., J. I. Kroschwitz, (Ed.), Wiley-Interscience, New York, 1985, p. 36.

55. M. D. Watson and K. B. Wagener, *Macromolecules*, **33**, 8963 (2000).

9 Nontraditional Step-Growth Polymerization: Transition Metal Coupling

Qiao-Sheng Hu

Department of Chemistry, City University of New York — College of Staten Island, Staten Island, New York 10314

9.1 INTRODUCTION

9.1.1 Historical Perspective

In 1977, Yamamoto and Yamamoto adopted the Kumada coupling,[1] the nickel-catalyzed coupling reaction of aryl halides with Grignard reagents (Scheme 9.1), for the synthesis of poly(*para*-phenylene) and related polymers (Scheme 9.2).[2] This marked an important step of using transition metal coupling for step-growth polymerization.* Although Grignard reagents and Ni(II) complexes are the commonly used combination for this coupling, other organometallic reagents such as organozincs and transition metal catalysts such as palladium complexes have also been used for this coupling polymerization later on (Scheme 9.3).[4,5]

$$R-X \quad + \quad R'MgX \xrightarrow{\text{Ni(II)}} R-R' \quad \text{Kumada coupling}$$

R = aryl, alkenyl

Scheme 9.1

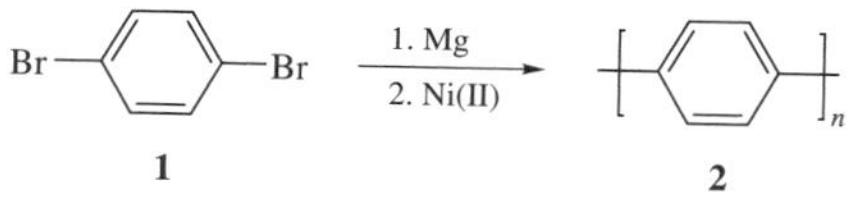

Br—⟨◯⟩—Br $\xrightarrow[\text{2. Ni(II)}]{\text{1. Mg}}$ ⟦⟨◯⟩⟧$_n$

     1   2

Scheme 9.2

---

* Cu-mediated Ullman reaction has been used for the polymerization of dihaloaryls. For example, see ref. 3. This type of polymerization as well as other transition-metal-mediated reactions that involve radicals in the polymerization process is not included in this chapter.

Synthetic Methods in Step-Growth Polymers. Edited by Martin E. Rogers and Timothy E. Long
© 2003 John Wiley & Sons, Inc. ISBN: 0-471-38769-X

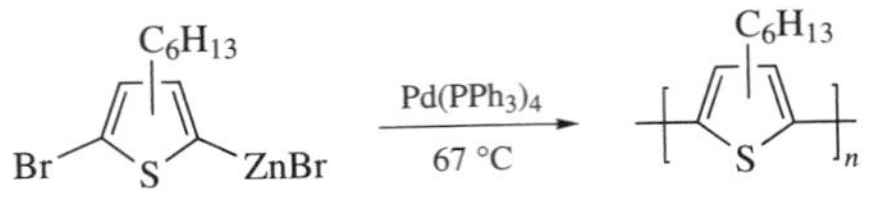

Scheme 9.3

The coupling reaction of aryl–alkenyl halides with alkenes in the presence of a palladium catalyst and a base is known as the Heck coupling (Scheme 9.4).[6] Since the early 1980s, this type of coupling reaction has been used for the synthesis of poly(arylenevinylene) and related polymers by polymerization of AB- or AA/BB-type of monomers (Scheme 9.5).[7]

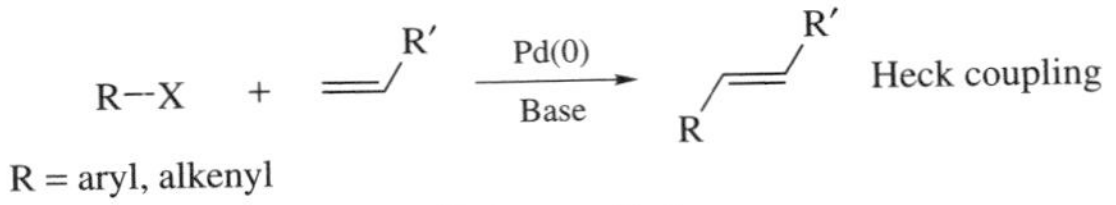

Scheme 9.4

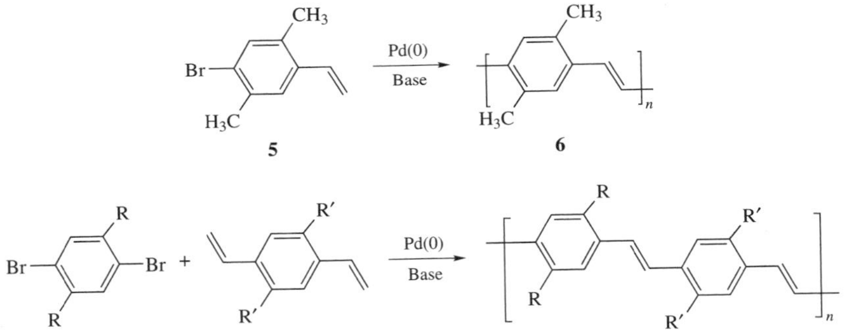

Scheme 9.5

By using a cocatalyst of Pd–Cu, aryl–alkenyl halides couple efficiently with alkynes to generate the disubstituted alkynes (Scheme 9.6).[8] This coupling reaction has been applied to the synthesis of polyphenyleneethynylenes by Yamamoto et al. in 1984 (Scheme 9.7).[9a] Due to the ready availability of diacetylenes and the mild coupling condition, this strategy has been widely used for the preparation of many poly(aryleneethynylene)s.[9]

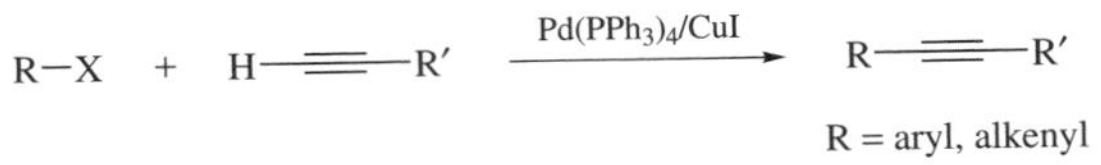

Scheme 9.6

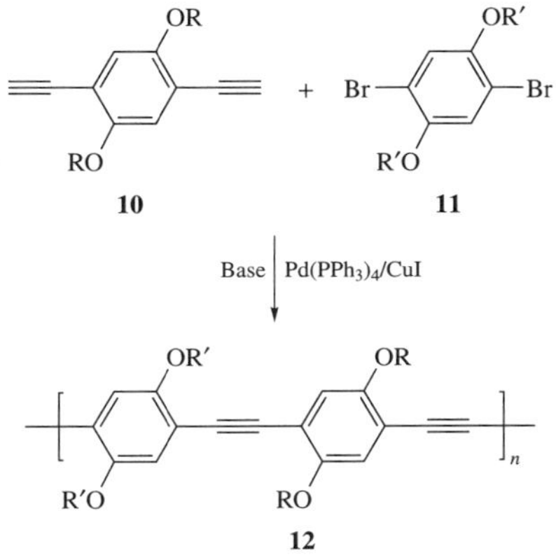

Scheme 9.7

In 1988, Yamamoto et al. reported the Ni(0)-mediated homo coupling of aryl halides for step-growth polymerization (Scheme 9.8).[10,11] Unlike the Kumada reaction that involves the highly nucleophilic Grignard reagents, this process can tolerate many functionalities and thus allows the synthesis of many highly functionalized polymers. Although a stoichiometric amount of Ni(0) complex was first introduced and is still very often used for the homo coupling of aryl halides, it is also possible to use a catalytic amount of Ni(0) with a stoichiometric amount of zinc for the polymerization (Scheme 9.9).[12]

$$\mathrm{Ar{-}X} \xrightarrow[\text{or Ni(II)/Zn}]{\text{Ni(0)}} \mathrm{Ar{-}Ar}$$

Scheme 9.8

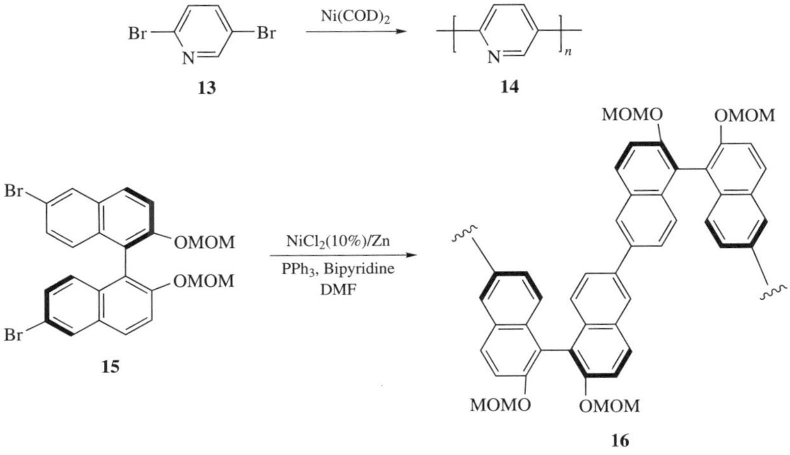

Scheme 9.9

Polymerization using the Stille coupling, the cross-coupling of aryl–alkenyl halides with organotins in the presence of palladium catalysts (Scheme 9.10),[13] appeared in 1989 (Scheme 9.11).[14] The low nucleophilicity of organotins makes it possible to use functionalized monomers for the polymerization.[15]

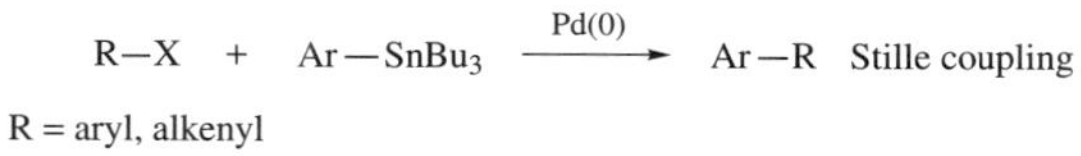

$$R{-}X \ + \ Ar{-}SnBu_3 \ \xrightarrow{Pd(0)} \ Ar{-}R \quad \text{Stille coupling}$$

R = aryl, alkenyl

Scheme 9.10

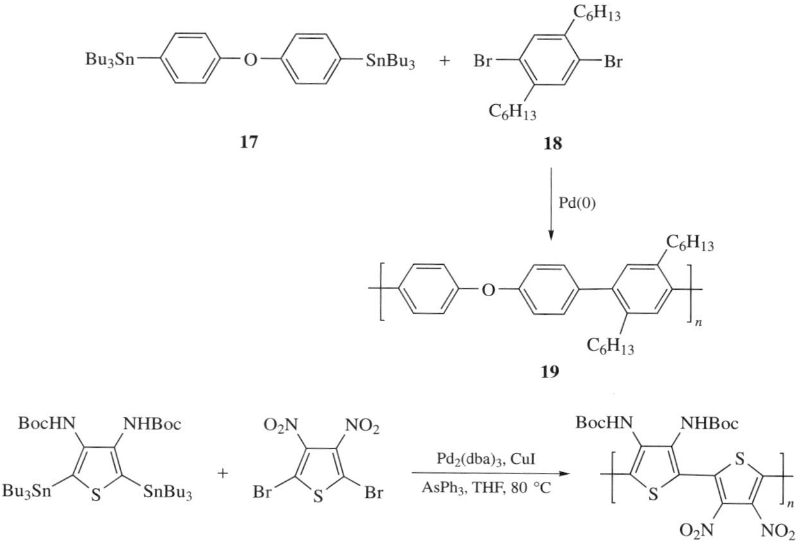

17   18

Pd(0)

19

20   21   22

Pd$_2$(dba)$_3$, CuI
AsPh$_3$, THF, 80 °C

Scheme 9.11

At about the same time, the application of the Suzuki coupling, the cross-coupling of boronic acids with aryl–alkenyl halides in the presence of a base and a catalytic amount of palladium catalyst (Scheme 9.12),[16] for step-growth polymerization also appeared. Schlüter et al. reported the synthesis of soluble poly(*para*-phenylene)s by using the Suzuki coupling condition in 1989 (Scheme 9.13).[17] Because aryl–alkenyl boronic acids are readily available and moisture stable, the Suzuki coupling became one of the most commonly used methods for the synthesis of a variety of polymers.[18]

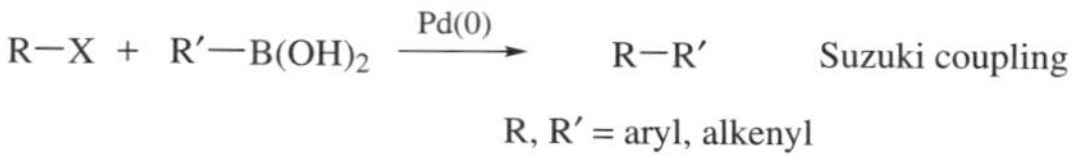

$$R{-}X \ + \ R'{-}B(OH)_2 \ \xrightarrow{Pd(0)} \ R{-}R' \quad \text{Suzuki coupling}$$

R, R' = aryl, alkenyl

Scheme 9.12

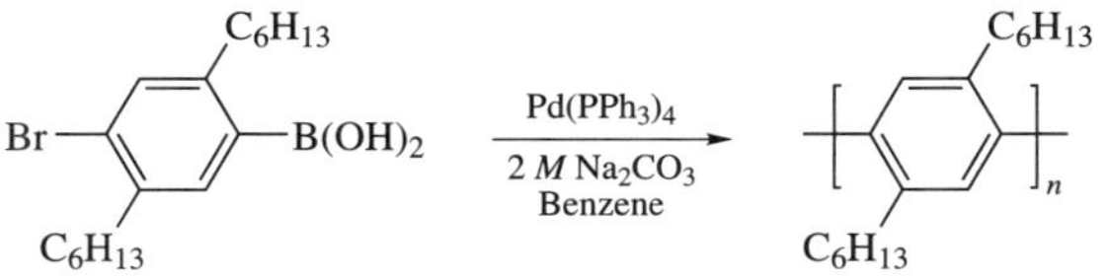

Scheme 9.13

Recently developed catalyst systems make it possible to construct carbon (sp^2)–carbon (sp^3) bonds and carbon–nitrogen bonds under mild conditions (Scheme 9.14).[19,20] These new developments have also been incorporated into step-growth polymerization. Kanbara et al. reported the synthesis of polyanilines and related polymers in 1996 (Scheme 9.15).[21] Wang and Wu reported the synthesis of polyketones in 1999 (Scheme 9.16).[22]

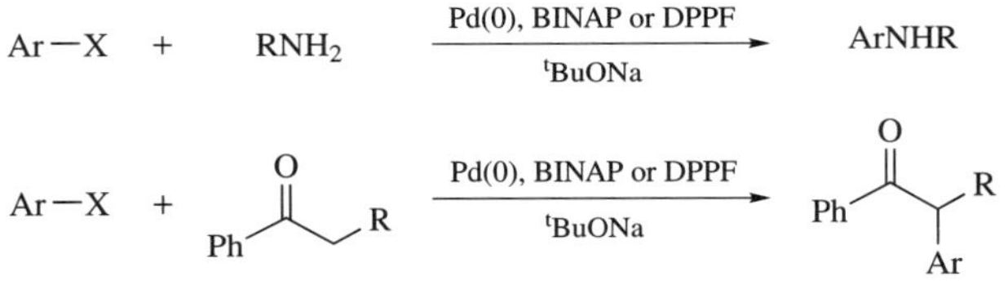

Scheme 9.14

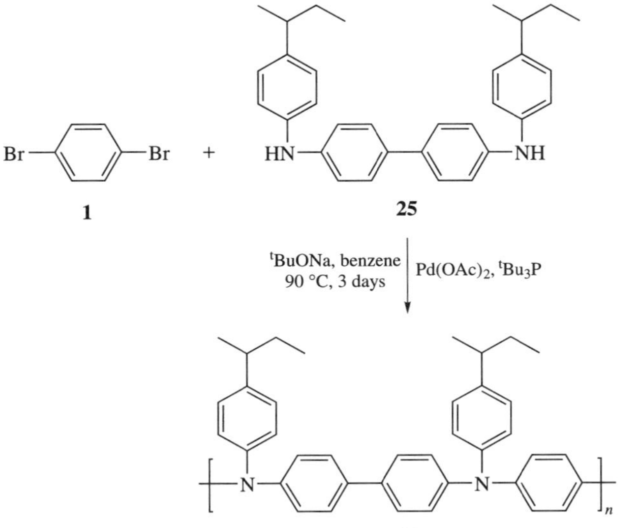

Scheme 9.15

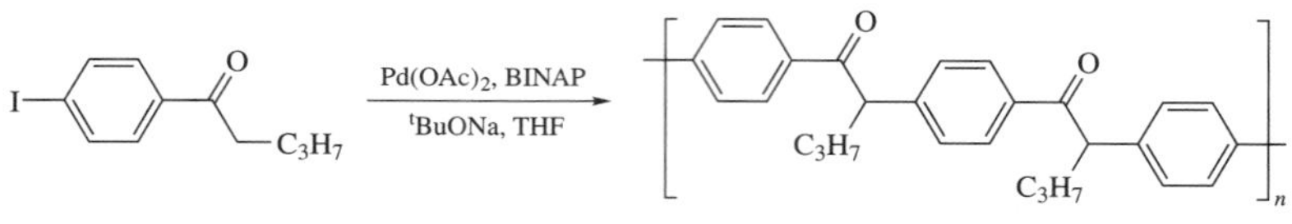

Scheme 9.16

9.1.2 Applications

In general, transition-metal-catalyzed coupling polymerizations have several advantages: (a) carbon–carbon bond or carbon–heteroatom bond formation can be achieved under mild conditions in the presence of many functionalities and (b) high regio- and chemoselectivities can be achieved to lead to well-defined structures. Because of these advantages, transition metal coupling polymerization has become a powerful tool for the synthesis of different types of polymers that cannot be easily accessible or are impossible by using other methods. These polymers can be generally divided into three classes: (1) linear polymers containing no chiral unit, (2) linear polymers containing chiral units, and (3) hyperbranched and dendritic polymers.

Linear polymers such as poly(*para*-phenylene)s, poly(*para*-phenylenevinylene)s, and polythiophenes have potential applications in the fields of organic conductors, nonlinear optics, electroluminescent materials, solar cells, high-energy density batteries, and modified electrodes.[23] Although these polymers have been prepared by other chemical and electrochemical methods, the nature of the polymerization processes casts doubts about the connection between each repeat unit and makes it difficult to prepare their analogues with desired functional groups. Early applications of transition-metal-catalyzed couplings for the step-growth polymerization have been in the synthesis of these polymers. Because of the high chemo- and stereoselectivity of the transition-metal-catalyzed reactions, polymers synthesized by transition metal coupling polymerization strategy have unambiguous bond connection between each repeat unit. It is also possible to introduce functional groups into the polymer either in the side chains or in the main chain to tune the properties of the polymers. This strategy has thus been widely used for the synthesis of different types of linear polymers with interesting optical and electrical properties. For example, they have been used for the construction of polymeric molecular sensors such as **29** and **30**[24,25]:

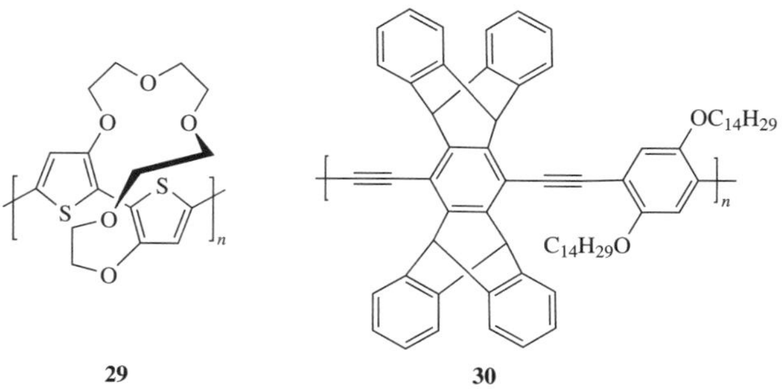

29   **30**

Optically active polymers are potentially very useful in areas such as asymmetric catalysis, nonlinear optics, polarized photo and electroluminescence, and enantioselective separation and sensing.[26] Transition metal coupling polymerization has also been applied to the synthesis of these polymers.[27] For example, from the Ni(II)-catalyzed polymerization, a regioregular head-to-tail polymer **32** was obtained (Scheme 9.17).[28] This polymer is optically active because of the optically active chiral side chains.

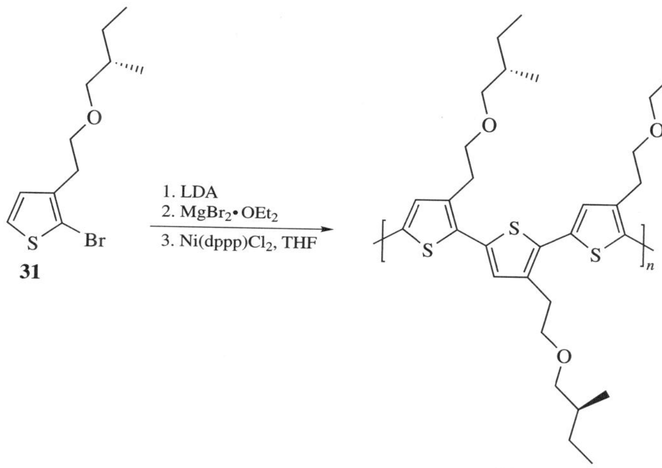

Scheme 9.17

Transition metal coupling polymerization has also been used to synthesize optically active polymers with stable main-chain chirality such as polymers **33**, **34**, **35**, and **36** by using optically active monomers.[29-31] These polymers are useful for chiral separation and asymmetric catalysis. For example, polymers **33** and **34** have been used as polymeric chiral catalysts for asymmetric catalysis. Due

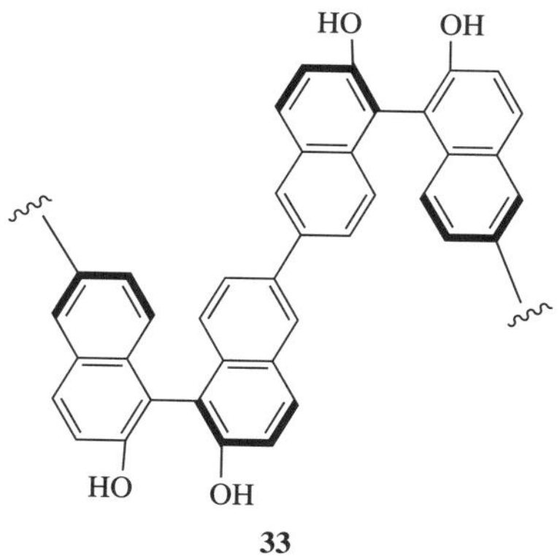

33

to the size and solubility difference between the polymers and monomers, using polymeric chiral catalysts not only can simplify the reaction work-up and allow the recovery and reuse of the catalysts by simple filtration or precipitation but also can make it possible to carry out reactions in a continuous mode. Polymer **39**, synthesized from the cross-coupling of **37** and **38**, has been found to be the best polymeric catalyst reported so far for the diethyl zinc addition to aldehydes (Scheme 9.18)[18a]:

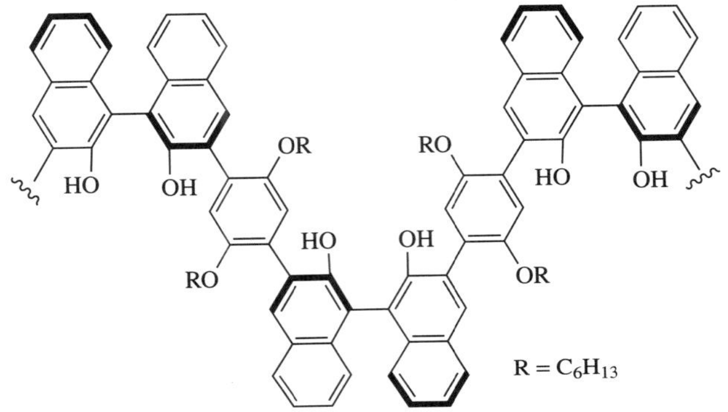

34

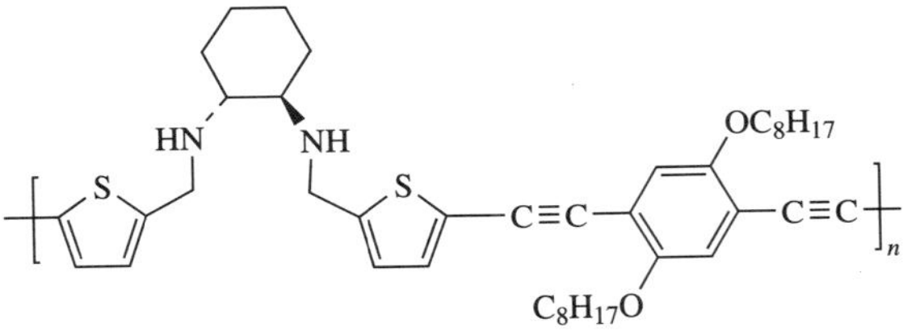

35

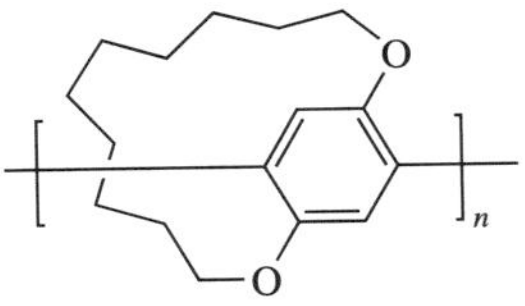

36

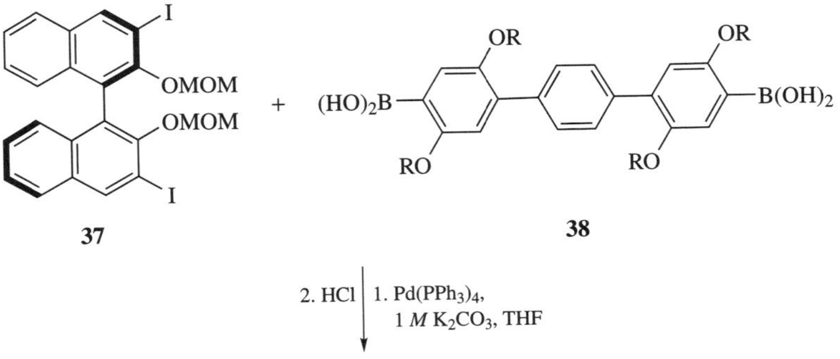

37 **38**

39

Scheme 9.18

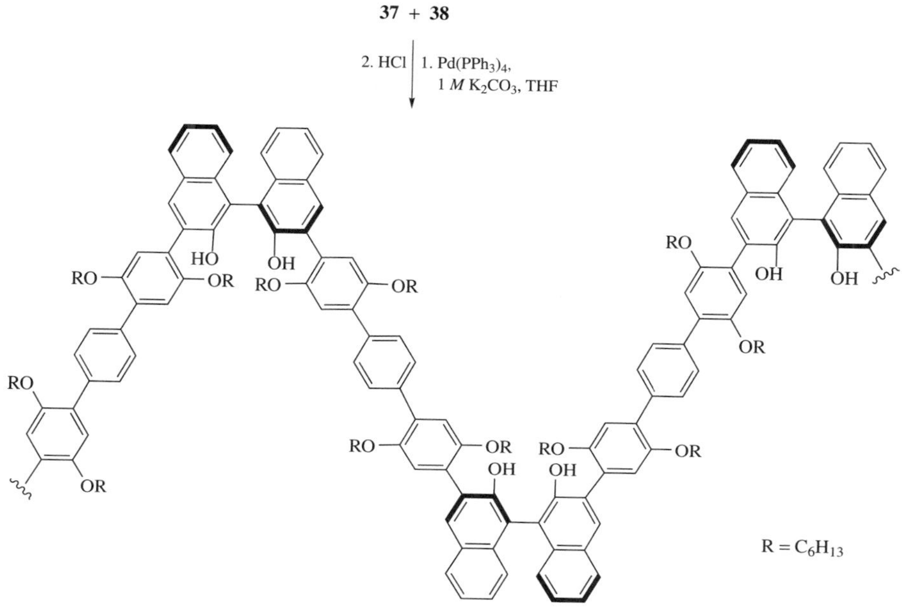

37 + 38

2. HCl | 1. Pd(PPh₃)₄,
          1 *M* K₂CO₃, THF

R = C₆H₁₃

39

Scheme 9.18   (*Continued*).

Hyperbranched and dendronized polymers such as **40**, **41**, and **42** have also been synthesized using the transition metal coupling strategies in recent years.[32] These polymers are fundamentally different from those traditional linear polymers. They possess dendritic arms within the polymer or along the polymer backbone. It is believed that they possess interesting properties and have potential applications in many fields such as nanotechnology and catalysis:

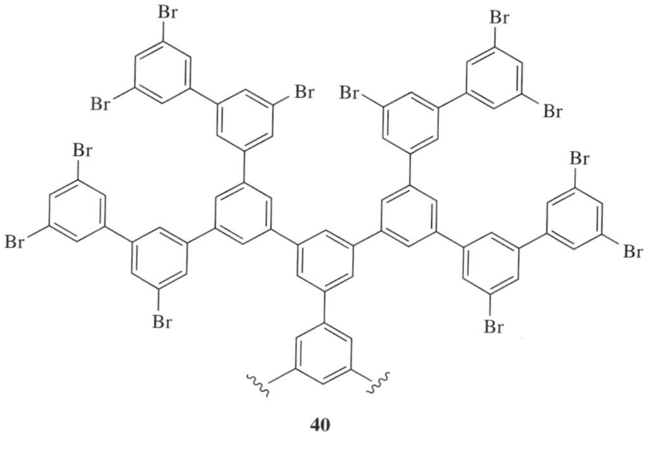

40

41

42

9.1.3 Research Trends and Critical Issues

Like other step-growth polymerization methods, factors such as the monomer purity, ratio of the monomers, conversion, temperature, and concentration will greatly influence the transition metal coupling polymerization. These factors have to be taken into account when higher molecular weight polymers need to be prepared.[33]

Because low-valent transition metals such as Ni(0) and Pd(0) are air and/or moisture sensitive,[34] the exclusion of oxygen and/or moisture is also crucial for the polymerization. Failure to exclude oxygen will deactivate the catalysts, thus causing the termination of the polymerization and influencing the polymerization degree.

The side reactions existing in the transition metal coupling reactions are sometimes responsible for the low molecular weight. These side reactions can be classified in two types: (1) reduction of monomer and (2) coupling of monomer with a nonreactive chain end. These side reactions can be minimized by proper choice of reaction temperature, catalysts, and catalyst loading.

Development of more efficient transition metal catalyst systems including using novel and efficient ligands has been one of the focuses in organometallic chemistry.[35] The developments in this area will allow not only to synthesize polymers under mild conditions with higher or desired molecular weights but also to use less expensive, more readily available materials for the polymerizations.

Conjugated polymers, including optically active polymers and dendronized polymers that are very useful in electrical and optical fields and asymmetric catalysis, will continue to attract interest from chemists and materials scientists. It is well anticipated that more and more polymers with interesting structures and properties will be synthesized from the transition metal coupling strategy.

9.2 STRUCTURE–PROPERTY RELATIONSHIPS

9.2.1 General Consideration

Because of the unambiguous reactive sites of monomers and the high chemo- and stereoselectivity of transition-metal-catalyzed coupling reactions, polymers prepared by transition metal coupling have predictable chemical structures. Functional groups can be easily and selectively introduced at the desired position within the polymer chains. Therefore, polymers with specific properties can be rationally designed and synthesized.

9.2.2 Role of Chemical Structure

Both the main chain and side chains can greatly influence the properties of polymers. By varying the side chains and using different linkers, properties of polymers such as the solubility and conjugation along the polymer backbone can be greatly altered.

Nonconjugated polymers containing flexible backbones such as **43** are generally more soluble in common organic solvents than the rigid, conjugated polymers[36]:

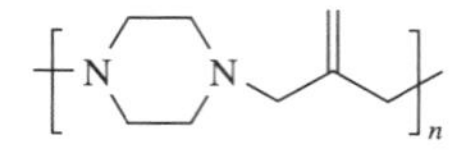

43

Conjugated polymers prepared from homo- and cross-coupling of aryl–alkenyl halides with organometallic reagent and its analogues usually have longer conjugation than their monomers with well-defined structures. The degree of the red shift depends on the steric hindrance around the bonds connecting the monomeric units. Polymers with inherently less steric hindrance in their intramonomer bonds showed larger red shift than those with more steric hindrance. For example, polythiophenes and polymethylthiozoles show larger red shifts from their corresponding monomers than poly(*para*-phenylene)s and polypyridines do (Fig. 9.1).[37–39]

Introduction of side chains will generally increase the solubility of the polymers but usually disrupt the conjugation along the polymer backbone. By properly choosing the side chains and linker, it is possible to prepare polymers with desired conjugation and solubilities.[40]

Copolymers composed of electron-donating and electron-withdrawing units are considered to have intramolecular charge transfer structure.[41] These copolymers show absorption bands at wavelengths longer than those of the corresponding homopolymers (Fig. 9.2).

Maximum conjugation can be achieved by converting a conjugated polymer to a ladder or fully conjugated planar polymer (Scheme 9.19).[42]

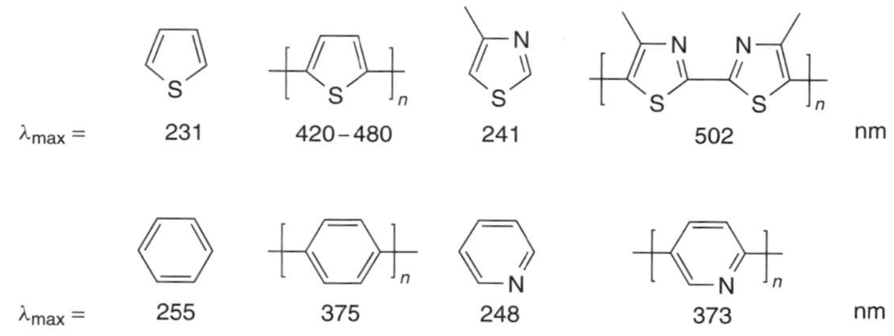

| $\lambda_{max} =$ | 231 | 420–480 | 241 | 502 | nm |

| $\lambda_{max} =$ | 255 | 375 | 248 | 373 | nm |

Figure 9.1

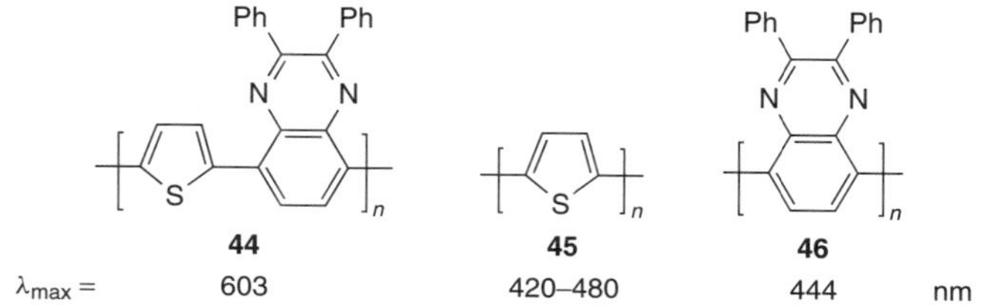

| | **44** | **45** | **46** | |
| $\lambda_{max} =$ | 603 | 420–480 | 444 | nm |

Figure 9.2

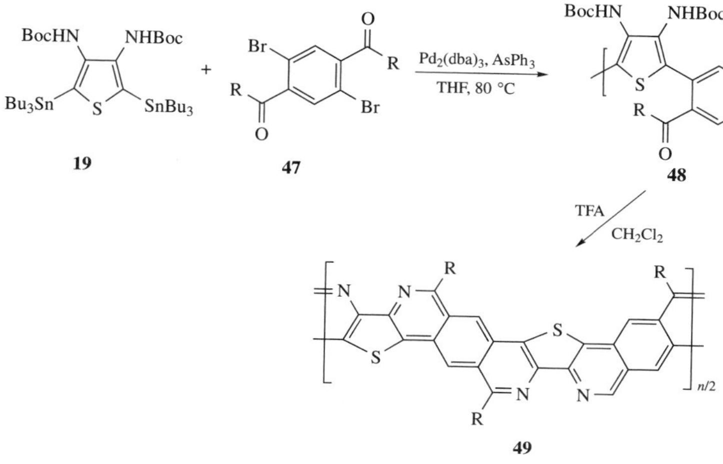

Scheme 9.19

Incorporation of chiral units into polymers generates optically active polymers.[27] Two types of optically active polymers could be obtained according to where the chiral units reside: optically active polymers with chirality derived from chiral side chains and optically active polymers with chirality derived from the chiral main chain. The circular dichroism (CD) measurement of **32**, an optically active polymer with chiral side chains, showed that the chiral substituents have induced main-chain chirality. The induced main-chain chirality disappeared at higher temperature and appeared upon cooling. This type of chiral conjugated polymer is potentially useful in reversing optical recording[28]:

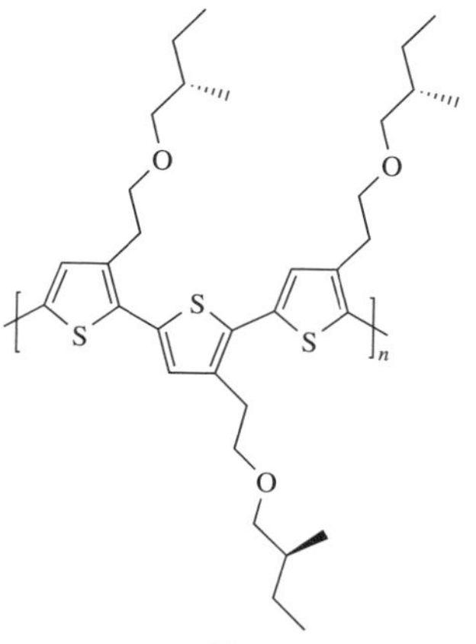

32

Optically active 1,1′-binaphthyl-based polymers incorporate chiral 1,1′-binaphthyl units into the polymer backbone and possess stable main-chain chirality.[43] Studies about this new class of optically active polymers showed that properties of the 3,3'-linked binaphthyl-based optically active polymers are very different from that of the 6,6'-linked polymers. For example, while 3,3'-linked polymers such as **34** are excellent polymeric chiral catalysts for asymmetric diethyl zinc addition to aldehydes and asymmetric hetero-Diels–Alder reactions, the 6,6'-linked polymers such as **33** only yield product with less than 40% enantiomeric excess (ee)[44,45]:

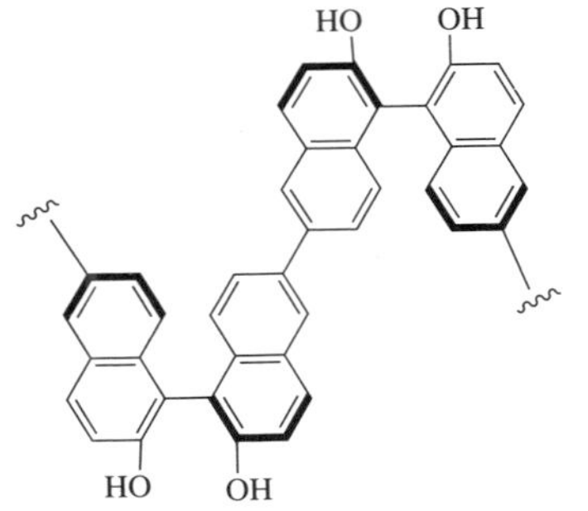

34

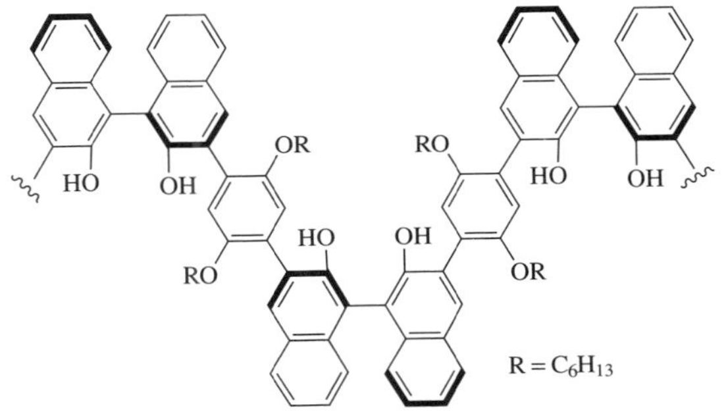

$R = C_6H_{13}$

35

   Optically active polymers containing planar chirality have also been synthesized and studied. Polymer **36** containing optically active cyclophanes exhibits chiraloptical activity in the region of $\pi - \pi^*$ transition with $\lambda_{max} = 332$ nm. A nonhelical chiral conjugated ladder polymer **50** displays extremely high Cotton effect in the long-wavelength $\pi - \pi^*$ absorption with $\lambda_{max} = 461$ nm[46]:

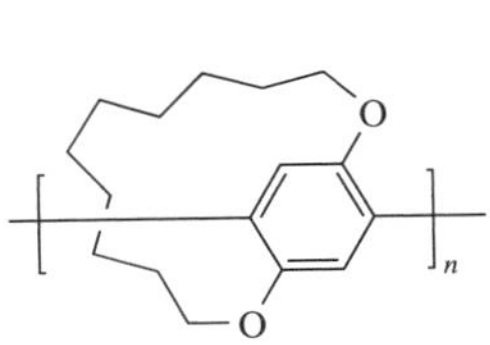

36

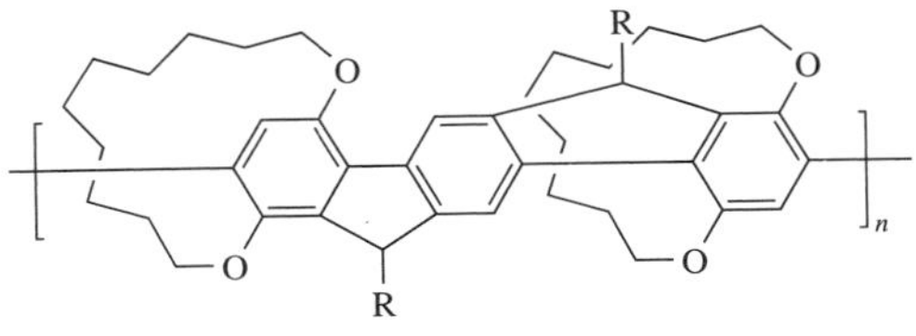

50

Due to the dendritic arms among or along the polymer main chain, hyper-branched and dendronized polymers are believed to have different solution and solid-state properties such as different solubility, viscosity, and shapes than that of their corresponding linear polymers. The unique structures of these polymers lead to unique properties. For example, it has been found that those monoden-drons attached to the polymer backbone can be light-harvesting antenna to harvest light which can migrate to the polymer main chain, thus making the dendronized polymers such as **51** very valuable for the light-emitting materials[47]:

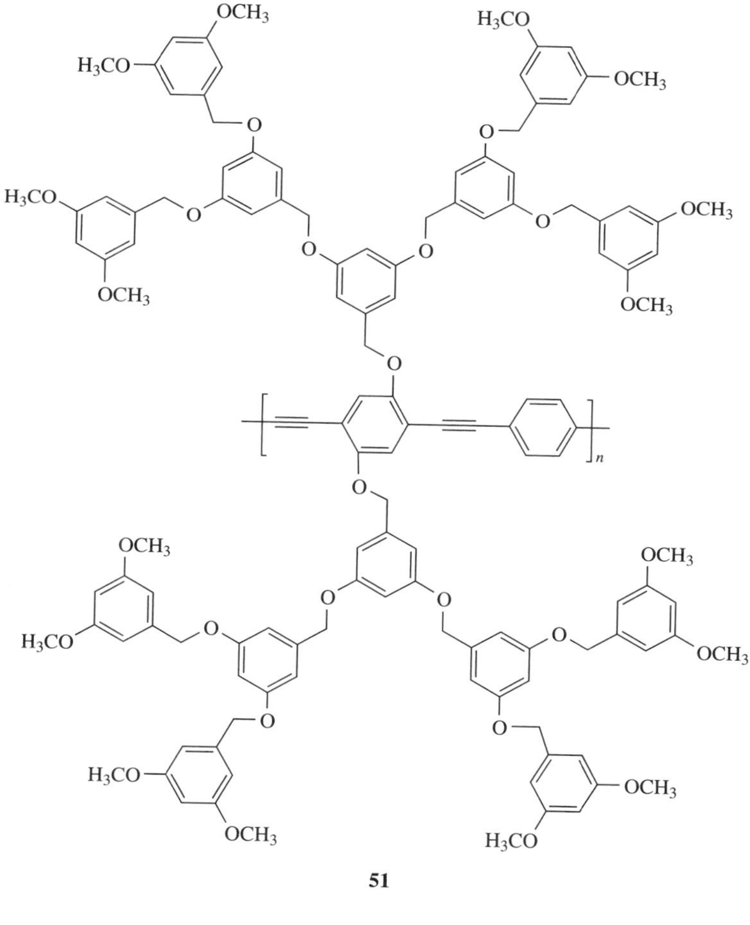

51

9.2.3 Role of Microstructure and Architecture

The microstructure and architecture of polymers can also greatly influence the properties of the polymers. For example, poly(3-substituted thiophene)s could have three microstructure joints: s-*trans* (head to tail), s-*trans* (head to head), and s-*cis* (head to tail) (Fig. 9.3). The regioregular head-to-tail poly(3-substituted thiophene)s exhibit higher electrical conductivity values and higher

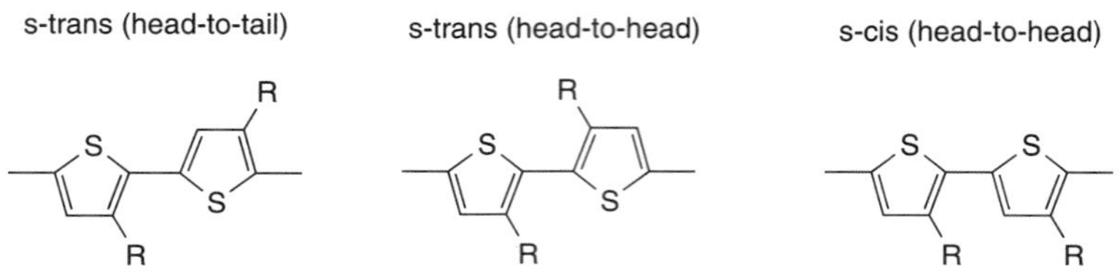
s-trans (head-to-tail)
s-trans (head-to-head)
s-cis (head-to-head)
R
S
R

Figure 9.3

crystallinity compared with those of regio-irregular polymers. They also form stacked structures.[48,49]

Conjugated polymers without sterically hindered side chains can usually line up to form linear materials. Fluorescence measurement reveals that this arrangement greatly reduced the quantum efficiency of the polymer. To reduce the interpolymer chain quenching, larger side chains have been introduced in the polymer backbone. For example, Zhou and Swager synthesized and studied a series of poly(phenyleneethynylene)s such as polymer **52** and **30**. Their study showed that the fluorescence of polymer **30** has increased significantly due to the introduction of the larger anthracene groups.[9c]

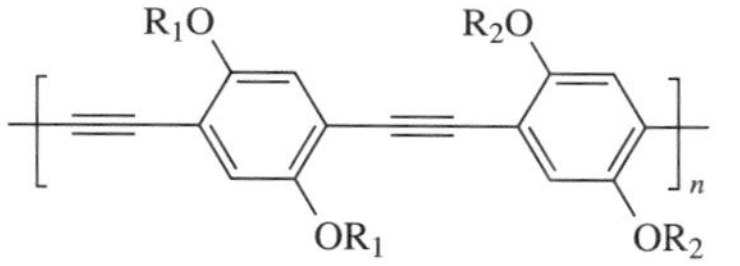
R_1O
R_2O
OR_1
OR_2
n

$R_1 = C_8H_{17}$, $R_2 = C_{16}H_{33}$

52

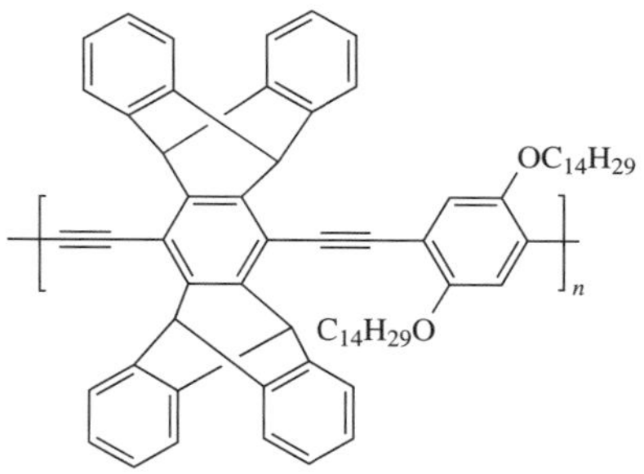
$OC_{14}H_{29}$
$C_{14}H_{29}O$
n

30

The main chain of dendronized polymers, due to the large size of the monodendrons, is usually forced to take a stretched shape; thus the whole molecule exists as a rigid rod architecture both in solution and in the solid state.[32d] Depending on the backbone stiffness, the degree of monodendron coverage, and the size of the monodendron, the architecture of these macromolecules is no longer a sphere but a cylinder; this dictates the properties of the dendronized polymers.

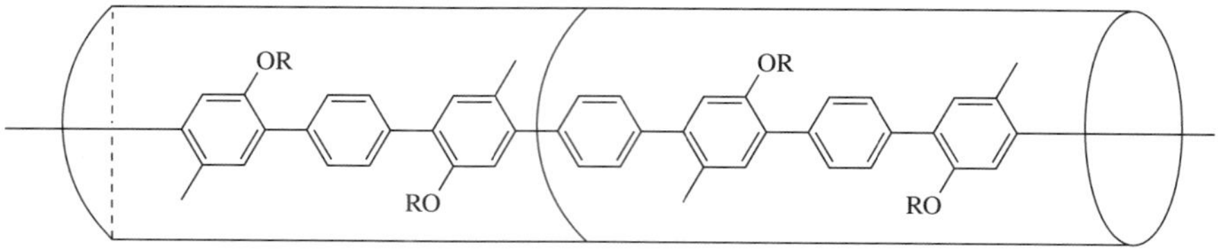
OR
RO

R = monodendron group

9.3 OVERVIEW OF CHEMISTRY AND ANALYTIC TECHNIQUES

9.3.1 Chemistry and Catalysis

Transition metal coupling reactions used for step-growth polymerization can be generally divided into the following four classes: (a) coupling reactions of aryl–alkenyl halides with organometallic reagents, (b) coupling reactions of aryl–alkenyl halides with alkenes, (c) homo-coupling reactions of aryl halides, and (d) coupling reactions of aryl halides with ketones and amines. Transition metal coupling polymerization is considered to follow similar reaction mechanisms.

9.3.1.1 Coupling Reactions of Aryl–Alkenyl Halides with Organometallic Reagents and Related Nucleophiles[50]

The general mechanism of coupling reactions of aryl–alkenyl halides with organometallic reagents and nucleophiles is shown in Fig. 9.4. It contains (a) oxidative addition of aryl–alkenyl halides to zero-valent transition metal catalysts such as Pd(0), (b) transmetallation of organometallic reagents to transition metal complexes, and (c) reductive elimination of coupled product with the regeneration of the zero-valent transition metal catalyst.

Transition metal complexes that are easy to handle and store are usually used for the reaction. The catalytically active species such as Pd(0) and Ni(0) can be generated in situ to enter the reaction cycle. The oxidative addition of aryl–alkenyl halides can occur to these species to generate Pd(II) or Ni(II) complexes. The relative reactivity for aryl–alkenyl halides is RI > ROTf > RBr > RCl (R = aryl–alkenyl group). Electron-deficient substrates undergo oxidative addition more readily than those electron-rich ones because this step involves the oxidation of the metal and reduction of the organic aryl–alkenyl halides. Usually

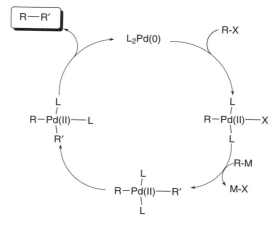

Figure 9.4

the slow step in the catalytic cycle is the transmetallation step. Therefore, side reactions may occur to oxidative product RPdX(L$_2$) prior to transmetallation.

A common side reaction in the phosphine-containing transition metal complexes involves the exchange of the organic group of Pd(II) species and the organophosphine ligands. The R group can also transfer to the P atom of the organophosphine ligands to form phosphonium salts (Scheme 9.20).

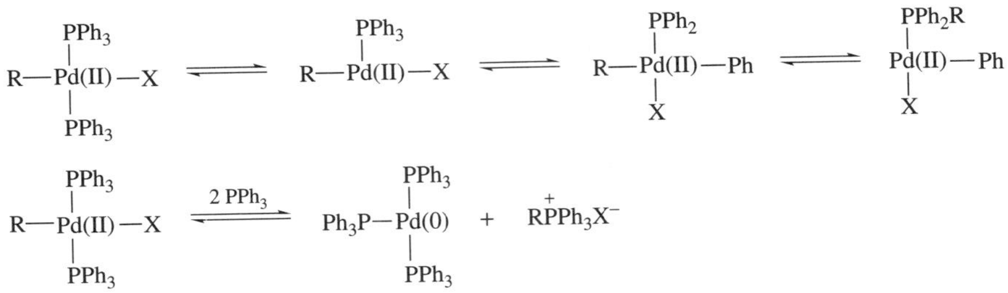

Scheme 9.20

The transmetallation reaction involves the transfer of the organic group from an organometallic species to a Pd(II) species and produces a trans Pd(II) species. Isomerization from the trans arrangement to a cis one is necessary prior to the reductive elimination step. Reductive elimination yields the coupled product and regenerates the transition metal catalyst. Because the reductive elimination is very fast, competing reactions leading to by-products are usually not observed.

Organometallic reagents and related nucleophiles used for step-growth polymerization include Grignard reagents, organozincs, organotins, organoboron reagents, and organocopper reagents[51,52]:

1. *Grignard Reagents and Organozincs.* Aryl, alkenyl, and alkyl Grignard reagents, arylzincs have been used for the polymerization. Besides Ni(II) complexes such as Ni(PPh$_3$)$_2$X$_2$ (X = Cl, Br, I) that are commonly used for the coupling, palladium complexes such as Pd(PPh$_3$)$_4$ and CuBr have also been employed as catalysts for the coupling. Aryl–alkenyl halides including chlorides and tosylates can be used as substrates for the reaction. The high nucleophilicity of Grignard reagents prevented the use of functionalized monomers that can accept the nucleophilic attack. Due to the moisture and air-sensitive nature of these reagents, the polymerization should be conducted under anhydrous, inert atmosphere conditions.

2. *Organotins.* The organotin reagents have much lower nucleophilicity than that of the Grignard reagents, thus allowing the use of a variety of functionalized monomers for the polymerization. Aryl–alkenyl iodides, bromides and tosylates have been used as substrates. Palladium complexes are commonly employed as catalysts for the reaction. Because the catalysts can be destroyed

upon exposure to O_2, the reaction is air-sensitive. However, it is much less sensitive to moisture than Grignard reagents.

3. *Organoboronic Acids or Borates.*   Although the nucleophilicity of organoboronic acids or borates is low, they have sufficiently high reactivity to undergo transmetallation to other metals. Both palladium and nickel complexes have been used for the coupling reaction. Aryl–alkenyl iodides, bromides, tosylates, and electron-deficient arylchlorides have been employed for step-growth polymerization. It has been demonstrated recently that aryl chlorides, including the electron-rich ones, can also be used for the reaction by using an electron-rich phosphine ligand for the palladium catalysts.[35] These developments provide tools for the synthesis of polymers by using cheap aryl–alkenyl chlorides. Both strong and weak bases can be employed for the reaction. The reaction is generally sensitive to oxygen and conducted in aqueous mixtures.

4. *Organocopper Reagents.*   In the presence of a base, the terminal alkynes react with CuI to form alkynylcopper species. The alkynylcopper species can undergo cross-coupling reaction with aryl–alkenyl halides by using a palladium catalyst. The base is usually a secondary or tertiary amine such as triethylamine. Aryl–alkenyl iodides, bromides, and tosylates and alkenyl chlorides are active enough to be used for the polymerization. The suggested mechanism is shown in Fig. 9.5.

9.3.1.2   *Coupling Reactions of Aryl–Alkenyl Halides with Alkenes*[53]

The Pd-catalyzed coupling reaction of aryl–alkenyl halides with alkenes in the presence of a base is known as the Heck reaction (Scheme 9.3).[5]

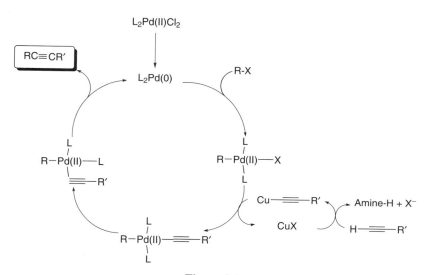

Figure 9.5

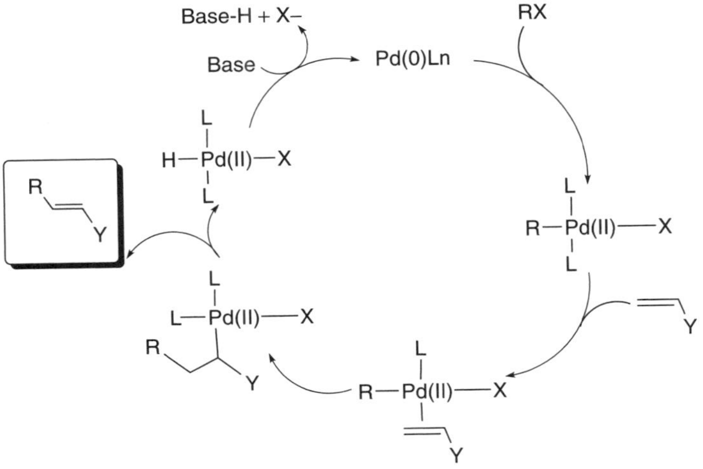

Figure 9.6

Tetrakis(triphenylphosphine)palladium(0) is often used for this reaction. However, Pd(II) complexes such as bis(triphenylphosphine)palladium dichloride or palladium acetate are also commonly employed for convenience, as they are stable in air. The base is typically a secondary or tertiary amine such as triethylamine. Weak bases such as sodium (potassium) acetate, bicarbonate, or carbonate are also used.

The general catalytic cycle for the coupling of aryl–alkenyl halides with alkenes is shown in Fig. 9.6. The first step in this catalytic cycle is the oxidative addition of aryl–alkenyl halides to Pd(0). The activity of the aryl–alkenyl halides still follows the order RI > ROTf > RBr > RCl. The olefin coordinates to the Pd(II) species. The coordinated olefin inserts into Pd–R bond in a syn fashion. β-Hydrogen elimination can occur only after an internal rotation around the former double bond, as it requires at least one β-hydrogen to be oriented syn perpendicular with respect to the halopalladium residue. The subsequent syn elimination yields an alkene and a hydridopalladium halide. This process is, however, reversible, and therefore, the thermodynamically more stable (E)-alkene is generally obtained. Reductive elimination of HX from the hydridopalladium halide in the presence of a base regenerates the catalytically active Pd(0), which can reenter the catalytic cycle. The oxidative addition has frequently assumed to be the rate-determining step.

9.3.1.3 Homo-Coupling Reactions of Aryl Halides

Early development of the homo-coupling reactions of aryl halides involves the use of stoichiometric amounts of air-sensitive Ni(0) complexes.[54] The reaction could also be realized with a catalytic amount of Ni(0) complexes formed in situ when a stoichiometric amount of Zn was present. Besides aryl iodides, tosylates,

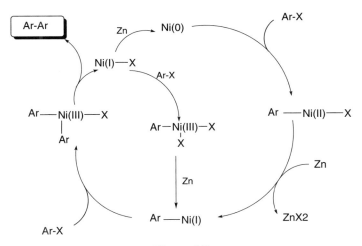

Figure 9.7

and bromides, aryl chlorides can also be used as substrates for the polymerization under optimized conditions. Ligands such as PPh₃ and bipyridine are used for the polymerization to enhance the yield.[55]

Figure 9.7 shows the mechanism suggested for the Ni(0)-mediated coupling reaction of aryl halides in the presence of excess of Zn. The first step of the mechanism involves the reduction of Ni(II) to Ni(0) by Zn. This is followed by the oxidative addition of aryl halides to Ni(0) to aryl Ni(II) halides. The Ni(II) species is then reduced to aryl Ni(I) species via one-electron reduction. Another aryl halide oxidatively adds to this Ni(I) species to generate the diaryl Ni(III) species. This diaryl Ni(III) species undergoes rapid reductive elimination to form the biaryl product. The generated Ni(I) species have two possible ways to reenter the catalytic cycle: Ni(I) is reduced by Zn to Ni(0), which can repeat the catalytic cycle, or ArX undergoes direct oxidative addition to Ni(I) followed by the reduction of Zn to form the aryl Ni(I) species once again.

The possible mechanism for the reactions involving stoichiometric amount of preformed Ni(0) complexes is shown in Fig. 9.8. The first step of the mechanism involves the oxidative addition of aryl halides to Ni(0) to form aryl Ni(II) halides. Disproportion of two aryl Ni(II) species leads to a diaryl Ni(II) species and a Ni(II) halide. This diaryl Ni(II) species undergoes rapid reductive elimination to form the biaryl product. The generated Ni(0) species can reenter the catalytic cycle.

The rate-determining step in the homo-coupling reaction of aryl halides could be the oxidative step or the reduction of Ni(II) to Ni(I) step.

9.3.1.4   Coupling Reactions of Aryl Halides with Ketones and Amines[19,20,56]

Extensive study by several groups showed that Ar—X (X = I, Ts, Br, and Cl) can undergo coupling reactions with amine and ketones in the presence of a

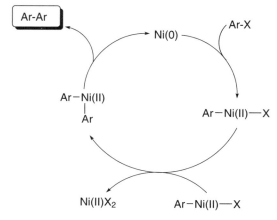

Figure 9.8

base. A chelated phosphine ligand such as BINAP or DPPF or a bulky, electron-rich monodentated phosphine ligand such as t-Bu$_3$P is essential for the reaction. NaO-t-Bu, K$_2$CO$_3$, and Cs$_2$CO$_3$ are commonly used as the base for the reaction. The enantioselective coupling of aryl halides with ketones is possible by using a chiral phosphine ligand. Figure 9.9 shows one of the suggested mechanisms for the chelated ligand involved aryl halide amination. Oxidative addition of an aryl halide to Pd(0) species generates an aryl Pd(II) halide complex. This arylpalladium complex undergoes substitution with NaO-t-Bu to generate an aryl-t-butoxylpalladium(II) complex which can react with amine to yield an aryl-lamidopalladium complex. Reductive elimination gives the coupled product and regenerates the Pd(0) species.

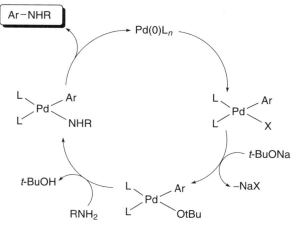

Figure 9.9

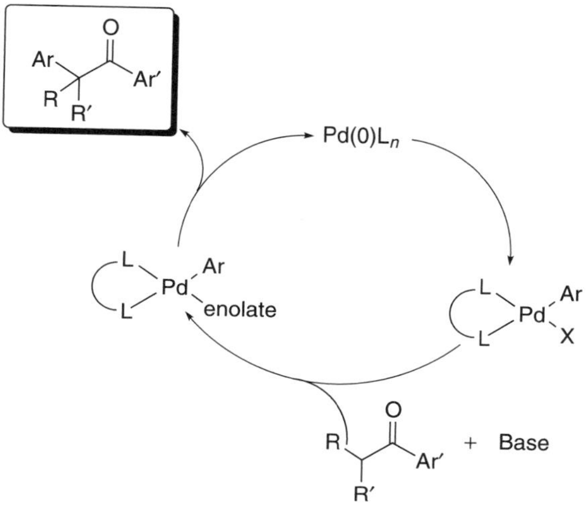

Figure 9.10

A general mechanism for the ketone arylation is shown in Fig. 9.10. Oxidative addition of an aryl halide to Pd(0) species generates an aryl-Pd(II) halide complex. This arylpalladium complex undergoes substitution with an enolate (in situ generated from a ketone and a base) to generate an arylenolatepalladium(II) complex. Reductive elimination gives the coupled product and regenerates the Pd(0) species.

Although only several reports appeared employing these reactions for polymerization, they could potentially be used for the synthesis of different types of polyamines and polyketones, including optically active ones with main-chain chirality.

9.3.2 Experimental Methods

Transition metal coupling polymerization, like its parent transition-metal-catalyzed coupling reaction, is usually conducted in solution phase, including homogeneous and heterogeneous conditions. Polymerization employing the Kumada coupling, the Heck coupling, the Stille coupling, and Ni(0)-mediated homo couplings starts from homogeneous solution with salts produced as the polymerization proceeds. A report using the heterogeneous condition for Heck coupling polymerization has been appeared.[57] Polymerizations applying the Suzuki coupling and Ni(II)–Zn coupling which start from two phases are conducted in heterogeneous conditions.[11,18]

9.3.3 Characterization of Polymers[33]

Because side-chain groups can be easily introduced in the polymer backbone, polymers synthesized via transition metal coupling are usually soluble in common organic solvents. Therefore, most of the methods used for the characterization of other types of step-growth or chain-growth polymers can be used for characterization of polymers synthesized via transition metal coupling.

Gel permeation chromatography (GPC) or size exclusion chromatography (SEC) has been routinely used to estimate the molecular weight of the polymers. The molecular weight measured by GPC is relative to a polymer standard, typically polystyrene; GPC is thus a relative method rather than an absolute one. For those polymers whose structure is very different from polystyrene, GPC molecular weight values could significantly differ from the real ones. In those cases, GPC values should only be regarded as a reference.

Light scattering is another technique frequently used in the measurement of polymer molecular weight. Because this method deals with the light scattering of the sample and is directly related to the size of the sample molecules, it is thus particularly good to measure high-molecular-weight polymers. Other methods such as vapor pressure osmometry, membrane osmometry, and matrix-assisted laser desorption/ionization (MALDI) mass spectrometry have also been used for the measurement of molecular weights of polymers.

Thermogravimetric analysis (TGA) and differential scanning calorimetry (DSC) are also very useful tools for the characterization of polymers. TGA and DSC provide the information about polymer stability upon heating and thermal behaviors of polymers. Most of the polymers synthesized via transition metal coupling are conjugated polymers. They are relatively stable upon heating and have higher T_gs.

Spectroscopy methods are routinely used for the characterization of the polymers. Nuclear magnetic resonance (NMR) is very useful for the characterization of these polymers. For example, from ^{1}H NMR, not only the molecular weight of the sample can be estimated based on the end group but also the structure regularity of the polymer can be obtained. For very high molecular weight polymers, because the endgroup is too dilute, this technique is no longer suitable for the estimation of the molecular weight. It will provide the information of the molecular structure.

Ultraviolet spectroscopy and photoluminescence are used to study the optical properties of polymers. These spectra provide information about the conjugation along the polymer backbone and the photo behavior of the polymers.

Optical rotation and circular dichroism have been used for the characterization of optically active polymers. They have been used to determine whether polymers are optically active and whether a secondary structure such as a helix exists.

Useful information such as the functionality and crystallinity of the polymers can be obtained by using infrared spectroscopy. Elemental analysis is also considered as one of the tools for the characterization of the polymers. Due to the endgroups and incomplete combustion of the carbon, it is common to observe the low-value carbon content than the theoretical one.

Other techniques such as cyclic voltammogram (CV), atomic force morphology (AFM), and scanning force morphology (SFM) have also been used for the characterization of polymers. CV has been used to study the electrochemical properties of the polymers. The surface properties of polymers have been characterized by using AFM and SFM.

9.4 SYNTHETIC METHODS

9.4.1 Linear Polymers without Chiral Units

9.4.1.1 Synthesis of Poly(p-phenylene) 24 by Using Kumada Coupling[58]

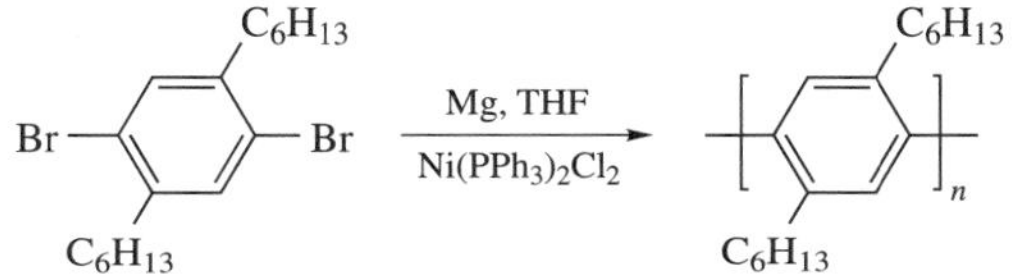

C_6H_{13}

Br — [benzene ring] — Br →(Mg, THF / Ni(PPh_3)_2Cl_2)→ [poly(p-phenylene)]_n

C_6H_{13}

18 **24**

Under nitrogen, freshly distilled 1,4-dibromo-2,5-dihexylbenzene **18** (12.0 g, 29.7 mmol) and magnesium (0.72 g, 29.7 mmol) were refluxed in THF (80 mL) for about 1 h. After being cooled to room temperature, Ni(PPh_3)_2Cl_2 (220 mg, 1.13 mmol) was added and the resulting dark solution was refluxed for 48 h. The reaction mixture was then poured into acetone (500 mL) and filtered. The solid was washed with dilute HCl (100 mL) and dried under vacuum (0.01 mm Hg). The resulting solid was extracted with toluene in a Soxhlet apparatus. The solution was then concentrated to about 10 mL. Upon adding acetone (500 mL) to the solution, the polymer was precipitated out. After freeze drying in benzene, polymer **24** was obtained in 62% yield (4.11 g).

9.4.1.2 Synthesis of Poly(p-phenylene) 24 by Using Suzuki Coupling[17]

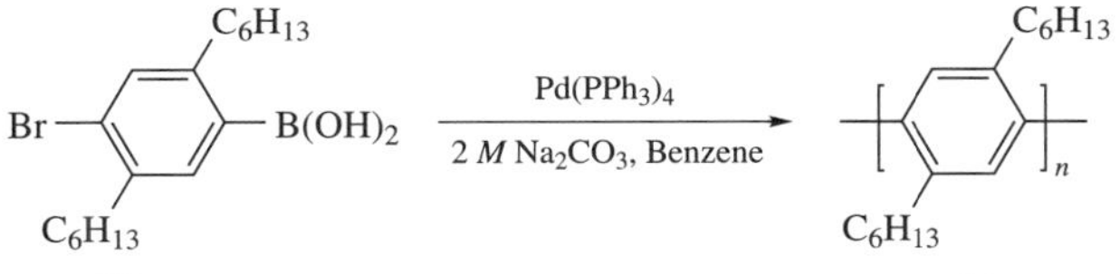

C_6H_{13}

Br — [benzene ring] — B(OH)_2 →(Pd(PPh_3)_4 / 2 M Na_2CO_3, Benzene)→ [poly(p-phenylene)]_n

C_6H_{13}

23 **24**

Under nitrogen, a mixture of compound **23** (2.88 g, 7.8 mmol), Pd(PPh_3)_4 (0.045 g, 0.039 mmol), and 2 M Na_2CO_3 (30 mL) in benzene (40 mL) was vigorously stirred and refluxed for 48 h. After being cooled to room temperature, the mixture was poured into acetone (200 mL). The precipitated solid material was recovered by filtration through a Büchner funnel, washed with dilute HCl (100 mL), and dried under vacuum (0.01 mm Hg) for 24 h. The resulting solid was extracted with toluene in a Soxhlet apparatus for ca. 20 h. The solution was concentrated to 20 mL and the product was precipitated out by adding acetone (200 mL). After freeze drying with benzene, polymer **24** was obtained in 96% yield (1.83 g).

9.4.1.3   Synthesis of Poly(p-phenylene) 55[59a]

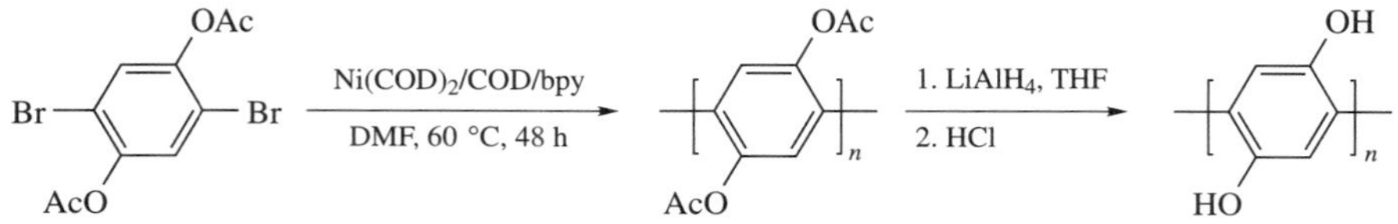

53   **54**   **55**

(1) Under nitrogen, a solution of 1,4-dibromo-2,5-diacetoxybenzene **53** (2.0 g, 5.7 mmol), Ni(COD)$_2$ (3.13 g, 11.4 mmol), 1,5-cyclooctadiene (1.23 g, 11.4 mmol), and bipyridine (Bpy) (1.77 g, 11.4 mmol) in DMF (50 mL) was stirred at 60°C for 48 h. The reaction mixture was concentrated to 10 mL and poured into dilute HCl. After filtration, the cream precipitate was washed with dilute HCl (twice), an aqueous solution of disodium ethylenediaminetetraacetate (twice) and dilute HCl (once) in this order and dried under vacuum. Polymer **54** was obtained in 72% yield. GPC (polystyrene standards): $M_w = 3400$ (polydispersity index (PDI) = 1.6). (2) A mixture of polymer **54** (0.30 g, 1.6 mmol) and LiAlH$_4$ (0.25 g, 6.5 mmol) in THF (50 mL) was stirred at 40°C for 48 h. After being cooled to room temperature, the solvent was removed by rotary evaporation. The residue was washed with dilute HCl repeatedly to give polymer **55** as a black solid in 100% yield. GPC (polystyrene standards): $M_w = 4800$, $M_n = 3400$.

9.4.1.4   Synthesis of Poly(2,5-Benzophenone) 57[59b]

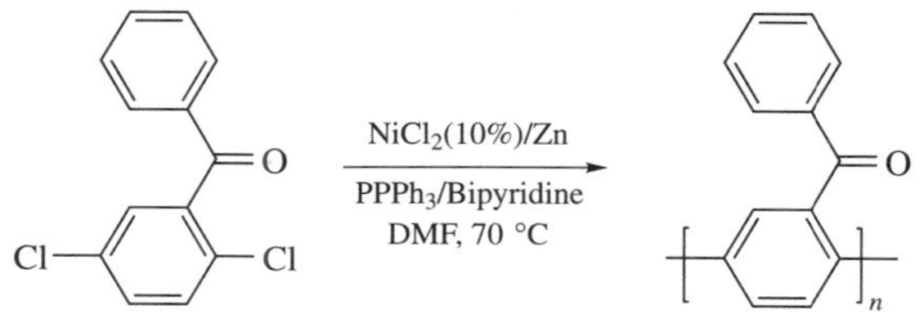

56   **57**

Under argon, to a mixture of 2,5-dichlorobenzophenone (2.5 g, 10 mmol), zinc (2.0 g, 30.6 mmol), NiCl$_2$ (0.13 g, 1 mmol), PPh$_3$ (1.04 g, 4 mmol), and bipyridine (01.6 g, 1.01 mmol) was added 30 mL DMF. The mixture was heated to 70°C and stirred at that temperature for 20 h. The reaction mixture was then dispersed into acetone. The solid was filtered and washed with 5% HCl (3 times), water (3 times), and acetone. Polymer **57** was obtained in quantitative yield as a yellowish powder. GPC (polystyrene standards): $M_w = 58,000$.

9.4.1.5 Synthesis of Poly(p-phenylene) Containing Azobenzene Unit 60[60]

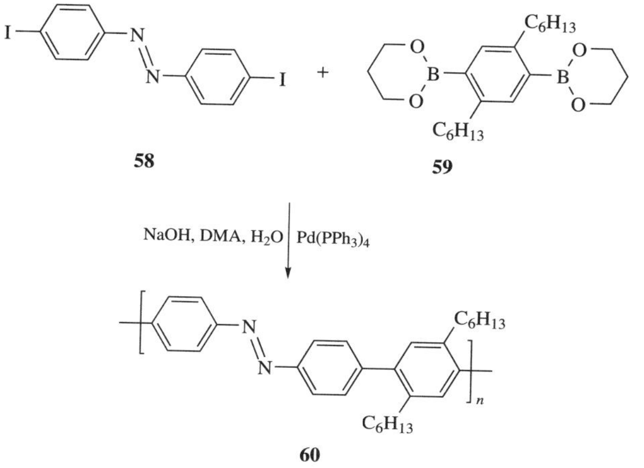

58 + **59**

NaOH, DMA, H$_2$O | Pd(PPh$_3$)$_4$

60

Under nitrogen, a mixture of **58** (0.174 g, 0.40 mmol), **59** (0.166 g, 0.40 mmol), Pd(PPh$_3$)$_4$ (0.0139 g, 0.012 mmol), NaOH (0.064 g, 1.6 mmol), H$_2$O (0.8 mL), and DMA (5.0 mL) was stirred at 110°C for 48 h. The reaction mixture was poured into MeOH (100 mL) and filtered. The solid was washed with MeOH–H$_2$O (1 : 1) and then with MeOH. After drying under vacuum, the polymer was redissolved in toluene and reprecipitated with MeOH in 100% yield. GPC (polystyrene standards): $M_n = 7700$ (PDI $= 1.8$).

9.4.1.6 Synthesis of Water-Soluble Poly(p-phenylene) 63[61]

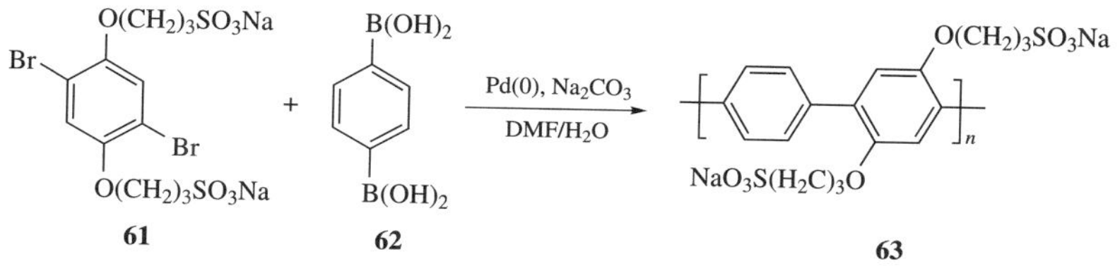

61   **62**   **63**

A mixture of 1,4-dibromo-2,5-bis(3-sulfonatopropoxy)benzene **61** (0.78 g, 1.39 mmol), **60** (0.23 g, 1.39 mmol), Na$_2$CO$_3$ (0.99 g) in doubly distilled water (47 mL), and DMF (20 mL) was heated at 85°C until the solids were completely dissolved. The resulting solution was cannulated to a 200-mL Schlenk flask with tris[(sulfonatophenyl)phosphine]palladium(0) (0.045 g) and the mixture was stirred at 85°C for 10 h. The reaction mixture was concentrated to 25 mL by boiling and filtered. The filtrate was added dropwise to cold acetone (250 mL) to precipitate out the polymer. The polymer was collected by filtration, redissolved in a minimum of hot water, and reprecipitated by cooling. After repeating this procedure twice, the polymer was redissolved in distilled water and dialyzed for 72 h in 3500 g·mol^{-1} cutoff membrane. After drying under vacuum, polymer **63** was obtained in 64% (0.42 g).

9.4.1.7 Synthesis of Poly(p-phenylene) 66[62]

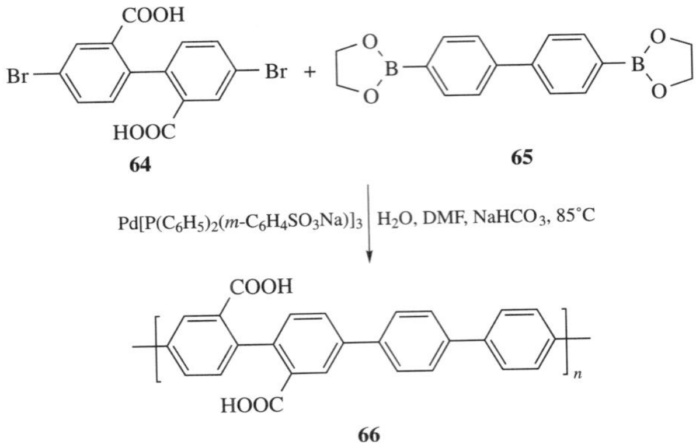

Under nitrogen, a mixture of **64** (0.080 g, 2.00 mmol), **65** (0.5878 g, 2.00 mmol), Pd catalyst (0.074 g, 0.06 mmol) in aqueous $NaHCO_3$–HOAc (42 mL, pH 8.0, 0.2 M), and DMF (18 mL) was stirred at 85°C for 10 h. The resulting mixture was poured into dilute HCl (0.5% w/v) and the mixture was stirred at room temperature for 2 days. The polymer was collected by centrifugation and washed four times with H_2O (20 mL). The crude polymer was dissolved in Na_2CO_3 (0.1 M), filtered, and reprecipitated with dilute HCl. The precipitation is repeated and followed by washing with EtOH (95%, 20 mL) twice and dried. Polymer **66** was obtained in 87% yield (0.681 g).

9.4.1.8 Synthesis of Polymer 69[63]

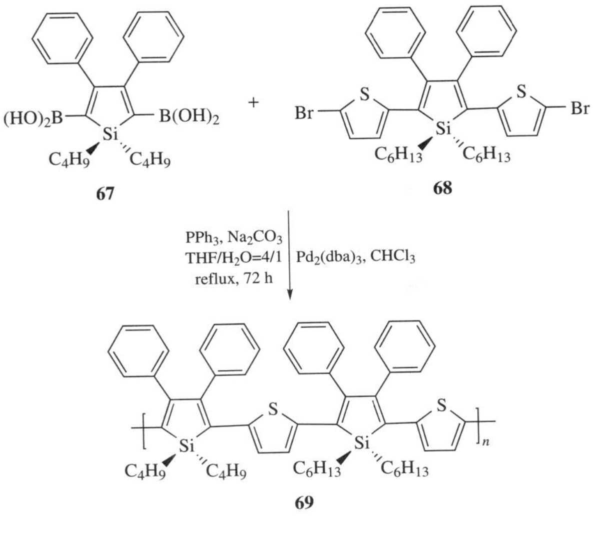

A mixture of **67** (0.868 g, 2.0 mmol), **68** (1.450 g, 2.0 mmol), [Pd$_2$(dba)$_3$]·CHCl$_3$ (0.104 g, 0.10 mmol), PPh$_3$ (0.105 g, 0.40 mmol), and Na$_2$CO$_3$ (0.424 g, 4.0 mmol) in THF–H$_2$O (4 : 1, 20 mL) was refluxed for 72 h. The reaction mixture was poured into hexanes (400 mL) and vigorously stirred. The precipitate was collected by filtration. The solid collected was suspended in MeOH–H$_2$O (1 : 1, 120 mL) and sonicated for 10 min. The insoluble part was collected and washed with MeOH (80 mL) and hexanes. Soxhlet extraction from the residue with THF gave the soluble polymer **69** in 98% yield (1.78 g) as a bronze solid. GPC (polystyrene standards): $M_w = 48,700$, $M_n = 18,400$.

9.4.1.9 Synthesis of Poly(phenylenevinylene) 72[53a]

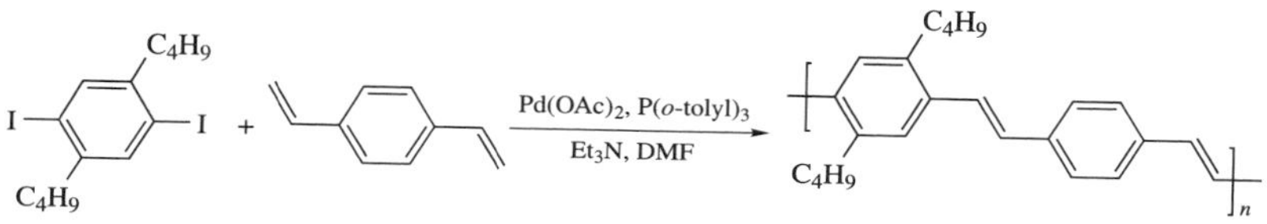

To a DMF (5 mL) solution of p-divinylbenzene (0.130 g, 1 mmol), 2,5-dibutyl-1,4-diiodobenzene (0.474 g, 1 mmol), Pd(OAc)$_2$ (0.009 g, 0.04 mmol), and tri-o-tolylphosphine (0.061 g, 0.2 mmol) was added Et$_3$N (0.35 mL, 2.5 mmol). The reaction mixture was heated at 100°C for 5 h under N$_2$ and then poured into MeOH (20 mL). The solid was collected by filtration. The solid was redissolved in a minimum amount of CHCl$_3$ and precipitated from acetone. The resulting polymer was extracted with MeOH for 24 h and then dried under vacuum at 40°C for 2 days. Yield: 87%. GPC (THF, polystyrene standards): $M_w = 32,000$, $M_n = 8300$ (PDI = 3.85).

9.4.1.10 Synthesis of Poly(phenylenevinylene) 74[53b]

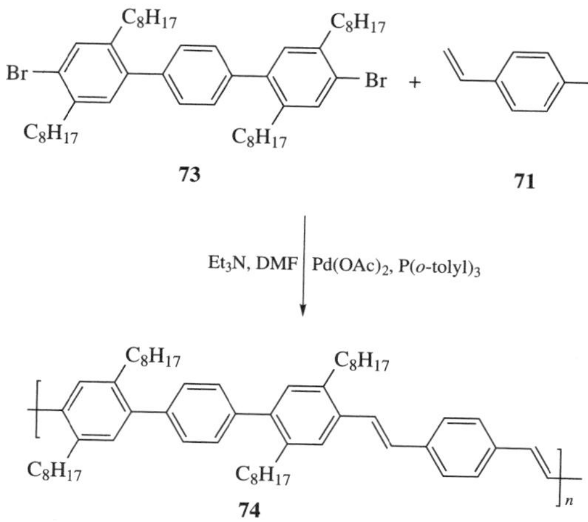

Under nitrogen, to a heavy-wall pressure tube were added Et$_3$N (2.0 mL), p-divinylbenzene (0.155 g, 1.2 mmol), **73** (1.0 g, 1.2 mmol), Pd(OAc)$_2$ (0.011 g, 0.05 mmol), tri-o-tolylphosphine (0.071 g, 0.23 mmol), and DMF (5 mL). The tube was degassed, sealed, and heated to 100°C. After 40 h, the reaction mixture was poured into methanol (75 mL). The precipitate was collected by filtration and dried under vacuum to give a brown-yellow polymer (0.88 g, conversion 91%). The crude polymer was redissolved in CHCl$_3$ and filtered through a small column of Kieselgel to remove traces of catalyst. The resulting solution was concentrated and precipitated from MeOH (75 mL). The yellow solid was collected by filtration and dried under vacuum. GPC (THF, polystyrene standards): $M_w = 4500$ (PDI = 1.7).

9.4.1.11   Synthesis of Poly(phenylenevinylene) 77[64]

Under nitrogen, n-Bu$_3$N (0.32 mL, 1.34 mmol) was added to a mixture of divinylbenzene (0.070 g, 0.545 mmol), **75** (0.5890 g, 0.5400 mmol), **76** (0.010 g, 0.0054 mmol), Pd(OAc)$_2$ (0.0049 g, 0.0217 mmol), and tri-o-tolylphosphine (0.0329 g, 0.108 mmol) in DMF (5 mL). The reaction mixture was stirred at 90°C overnight and then poured into methanol. The precipitate was collected and redissolved in CHCl$_3$. After filtration to remove the catalyst residue, the filtrate was concentrated and precipitated repeatedly. The resulting polymer was further purified by extraction in a Soxhlet extractor with MeOH for 24 h and then dried under vacuum at 40°C for 24 h. GPC (polystyrene standards): $M_n = 18,000$ (PDI = 1.9).

9.4.1.12   Synthesis of Poly(phenyleneethynylene) 79[52a]

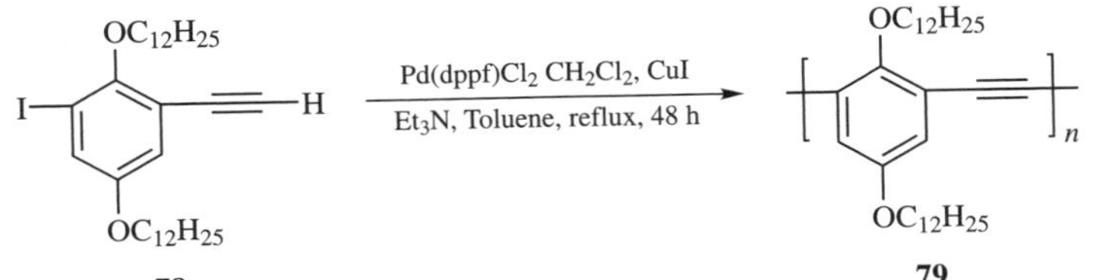

Under nitrogen, a mixture of **78** (4.26 g, 10 mmol), Pd(dppf)Cl$_2$·CH$_2$Cl$_2$ (0.16 g, 2 mol %), CuI (0.11 g, 6 mol %) in Et$_3$N (10 mL) and toluene (50 mL) was stirred at reflux for 48 h. The reaction mixture was poured into MeOH and the solid was centrifuged and dried under vacuum. The crude polymer was dissolved in CH$_2$Cl$_2$ and passed through a short column of silica gel to remove trace amount of catalyst. The resulting solution was concentrated and precipitated with MeOH to give polymer **79** in 87% yield. GPC (polystyrene standards): $M_n = 8399$ (PDI = 3.82).

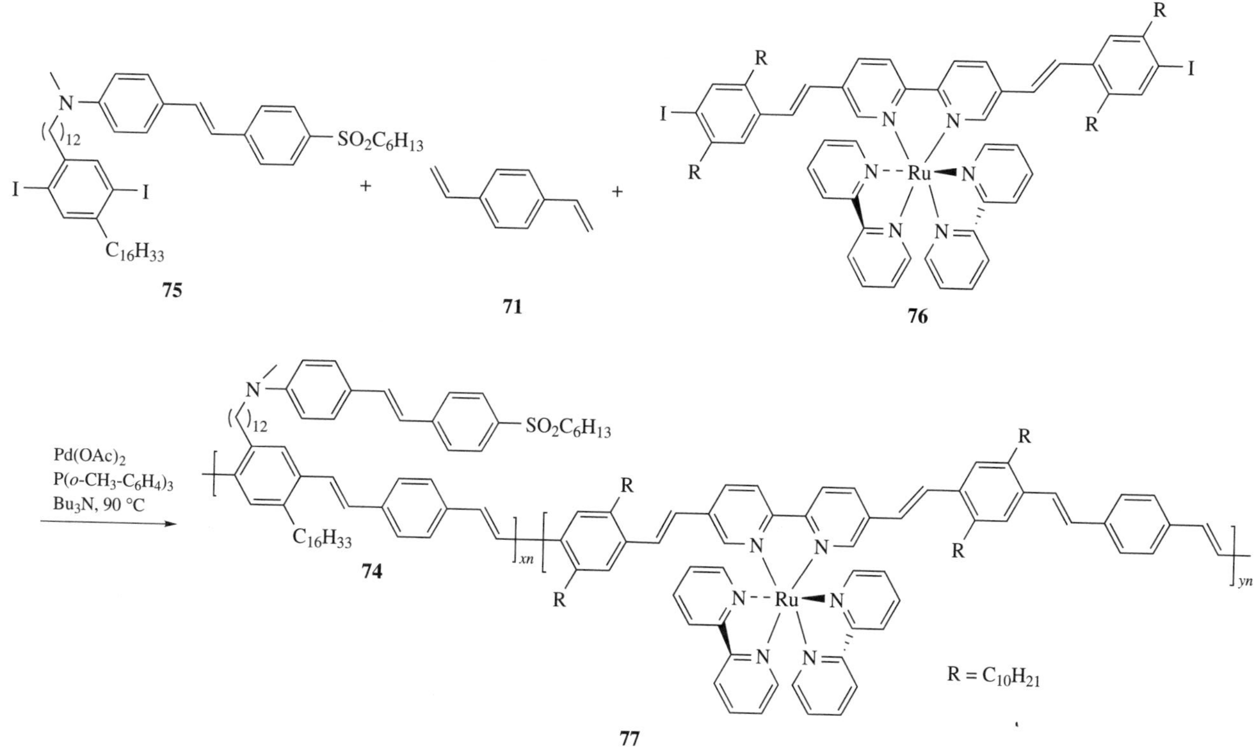

76

71

75

74

77

R = C$_{10}$H$_{21}$

Pd(OAc)$_2$
P(o-CH$_3$-C$_6$H$_4$)$_3$
Bu$_3$N, 90 °C

SO$_2$C$_6$H$_{13}$

SO$_2$C$_6$H$_{13}$

C$_{16}$H$_{33}$

C$_{16}$H$_{33}$

9.4.1.13 Synthesis of Poly(aryleneethynylene) 82[52b]

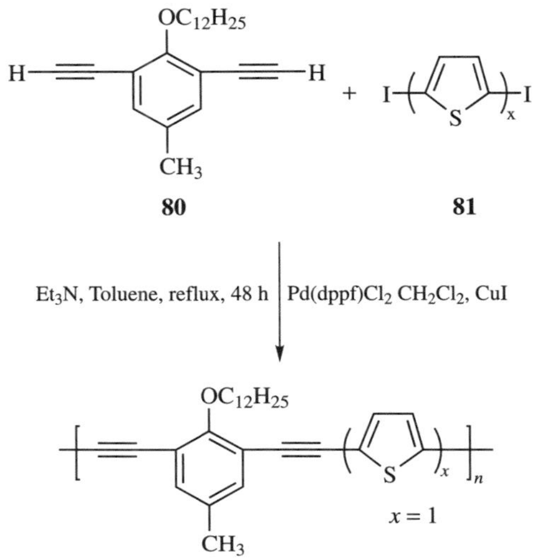

80 **81**

Et$_3$N, Toluene, reflux, 48 h | Pd(dppf)Cl$_2$ CH$_2$Cl$_2$, CuI

$x = 1$

82

Under nitrogen, a mixture of **80** (3.24 g, 10 mmol), **81** (10 mmol), Pd(dppf)Cl$_2$·CH$_2$Cl$_2$ (0.16 g, 2 mol %), and CuI (0.11 g, 6 mol %) in Et$_3$N (10 mL) and toluene (50 mL) was stirred at reflux for 48 h. The reaction mixture was poured into MeOH and the solid was centrifuged and dried under vacuum. The crude polymer was dissolved in CH$_2$Cl$_2$ and passed through a short column of silica gel to remove the trace amount of catalyst. The resulting solution was concentrated and precipitated with MeOH to give polymer **82** in 81% yield. GPC (polystyrene standards): $M_n = 3563$ (PDI = 2.35).

9.4.1.14 Synthesis of Poly(phenyleneethynylene) 85[52c]

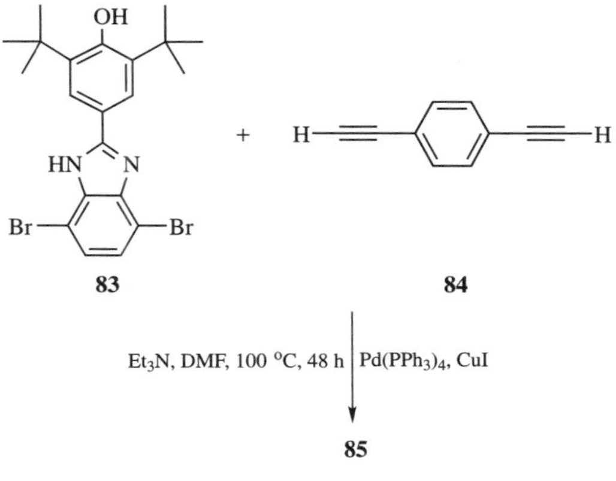

83 **84**

Et$_3$N, DMF, 100 °C, 48 h | Pd(PPh$_3$)$_4$, CuI

85

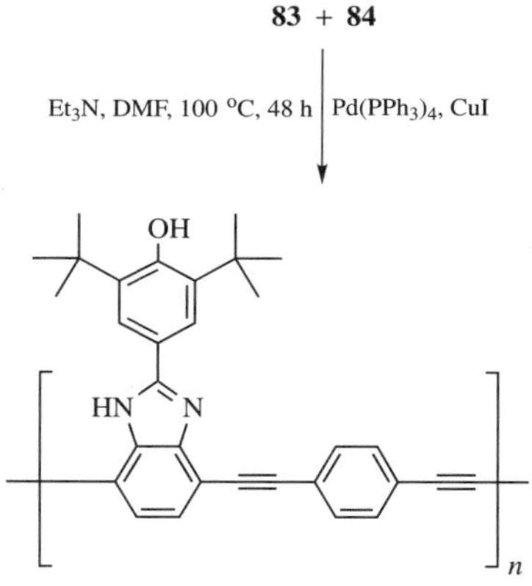

83 + 84

Et₃N, DMF, 100 °C, 48 h | Pd(PPh₃)₄, CuI

85

Under nitrogen, anhydrous DMF (10 mL) was added to a mixture of **83** (0.240 g, 0.5 mmol), **84** (0.063 g, 0.5 mmol), and Et₃N (1 mL). Pd(PPh₃)₄ (0.027 g, 0.025 mmol) and CuI (0.005 g, 0.025 mmol) were then added and the reaction mixture was stirred at 100°C for 48 h. After being cooled to room temperature, the reaction mixture was poured into MeOH and filtered. The solid was washed with MeOH and dried under vacuum. The repetition of the precipitation procedure gave polymer **85** as orange powder in 96% yield (0.212 g). GPC (polystyrene standards): $M_n = 15,100$.

9.4.1.15 Synthesis of Poly(aryleneethynylene) 87[65]

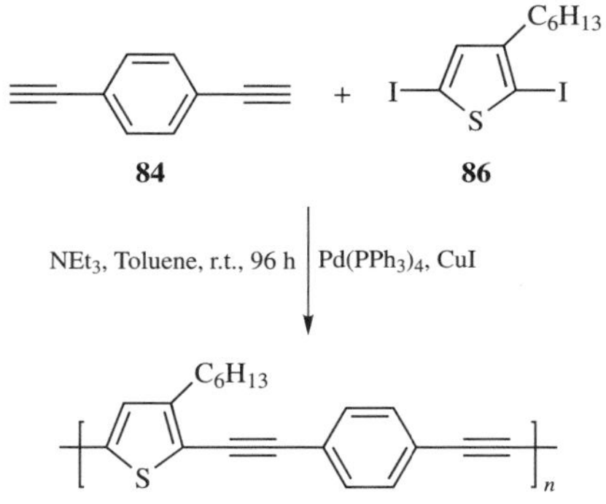

84   **86**

NEt₃, Toluene, r.t., 96 h | Pd(PPh₃)₄, CuI

87

Under nitrogen, a mixture of **84** (2.15 mmol), **86** (2.15 mmol), Pd(PPh₃)₄ (0.02 mmol), and CuI (0.04 mmol) in Et₃N (1.5 mL) and toluene (12 mL) was stirred at 110°C for 2 h. After cooling to room temperature, the reaction mixture was poured into MeOH and filtered. The solid was washed with MeOH repeatedly and dried under vacuum to give polymer **87** in 93% yield. Light-scattering method gave $M_w = 9600$.

9.4.1.16   Synthesis of Poly(phenyleneethynylene) 89[66]

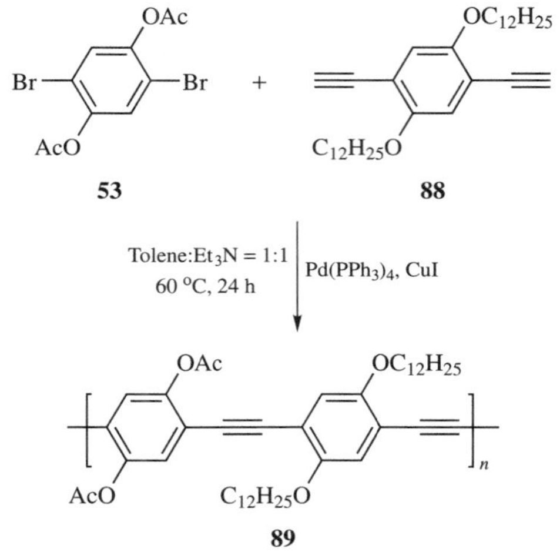

Under nitrogen, a mixture of **53** (0.200 g, 0.57 mmol), **88** (0.280 g, 0.57 mmol), Pd(PPh$_3$)$_4$ (0.033 g, 0.028 mmol), and CuI (0.0054 g, 0.028 mmol) in 1 : 1 toluene and Et$_3$N was stirred at 60°C for 24 h. The polymer **89** was extracted with CHCl$_3$ and precipitated with MeOH in 89% yield. GPC (polystyrene standards): $M_n = 7700$, $M_w/M_n = 1.7$.

9.4.1.17   Synthesis of Ferrocenyl-Containing Poly(aryleneethynylene) 92[67]

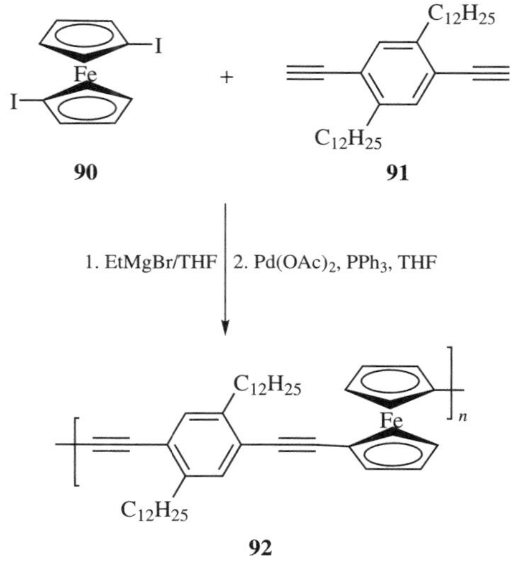

Under nitrogen, to a solution of **90** (0.15 g, 0.32 mmol) in THF (5 mL) was added EtMgBr (0.670 mL, 0.96 M in THF) and the mixture was stirred at room temperature for 1.5 h. The reaction mixture was then added to a mixture of **91** (0.14 g, 0.32 mmol), Pd(OAc)$_2$ (0.03 g, 0.01 mmol), and PPh$_3$ (0.03 g, 0.01 mmol). The resulting mixture was stirred at reflux for 2 h. Diluted HCl was then added and the product was extracted with CHCl$_3$. After concentration, the residue was washed with MeOH repeatedly and dried under vacuum to give polymer **92** in 100% yield (0.206 g). GPC (polystyrene standards): $M_w = 6900$, $M_n = 4700$.

9.4.1.18   Synthesis of Poly(phenyleneethynylene) 30[68]

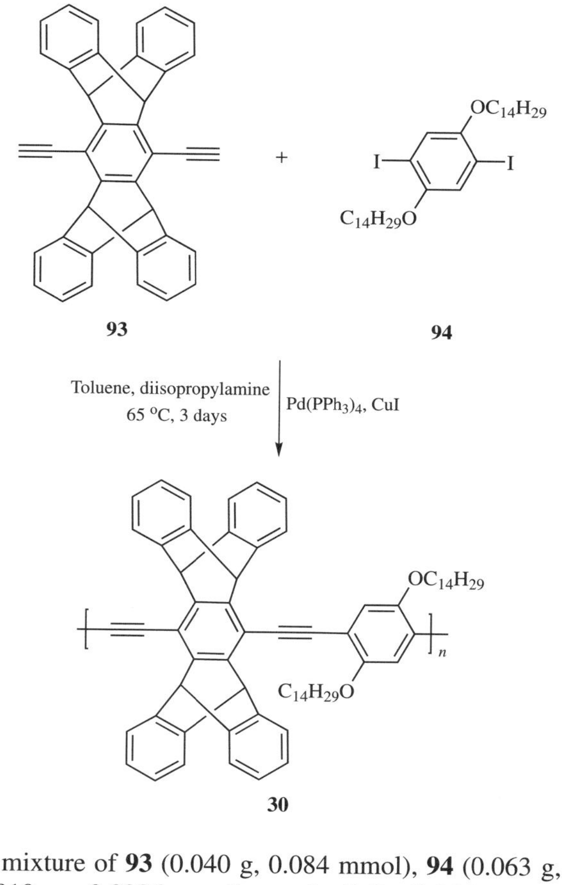

93   **94**

Toluene, diisopropylamine
65 °C, 3 days   Pd(PPh$_3$)$_4$, CuI

30

Under argon, a mixture of **93** (0.040 g, 0.084 mmol), **94** (0.063 g, 0.084 mmol), Pd(PPh$_3$)$_4$ (0.010 g, 0.0086 mmol), and CuI (0.010 g, 0.053 mmol) in diisopropylamine–toluene (2 : 3, 2.5 mL) was stirred at 65°C for 3 days. The reaction mixture was extracted with CHCl$_3$. The organic solution was washed with water and aqueous NH$_4$Cl and dried (MgSO$_4$). After concentration, the residue was precipitated from MeOH. Repetitive precipitation–centrifugation using THF and MeOH (three times) gave polymer **30** in 75% yield (76 mg). GPC (polystyrene standards): $M_n = 144{,}000$ (PDI $= 2.6$).

9.4.1.19   Synthesis of Poly(thiophenepyridine) 96[54a]

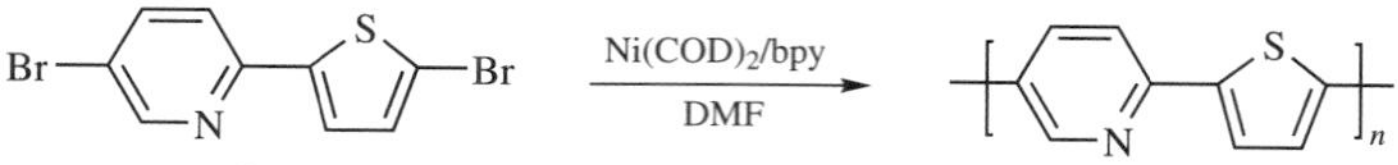

Under nitrogen, 2,2'-bipyridine (0.322 g, 2.07 mmol) was added to a mixture of Ni(COD)$_2$ (0.570 g, 2.07 mmol) in anhydrous DMF (18 mL) and the mixture was stirred at room temperature for 1 h. Then **95** (0.510 g, 1.60 mmol) was added and the reaction mixture was stirred at 60–70°C for 28 h. After cooling to room temperature, HCl (100 mL, 2 M) was added and the mixture was stirred overnight. After filtration, the solid was washed with a mixture of MeOH (60 mL) and concentrated HCl (10 mL), a warm aqueous solution of ethylenediaminetetraacetic acid (EDTA) (pH 9, then pH 3), dilute ammonia, water, and MeOH repeatedly and dried under vacuum for 20 h at 80°C. Polymer **96** was obtained in 100% yield (0.261 g). Light scattering gave $M_w = 5400$.

9.4.1.20   Synthesis of Poly(phenylenethiophene) 99[69]

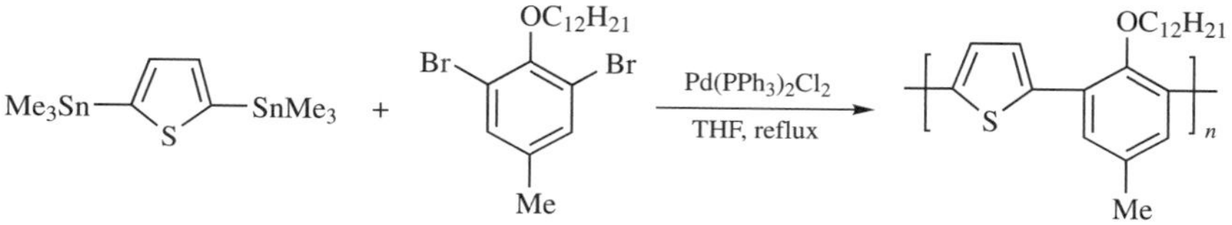

Under nitrogen, a mixture of **97** (2 eq.), 2,6-dibromo-1-dedocyloxy-4-methylbenzene **98** (1 eq.), and Pd(PPh$_3$)$_2$Cl$_2$ in THF was stirred at reflux for 24 h. Polymer **99** was purified by repetitive precipitation–centrifugation using THF and MeOH. GPC (polystyrene standards): $M_n = 1700$, PDI $= 1.4$.

9.4.1.21   Synthesis of Poly(thiophene) 22[15]

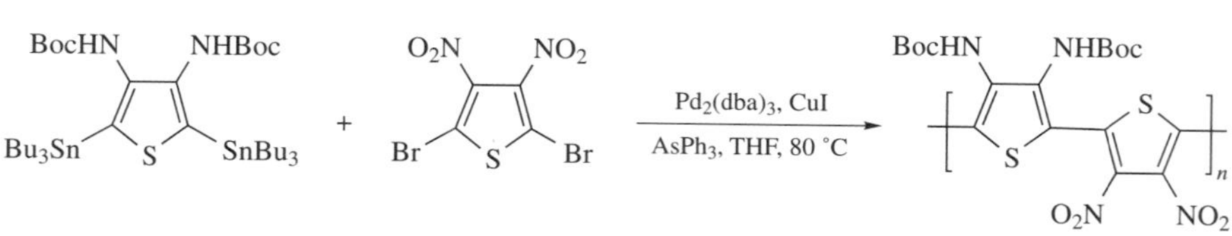

Under nitrogen, Pd$_2$(dba)$_3$ (0.0377 g, 0.0412 mmol), CuI (0.0393 g, 0.2063 mmol), and AsPh$_3$ (0.0252 g, 0.0824 mmol) were added to a mixture of **20** (1.8392 g,

2.06 mmol) and **21** (0.6978 g, 2.10 mmol) in THF (5.0 mL). The reaction mixture was stirred at 80°C for 72 h. After being cooled to room temperature, the mixture was poured into a solution of KF (10 mL, 1 M in H$_2$O) and stirred for 30 min followed by filtration. The filtrate was extracted with CH$_2$Cl$_2$ and dried over Na$_2$SO$_4$. After concentration, the residue was dissolved in acetone (5 mL) and precipitated with hexane (200 mL). Repeating the dissolvation and precipitation sequence gave polymer **22** in 86% yield (0.86 g). GPC (polystyrene standards): $M_w = 10,200$, $M_n = 7000$.

9.4.1.22 Synthesis of Ladder Polymer 102[42b]

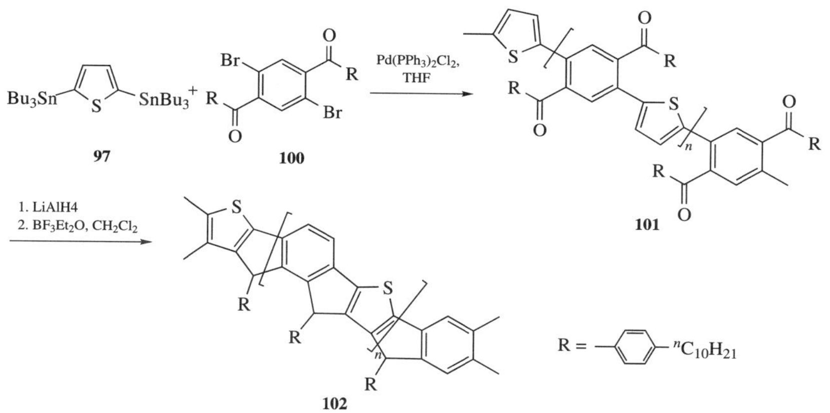

Under nitrogen, a mixture of **97** (0.955 g, 1.5 mmol) and **100** (1.088 g, 1.5 mmol) in 20 mL of THF was added Pd(PPh$_3$)$_2$Cl$_2$ (0.050 g) in THF (5 mL). The mixture was stirred at reflux for 3 days. After cooling to room temperature, CH$_2$Cl$_2$ (50 mL) was added and the organic layer was separated and washed with H$_2$O and 2 N HCl. After concentration, the polymer was precipitated with MeOH (1 : 10). Repeating the dissolution and precipitation procedure gave polymer **101** in 73% yield. GPC (polystyrene standards): $M_n = 15,500$, PDI = 1.8. The resulting polymer was then treated with lithium aluminum hydride (LAH) in THF followed by the treatment of BF$_3$·Et$_2$O to give ladder polymer **102**. GPC (polystyrene standards): $M_n = 17,000$ (PDI = 2.0).

9.4.1.23 Synthesis of Poly(pyridine) 106[70]

Under nitrogen, THF (20 mL) and NMP (10 mL) were added to a mixture of **103** (1.81 g, 2.04 mmol), **104** (0.81 g, 2.00 mmol), Pd(PPh$_3$)$_2$Cl$_2$ (0.028 g, 0.04 mmol), and CuI (0.008 g, 0.04 mmol). The reaction mixture was stirred at 90°C for 2 days. After cooling to room temperature, CH$_2$Cl$_2$ was added and the mixture was filtered through a pad of Celite. The filtrate was concentrated to ca. 10 mL and acetone (150 mL) was added dropwise. The precipitate was collected by filtration and dissolved in a minimum amount of CHCl$_3$ to repeat the precipitation procedure. Polymer **105** was obtained as a brown solid (0.695 g). GPC

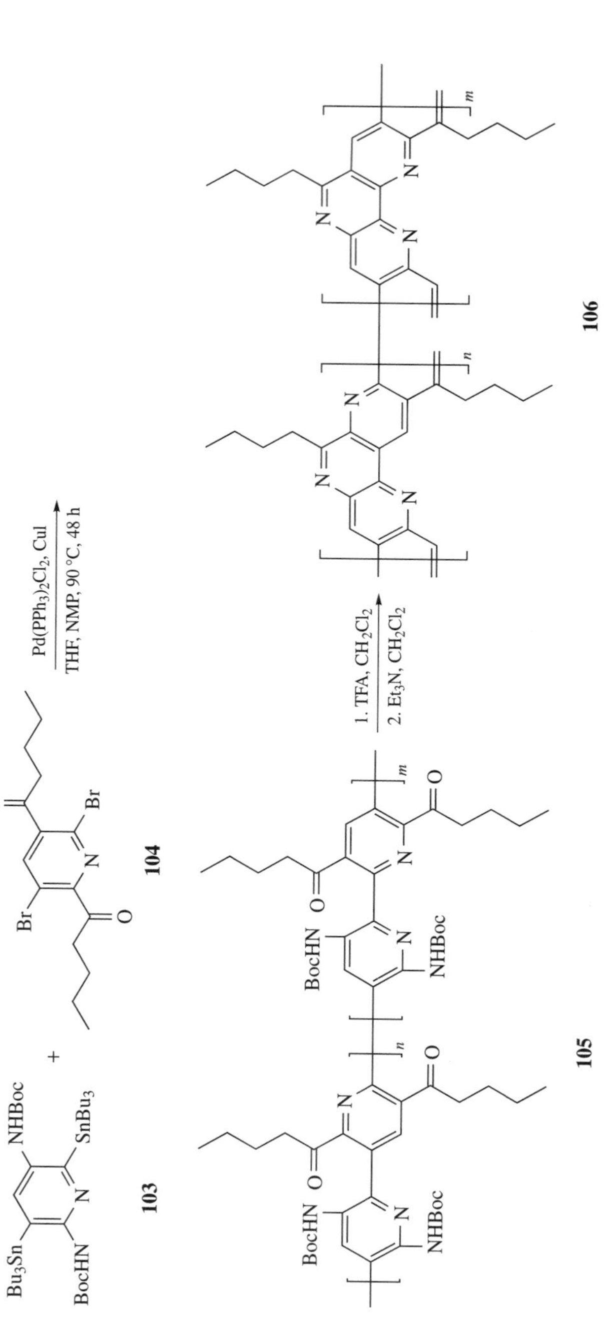
Bu3Sn
BocHN
NHBoc
SnBu3
103
+
Br
O
Br
O
104
Pd(PPh3)2Cl2, CuI
THF, NMP, 90 °C, 48 h
BocHN
NHBoc
O
n
m
105
1. TFA, CH2Cl2
2. Et3N, CH2Cl2
N
n
m
106

(polystyrene standards): $M_n = 15,250$, $M_w = 65,700$. To a solution of polymer **105** (0.565 g) in CH_2Cl_2 (20 mL) was added trifluoroacetic acid (10 mL). The mixture was refluxed overnight. After cooling to room temperature, the solution was added slowly and carefully to a mixed solvent of CH_2Cl_2 (35 mL) and Et_3N (35 mL). The precipitate was collected. The solid obtained was suspended in Et_3N (15 mL) and heated to 100°C for 10 h and 180°C for 1 day. Filtration followed by washing with CH_2Cl_2 and Et_2O yielded polymer **106** as a dark brown solid in 74% yield (0.380 g).

9.4.1.24 Synthesis of Planar Polymer 109[71]

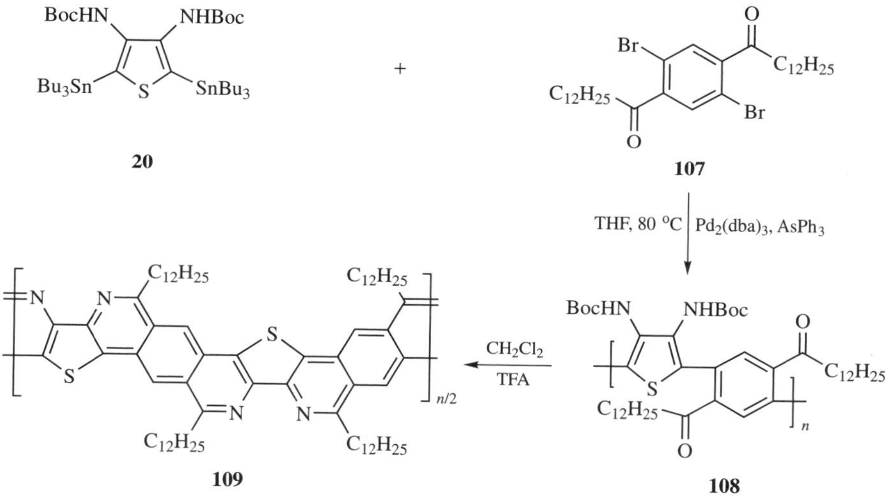

(1) Under nitrogen, $Pd_2(dba)_3$ (0.00927 g, 0.0100 mmol) and $AsPh_3$ (0.0061 g, 0.0200 mmol) were added to a mixture of **20** (0.4374 g, 0.4900 mmol) and **107** (0.3140 g, 0.5000 mmol) in THF (5.0 mL). The reaction mixture was stirred at 80°C for 72 h. After cooling to room temperature, the mixture was poured into a solution of KF (10 mL, 1 M in H_2O) and stirred for 30 min followed by filtration. The filtrate was extracted with CH_2Cl_2 and dried over Na_2SO_4. After concentration, the residue was dissolved in acetone (5 mL) and precipitated with hexane (200 mL). Repeating the dissolvation and precipitation gave polymer **108** in 51% yield (0.20 g). GPC (polystyrene standards): $M_n = 9.400$, $M_w = 15,000$. (2) Under nitrogen, trifluoroacetic acid (1.5 mL) was added to the mixture of polymer **108** (0.1302 g, 0.1666 mmol) in CH_2Cl_2 (1.5 mL) and the reaction mixture was stirred at room temperature for 12 h and then poured into a solution of NaOH (10 mL, 3 M). The mixture was stirred for another 2 h and heated to evaporate CH_2Cl_2. After filtration, the solid was washed with H_2O, CH_2Cl_2, and ether. To this solid was added Et_3N and the mixture was stirred at 80°C for 12 h. After cooling to room temperature, the solid was collected by filtration and washed with H_2O, ether, and CH_2Cl_2 and dried to give polymer **109** in 91% yield (0.0847 g).

9.4.1.25 Synthesis of Poly(arylamine) 26[21b]

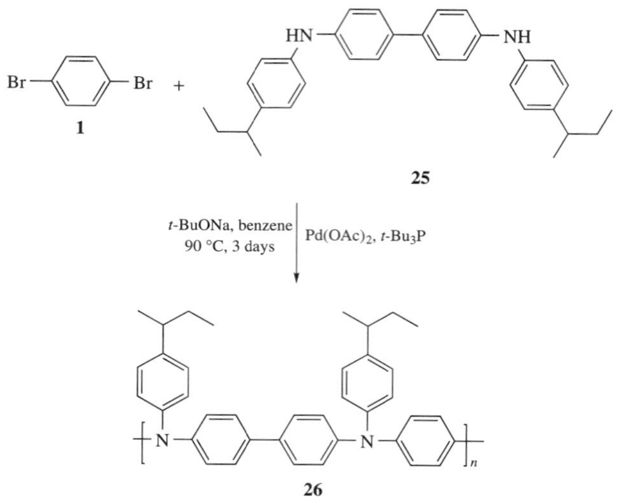

Under nitrogen, a mixture of 1,4-dibromobenzene (0.25 mmol), diamine **25** (0.25 mmol), NaO-t-Bu (72 mg, 0.75 mmol), and 0.25 mL (0.5 mmol%) solution of benzene that is 0.01 mM in Pd(OAc)$_2$ and 0.03 mM in-t-Bu$_3$P was stirred in a sealed vial at 90°C for 3 days. After 2 mL of 0.5 M citric acid was added, the reaction mixture was extracted with toluene and washed with 5% KCN. After concentration, the solid was dissolved in a minimum of toluene. The polymer was obtained by precipitation from MeOH (50 mL) in 95% yield. GPC (polystyrene standards): $M_w = 102,100$ and $M_n = 13,600$.

9.4.1.26 Synthesis of Poly(m-arylamine) 112[72]

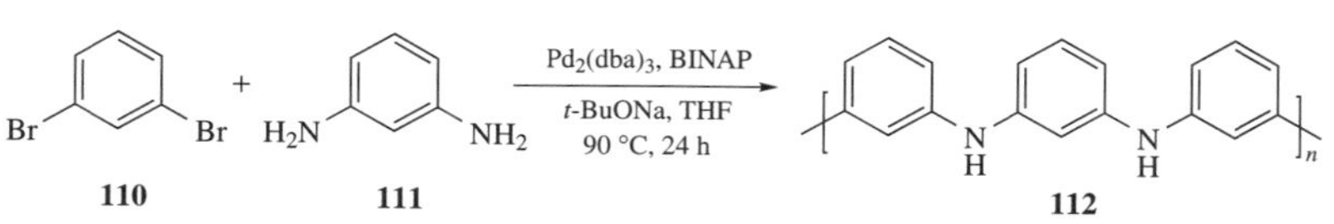

To a mixture of 1,3-dibromobenzene (4.36 g, 18.49 mmol), 1,3-phenylene diamine **111** (2.00 g, 18.49 mmol), NaO-t-Bu (3.73 g, 38.84 mmol), Pd$_2$(dba)$_3$ (0.339 g, 0.37 mmol), and BINAP (0.691 g, 1.11 mmol) in a heavy-walled flask equipped with a Teflon valve was added THF (15 mL) under inert atmosphere. The flask was sealed and heated to 90°C. After 24 h, the reaction mixture was cooled to room temperature and neutralized using 0.2 mol equivalents of 2.4 N HCl in MeOH. The polymer was precipitated from hexanes, filtered, and dried under vacuum. The dried polymer was redissolved in THF, filtered through Celite, and reprecipitated. The solid was collected by filtration and then purified by Soxhlet extraction with CH$_2$Cl$_2$ overnight. After drying under vacuum, a green-tan solid was obtain in ~106% yield. GPC (NMP, polystyrene standards): $M_w = 39,000$;

$M_n = 11,000$; PDI $= 3.6$. GPC (THF, polystyrene standards): $M_w = 8800$, $M_n = 5600$ (PDI $= 1.6$).

9.4.1.27   Synthesis of Polymer 43[36b]

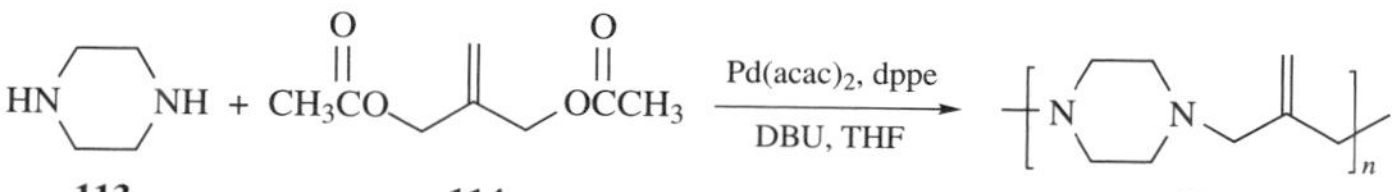

Under nitrogen, a mixture of **113** (0.344 g, 2 mmol), 1,8-diazabicyclo[5,4,0] undec-7-ene (DBU) (0.913 g, 6 mmol), Pd(acac)$_2$ (0.0061 g, 0.02 mmol), and 1,2-bis(diphenylphosphino)ethane (dppe) (0.0159 g, 0.04 mmol) in THF (1 mL) was stirred at 50°C to produce a homogeneous mixture. Then **114** (0.172 g, 2 mmol) in THF (4 mL) was added and the mixture was stirred at 50°C for 5 h. The resulting reaction mixture was poured into MeOH (50 mL) and the polymer **43** was collected by filtration which was further purified by reprecipitation from CHCl$_3$ with adding CH$_3$OH. After drying under vacuum, polymer **43** was obtained in 86% yield (273 mg). Vapor pressure osmometry (40°C, CHCl$_3$): $M_n = 2100$.

9.4.2 Linear Polymer with Chiral Units

9.4.2.1   Synthesis of Poly(thiophene) 117[24a]

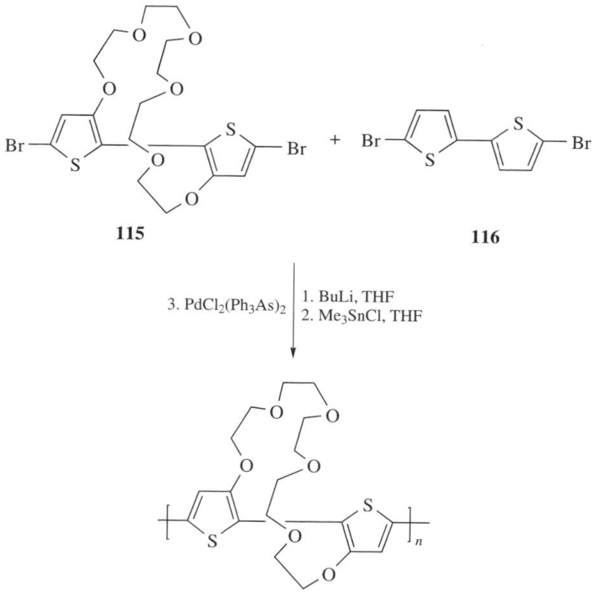

Under nitrogen, **115**(1 eq.) was reacted with BuLi (2 eq.) at 0°C for 1 h followed by treatment with trimethyltin chloride (2.5 eq.) in THF at room temperature for 30 min. Then a solution of **116** (1 eq.) and PdCl$_2$(Ph$_3$As)$_2$ (0.02 eq.) in THF was combined and the mixture was stirred at reflux overnight. The polymer was precipitated with MeOH followed by filtration and dried under vacuum. GPC (polystyrene standards): $M_n = 2000$.

9.4.2.2   Synthesis of Polycyclophane 36[38]

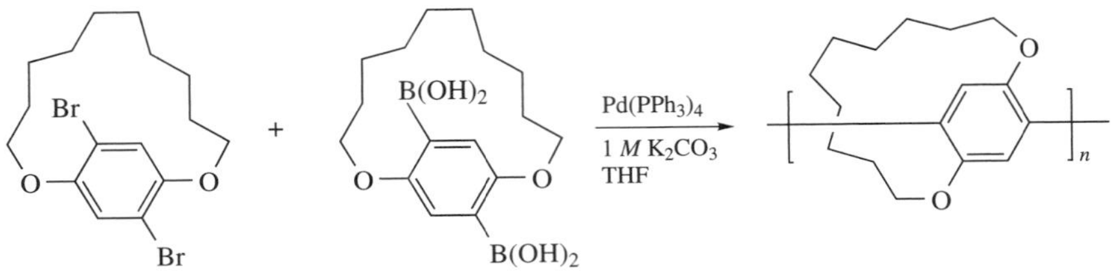

Under nitrogen, a solution of **118** (0.606 g, 1.5 mmol) and **119** (0.5 g, 1.5 mmol) in THF (20 mL) was added to 1 M K$_2$CO$_3$ (20 mL). Pd(PPh$_3$)$_4$ (26 mg, 1.5 mol %) in THF (5 mL) was added and stirred at reflux for 2 days. The reaction mixture was then poured into MeOH (200 mL). The precipitated polymer was collected by filtration, washed with dilute HCl (50 mL), and redissolved in CHCl$_3$. The resulting solution was dried and concentrated and MeOH was added to precipitate out polymer **36** in 48% yield (0.533 g). GPC (polystyrene standards): $M_w = 12,700$, $M_n = 3000$.

9.4.2.3   Synthesis of Polyketone 28[22]

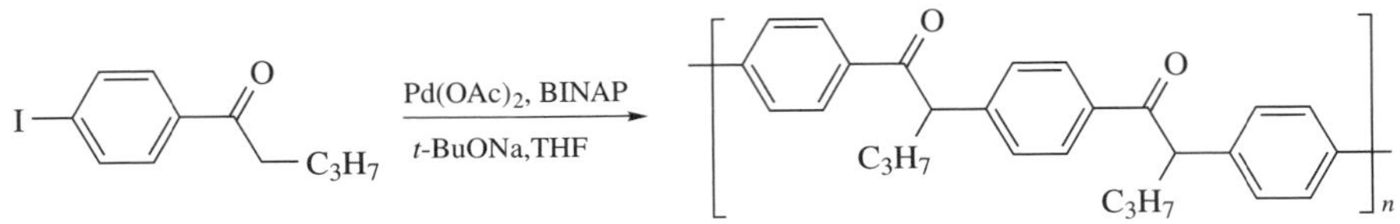

Under nitrogen, a mixture of **27** (1 mmol), Pd(dba)$_3$ (1.5 mmol %), BINAP (3.6 mmol %), and t-BuONa in 3 mL THF was stirred at 60–90°C for 10 h. After cooling to room temperature, MeOH was added to precipitate out the polymer in 65% yield. GPC (polystyrene standards): $M_w = 13,200$, $M_n = 10,100$.

9.4.2.4 Synthesis of Optically Active Polymer 35[57]

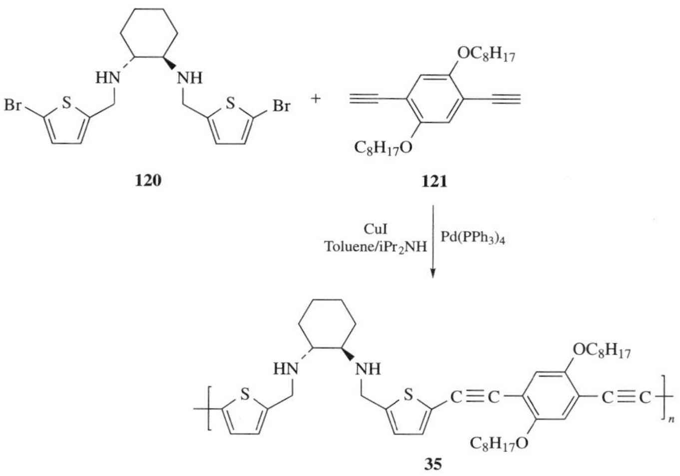

In a toluene–i-Pr$_2$NH mixture (25 mL/10 mL), **120** (0.474 g, 1.02 mmol) was reacted with **121** (0.390 g) in the presence of Pd(PPh$_3$)$_4$ (4 mol %) and CuI (4 mol %). Polymer **35** was obtained in 90% yield. GPC (polystyrene standards): $M_w = 11,900$ (PDI = 1.7); $[\alpha]_D = -65°$ ($c = 0.9$, THF).

9.4.2.5 Synthesis of Optically Active Poly(1,1′-bi-2-naphthyl) 16[54c]

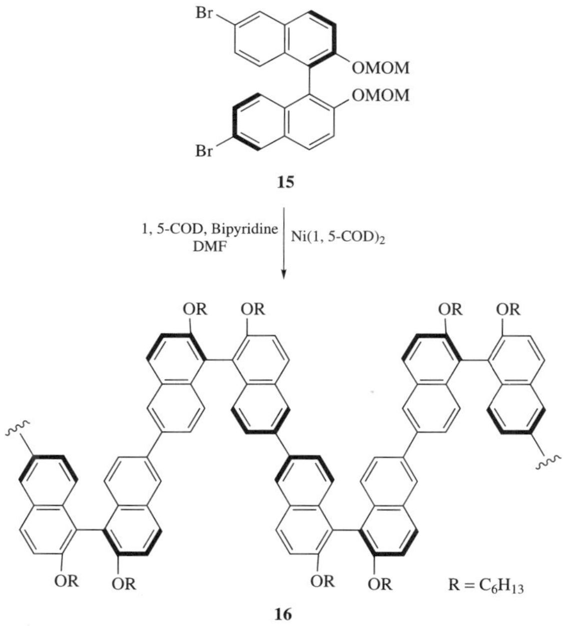

Under nitrogen, to a 50-mL Schlenk flask were added (R)-**15** (0.855 g, 1.42 mmol), bis(cyclooctadiene)nickel(0) [(1,5-COD)$_2$Ni] (0.489 g, 1.76 mmol), 2,2′-bipyridine (0.330 g, 2.09 mmol), and 1,5-cyclooctadiene (1,5-COD) (0.6 mL, 4.9 mmol) in anhydrous DMF (9 mL), and the reaction mixture was heated at 70°C for 21 h. Aqueous HCl (1 N, 25 mL) was then added to the reaction mixture at room temperature and was extracted with CH$_2$Cl$_2$ (2 × 25 mL). The organic layer was passed through a plug of silica gel and concentrated under vacuum. The polymer was precipitated out with MeOH and isolated after centrifugation. The yield of (R)-**16** is 58%. GPC (polystyrene standards): M_w = 15,000, M_n = 6800 (PDI = 2.2); [α]$_D$ = −215° (c = 0.52, CH$_2$Cl$_2$).

9.4.2.6 Synthesis of Optically Active Poly(phenylenevinylene) 124[73]

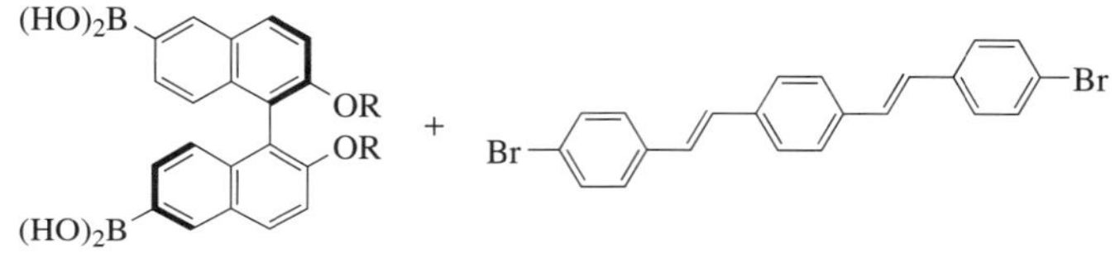

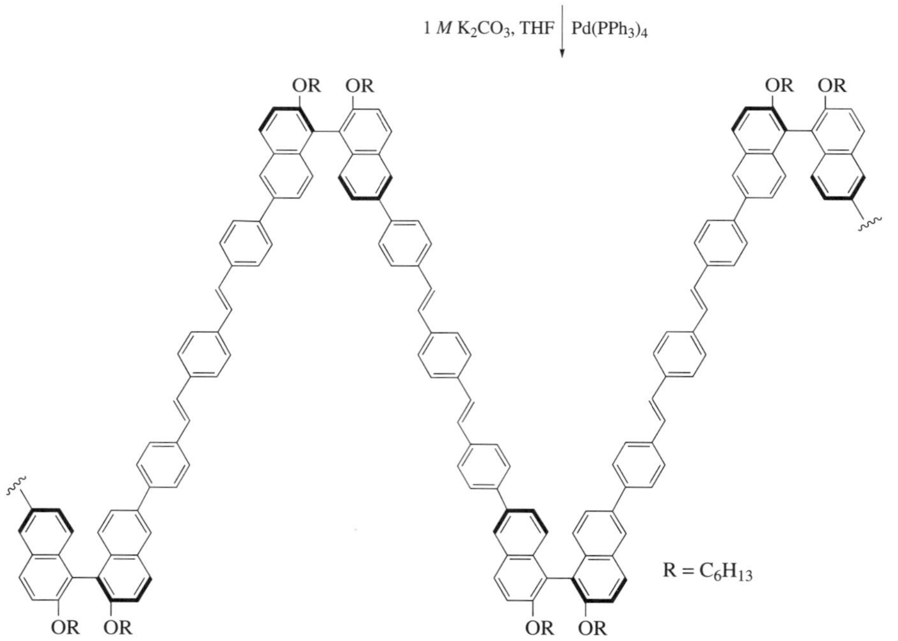

Under nitrogen, a solution of Pd(PPh$_3$)$_4$ (0.058 g, 0.050 mmol) in THF (5 mL) was added to the mixture of (R)-**122** (0.544 g, 1 mmol) and **123** (0.440 g,

1 mmol) in THF (10 mL). The reaction mixture was stirred at reflux for 48 h. After cooling to room temperature, CH_2Cl_2 (100 mL) was added. The organic layer was washed with 1 N HCl (30 mL) and brine. The solvent was removed by evaporation under vacuum. The residue was dissolved in THF and precipitated with MeOH. This procedure was repeated twice. Polymer (R)-**124** was obtained in 95% yield (695 mg). GPC (polystyrene standards): $M_w = 67{,}000$, $M_n = 20{,}000$ (PDI = 3.4); $[\alpha]_D = -351°$ (c = 0.38, THF).

9.4.2.7   Synthesis of Optically Active Poly(binaphthol) 34[44]

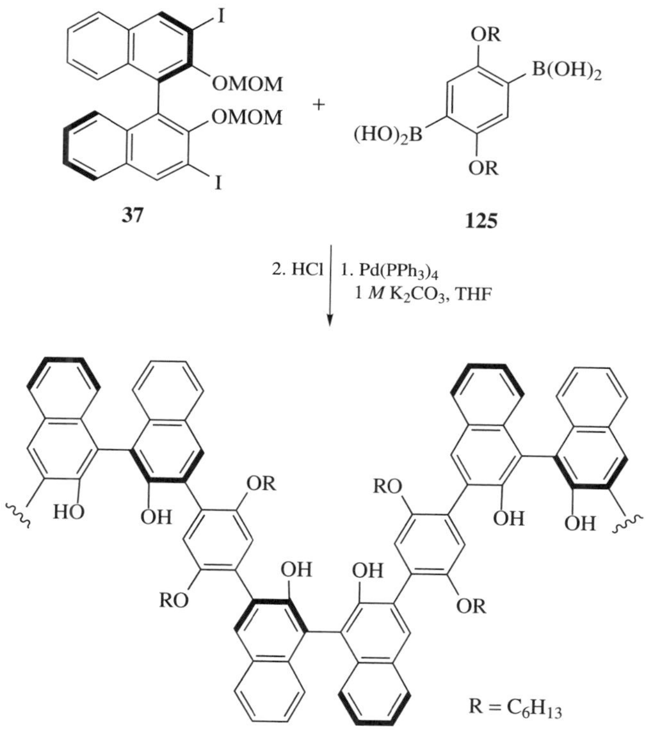

37   **125**

2. HCl | 1. Pd(PPh$_3$)$_4$
       | 1 M K$_2$CO$_3$, THF

R = C$_6$H$_{13}$

34

To a flask containing (R)-**37** (13.70 g, 22.0 mmol), **125** (8.0 g, 22.0 mmol), THF (75 mL), and 1 M K$_2$CO$_3$ (100 mL) was added Pd(PPh$_3$)$_4$ (0.5 g in 25 mL THF), and the reaction mixture was heated at reflux under nitrogen for 36 h. EtOAc was then added. The organic layer was washed with H$_2$O and filtered. After removal of EtOAc with rotary evaporation, the residue was redissolved in CH_2Cl_2 and precipitated with MeOH. This procedure was repeated three times. After being dried under vacuum, the resulting polymer was dissolved in THF (30 mL) to which 6 N HCl (20 mL) was added subsequently. After the mixture was heated at reflux for 16 h, CH_2Cl_2 was added. The organic layer was separated and washed with H_2O. Removal of the solvent with rotary evaporation gave a polymer residue which was redissolved in CH_2Cl_2 and precipitated with MeOH. This procedure was repeated three times. After being dried under vacuum, (R)-**34** was obtained as a yellow solid in 88% yield (two steps). GPC (polystyrene standards): $M_w = 24,300$, $M_n = 9900$ (PDI = 2.5); $[\alpha]_D = -16.6°$ ($c = 0.5$, THF).

9.4.2.8   Synthesis of Optically Active Poly(binaphthol) 39[18a]

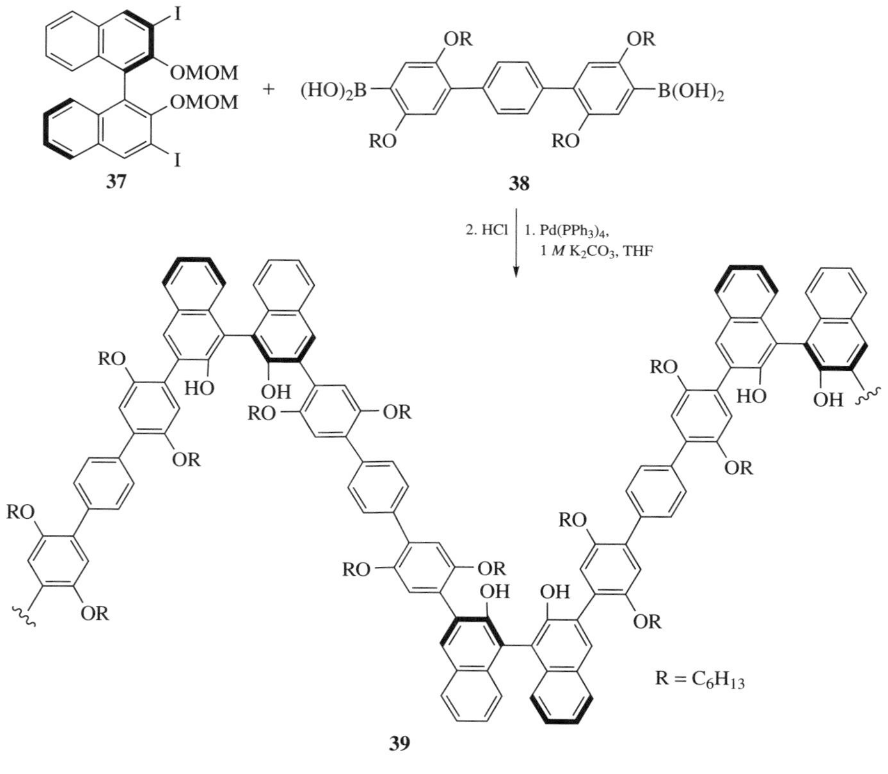

37   **38**

2. HCl   1. Pd(PPh$_3$)$_4$,
          1 M K$_2$CO$_3$, THF

$R = C_6H_{13}$

39

To a mixture of (R)-**37** (5.00 g, 8.0 mmol), **38** (6.20 g, 8.6 mmol), THF (50 mL), and 1 M K$_2$CO$_3$ (50 mL) under nitrogen was added Pd(PPh$_3$)$_4$ (0.060 g). After the mixture was heated at reflux for 36 h until the ^{1}H NMR spectrum showed the disappearance of the iodide endgroup signals, 4-*tert*-butylbromobenzene (0.5 mL, 2.9 mmol) was added to the mixture to cap the polymer chain by reaction with the boronic acid endgroups. The mixture was heated at reflux for another 5.5 h and then cooled to room temperature. Ethyl acetate was added and the organic layer was washed with brine. Evaporation of the solvent gave a yellow residue that was redissolved in CH_2Cl_2 and precipitated with methanol. This procedure was repeated three times. After the residue was dried under vacuum, a polymer was obtained as a yellow solid in 94% yield (7.55 g). GPC (polystyrene as standard): $M_w = 24,000$, $M_n = 12,200$ (PDI = 1.97); $[\alpha]_D = -140.98°$ ($c = 1.00$, CH_2Cl_2). Hydrolysis of this polymer in 6 N HCl–THF at reflux under nitrogen for 17 h gave (R)-**39** in 96% yield. GPC (polystyrene standards): $M_w = 25,800$, $M_n = 14,300$ (PDI = 1.80); $[\alpha]_D = -92.9°$ ($c = 1.01$, CH_2Cl_2).

9.4.2.9 Synthesis of Optically Active Poly(BINAP) 127[74]

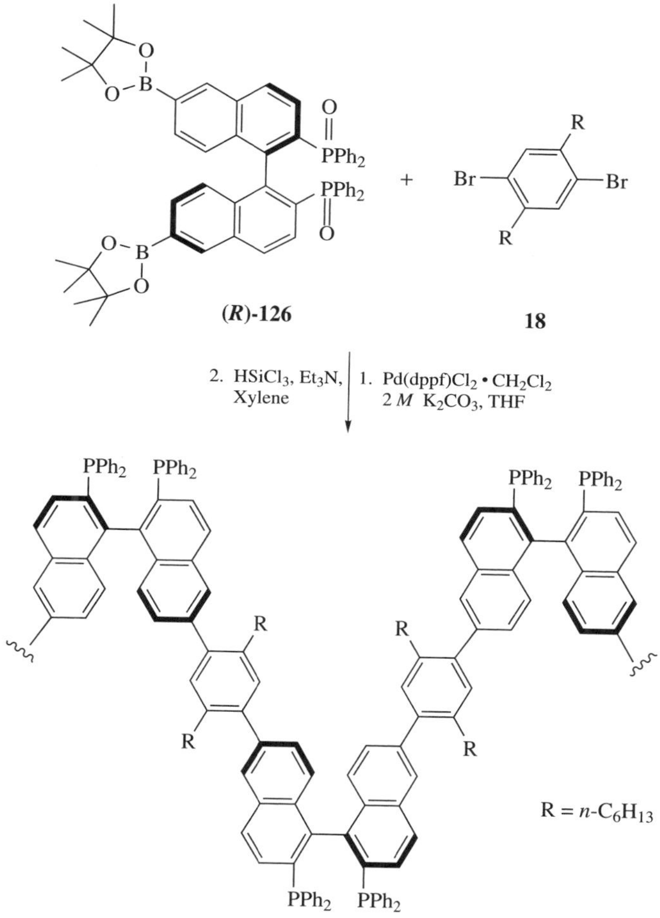

(R)-**126**   **18**

2. $HSiCl_3$, Et_3N,   1. $Pd(dppf)Cl_2 \cdot CH_2Cl_2$
   Xylene   2 M K_2CO_3, THF

$R = n\text{-}C_6H_{13}$

127

Under nitrogen, $Pd(dppf)Cl_2 \cdot CH_2Cl_2$ (0.024 g, 0.03 mmol) and 2 M K_2CO_3 (8 mL) were added to a solution of (R)-**126** (0.092 g, 1.02 mmol) and 2,5-dihexyl-1,4-dibromobenzene (0.040 g, 1.00 mmol) in THF (10 mL). After the reaction mixture was stirred at reflux for 51 h, 4-butyl bromobenzene (100 μL) was added. The mixture was refluxed for another 18 h to cap the polymer chain end. The organic layer of the mixture was washed with 1 N HCl (5 mL × 2) and brine (5 mL × 2). The solvent was removed under vacuum. MeOH (20 mL) was added to the residue to precipitate out the polymer. The dissolution with CH_2Cl_2 and precipitation with MeOH were repeated three times to give a polymer as a dark brown solid. To the resulting polymer (0.60 g, 0.67 mmol) in o-xylene (50 mL) and Et_3N (7 mL) was added $HSiCl_3$ (5 mL). The reaction mixture was heated at reflux for 48 h. After cooling to room temperature, NaOH (30%, 30 mL) was added carefully and the mixture was stirred for another 10 min. The resulting mixture was quickly extracted with degassed CH_2Cl_2 (50 mL × 3). The extract was concentrated to 20 mL and filtered. The filtrate was further concentrated and degassed MeOH was added to precipitate out the polymer. Quickly repeating the dissolution with CH_2Cl_2 and precipitation with methanol three times gave (R)-**127** as a yellow solid in 75% yield (two steps). GPC (polystyrene standards): $M_w = 5800$, $M_n = 4300$ (PDI = 1.35); $[\alpha]_D = 36.6°$ ($c = 0.12$, THF).

9.4.2.10 Synthesis of Optically Active 130[75]

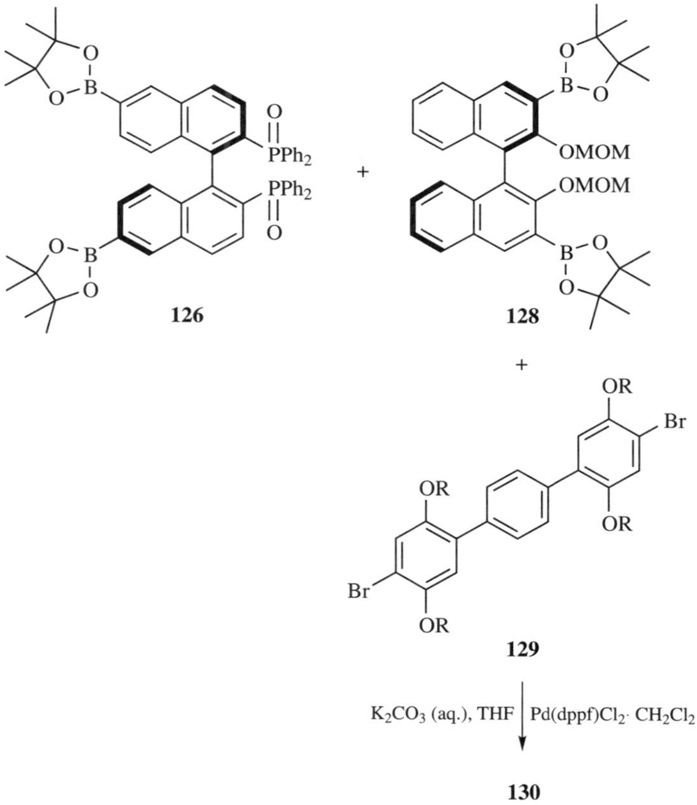

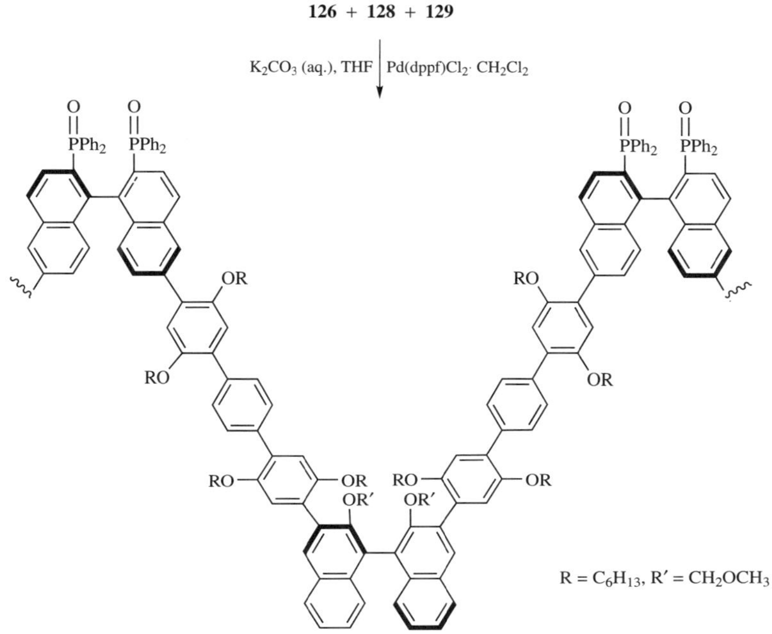

126 + 128 + 129

K$_2$CO$_3$ (aq.), THF | Pd(dppf)Cl$_2$· CH$_2$Cl$_2$

R = C$_6$H$_{13}$, R′ = CH$_2$OCH$_3$

130

Under nitrogen, a mixture of (R)-**128** (0.13 g, 0.20 mmol), **129** (0.32 g, 0.40 mmol), and (R)-**126** (0.18 g, 0.20 mmol) in THF (20 mL) was degassed for 20 min. The resulting mixture was combined with Pd(dppf)Cl$_2$·CH$_2$Cl$_2$ complex (0.032 g, 0.04 mmol) and 2 M aqueous K$_2$CO$_3$ (12 mL, degassed). The reaction mixture was stirred at reflux for 46 h. 4-*tert*-Butyl bromobenzene (0.100 mL) was then added and the mixture was then stirred at reflux for another 10 h in order to cap the polymer chain end. After cooling to room temperature, the organic layer was concentrated. The polymer was precipitated out with MeOH. Repeating the dissolution with CH$_2$Cl$_2$ and precipitation with MeOH three times gave polymer **130** as a gray solid in 93% yield (0.42 g).

9.4.2.11   Synthesis of Optically Inactive Poly(phenylene-thiophene) 132[40]

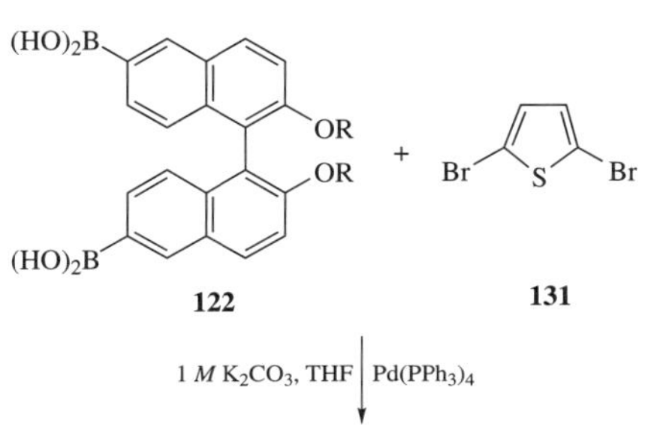

122   **131**

1 M K$_2$CO$_3$, THF | Pd(PPh$_3$)$_4$

132

122 + 131

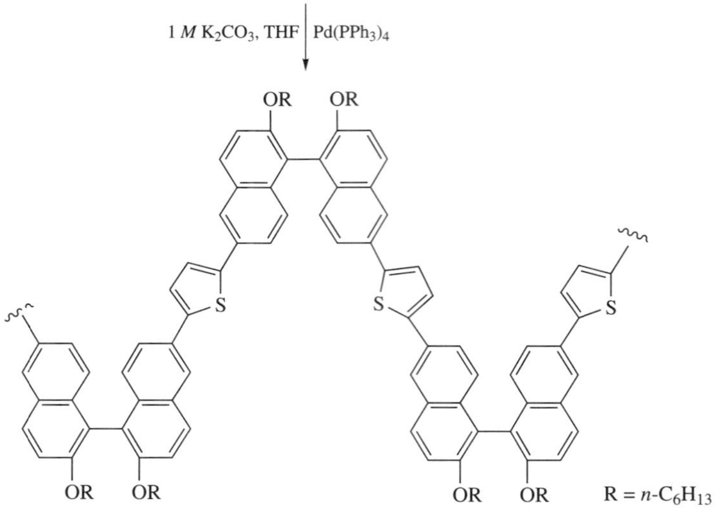

Under nitrogen, Pd(PPh₃)₄ (0.027 g, 0.024 mmol) and 1 M K₂CO₃ (7 mL) were added to the solution of racemic **122** (0.46 mmol) and **131** (0.46 mmol) in THF (8 mL). The reaction mixture was stirred at 60°C for 48 h. After cooling to room temperature, CH₂Cl₂ (20 mL) was added and the organic layer was separated. The aqueous layer was extracted with CH₂Cl₂ (20 mL × 2). The combined organic layer was washed with 1 N HCl and brine and then was concentrated by rotary evaporation under vacuum. The residue was dissolved in a minimum amount of THF and MeOH was added to precipitate out the polymer. This procedure was repeated twice to give **132** as a greenish-yellow solid in 93% yield (0.230 g). GPC (polystyrene standards): $M_w = 35,500$, $M_n = 13,900$ (PDI = 2.6).

9.4.2.12 Synthesis of Optically Active Poly(phenyleneethynylene) 135[76]

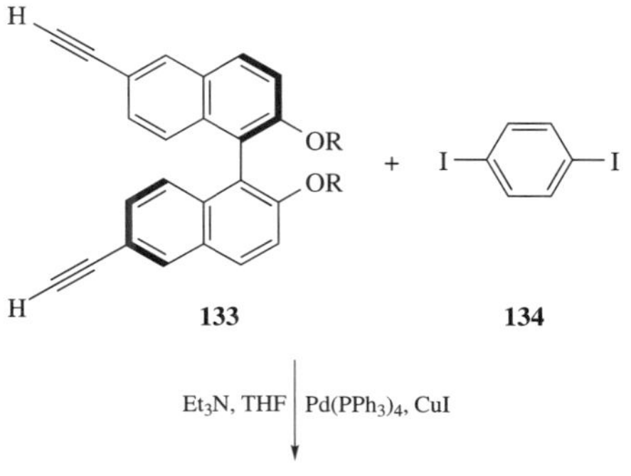

135

133 + **134**

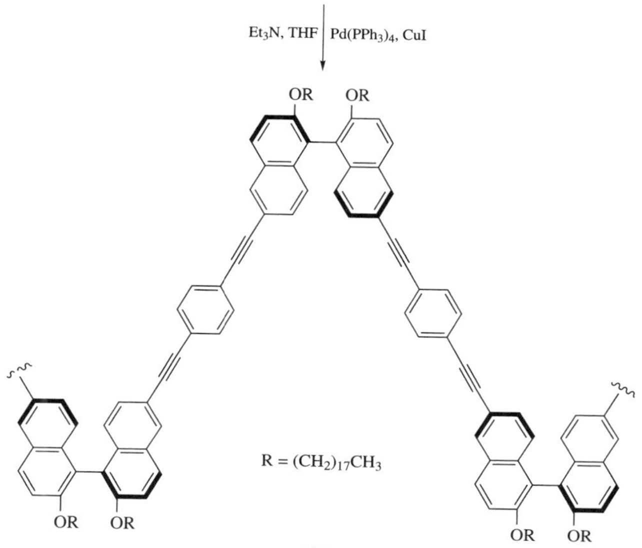
Et3N, THF
Pd(PPh3)4, CuI
OR
OR
R = (CH2)17CH3

135

Under nitrogen, to a 50-mL flame-dried Schlenk flask were loaded (*R*)-**133** (0.210 g, 0.25 mmol), 1,4-diiodobenzene (0.083 g, 0.25 mmol), Et_3N (2 mL), and toluene (8 mL). The resulting solution was degassed for 30 min and then $Pd(PPh_3)_4$ (0.0144 g, 0.013 mmol) and CuI (0.0024 g, 0.013 mmol) were added in the drybox. After this reaction mixture was refluxed for 48 h, it was filtered to remove the Et_3NHBr precipitate. The salt was rinsed with Et_2O. The combined filtrate was evaporated to dryness and the residue was dissolved in a minimum amount of CH_2Cl_2 and precipitated with MeOH (75 mL) twice. After filtration and drying under vacuum, (*R*)-**135** was obtained as a yellow solid in 96% yield (0.220 g). GPC (THF, polystyrene standards): $M_w = 29{,}000$, $M_n = 12{,}000$ (PDI = 2.4); $[\alpha]_D = -272.2°$ ($c = 0.5$, CH_2Cl_2).

9.4.2.13 Synthesis of Propeller-Like Polymer 137[77]

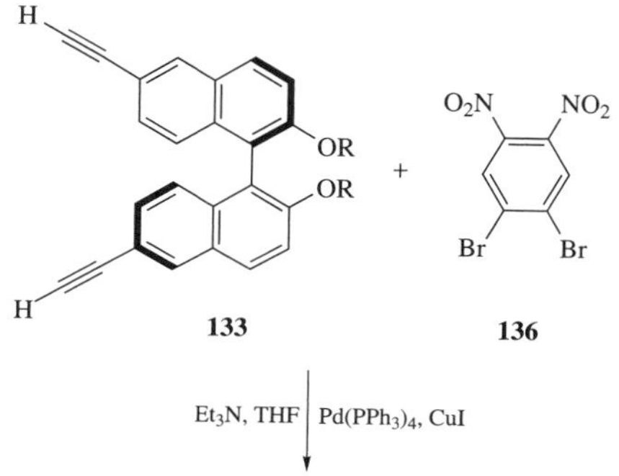
H
OR
OR
H
+
O2N
NO2
Br
Br
133
136
Et3N, THF
Pd(PPh3)4, CuI

137

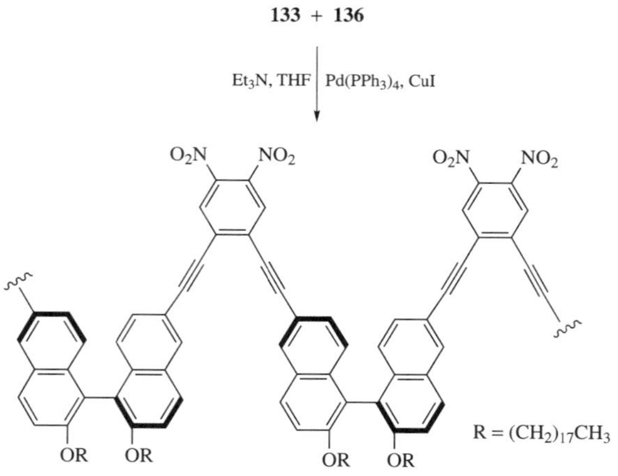
133 + 136
Et3N, THF
Pd(PPh3)4, CuI
O2N
NO2
OR
R = (CH2)17CH3

137

Under nitrogen, to a 50-mL flame-dried Schlenk flask were loaded (*R*)-**133** (0.210 g, 0.25 mmol), **136** (0.082 g, 0.25 mmol), Et_3N (1 mL), and THF (4 mL). The resulting solution was degassed for 30 min and then $Pd(PPh_3)_4$ (0.0144 g, 0.013 mmol) and CuI (0.0024 g, 0.013 mmol) were added in the drybox. After this reaction mixture was stirred at room temperature for 2 days, it was filtered to remove the Et_3NHBr salt. The salt was rinsed with Et_2O until the filtrate was clear. The combined filtrate was evaporated to dryness to give a brown residue which was dissolved in a minimum amount of CH_2Cl_2 and precipitated with MeOH (75 mL) twice. After filtration and drying under vacuum, (*R*)-**137** was obtained as an orange solid in 83% yield (0.208 g). GPC (THF, polystyrene standards): $M_w = 18{,}000$, $M_n = 12{,}000$ (PDI = 1.5); $[\alpha]_D = -163.9°$ ($c = 0.16$, CH_2Cl_2).

9.4.2.14 Synthesis of Optically Active Poly(arylene) 139 [54b]

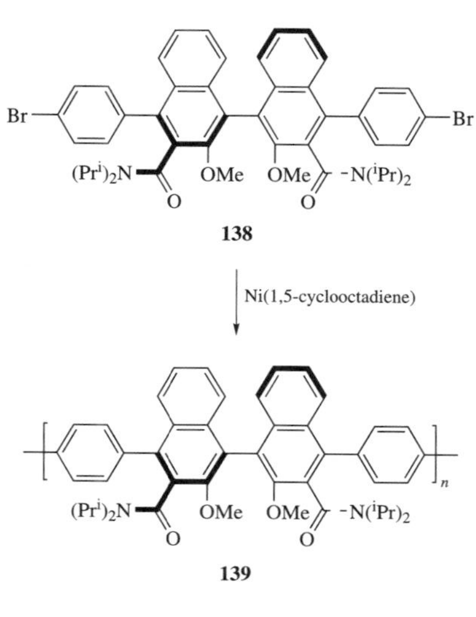
Br
(Pri)2N
OMe
O
-N(iPr)2
138
Ni(1,5-cyclooctadiene)
n

139

To a dry, N$_2$-purged test tube were added Ni(0)(1,5-cyclooctadiene) (0.11 g, 0.4 mmol), 2,2'-bypyridine (0.06 g, 0.4 mmol), (−)-(R, S, R)-**138** (0.20 g, 0.25 mmol), DMF (4 mL), and 1,5-cyclooctadiene (0.05 mL, 0.4 mmol) sequentially. The mixture was then heated to 80°C overnight. The mixture was then poured into 3 N HCl and the solid was collected by filtration to give polymer **139** in 80% yield (0.14 g). GPC (polystyrene standards): $M_n = 6800$ (PDI = 2.29). $[\alpha]_D^{23} = -63$ (c = 0.16, THF).

9.4.3 Hyperbranched and Dendritic Polymers

9.4.3.1 Synthesis of Phenyl-Based Hyperbranched Polymer 40[32a]

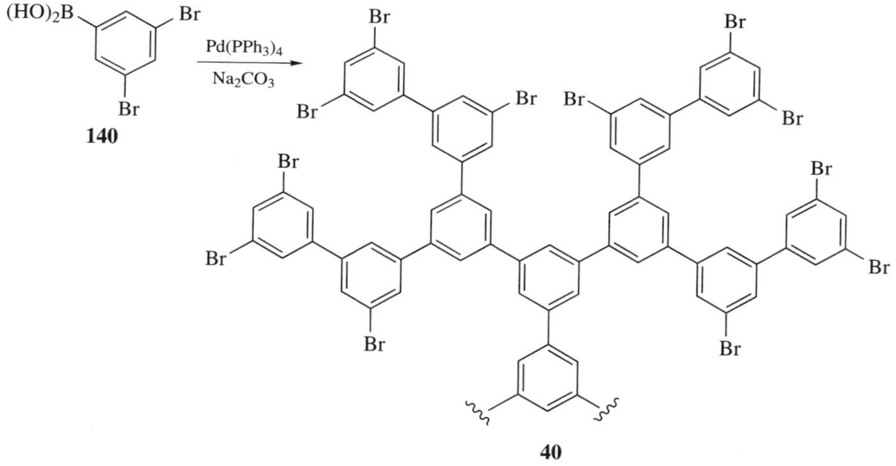

140

40

Under nitrogen, 3.0 g of **140**, 0.030 g of Pd(PPh$_3$)$_4$ in 50 mL organic solvent, and 20 mL of aqueous Na$_2$CO$_3$ were reacted to give polymer **40** in 80–95% yield. GPC (polystyrene standards): $M_w = 5750$, $M_n = 3820$.

9.4.3.2 Synthesis of Hyperbranched Polyamine 142[78]

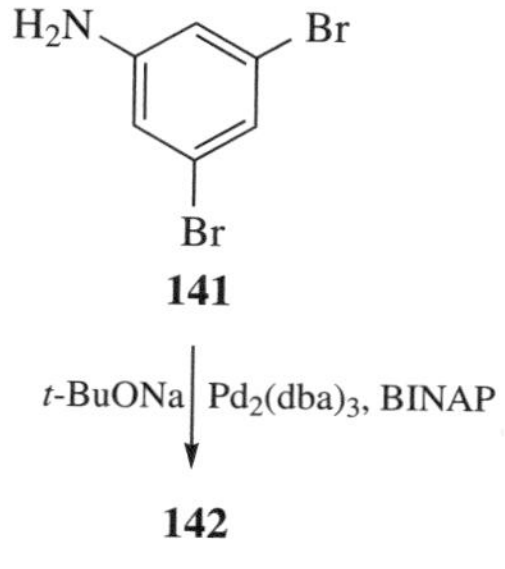

141

t-BuONa | Pd$_2$(dba)$_3$, BINAP

142

141

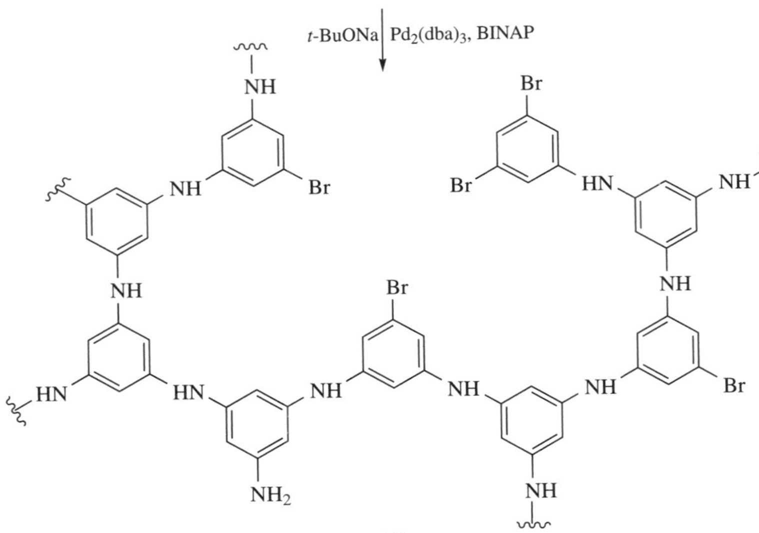

142

To a mixture of 3,5-dibromoaniline (0.9947 g, 3.96 mmol), NaO-t-Bu (0.4382 mg, 4.56 mmol), Pd$_2$(dba)$_3$ (0.726 g, 0.08 mmol), and BINAP (0.1481 g, 0.24mmol) in a heavy-walled flask equipped with a Teflon valve was added THF (15 mL) under inert atmosphere. The flask was sealed and heated to 90°C. After 24 h, the reaction mixture was cooled to room temperature and was neutralized using 0.2 mol equivalents of 2.4 N HCl in MeOH. The hyperbranched polymer was precipitated from hexanes, filtered, and dried under vacuum. The dried polymer was redissolved in THF, filtered through Celite, and reprecipitated. The solid was collected by filtration and then purified by Soxhlet extraction with CH$_2$Cl$_2$ overnight. After drying under vacuum, a green-tan solid (**142**) was obtained in 39% yield. GPC (NMP, polystyrene standards): $M_w = 7000$, $M_n = 2200$ (PDI = 3.2).

9.4.3.3 Synthesis of Dendronized Poly(p-phenylene) 41[32b]

Under nitrogen, tris[tri(p-tolyl)phosphine]-palladium(0) (0.012 g, 1.1 mol %) was added to a mixture of **143** (1.87 g, 1.01 mmol), **144** (diboronic acid) (0.342 g, 1.03 mmol) in 1 M Na$_2$CO$_3$ (100 mL) and toluene (30 mL). The resulting mixture was stirred at reflux for 48 h. The aqueous phase of the reaction mixture was washed with toluene (50 mL) and the combined organic layer was dried with MgSO$_4$. After concentrating to ca. 10 mL, the mixture was poured into MeOH (300 mL) and polymer **41** was recovered by centrifugation (96% yield). GPC (polystyrene standards): $M_w = 275,500$, $M_n = 52,400$.

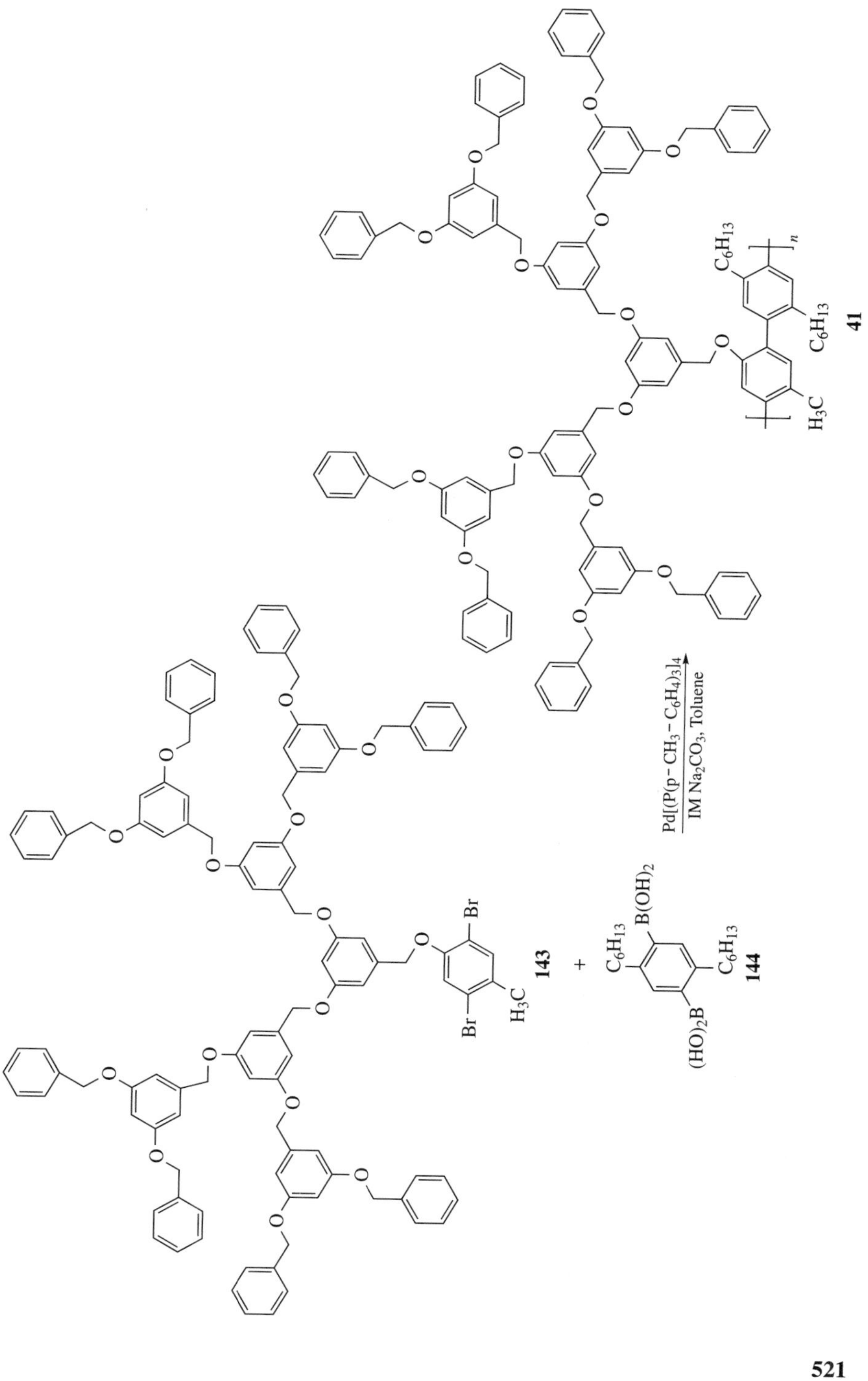

Br
Br
H_3C
143
+
C_6H_{13}
$B(OH)_2$
$(HO)_2B$
C_6H_{13}
144
$Pd[(P(p-CH_3-C_6H_4)_3]_4$
IM Na_2CO_3, Toluene
H_3C
C_6H_{13}
C_6H_{13}
n
41

9.4.3.4 Synthesis of Dendronized Poly(phenyleneethynylene) 42[32c]

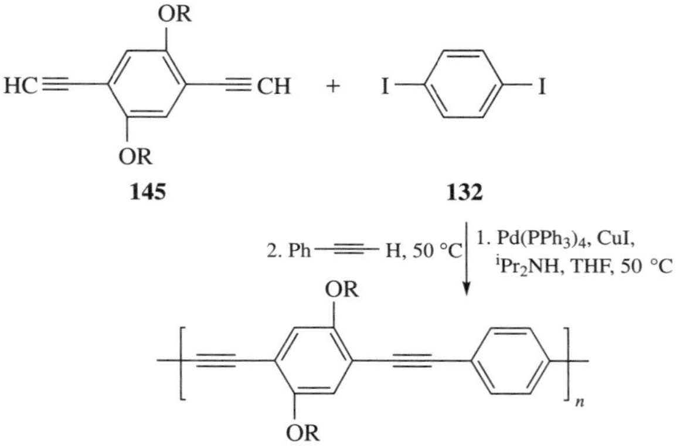

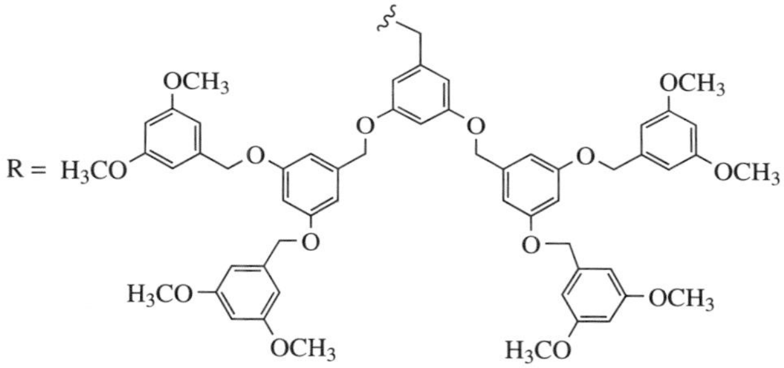

Under argon, a mixture of **145** (0.046 mmol), **132** (0.046 mmol), Pd(PPh$_3$)$_4$ (0.2 μmol), CuI (0.2 μmol), and diisopropylamine (0.015 mmol) in THF (4 mL) was stirred in the dark at 50°C for 2 days. Ethynylbenzene (0.92 mmol) was then added and stirred at 50°C overnight. After concentration, the residue was dissolved in CHCl$_3$ and filtered. The filtrate was subjected to preparative SEC with CHCl$_3$ as eluent in order to remove catalyst residues and unreacted starting materials. Polymer **42** was obtained as a yellow solid in 85% yield. SEC analysis (THF, polystyrene standards): $M_w = 280,000$ (PDI = 6.5).

Abbreviations Used in Structures

BINAP	2,2′-Bis(diphenylphosphino)-1,1′-ninaphthyl
bpy	2,2′-bipyridine
COD	1,5-cyclooctadiene
DBU	1,8-diazabicyclo[5,4,0]undec-7-ene
DMA	N,N'-dimethylacetamide
DMF	N,N'-dimethylformamide
DPPF	1,1′-bis(diphenylphosphino)ferrocene

dppp	1,3-bis(diphenylphosphino)propane
EDTA	Ethylenediaminetetraacetic acid
NMP	1-methyl-2-pyrrolidinone
LDA	Lithium diisopropylamide
Pd(acac)$_2$	Palladium(II) acetylacetonate
Pd$_2$(dba)$_3$	tris(dibenzylideneacteone)dipalladium
P(o-tolyl)$_3$	tri-o-tolylphosphine
TFA	trifluoroacteic acid
THF	Tetrahydrofuran

ACKNOWLEDGMENT

I would like to thank Lin Pu at the University of Virginia for encouraging me to work in this fascinating area and for his guidance. I would also like to thank Hong-Bin Yu for her helpful discussion and editorial assistance.

REFERENCES

1. (a) K. Tamao, K. Sumitani, and M. Kumada, *J. Am. Chem. Soc.*, **94**, 4374 (1972); (b) M. Kumada, *Pure. Appl. Chem.*, **52**, 669 (1980).
2. T. Yamamoto and A. Yamamoto, *Chem. Lett.*, 353 (1977).
3. G. K. Noren and J. K. Stille, *Makromol. Rev.*, **5**, 385 (1971).
4. (a) T. Yamamoto, *Bull. Chem. Soc. Jpn.*, **72**, 621 (1999); (b) L. Pu, in *Electrical and Optical Polymer Systems: Fundamentals, Methods, and Applications*, D. L. Wise, G. E. Wnek, D. J. Trantolo, T. M. Cooper, and J. D. Gresser (Eds.), Marcel Dekker, New York, 1998, Chapter 24; (c) V. Percec and D. H. Hill, in *Step-Growth Polymers for High-Performance Materials*, J. L. Hedrick and J. W. Labadie (Eds.), ACS Symposium Series 624, American Chemical Society, Washington, DC, 1996, Chapter 1.
5. (a) T. A. Chen and R. D. Rieke, *J. Am. Chem. Soc.*, **114**, 10087 (1992); (b) M. J. Marsella, P. J. Carroll, and T. M. Swager, *J. Am. Chem. Soc.*, **116**, 9347 (1994).
6. R. F. Heck, in *Comprehensive Organic Synthesis*, Vol. 4, B. M. Trost (Ed.), Pergamon, New York, 1991, p. 833.
7. (a) Jpn. Pat. 57,207,618, 1981, Asahi Glass Co.; *Chem. Abstr.*, **99**, 23139k (1983); (b) A. Greiner and W. Heitz, *Macromol. Chem. Rapid Comm.*, **9**, 581 (1988).
8. S. Takahash, Y. Kuroyama, K. Sonogashira, and N. Hagihara, *Synthesis*, 627 (1980).
9. (a) K. Sanechika, A. Yamamoto, and T. Yamamoto, *Bull. Chem. Soc. Jpn.*, **57**, 752 (1984); (b) T. Yamamoto, M. Takagi, K. Kizu, T. Maruyama, K. Kubota, H. Kanbara, T. Kurihara, and T. Kaino, *J. Chem. Soc. Chem. Comm.*, 797 (1993); (c) Q. Zhou and T. M. Swager, *J. Am. Chem. Soc.*, **120**, 5321 (1998).
10. (a) M. Tiecco, L. Testaferri, M. Tingoli, D. Chianelli, and M. Mantanucci, *Synthesis*, 736 (1984); (b) I. Colon and D. R. Kelsey, *J. Org. Chem.*, **51**, 2627 (1986).
11. (a) T. Yamamoto, T. Ito, and K. Kubota, *Chem. Lett.*, 153 (1988); (b) V. Percec, S. Okita, and R. Weiss, *Macromolecules*, **25**, 1816 (1992).

12. (a) T. Yamamoto, A. Kashiwazaki, and K. Kato, *Makromol. Chem.*, **190**, 1649 (1989); (b) I. Colon and G. T. Kwiakowski, *J. Poly. Sci. Part A: Poly. Chem. Ed.*, **28**, 367 (1990); (c) Q.-S. Hu, X.-F. Zheng, and L. Pu, *J. Org. Chem.*, **61**, 5200 (1996).

13. J. K. Stille, *Pure. Appl. Chem.*, **57**, 1771 (1985).

14. M. Bochmann and K. Kelly, *J. Chem. Soc. Chem. Comm.*, 532 (1989).

15. Q. T. Zhang and J. M. Tour, *J. Am. Chem. Soc.*, **120**, 5355–5362 (1998).

16. (a) N. Mitaura, T. Yanagi, and A. Suzuki, *Synth. Comm.*, **11**, 513 (1981); (b) N. Mitaura, T. Yanagi, and A. Suzuki, *Chem. Rev.*, **95**, 2457 (1995).

17. M. Rehahn, A.-D. Schlüter, G. Wegner, and W. Feast, *J. Polymer*, **30**, 1060 (1989).

18. (a) Q.-S. Hu, W.-S. Huang, and L. Pu, *J. Org. Chem.*, **63**, 2798 (1998); (b) A. D. Schluter and J. P. Rabe, *Angew. Chem. Int. Ed. Engl.*, **39**, 864 (2000); (c) J. J. S. Lamba and J. M. Tour, *J. Am. Chem. Soc.*, **116**, 11723 (1994); (d) M. B. Goldfinger and T. M. Swager, *J. Am. Chem. Soc.*, **116**, 7895 (1994).

19. (a) M. Kawatsura and J. F. Hartwig, *J. Am. Chem. Soc.*, **121**, 1473 (1999); (b) M. Palucki and S. L. Buchwald, *J. Am. Chem. Soc.*, **119**, 11108 (1997).

20. (a) J.-F. Marcoux, S. Wagaw, and S. L. Buchwald, *J. Org. Chem.*, **62**, 1568 (1997); (b) M. S. Driver and J. F. Hartwig, *J. Am. Chem. Soc.*, **118**, 7217 (1996).

21. (a) T. Kanbara, A. Honma, and K. Hasagawa, *Chem. Lett.*, 1135 (1996); (b) T. Kanbara, K. Izumi, T. Narise, and K. Hasagawa, *Polym. J.*, **30**, 66 (1998); (c) F. E. Goodson, S. I. Hauck, and J. F. Hartwig, *J. Am. Chem. Soc.*, **121**, 7527 (1999); (d) N. Spetseris, R. E. Ward, and T. Y. Meyer, *Macromolecules*, **31**, 3758 (1998).

22. D. Wang and Z. Wu, *Chem. Commun.*, 529 (1999).

23. (a) T. A. Skotheim (Ed.), *Handbook of Conducting Polymers*, Vols. 1 and 2, Marcel Dekker, New York, 1986; (b) H. G. Kiess (Ed.), *Conjugated Conducting Polymers*, Springer-Verlag, New York, 1992.

24. (a) M. J. Marsella and T. M. Swager, *J. Am. Chem. Soc.*, **115**, 12214 (1993); (b) M. J. Marsella, P. J. Carroll, and T. M. Swager, *J. Am. Chem. Soc.*, **117**, 9832 (1995).

25. J.-S. Yang and T. M. Swager, *J. Am. Chem. Soc.*, **117**, 7017 (1995).

26. J. L. Bredas and R. Silby, *Conjugated Polymers: The Novel Science and Technology of Highly Conducting and Nonlinear Optically Active Materials*, Kluwer Academic, Boston, 1991.

27. L. Pu, *Acta Polymer.*, **48**, 116 (1997).

28. M. M. Bouman and E. W. Meijer, *Advanced Mater.*, **7**, 385 (1995).

29. L. Pu, *Chem. Eur. J.*, **5**, 2227 (1999).

30. J. Huber and U. Scherf, *Macromol. Rapid Commun.*, **15**, 897–902 (1994).

31. J. P. Lere-Porte, J. J. E. Moreau, F. Serein-Spirau, and S. Wakim, *Tetrahedron Lett.*, **42**, 3073 (2001).

32. (a) Y. H. Kim and O. W. Webster, *J. Am. Chem. Soc.*, **112**, 4592 (1990); (b) B. Karakaya, W. Claussen, K. Gessler, W. Saenger, and A.-D. Schlüter, *J. Am. Chem. Soc.*, **119**, 3296 (1997); (c) T. Sato, D.-L. Jiang, and T. Aida, *J. Am. Chem. Soc.*, **121**, 10658 (1999); (d) A. D. Schlüter and J. P. Rabe, *Angew. Chem. Int. Ed. Engl.*, **39**, 864 (2000).

33. G. Odian, *Principles of Polymerization*, 3rd ed., Wiley, New York, 1991.

34. J. P. Collman, L. S. Hegedus, J. R. Norton, and R. G. Finke, *Principles and Applications of Organotransition Metal Chemistry*, University Science Books, Mill Valley, CA, 1987.

35. (a) R. Sturmer, *Angew. Chem. Int. Ed. Engl.*, **38**, 3307 (1999). Also see: (a) A. F. Littke and G. C. Fu, *Angew. Chem. Int. Ed. Engl.*, **37**, 3387 (1998); (b) J. P. Wolfe, R. A. Singer, B. H. Yang, and S. L. Buchwald, *J. Am. Chem. Soc.*, **121**, 9550 (1999).

36. (a) T. Yamamoto, T. Taguchi, K. Sanechika, Y. Hayashi, and A. Yamamoto, *Macromolecules*, **16**, 1555 (1983); (b) M. Suzuki, J.-C. Lim, M. Oguni, A. Eberhardt, and T. Saegusa, *Polymer J.*, **22**, 815 (1990).

37. T. Yamamoto, H. Suganuma, T. Maruyama, and K. Kubota, *J. Chem. Soc. Chem. Comm.*, 1613 (1995).

38. T. Yamamoto, T. Kanbara, C. Motr, T. Wakayama, T. Fukuda, T. Inoue, and S. Sasaki, *J. Phys. Chem.*, **30**, 12631 (1996).

39. T. Yamamoto, T. Maruyama, Z.-H. Zhou, T. Ito, T. Fukuda, Y. Yoneda, F. Begum, T. Ikeda, S. Sasaki, H. Takezoe, A. Fukuda, and K. Kubota, *J. Am. Chem. Soc.*, **116**, 4832 (1994).

40. K. Y. Musick, Q.-S. Hu, and L. Pu, *Macromolecules*, **31**, 2933 (1998).

41. T. Yamamoto, Z.-H. Zhou, T. Kanbara, M. Shimura, K. Kizu, T. Maruyama, Y. Nakamura, T. Fukuda, B.-L. Lee, N. Ooba, K. Kubota, and S. Sasaki, *J. Am. Chem. Soc.*, **118**, 10389 (1996).

42. (a) U. Sherf and K. Mullen, *Macromol. Chem. Rapid Comm.*, **12**, 489 (1991); (b) M. Froster, K. O. Annan, and U. Sherf, *Macromolecules*, **32**, 3159 (1999).

43. L. Pu, *Chem. Rev.*, **98**, 2405 (1998).

44. Q.-S. Hu, W.-S. Huang, D. Vitharana, X.-F. Zheng, and L. Pu, *J. Am. Chem. Soc.*, **119**, 12454 (1997).

45. K. B. Simonsen, K. A. Jorgensen, Q.-S. Hu, and L. Pu, *J. Chem. Soc. Chem. Comm.*, 811 (1999).

46. R. Fiesel, J. Huber, and U. Scherf, *Angew. Chem. Int. Ed. Engl.*, **35**, 2111 (1996).

47. T. Sato, D.-L. Jiang, and T. Aida, *J. Am. Chem. Soc.*, **121**, 10658 (1999).

48. T. Yamamoto, K. Sanechika, and A. Yamamoto, *Bull. Chem. Soc. Jpn.*, **56**, 1497, 1503 (1983).

49. T. Yamamoto, D. Komarudin, M. Arai, B.-L. Lee, H. Suganuma, N. Asakawa, Y. Inoue, K. Kubota, S. Sasaki, T. Fukuda, and H. Matsuda, *J. Am. Chem. Soc.*, **120**, 2047 (1998).

50. (a) M. Beller and C. Bolm, *Transition Metals for Organic Synthesis*, Wiley-VCH, Weinheim, (1998); (b) F. Diederich and P. J. Stang (Eds.), *Metal-Catalyzed Cross-Coupling Reactions* Wiley-VCH, Weinheim (1997).

51. (a) T. A. Chen, R. A. O'Brien, and R. D. Rieke, *Macromolecules*, **26**, 3462 (1993); (b) M. J. Marsella, P. J. Carroll, and T. M. Swager, *J. Am. Chem. Soc.*, **116**, 9347 (1994).

52. (a) B. S. Kang, D. H. Kim, S. M. Lim, J. Kim, M.-L. Seo, K.-M. Bark, and S. C. Shin, *Macromolecules*, **30**, 7196–7201 (1997); (b) B. S. Kang, D. H. Kim, S. M. Lim, J. Kim, M.-L. Seo, K.-M. Bark, and S. C. Shin, *Macromolecules*, **30**, 7196–7201 (1997); (c) H. Hayashi and T. Yamamoto, *Macromolecules*, **31**, 6063–6070 (1998).

53. (a) Z. Bao, Y. Chen, R. Cai, and L. Yu, *Macromolecules*, **26**, 5281 (1993); (b) A. Hilberer, H.-J. Brouwer, B.-J. van der Scheer, J. Wildeman, and G. Hadziioannou, *Macromolecules*, **28**, 4525 (1995); (c) A. Greiner and W. Heitz, *Makromol. Chem. Rapid. Commun.*, **9**, 581 (1988).

54. (a) T. Yamamoto, Z.-H. Zhou, T. Kanbara, M. Shimura, K. Kizu, T. Maruyama, Y. Nakamura, T. Fukuda, B.-L. Lee, N. Ooba, S. Tomaru, T. Kurihara, T. Kaino, K. Kubota, and S. Sasaki, *J. Am. Chem. Soc.*, **118**, 10389–10399 (1996); (b) P. V. Bedworth and J. M. Tour, *Macromolecules*, **27**, 622 (1994); (c) Q.-S. Hu, D. Vitharana, X.-F. Zheng, C. Wu, C. M. S. Kwan, and L. Pu, *J. Org. Chem.*, **61**, 8370–8377(1996).

55. (a) M. Ueda and T. Ito, *Polym. J.*, **23**, 297 (1991); (b) M. Ueda and F. Ichikawa, *Macromolecules*, **23**, 926 (1990).

56. (a) J. P. Wolfe, S. Wagaw, J.-F. Marcoux, and S. L. Buchwald, *Acc. Chem. Res.*, **31**, 805 (1998); (b) J. F. Hartwig, *Acc. Chem. Res.*, **31**, 852 (1998).

57. M. Jikei, M. Miyauchi, Y. Ishida, M. Kakimoto, and Y. Imai, *Macromol. Chem. Rapid Comm.*, **15**, 979 (1994).

58. M. Rehahn, A.-D. Schlüter, G. Wegner, and W. J. Feast, *Polymer*, **30**, 1054 (1989).

59. (a) T. Yamamoto, T. Kimura, and K. Shiraishi, *Macromolecules*, **32**, 8886–8896 (1999); (b) Y. Wang and R. P. Quirk, *Macromolecules*, **28**, 3495–3498 (1995).

60. A. Izumi, M. Teraguchi, R. Nomura, and T. Masuda, *Macromolecules*, **33**, 5347–5352 (2000).

61. A. D. Child and J. R. Reynolds, *Macromolecules*, **27**, 1975 (1994).

62. T. I. Wallow and B. M. Novak, *J. Am. Chem. Soc.*, **113**, 7411–7412 (1991).

63. S. Yamaguchi, T. Goto, and K. Tamao, *Angew. Chem. Int. Ed. Engl.*, **39**, 1695 (2000).

64. Q. Wang, L. Wang, and L. Yu, *J. Am. Chem. Soc.*, **120**, 12860 (1998).

65. T. Yamamoto, W. Yamada, M. Takagi, K. Kizu, and T. Maruyama, *Macromolecules*, **27**, 6620 (1994).

66. T. Yamamoto, T. Kimura, and K. Shiraishi, *Macromolecules*, **32**, 8886–8896 (1999).

67. T. Yamamoto, T. Morikita, T. Maruyama, K. Kubota, and M. Katada, *Macromolecules*, **30**, 5390 (1997).

68. J.-S. Yang and T. Swager, *J. Am. Chem. Soc.*, **120**, 11864 (1998).

69. B. S. Kang, M.-L. Seo, Y. S. Jun, C. K. Lee, and S. C. Shin, *Chem. Commun.*, 1167 (1996).

70. Y. Yao, J. J. S. Lamba, and J. M. Tour, *J. Am. Chem. Soc.*, **120**, 2805 (1998).

71. Q. T. Zhang and J. M. Tour, *J. Am. Chem. Soc.*, **119**, 9624 (1997).

72. N. Spetseris, R. E. Ward, and T. Y. Meyer, *Macromolecules*, **31**, 3158 (1998).

73. Q.-S. Hu, D. Vitharana, G.-Y. Liu, V. Jain, M. W. Wagaman, L. Zhang, T. R. Lee, and L. Pu, *Macromolecules*, **29**, 1082 (1996).

74. H.-B. Yu, Q.-S. Hu, and L. Pu, *Tetrahedron Lett.*, **41**, 1681 (2000).

75. H.-B. Yu, Q.-S. Hu, and L. Pu, *J. Am. Chem. Soc.*, **122**, 6500 (2000).

76. L. Ma, Q.-S. Hu, K. Y. Musick, D. Vitharana, C. Wu, C. M. S. Kwan, and L. Pu, *Macromolecules*, **29**, 5083 (1996).

77. L. Ma, Q.-S. Hu, D. Vitharana, C. Wu, C. M. S. Kwan, and L. Pu, *Macromolecules*, **30**, 204 (1997).

78. N. Spetseris, R. E. Ward, and T. Y. Meyer, *Macromolecules*, **31**, 3158 (1998).

10 Depolymerization and Recycling

Malcolm B. Polk

Georgia Institute of Technology, School of Textile and Fiber Engineering,
Atlanta, Georgia 30332

10.1 INTRODUCTION

10.1.1 Historical Perspective

Poly(ethylene terephthalate) (PET) is one of the most extensively recycled polymeric materials. In 1995, 3.5×10^4 tons of PET were recycled in Europe.[1] The main reason for the widespread recycling of PET is its extensive use in plastic packaging applications, especially in the beverage industry as plastic bottles. The consistency in terms of volume and availability of postconsumer bottles from sorting facilities and its high material scrap value create an excellent economic environment for PET recycling.

The first laboratory samples of PET were prepared by J. R. Whinfield and J. T. Dickson at a small English company in 1941.[2] Development of the new fiber was delayed by World War II, and public announcement of the invention of the fiber was not made until 1946. Imperial Chemical Industries purchased the rights for manufacture of the fiber for all countries except the United States, where DuPont obtained the manufacturing rights. DuPont introduced its polyester fiber in 1951 as Dacron. In 1962, Goodyear introduced the first polyester tire cord, and in the late 1960s and early 1970s, polyesters were developed specifically for packaging as film, sheet, coatings, and bottles.[3]

PET is the polyester of terephthalic acid and ethylene glycol. Polyesters are prepared by either direct esterification or transesterification reactions. In the direct esterification process, terephthalic acid is reacted with ethylene glycol to produce PET and water as a by-product. Transesterification involves the reaction of dimethyl terephthalate (DMT) with ethylene glycol in the presence of a catalyst (usually a metal carboxylate) to form bis(hydroxyethyl)terephthalate (BHET) and methyl alcohol as a by-product. In the second step of transesterification, BHET

Synthetic Methods in Step-Growth Polymers. Edited by Martin E. Rogers and Timothy E. Long
© 2003 John Wiley & Sons, Inc. ISBN: 0-471-38769-X

undergoes polycondensation in the presence of a catalyst (usually antimony tri-oxide) to form PET and ethylene glycol as a by-product. Since polyesterification reactions are equilibrium processes, depolymerization may involve the simple reversal of the polyesterification reactions by using by-product molecules such as water, methanol, and ethylene glycol to react with the waste PET. Chemical recy-cling of PET was initiated nearly parallel to the manufacture of the polymer on a commercial scale, as demonstrated by patents beginning in the 1950s. Initially, recycling was used to recover wastes generated during the production stages. More recently, depolymerization has been directed to the chemical recycling of postconsumer PET.

However, recycling of bottle-grade PET is more complicated because it is generally manufactured from a mixture of glycols and dicarboxylic acids. The glycols may include diethylene glycol and cyclohexanedimethanol and isophthalic acid. Most PET bottles are based on isophthalic-acid-modified PET[1] because of improved clarity, ductility, and processibility as well as lower crystallinity compared to homopolymeric PET. In the 1970s, a new technology referred to as "solid stating" was used to produce higher molecular weight, high-strength PET which could be melt blown into containers and used as beverage containers.[4] PET used in beverage containers has a crystallinity of ca. 25%, although the maximum crystallinity of PET is 55%. Since the dimensions of the crystallites are important for clarity, the crystallinity and rate of crystallization are controlled during the polymerization process by adding a comonomer and during processing by adding a nucleating agent.

In 1930, DuPont launched the synthetic fiber industry with the discovery of nylon-6,6.[2] In 1938, a pilot plant for nylon-6,6 production was put into opera-tion, and in 1939, production was commenced at a large-scale plant in Seaford, Delaware. The classical method for the synthesis of nylon-6,6 involves a two-step process. In the first step, hexamethylene diamine (HMDA) is reacted with adipic acid (AA) to form a nylon salt. Polymerization of the aqueous salt solution is carried out at temperatures in the range of about 210–275°C at a steam pressure of about 1.7 MPa. When 275°C is reached, the pressure is reduced to atmospheric pressure and heating is continued to drive the reaction to completion.

Polycaprolactam (PCA) is a crystalline polymer obtained by the ring-opening polymerization of caprolactam brought about by anionic or cationic initiators or with water. Nylon-6 is the product of the fiber spinning, extrusion, and injec-tion molding of PCA. In the manufacture of yarn, cord, rod, blocks, and molded nylon-6 parts, waste is produced in the form of polymeric lumps, cables, entan-gled masses of filaments, drawn fibers, rejected bobbins, out-of-specification batches of polymers, and low molecular weight oligomers formed during the polymerization process.[5]

The majority of polyamides used commercially are nylon-6,6 or nylon-6. The largest supply of waste for recycling of nylons is obtained from used carpets. Approximately 30–40% of the nylon produced in the world is used as carpet face fibers.[1] Carpets consist of a multicomponent construction of face fibers and a primary and secondary backing. The face fibers are generally made of nylon-6, and nylon-6,6. The backing is generally made from polypropylene filled with calcium carbonate. Styrene–butadiene rubber latex is generally used as an adhesive. The waste carpets are collected and the carpet is subjected to a mechanical shredding process for removing the face fibers from the backing, adhesive, and filler.

A process for depolymerizing nylon-6 and polyester–nylon-6 mixed scrap was patented by Allied Chemical Corporation in 1965[6] and 1967.[7] Ground scrap was dissolved with high-pressure steam at 125–130 psig (963–997 kPa) pressure and 175–180°C for 0.5 h in a batch process and then continuously hydrolyzed with superheated steam at 350°C and 100 psig (790 kPa) to form ε-caprolactam at an overall recovery efficiency of 98%. The recovered monomer could be repolymerized without additional purification.

In a patent granted to the DuPont Company in 1946, Myers[8] described the hydrolysis of nylon-6,6 with concentrated sulfuric acid which led to the crystallization of AA from the solution. HMDA was recovered from the neutralized solution by distillation. In a later patent assigned to the DuPont Company by Miller[9], a process was described for hydrolyzing nylon-6,6 waste with aqueous sodium hydroxide in isopropanol at 180°C and 305 psi pressure. After distillation of the residue, HMDA was isolated, and on acidification of the aqueous phase, AA was obtained in 92% yield.

Polyurethanes (PURs) are usually described as being prepared by the reaction of diols with diisocyanates. However, this is an oversimplification because often water is deliberately added in the production of flexible polyurethane foams. Unreacted isocyanate groups react with water to form carbon dioxide and urea groups in the polymer chain. The carbon dioxide acts as a "blowing" agent in the production of PUR foams. Also, polyurethanes can be formed by the reaction of bischloroformates with diamines.

In 1976, Meluch and Campbell at General Motors obtained a patent[10] for a steam hydrolysis depolymerization of PUR. High-pressure steam hydrolyzes flexible PUR foams rapidly at temperatures of 232–316°C to form diamines and polyols. The diamines are distilled and extracted from the steam and the polyols are obtained from the hydrolysis residue. In 1977, Bayer AG obtained a patent for a continuous PUR hydrolysis process using a specially designed extruder.[11]

10.1.2 Applications

10.1.2.1 Applications of Recycled and Depolymerized PET Products

PET has the second highest scrap value for recycled materials, second only to that of aluminum.[1] A typical PET beverage bottle consists of PET (60 g), polyethylene (1 g), and label and glue (5 g). Therefore, the PET bottle scrap must be separated before the PET can be recycled. The bottles are sorted at a material recovery facility and compressed. The compressed PET bottles are then washed and converted to flake by grinding. The flakes may be converted to pellets in an extruder. The pellets are more suitable for material handling. Currently recycled PET is being used to make food and nonfood containers, straps, sheeting, and textile fibers.[4] Allied Signal has developed a high-performance material, PETRA, from recycled PET. PETRA has a high heat deflection temperature and excellent chemical resistance and mechanical properties. PETRA is used in the automotive industry for such applications as panels and headlight brackets. General Electric and NCR Polymer in Chicago have produced an engineering resin from recycled PET by forming a polymeric alloy with polycarbonate. The polymeric alloys are used in the automobile industry as decorative body panels and trim.

The major PET manufacturers are depolymerizing scrap PET with glycols (glycolysis) or methanol (methanolysis) to form low-molecular-weight polyester diols (and BHET) and dimethyl terephthalate.[3] The purified products are then used to make new products. Goodyear uses glycolysis to make REPETE, a new product which contains 10–20% recycled PET. Hoechst Celanese used methanolysis to produce DMT for repolymerization. Eastman Chemicals uses depolymerization of PET to recover used X-ray scrap.

Recycled bottles are regulated by the Food and Drug Administration. Chemically recycled bottles are allowed to come into contact with both liquid and solid food products.

By depolymerizing PET waste with a polyol and subsequently condensing the oligomeric product with a polycarboxylic acid or anhydride, polyester resins are produced which have wide industrial applications. Depending on the polyol and polycarboxylic acid or anhydride used, saturated resins, alkyd resins, or unsaturated resins are obtained. PET wastes have been used for the production of alkyd resins in water thinnable paints. The materials obtained from the reaction of PET with a mixture of fatty acids high in linoleic acid content and trimethylolethane have been used in the preparation of water-dispersible coatings. Products of the depolymerization of PET with trimethylolpropane and pentaerythritol are used in the manufacture of high-solids paints. In the first step, PET is depolymerized with trimethylopropane and pentaerythritol at temperatures of 230–240°C. The final paint compositions contain 30–50% of PET depolymerization products.[12]

PET waste has been used in the manufacture of terephthalic electroinsulation lacquers. Here PET is heated in a mixture of a triol and glycol at temperatures of 230–260°C followed by catalytic transesterification and distillation of low-molecular-weight products of degradation until a polyester with the required softening temperature (ca. 60–100°C) is obtained.

The transesterification of PET with 2-ethylhexanol produces dioctyl terephthalate and small amounts of octyl(2-hydroxyethylene)terephthalate and terephthalate oligoesters. This mixture is used as a plasticizer for polyvinyl chloride.[12]

10.1.2.2 Application of Depolymerized Nylon-6

Chemical recycling of nylon-6 carpet face fibers has been developed into a closed-loop recycling process for waste nylon carpet.[5] The recovered nylon-6 face fibers are sent to a depolymerization reactor and treated with superheated steam in the presence of a catalyst to produce a distillate containing caprolactam. The crude caprolactam is distilled and repolymerized to form nylon-6. The caprolactam obtained is comparable to virgin caprolactam in purity. The repolymerized nylon-6 is converted into yarn and tufted into carpet. The carpets obtained from this process are very similar in physical properties to those obtained from virgin caprolactam.

10.1.2.3 Applications of Depolymerized Nylon-6,6 Products

Ammonolysis is the preferred route currently in use at the DuPont Company for the depolymerization of nylon-6,6 carpet waste. McKinney[13] has described the reaction of nylon-6,6 and nylon-6 mixtures with ammonia at temperatures in the range of 300–350°C at a pressure of about 68 atm in the presence of an ammonium phosphate catalyst to form a mixture of nylon-6,6 and nylon-6 monomers (HMDA, AA, and ε-caprolactam) and adiponitrile, 5-cyanovaleramide, 6-aminocapronitrile, and 6-aminocaproamide.

10.1.2.4 Applications of Depolymerized Polyurethane Products

The Upjohn Company has obtained a patent[14] for the glycolysis of rigid and flexible polyurethane foams. The polyols obtained were those used in the preparation of rigid foams. General Motors obtained a patent[10] for a PUR steam hydrolysis process. High-pressure steam hydrolyzes flexible PUR foams rapidly to form the diamines and polyols. Excellent results were obtained by using the reclaimed polyols in flexible-foam recipes at the 5% level. Rigid-foam, flexible-foam, reaction injection molding (RIM), and microcellular elastomer scrap have been treated with an equal weight of a 90/10 mixture of di(alkylene) glycol–diethanolamine at 190–210°C for several hours. The residue is cooled and treated with propylene oxide. The resulting polyols are most suitable for the production of rigid foams. In some rigid-foam formulations, the polyols produced may be substituted for up to 40% of the virgin polyol.[5]

10.1.3 Research Trends and Critical Issues

PET recycling is one of the most extensive and successful examples of polymer recycling. In 1995, 3.5×10^4 tons of PET were recycled in Europe, compared to 2.2×10^4 tons in 1993 and 1.3×10^4 tons in 1991.[1] The main reason for the increase in recycling of postconsumer PET is its increased use, particularly in beverage bottles, making it the primary target of plastic recycling. Critical issues include the following:

1. When used in nontraditional PET applications, recycled PET must compete with less expensive resins.
2. Trace remnants of label additives lead to discoloring of recycled PET and loss of clarity.
3. Waste PET must be carefully dried or it decomposes rapidly on reprocessing in the presence of water.

4. Collection costs are quite high because of the low bulk density of PET bottles.

In recent months, three nylon producers (DMS, DuPont, and Honeywell) have developed closed-loop recycling processes for nylon carpet,[15] thereby joining companies like BASF, Allied, and Rhodia, which have been recycling nylon on a modest level for years. DuPont is building a demonstration plant in Maitland, Ontario, which will be dedicated to the chemical recycling of nylon-6,6 and nylon-6. The newly developed ammonolysis process invented by DuPont can be used to depolymerize both nylon-6 and nylon-6,6. However, the cost of recycled nylon is estimated to exceed that of virgin nylon by ca. 25%.

Glycolysis is the most promising approach for the chemical recycling of polyurethanes.[1] The chemistry of PUR depolymerization is complicated by the presence of other chemical groups in the polymer, such as ureas, allophanates, and biurets.

10.2 STRUCTURE–PROPERTY RELATIONSHIPS

10.2.1 General Considerations

The preparation of many step-growth polymers involves equilibrium reactions. Polyesterifications are equilibrium reactions. In order to obtain high yields of high-molecular-weight polyesters, such polymerizations are conducted in such a manner so as to continually shift the equilibrium in the direction of the polymeric product. This is easily accomplished by the removal of water in the case of the reaction of terephthalic acid with ethylene glycol and the removal of methanol in the transesterification of dimethyl terephthalate with ethylene glycol. Depolymerization would involve shifting the equilibrium toward monomers by reacting PET with water, methanol, ethylene glycol, and other glycols. Polyamidations are equilibrium reactions, which have K values much greater than that of polyesterifications. Typically, the equilibrium constant K for a polyesterification is no greater than $1-10$, while K for a transesterification is in the range of $0.1-1$ and K for a polyamidation is in the range of 10^2-10^3.[16] Therefore, chemical depolymerization of polyamides occurs under more stringent conditions. Depolymerization of nylon-6 occurs more readily than that of nylon-6,6 because the thermodynamic stability of caprolactam acts as a driving force for depolymerization. Polyurethanes are formed by the irreversible, nonequilibrium reaction of diols with diisocyanates. Polyurethane formation does not necessarily involve a condensation reaction; that is, there may be no release of small molecules such as water. As mentioned earlier, depolymerization of polyurethanes is also complicated by the presence of ureas, allophanates, and biurets in the polyurethane structure.

Therefore, since step-growth polymers are often prepared by reversible reactions, it is feasible to convert them back to their monomers or oligomers/chemicals by solvolytic processes such as hydrolysis, glycolysis, methanolysis, aminolysis, ammonolysis, transesterification, alcoholysis, hydroglycolysis, acidolysis, and transamidation.

The principal solvolysis reactions for PET are methanolysis with dimethyl terephthalate and ethylene glycol as products, glycolysis with a mixture of polyols and BHET as products, and hydrolysis to form terephthalic acid and ethylene glycol. The preferred route is methanolysis because the DMT is easily purified by distillation for subsequent repolymerization. However, because PET bottles are copolyesters, the products of the methanolysis of postconsumer PET are often a mixture of glycols, alcohols, and phthalate derivatives. The separation and purification of the various products make methanolysis a costly process. In addition to the major product DMT, methanol, ethylene glycol, diethylene glycol, and 1,4-cyclohexane dimethanol have to be recovered to make the process economical.[1]

In the case of glycolysis, in addition to BHET, higher oligomers are also produced. Also, the reaction product obtained by glycolysis is difficult to purify by techniques such as distillation or crystallization, so BHET is difficult to isolate. Glycolysis is more appropriate for the recycling of postindustrial scrap. In order for glycolysis to be used successfully for postconsumer scrap, careful control of the purity of the bottle scrap is required and the recovered monomer must be mixed with virgin monomer for repolymerization.[1]

Hydrolysis, although a simple method in theory, yields terephthalic acid (TPA), which must be purified by several recrystallizations. The TPA must be specially pretreated to blend with ethylene glycol to form premixes and slurries of the right viscosities to be handled and conveyed in modern direct polyesterification plants. The product of the alkaline hydrolysis of PET includes TPA salts, which must be neutralized with a mineral acid in order to collect the TPA. That results in the formation of large amounts of inorganic salts for which commercial markets must be found in order to make the process economically feasible. There is also the possibility that the TPA will be contaminated with alkali metal ions. Hydrolysis of PET is also slow compared to methanolysis and glycolysis.[1]

Important solvolysis reactions for nylons are hydrolysis, methanolysis, glycolysis, aminolysis, ammonolysis, transamidation, and acidolysis.[17] Hydrolysis of nylon-6 with steam in the presence of an acid catalyst to form caprolactam is the preferred depolymerization approach. However, when recycling carpet face fibers, the fillers in the polymer may react with the acid catalyst and lower the efficiency of the catalyst.

Hydrolysis of nylon-6 in a high-pressure steam reactor can be carried out to give high yields of ε-caprolactam; however, removing the water from the caprolactam requires a distillation and the process is extremely expensive. Research at DuPont has identified ammonolysis as the best method for the recycling of scrap carpet. Ammonolysis may also be used for the recycling of nylon-6. The process, however, is expensive and the cost of recycled monomer is approximately 25% greater than virgin monomer.

The purpose of the depolymerization of polyurethanes is the isolation of polyols, which may be reused in the production of PUR. Important solvolysis reactions for PURs are hydrolysis,[18] alcoholysis, glycolysis, aminolysis, and interchange reactions. Glycolysis is the most promising route for the depolymerization of PURs.

10.2.2 Role of Chemical Structure

When a carbonyl group is bonded to a substituent group that can potentially depart as a Lewis base, addition of a nucleophile to the carbonyl carbon leads to elimination and the regeneration of a carbon–oxygen double bond. Esters undergo hydrolysis with alkali hydroxides to form alkali metal salts of carboxylic acids and alcohols. Amides undergo hydrolysis with mineral acids to form carboxylic acids and amine salts. Carbamates undergo alkaline hydrolysis to form amines, carbon dioxide, and alcohols.

Ingold[19] classified the possible mechanisms of ester hydrolysis into eight categories and a ninth was subsequently added.[20] The primary division is made concerning whether the acyl–oxygen or alkyl–oxygen bond is broken. The mechanisms may be elucidated by isotope tracer studies using ^{18}O or by hydrolyzing an ester with an asymmetric carbon in the alcohol portion bonded to the oxygen. The reactions occurring by alkyl–oxygen cleavage are nucleophilic substitutions at the saturated carbon with carboxylic acid or carboxylate ion leaving groups. The most common ester hydrolysis processes (Fig. 10.1) involve acyl–oxygen cleavage and occur by the A_{AC}^2 mechanism for acid catalysis and the B_{AC}^2 mechanism for alkaline hydrolysis.[20] These are hydrolyses that occur by the tetrahedral intermediate mechanism.

In general, the solvolytic reactions of polymers involve the cleavage of the C–X bonds of the polymer chain (Fig. 10.2), where X is a heteroatom (O, N,).[12]

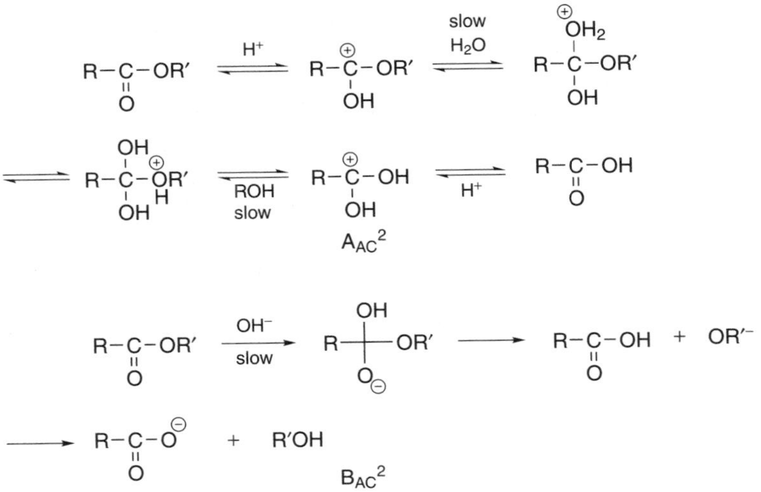

Figure 10.1 A_{AC}^2 and B_{AC}^2 mechanisms.

$$-\overset{|}{\underset{|}{C}}-X-\overset{|}{\underset{|}{C}}- \quad + \quad YZ \quad \longrightarrow \quad -\overset{|}{\underset{|}{C}}-XZ \quad + \quad Y-\overset{|}{\underset{|}{C}}-$$

Figure 10.2 Solvolysis reactions of polymers.

The mechanism of the solvolytic process is shown in Fig. 10.2, where YZ is the solvolytic agent (e.g. water, alcohol, acid, alkali, or amine). The general mechanism for the degradation of polyester in an acid and basic solution is shown in Fig. 10.3. In an acid environment, the mechanism is the A_{AC}^2 mechanism. A hydrogen ion adds to the carbonyl oxygen to form a resonance-stabilized cation with the positive charge shared by the O and the carbon. Attack by water reversibly on the carbon cationic center leads to the formation of a carboxylic acid group and an alcohol group. This mechanism is different than that of alkaline hydrolysis of polyesters. In an alkaline environment, the mechanism is the B_{AC}^2 mechanism. The alkaline hydrolysis of polyester involves reversible attack by a hydroxide ion on the carbon atom of the carbonyl to form an ion with the negative charge on oxygen followed by reversible chain cleavage and formation of carboxylate anions and alcohols. The four major polyester depolymerization processes (hydrolysis, alcoholysis, acidolysis, and aminolysis) are shown in Fig. 10.4.

The chemistry of the glycolysis of polyurethanes is complicated by the fact that there are additional groups in the polymer such as ureas, allophanates, and biurets, and the PURs may be crosslinked. In the presence of the appropriate glycols and at about 200°C, PURs undergo transesterification to form polyols. Under the same conditions, ureas undergo glycolysis to form urethanes and amines (Fig. 10.5).

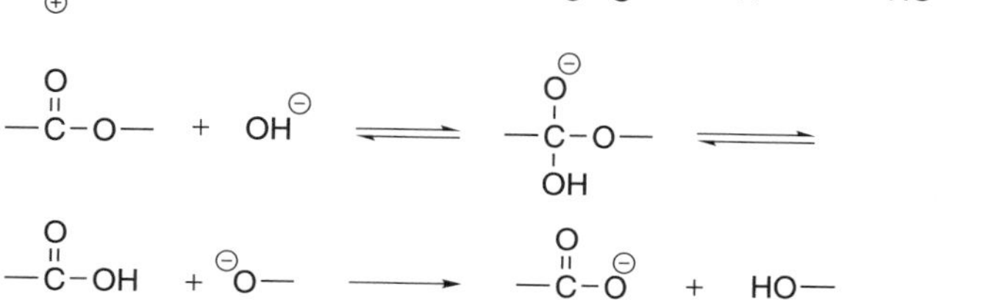

Figure 10.3 General mechanisms for degradation of polyester in acid and basic solutions.

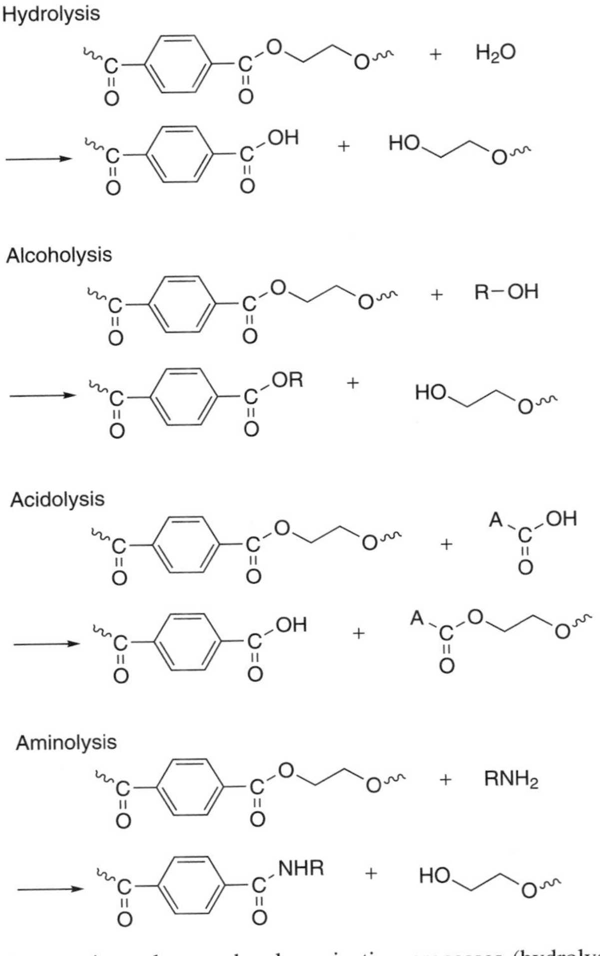

Hydrolysis

Alcoholysis

Acidolysis

Aminolysis

Figure 10.4 Four major polyester depolymerization processes (hydrolysis, alcoholysis, acidolysis, and aminolysis).

10.2.3 Role of Microstructure and Architecture

PET fibers consist of crystalline, oriented semicrystalline, and noncrystalline (amorphous) regions. In the crystalline state the molecules are well organized and form crystallites, which are crystalline regions with dimensions of a few hundred angstrom units. The maximum crystallinity is about 55%. The unit cell for PET was determined to be triclinic by X-ray diffraction techniques; it contains one repeating unit. The aromatic, carboxyl, and aromatic groups are practically planar and lie adjacent to one another. Distances between atoms in neighboring molecules are within normal van der Waals distances, and there is no evidence for unusually strong forces between the molecules. This agrees with the fact

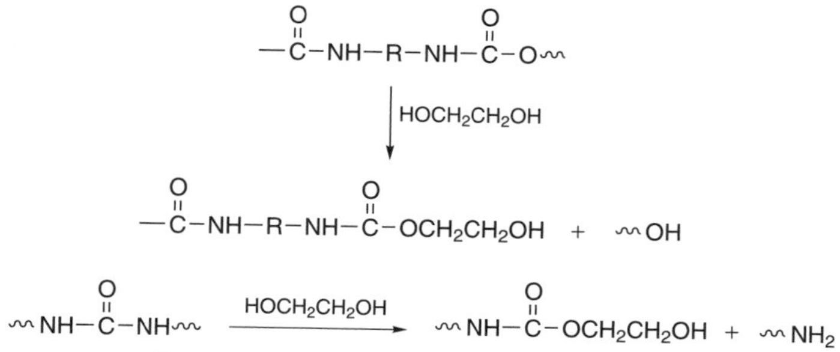

Figure 10.5 Glycolysis of PURs and ureas.

that the calculated PET cohesive density[21] is comparable to that of an aliphatic polyester, poly(ethylene adipate). Conjugation of the ester groups with the benzene ring increases their polarity and facilitates hydrogen bond formation. The intermolecular interactions create a relatively inflexible macromolecule with high modulus, strength, and resistance to moisture, dyestuffs, and solvents.

The crystallinity in PET soft drink bottles is about 25%. Because a more crystalline state is normal for PET, the amorphous content is increased intentionally by copolymerization and rapid cooling for the molten PET from the melt to a temperature below the glass transition temperature. Companies which perform high-speed blow molding of PET prefer PET resins made with small amounts of glycol and diacid comonomers.

Nylon fibers are semicrystalline, that is, they consist of crystallites separated by amorphous regions. Hydrogen bonding is an important secondary valence interaction in nylon-6 and nylon-6,6. Individual chains in the microcrystalline regions of nylons are held together by hydrogen bonds. Nylons are resistant to aqueous alkali but deteriorate more readily on exposure to mineral acids.

Polyurethane materials are extremely versatile in that it is possible to produce a large variety of structures which range in properties from linear and flexible to crosslinked and rigid. The crosslinked PURs are thermosets, which are insoluble and infusible and therefore cannot be reprocessed by extrusion without suffering extensive thermal degradation. At present, the main sources of recyclable waste are flexible PUR foams and automobile waste. Waste and scraps of these materials may consist of 15–25% by weight of total PUR foam production.

10.3 FACTORS AFFECTING THE USE OF RECYCLED MONOMERS OR OLIGOMERS

10.3.1 Recycling of PET Monomers

Contamination problems act as a barrier to the recycling of PET bottle waste. The presence of impurities that generate acid compounds at the high temperatures reached during the extrusion process prior to blow molding is a major problem in the reprocessing of PET because chain cleavage reactions are acid catalyzed. EVA (ethylene–vinyl acetate copolymer) from cap liners produces acetic acid. Rosin adhesives from the label produce abietic acid. PVC (polyvinyl chloride) flakes from PVC bottles produce hydrogen chloride. Even small amounts of PVC contamination cause a drastic decrease in PET melt viscosity and therefore molecular weight because of acid-catalyzed chain scission at the ester groups. This results in a decrease in intrinsic viscosity of the PET. For each PET application, a specific intrinsic viscosity range is required. The range for PET-carbonated soda bottles is 0.73–0.80 and solid stating may be required to achieve that goal.[1]

PVC contamination can also increase the amount of undesirable cyclic oligomers formed during reprocessing. The formation of low-molecular-weight oligomers, particularly cyclic oligomer, is an unavoidable part of the processing of PET. Among the various oligomers in PET fibers, cyclic trimers occur in the largest amounts. PVC liners in PET waste are a significant problem because, after shredding, the PVC particles are difficult to separate from PET because the densities are very similar. Also, during the baling process, if PVC bottles are not rigorously separated, the PVC will decompose on exposure to the processing temperatures used, leading to the formation of color and black specks in the PET bottles. Levels of PVC contamination greater than 1 ppm cause the discoloration of PET bottles. A quality control method has been developed for quantitatively determining the amount of PVC contamination in recycled PET by heating 500 g of PET at 180°C for 4 h and then weighing the collected darkened particles of degraded PVC. Moisture contamination of PET must be below 0.02%[1] in order to avoid the deleterious hydrolysis of PET resulting in molecular weight reduction. It has been shown that the rate of hydrolysis of PET is several orders of magnitude greater than the rate of thermo-oxidative degradation. In fact, the degradation of wet PET has been shown to occur in two distinct kinetic steps. The initial fast rate of degradation has been attributed to hydrolysis, caused by the residual water present in recycled PET. The second process occurring at a slower rate involves thermo-oxidative chain cleavage.

A surprisingly low concentration of water can reduce the viscosity such that reclaimed PET cannot be used for the blow molding of bottles with acceptable physical properties. The established solution to the moisture problem is to dry the recycled PET in special dryers prior to use. However, the drying process is both time and energy intensive. Paper labels can cause problems in PET recycling if they decompose during washing and removal. The paper fibers formed can produce cellulose fibers that are difficult to remove from the reprocessed PET.[1]

There are three types of adhesives used on paper labels for PET bottles. Synthetic adhesives in water-based emulsions are based on polyvinyl acetate or an ethylene–vinyl acetate copolymer plasticized with dibutyl phthalate. This adhesive does disintegrate to some extent in water, but some residue may remain on the bottle and become involved in the recycling process. Also used are thermofusible (hot-melt) adhesives, which are based on ethylene–vinyl acetate copolymers. These glues remain insoluble during the washing process; however, because of their low melting points, they do soften during mechanical action to release the paper label but most of the glue remains on the bottle. The third type of glue is the preferred alkali-soluble glue. These glues will separate and totally disintegrate on exposure to a 2 wt % solution of sodium hydroxide. However, the alkali soluble glues are 20–30% more expensive than the traditional adhesive and this has prevented their extensive use.[1]

Color contamination can occur as a result of the dyes used on colored bottles and, often, the printed ink used on the labels can lead to discoloration by staining on exposure to water during the washing process.

Acetaldehyde is formed during the degradation of PET. Vinyl ester endgroups formed during thermal degradation of PET liberate vinyl alcohol on transesterification with hydroxyethylterephthalate polymeric endgroups (Fig. 10.6). The vinyl alcohol tautomerizes to form acetaldehyde, which can affect the taste of foods in PET food contact applications.[1]

Methanolysis products are separated and purified by distillation. BHET, the monomer obtained by PET glycolysis, is normally purified by melt filtration under pressure. One of the problems encountered in neutral hydrolysis of PET is that the terephthalic acid isolated contains most of the impurities initially present in the PET waste. Hence very elaborate purification processes are required to obtain terephthalic acid of commercial purity.

10.3.2 Recycling of Caprolactam

Approximately 10–12% by weight of oligomers is formed in the synthesis of polycaprolactam (nylon-6). These oligomers may be removed by extraction with water or by distillation under vacuum. In the process, two types of liquid wastes are formed: (1) a 4–5% aqueous solution of low-molecular-weight compounds consisting of ca. 75% by weight of caprolactam and ca. 25% by weight of a mixture of cyclic and linear caprolactam oligomers and (2) a caprolactam–oligomer melt containing up to 98% caprolactam and small amounts of dimer, water, and organic contaminants. The recycle of caprolactam involves two different stages: depolymerization of polymeric waste and purification of the caprolactam and oligomers obtained.[5,22]

A typical recovery of caprolactam from liquid waste generates 20–25% oligomers along with organic and inorganic compounds as impurities. The distillation of caprolactam under reduced pressure produces a residue which consists of inorganic substances such as permanganates, potassium hydrogen sulfate, potassium

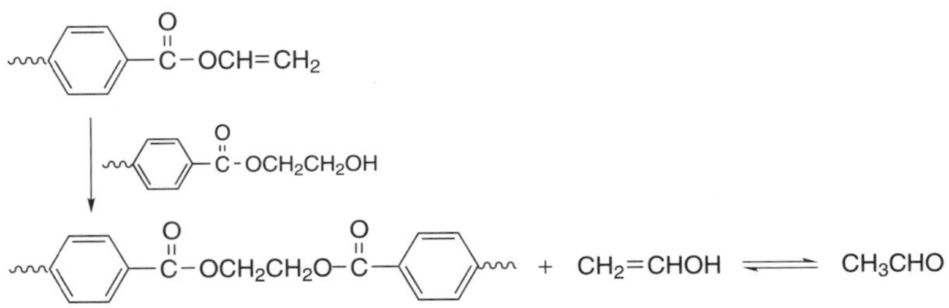

Figure 10.6 Vinyl ester endgroup reaction.

sulfate, sodium hydrogen phosphate, and sodium phosphate. The larger portion of the residue contains cyclic and linear chain oligomers plus 8–10% of caprolactam (CL). The types and exact amounts of impurities depend on the method used for the purification and distillation of caprolactam.

The cyclic oligomers are only slightly soluble in water and dilute solutions of caprolactam. They tend to separate out from the extracted waste during the process of concentration and chemical purification of the caprolactam. The cyclic oligomers tend to form on the walls of the equipment used in the process equipment. 6-Aminocaproic acid or sodium 6-aminocaproate may also be found in the oligomeric waste, especially if sodium hydroxide is used to initiate the caprolactam polymerization.

Many impurities are present in commercial caprolactam which pass into the liquid wastes from PCA manufacture from which caprolactam monomer may be recovered. Also, the products of the thermal degradation of PCA, dyes, lubricants, and other PCA fillers may be contained in the regenerated CL. Identification of the contaminants by IR spectroscopy has led to the detection of lower carboxylic acids, secondary amines, ketones, and esters. Aldehydes and hydroperoxides have been identified by polarography and thin-layer chromatography.

Caprolactam is a thermally unstable compound which on distillation may form methyl-, ethyl-, propyl-, and n-amylamines. Also, at high temperatures, CL reacts with oxygen to form hydroperoxides which in the presence of iron or cobalt ions are converted into adipimide. N-alkoxy compounds are also formed by the reaction of CL with aldehydes during storage.

Therefore, CL and the depolymerized product from which CL is regenerated contain various impurities which are present in widely fluctuating amounts depending on the reclamation processes involved. In particular, the presence of cyclohexanone, cyclohexanone oxime, octahydrophenazine, aniline, and other easily oxidized compounds affects the permanganate number. Also volatile substances such as aniline, cyclohexylamine, cyclohexanol, cyclohexanone, nitrocyclohexanone, and aliphatic amines may also be present in the CL.[22]

CL must be very carefully purified to exclude small concentrations of (1) ferric ions which would catalyze the thermal oxidative degradation of polycaprolactam and (2) aldehydes and ketones which would markedly increase oxidizability of CL. The impurities in CL may retard the rate of CL polymerization as well as having a harmful effect on the properties of the polymer fiber. In the vacuum depolymerization of nylon-6, a catalyst must be used because in the absence of a catalyst by-products such as cyclic olefins and nitriles may form, which affects the quality of the CL obtained.[1]

The caprolactam obtained must meet the specifications of permanganate number, volatile bases, hazen color, UV transmittance, solidification point, and turbidity in order to be used for repolymerization alone or in combination with virgin CL.[5] Reported CL purification methods include recrystallization, solvent extraction, and fractional distillation. One solvent extraction technique involves membrane solvent extraction. Ion exchange resins have been shown to be effective in the purification of aqueous caprolactam solutions. In one such process, the oily impurities are removed by extraction with organic solvents followed by treatment with carbon at 60–80°C. Cationic and anionic exchange resins are then used to complete the purification process. Ion exchange resins remove all ionic impurities as well as colloidal and floating particles; that is, alkali metal salts formed in permanganate treatment are removed during ion exchange treatment. Also, the treatment of aqueous solutions of CL with ion exchange resins helps to remove the distillation residue. Treatment of caprolactam with activated carbon helps to remove anionic and cationic impurities.

Impurities in CL have also been destroyed by oxidation with ozone[22] followed by distillation. Ozonation treatment of waste CL leaves no ionic impurities. However, the most commonly used oxidizing agents are potassium permanganate, perboric acid, perborate, and potassium bromate. Treatment of CL with these oxidizing agents is carried out in a neutral medium at 40–60°C. Strongly alkaline or acidic conditions accelerate the oxidation of CL to form isocyanates. The undesirable oxidation reaction is fast above pH 7 because of the reaction with isocyanate to form carbamic acid salts, which shifts the equilibrium to form additional isocyanate.

In a typical process, potassium permanganate is used to treat the cracked liquor exiting the depolymerization plant without any pH adjustment. The liquor is usually acidic because it contains some of the phosphoric acid depolymerization catalyst. The $KMnO_4$ treatment is followed by treatment of the CL aqueous solution with carbon followed by filtration. Next the filtered 20–30% CL aqueous solution is concentrated to 70% and the pH is adjusted to 9–10 by addition of sodium hydroxide. The caprolactam alkaline concentrate is treated with $KMnO_4$ followed by distillation under reduced pressure to remove water and low-boiling impurities.

Also, the CL aqueous solution may be hydrogenated at 60°C in the presence of 20% sodium hydroxide and 50% palladium absorbed on carbon to provide caprolactam of very high purity after distillation. Treatment with an ion exchange resin before or after the oxidation or hydrogenation process also improves the quality of the CL obtained after distillation. CL has also been purified by treatment with alkali and formaldehyde followed by fractional distillation to remove aromatic amines and other products.

Also, nylon-6 waste may be hydrolyzed in the presence of an aqueous alkali metal hydroxide or acid[5] to produce an alkali metal or acid salt of 6-aminocaproic acid (ACA). The reaction of nylon-6 waste with dilute hydrochloric acid is rapid at 90–100°C. The reaction mixture is poured into water to form a dilute aqueous solution of the ACA salt. Filtration is used to remove undissolved impurities such as pigments, additives, and fillers followed by treatment of the acid solution with a strong cation exchange resin. A sulfonic acid cationic exchanger absorbs ACA salt and pure ACA is eluted with ammonium hydroxide to form a dilute aqueous solution. Pure ACA is obtained by crystallization of the solution.

Alternatively, nylon-6 waste may be hydrolyzed with aqueous sodium hydroxide and the sodium salt of ACA converted into pure ACA by passing the aqueous solution through an anion exchange resin.

10.3.3 Recycling of Nylon-6,6 Monomers

Adipic acid and HMDA are obtained from nylon-6,6 by the hydrolysis of the polymer in concentrated sulfuric acid (Fig. 10.7). The AA is purified by recrystallization and the HMDA is recovered by distillation after neutralizing the acid. This process is inefficient for treating large amounts of waste because of the required recrystallization of AA after repeated batch hydrolyses of nylon-6,6 waste. In a continuous process,[5] nylon-6,6 waste is hydrolyzed with an aqueous mineral acid of 30–70% concentration and the resulting hydrolysate is fed to a crystallization zone. The AA crystallizes and the crystals are continuously removed from the hydrolysate. Calcium hydroxide is added to neutralize the mother liquor and liberate the HMDA for subsequent distillation.

Continuous recovery requires AA crystals having an average diameter of ca. 40–50 nm.[5] Such crystals are obtained by continuously introducing the hot hydrolysate containing 10–20% AA into an agitated crystallization vessel while maintaining an average temperature of 20–30°C. The slurry obtained from the crystallization vessel is filtered to collect the AA crystals, and the filtrate which contains the HMDA acid salt is continuously neutralized with calcium hydroxide. The calcium salt formed is removed by filtration, and the HMDA in the filtrate is isolated by distillation.

In the case of nylon-6,6 waste recycled by ammonolysis, nylon is treated with ammonia in the presence of a phosphate catalyst. Reaction occurs at 330°C and 7 MPa. Distillation of the reaction mixture produces ammonia which is recycled and three fractions containing (a) caprolactam, (b) HMDA and aminocapronitrile, and (c) adiponitrile. Aminocapronitrile and adiponitrile are hydrogenated to yield pure HMDA, and the caprolactam is either converted to aminocapronitrile by further ammonolysis or distilled to produce pure caprolactam. The HMDA produced by this process is extremely pure (>99.8).[1] The main impurities are aminomethylcyclopentylamine and tetrahydroazepine, which are expected to be removed more effectively in the larger distillation columns employed in the larger plants.

10.3.4 Recycling of Polyols from Polyurethanes

The glycolysis of rigid polyurethane foams produces polyol products which can be reintroduced into the production cycle of PUR insulation materials to form materials with properties practically equivalent to those of materials produced using virgin polyols. Aromatic amines produced as by-products in the glycolysis process are toxic and therefore undesired side products. The most frequently observed side product is diphenylmethanediamine (DMDA), which is formed

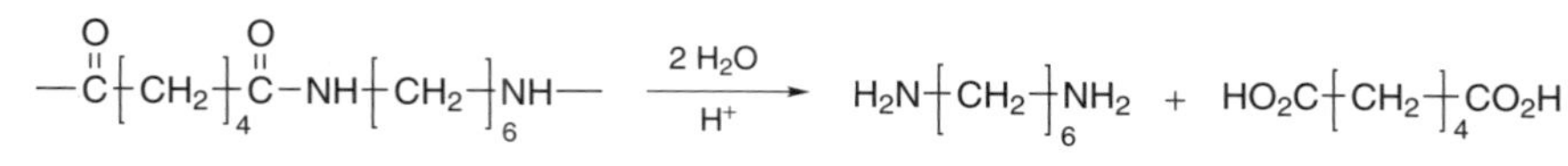

Figure 10.7 Hydrolysis of nylon-6,6.

when water present in the system decomposes the carbamic acid formed. In addition to being highly toxic, DMDA lowers the reactivity of the polyol obtained. A deamination agent may be added to remove DMDA from the reaction mixture. Deamination agents such as ethylene oxide or propylene oxide react with DMDA to form polyoxyalkylene adducts which phase separate from the polyol reaction product.[1]

10.4 CHEMISTRY AND CATALYSIS

10.4.1 PET Chemistry and Catalysis

Glycolysis is a method for the depolymerization of PET which involves heating PET under pressure with an excess of glycol, usually ethylene glycol, to form BHET. Glycolysis is carried out in the presence of a transesterification catalyst, usually zinc, lead, manganese, cobalt, sodium, or lithium acetate. The formation of diethylene glycol can be minimized by the use of a combination of lithium and zinc acetates and antimony trioxide as catalyst.[23] Other products formed during PET glycolysis in addition to diethylene glycol include ethylene terephthalate, dioxane, aldehydes, and cyclic trimers.[24] Campanelli et al.[25] demonstrated that although zinc salts catalyze glycolysis of PET at temperatures below 245°C, there is no effect at higher temperatures. This was used as evidence that the catalysis is an interfacial phenomenon since 245°C is the melting point of PET. Another glycolysis method for the depolymerization of PET involves the reaction of PET with an alkene oxide, usually ethylene or propylene oxides, with heating to form polyols. Basic compounds such as sodium hydroxide, potassium hydroxide, and tertiary aminoalkyl phenols were used to catalyze the reaction.

PET methanolysis involves the reaction of PET with methanol at high temperatures and pressures in the presence of transesterification catalysts such as magnesium acetate, cobalt acetate, and lead dioxide.

Acids such as sulfuric or nitric acids or bases such as sodium hydroxide may catalyze the hydrolysis of PET. It has been demonstrated that the rate of alkaline PET hydrolysis increases in the presence of quaternary ammonium compounds.[26,27] Niu et al.[26] reported an increase in the rate of alkaline PET degradation in the presence of dodecylbenzyldimethylammonium chloride at 80°C. Polk et al.[27] reported increases in the rate of sodium hydroxide depolymerization of PET in the presence of trioctylmethylammonium chloride, trioctylmethylammonium bromide, and hexadecyltrimethylammonium bromide at 80°C.

Neutral PET hydrolysis usually takes place under high temperature and pressure in the presence of alkali metal acetate transesterification catalysts.[28] It is thought that the catalytic effect observed on the part of zinc salts is the result of electrolytic changes induced in the polymer–water interface during the hydrolysis process. The catalytic effect of zinc and sodium acetates is thought to be due to the destabilization of the polymer–water interface in the hydrolysis process.

Ammonolysis of PET involves the reaction of PET and ammonia with heating and usually in the presence of ethylene glycol to form terephthaldiamide. The reaction is catalyzed by zinc acetate.

10.4.2 Nylon-6 and Nylon-6,6 Chemistry and Catalysis

The primary method of nylon-6 depolymerization is hydrolysis. Acid hydrolysis of nylon-6 generally takes place in the presence of H_3PO_4 or H_3BO_3 at high temperature.[29] Nylon-6 may be depolymerized in vacuo to produce ε-caprolactam in good yields (>80%). A catalyst such as potassium carbonate is needed to prevent the formation of by-products such as cyclic olefins and nitriles and lower quality ε-caprolactam.[1] Ammonolysis is a relatively new method for the depolymerization of nylon-6 and nylon-6,6. It involves the reaction of nylon-6 and nylon-6,6 with ammonia at high temperatures and pressures in the presence of ammonium phosphate to form HMDA, adiponitrile, 5-cyanovaleramide, ε-caprolactam, 6-aminocapronitrile, and 6-aminocaproamide.[13] McKinney[30] has also reported the use of Lewis acid catalysts such as Zn, Co, Pd, Mn, Ti, and W chlorides to enhance the depolymerization conversions.[30] Polk et al.[27] have demonstrated the alkaline hydrolysis of nylon-6,6 in the presence of benzyltrimethylammonium bromide to form adipic acid after acidification.

10.4.3 Chemistry and Catalysis of Polyurethanes

The most important methods for the depolymerization of PURs are glycolysis and hydrolysis. Glycolysis of polyurethanes involves the reaction of PURs with ethylene glycol, diethylene glycol, propylene glycol, and dipropylene glycol at high temperatures and in many cases at atmospheric pressure.[31,32] Although catalysis may not be necessary, organometallic compounds such as $Ti[O(CH_2)_3CH_3]_4$ and organic compounds based on tin appear to be selective catalysts for glycolysis.[1] Amines and Lewis acids have also been used as transesterification catalysts. When potassium acetate is employed as a catalyst for rigid PUR foam glycolysis, large amounts of amines are formed as part of the glycolysis product.

The Ford hydroglycolysis process is an example of a combined approach for the depolymerization of PURs. In a reactor, polyurethane foam is reacted with a mixture of water, diethylene glycol, and alkali metal hydroxides at high temperature to form polyols. When sodium hydroxide is added as a catalyst, a cleaner polyol is formed because all of the carbamates and ureas in the product are converted into amines and alcohols by hydrolysis.[33]

10.5 EXPERIMENTAL METHODS

10.5.1 Glycolysis of PET

Glycolysis is the oldest method of PET depolymerization and is used widely on a commercial scale. Glycolysis (Fig. 10.4) involves the reaction of PET with excess glycol, usually ethylene glycol, at temperatures of 180–240°C and high pressures to form BHET.[22] PET dissolves in boiling ethylene glycol at atmospheric pressure and even more readily when heated under pressure. Oligomers of PET are formed on cooling the solution. There is an equilibrium formed between hydroxyethyl-terephthalate, BHET, and low-molecular-weight oligomers. The equilibrium may be shifted to the left by addition of ethylene glycol to the glycolysis product. Cooling of the solution causes BHET to crystallize out. Glycols other than ethylene glycol may be used: that is, PET reacts with propylene glycol at 200°C in the presence of zinc acetate to yield bis(hydroxypropyl) terephthalate, BHET, and hydroxypropyl-hydroxyethyl terephthalate.[34,35] BHET[36] may be added to promote the solubility of the PET waste and facilitate depolymerization. If ethylene glycol and polyester scrap are mixed in the absence of BHET in conventional depolymerization and heated, they cannot be agitated because of the resulting solidification and depolymerization time increases in the inhomogeneous reaction system. Rapid solution of the polyester occurs with a weight ratio of BHET to waste PET of about 3.1 to about 3.4 : 1. The reaction occurs more rapidly in the presence of an excess of ethylene glycol; however, too large an excess may lead to the formation of diethylene glycol. Sodium acetate trihydrate and water are added to inhibit diethylene glycol formation.

A proposed mechanism of PET glycolysis would be (1) glycol diffusion into the polymer; (2) swelling of the polymer, which increases the rate of diffusion; and (3) reaction (transesterification) of the glycol hydroxy group at an ester group in the polyester chain.[24] Since it is a transesterification process, metal acetate salts are effective catalysts.

Polybutylene terephthalate (PBT) has been produced from PET scrap by transesterification with 1,4-butanediol.[1] In the process, classified and cleaned polymer flake from postconsumer PET bottles is reacted with 1,4-butanediol in an extruder. PBT is used as an engineering plastic. Ethylene glycol and tetrahydrofuran produced as by-products are recovered by distillation.

Glycolysis may be applied to the conversion of PET scrap to polyols for the production of important industrial materials such as polyurethanes, saturated resins, alkyd resins, and unsaturated polyesters.[1,5] The conversion of scrap PET to polyols may be achieved by reaction with diols such as 1,6-hexanediol and propylene glycol, polyglycols such as polyethylene glycol, or ethylene oxide. The reaction of PET with polyglycols incorporates TPA in the resulting polyols produced. The polyols may then be reacted with diisocyanates such as methylene-4,4'-diphenyldiisocyanate, toluene-2,4-diisocyanate, and hexamethylene diisocyanate to form polyurethanes. Unsaturated polyesters are obtained by the reaction of PET-based polyols with substances such as maleic anhydride. The PET-based polyols may be reacted with acids such as phthalic acid, isophthalic acid, and adipic acid to form saturated resins or a fatty acid together with phthalic anhydride to produce alkyd resins.[1]

10.5.2 Methanolysis of PET

PET methanolysis is based on the reaction of PET with methanol at high temperatures (180–280°C) and pressures (20–40 atm) to form DMT and ethylene glycol, which are raw materials for the production of PET as the main products.[37] The reaction occurs in the presence of transesterification catalysts such as zinc, magnesium, and cobalt acetates and lead dioxide. The reaction temperature can be achieved by heating and melting PET in the first step, followed by exposure of the hot polymer to methanol or heat necessary to melt the PET polymer may be supplied by superheated methanol vapors. In other instances, a solvent may be used to dissolve the PET, as in the method used by Naujokas and Ryan,[38] who invented a process in which the scrap PET is dissolved in oligomers of DMT and ethylene glycol followed by passing superheated methanol vapors through the solution. The methanolysis process has been used by such large manufacturers of PET as Hoechst, DuPont, and Eastman. One of the opportunities offered by methanolysis is the possibility of locating the methanolysis unit in the polymer production line, thereby enabling the waste generated in the process to be utilized as part of the process.[12]

Both continuous and batch methods may be used in methanolysis. The batch method requires an autoclave, crystallizer, and centrifuge and a system for the melting and distillation of the DMT obtained. In the two-stage Hoechst continuous process, waste PET is melted and fed to a reactor. Preheated methanol is added to the autoclave, which is equipped with a mixer. The conversion reaches 70–90% in the first reactor, after which the reaction stream is introduced into a second autoclave at a lower temperature near the bottom, where it rises slowly and the higher density impurities settle at the bottom. The reaction stream leaves the second autoclave and its pressure is reduced to 0.3 MPa. On further reduction of the pressure and cooling, DMT precipitates and is subsequently purified.[12]

10.5.3 Hydrolysis of PET

10.5.3.1 Acid Catalysis

PET reacts with water to produce TPA and ethylene glycol (EG). The process can be carried out under acidic, basic, or neutral conditions. Polyester waste may be hydrolyzed with strong acids such as sulfuric, phosphoric, and nitric acids to produce terephthalic acid and ethylene glycol in high yields without the use of high temperatures and pressures. However, low temperatures and low acid concentrations result in slow rates of hydrolysis. Several patents describe the acid hydrolysis of PET with sulfuric acid. The process occurs at temperatures of $25–100°C$ in a few minutes at atmospheric pressure. The hydrolysis product is neutralized with base to produce the TPA salt. After treatment with ion exchange columns, the solution is acidified to produce TPA.[39,40] A major problem with this process is the corrosion caused by the reaction mixture and the production of large amounts of liquid wastes containing inorganic salts and sulfuric acid. Yoshioka et al.[1] describe a process for acid-catalyzed depolymerization of PET in which PET bottles are digested in $7–13\ M$ nitric acid at temperatures ranging from 70 to 100°C for 72 h to form TPA, EG, and oxalic acid. The oxalic acid is produced by the oxidation of EG by nitric acid. The NO is recycled by oxidation and the addition of water to form nitric acid. Since the oxalic acid is more valuable than TPA or EG, the economics of the process are improved compared to acid-catalyzed hydrolysis to produce TPA and EG only.

10.5.3.2   Alkaline Hydrolysis

The alkaline hydrolysis of PET involves reacting PET with a 4–20 wt% solution of NaOH under pressure at temperatures in the range of 200–250°C for several hours to form the sodium salt of TPA. Acidification of the sodium salt of TPA yields the TPA on precipitation from solution. Niu[26] has demonstrated the use of quaternary ammonium salts to accelerate the alkaline hydrolysis of PET at 80°C and atmospheric pressure. Collins and Zeronian[41] have reported that the reaction of PET with methanolic sodium hydroxide accelerates the alkaline degradation of PET compared to aqueous systems. Oku et al.[42] quantitatively converted PET to disodium terephthalate by reacting the polyester with sodium hydroxide dissolved in anhydrous ethylene glycol at 150°C. Disodium terephthalate precipitated from the ethylene glycol solution. Yoshioka et al.[43] described a process for the conversion of postconsumer PET to TPA and oxalic acid in concentrated sodium hydroxide solution. In the process, PET is completely converted to TPA and ethylene glycol in aqueous sodium hydroxide at 250°C followed by conversion of ethylene glycol to oxalic acid and carbon dioxide by base-catalyzed oxidation with oxygen. Conditions for the base-catalyzed oxidation were a base concentration of 27.5 M NaOH, a temperature of 250°C, a reaction time of 5 h, and a partial pressure of oxygen of 5 MPa. Disodium terephthalate is stable to oxidation and is formed in 100% yield. The production of oxalic acid gives the process an economic boost because oxalic acid has a higher value than ethylene glycol. PET has also been treated with ammonium hydroxide and steam at 200°C to form diammonium terephthalate and EG. TPA was obtained by acidification with sulfuric acid.[5]

Polk et al. reported[27] that PET fibers could be hydrolyzed with 5% aqueous sodium hydroxide at 80°C in the presence of trioctylmethylammonium bromide in 60 min to obtain terephthalic acid in 93% yield. The results of catalytic depolymerization of PET without agitation are listed in Table 10.1. The results of catalytic depolymerization of PET with agitation are listed in Table 10.2. As expected, agitation shortened the time required for 100% conversion. Results (Table 10.1) for the quaternary salts with a halide counterion were promising. Phenyltrimethylammonium chloride (PTMAC) was chosen to ascertain whether steric effects would hinder catalytic activity. Bulky alkyl groups of the quaternary ammonium compounds were expected to hinder close approach of the catalyst to the somewhat hidden carbonyl groups of the fiber structure. The results indicate that steric hindrance is not a problem for PET hydrolysis under this set of conditions since the depolymerization results were substantially lower for PTMAC than for the more sterically hindered quaternary salts.

Hexadecyltrimethylammonium bromide (HTMAB) was included in the investigation because it had been employed previously. The past success of HTMAB was attributed to its long-chain alkyl group. Mimicking a long polymer chain, the hexadecyl group probably aided in the solvation of PET.

TABLE 10.1   Results of Catalytic Depolymerization of PET without Agitation

Catalyst	30 min	60 min	90 min	150 min	240 min
Control (wt % PET reacted)	3.99	4.69	12.7	22.6	32.8
TOMAC (wt % PET reacted)	21.84	44.09	52.78	66.07	69.79
TOMAB (wt % PET reacted)	24.81	44.71	54.44	64.58	90.15
HTMAB (wt % PET reacted)	7.08	32.87	44.66	66.22	81.72
TEAOH (wt % PET reacted)	3.28	5.11	8.01	8.86	22.52
PTMAC (wt % PET reacted)	1.09	3.58	15.15	30.51	46.76

TABLE 10.2   Results of Catalytic Depolymerization of PET with Agitation

Catalyst	30 min	60 min	90 min	150 min	240 min
Control	3.5	11.33	13.71	26.05	41.74
TOMAC	79.8	92.41	99.52	100	
TOMAB	85.71	98.55	100		
HTMAB	26.31	50.97	60.68	88.96	100

As expected, HTMAB made a respectable showing in these experiments. Trioctylmethylammonium chloride (TOMAC) and trioctylmethylammonium bromide (TOMAB) outperformed all other catalysts. It was postulated that the three octyl groups were the proper length for solvation of the polymer while at the same time small enough to avoid sterically hindering the reaction. In order to determine if TOMAB could be used to catalyze PET depolymerization for more than one treatment cycle, the catalyst was recovered upon completion of one treatment and added to a second run for 60 min. Tetraethylammonium hydroxide (TEAOH) was studied as a catalyst in order to demonstrate the effect of hydroxide ion as a counterion. The percent PET conversion for the second cycle was 85.7% compared to a conversion of 90.4% for the first treatment cycle.

10.5.3.3 *Neutral Hydrolysis*

Neutral hydrolysis of PET is usually carried out under pressure (1–4 MPa) at temperatures of 200–300°C.[12] High-purity TPA and EG may be obtained by the hydrolytic depolymerization of PET in an autoclave with excess water. PET hydrolysis occurs faster in the molten state than as a solid; therefore, it is advantageous to carry out the process at temperatures greater than 245°C. A study of the factors involved in neutral hydrolysis of PET showed that depolymerization occurred in 2 h at 265°C with an initial ratio of 5.1 : 1 (w/w) water to PET.[44] Alkali metal acetates typically catalyze the process. The major problem with this process is that the impurities present in the waste PET are carried over with the TPA product, thereby requiring extensive purification of the TPA obtained.

10.5.4 Ammonolysis and Aminolysis of PET

Ammonolysis of PET involves the reaction of PET with ammonia at temperatures of 70–180°C, usually under pressure in EG. The ammonolysis of PET postconsumer bottles has been carried out at temperatures in the range of 120–180°C and a pressure of ca. 2 MPa for 1–7 h. The TPA diamide formed may be converted to terephthalonitrile. Terephthalonitrile may be hydrogenated to form *p*-xylylenediamine and 1,4-bis(aminomethyl)cyclohexane.[12] A low-pressure PET ammonolysis process in EG has been developed. The process is catalyzed by 0.5 wt% zinc acetate at a temperature of 70°C and a PET–NH_3 ratio of 1 : 6 (w/w). The yield of TPA diamide was 87%.

Aminolysis of PET involves the reaction of PET with primary amines such as methylamine, ethylamine, and ethanolamine[37] at temperatures in the range of 20–100°C to form the mono- and diamides of TPA.

10.5.5 Combined Methods of Depolymerization of PET

The combined depolymerization methods consist of two or more steps involving different treatment regimes. Examples include glycolysis–hydrolysis, methanolysis–hydrolysis, and glycolysis–methanolysis. Doerr[45] has developed a glycolysis–hydrolysis process for the chemical recycling of PET. In the first stage, PET is reacted with EG in a high-pressure screw extruder at 280–290°C (residence time 2 min) in order to reduce the molecular weight by at least 50%. At this stage the product consists of BHET and hydroxy-terminated oligomers. In the second step, the initial product mixture undergoes neutral hydrolysis by reaction with water to form TPA and EG. The advantage of the two-step approach is that the time required for the two-step process is much less than that for a single-step neutral hydrolysis.

A two-step methanolysis–hydrolysis process[37] has been developed which involves reaction of PET with superheated methanol vapors at 240–260°C and atmospheric pressure to produce dimethyl terephthalate, monomethyl terephthalate, ethylene glycol, and oligomeric products in the first step. The methanolysis products are fractionally distilled and the remaining residue (oligomers) is subjected to hydrolysis after being fed into the hydrolysis reactor operating at a temperature of ca. 270°C. The TPA precipitates from the aqueous phase while impurities are left behind in the mother liquor. Methanolysis–hydrolysis leads to decreases in the time required for the depolymerization process compared to neutral hydrolysis; for example, a neutral hydrolysis process that requires 45 min to produce the monomers is reduced to 15 min for the two-stage process.[37] This process is effective even when the waste PET is contaminated by metals, dyes, paper, film, and other resins. This process has been used successfully to depolymerize polybutylene terephthalate with 1,4-butanediol used as the depolymerizing agent in the glycolysis step.

Gamble et al.[46] devised a glycolysis–methanolysis process which involves first dissolving PET in a mixture of EG, TPA, and dimethyl terephthalate oligomers (this stream actually is obtained from the molten PET formed in the polymerization process) at $230-290°C$. The product stream is then reacted with superheated methanol at temperatures between 250 and $290°C$ to form a product mixture which is rich in dimethyl terephthalate. Most of the contaminants remain in the first reactor, which enables the process to produce high-purity dimethyl terephthalate.

10.5.6 Hydrolysis of Nylon-6

A process for depolymerizing nylon-6 scrap using high-pressure steam was patented by the Allied Chemical Corporation in 1965.[6] Ground scrap was dissolved in high-pressure steam at 125–130 psig (963–997 kPa) and 175–180°C for 0.5 h in a batch process and then continuously hydrolyzed with superheated steam at 350°C and 100 psig (790 kPa) to form ε-caprolactam at an overall recovery efficiency of 98%. The recovered monomer could be repolymerized without additional purification. Recently, Braun et al.[47] reported the depolymerization of nylon-6 carpet in a small laboratory apparatus with steam at 340°C and 1500 kPa (200 psig) for 3 h to obtain a 95% yield of crude ε-caprolactam of purity 94.4%.

Acid hydrolysis of nylon-6 wastes[29,48] in the presence of superheated steam has been used to produce aminocaproic acid which under acid conditions is converted to ε-caprolactam. Acids used for the depolymerization of nylon-6 include inorganic or organic acids such as nitric acid, formic acid, benzoic acid, and hydrochloric acid.[5] Orthophosphoric acid and boric acid are typically used as catalysts at temperatures of 250–350°C. In a typical process, superheated steam was passed through the molten nylon-6 waste at 250–300°C in the presence of phosphoric acid. The resulting solution underwent a multistage chemical purification before concentration to 70% liquor, which was fractionally distilled in the presence of base to recover pure ε-caprolactam. Boric acid (1%) was used to depolymerize nylon-6 at 400°C under ambient pressure. A recovery of 93–95% ε-caprolactam was obtained by passing superheated steam through molten nylon-6 at 225–350°C in the presence of 0.3–1.0 wt% mixed phosphoric acid–boric acid catalyst.

Sodium hydroxide has been used successfully as a catalyst for the base-catalyzed depolymerization of nylon-6. At 250°C, a pressure of 400 Pa, and a sodium hydroxide content of 1%, the yield of ε-caprolactam was 90.5%.[49]

10.5.7 Hydrolysis of Nylon-6,6 and Nylon-4,6

The depolymerization of nylon-6,6 and nylon-4,6 involves hydrolysis of the amide linkages, which are vulnerable to both acid- and base-catalyzed hydrolysis. In a DuPont patent,[9] waste nylon-6,6 was depolymerized at a temperature of at least 160°C in the presence of a propyl or butyl alcohol with an aqueous solution of sodium hydroxide in the amount of at least 20% excess equivalents of the acid component to be recovered. Craig[50] depolymerized nylon-6,6 fibers in an inert atmosphere at what was reported to be a superatmospheric pressure of up to 1.5 MPa and at a temperature in the range of 160–220°C in an aqueous solution containing at least 20% excess equivalents of sodium hydroxide.

Polk[27] reported the depolymerization of nylon-6,6 and nylon-4,6 in aqueous sodium hydroxide solutions containing a phase transfer catalyst. Benzyltrimethylammonium bromide was discovered to be an effective phase transfer catalyst in 50% sodium hydroxide solution for the conversion of nylon-4,6 to oligomers. The depolymerization efficiency (percent weight loss) and the molecular weight of the reclaimed oligomers were dependent on the amount and concentration of the aqueous sodium hydroxide and the reaction time. Table 10.3 exhibits the effects of experimental conditions on the depolymerization efficiency and the average molecular weight of the oligomers. The viscosity-average molecular weight was calculated from the Mark–Houwink equation: $[\eta] = K M_v^a$, where M_v is the viscosity-average molecular weight, $K = 4.64 \times 10^{-2}$ dL/g, and $a = 0.76$ at 25°C in 88% formic acid. Nylon-4,6 fibers ($M_v = 41{,}400$ g/mol) did not undergo depolymerization on exposure to 100 mL of 25 wt % sodium hydroxide solution at 165°C. Out of 6.0 g of nylon fibers fed for depolymerization, 5.95 g was unaffected. When the concentration of sodium hydroxide was increased to 50 wt %, the depolymerization process resulted in the formation of low-molecular-weight oligomers. Hence, even in the presence of a phase transfer agent, a critical sodium hydroxide concentration exists between 25 and 50 wt %, which is required to initiate depolymerization under the conditions used. Soluble amine salts were also obtained.

In order to establish the feasibility of alkaline hydrolysis with respect to recycling of nylon-4,6, it was necessary to determine whether the recovered oligomers could be repolymerized to form nylon-4,6. For this purpose, solid-state polymerization was performed on nylon-4,6 oligomers formed via alkaline hydrolysis with 50 wt % NaOH at 165°C for 24 h. The solid-state polymerization process

TABLE 10.3 Effects of Experimental Conditions on the Depolymerization Efficiency of Nylon-4,6

Weight BTEMB (g)	Weight Percent of NaOH	Volume of NaOH (mL)	Weight Percent of Fibers Hydrolyzed	Weight of Oligomers Formed (g)	Intrinsic Viscosity (dl/g)	M_v (g/mol)
0.2	25	100	0			
0.2	50	100	38.4	3.7	0.14	1846
0.2	50	50	32.0	4.1	0.34	5946
0.2	50	100	39.3	3.6	0.13	1679
0.0	50	100	8.7	5.5	0.29	4704

was carried out in a round-bottom flask at 210°C for 16 h under vacuum. Solid-state polymerization of the nylon-4,6 oligomers resulted in an increase in intrinsic viscosity from 0.141 to 0.740 dL/g. That corresponds to an increase in viscosity-average molecular weight from 1846 to 16,343 g/mol. In theory, higher molecular weights would be obtained by heating for a longer time interval.

The product of the depolymerization of nylon-6,6 with 50% aqueous sodium hydroxide solution was relatively low molecular weight oligomers. A series of experiments were run in order to examine the applicability and efficiency of benzyltrimethylammonium bromide (BTEMB) as a phase transfer catalyst in the depolymerization of nylon-6,6. Table 10.4 shows the effect of the feed ratio of the nylon-6,6 to BTEMB on the viscosity-average molecular weight of the depoly-merized nylon. The product of the run with no phase transfer agent showed a 15.9% increase in weight compared to the weight of the original nylon-6,6. The calculated percent increase in weight for a 19-fold decrease in molecular weight (due to the addition of water) would be ca. 1%. Therefore, a large part of the increase must be due to leaching of silicates of the glass container (resin reaction kettle) by the strong alkali (50 wt%) at the temperature of the reaction (130°C) over 24 h. The oligomer obtained had a viscosity-average molecular weight of 1644 g/mol (the original nylon-6,6 had a molecular weight of 30,944 g/mol). The runs with phase transfer agent produced oligomers with decreases in weight of 40–50%. Although the occurrence of leaching of silicates from the glass container made quantitative assessment difficult, these results suggested that in the absence of phase transfer agent only oligomers are formed; however, soluble low-molecular-weight products are formed in the presence of phase transfer agent. The oligomers obtained were repolymerized in the solid state by heating at 200°C in a vacuum. The viscosity-average molecular weight of the solid-state polymerized nylon-6,6 obtained was ca. 23,000 g/mol (the molecular weight of the oligomeric mixture was 1434 g/mol).

In order to isolate adipic acid, nylon-6,6 fibers were depolymerized under reflux with a 50% NaOH solution in the presence of catalytic amounts of ben-zyltrimethylammonium bromide. The oligomers formed in successive steps were depolymerized under similar conditions. The yields in steps 1, 2, and 3 were

TABLE 10.4   Effect of Feed Ratio of Nylon-6,6 to BTEMB

Experiment	Experiment 1	Experiment 2	Experiment 3	Experiment 4	Experiment 5	Experiment 6
Feed ratio of nylon/BTEMB (wt/wt)	5	10.3	20.6	29.4	59.8	no PTA
Decrease in weight of oligomers	40.3	49.5	55.5	42.5	40.8	−15.9
Oligomer M_v	1556	—	1912	1697	2396	1644

57.8, 38.7, and 100+% theoretical. However, the researchers were unable to isolate hexamethylene diamine. The overall yield of adipic acid was 59.6%.

10.5.8 Ammonolysis of Nylon-6,6

McKinney[13] has described the reaction of nylon-6,6 and nylon-6,6–nylon-6 mixtures with ammonia at temperatures between 300 and 350°C and a pressure of about 68 atm in the presence of an ammonium phosphate catalyst to yield a mixture of the following monomeric products: HMDA, adiponitrile, and 5-cyanovaleramide from nylon-6,6 and ε-caprolactam, 6-aminocapronitrile, and 6-aminocaproamide from nylon-6. The equilibrium is shifted toward products by continuous removal of water formed. Most of the monomers may be transformed into HMDA by hydrogenation. Kalfas[51] has developed a mechanism for the depolymerization of nylon-6,6 and nylon-6 mixtures by the ammonolysis process. The mechanism includes the amide bond breakage and amide end dehydration (nitrilation) reactions plus the ring addition and ring-opening reactions for nylon-6 cyclic lactams. On the basis of the proposed mechanism, a kinetic model was developed for the ammonolysis of nylon mixtures.

10.5.9 Glycolysis of Polyurethanes

Polyurethane glycolysis is usually carried out at temperatures of about 200°C with excess glycol concentration in many instances at atmospheric pressure. Ulrich et al.[14] reported the glycolysis of rigid and flexible polyurethane foams to form polyols which in a 50 : 50 blend with virgin polyols could be utilized in the preparation of new rigid and semirigid foams. Rigid-foam, flexible-foam, RIM, and microcellular elastomer scrap has been treated with an equal weight of a 90/10 mixture of dialkylene glycol–diethanolamine at 190–210°C for several hours. The cooled residue was reacted with propylene oxide. The resulting polyols could be substituted for up to 40% of virgin polyols in some rigid-foam formulations. In a recent patent, Munzmay et al.[52] used glycolysis for the chemical recycling of polyurethane–polyurea or polyurea wastes. The polymeric wastes were reacted with a diol or polyol at 200°C followed by reaction of the alcoholysis product with a carbamic acid ester. The second reaction reduces the concentration of amino-capped products, which makes the final product suitable for reuse via the reaction with diisocyanates.

10.5.10 Hydrolysis of Polyurethanes

Meluch et al.[10] reported that high-pressure steam hydrolyzes flexible polyurethane foams rapidly at temperatures of 232–316°C. The diamines are distilled and extracted from the steam and the polyols are isolated from the hydrolysis residue. Good results were obtained by using reclaimed polyol in flexible-foam recipes at the 5% level. Mahoney et al.[53] reported the reaction of polyurethane foams with superheated water at 200°C for 15 min to form toluene diamines and polypropylene oxide. Gerlock et al.[54] studied the mechanism and kinetics of the reaction of polyurethane foam with steam using a polyether–toluene diisocyanate-based polyurethane as starting material. Reaction with steam at temperatures between 190 and 250°C led to the hydrolysis of all urea and urethane linkages to form a high-quality polyol product. The authors observed that the urethane bonds are rapidly broken by hydrolysis according to the reaction scheme shown in Fig. 10.8, while the urea bonds are slowly thermally converted to the corresponding iso-cyanate and amine. The polyols formed by this approach were used in a 20/80 ratio with virgin polyol to react with diisocyanate to form a high-quality flexi-ble foam.

10.5.11 Ammonolysis and Aminolysis of Polyurethanes

Sheratte[55] reported the decomposition of polyurethane foams by an initial reaction with ammonia or an amine such as diethylene triamine (at 200°C) or ethanolamine (at 120°C) and reacting the resulting product containing a mixture of polyols, ureas, and amines with an alkylene oxide such as ethylene or propylene oxide at temperatures in the range of 120–140°C to convert the amines to polyols. The polyols obtained could be converted to new rigid foams by reaction with the appropriate diisocyanates.

10.5.12 Combined Chemolysis Methods for the Depolymerization of Polyurethanes

Polyurethane[33] is degraded by a mixture of water, diethylene glycol, and alkali metal hydroxides at 200°C. When sodium hydroxide is used as a catalyst, carbamates and ureas in the product are converted to amines and alcohols by hydrolysis. After 4 h, the polyurethane is completely decomposed to form polyols, amines, isocyanates, and carbon dioxide. Kondo et al.[56] have reported the degradation of polyurethane and polyisocyanurate foams by reaction with a mixture of a glycol and ethanolamine at temperatures of about 200°C for 30 min. The resulting product was used for the production of rigid polyurethane foams. Van der Wal[57] describes the combined aminolysis and hydrolysis of

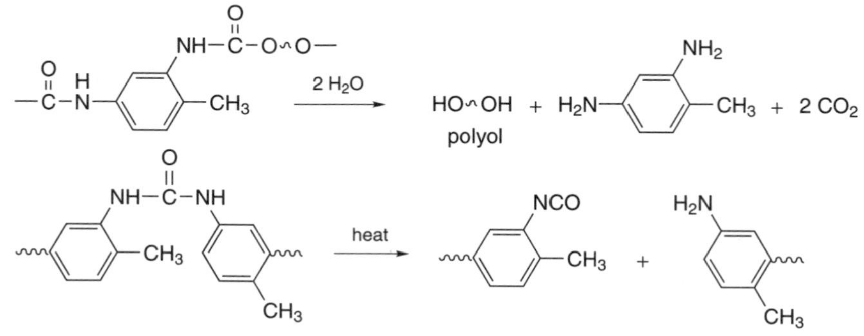

Figure 10.8 Polyurethane hydrolysis.

polyurethane scrap by reaction with a mixture of diethanolamine and aqueous sodium hydroxide. The reaction product is treated with propylene oxide to form a polyol, which is essentially free of amino endgroups.

10.5.13 Characterization/Monitoring of Depolymerization Process

10.5.13.1 Polyesters

The hydrolytic depolymerization of PET is followed by obtaining the initial and final weights of the PET and applying the formula:

$$\text{Percent degradation of PET} = \frac{\substack{\text{weight of PET} \\ \text{before reaction}} - \substack{\text{weight of PET} \\ \text{after reaction}}}{\text{weight of PET before reaction}} \times 100\%$$

The extent of depolymerization of PET as a result of glycolysis requires the determination of the hydroxyl value of the glycolyzed PET:

$$\text{Extent of depolymerization} = \frac{\text{hydroxyl value of glycolyzed PET}}{\text{hydroxyl value of BHET}} \times 100\%$$

The hydroxyl value[58] is calculated from the formula:

$$\text{Hydroxyl value} = \frac{56(v_1 - v_2)n}{m},$$

where v_1 and v_2 are the volumes in milliliters of 0.5 N KOH used for titration of the blank and the sample, respectively; n is the normality of the KOH solution, and m is the weight of the polyol sample in grams. To obtain the hydroxyl value, about 2 g of polyol is weighed into a 250-mL flask and 20 mL of the acetylation mixture of anhydrous pyridine (17.6 mL) and acetic anhydride (2.4 mL) is added. A reflux condenser is affixed to the flask and the contents of the flask are heated on a boiling water bath with occasional shaking for 1 h or more to cause complete dissolution of the PET polyol. When the polyol has completely dissolved, 25 mL of dry dichloroethane or benzene is added, the flask is stoppered, and the contents of the flask are vigorously shaken. The solvent addition should help to dissolve or at least finely divide the material. Next, 100–150 mL of water is added and after vigorous shaking the solution is titrated with 0.5 N KOH, with phenolphthalein added as an indicator.

10.5.13.2 Nylon-6[48]

The hydrolytic depolymerization of nylon-6 was followed by gel permeation chromatography (GPC), viscometry, and gravimetry. GPC determinations were performed on a Waters 150C chromatography system using benzyl alcohol as the eluant, two Plgel 10-μm crosslinked polystyrene columns, and a differential refractometer detector. The flow rate was 1 mL/min. The concentration of the polymer solutions was 0.5 wt % and dissolution was accomplished at 130°C.

The molecular weights were estimated by applying the Mark–Houwink relationship $[\eta] = K M^a$. The intrinsic viscosity $[\eta]$ was determined in m-cresol at 20°C. The values for K in the literature vary from 1.35 to 18.10 × 10^{-4} and the a values vary from 0.654 to 0.96 for polyamide–cresol systems.

Disc samples for gravimetry which were 14 mm in diameter and 2 mm thick were dried at 80°C for 2 days. The dried samples were weighed (m_0) and then placed in an acid solution. After an appropriate immersion period, they were vacuum dried at 80°C for 1 week and reweighed (m_1). The weight variations during hydrolysis are represented by $[(m_0 - m_1)/18m_0] \times 100\%$

10.5.13.3   *Polyurethane*

The glycolysis of polyurethane with diethylene glycol[59] was followed by gel permeation chromatography on a Waters Associates model 150-C GPC system equipped with a refractive index detector, a model 730 data module with GPC calculation capability, and a μ-Styragel column set consisting of five columns with permeability limits of 10^3, 10^4, 10^5, 500, and 100 Å. Tetrahydrofuran was the eluant at 40°C and a flow rate of 1.5 mL/min. The GPC was calibrated with commercially available polypropylene glycols and dipropylene glycol in order to calculate average molecular weights of glycolysis products.

Mahoney et al.[53] followed the hydrolysis of polyurethane by UV analysis of spectra recorded by a Cary model H spectrophotometer from 2500 to 3300 Å. Reference samples of pure 2,4-toluenediamine and pure 2,6-toluenediamine were shown to obey Beer's law in the range 2500–3300 Å. For the 2,4-isomer, ε_{max} at 2930 Å was 2300 ± 50 $M^{-1} \cdot cm^{-1}$ and for the 2,6-isomer, ε_{max} at 2940 Å was 1170 ± 20 $M^{-1} \cdot cm^{-1}$. The ratio of the 2,4-isomer to the 2,6-isomer in the foam sample was 4 : 1 with a theoretical yield equal to 18.7 mg/100 mg of foam. Therefore, for complete reaction, the calculated absorption at 2940 Å, due to the 2,4-isomer/100 mg of foam/mL of water and 2,6 isomer/100 mg foam/100 mg water would be 278 and 33, respectively. The yields of diamine as a function of time were calculated by the equation

$$\text{Percent yield} = \frac{\text{absorption/100 mg foam/mL of water}}{311} \times 100\%$$

10.6 SYNTHETIC METHODS

10.6.1 Depolymerization of Polyesters

10.6.1.1   *Glycolysis of PET to Obtain BHET*[36]

To a stainless steel reactor equipped with a heating mantle, a charging port, a condenser for removing ethylene glycol, an inert gas inlet, and a sampling valve were added 400 g of bis(2-hydroxyethyl)terephthalate, 136 g of ethylene glycol, and 0.035 g (or 0.225 g) of sodium acetate trihydrate. The temperature was raised to between 190 and 200°C in 1 h and then 454 g of waste polyester was added. The waste polyester consisted of polyester fibers which contained, on average, about 0.5% spin finish with 22% of the waste being of an ethylene terephthalate copolyester containing 2 mol % of ester units from the addition of 3,5-di(carbomethoxy)benzenesulfonate. The temperature of the material in the reactor was raised over a period of 1.25 h to allow the polyester to undergo glycolysis. In four of the runs, the temperature was raised to 225°C and in the remaining two runs to 235°C; in half of the runs at each temperature, water was added. Following the glycolysis interval, unreacted ethylene glycol was removed over a period of 4.75 h.

The BHET obtained by depolymerization may be conveyed to a polymerizing reactor through a transfer valve for repolymerization.

10.6.1.2 Polyester Polyols from Glycolyzed PET Waste

Example 1. Depolymerization with EG, Propylene Glycol, or Diethylene Glycol.[35] *The PET waste was depolymerized to form oligomeric diols by refluxing it with 62.5% (w/w) ethylene glycol, propylene glycol, or diethylene glycol. The depolymerization process was carried out in the presence of 0.5% (w/w) zinc acetate based on the weight of PET at the reflux temperature of the glycol for 8 h. Extents of depolymerization varied form 0.8461 to 0.9955.*

Example 2. Depolymerization with Diethylene Glycol in the Presence of Tetrabutyl Titanate.[60] *Diethylene glycol (25,300 g) and 10 g of tetrabutyl titanate were charged to a 50-L reactor fitted with a mechanical stirrer, a thermometer, a reflux condenser, an inlet and outlet for nitrogen, and a heating mantle. The reaction mixture was heated to 220°C and 18,800 g of PET scrap pellets was added to the reactor over a period of 2.5 h. Phthalic anhydride (4500 g) was added to the resulting solution, and the ingredients were heated at 240°C for 3 h. The water formed was removed from the reactor as the reaction proceeded. The polyol mixture obtained had an acid number of 1.4, a viscosity of 1250 cps at 25°C, a free diethylene glycol content of 19.06%, a free ethylene glycol content of 3.66%, and a hydroxyl number of 439.5.*

Example 3. Depolymerization Followed by Repolymerization.[61] *BHET oligomer (2142 parts) having an EG-to-TPA equivalent ratio of 1.5 was added to a depolymerizing reactor containing a rectifier, an agitator, and a heater. The BHET oligomer was melted by heating to a temperature of 220°C, and 1850 parts of PET scrap pellets were added to the molten BHET oligomer through a scrap delivery pipe under nitrogen. Simultaneously, 292 parts of EG were added continuously for 40 min, the moisture was distilled from the reactor by using the rectifier equipment, and depolymerization was completed in 65 min while the EG was refluxing. The reaction mixture was maintained at a temperature of 230°C during the course of the depolymerization process. Next 2142 parts of the product obtained were filtered and transferred to a polymerization reactor. Phosphoric acid (0.37 part) and 0.555 part of antimony trioxide were added to the polymerization reactor. The temperature and pressure in the polymerization reactor were raised to 290°C* *and 0.05 mm Hg, respectively, over a period of 60 min, and the polymerization was continued for 2 h to produce a polyester having an intrinsic viscosity of 0.650 (in o-chlorophenol at 25°C) and a softening point of 256°C.*

Example 4. Depolymerization under Pressure.[62] *PET resin was depolymerized at pressures which varied from 101 to 620 kPa and temperatures of 190–240°C in a stirred laboratory reactor having a bomb cylinder of 2000 mL (Parr Instrument) for reaction times of 0.5, 1, 2, and 3 h and at various ratios of EG to PET. The rate of depolymerization was found to be directly proportional to the pressure, temperature, and EG–PET ratio. The depolymerization rate was proportional to the square of the EG concentration at constant temperature, which indicates that EG acts as both a catalyst and reactant in the chain scission process.*

Example 5. Glycolysis to Yield Polyols Used to Synthesize Polyurethanes.[63] *A three-necked flask equipped with a condenser and stirrer was charged with 40 g of PET, 66 g of ethylene glycol, and 0.53 g of zinc acetate. The reaction mixture was heated at 190°C for 6 h. After cooling the mixture was mixed with 1 L of water and filtered. The white product obtained was dried under a vacuum at 80°C for 24 h.*

A three-necked flask equipped with a condenser and stirrer was charged with the PET depolymerization product (0.05 mol of BHET and dimer in the ratio of 80 to 20 wt%), 0.05, 0.10, and 0.15 mol of ε-caprolactone (in separate experiments), and 0.1 wt% of dibutyltin dilaurate. The reaction mixture was heated at 150°C for 2 h. The resulting co-oligomer (0.01 mol) was dissolved in 500 mL of tetrahydrofuran in a three-necked flask equipped with a condenser and a stirrer. After the temperature was raised to 67°C, a solution of 0.01 mL of hexamethylene diisocyanate in 50 mL of tetrahydrofuran was added dropwise. After heating and stirring the reaction mixture for 12 h, it was cooled and precipitated in ether. The polyurethane precipitate was collected by filtration and dried at 70°C for 12 h.

10.6.1.3 Unsaturated Polyesters from PET Waste[34]

PET waste was glycolyzed at different weight ratios of PET to propylene glycol in the presence of 0.5% (w/w) zinc acetate, based on the weight of PET, as the catalyst. The reaction was carried out at about 200°C under reflux in a nitrogen atmosphere for 4 h in a four-necked round-bottom flask fitted with a reflux condenser, gas bubbler, thermometer, and stirrer.

Unsaturated polyesters were obtained by reacting the glycolyzed product with maleic anhydride at a hydroxy-to-carboxyl ratio of 1 : 1. The hydroxyl number was determined without separation of the free glycol. The polyesterification reaction was conducted in a 2-L round-bottom flask equipped with a condenser, a gas bubbler, a thermowell, and a stirrer. The reaction mixture was heated from room temperature to 180°C in about 1–1.5 h. The temperature was maintained at 180°C for about 3 h, then raised to 200°C and maintained until the acid value reached 32 mg KOH/g.

10.6.1.4   Methanolysis of PET[64]

A vessel heated with diphenyl to 285°C and equipped with a pump was charged over a 24-h period with 15 tons of polyester waste in the form of ribbons, fibers, foils, and chips. The melted PET waste was pumped under nitrogen into a second vessel with a stirrer and maintained at an internal temperature of 250–270°C. The polyester melt was then transferred into an autoclave into which methanol heated to 200°C was simultaneously metered. Molten PET (400 kg) and 1600 kg of methanol were added to the autoclave over a period of 1 h. A temperature of 210°C and a pressure of 3–4 MPa were maintained in the autoclave. The vessel contained an overflow such that after 10 min the reaction mixture flowed through an immersion tube into a second autoclave. The second vessel also had an overflow so that the reaction mixture flowed upward at a temperature of 200°C and under a pressure of 3–4 MPa. After leaving the second autoclave, the pressure of the reaction products was reduced to 0.3 MPa and the products were passed into a stirred vessel having an internal temperature of 100°C. Next the DMT product in methanol was conveyed through heated tubes into stirred vessels and the pressure was decreased to atmospheric pressure, the mother liquor was filtered off, and the crystallized reaction product was washed with fresh methanol, suspended, and separated by centrifugation. A conversion of 70–90% of the PET to DMT was achieved in the first stage over a residence time of 7–13 min.

10.6.1.5   Methanolysis of PET and Repolymerization of EG and DMT Obtained[65]

DMT (72.75 g, 0.375 mol), 93 g of EG (1.5 mol), 1 mL of an EG solution containing 1.588 g of zinc acetate dihydrate catalyst per 100 mL of EG, and 2 mL of an EG solution containing 2.18 g of antimony trioxide catalyst per 200 mL of EG were added to a 500-mL three-necked round-bottom flask equipped with a ground-glass distillation head, a stirrer, and a nitrogen inlet. Commercial post-consumer PET flake (72 g, 0.375 mol, intrinsic viscosity (IV) 0.70) obtained by cleaning and grinding PET soft-drink bottles is added to the mixture in the flask. Next the flask is immersed in a Belmont metal bath and heated for 1 h at 200°C and about 2 h at 210°C. The theoretical amount of methanol was collected. The mixture was a clear, transparent solution. After removing any solid with a glass tube siphon (while keeping a nitrogen blanket over the molten solution) with a vacuum line attached, the contents of the flask were transferred to a second poly-condensation flask equipped as before for the transesterification flask. The molten liquid at 210°C was filtered to remove any solids and the polycondensation flask was immersed in a Belmont metal bath that has been preheated to 210°C. The temperature was increased to 280°C and the pressure in the flask was reduced to 13.3 to 66.7 Pa for 20 min. The vacuum was then discontinued and nitrogen was passed into the flask. The flask containing the molten polymer was removed and the molten polymer was poured into a 2-L stainless steel beaker containing ice water. The water was stirred continuously and the molten polymer was quenched to form a continuous rod. The amorphous rod had an IV of 0.36. The rod was ground to pass a 3-mL screen and mixed with similarly ground virgin PET of IV 0.36 to give a 25% or more content of recycled PET. The blend was then allowed to undergo solid-state polymerization at about 215°C in a fluidized bed with a stream of dry nitrogen until the IV became 0.76.

10.6.1.6   Hydrolysis of Waste PET by Sulfuric Acid[43]

PET powder was prepared from commercial PET bottles and had a number-average molecular weight of 30,000. PET powder (0.2 g) was placed in 30 mL of relatively dilute sulfuric acid (10 M) in a sealed Pyrex tube and heated to 150°C for 1–6 h in an oven. After the reaction mixture was allowed to cool, the precipitate of TPA and PET residue deposited was filtered through a 1G-5 glass filter. TPA was separated as the ammonium salt by dissolving it in 12 mL of 5 M NH$_3$. The TPA was reprecipitated by sulfuric acid solution. The yield of TPA was 100% in 10 M sulfuric acid solution.

10.6.1.7   Hydrolysis of PET with Sodium Hydroxide in Ethylene Glycol

Example 1. Test   Tube   Experiment.[42] *PET pellets (0.48–1.92 g), sodium hydroxide (0.21–0.84 g), and ethylene glycol (5 mL) were added to a test tube under a nitrogen atmosphere, and the mixture was heated to 150–180°C with stirring. After 80, 50, 30, and 15 min at 150, 160, 170, and 180°C, respectively, the test tube was cooled quickly in cold water, and the reaction mixture was poured into 50 mL of distilled water and titrated with standard 0.2 N HCl solution up to pH 7 to measure the amount of NaOH consumed (which is equivalent to the molar amount of sodium carboxylate formed). In preliminary experiments, the titration results for the filtrates, which were obtained after removing disodium terephthalate and unreacted pellets, were found to be the same as those for unfiltered gross mixtures; therefore, all titrations were carried out on the gross mixtures. After titration, the unreacted pellets were removed and the filtrate acidified with concentrated hydrochloric acid to form terephthalic acid. The terephthalic acid obtained was dried and weighted. The reaction occurring at 180°C for 15 min gave a quantitative yield of TPA.*

Example 2. Reactor Experiment.[66] *Waste PET (110 g), 800 g of ethylene glycol, and 93 g of 50% aqueous sodium hydroxide were introduced into a reactor. The reaction mixture was heated to 170°C with agitation while collecting distillate (mostly water with some ethylene glycol). The slurry, which consisted of disodium terephthalate in ethylene glycol, was filtered at a temperature of 170°C in a vacuum filter. The disodium terephthalate obtained was pressed as dry as possible and the ethylene glycol was recovered. The filter cake was washed with room temperature EG to remove impurities and to cool the disodium terephthalate to less than 100°C, followed by washing with a saturated solution of disodium terephthalate in water (maintained at 90–100°C).*

The disodium terephthalate was dissolved in water (15 g of disodium terephthalate is soluble in 100 mL of cold water). The solution was filtered to remove any insoluble impurities (e.g., pigments, polyethylene, vinylidene-chloride-type degradation products). The solution was then neutralized with 100% sulfuric acid. The resulting slurry was heated at 90–95°C for 15 min, followed by filtration. The filter cake was then washed with seven 300-mL portions of distilled water to remove the inorganic sulfates formed during neutralization. The terephthalic acid obtained was dried in a dryer for 3 h at a temperature of about 125°C.

Immediately prior to the neutralization with sulfuric acid, the solution may be heated to about 90°C with about 3% (based on the total weight of terephthaloyl content) of pulverized activated carbon added while the solution was being agitated. The mixture was heated for 15 min at 90–95°C followed by removal of the carbon by filtration. This treatment substantially improved the color of the solution. The purified terephthalic acid obtained after neutralization was recycled back into the PET polymerization process by adding it directly to the BHET (formed by the ester interchange reaction between DMT and EG) prior to substantial polymerization having taken place in up to 20% by weight based on the weight of DMT. Under these conditions, the resulting polymer had excellent color and properties equivalent to control polymer.

10.6.1.8 Hydrolysis of PET by Sodium Hydroxide in the Presence of Phase Transfer Catalysts[27]

In a 300-mL round-bottom flask, a 5% sodium hydroxide solution (250 mL) was heated to 80°C in a constant-temperature bath. The catalysts were added in the following amounts in separate experiments: trioctylmethylammonium chloride (TOMAC) (0.04 g, 0.0001 mol); trioctylmethylammonium bromide (TOMAB) (0.045 g, 0.0001 mol); hexadecyltrimethylammonium bromide (HTMAB) (0.045 g, 0.0001 mol); tetraethylammonium hydroxide (TEAOH) (0.015 g, 0.0001 mol); and phenyltrimethylammonium chloride (PTMAC) (0.02 g, 0.0001 mol). PET fibers (1.98 g, 0.01 mol) were added to the mixture and allowed to react for 30, 60, 90, 150, and 240 min. Upon filtration, any remaining fibers were washed several times with water, dried in an oven at 130–150°C, and weighed. The results are shown in Table 10.1.

The agitation studies for PET depolymerization were performed in the Atlas "Launder-ometer." The Launder-ometer is a device for rotating closed containers in a thermostatically controlled water bath. The procedure used in these experiments was adapted from an American Association of Textile Chemists and Colorists (AATCC) standard test method. The 5% sodium hydroxide solution (250 mL) was preheated to 80°C in a 1-pint stainless steel jar. The catalysts were added in the following amounts in separate experiments: TOMAC (0.04 g, 0.0001 mol); TOMAB (0.045 g, 0.0001 mol); and HTMAB (0.045 g, 0.0001 mol). The PET fiber specimens (1.98 g, 0.01 mol) were placed in the containers along with ten $\frac{1}{4}$-in. stainless steel balls to aid in the agitation process. The jars were sealed in the Launder-ometer, whose bath was at the desired temperature (80°C). The machine was allowed to run for the allowed treatment times (i.e., 30, 60, 90, 150, and 240 min) at 42 rpm. Upon decanting, any residual fibers were washed several times with water, dried in an oven at 144°C, and weighed. The results are shown in Table 10.2.

Alkaline depolymerization of PET at 80°C in the presence of phase transfer agents resulted in the isolation of terephthalic acid in yields as high as 93%.

10.6.1.9   Neutral Hydrolysis of PET

Example 1. Hydrolysis of Bottle Waste.[67] *The PET used in this experiment was postconsumer uncolored bottle polymer which was chopped into flakes. Analysis showed that the PET contained 29.37 wt % ethylene glycol, 1.77 wt % diethylene glycol, 0.92 wt % triethylene glycol, and 2.63 wt % cyclohexanedimethanol. Waste PET (200 g) and distilled water (800 g) were added to a stirred 1.8-L Hastelloy C PARR TM autoclave. The autoclave was pressurized with nitrogen to 250 psig, stirring was commenced, and the autoclave was heated to 220°C. The reaction was allowed to continue with stirring under autogenous pressure at 220°C for 2 h. The autoclave was cooled to ambient temperature followed by venting. The reaction mixture was removed from the autoclave and vacuum filtered. Analysis of the aqueous filtrate (670 g) by high-pressure liquid chromatography and gas chromatography revealed the presence of the following substances: 0.2 wt % 2-hydroxyethyl terephthalic acid, 6.87% wt % ethylene glycol, 0.43 wt % diethylene glycol, 0.015 wt % triethylene glycol, 0.121 wt % cyclohexanedimethanol, 0.023 wt % acetaldehyde, and 0.006 wt % 2-methyl-1, 3-dioxolane. The solid precipitate contained 96.4 wt % TPA and 2.4% wt % 2-hydroxyethyl terephthalic acid.*

Example 2. Hydrolysis of PET Chips.[68] *PET chips (10 g) and 100 g of water were charged to a Parr 4560 minireactor equipped with a 300-mL bomb and a variable-speed motor for stirring and the reaction mixture was heated to 265°C. The temperature was maintained for 10 min at a rate of stirring of 400 rpm and autogenous pressure. The vessel was then quickly removed from the heating mantle, immersed in an ice bath, and quenched to ambient conditions. The mixture in the bomb was filtered through a sintered glass filter to yield a solid and a liquid phase. The liquid phase composed of water, ethylene glycol, and diethylene glycol was analyzed by gas chromatography. The solid phase, which mainly consisted of unreacted PET, oligomers, and terephthalic acid, was washed by deionized water, dried at 80°C, and ground to form a fine powder. The carboxylic acid concentration was determined by titration. The yield of EG was 98.1% and diethylene glycol 1.15%.*

Example 3. Hydrolysis in an Extruder.[69] *PET reactive extrusion experiments were carried out on a 25-mm Berstorff ZE25 corotating twin-screw extruder with a barrel length-to-screw ratio of 28 : 1. The extruder consisted of six barrel sections equipped for heating, cooling, and controlling the temperature of each section of the extruder. Initially reaction extrusion of PET and water was performed with cold water at room temperature injected into the extruder. Typical operating conditions were reaction temperatures of 230–265°C, extruder speeds* *of 10–50 rpm, PET feed rate of 0.62–2.67 kg/h, and water injected at 21–81% of the weight of PET. The highest degree of depolymerization occurred at the lowest PET feed rate and screw speed and highest reaction temperature. A carboxyl group concentration of 0.1248 mmol/g was obtained for a ratio of 24% weight of water to PET; the carboxyl group concentration for virgin unprocessed PET was 0.0137 mmol/g.*

Experiments were also carried out with saturated steam at reaction and steam temperatures of 260, 280, and 300°C with a low screw speed and feed rate. When extruding polymer melts alone, the lowest residence time measured was about 18 min. As a result of the depolymerization of PET decreasing the viscosity, steam injection cut the residence times approximately in half. At a temperature of 260°C and steam injected at 1.52 times the weight of PET, the carboxyl group concentration of the extrudate was 0.1611 mmol/g. With approximately the same steam feed rate at 280°C, the carboxyl group concentration increased to 0.2123 mmol/g, and at 300°C, the carboxyl group concentration of the extrudate was 0.249 mmol/g.

Experiments were also carried out at high pressures generated in the reaction zone by using a throttling device to control extruder output. When the pressure in the extruder was raised to 1103 kN/m² (160 psig) at a temperature of 270°C, a PET feed rate of 0.58 kg/h, and a water–PET ratio of 1.73 : 1, the extrudate carboxyl group concentration was 0.9902. The carboxyl group concentration of the extrudate increased under similar conditions when the temperature was raised to 300°C and the pressure in the reaction zone was 965 kN/m² (140 psig).

10.6.1.10 Ammonolysis of PET[70]

In a continuous process, ground PET bottles (830 parts) in an aqueous slurry were pumped into an autoclave equipped with a stirrer and maintained at 450–550 psig pressure and 191–232°C. Ammonium hydroxide (300 parts) solution consisting of water (7857 parts), ethylene glycol (493 parts), and ammonium sulfate (918 parts) was introduced into the reactor. The retention time in the reactor varied from 5 to 45 min. The aqueous diammonium terephthalate and ethylene glycol solution was withdrawn from the reactor and filtered while hot to remove solid impurities such as pigments, pieces of metal caps, labels, and cap liners. The filtrate was acidified with sulfuric acid solution to liberate the TPA product. The recovered TPA usually had a purity of 99% or higher.

10.6.1.11 Aminolysis of PET[71,72]

Most degradation experiments carried out on polyesters have involved treating the PET with aqueous primary amines, usually 40% aqueous methylamine at 20°C (or room temperature). In most cases, 30–50 mg of vacuum-dried PET was exposed to excess aqueous amine contained in an airtight tube to prevent the loss of gaseous methylamine from the aqueous solution. The reactions were carried out with or without agitation over various periods of time. After the degradation experiment, the sample was filtered, carefully washed, and dried at 60°C for 3 h to remove water and residual methylamine. After drying, the weight loss and extent of chain scission were determined (by viscometry). In some experiments, the residues after aminolysis were separated by centrifugation. Considerable weight losses and viscosity decreases have been observed.

10.6.1.12   Glycolysis–Hydrolysis of PET[45]

PET was passed into an extruder where it was melted. Ethylene glycol was also pumped to the extruder. The molten PET and ethylene glycol were mixed in the extruder at a temperature in the range of 280–285°C and a pressure of about 6.9 MPa. The total residence time of the EG and PET in the pre–molecular weight reduction zone was about 2 min. The lower molecular weight PET was then passed into a hydrolyzer, in which the molten lower molecular weight PET was exposed to water at a high temperature at a pressure of about 4.1 MPa. The polyester was hydrolyzed to EG and TPA in the hydrolyzer. The hydrolysis reaction was complete in approximately 15 min. In contrast, if the PET were passed to the hydrolyzer operated at 4.1 MPa without undergoing preglycolysis, about 45 min would be required to complete the hydrolysis reaction.

10.6.1.13   Glycolysis–Methanolysis of PET[46]

A dissolver equipped with an agitator was charged with 2000 parts of PET oligomers, containing a mixture with degrees of polymerization varying from 2 to 20 and heated to about 250°C to melt the material. Ground scrap PET bottles containing polyolefin bottom cups, aluminum bottle caps, labels, and adhesives were fed to the dissolver at a rate of 2.5 parts per minute (mixing time is about 5–60 min). The dissolver melt obtained was fed to a reactor (operated at a pressure of about 30–50 psig and a temperature of about 250–290°C) and superheated methanol vapor was fed to the reactor at a rate of 20 parts per volume and circulated through the molten mass to form a melt consisting of low-molecular-weight polyesters, monohydric alcohol-ended oligomers, glycols, and DMT. Methanol, glycols, and DMT were collected by distillation. A light layer of polyolefins could be skimmed from the mass and aluminum collected from the bottom of the melt for removal from the dissolver. The reactor melt was returned to the dissolver as ground scrap PET was continuously fed to the dissolver.

10.6.2 Depolymerization of Nylons

10.6.2.1   Neutral Hydrolysis of Nylon-6 in Carpet Waste

Example 1. Neutral Hydrolysis at 410–450 kPa.[47] *A Waters 590 high-performance liquid chromatography (HPLC) pump was used to pump water at a rate of 2–6 g/min through two steam generators in series. Heating tape was used to heat transfer lines and valves. Reactor heating was achieved by heating three zones with individual 400-W cable heaters for each zone. High-temperature needle valves were situated at the reactor overheads and bottoms exit points.*

In a typical reactor, 180 g of chipped carpet extrudate was charged to a 43 × 1 -in.-diameter stainless steel reactor while the reactor was purged with nitrogen. Analysis of the extrudate indicated a 57.6% nylon content. The system was sealed and all zones except that in contact with the carpet extrudate itself were brought up to temperature. Next the reactor body was heated and steam flow started when the reactor's internal temperature reached ~120°C (~20 min). The reactor was maintained at a pressure in the range of 410–450 kPa (45–50 psig) during heatup and increased to the desired pressure when the temperature reached ~180°C. Samples of overhead condensate were collected throughout the experiment. At the end of the reaction, the steam flow was terminated and the overhead exit valve was closed. The nitrogen inlet valve at the top was opened and the pressure in the system was allowed to increase slightly. The residue was removed by carefully opening the valve at the bottom of the reactor and collecting the expelled residue in a metal beaker. After sampling the overhead fractions for GC and HPLC analysis, 50% of the combined overhead samples were evaporated to dryness on a rotary evaporator. The solid obtained was analyzed by GC and HPLC. The permanganate number and color were also determined for the solid ε-caprolactam. The solid sample collected from the bottom of the reactor was analyzed for residual nylon by nitrogen analysis. At temperatures between 300 and 340°C, 3 or 6 g/min steam, and pressures of 620–1500 kPa (75–200 psig), crude ε-caprolactam yields of 87–95% were obtained.

Example 2. Neutral Hydrolysis of Nylon-6 at 932 kPa.[73] *A shredded carpet having nylon-6 face fibers and a backing of polypropylene and calcium-carbonate-filled styrene-butadiene rubber latex (SBR) was extruded and the extrudate ground to 5 mesh chips. The resulting whole carpet feedstock (178.8 g) containing 57.6% by weight of nylon-6 was placed in a cylindrical stainless steel reactor of 24.5 mm diameter and 1070 mm height. The reactor was connected to a condenser which was equipped with a back-pressure valve at the exit set at 932 kPa (9.2 atm). Superheated steam was passed through the bottom of the reactor while the temperature of the reactor was maintained at 300°C. Overhead cuts (distillate) were taken frequently and analyzed for caprolactam, caprolactam oligomers, and ammonia. In total, 1094 g of distillate was collected in 6.0 h containing 92.5 g caprolactam, 0.54 g caprolactam cyclic dimer, and 0.126 g ammonia. The molar yield of ε-caprolactam obtained from the nylon-6 present in the carpet was 89.8%. The relative moles of cyclic dimer (expressed as caprolactam equivalents) and relative moles of ammonia compared to lactam produced were 0.58 and 0.91%, respectively.*

10.6.2.2 Acid-Catalyzed Hydrolysis of Nylon-6 in Carpet Waste[74]

Nylon-6 (108 g) carpet backed with calcium-carbonate-filled latex and polypropylene was charged to a 1000-mL three-neck round-bottom flask (equipped with a condenser) with 6 mL of 85% phosphoric acid. Superheated steam was injected continuously during a 45-min period. The vapor temperature of the reaction medium was 250–300°C. The volume of distillate collected was 1065 mL. The distillate contained 1.9% ε-caprolactam (as determined by GC), which corresponded to a crude yield of 37.5%. The distillate was fractionated in a distillation column and the nonaqueous phase removed. The remaining aqueous phase was treated with 2% potassium permanganate at 40–50°C for 2 h. Evaporation of the water produced solid ε-caprolactam. The crude ε-caprolactam obtained was distilled at about 1 mm Hg to produce pure ε-caprolactam.

10.6.2.3   Acid-Catalyzed Hydrolysis of Nylon-6,6[8]

Dyed waste nylon-6,6 stockings were thoroughly washed and dried. The dried stockings were treated with sulfuric acid in water in the ratio of 1 : 1 : 1 at 115–120°C at reflux for 5 h. The reaction mixture was cooled and the solid adipic acid, which precipitated on cooling, was collected by filtration. The filtrate was hydrolyzed for an additional 5-h period and the adipic acid collected by filtration. The filtrate, which contained sulfuric acid, HMDA sulfate, and a small quantity of adipic acid was slowly added to a slurry of lime in water (1 : 7) maintained at a temperature of 70°C to neutralize the sulfuric acid and obtain an aqueous solution of the free amine. The solid calcium sulfate and unreacted lime were collected by filtration and the calcium sulfate cake was reslurried with water two or three times followed by adding the aqueous washings to the free amine solution. Pure HMDA was obtained by first distilling water from the aqueous washings at atmospheric pressure and them distilling HMDA at 35 mm Hg and a temperature of 108°C. The adipic acid was purified by treatment with decolorizing carbon followed by two recrystallizations from water. The HMDA and adipic acid obtained in the process were suitable for making new nylon products which compared favorably with those made from virgin monomers.

10.6.2.4   Acidolysis of Nylon-6,6[75]

Waste nylon-6,6 was washed in a diluted commercial detergent solution at 100°C for 0.5 h and then rinsed twice with water to remove any finishes present. The washed nylon-6,6 was then reacted with molten adipic acid for 1.5 h or more at a temperature of 175°C with a weight ratio of nylon-to-adipic acid of 0.15 : 1. The molten product was then exposed to steam at a temperature of 230–233°C to remove any stabilizers present. The acidolysis product was then hydrolyzed with water at a temperature of 204°C under autogenous pressure for 0.5 h or longer with a ratio of water to acidolysis product of 0.50 : 1 (w/w). The hot solution was then filtered at 100°C to remove any titanium dioxide present. The filtered product was then mixed with HMDA to neutralize any excess acid present. The solution was then filtered to remove any solids. A 50% by weight aqueous solution of HMDA was added to the filtrate, and under standard polymerization conditions, polyhexamethylene adipamide (nylon-6,6) was produced.

10.6.2.5   Alkaline Hydrolysis of Nylon-6,6[9]

Example 1. Alkaline Hydrolysis of Nylon-6,6 in Isopropanol.[9] *Isopropanol (209 parts), 267 parts of water, 54 parts of 97% sodium hydroxide, and 100 parts of nylon-6,6 were charged to a 1-gal stainless steel autoclave. Under an atmosphere of nitrogen, the charge, which weighed 1060 g, was heated in the autoclave to 180°C for 1.5 h with constant agitation and then cooled. A clear two-phase liquid was obtained which had an alcoholic upper phase and an* *aqueous lower phase. The alcohol phase was decanted and the aqueous phase was washed with a small amount of fresh isopropanol. The combined extracts were distilled at atmospheric pressure to yield isopropanol at 86°C in 98% yield, which was appropriate for reuse. The residue underwent vacuum fractionation at 30 mm Hg. HMDA was obtained in the fraction distilling at 105–106°C in 92.5% yield (80.4 g). The aqueous phase was acidified with 50% sulfuric acid to a pH of 1.0, giving a precipitate of adipic acid in 92% yield (105.3 g).*

Example 2. Alkaline Hydrolysis in *n*-Butanol.[50] *Water (600 parts), sodium hydroxide (120 parts), and nylon-6,6 (220 parts) were charged to a 4-L stainless steel autoclave. Under an atmosphere of nitrogen and with agitation, the reaction mixture was heated at 190°C for 3 h, then cooled to 100°C, and discharged from the reactor as a clear, light brown solution. The solution was cooled to 40°C and extracted with five 105-part portions of n-butanol containing 20% water. The n-butanol extracts were combined and the n-butanol distilled off at atmospheric pressure through a fractionating column. Distillation was terminated when the temperature in the still head exceeded 120°C, then the column was washed down with a small amount of water, the washings being added to the residue in the distillation flask. The residue in the distillation flask was then distilled at a pressure of 60 mm Hg without any separation of fractions (the fraction boiling at 122–124°C at this pressure corresponds to HMDA) to give an aqueous solution of HMDA. The yield of HMDA, estimated by titration, was 93.1%.*

The aqueous layer remaining after extraction with n-butanol was acidified (to pH 1) by the addition of 50% sulfuric acid, giving a precipitate of adipic acid which was collected by filtration, washed with 120 parts of water in two equal portions, and dried at 110°C. The crude adipic acid obtained was recrystallized from twice its weight of water to provide adipic acid in 90.2% yield, which was pure enough to be used in the synthesis of adiponitrile.

10.6.2.6   *Phase-Transfer-Catalyzed Alkaline Hydrolysis of Nylon-6,6*[27]

In a 500-mL resin reaction kettle fitted with a mechanical stirrer and a reflux condenser were placed 6 g (0.025 mol) of nylon-6,6 fibers and 200 mL of 50% aqueous sodium hydroxide solution. Benzyltrimethylammonium bromide (0.2 g, 0.0008 mol) was added to the reaction mixture. The reaction mixture was heated for 24 h under reflux (130°C). Next, the unreacted nylon-6,6 fibers were removed by filtration and dried in a vacuum oven at 100°C for 24 h. The remaining aqueous solution was evaporated and the residue was extracted with isopropanol in a Soxhlet apparatus for 24 h.

The mixture of oligomers was heated in a flask to 200°C in a mineral bath (the DSC peak melting temperature of the oligomers was 230°C) while stirring and removing water formed with a vacuum pump to form high-molecular-weight nylon-6,6.

Nylon-6,6 fibers (6.0 g) were depolymerized under reflux with 200 mL of 50% NaOH solution in the presence of 0.20 g BTEMB to form 0.55 g of adipic acid (after acidification) and 4.52 g of oligomer. In a second step, the oligomer obtained in step 1 was depolymerized under the same conditions to form 0.72 g of adipic acid and 1.62 g of oligomer. In the third step, the 1.62 g of oligomer formed in step 2 were depolymerized with 50 mL of 50% NaOH solution and 0.54 g of BTEMB to form 1.05 g of adipic acid after initial exposure of the oligomer to heat in an attempt to obtain hexamethylenediamine by distillation. The overall yield of adipic acid was 59.6%.

10.6.2.7 Phase-Transfer-Catalyzed Alkaline Hydrolysis of Nylon-4,6[27]

A 500 mL round-bottom flask containing aqueous sodium hydroxide (25 or 50 wt%) and benzyltrimethylammonium bromide as the phase transfer agent (0.2 g, 0.0008 mole) was placed in a constant-temperature oil bath and heated to the reflux temperature of 165°C. Chopped nylon-4,6 fibers (6 g, 0.030 mol) were placed in the reaction flask. The reaction mixture was constantly stirred with a magnetic stirrer and the reaction was carried out at atmospheric pressure for a period of 24 or 36 h (see Table 10.3 for results). The aqueous sodium hydroxide solution containing products of depolymerization was concentrated by evaporation and 30 mL of 35% aqueous hydrochloric acid was added to the concentrate. The precipitate was filtered and washed with water. The product was dried in a vacuum oven at 120°C for 24 h.

Oligomers (2.0 g) obtained from the alkaline hydrolysis of nylon-4,6 were charged to a round-bottom flask. The flask containing the oligomers was heated in a mineral bath at 210°C for 16 h under vacuum.

10.6.2.8 Ammonolysis of Nylon-6,6[13]

Nylon-6,6 (15.0 g) and 0.25 g of diammonium hydrogen phosphate were charged into a vertical cylindrical reactor (72 cm^3 inner volume) with a 5-μm fritted disk at the bottom. The reactor was sealed and purged with nitrogen. Liquid ammonia was fed into the reactor at a rate of 2.0 mL/min through a preheater at a temperature of 320° and the fritted disk. The reactor was heated by a band heater set at 320°. Constant pressure was maintained by a back-pressure regulator set at 1000 psig throughout the 90 min reaction time. Under the reaction conditions, monomeric products were volatilized and transported from the reactor through the regulator and condensed out of the ammonia stream into a cooled receiver. The ammonia passed out of the receiver into a water scrubber. Monomeric product yields (by GC) based on nylon-6,6 charged were HMDA (18%), adiponitrile (17%), and 5-cyanovaleramide (2%). When the same apparatus and conditions were applied to a mixture of nylon-6,6 (7.5 g), nylon 6 (7.5 g), and ammonium phosphate (0.25 g), monomeric product yields based on nylon-6,6 charged were HDMA (56%), adiponitrile (38%), and 5-cyanovaleramide (3%) and based on nylon-6 charged were 6-aminocapronitrile (50%), ε-caprolactam (37%), and 6-aminocaproamide (1%).

10.6.2.9 Lewis-Acid-Catalyzed Ammonolysis of Nylon[30]

Nylon-6,6 (7.5 g), nylon-6 (7.5 g), and 0.25 g of Lewis acid catalysts such as zinc chloride, cobalt chloride, calcium chloride, and barium carbonate were charged into a vertical cylindrical reactor (72 cm^3 inner volume) with a 5-μm fritted disk at the bottom. The reactor was sealed and purged with nitrogen. Liquid ammonia was fed into the reactor at a rate of 1.8 g/min through a preheater at a temperature of 300° and the fritted disk. The reactor was heated by a band heater set at 300°. Constant pressure was maintained by a back-pressure regulator set at 1000 psig throughout the 30 min reaction time. Under the reaction conditions, monomeric products were volatilized and transported from the reactor through the regulator and condensed out of the ammonia stream into a cooled receiver. The ammonia passed out of the receiver into a water scrubber. Monomeric product yields (by GC) were HMDA (7.01 mmol), adiponitrile (1.04 mmol), 5-cyanovaleramide (0.40 mmol), 6-aminocapronitrile (8.5 mmol), ε-caprolactam (10.4 mmol), and 6-aminocaproamide (0.59 mmol) for the zinc-chloride-catalyzed case. The total yield of monomers was 21%. Monomeric yields for the control (without catalyst) were HMDA (2.437 mmol), adiponitrile (0.54 mmol), 5-cyanovaleramide (0.04 mmol), 6-aminocapronitrile (2.58 mmol), ε-caprolactam (3.84 mmol), and 6-aminocaproamide (0.13 mmol). The total yield of monomers was 7%.

10.6.3 Depolymerization of Polyurethanes

10.6.3.1 Glycolysis of Polyurethanes

Example 1. Glycolysis of Polyurethanes with Diethylene Glycol.[59] *In a typical experiment, 300 g of vacuum-distilled diethylene glycol (DEG) was weighed into a 1-L round-bottom flask equipped with an electrically driven stirrer, a reflux condenser, and nitrogen purge and degassed with hot copper (450°C) scrubbed nitrogen for 1 h. The DEG was brought to the reaction temperature (190–220°C) by heating the flask with an electric heating mantle controlled by a variac. Temperatures were monitored by a stainless steel sheathed type-K calibrated thermocouple immersed in the reactor and followed with a thermocouple reader. The desired amount of foam (1–5 g) was quickly added to the reactor. Foam was added through the nitrogen purge neck by briefly substituting the tube with a DEG lubricated tube through which the foam was pushed. Samples of the reaction mixture were withdrawn through the sampling port with a 5-cm^3 glass syringe equipped with an 8-in.-long, 16-gauge stainless steel needle for GPC (gel permeation chromatography) analysis. GPC analysis revealed the presence of seven major components. Among the components were isomeric toluene diamine (TDA), TDA monocarbamates, 2,4- and 2,6-TDA dicarbamates, and urea-linked TDA oligomers.*

Example 2. Glycolysis of Polyurethanes with Diethylene Glycol in the Presence of Superheated Steam.[76] *Waste polyurethane foam (600 g) was dissolved in 600 g of diethylene glycol (DEG) by adding 100-g amounts of foam every 10 min to the DEG at 200°C with stirring. Then superheated steam was bubbled through the reaction mixture. The overhead gases passed through a condenser. At 30-min intervals, samples of the reaction mixture were withdrawn from the reaction vessel and analyzed by HPLC to identify the presence of mono- and bicarbamates of diethylene glycol and the isocyanates used to make the foam. It was discovered that the carbamates do not disappear from the reaction mixture until most of the* *amines have been distilled from the reactor by the flowing superheated steam (a period of about 8 h at a rate of steam addition of 300 cm³/min). Since diethylene glycol was also distilling with the steam, additions of diethylene glycol to the reactor every 15 min were required to maintain the reaction volume.*

After 8 h of reaction, the reactor was allowed to cool. A two-layer liquid formed. The top layer was found to contain mostly polypropylene ether triols with about 20% by weight diethylene glycol and 5% by weight toluene diamines. The top layer was purified by vacuum distillation at 2 mm Hg and 200°C to produce 320 g of a light brown liquid residue. This residue (polyols) was used as a replacement for 5% by weight of the Pluracol 535 polyol in the formulation of a flexible polyurethane foam. A flexible foam which had good resiliency and a density of 2.2 lb/ft³ was obtained. At higher replacement levels, lesser quality foams were obtained.

Example 3. Glycolysis of Polyurethanes with Propylene Glycol.[32] *Propylene glycol (487 g) was added to a 1-L flask equipped with a stirrer, thermometer, and reflux condenser. The flask was immersed in an oil bath and the propylene glycol was heated to reflux at 200°C. To the flask was added 230 g of flexible polyurethane foam. The foam went into solution. As the polyurethane foam dissolved, the color of the reaction mixture changed to light yellow, yellow, and brown and the foam was completely dissolved after 5 h. The reaction mass was transferred to a separatory funnel to be stored overnight while cooling. It separated into two layers with the lower layer occupying about five-sixths of the volume. The lower layer was recovered and the excess propylene glycol was removed by distillation at 2 mm Hg and 63°C to leave 120 g of a black paste. The paste recovered was washed with water and dried. Infrared spectral analysis indicated that the paste was an amine compound containing an amino group and a urethane linkage.*

The top layer was washed with water and ethylene glycol and dried to give 135 g of a yellow oily material that from infrared spectral analysis was identified to be the same triol used to prepare the flexible polyurethane foam.

Example 4. Glycolysis of Polyurethanes with Propylene Oxide after Pretreatment with a Mixture of Diethanolamine and Potassium Hydroxide.[57] *Polyurethane scrap was treated with a mixture of diethanolamine and potassium hydroxide at a temperature between about 80 and 140°C with stirring to form an intermediate product. The weight ratio of the scrap PUR polymer to the mixture of diethanolamine and potassium hydroxide was from about 15 : 1 to 30 : 1. The intermediate product was reacted with propylene oxide at a temperature of from about 100 to 120°C in a closed reaction vessel to form a polyol. The propylene oxide was added at a rate to maintain a pressure of from about 2 to 5 atm (29–73 psi). The progress of the reaction was followed by following the change of pressure with time. When the pressure remained constant, the reaction of the intermediate product with propylene oxide was considered to be complete. The crude polyol obtained was treated with 10 mol % excess of dodecylbenzene sulfonic acid to remove the potassium hydroxide.*

Example 5. Glycolysis of Polyurethanes with Propylene Oxide after Pretreatment with Ethanolamine.[55] *A rigid polyurethane foam (ca. 100 g) was dissolved in 30 g ethanolamine by heating. Excess ethanolamine was stripped, leaving a clear solution. Infrared and GPC analysis indicated that the clear solution obtained contained some residual polyurethane, aromatic polyurea, aliphatic polyols, aromatic amines, and N,N'-bis(β-hydroxyethyl)urea. Next the mixture was dissolved in 45 g propylene oxide and heated at 120°C in an autoclave for 2 h. The pressure increased to 40 psi and then fell to 30 psi at the end of the 2-h heating period. The product was a brown oil with a hydroxyl number of 485.*

10.6.3.2 Hydrolysis of Polyurethanes

Example 1. Hydrolysis of Polyurethane in a Pyrex Hydrolysis Reactor.[54] *A 100-mg sample of polyurethane foam was placed in a Pyrex hydrolysis reactor. The hydrolysis reactor consists of a stainless steel wire mesh sample basket which was placed in a reactor tube that was equipped with a nitrogen inlet, a heated sidearm for steam entry, and a sample holding zone. The Pyrex hydrolysis reactor was heated by a Lindberg heavy-duty furnace. The stainless steel wire mesh basket was then attached to the Pyrex sample positioning tube and placed in the holding zone of the reactor. A thermocouple was positioned so that it ran through the sample positioning tube to touch the foam in the sample basket. The exposed thermocouple allowed the temperature of the sample to be determined with little heat transfer delay. When the hydrolysis zone of the reactor reached the reaction temperature and a constant flow of steam was achieved, the sample basket was pushed into the preheat zone of the reactor. Next the sample was rapidly warmed (in less than 4 min) with dry, oxygen-free, 200°C nitrogen to 185°C. When the temperature of the foam reached 185°C, the sample basket was introduced into the hydrolysis zone of the reactor. Preheating the sample prevents steam condensation when the sample enters the hydrolysis zone. For reactions conducted at temperatures of 210–230°C, the total yields of toluene diamine (TDA) approached 98% theoretical. Use of sodium hydroxide as a catalyst significantly increased the rate of the hydrolysis process.*

Example 2. Hydrolysis of Polyurethane with Saturated Steam.[77] *A polyurethane (1200 g) which was produced by the reaction of polytetrahydrofuran diol with toluene diisocyanate followed by curing with 4,4'-methylenediamine was heated under nitrogen to 200°C in saturated steam at 225 psig and held at this temperature for 12 h with the pressure reaching 550 psig. The reactor was gradually cooled to 75°C, and the hydrolyzed mass was transferred to a vessel which contained 100 mL of tetrahydrofuran which contained 1 g of the antioxidant 2,6-di-t-butyl-p-cresol. The product was two incompletely separated red-brown phases. Water and tetrahydrofuran were removed under vacuum to give a red-brown oil. The oil was kept under vacuum of nitrogen because on exposure to air it darkened. After several days, a dark solid deposit of sodium chloride and a tarry material separated. The oil (1149 g) was dissolved in 1200 mL of toluene. Gaseous hydrogen chloride was passed through the solution until no more amine* *salts precipitated. The salts were isolated by filtering through Celite. The solvent present in the almost colorless filtrate was removed under reduced pressure to give 624 g (76% yield based on hydrolyzed polymer) of a straw-colored oil. The oil consisted of approximately 79% of polytetramethylene ether glycol, 21% bis(2-ethylhexyl)phthalate, and trace amounts of 2,4- and 2,6-toluene diamines and 4,4'-methylenedianiline.*

Example 3. Hydrolysis of Polyurethane with Superheated Steam.[53] *A 300-mL stainless steel pressure reactor was charged with 10.0 g of polyurethane foam (shredded or in single pieces) and 100 mL of water. The assembly was lowered via a pulley system into a large constant-temperature oil bath. The contents of the reactor were heated for 15 min with superheated steam at 200°C. The reactor was removed from the bath and air cooled to room temperature. On opening the reactor, a two-phase liquid system was collected. The top phase was an aqueous solution containing toluene amines while the lower phase was a dark-colored, water-insoluble liquid of density 1.01 which by gel permeation chromatography was found to be pluracol polyol 443 (the polyol utilized in foam preparation). Vacuum distillation of the aqueous phase yielded a liquid fraction at 120°C and 0.1 mm Hg which solidified on cooling to form a pale yellow solid. The melting point and infrared and ultraviolet spectra were identical to that of a reference sample of 80% 2,4-toluene diamine and 20% 2,6-toluene diamine.*

REFERENCES

1. J. Scheirs, *Polymer Recycling*, Wiley, New York, 1998.

2. M. J. Joseph, *Introductory Textile Science*, 5th ed., Holt, Rinehart, and Winston, New York, 1986.

3. J. Milgrom, "Polyethylene Terephthalate (PET)," in *Plastics Recycling*, R. J. Ehrig (Ed.), Hanser, New York, 1992, p. 47.

4. D. Mangaraj, "Overview of Polymer Recycling Technologies," in *Comprehensive Polymer Science*, Second Supplement, G. Allen (Ed.), Pergamon, New York, 1989, p. 622.

5. P. Bajaj and N. D. Sharma, "Reuse of Polymer and Fibre Waste," in *Manufactured Fibre Technology*, V. B. Gupta, and V. K. Kothari (Eds.), Chapman & Hall, New York, 1997, p. 615.

6. J. H. Bonfield, R. C. Hecker, O. E. Snider, and B. G. Apostle, U.S. Patent 3,182,055, 1965.

7. S. D. Lazarus, I. C. Twilley, and O. E. Snider, U.S. Patent 3,317,519, 1967.

8. C. D. Myers, U.S. Patent 2,407,896, 1946.

9. B. Miller, U.S. Patent 2,840,606, 1958.

10. W. C. Meluch and G. A. Campbell, U.S. Patent 3,978,128, 1976.

11. E. Geigat and H. Hetzel, U.S. Patent 4,051,212, 1977.

12. D. Paszun and T. Spychaj, *Ind. Eng. Chem. Res.*, **36**, 1373 (1997).

13. R. J. McKinney, U.S. Patent 5,302,756, 1994.

14. H. Ulrich, A. Odinak, B. Tucker, and A. A. R. Sayigh, *Poly. Soc. Eng.*, **18**(11), 844 (1978).

15. C & EN, January 24, 2000, p. 23.

16. G. Odian, *Principles of Polymerization*, 3rd ed., Wiley, New York, 1991, p. 69.

17. M. Xanthos and S. H. Patel, "Solvolysis" in *Frontiers in the Science and Technology of Polymer Recycling*, G. Akovali, C. A. Bernardo, J. Leidner, L. A. Utracki, and M. Xanthos (Eds.), Kluwer Academic, Netherlands, 1998, pp. 425–436.

18. W. Schnabel, *Polymer Degradation, Principles and Practical Applications*, Macmillan, New York, 1981, pp. 184–186.

19. C. K. Ingold, *Structure and Mechanism in Organic Chemistry*, 2nd ed., Cornell University Press, Ithaca, NY, 1969, p. 1131.

20. T. H. Lowry and K. S. Richardson, *Mechanism and Theory in Organic Chemistry*, Harper & Row, New York, 1976, pp. 439–449.

21. G. W. Davis and J. R. Talbot, "Polyesters, Fibers," in *Polymers: Fibers and Textiles, A Compendium*, J. I. Kroschwitz (Ed.), Wiley, New York, 1990, p. 672.

22. L. N. Dmitrieva, A. A. Speranskii, S. A. Krasavin, and Y. N. Bychkov, *Fibre Chem.*, **17**, 229 (1986).

23. K. T. Barkley, U.S. Patent 3.830,759, 1974.

24. D. Gintis, *Makromol. Chem. Macromol. Symp.*, **57**, 185 (1992).

25. J. R. Campanelli, M. R. Kamal, and D. G. Cooper, *J. Appl. Polym. Sci.*, **54**, 1731 (1994).

26. S. Niu, T. Wakida, S. Ogasawara, H. Fujimatsu, and S. Takekoshi, *Textile Res. Res. J.*, **65**(12), 771 (1995).

27. M. B. Polk, L. L. LeBoeuf, M. Shah, C.-Y. Won, X. Hu, and Y. Ding, *Polym.-Plast. Technol. Eng.*, **38**(3), 459 (1999).

28. J. R. Campanelli, D. G. Cooper, and M. R. Kamal *J. Appl. Polym. Sci.*, **53**, 985 (1994).

29. H. V. Datye, *Indian Fibre Textile Res.*, **16**(1), 46 (1991).

30. R. J. McKinney, U.S. Patent 5,395,974, 1995.

31. W. R. McElroy, U.S. Patent 3,300,417, 1967.

32. O. Kinoshita, U.S. Patent 3,632,530, 1972.

33. J. Braslaw and J. L. Gerlock, *Ind. Eng. Chem. Process Des. Dev.*, **23**, 552 (1984).

34. U. R. Vaidya and V. M. Nadkarni, *Ind. Chem. Eng. Res.*, **26**, 194 (1987).

35. U. R. Vaidya and V. M. Nadkarni, *J. Appl. Polym. Sci.*, **38**, 1179 (1989).

36. A.-I. Malik and E. E. Most, U.S. Patent 4,078,143, 1978.

37. J. Aguado and D. Serrano, *Feedstock Recycling of Plastic Wastes*, Royal Society of Chemistry, Cambridge, 1999 p. 34.

38. A. Naujokas and K. M. Ryan, U.S. Patent 5,051,528, 1991.

39. G. E. Brown and R. C. O'Brien, U.S. Patent 3,952,053, 1976.

40. S. F. Pustaszeri, U.S. Patent 4,355,175, 1982.

41. M. J. Collins and S. H. Zeronian, *J. Appl. Polym. Sci.*, **45**, 797 (1992).

42. A. Oku, L.-C. Hu, and E. Yamada, *J. Appl. Polym. Sci.*, **52**, 1353 (1994).

43. T. Yoshioka, T. Sato, and A. Okuwaki, *J. Appl. Polym. Sci.*, **52**, 1353 (1994).

44. J. R. Campanelli, M. R. Kamal, and D. G. Cooper, *J. Appl. Polym. Sci.*, **48**, 443 (1993).

45. M. L. Doerr, U.S. Patent 4,620,032, 1986.

46. W. J. Gamble, A. A. Naujokas, and B. R. DeBruin, U.S. Patent 5,298,530, 1994.

47. M. Braun, A. B. Levy, and S. Sifniades, *Polym.-Plast. Technol. Eng.*, **38**(3), 471 (1999).

48. N. Chaupart, G. Serpe, and J. Verdu, *Polymer*, **39**(6-7), 1375 (1998).

49. A. R. Mikherjee and D. K. Goel, *J. Appl. Polym. Sci.*, **22** (2), 361 (1978).

50. J. T. Craig, U.S. Patent 3,223,731, 1965.

51. G. A. Kalfas, *Polymer React. Eng.*, **6**(1), 41 (1998).

52. T. Munzmay, H. Nefzger, W. Rasshofer, and W. Meckel, U.S. Patent 5,508,312, 1996.

53. L. R. Mahoney, S. A. Weiner, and F. C. Ferris, *Environ. Sci. Technol.*, **8**, 135 (1974).

54. J. L. Gerlock, L. R. Braslaw, L. R. Mahoney, and F. C. Ferris, *J. Polym. Sci.*, **18**, 541 (1980).

55. M. B. Sheratte, U.S. Patent 4,110,266, 1978.

56. O. Kondo, T. Hashimoto, and H. Hasegawa, U.S. Patent 4,014,809, 1977.

57. H. R. Van der Wal, U.S. Patent 5,274,004, 1993.

58. J. Urbanski, W. Czerwinski, K. Janicka, F. Majewska, and H. Zowall, *Handbook of Synthetic Polymers and Plastics*, 1st ed., Wiley, New York, 1977, pp. 52-53.

59. J. Gerlock, J. Braslaw, and M. Zindo, *Ind. Eng. Chem. Process Des. Dev.* **23**, 545 (1984).

60. R. K. Hallmark, M. J. Skowronski, and W. D. Stephens, U.S. Patent 4,873,268, 1989.

61. A. Fujita, M. Sato, and M. Murakami, U.S. Patent 4,609,680, 1986.

62. J. Y. Chen, C. F. Ou, Y. C. Hu, and C. C. Lin, *J. Appl. Polym. Sci.*, **42**, 1501 (1991).

63. J. M. Lusinchi, R. Pietrasanta, and J. J. Boutevin, *J. Appl. Polym. Sci.*, **69**, 657 (1998).

64. H. Grushke, N. Taunus, W. Hammerschick, B. Nauheim, and H. Medem, U.S. Patent 3,403,115, 1965.

65. B. J. Sublett and G. W. Connell, U.S. Patent 5,559,159, 1996.

66. R. J. England, U.S. Patent 3,544,622, 1970.

67. G. C. Tustin, T. M. Pell, D. A. Jenkins, and M. T. Jernigan, U.S. Patent 5,413,681, 1995.

68. C.-Y. Kao, B.-Z. Wan, and W.-H. Cheng, *Ind. Eng. Chem. Res.*, **37**, 1228 (1998).

69. M. R. Kamal, R. A. Lai-Fook, and T. Yalcinyuva, *SPE ANTEC*, San Francisco, May 1-5, 1994, 2896 (1994).

70. B. A. Lamparter, B. A. Barna, and D. R. Johnsrud, U.S. Patent 4,542,239, 1985.

71. V. A. Popoola, *J. Appl. Polym. Sci.*, **36**, 1677 (1988).

72. Y. M. Awordi, A. Johnson, and A. V. Popoola, *J. Appl. Polym. Sci.*, **33**, 2503 (1987).

73. S. Sifniades, A. B. Levy, and J. A. J. Hendrix, *SPE ANTEC, Technical Papers*, U.S. Patent 5,681,952, 1997.

74. T. F. Corbin, E. A. Davis, and J. A. Dellinger, U.S. Patent 5,169,870, 1992.

75. U.S. Patent Specification 921,667, 1960.

76. J. L. Gerlock, J. Braslaw, and J. Albright, U.S. Patent 4,316,992, 1982.

77. D. F. Lohr and E. L. Kay, U.S. Patent 4,035,314, 1977.

Index

Synthetic Methods in Step-Growth Polymers.   Edited by Martin E. Rogers and Timothy E. Long
© 2003 John Wiley & Sons, Inc.   ISBN 0-471-38769-X